W0193202

Artificial Particle
Beams in Space
Plasma Studies

NATO ADVANCED STUDY INSTITUTES SERIES

A series of edited volumes comprising multifaceted studies of contemporary scientific issues by some of the best scientific minds in the world, assembled in cooperation with NATO Scientific Affairs Division.

Series B: Physics

Recent Volumes in this Series

This series is published by an international board of publishers in conjunction with NATO Scientific Affairs Division

A Life Sciences

B Physics

C Mathematical and
   Physical Sciences

D Behavioral and
   Social Sciences

E Applied Sciences

Plenum Publishing Corporation
London and New York

D. Reidel Publishing Company
Dordrecht, The Netherlands
and Hingham, Massachusetts, USA

Martinus Nijhoff Publishers
The Hague, The Netherlands

Artificial Particle Beams in Space Plasma Studies

Edited by

Bjørn Grandal

Norwegian Defence Research Establishment
Kjeller, Norway

SPRINGER SCIENCE+BUSINESS MEDIA, LLC
Published in cooperation with NATO Scientific Affairs Division

Library of Congress Cataloging in Publication Data

NATO Advanced Research Institute on Artificial Particle Beams in Space Plasma Studies
(1981: Geilo, Norway)
Artificial particle beams in space plasma studies.

(NATO advanced study institutes series. Series B, Physics; v. 79)
"Proceedings of a NATO Advanced Research Institute on Artificial Particle Beams in
Space Plasma Studies, held April 21-26, 1981, in Geilo, Norway" — T.p. verso.
"Published in cooperation with NATO Scientific Affairs Division."
Includes bibliographical references and index.
1. Space plasmas — Congresses. 2. Particle beams — Congresses. I. Grandal, Bjørn,
1948-   . II. North Atlantic Treaty Organization. Division of Scientific Affairs. III.
Title. IV. Series.
QC809.P5N27   1981   530.4′4   82-470
                                                                             AACR2

ISBN 978-1-4684-4225-0   ISBN 978-1-4684-4223-6 (eBook)
DOI 10.1007/978-1-4684-4223-6

Proceedings of a NATO Advanced Research Institute on Artificial Particle Beams
Utilized in Space Plasma Studies, held April 21 — 26, 1981, in Geilo, Norway

© 1982 by Springer Science+Business Media New York
Originally published by Plenum Press New York in 1982
Softcover reprint of the hardcover 1st edition 1982

All rights reserved

No part of this book may be reproduced, stored in a retrieval system, or transmitted,
in any form or by any means, electronic, mechanical, photocopying, microfilming,
recording, or otherwise, without written permission from the Publisher

PREFACE

These proceedings are based upon the invited review papers and the research notes presented at the NATO Advanced Research Institute on "Artificial Particle Beams in Space Plasma Studies" held at Geilo, Norway April 21-26, 1981.

In the last decade a number of research groups have employed artificial particle beams both from sounding rockets and satellites in order to study various ionospheric and magnetospheric phenomena.

However, the artificial particle beams used in this manner have given rise to a number of puzzling effects. Thus, instead of being just a probe for studying the ambient magnetosphere, the artificial particle beams have presented a rich variety of plasma physics problems, in particular various discharge phenomena, which in themselves are worthy of a careful study.

The experimental studies in space using artificial particle beams have in turn given rise to both theoretical and laboratory studies. In the laboratory experiments special attention has been paid to the problem of creating spacelike conditions in the vacuum chamber. The theoretical work has addressed the question of beam-plasma-neutral interaction with emphasis on the wave generation and the modified energy distributions of the charged particles. Numerical simulations have been used extensively.

With the advent of the Space Shuttle in which several artificial particle beam experiments are planned for the 1980's, there is a growing interest in such experiments. Furthermore, there is a need for coordinating these studies, both in space and in the laboratory.

This book contains a broad review of recent experiments both in space, Chapter 1, and in the laboratory, Chapter 3. Chapter 2 is devoted to a brief discussion of the natural particle beams in space. Recent theoretical work, with emphasis on the beam-plasma-discharge (BPD) is given in Chapter 4. The present understanding of the charging and neutralization processes in the vicinity of a space vehicle employing particle accelerators is outlined in Chapter 5.

At the end of chapters 1, 3 and 5, summaries of the general discussions at the respective sessions of the meeting are given. The subject matter covered in these general discussions does overlap and no attempt has been made to sort out the related topics. These summaries, included to give the reader an impression of the issues of interest, are written by the editor, based on notes taken by T Hallinan, N Kawashima and himself.

Chapter 6 includes a summary of the final discussions with future plans, written by B N Mæhlum, and a few of the contributed papers outlining future work.

At the end of most of the papers, selected questions of general interest are included.

The responsibility for the uneven coverage of some of the topics lies with the editor and the program committee, rather than with the individual contributors.

The editor would like to express his gratitude to B N Mæhlum for his constant support.

Kjeller, July 1981

Bjørn Grandal

DIRECTOR'S PAGE

Bernt N Maehlum, Program Director

Several studies of beam-plasma interactions have
been conducted in space and in laboratory plasma
chambers in recent years. Also, particle accelerators
on rockets and satellites have been used for probing
the ionosphere and magnetosphere.

This work has resulted in several publications,
but it is well known that many results from similar
studies have not yet been presented in the literature
due to apparent lack of consistency.

During the NATO-ARI several unexplained and
unpublished observations were presented, and I per-
sonally feel that a somewhat coherent picture is now
emerging from the apparent confusion. It is hoped that
the results from the meeting may serve as a platform
for future activity in the field.

Let me acknowledge the assistance given me by
the program committee: Drs C Beghin, W Bernstein,
A Johnstone, L M Linson and J R Winckler. Without their
help we could not have organized the meeting.

Also, an active organizing committee, consisting
of Drs B Grandal, J Holtet and T A Jacobsen, have taken
their share of the work.

Finally, let me thank the secretary, Mrs E Kurland,
who kept track of all the loose papers before, during
and after the conference.

In addition to the NATO grant which covered most of
the expenses, some financial support was obtained from
the Royal Norwegian Council for Scientific and
Industrial Research and the Norwegian Defence Research
Establishment.

OPENING SPEECH

Bjørn Landmark

Royal Norwegian Council for Scientific and Industrial
Research
Oslo, Norway

Twenty years ago, in 1961, a group of ionospheric physicists at the Norwegian Defence Research Establishment arranged the first NATO Advanced Study Institute in Norway at a remote mountain hotel. The title of the Institute was "Electron Density Profiles in the Ionosphere and Exosphere", and the meeting was successful partly because the right people were present and partly because it was arranged at a secluded place with ample opportunities for the participants to meet during leisure hours at the hotel and on the ski slopes for informal discussions.

Two years later a new NATO meeting was arranged at the same hotel. This time the topic of the conference was "Electron Density Profiles in the Ionosphere and Exosphere". This was also a successful meeting, and it was followed by a third meeting in 1965, which again was devoted to "Electron Density Profiles in the Ionosphere and Exosphere".

After these three initial meetings with the same "winning title", six more NATO Advanced Study Institutes have been arranged on Upper Polar Ionosphere and Space Plasma problems. All of these have had a fairly broad scope. In fact, the one last year was entitled "Exploration of the Polar Upper Atmosphere".

The meeting which starts here today is therefore number ten in the series of NATO conferences in Norway, which calls for a celebration. In contrast to the previous Institutes the format of this meeting is so specialized that the NATO Science Committee initially declined to fund it under the Advanced Study Institute program. However, the committee turned around and established a new series of conferences, named NATO Advanced Research Institutes. As we understand it, the future of the Research Institute program depends largely on the success of the present meeting. Therefore, we need to work hard to keep this program going.

All the previous NATO Advanced Institutes in Norway on upper atmosphere and magnetospheric phenomena are based on passive observations of the natural phenomena. Also, due to the geographical location of Norway emphasis has been placed on high latitude, auroral problems.

The present meeting on the other hand, will mainly be concerned with various methods for artificial modification of the upper atmosphere in order to study, under controlled conditions, the various beam-plasma interactions in space, such as those taking place in the auroral ionosphere. The first attempt made for producing artificial auroras in the laboratory was conducted early this century by the Terrellas, and as you know - my countryman Kr Birkeland - was one of the pioneers in this field. His Terrellas are on display in the Technical Museum in Oslo if you would like to see them. Some 50 years later the first artificial aurora in the upper atmosphere was created on a rocket launched some ten years ago by Wilmot Hess and his colleagues.

Strangely enough, during the last 5-6 years, when this field has been rapidly expanding, the interest for generation for artificial auroras is no longer the main driving force. The particle beams are now mostly used as diagnostic tools for studying ionospheric and magnetospheric plasma processes, such as plasma heating by fast particle beams, scattering or fast electrons in a plasma, tracing of geomagnetic field lines, enhancements in the plasma density, wave generation through beam-plasma interactions, detection of electric double layers etc. It is not quite clear yet how many of these studies that will contribute to a deeper understanding of the auroral processes, since the current density in most of the particle beam experiments generally exceeds the natural auroral particle fluxes by orders of magnitude. However, the studies of upper atmosphere beam-plasma interactions per se give important results from a plasmaphysics point of view. I believe that also some astrophysical problems may be elecidated by these experiments.

As a complement to the space plasma studies by particle beams, several laboratory experiments have been conducted in recent years. These have given valuable information on the energy dissipation processes. It has been demonstrated how the instabilities which occurs for certain beam currents and plasma densities may enhance the transfer rate of energy from the beam to the plasma. We will hear more about results from these studies at the conference.

The new observations of the beam-plasma interaction in the laboratory and in space have created some interest among theoreti-

cians, and there is now a fruitful exchange of ideas between ex-
perimentalists and the theoreticians in the field. A session at
this conference is devoted to theoretical studies of the beam-
plasma interaction.

Among the applied aspects of this field are the problems related to space vehicle neutralization in the ionosphere. Several
groups are involved in studies of this problem, and a series of
papers will present us with the state of art.

One of the main aims of the meeting is to bring workers from
different but related fields together for informal discussions of
problems of joint interest. Also, during the last session future
plans and possible recommendations for new and cooperative experiments will be discussed. I feel that both from a scientific and
from an economic point of view international cooperation is important, and it is my hope that this meeting will contribute to
a continued growth in the field. Future Spacelabs and rocket
flights open up new opportunities, and it is hoped that the plans
for closing the large plasma chamber at Johnson Space Center will
not spread to other facilities. There are still a number of problems to be solved in the laboratory.

CONTENTS

Chapter 1:   ACCELERATOR EXPERIMENTS IN SPACE

CONTENTS

CONTENTS

Chapter 1:   ACCELERATOR EXPERIMENTS IN SPACE

THE USE OF ARTIFICIAL ELECTRON BEAMS AS PROBES OF THE DISTANT MAGNETOSPHERE

John R. Winckler

School of Physics and Astronomy, University of Minnesota
Minneapolis, Minnesota, 55455

Abstract and Summary

Electron beams have been used as analytical tools in laboratory plasmas, particularly for diagnosing fluctuating fields. This paper describes experiments in which electron beams have been injected into the magnetosphere to diagnose plasma processes at great distance by measurements made in the ionosphere. In some of these experiments the conjugate region atmosphere was used to detect the electron beams. In others conjugate "echoes" were detected near the injection region. Conjugate locations, field line lengths, electric and magnetic drifts, field fluctuations and electron scattering and diffusion have been analyzed. Beams intended for probing have been injected with up to 40 kev energy and at currents up to 0.8 A. The injecting vehicle constitutes a complex system which produces major perturbations of the local ionosphere including a large space charge. These effects may propagate to large distances and affect the regions under study. The probing beams may also conceivably alter the distant plasmas. Experiences with echo detection by particle counters on some of the ECHO rocket series are discussed in some detail. The echoes are seen to respond to changes in the convection fields and to reflect auroral zone activity. Theoretical and experimental echo patterns are discussed. Evidence for beam pitch angle scattering and altered mirror heights is presented and may occur even in the time of one bounce period. The use of the atmospheric response to electron beams in the loss cone as a detector has been achieved using optical, x-ray and radar techniques and examples are given. Future programs will search for acceleration processes both for the trapped radiation and for auroral-producing precipitation. Beams injected from a long-range rocket or polar orbiting satellite may delineate the regions of open and closed field lines. Beams injected from geostationary orbit may be a powerful means of analyzing magnetospheric electrodynamics if detected and analyzed in the conjugate region in the polar ionosphere.

3

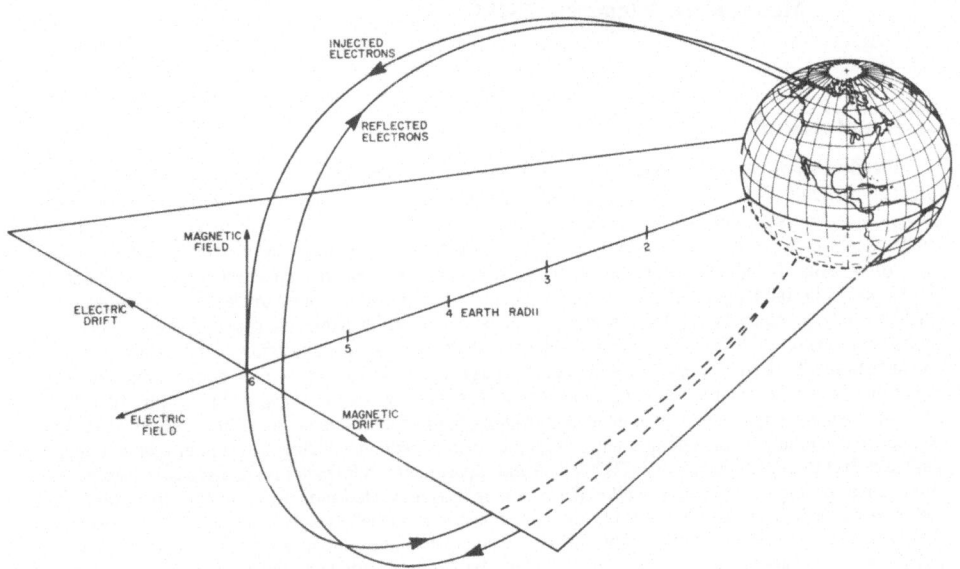

Frontespiece. The geometry of the near tail region of the magnetosphere and field lines traced by ECHO electron beams injected from Alaska.

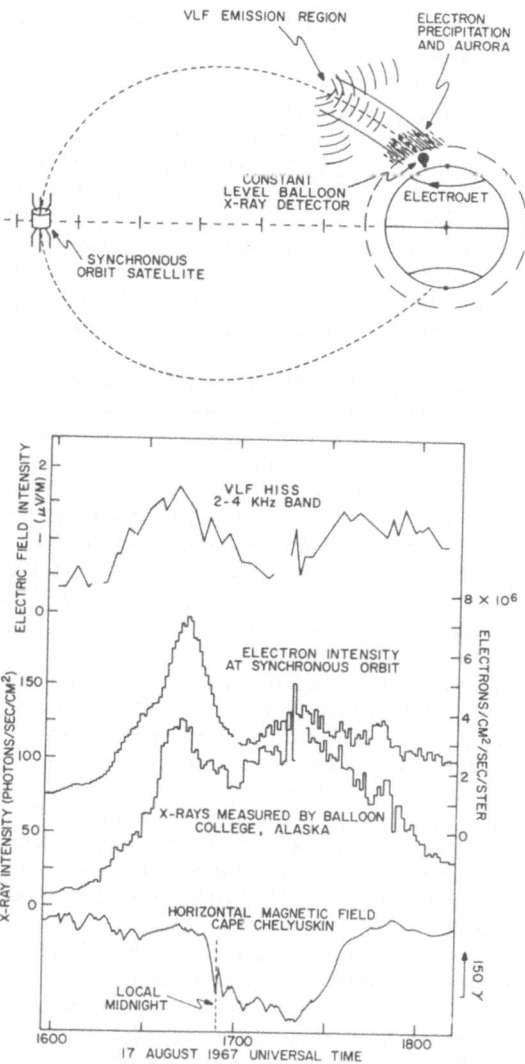

Figure 1.   (Top) The geometry of synchronous orbit and the auroral zone. (Bottom) An injection-precipitation
event observed at synchronous orbit and simultaneously by balloon at the conjugate region in early
morning, accompanied by a major substorm (after Parks et. al., 1968).

1. THE SCIENTIFIC PURPOSES OF THE EXPERIMENTS

A beam of electrons, if injected into a plasma under controlled conditions, can be used as a
diagnostic tool to explore the time and space structure of the electromagnetic fields in the plasma. The
small mass of the electron make it very sensitive to a wide range of AC and DC electric and magnetic
fields. Further, beams of electrons are easy to generate in a controlled manner, and detection and
analysis in the laboratory, at least, has been successful (Cannara et. al., 1967, Mills et. al., 1967)

This paper discusses the application of electron beams as probes and some significant results which
have been obtained in a number of areas. A concise discussion of beams as probes for studying the
distant electric convection fields was given recently (Winckler, 1981) as well as an extensive review of
many aspects of electron beams in space (Winckler, 1980). Electron beams also constitute a type of

"active" experiment, and the process of beam injection in the ionosphere produces large plasma disturbances. The extent to which the natural magnetosphere or the initial state of the probing beams are modified by these disturbances deserves careful consideration, as their effects are undesirable for the present purposes and should be minimized.

In the ARAKS (Cambou et. al., 1980) and the Hess "Kaui" (Davis et. al., 1980) experiments electron beams have been used to locate by direct experiment conjugate points between hemispheres at the 100 km level in the ionosphere, and the results are in agreement with the predictions of acceptable field models. The very important question of how the magnetic equatorial plane maps along the field lines into the ionosphere is not answered by conjugate measurements. The shape of the distant field predicted by models can, however, be compared with the bounce times and gradient and curvature drifts of injected beams. Experience shows these measurements to be difficult or impossible with conjugate type experiments but are well suited, and produce first order effects in "echo" type experiments.

The mapping of the equatorial plane to the earth's surface magnetically may be explored best by beams injected from a vehicle at great distance with observations made in the conjugate atmosphere. A satellite in geostationary orbit could investigate a particularly interesting region of the magnetic field conjugate to the auroral zone, and containing the very active inner edge of the plasma sheet. The instabilities induced by convective motion of the plasma sheet into the inner magnetosphere and their importance in precipitation and acceleration were pointed out long ago by Kennel (1968), and have been since much discussed from the standpoint of electrostatic mode wave generation and their affects on particles of all energies (Lyons, 1974, Ashour-Abdalla and Kennel, 1978). Phenomena occurring in this region are shown schematically in Figure 1, upper panel. The lower panel shows an actual comparison of an acceleration-injection event at synchronous orbit with auroral zone conjugate observations of various types during a substorm event. Since the geostationary satellite would inject beams from a precisely known, fixed location, the observation of these beams by atmospheric fluorescence or other means in the conjugate region in the polar ionosphere would be a powerful tool for studying magnetospheric dynamics. The experiment would function like a laboratory cathode ray oscillograph and write a wave form in the polar night sky of substorm dynamics and many other effects.

An example of large-scale dynamics accessible to these methods is shown in Figure 2, which displays by super-posed epoch analysis the typical growth-phase inflation and expansion-phase collapse of the magnetosphere near synchronous orbit (Swanson, 1978). A simple measurement of the progression of field line lengths by a series of ECHO rockets or by beams from synchronous orbit injected at the times indicated near the bottom of Figure 2 could directly display the inflation and collapse of the magnetosphere tail region associated with the changes of distant currents during a substorm, a result probably unobtainable by other means.

Several programs have been instituted to search for electric fields parallel to B by injecting beams upwards on auroral field lines and detecting echoes returning downwards from within one earth radius. The results of this very difficult experiment are preliminary. Nevertheless if the techniques could be improved, pressing questions about the reality of double layers, shocks, and basic auroral physics could be answered (Maehlum, 1980, Wilhelm, 1980). Where the ECHO technique is carried out between hemispheres by using conjugate mirror point reflection the effects of parallel electric fields may be second order and difficult to identify.

Electron beams injected from an ordinary research rocket in the polar ionosphere can traverse the magnetic equatorial plane and interact with the active plasma sheet or the plasmapause. Here there is the fascinating possiblilty to duplicate and measure the natural acceleration and injection processes for the trapped radiation and the precipitation of the auroral and higher energy particles. This is the particular goal of the ECHO experiments, and has also been studied in the ARAKS flights (Pyatsi, 1980, Gringauz, 1980, Roeder, 1980). Evidently strong pitch angle diffusion and strong fluctuating electric field effects have been noted. The first ECHO experiment and the ARAKS experiment, were conducted in the inner magnetosphere at L-values of 2.6 and 3.7, respectively. This is a region of stable magnetic configuration but interesting plasma phenomena. The present paper will detail higher latitude measurements typified by the ECHO 3, 4, and 5 experiments. The regions accessible to electron beams from these rockets is shown in the frontespiece.

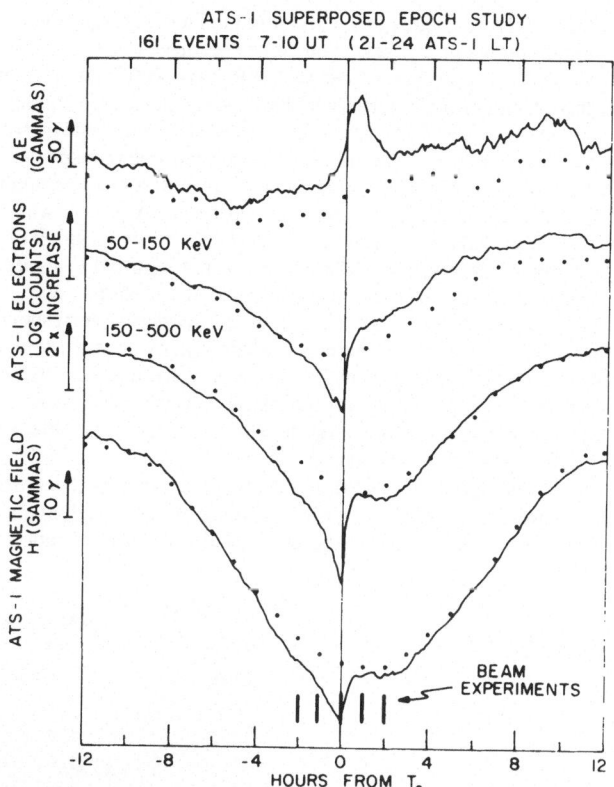

Figure 2. Superposition of 161 traces of energetic electrons, AE index and magnetic field (from Swanson, 1978) Zero epoch time was indexed to a substorm expansion phase shown by a particle increase. The substorm inflation-collapse sequence should be a prime subject for study by a series of ECHO experiments through the growth-expansion period.

2. THE BASIC ELEMENTS OF THE EXPERIMENTAL SYSTEM

a. The Beam Injecting Rocket System in the Ionosphere

If Electron beams are to be useful probes, they must enter the undisturbed magnetosphere with a known configuration, e.g. with known pitch angle, energy and spatial distribution. The effects of vehicle potential, near-vehicle discharges and heating due to the neutralizing processes, beam-plasma discharges, self effects of beam space charge, etc., may be bothersome and alter the beam from the near-vehicle configuration determined by the accelerator.

A few comments of technical nature are in order, concerning high-power, high-current electron guns for space use. Triode planar geometries have been used in several experiments, (Hess et. al., 1971, Davis et. al., 1980, Paton et. al., 1978). Space-charge grid control in these systems permits easy programming and high-frequency modulation capability. Sellen (1976) has discussed high-powered gun systems for space use, considering how to achieve maximum perveance. These systems have used oxide cathodes, which must be vacuum sealed until exposed in space, and are accordingly difficult to test. In the Norwegian "Polar" series (Maehlum et. al., 1980), the PRECEDE-EXCEDE series (O'Neil et. al., 1978) and the ECHO series (Winckler, 1980) diode guns with relatively indestructable tungsten or tantulum cathodes were used. A lower perveance per gun, but at higher voltage was achieved. The gun current and voltage programming for diodes must be applied to the primary power drive, which limits the frequency range of modulation to a few KHz. This frequency range suffices, however, for conjugate experiments or distant echo observations. Figure 3 shows the voltage-current characteristic of one of the ECHO 5 diode guns, similar to the Pierce design (Pierce, 1949) The beam angular profile of this gun is shown in Figure 4. The diode was driven with a voltage profile as shown in Figure 5, derived from a 500 Hz power converter with full-wave rectification but no filtering. Figure 5 shows the 0.5 msec turn-on and turn-off pulses and three 1 msec pulses, providing a 4 msec injection repeated every 20 msec in a 90 pulse "fast" series. A superposed epoch analysis of the "fast" series pulses has been carried out by Swanson in searching for weak atmospheric responses (see below). Four such diode guns were built into the ECHO 5 rocket payload, resulting in a maximum power of 30KW (0.8 amperes at 37KV). The perveance for a single gun defined as $A = I/V^{3/2}$ was 4×10^{-8} with I in amperes and V in volts. These ECHO electron beams are focussed into a small divergence and well defined pitch angle, and have a perveance probably below that required to ignite the beam-plasma discharge in space. These qualities are especially desirable for probing.

It seems impossible to inject beams of the order of 0.5 amperes, corresponding to 1×10^{19} electrons/second without producing high fluxes of beam-energy electrons in on-board particle counters. Electrons may scatter from the gun anode or exit port, from gas emitted by the rocket or by the ambient atmosphere, and scatter further on the metallic skin of the rocket. Such a halo may sometimes be misinterpreted as evidence for a large rocket potential, as if ambient electrons had been accelerated from thermal energies back to the rocket.

The following is a brief description of conditions near the ECHO 5 rocket derived from particle and xray detectors and photometers by Swanson (1981). Figure 6 shows schematically the ECHO 5 payload in flight from the Poker Flat range in Alaska. This payload was attitude controlled and flew

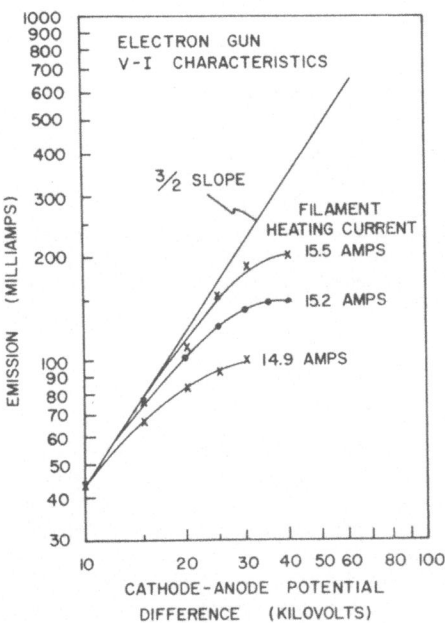

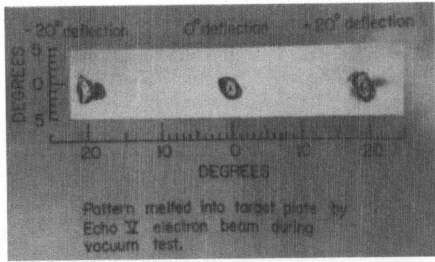

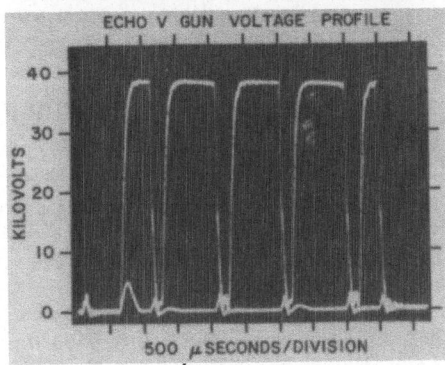

Figure   3.   The voltage-current characteristic of the Echo 5 diode gun shown for various filament temperatures. A current of 250ma was achieved.

Figure   4.   The beam pattern formed by melting the surface of a metal plate located one meter from gun anode. The deflections were produced as in flight by electromagnets and would normally correspond to pitch angles of 70, 90 and 110 degrees.

Figure   5.   Voltage profile applied to the ECHO 5 gun for one of the 4 msec "fast series" injections. There is a definite off time of 10-50 micro-secs duration between each 1 msec square-wave drive pulse. All beam injections followed this profile regardless of length, but with a peak voltage of either 25 or 37KV.

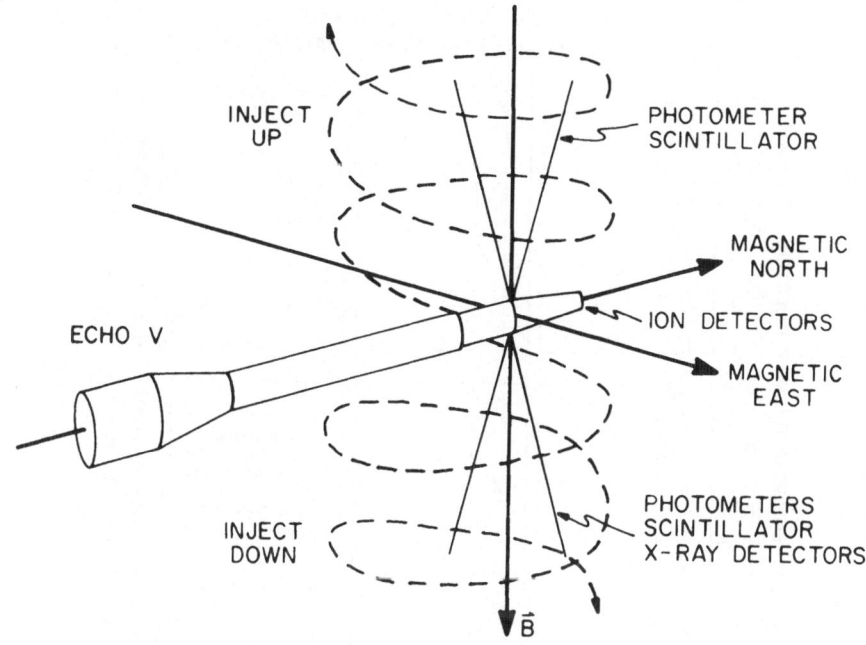

Figure 6. The ECHO 5 payload in flight aspect showing the beam Larmor spirals for up and down injections. in relation to the view directions of various detectors.

with the long axis magnetically horizontal and aligned north-south. The beams from the four guns thus had access to a full range of pitch angles from zero to 180 degrees. The pitch angle of beam injection was set by deflection electromagnets.

As shown in Figure 6, there was a photometer which looked directly up the magnetic field filtered for a neon line (540.1 nm) but which also responded to beam-generated continuum in its 5 nm bandwidth. The field of view was 16 degrees FWHM. A 60 degree FWHM scintillation counter sensitive to electrons of E>5kev as in the ECHO 4 experiment was also aimed upward. An identical photometer and scintillator looked directly down the field as shown in Figure 6. Also looking down the field was a battery of x-ray detectors and other wavelength photometers designed to search for responses at the 100 km level in the atmosphere.

The responses of the up- and down-looking scintillators and photometers and the down-looking x-ray detectors have been processed by the superposed epoch technique using the "fast" series pulses which were designed specially for this purpose. The zero epoch was set at the start of each 4 msec duration pulse and the data averaged over the 20 msec interval between pulses in the 90-pulse series. Beam injections upwards at 110 degree pitch angle and 24 kev energy gave responses as shown in the top panels of Figure 7. Beam injections downward were at higher energy (38 kev) with 70 degree pitch angle and in the same "fast" mode as the upwards injections. In addition, a series of 2 msec duration downward pulses are shown during the release of 2 moles/sec of neon gas from orifices symmetrically directed up and down the field. All responses have been background corrected and normalized to the same electron flux.

The results summarized in Figure 7 are of interest because they give insight into the rapidity of dispersal of the energetic electrons following beam injection for the rather high power (30 kw) beams of ECHO 5, as measured both in terms of electrons directly and in terms of visible and x-ray photons. Also, the configuration of the photometers looking directly into the Larmor spiral of the injected beams for several gun currents, pressures and beam energies can definitively identify the "beam-plasma discharge" if it occurs in space for ECHO 5, as the geometry is precisely the same as that described in the large tank experiments of Bernstein et. al. (1979). The question of the "BPD" is discussed in Sec. 2.d.

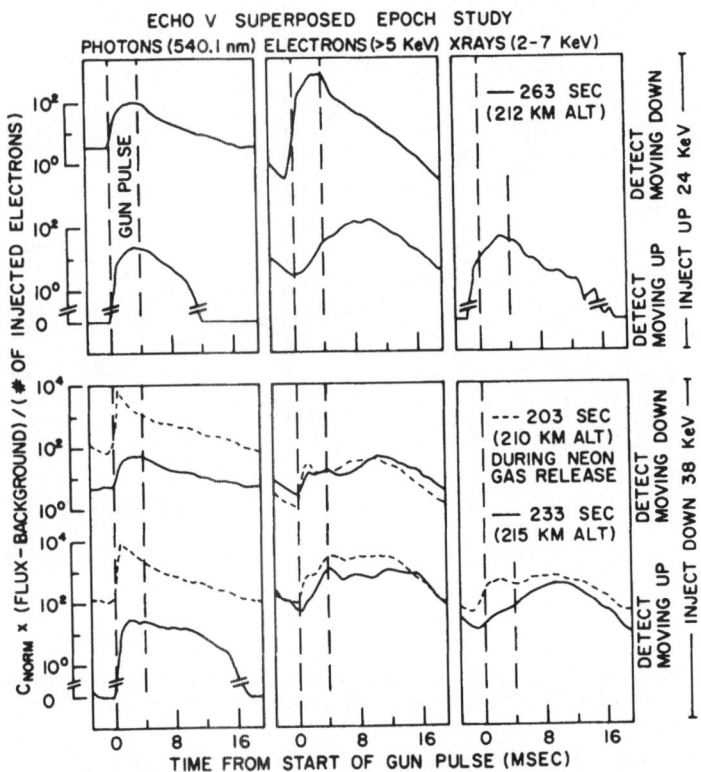

Figure 7.   Result of a superposition of various sets of 90 "fast series" pulses for up and down injections and with and without gas release. The time span is the 20 msec interval between pulses. Each pulse had a profile as shown in Figure 5.

The results shown in Figure 7 may be summarized as follows:

Electrons in the center column show vigorous scattering by the ambient atmosphere and magnetically; upward injections are detected moving back down by scattering and are seen moving up with a 10 msec delay characteristic of "quick echoes" magnetically mirrored below; downward injections give a much larger up flux which nearly saturated the down looking scintillator, but with a "quick echo" maximum still visible; the gas injection increased the upward flux appraciably; downward injections likewise gave a small down moving flux coming from a second scattering above the payload.

X-rays are produced below the rocket by upward injections, obviously by back-scattering downward past the detectors; downward injections produce a larger x-ray flux below the rocket which shows a 10 msec delay characteristic of "quick echoes"; finally, x-ray production is enhanced by the neon gas; the x-ray signal rises before the 0 mark due to the long 3.5 msec data sample.

Visible photons are produced above and below the payload during beam injections with about the same intensity, indicating that they are produced in the vehicle neutralizing current discharge; after the 4 msec beam pulse the photons die away, generally more rapidly than the electron residual; The effect of the neon gas injection is spectacular as seen in the well-known orange line 540.1 nm, and is two orders of magnitude; this is equivalent to an increase of the gas pressure near the rocket of 1000x or a decrease in altitude to 110 km.

All of the above high-energy associated phenomena are ended by the next beam pulse, i.e. in about

20 msec. Thus there is a sort of "quasi-trapping" on the injection tube of magnetic force; particularly for downward injections, but even for upward, the space between the 100 km. level and a region extending far above the rocket contains electrons of beam origin which finally follow the main beam away to infinity with a delay of some tens of milliseconds.

b. Beam Detection and Analysis

b.1. Echo Detection by Particle Counters

All cases of conjugate echoes reported so far have been under the ECHO program and have been accomplished with particle detectors carried on the beam-emitting rockets. In considering this seemingly impossible result one tends to forget that the rocket is located initially on the precise magnetic line of force where the beam is moving. Thus to encounter the beam again after one bounce the rocket motion has only to match the bounce-integrated transverse displacement of the beam. The results show that this is possible for regions up to at least $L = 6$ over a considerable range of rocket motions and geophysical conditions (See Table I). The outcome of an echo-type experiment is quite predictable if the first and second invariants are preserved. Thus the observation of echoes with the ECHO 1 experiment, and their non-observation with the ECHO 5 and ARAKS experiments is well understood. Probably the most amazing feature is the validity of invariant motion over such vast distances, yet it is the departures from invariant motion which result in particle acceleration, precipitation and injection, and are therefore of paramount interest.

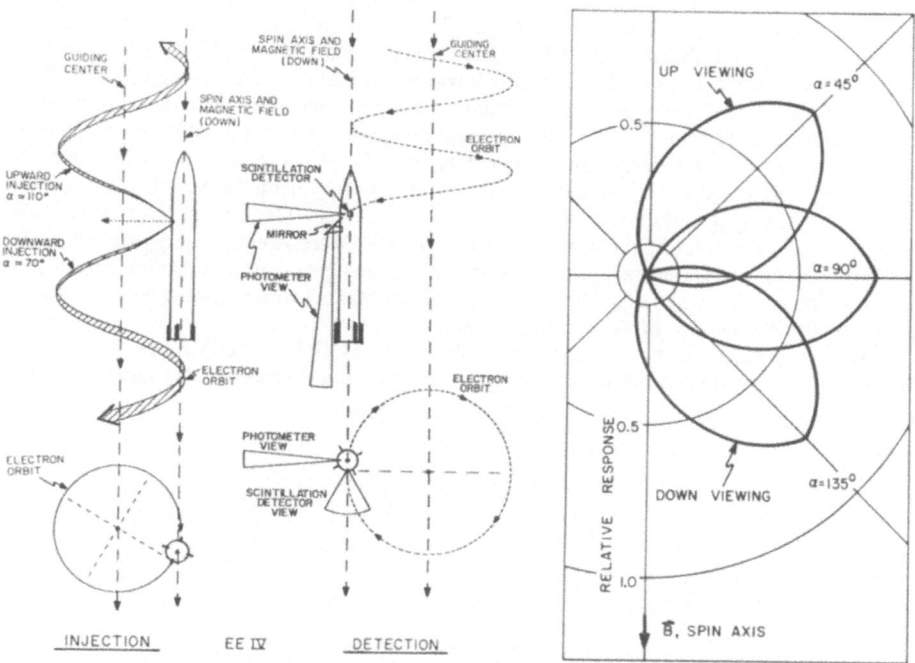

Figure   8.   Beam injection and echo detection geometry for the ECHO 4 experiment. The payload remained attached to the rocket motor throughout the flight to increase the return current collection area. The rocket axis was aligned along the B vector and spun at 1 RPS. The electron gyro orbits shown in the lower panels rotated with the spinning rocket. The beam pitch angle was controlled by electromagnets over a range from 30 to 160 degrees.

Figure   9.   Meridian profile of the ECHO 4 Scintillation counter response in pitch angle. At 90 degrees pitch angle the azimuth coverage was obtained with six detectors spaced each 60 degrees. At 45 and 135 degrees pitch angle three units each were used spaced 120 degrees in azimuth.

TABLE I

Flight	Date	Location	L	Dir.	Echoes	Details
Echo 1	Aug 70	Wallops	2.6	East	Yes	Remote Scattering
Echo 2	Sep 72	Churchill	8	East	No	Local Scattering
Echo 3	Apr 74	Poker	6	East	Yes	E, B Models
Echo 4	Jan 76	Poker	6	East	Yes	E, TV, Gas
Echo 5	Nov 79	Poker	6	North	?	Hi Pwr, TV, Gas
Echo 6	Nov 82	Poker	6	North	In Progress - TADS, Echoes

The Echo 3 and 4 experiments are the most interesting for the purposes of this paper since they were launched from the auroral zone region, connected magnetically with the plasma sheet as shown in the frontespiece. The Echo 4 was particularly well suited for on-board echo detection since the payload axis was set along the B-vector by an attitude control system and was rotated at 1RPS, as shown in Figure 8. An attempt was made to measure the energy of the echoes accurately by equipping the ECHO 4 with octospherical electrostatic analysers, supplied by the U. of New Hampshire group, of maximum geometry factor (0.01 ster-cm^2). The echoes were also studied with 12 scintillation counters of 8 str-cm^2 operated in a current mode, sensitive to electrons of >5 kev and pointing in various azimuths and pitch angles as shown in Figure 9. ECHO 4 also carried groups of photometers which have been used to observe the glow discharge near the beam and in the rocket sheath (Israelson and Winckler, 1979) but which will not be discussed here.

The 80 ma beam had a narrow divergence similar to the Echo 5 experiment, and the pitch angle was controlled by electromagnets. The complete gun sequence is shown in Figure 10. The 0.5 sec down marker was for producing reference atmospheric fluorescent streaks to be detected by television, which was successful (Hallinan et. al., 1978). The 0.5 sec up marker immediately after was intended for conjugate echo detection by television which was not successful. The up marker as well as the sequences of pickets and spikes as shown in Figure 10 were intended for echo detection by particle detectors and were successful. "Spikes" were injected at maximum voltage and at a pitch angle of 150 degrees to enter the loss cone in the conjugate hemisphere. Echoes were detected, from atmospheric back scattered electrons with an inherent large energy spread. In general the particles detected passed to the southern hemisphere at the maximum energy (approx. 38 kev) but returned with lower energy, depending on the detection criteria at the moment. "Pickets" were swept in energy from 8 to 18 kev over a 100 msec interval and were injected to mirror at the conjugate region. It was intended that if the echo energy lay in that range that a double echo would be obtained with a spacing between 0 and 100 msec to identify the energy of the echoing particles.

Two distinct groups of echoes were observed during the ECHO 4 experiment as shown in Figure 11, lower section, in which a portion of the rocket trajectory has been projected into the "injection

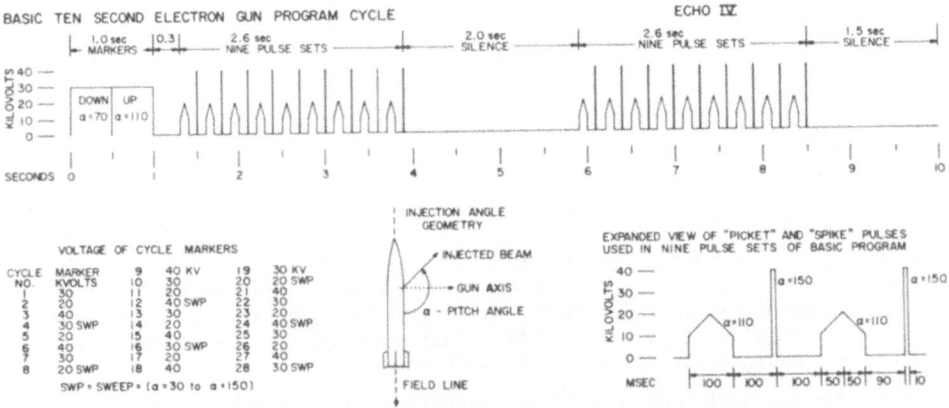

Figure 10.   Details of the ECHO 4 gun sequence. See text for explanation.

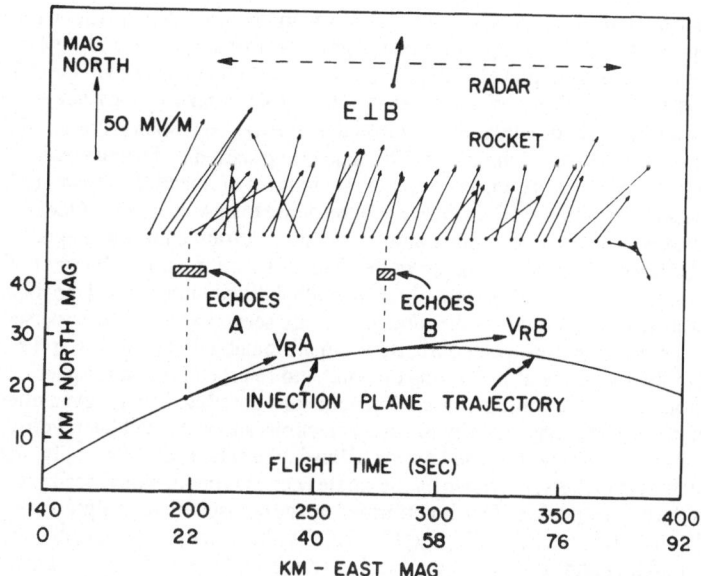

Figure 11. The ECHO 4 trajectory projected into the injection plane. The rocket was launched generally eastward to follow the magnetic drifts. Groups of echoes were observed at two times during the flight, corresponding to two directions of the distant transverse electric field. Neither of these groups corresponded to "L-Apogee" at 300 secs time. The on-board field vectors are shown just above the trajectory, and the flight average field measured by the Chatanika Radar at top.

plane'', a plane perpendicular to B and containing the rocket as beam origin. In this view the trajectory is still a parabola but apogee is now in magnetic latitude rather than altitude. This has been termed ''L-Apogee'' and at that instant the rocket was moving parallel to the electron magnetic drift sells (See discussion in Winckler, 1980, 1981). Echoes ''B'' were close to L-Apogee but echoes ''A'' occurred when the rocket had a considerable northward motion magnetically. The explanation for the ''A'' group and in fact the details of all echoes for high latitude situations like ECHO 4 must include the bounce-integrated effect of distant electric field drifts, which may vary on a time scale shorter than the rocket flight duration. The local electric field vectors sampled every 15 secs are shown in Figure 11, upper section and will be discussed in detail below. The deviation of the electric field eastward at 200 secs is very clear, and corresponds to the appearance of echoes ''A''.

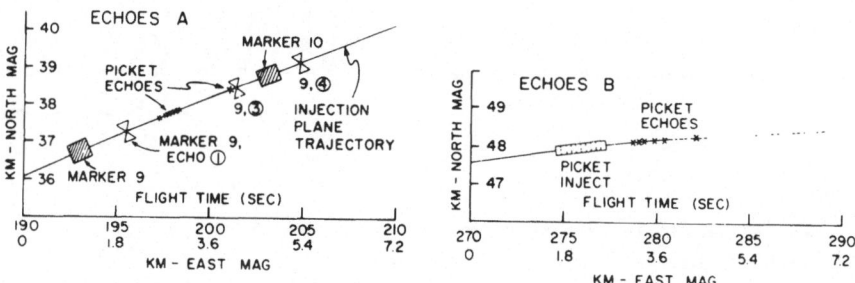

Figure 12. Two short segments of the tragectory in the injection plane showing echo details. Echoes ''A'', left panel, include 3 echoes from the single long marker 9, injected at 40 kev. The echoes, labeled 1, 3, and 4 with 2 missing apparently came from the downward part of the marker which was then backscattered on the atmosphere. Other echoes in this rather long period originated with pickets and spikes. Echoes ''B'', right panel originated with picket and spike injections. (See Figure 13)

Figure 12 shows two segments of the trajectory in the injection plane with echo details schematically represented. In the left panel marker 9 gave rise to echoes 1, 3, and 4, (2 missing), and in addition there were echoes from picket and spike pulses (injection not shown) in the "A" group. The right panel shows picket echoes and injection in "B" group (spike echoes not shown).

Figure 13, as an example of actual data, shows the superposed responses of all of the 90 degree pitch angle detectors for the "B" echo group. The echo group started in the quiet period before a set of picket and spike injections and originated with the previous set of injections shown in the line below. After 279.2 secs identification is difficult because of response to new injections. One sequence labeled P1, P2,—P6 appeared suddenly with large amplitude and is attributed to the magnetically mirrored picket pulses. The large amplitude variation in the detectors for the same echo (more than order-of-magnitude, see log scale on left) reflects the sharp azimuthal distribution around the rocket. The entire echo pattern is transient, and not all of the nine pickets are seen as echoes. Interspersed between the pickets are small broad responses labeled S1, S2,—S6 attributable to the 38 kev spike pulses, which were injected into the loss cone at the conjugate point and returned after atmospheric scattering and energy degradation. There are some immediate conclusions possible: Firstly, even after a round-trip distance of 38 earth radii the magnetically mirrored electrons showed small lateral diffusion, as large intensity variations were seen within one Larmor radius in the echo beam. Secondly, the echo energy was about 15 kev for the pickets, as shown by the single echo maximum which occurred therefore near the maximum of the energy profile. This is confirmed by the octospherical analysers which gave about 18 kev energy for the intense echo P3 (Figure 13). Thirdly, the electrons responsible for spike echoes travelled south at 38 kev and returned at a much lower energy (6 kev) but still satisfied the net drift requirements. No multiple bounces were observed for the B echoes, indicating that the echo region passed over the rocket rather briefly. The most consistant bounce time for the spike echoes was 3.04 seconds. The six observed picket echoes were delayed a minimum of 3.07 seconds.

A remarkable feature of the echoes A shown schematically in the left side of Figure 12 was a multiple-bounce event in which four at least, and possibly five echoes were observed. The event is shown quantitatively in Figure 14, which gives the time record of intensity during four bounces. The injection consisted of an 80 ma 38 kev marker pulse (no. 9) directed downward at a loss cone pitch angle of 70 degrees for 0.5 seconds, followed immediately by an upward injection at 110 degrees also for 0.5 seconds. The initial high value in Figure 14 is calculated by assuming that the beam current was injected over the area of a gyro circle (15 m radius), which gives the value $5 \times 10^{10} m^{-2} sec^{-1}$ at 193 seconds flight time. The echoes are due to the backscattered spectrum from the loss cone, which

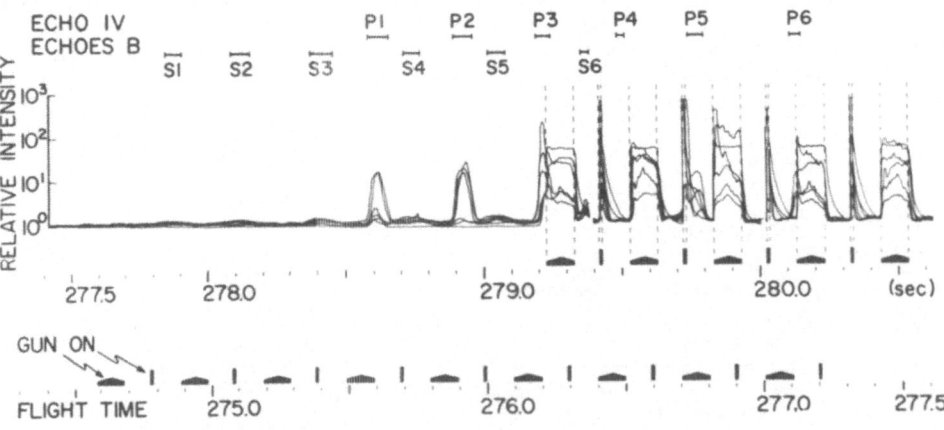

Figure 13.   The superposed responses of all 12 of the Echo 4 scintillators during the echoes "B" series. The responses beginning at 278.6 secs labeled P are attributed to pickets, which were injected at 110 degree pitch intended to mirror at the conjugate hemisphere at 300 km. altitude. The broad, low responses between labeled S are due to spike pulses injected at 150 degrees and back-scattered from the conjugate loss cone region. After 279.2 secs the echoes are obscured by the next injection series.

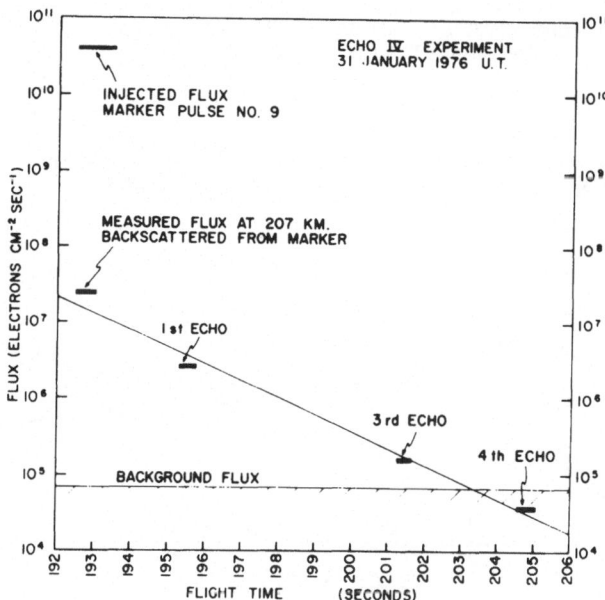

Figure 14.   Time-Intensity history of the multiple-bounce event echoes a.

should have been about 10 percent of the downward flux. The actual measured upward moving backscattered flux at the rocket which constitutes the "zero echo" in Figure 14 and which lies nicely on the echo decay curve was only 0.001 of the downward injection. It is likely that the rocket was not in the center of the "zero echo" pattern, considering its 20 meter motion during the backscatter process, and in view of the sharp lateral distributions obtained from the ECHO 1 experiment for a similar backscatter process. ECHO 2 has confirmed this picture (Winckler et. al., 1975), but the factor of 0.001 seems difficult to explain by this means.

The echo flux decays with a decrement of factor 10 per bounce, and the mean bounce time is 2.88 seconds. The long time of observation must be due to a chance variation in the electric convection field which in effect kept the echo locus on the rocket trajectory. The backscatter also facilitated echo detection by injecting a broad spectrum of electrons with the maximum flux at about half the beam energy, as shown by analysis of the ECHO 2 (Winckler et. al., 1975). The echo energy is not known directly for echoes A, but 15 kev may be inferred using the bounce time and Figure 15.

b.2. Echo Analysis

In the present analysis the magnetic effects will be derived from models, and the transverse electric field effects will be solved for as unknowns. In experiments where more details of echo patterns and energies are measured, both magnetic and electric effects may be extracted from the data without such limiting assumptions. Figure 15 displays in convenient form the bounce times and gradient and curvature drifts based on a recent model by Olson and Pfitzer (1977). (See also description in Walker, 1979). The computations were carried out by Malcolm (U. of Minnesota, private communication) for the $L = 6$ region for the ECHO 3, 4 and 5 experiments.

In Figure 16 are shown echo loci corresponding to the Echo IV injection plane trajectory pictured in Figure 11 and Figure 12, right, for echoes "B". These loci were constructed as follows:

1) A coordinate system lying in the injection plane and fixed in the rocket rest frame was aligned magnetically north-south and east-west.

2) In this reference frame the earth system was moving with a velocity $-\vec{V}_R$. The bounce-averaged electric field drift of the beam in the earth reference frame projected along field lines into the injection plane was designated $\vec{V}_E$, and the beam magnetic gradient and curvature drifts combined also bounce-averaged in the earth reference frame were designated by $\vec{V}_{GC}$. The net transverse

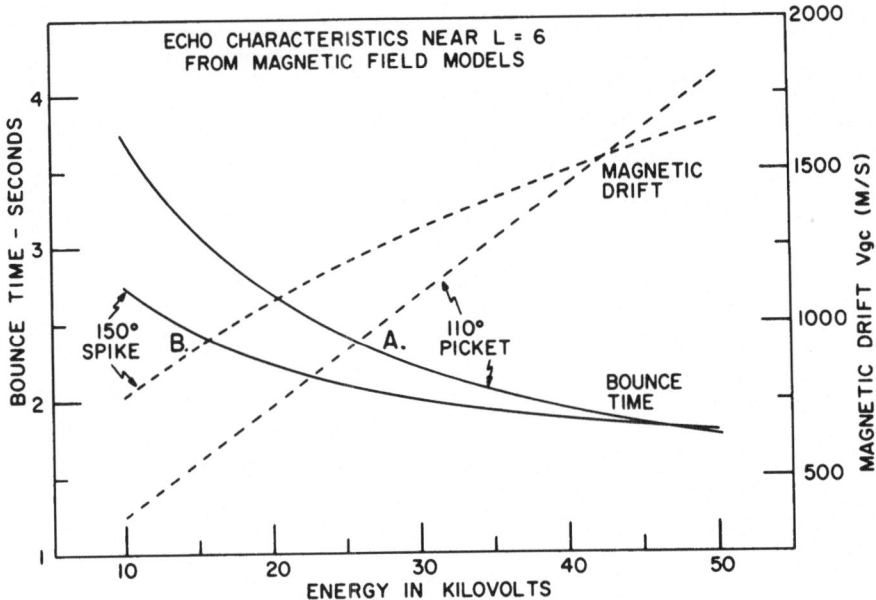

Figure 15.   Total Magnetic drift as function of bounce time for beams injected near L = 6. Curve A is for
injection at 110 degree pitch angle and 200 km. altitude, based on a model of Olson and Pfitzer
(1977) which theoretically should produce magnetic mirroring at the conjugate point and return at
70 degree pitch angle, with beam energy as parameter along the curve, characteristic of picket
pulses, for example. Curve B is for spike type pulses injected at 150 degrees and 40 kev energy, and
which return with the energy indicated after atmospheric backscattering from the loss cone at the
conjugate region. Curvature drift dominates for low altitude injections, and the bounce time is
insensitive to the injection pitch angle.

drift velocity with respect to the rocket of the echo beam was obtained as the vector sum of these
three components.

3) The vector position of points along the echo locus in the injection plane for any given time and
   energy is obtained by multiplying the net transverse vector drift by the bounce time of the echo.

Physically, the echo locus represents the position where echo electron guiding centers arrive back in
the injection plane from a continuously injected DC beam containing a complete spread of electron
energies. At very low energies where $\tau_B \to \infty$ and $\vec{V}_{GC} \to 0$ the locus goes off towards infinite distance
asymptotic to the direction $\vec{V}_E + (-\vec{V}_R)$. At very high energies the locus converges with the $\vec{V}_{GC}$
direction. Interception of echoes by the rocket occurs when the locus passes over the rocket and
corresponds to the equation

$$\vec{V}_E + (-\vec{V}_R) + \vec{V}_{GC} = 0.$$

In the example given in Figure 16 the vector diagram at top is drawn to correspond to locus 2 which
is shown intersecting the rocket and corresponds to the detection of echoes B at an energy on the locus
of 15 keV. Locus 1 occurred earlier when the rocket had a magnetic north component of velocity and
locus 3 occurred after L-apogee when the rocket had turned southward. The solution for $\vec{V}_E$ was
derived as follows:

1) The components of $-\vec{V}_R$ were determined at the time of echoes and are independent of energy.
2) The energy on the locus of echoing electrons intercepting the rocket was determined to be about 15
   kev as discussed above.
3) Knowing the energy, $\vec{V}_{GC}$ was determined, also from the field model.
4) $\vec{V}_E$ was determined from equation (1).

The detailed values of the resultant "remote" electric fields for this and the other solutions that follow are discussed in section 3.a.

In this first example of echoes "B" on the Echo IV flight the motion of the echo locus across the rocket was caused by the changing $-V_R$ with V_E assumed fixed. This seems justified by examining the E-vectors plotted in Figure 11 from the rocket measurements (field booms), which remain relatively constant around 273 seconds when the echoes B were observed. In contrast the earlier echoes A occurred at a time of changing local E-fields, and in fact the sharp eastward turning of the rocket field in Figure 11 at 200 seconds seems responsible for bringing the locus over the rocket provided the remote field behaves similarly. As shown in Figure 17, and using this assumption, the loci moved from south of the rocket (no. 1) up to the rocket (no. 2), and then south again (no. 3). The larger south component of $-V_R$ during this early part of the flight is balanced by an enhanced

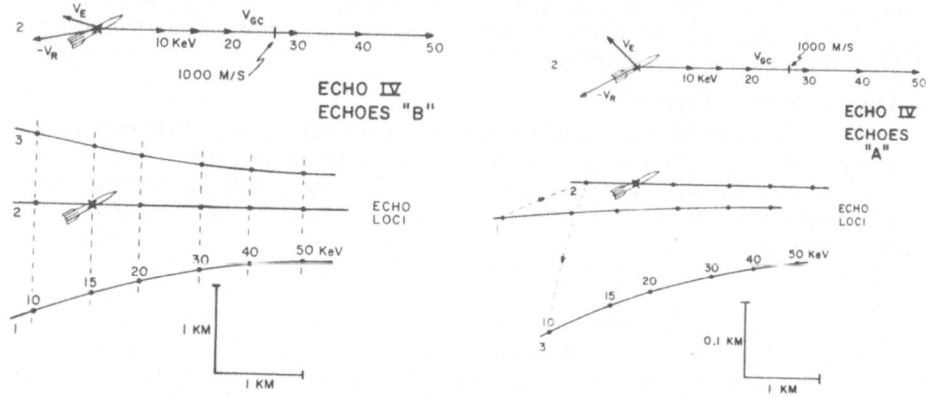

Figure 16.   Echo loci before (1), during (2), and after (3) echo detection. The sweeping of the locus over the rocket is caused by rocket motion as the electric fields were held constant in this example. The vector solution is shown in the upper panel. Fixed energy points on the locus move essentially northward at this time.

Figure 17.   The motion of the echo locus in this case for echoes "A" is due to the changing electric field which moved the locus to the rocket and then away again. A point of fixed energy on the locus may have east-west motion components. Echo detection corresponds again to the locus intercepting the rocket (locus 2). The vector diagram is in the upper panel.

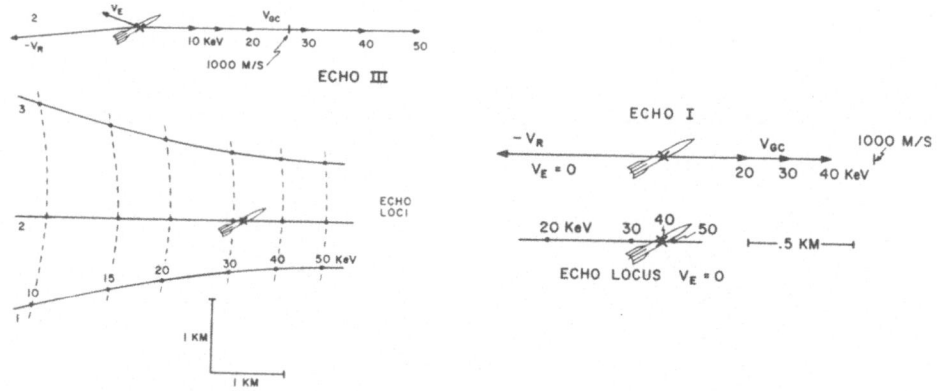

Figure 18.   Loci for the Echo 3 experiment. The locus motion is due to the motion of the rocket. Geophysical conditions were very quiet and stable for this flight.

Figure 19.   Loci for Echo 1, where electric fields were negligible.

northward $\tilde{V}_E$. An echo energy of about 15 kev is obtained from the 2.88 sec. bounce time and Figure 15.

The ECHO 3 experiment has been discussed in considerable detail (Hendrickson et. al., 1976), but not in the framework of echo loci as shown in Figure 18. The ECHO 3 rocket had a very large eastward velocity and well-defined beam energy. The echoes occurred at 31.5 keV on locus 2, just at the lower end of the injected beam energy interval. The resulting $\tilde{V}_{GC}$ was 1180 m sec^{-1}, which was mostly balanced by the $-\tilde{V}_R$ west component. The Echo III experiment was conducted during quiet times and has the least uncertainty of all of the flights in comparing remote and local fields (see Section 3.a.) The loci numbers 1 and 3 represent the effects of rocket motion north and south before and after L-apogee assuming E = Constant.

For completeness, Figure 19 shows the vector echo solution and the echo locus during echo detection for the Echo I experiment, where $\tilde{V}_E = 0$ at L = 2.6. The echo energy was 40 keV, at the upper end of the energy range of the accelerator. Echoes were detected accurately at L-apogee (see Winckler, 1980 and references therein).

b.3. The Structure of Echoes

The simple echo locus just described was approached in practice only by the ECHO 1 experiment, and without appreciable electric field effects. The ECHO 1 locus was extremely stable, had a well-defined width with e-folding intensity fall-off of about one Larmor radius (15 m.) determined entirely by atmospheric diffusion at the conjugate region, and was swept across the rocket by orbital motion at L = 2.6. In contrast, at L = 6 in the auroral zone the echoes are clearly caused to sweep rapidly and irregularly as much by time-varying fields as rocket motion. For example, some echoes in a regular series would be missing or appear at other than "L-Apogee" as just described. These effects must be produced at great distance, yet they are reminiscent of the motion of auroral rays both in time and space. Like auroral rays, the echoes maintain their sharp boundaries at high latitude as in ECHO 1, as shown by the large intensity variations in one Larmor radis. Another particularly good example is the third echo in the echoes A group (refer to Figure 12, left) with a time profile as shown in Figure 20, upper. The echo electron intensity circulating around guiding centers located one Larmor radius from the rocket at the magnetic azimuth shown is plotted on a linear radial scale for selected sampling times during the echo as shown. The echo swings over the rocket from the south and recedes to the north, moving approximately perpendicular to the rocket velocity vector. The sweeping action is presumed

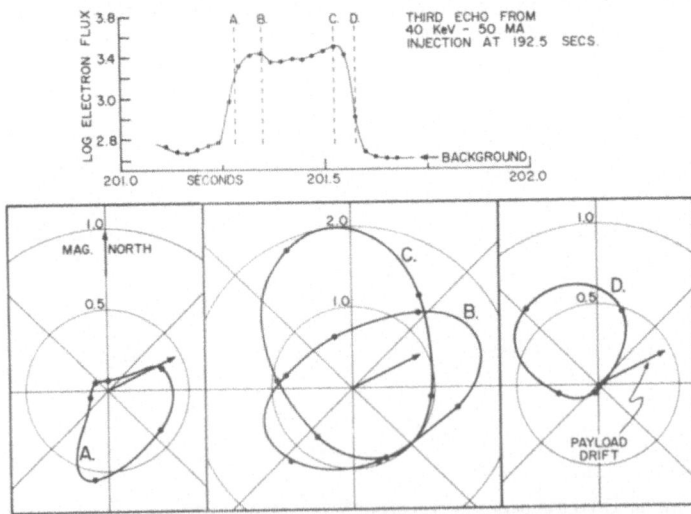

Figure 20.   The time history and guiding center density around the rocket at selected times during the event. The radial scale is labeled in units of 10^6 cm^{-2}, sec^{-1} of echo electron flux.

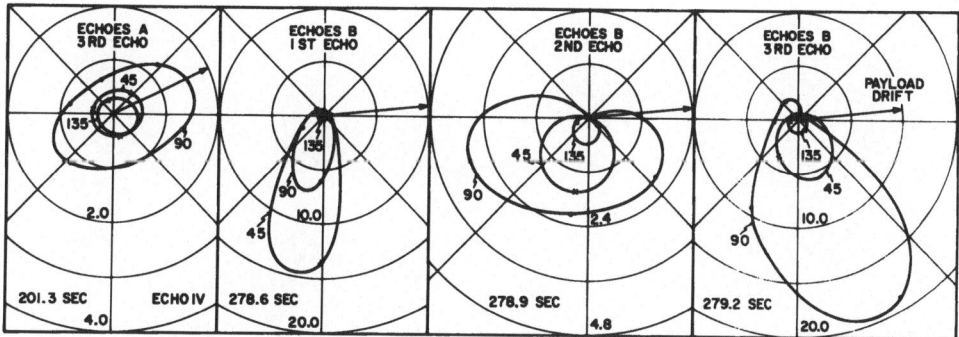

Figure 21.   Echo electron flux in the three pitch angle ranges of Figure 9 as a function of the guiding center azimuth around the rocket. Units same as Figure 20.

due to the electric field drift variation discussed earlier and summarized in the motion of the echo locus in Figure 16. The lateral sharpness of the echo is shown by the order-of-magnitude changes of flux in one Larmor diameter (17 m.).

Some echo pitch angle distributions measured by the three sets of scintillators described in Figure 9 are shown in Figure 21 in the same format as Figure 20. The third A echo, measured at time B in Figure 20, shows predominately a mirror distribution with the maximum in the 90 degree range. In contrast, the first B echo has a loss-cone distribution peaked at 45 degrees, with less intensity at 90 degrees and essentially none moving upward at 135 degrees. This situation shifts to a mirror distribution in the second and third echoes. The echo guiding centers are predominately on the south side of the rocket. The pitch angle distribution seems to be a function of the lateral position in the echo, indicating a very complex structure.

Since bounce and drift times depend on energy, a short injected pulse containing a wide energy spread such as often used in ECHO flights produces an echo whose high energy parts arrive first and at greater distance and whose low energy parts arrive near and later. The echo descends like a curved guillotine knife along the locus. The picture of a real-life high latitude echo locus that emerges from the above measurements is a writhing snake-like structure of variable width - possibly resembling a section through an auroral drapery.

b.4. The Atmosphere as a Detector

Beginning with the first Hess experiment in 1969 (Hess et. al., 1971) low light level TV systems have been successful in detecting the sub-visual artificial auroral streaks from rocket-borne accelerators (see summary in Winckler, 1980). The Hess Kaui experiment (Davis et. al., 1980) was remarkable in that an auroral streak was observed at the conjugate point by a high-flying aircraft using TV, after a distance of several thousand kilometers from the injection point. In the ZARNITZA experiments (Cambou et. al., 1975), and the O'Neil experiments (O'Neil et. al., 1979) the luminosity occurred in the region near the rocket, showing a field-aligned streak attached to the rocket as well as a rocket discharge glow and various chemi-luminescence effects. In the Hess and ECHO experiments the auroral streaks were produced in the 90-100 km region by a rocket at much higher altitude. A montaged presentation of the auroral streaks from the ECHO 5 flight is given in Figure 22. The photo vignettes were photographed from the acreen of a TV monitor with a camera hand synchronized to the injected electron beam pulses. Photographic integration on the screen of a video flight playback is very effective and gives much more flexibility than integration in the image tube before recording. Experience shows that the best sensitivity for electron streaks is obtained by including the entire white-light spectrum, as much continuum radiation is produced by this relatively high energy excitation. All the streaks shown in Figure 22 were produced by direct downward injections from the rocket, and were viewed from the ground at Ft. Yukon as shown schematically. The images were obtained by the Hallinan group of the University of Alaska using the original Davis image tubes (see Davis et. al., 1980).

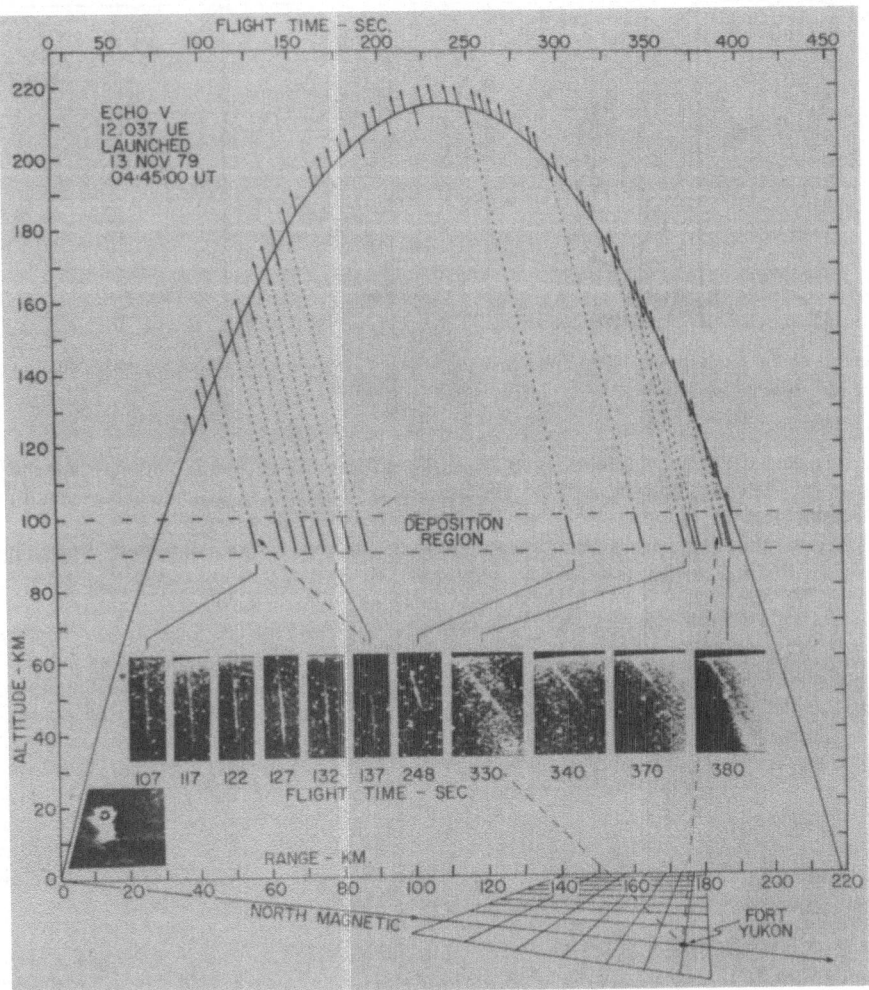

Figure 22.   Photo montage of the artificial auroral streaks produced by ECHO 5, viewed by TV at Fort Yukon, Alaska. The rocket was launched from Poker Flat. Upward injections are also shown. Note the bright 30 msec duration streaks at 330 secs and after, visible against an increasing auroral background.

The observation of the artificial auroral streaks from the first Hess experiment at the expected atmospheric depth region for the injected beam energy and with the expected brightness showed immediately that such beams were not catastrophically thermalized by instabilities, nor were they trapped by severe vehicle charging. The second Hess experiment further showed that most of the beam energy could propagate for large distances at least in the low-latitude inner magnetic field regions. The ECHO 4 downward marker pulses were recorded by the Davis-Hallinan TV system, and a careful analysis and comparison with the Hess flights showed that with excellent sky conditions, so that the cameras could detect 10th magnitude stars, the threshhold for detectability was 1kw. beam power level. The 40 kev energy beams from ECHO 4 produced streaks extending between 90 and 100 km., about as expected and somewhat deeper than the 20 kev beams in the Hess experiments. The purpose of the ECHO streaks was to serve as markers to facilitate the location of conjugate echoes by optical means. This technique has not so far been successful.

Although the ECHO 4 beams travelled from the rocket at 200 km. altitude to the 100 km. atmosphere without gross changes, the injections at the highest altitudes produced a more diffuse streak laterally, extending to a diameter of 150m. The sharp streaks had 50m widths, close to the resolution of the TV, which is about twice the Larmor diameter of the beam at injection. The ECHO 5 marker streaks shown in Figure 22 were about 10 times the power of ECHO 4. The ECHO 5 power was increased to the maximum possible to test the feasibility of optical echo detection using both on-board and ground-based detectors. The echo streaks were expected to be parallel to, and displaced from the markers by drifting, but none have been unambiguously identified. The ECHO 5 on-board equipment included banks of photometers and large-area x-ray counters collimated to view the atmosphere directly down the field line from the rocket. These detectors gave a large response to photons generated near the rocket (see discussion by Swanson, Sec. 2.a. above) but were not able to record the atmospheric fluorescence lying 100 km. below, even though bright images were obtained by the TV system.

The absence of identified optical images of echoes in the ECHO 5 experiment means that the echo brightness was 30 times less than the marker pulses, since the ECHO 4 experience showed that 1kw was clearly visible. Since the beams were sharply defined and injected upward so as to return in the loss cone nearby if invariant motion was followed, one must assume that a strong interaction with the distant magnetosphere dispersed the beam. This matter is discussed in detail below (Sec. 3.b.)

In Figure 22 the streaks after 330 secs. were due to very short injections of only 30 msec., which is only the length of two TV frames. Nevertheless they were brilliant and easy to detect even in the presence of strong aurora. The viewing direction nearly parallel to the magnetic field greatly increased the optical depth of the streak in the later parts of the flight, but there appears to be some other type of enhancement which makes the streak flash out with such brilliance, even against the background of a bright auroral ray.

The ECHO 5 flight was a major test of the use of optical and x-ray techniques for echo beam detection and analysis, and the results, for whatever reasons, even including large intensity losses due to the effects under investigation in the distant magnetosphere, show this not to be a viable technique. One recalls that the observation of conjugate echoes in the ARAKS program was thwarted by bad weather and background light (Cambou et. al., 1980). Thus the Hess Kaui result constitutes the only observation to date of an electron beam optically at a large distance from the injection point.

Artificial auroral streaks in the lower ionosphere represent ionization enhancements which can be detected by sensitive auroral radars. This is discussed in the next section.

b.5. Radio and Radar Methods

Beam injection in the ionosphere gives rise to a rich variety of plasma and electromagnetic mode wave emissions observed by many groups (See for example the review by Winckler, (1980), Sec. 3, and the special issue of Annales de Geophysique, Vol. 36, No. 3 (1980) on the ARAKS experiment). Plasma frequency emission up to 60MHz was observed both in ARAKS Dechambre et. al., 1980 (Mishin and Ruzhin, 1980), and in ZARNITZA (Mishin and Ruzhin, 1978) by ground-based antennas. This emission corresponded to the much enhanced plasma density created by beam ionization or by the rocket return current discharge. HF electromagnetic waves have been observed by the Kellogg group with ground receivers from the ECHO 4 and 5 experiments. An example is shown in Figure 23 from ECHO 4 recorded by a narrow-band receiver at twice gyro frequency (2.96 MHz.) during a marker pulse injection. This emission occurred during a strong cavity-mode oscillation involving the ion gyro frequency around the rocket shown in Figure 23 by the periodic oscillations in light and return current (Israelson and Winckler, 1979). Both the radio emission and the oscillations occurred during the 0.5 sec. down injection at 70 degrees pitch angle, and disappeared at 110 degrees. This event occurred during a nitrogen gas injection. A similar event occurred on the ECHO 5 flight. In ECHO 4, every beam injection was accompanied by HF emission at ground level (Monson and Kellogg, 1978). Despite the ten-fold power increase in ECHO 5 over ECHO 4, only one weak emission was observed by the Kellogg group (Private communication). The presence of strong aurora during the ECHO 5 flight evidently introduced strong ionospheric absorption, although the details are complex. It appears that all emissions observed so far are produced by the large plasma disturbances near the rocket. The Kellogg group has not reported any certain evidence for echo detection by wave emission, even with

ECHO IV   OSCILLATION EVENT

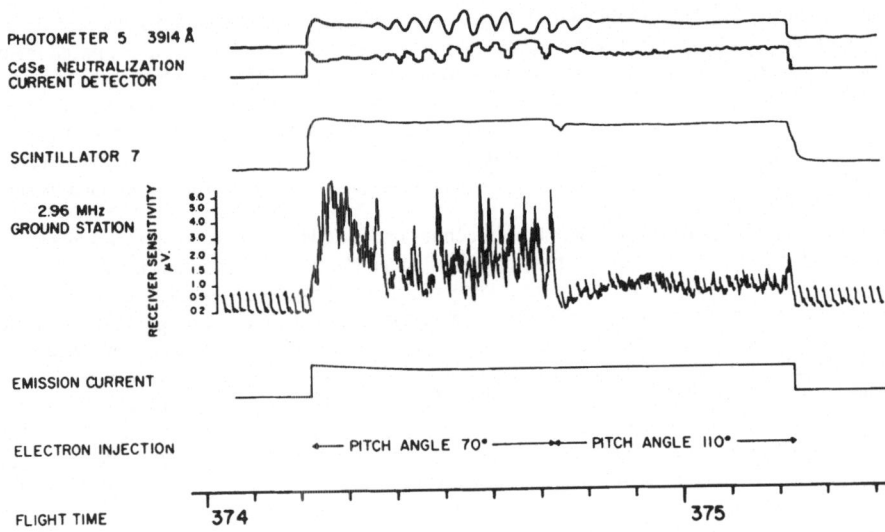

Figure 23.   Center panel - ground level radio signal from an ECHO 4 40 kev injection (courtesy P. J. Kellogg) and other related data. The top lines show that the discharge around the rocket was oscillatory in light emission and return current (note opposite phase). The high energy electrons (scintillator) were not affected.

plasma wave detectors flying in the ionosphere near the region where echoes were detected by particle counters.

Auroral radars in Siberia were successful in detecting ionization streaks from the ARAKS experiment both in the pulsed mode (Pyatsi and Zarnitsky, 1980) and in the doppler bi-static mode (Uspensky et. al., 1980). The frequencies used were 23 and 44 MHz, and the installations were located 433 km slant range to the south of the conjugate region to insure normal reflection of the echoes from the field-aligned ionization streaks. The artificial streaks would be expected to resemble meteor trails in some respects, or faint aurora. Figure 24, left panel (from Pyatsi and Zarnitsky, 1980) shows the southern progression of the echoes as the first ARAKS rocket moved northward at the conjugate region at Kerguelen Is. They correspond sequentially to the 15 kev. 3.84 sec. long injections, but with a delay of between 2.75 and 3.56 seconds from the start of the upward injection at Kerguelen. The expected delay for 15 kev. was 0.8 secs. One notes that the ionization trails were seen to perceptively move south during the injection period, but also long-lasting echoes of 10 seconds remained fixed in the ionosphere after injection ceased. The right panel of Figure 24 refers to a power spectrum analysis of the echoes which shows peaks at the bounce frequency, indicating beam trapping. This remarkable observation will be discussed in Sec. 3.b., below. Sensitive auroral type coherent scatter radars would thus seem to provide the "missing link" in ground-based systems, and to be promising for echo detection.

2.c. Perturbation of the Distant Magnetosphere by Beam Injection

The injection of powerful electron beams into the ionosphere produces strong local plasma heating, plasma waves and instabilities, and electric fields and space charge effects including vehicle charging. This subject has been much discussed (See for example Winckler, 1980, Sec. 2, and Arnoldy and Winckler,1981). The concern is that these ionospheric level effects may propagate out the magnetic field lines in the precise region under study and alter the natural magnetosphere. Previous

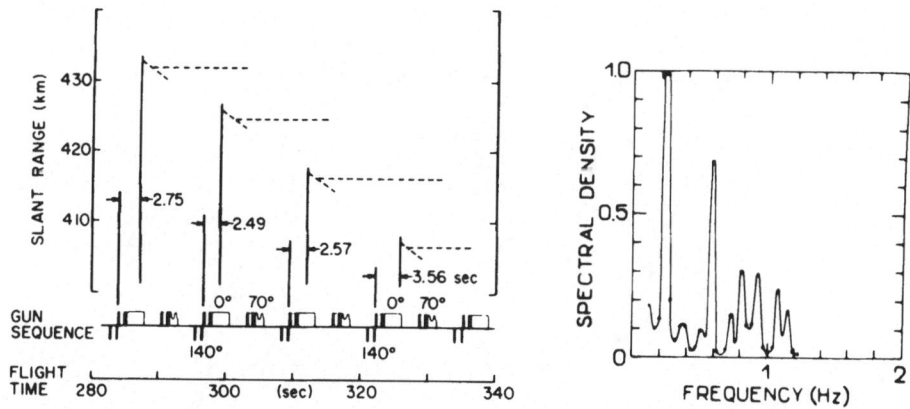

Figure 24. Left panel, pulsed auroral coherent scatter reflection from the ARAKS conjugate ionization trails. Right panel, power spectrum analysis of a long pulse showing evidence for the bounce frequency for 15 kev electrons between Kerguelen and Siberia. (Courtesy Pyatsi and Zarnitsky).

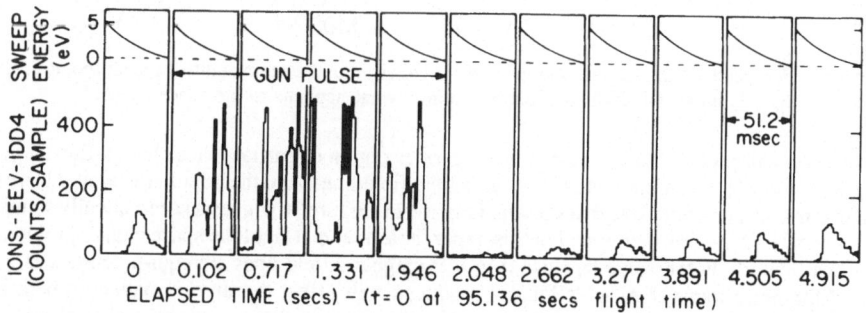

Figure 25. The attenuation and recovery of atmospheric ions following beam injection. ECHO 5 (courtesy R. L. Arnoldy).

estimates of the size of the disturbed region laterally were in the range 50-100 meters (Cartwright and Kellogg, 1978, Winckler, 1980). That the region can be much larger is indicated in Figure 25, derived from the thermal ion measurements of Arnoldy on the ECHO 5 flight (Arnoldy, private communication, see also Arnoldy and Winckler, 1981). The electrostatic analyzer swept from zero to five volts in 51.2 msec, as indicated at the top of the Figure. Eleven selected spectra extend from just before a "fast mode" 2 sec. pulse through the injection period and out to 3 sec. after injection. The spectrum before the injection at t = 0 is normal for the auroral ionosphere at 150 km. altitude. Enhanced, sporadic ion fluxes are seen during injection. Since these ions could not have reached the payload if the potential were positive the counts must have been accumulated during the very short (50 micro-sec.) off periods between each 1 msec pulse as shown in the pattern of Figure 5. As shown in Arnoldy and Winckler, (1981), the payload potential would be expected to snap negative with great rapidity to reach the hot plasma floating potential just after injection. The most important feature of Figure 25 is the depletion of ions after the injection, and the slow recovery requiring three seconds to full intensity. In this period the rocket motion would carry it laterally 1.2 km., which sets a new much larger size to the disturbed region. The ion depletion may be reflecting the large positive space charge formed around the rocket during injection. The pulse shown sent 0.3 coulombs of negative charge off to infinity, a relatively huge amount in terms of electric effects. This charge is equivalent to the normal electron content of a sphere of several hundred meters radius in the ionosphere, and the violent departure from neutrality of

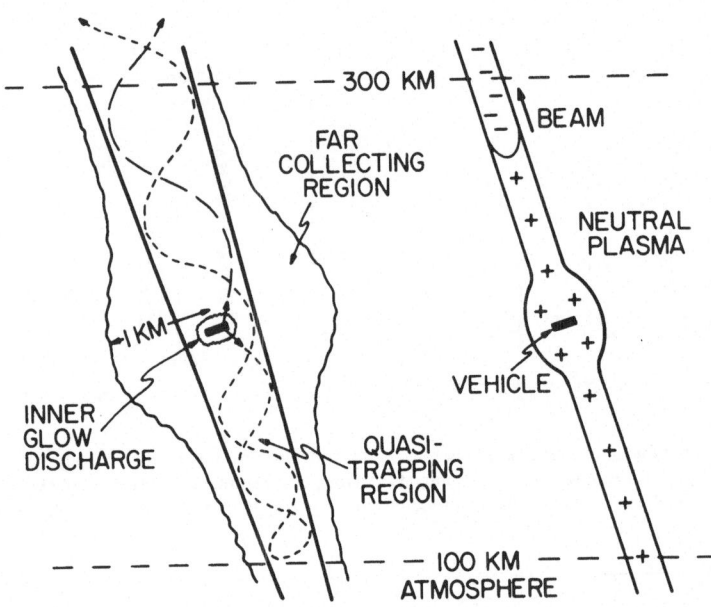

Figure 26.   Schematic of the various regions of disturbance around a beam emitting rocket (not to scale). The right hand sketch addresses the question, "what happens to the space charge?".

such a region will produce in actuality a much greater region of partial space charge. Such a positive region would repell the ambiant ions away along the field lines, and the population would be restored by diffusion with a recovery like that shown in Figure 25. Figure 26 suggests schematically some of the types of disturbed regions mentioned in this paper, at left, and at right the space charge problem. The latter is particularly bothersome because if in fact magnetic field lines are equi-potentials as is often assumed the large transverse fields locally produced will affect the distant magnetosphere after a suitable propagation time, determined perhaps by the electron thermal speed. Large electric fields have been measured near the Polar 5 vehicle during beam injection, reflecting this space charge. The study of such vector fields by free-flyers is extremely important.

Another possible source of perturbation is the beam probe itself which may excite instabilities at large distance, or produce effects from the coulombic charge along the trajectory. A beam-beam interaction was possibly observed in the ECHO 1 experiment by McEntire (1972) in which, during continuous, rapid beam injections an additional small transverse displacement was noted in the echo pattern. This effect could alternatively be interpreted as due to payload potential effects propagating to great distance.

2.d. Changes in the Injected Beam in the Near-Rocket Region

The possibility has long been considered that a high-powered beam like that shown in Figures 4 and 5 for the ECHO 5 mission, although well defined in energy and divergence when leaving the accelerator might become unstable after a few tens of meters and partially thermalize, producing a beam of undesirable characteristics. Attention has recently focussed on the "Beam Plasma Discharge" discovered by Getty and Smullen (1963) and recently observed by Bernstein et. al., (1979) in large vacuum facilities. Television images obtained by Hallinan of the Bernstein experiments in the Johnson Space Center facility are shown in Figure 27. The left panel shows a well-focussed beam of 50 ma and 4 kev spiralling away from the camera and observable by faint collisional luminosity. In the right panel the BPD has been ignited by reducing the energy to 2 kev with the same current. The Larmor spiral, now smaller, is filled with luminosity giving a total photometer signal 50 times the previous value. The particle measurements of Jost et. al., (1980) show that the BPD spreads the beam energy into a tail about 20% above the original value, and a major continuum extending to

TABLE II

Experiment	Perveance - Amps/volts$^{3\,2}$	BPD ?
ECHO 1 - 4	1×10^{-8}	No
ECHO 5	$4\text{-}10 \times 10^{-8}$	No
Hess 1, 2	18×10^{-8}	Not known
Polar 5	13×10^{-8}	Possible
E∥B	$2\text{-}90 \times 10^{-8}$	Possible
Chamber Exps.	same	Yes
Zarnitza	50×10^{-8}	Yes
ARAKS	$11\text{-}27 \times 10^{-8}$	Possible

In the recent ECHO 5 experiment a search for the BPD was made by photometers which looked along the Larmor spiral of the beam, as shown in Figure 6, with closely the same geometry as the chamber views in Figure 27. Figure 28, prepared by Swanson (U. of Minnesota, private communication), shows a linear relation between beam current controlled by the number of guns firing, at a low energies. It is clear that the BPD ignites when the acceleration of plasma or secondary electrons by the fluctuating electric fields of the instability is enough to reach the ionization threshhold for the residual gas (10 volts for nitrogen).

The critical current for BPD ignition scales according to the equation

$$I_c = C \frac{V^{3\,2}}{B^{0\,7}PL}$$

where V is the accelerating voltage, B the magnetic field, P the pressure and L the scale length of the experiment. The $V^{3/2}$ relation is (by chance) the same as for a space charge limited diode such as the ECHO gun shown in Figure 3. The perveance of such a diode (also used for triodes) is defined as $A = I/V^{3/2}$. In the Bernstein experiments a gun perveance of 50×10^{-8} amperes/volt$^{3/2}$ was sufficient to ignite the discharge with magnetic fields and pressures very similar to rocket altitudes of 150 km. The scale length L however is not known in the space environment. But if the required perveance to produce the BPD in space were known, then L could be determined from the vacuum chamber values by the relation $L_s = L_t A_t/A_s$ where the subscripts t and s refer to lab and space. Table II lists estimates from the various beam experiments which have been described in the literature for the perveance and the presence or absence of the BPD. The threshhold perveance in space seems to be about 12×10^{-8} amps/volts$^{3/2}$. This would therefore set the space L of order $10 \times$ that in the lab.

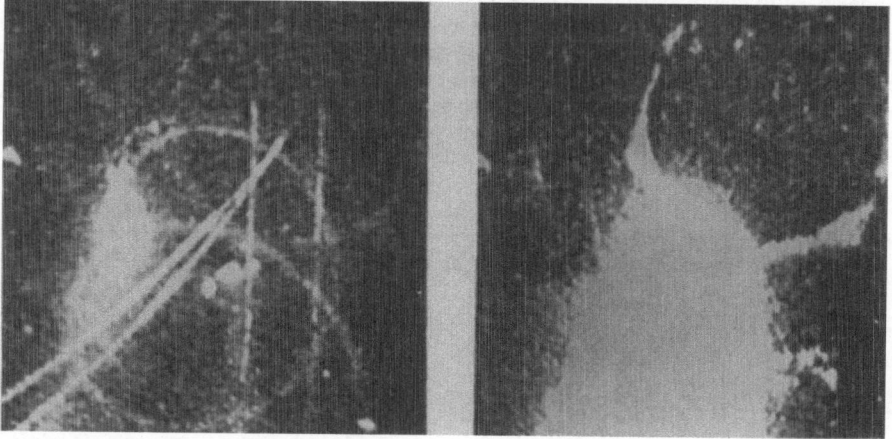

Figure 27.   Left panel, 4 kev 50 ma beam, no BPD. Right panel, same, except 2 kev, showing BPD ignition. The geometry of this figure is duplicated by the ECHO 5 in flight (see Figure 6).

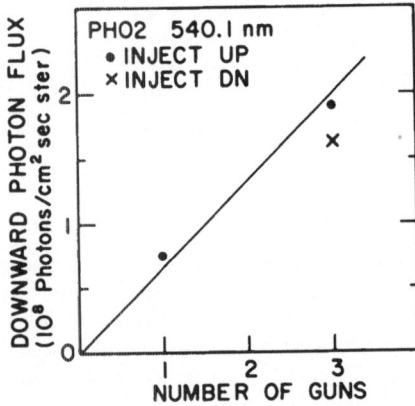

ECHO V 38 KeV INJECTIONS
260 mA/GUN, 160 KM. ALT.

Figure 28.   Response of a wide-angle upward looking photometer during beam injection upward with several values of beam current. Even with a downward injection as shown almost an equal response is obtained on the upward looking photometer.

current of 250 ma per gun, and the photometric output where the dots refer to an upward looking photometer and a beam injected upward as in Figure 6. One notes that a down injection produces almost as large a photon flux coming down from above the rocket as the up injections. One concludes that the response indicated is probably from a glow discharge near the rocket associated with the neutralizing return current and that very little effect is produced by the beam itself. Conclusions regarding the presence or absence of a beam plasma discharge are difficult to make. However, the results are probably not in accord with the presence of a large bright discharge filling the Larmor spiral of the beam as shown in Figure 27 for a laboratory experiment.

3. SOME OBSERVATIONS OF THE DISTANT MAGNETOSPHERE BY BEAMS

3.a. The Comparison of Distant and Local Electric Fields

The three examples of remote electric field drifts derived from high-latitude ECHO measurements are summarized in Figure 29 with numerical E-values listed in Table III. In each case an incoherent back-scatter radar value of ionosphere drift was available from the Chatanika radar (Leadabrand et. al., 1973) located within 50 km of the rocket and generally as an average during the five minutes of rocket flight above 150 km altitude (as shown for example in Figure 11, upper part). The radar and remote values for Echo III are in reasonable agreement considering the uncertainties in both measurements. The $\bar{V}_E$ local is derived from Morgan and Arnoldy (1978) from on-board ion drift

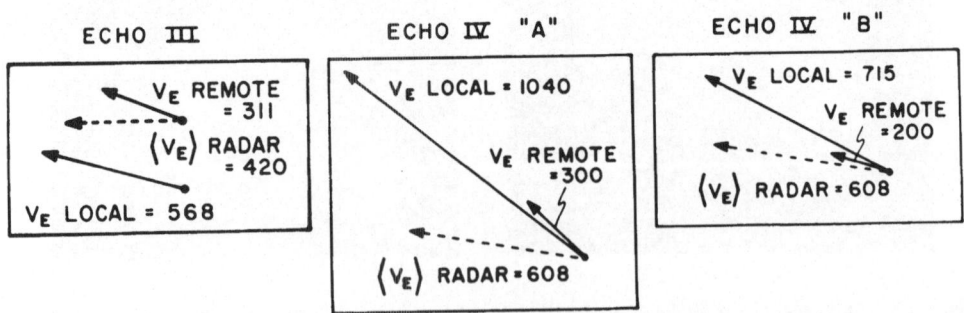

Figure 29.   Comparison of local and distant electric field vectors derived from ECHO experiments.

TABLE III

		Echo III	Echo IV "A"	Echo IV "B"
Injected Energy (KeV)		32-37	38	8-18
Pitch Angle (DEG)		73-117	110	110
Measured Bounce Time (Sec)		2.2-2.02	2.88	3.07-3.98
		(2.03)		(2.88)
Corresponding Energy (KeV)		26.32	15	15
(O-P Model)		(31.5)		
$\bar{V}_{GC}$ (m/s)		990-1200	570	570
(O-P Model)		(1180)		
$\bar{V}_R$ (m/s)	North	84	195	58
	East	881	365	361
$\bar{V}_E$ Remote	North	84	195	58
(m/s)	East	−299	−205	−209
$\bar{E}$ Remote	North	15.0	10.3	10.5
(mV/m)	East	4.2	9.8	2.9
$\bar{E}$ Radar	North	21	30	30
(mV/m)	East	0	5	5
$\bar{E}$ Local	North	25±4	41.2	31.6
(mV/m)	East	6±4	32.0	16.8

measurements (revised from Hendrickson et. al., 1976). The direction of this latter measurement is in good agreement but the magnitude substantially exceeds the other values. Nevertheless, it constitutes an on-board measurement and is therefore spatially close to the echo field line.

In the case of Echo IV, echoes B, the radar and local values are similar, but the remote value, while directionally similar, is definitely much smaller. The discrepancy seems outside the errors for echoes B. The Echo IV local measurements were made by electric field booms on the injecting rocket. The booms showed a large and long-lasting effect following beam injection which was not of instrumental origin, but seemed due to a plasma disturbance in space due to the injection process. It is felt that the boom E-field values are too high in magnitude due to this effect (Cahill, private communication), but directionally satisfactory. It seems no coincidence that the field deviation at 200 seconds coincided in the proper sense with the appearance of echoes A. The radar values for echoes A cannot reflect the rapid directional changes at that time. One tentatively concludes that for the ECHO 4 expeeriment, flown in the recovery phase of a small substorm, the bounce-averaged transverse electric drifts are reduced from their conjugate ionospheric values. The general agreement during the quiet time case of Echo III, on the contrary, leads to the conclusion that the remote and ionospheric fields were very similar and that magnetic field lines were possibly equi-potential loci. A theorem of Roederer (1970) is used in this connection.

The measurement of electric fields in the magnetosphere of all varieties is one of the best uses of electron beam probes. In the preceding paragraphs distant fields were related to local fields, and the equipotential nature of magnetic lines in space due to the high parallel conductivity was used as a guiding principle. In fact, the mapping of static electric fields into the ionosphere certainly does not follow for the case of fluctuating fields, where the magnetosphere may impose a frequency-dependent filter between the ionosphere and distant regions traversed by probes. Certainly a wide frequency range of fluctuations is observed in the motion of echoes. If the time scale of fluctuations is longer than the beam electron bounce time, and if the space scale is also large, then a simple interpretation may be made of echo motion. If the fluctuations have a smaller scale, then the echo displacements represent an average over the space and time variations. The problem is then one of diffusion, and the observations must be so interpreted. There is considerable evidence for the diffusion process, which will be discussed in the next section.

3.b. Electron Diffusion

The most familiar type of diffusion for magnetospheric particles is multiple coulomb scattering.

which formerly was much discussed in trapped radiation studies. A thorough study of this by Monte-Carlo techniques was made by McEntire (1972) for ECHO 1 beams which entered the loss cone at the conjugate region and which therefore gave echoes by virtue of both coulomb and magnetic scatter. The results agreed with known cross sections for multiple scattering and produced echoes which laterally were still rather narrow, with an e-folding distance of about the Larmor radius for 40 kev electrons. Coulomb scattering has been observed for all downward directed loss-cone pulses from rockets near the atmosphere, for example the ECHO 2 flight where the back-scattering for "quick echoes" closely resembled the scattering of conjugate echoes for ECHO 1.

The lateral diffusion of a beam even when backscattered from the dense atmosphere is thus seen to be moderate, and the lateral dimension will grow only about a Larmor radius in one loss cone interaction. This is not true for energy diffusion. The 40 kev beam of ECHO 2 backscattered with a broad energy spectrum peaked near 20 kev, and reaching to very low energies. The auroral streaks recorded in the ECHO 4 TV observations showed a different type of lateral diffusion. "Diffuse" streaks appeared from the injections near 200 km. of 150 m. diameter, while "sharp" streaks of 50 m size predominated from the 120-150 km. range (Hallinan et al., 1978). The authors suggested a collisionless interaction in the ionosphere as the underlying mechanism, but the details are lacking.

From the descriptions given earlier in this paper it will be clear that echoes are observed after traversing large distances (38 earth radii/bounce) to retain a surprisingly sharp lateral profile, of the order of a Larmor radius. Although the entire beam is seen to displace laterally in response to varying fields, this should more properly be called an interchange than diffusion.

There are several lines of evidence that pitch angle scattering and diffusion are strong processes in the distant regions which involve the loss of particles from echo beams. Any beam injections in the ionosphere even at 90 degree pitch angle correspond to a very small range of equatorial pitch angles in the loss cone near zero degrees. The remainder of phase space is empty of trajectories. Thus even very small angular scattering will spread the beam towards isotropicity and will tend to empty the loss cone. This is a violation of the second invariant of the motion, and the electron mirror heights will rapidly rise out of the ionosphere making the echoes inaccessable to detection by atmospheric streaks or detectors on the injecting vehicle. The missing particles become permanently trapped and drift in a shell around the earth. In the ECHO experiments there has always been an unaccountable loss of particles; for example in ECHO 1 Hendrickson (1972) after a careful spatial integration over the echo pattern could account for only 0.1 of the intensity predicted by the backscatter calculations at the conjugate regions. The absence of optical echoes visible by TV in the ECHO 5 experiment strongly indicates scattering out of the atmospheric loss cone. Similar evidence was obtained in ECHO 4 (Hallinan et. al., 1978). In the case of ARAKS the radar observations of Pyatsi and Zarnitsky shown in Figure 24 above indicate that the beam injected parallel to the B-vector at Kerguelen did not reach the atmosphere over Siberia in the expected 0.8 secs, but instead became trapped and after several bounces scattered enough electrons into the loss cone each bounce to produce a periodic enhancement of the radar echoes with the bounce frequency. It was known that magnetic conditions were disturbed and a dawn chorus was observed in VLF. This description for ARAKS resembles the multiple bounce echo observed with ECHO 4 shown in Figure 14, which also occurred during a disturbed period following a small substorm. The large loss of intensity with ECHO 5 may indicate even stronger pitch angle scattering as the experiment was conducted in the presence of strong aurora. The failure of optical observations in Siberia during ARAKS may be due as much to pitch angle scattering of the beam as to poor optical conditions. It is probably that the Hess Kaui experiment avoided strong scattering because of its very low latitude in the inner magnetic field.

There is a growing realization of the importance of electrostatic mode waves above the electron gyro frequency in producing scattering and precipitation of the natural magnetospheric particles (See summary in Ashour-Abdalla and Kennel, 1978). Electron diffusion by electrostatic waves has been treated theoretically by Lyons (1974) based on earlier work of Kennel and others. The electrostatic wave amplitude required for strong diffusion, i.e., when the pitch angle diffusion coefficient at the edge of the loss cone exceeds α/τ where α is the equatorial pitch angle at the edge of the loss cone and τ is the quarter bounce period of the particles, is shown in Figure 30, from Lyons (1974). The density and energy of the thermal background is critical for this process, but the required wave amplitudes are well within those measured. The injection-precipitation event shown in Figure 1 is an example which

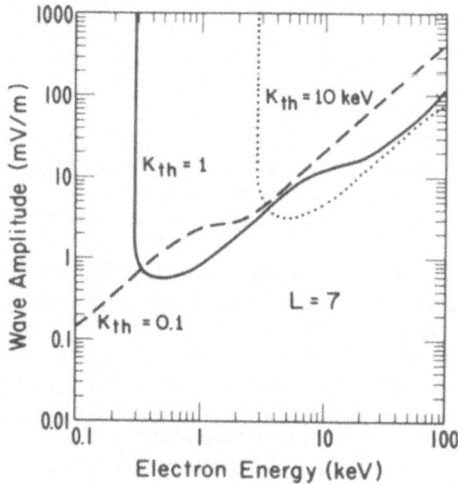

Figure 30.   The minimum equatorial wave amplitude required for strong pitch angle diffusion at L = 7 as a function of electron energy for thermal energies of 0.1, 1, and 10 kev. (Courtesy W. Lyons).

is likely very strong electrostatic wave effects in the inner plasma sheet region which produces electron acceleration and injection into the trapping region during a substorm.

3.c. The Distant Magnetic Field

As shown in the Hess and ARAKS experiments electron beams can make good tests of the conjugate features of magnetic field models, and generally good agreement is achieved if quiet conditions prevail. More dynamical models which include particle loading as well as distant currents such as the Olson-Pfitzer model mentioned earlier are crucial in describing the distant field, and eventually may be used during variations like substorms. In the ECHO 3 experiment where good bounce times were measured for known energy particles it was possible to compare theoretical and experimental field line lengths directly (Hendrickson et al., 1976). The results show significant differences with the models, giving the hope that improvement in echo type measurements can be a guide to better dynamical models. The magnetic drifts are also derivable from models, but are not so far subject to direct check as electric and magnetic drifts are not uniquely seperated in the experiments. Rather, the magnetic drifts derived from models are used to compute the electric drifts as described earlier. It is important to study simultaneously the echo displacements and bounce times for a number of energy beam particles to make significant advances.

4. FUTURE POSSIBILITIES

The Minnesota group is presently engaged in the design of a new remote probe designated ECHO 6, which is an attempt to improve the detection and analysis of echoes. The proposed flight plan is shown in Figure 31. Like ECHO 5, the flight trajectory will be magnetically northward over auroral arcs during evening hours (corresponding to t - 1 or 2 hours as shown in Figure 2). The payload will be oriented magnetically horizontal at an azimuth suitable for ejecting a number of TADS (throw-away detectors) in pairs across the echo loci as shown in Figure 32. The beam injection program will be designed for bounce time, drift and energy analysis, which will be carried by particle detectors on the tads remote from the injecting rocket. A "daughter" nose experiment will measure ambient electric fields as well as those produced by beam injection, to further investigate possible perturbations of the distant regions. The experiment is scheduled for winter, 1982-83.

Another imporatant science objective which justifies further work is the study of beam energy dispersion by fluctuating fields. An injected beam of sharply defined energy may show straggling both towards higher and lower energies after passing through regions like the active plasma sheet. Radial and pitch angle diffusion will problbly accompany the energy changes and are observable by echo anaylsis. Further work might include the injecting of electron beams from a polar-moving rocket or a polar orbiting satellite.This would be a very powerful tool if such beams could be captured and analyzed successfully. The delineation of the open and closed field line regions is a major unknown in the magnetosphere, and the disappearance of echoes would be a measure of this region. The role in particle processes played by closed field lines is often unknown. For example, if discrete auroras are produced by nearby "double layers" or E∥B fields, how do these produce conjugate auroral patterns in the two hemispheres?

A final fascinating area for exploration is to duplicate the role played by natural electron beams in space. Besides auroral precipitation, electron beams are well known to be produced in solar active regions and are thought to be responsible for type III "fast drift" bursts of radio emission first discovered and interpreted by Wild (Wild and Smerd, 1972). Radio emissions have been widely observed from electron beam experiments including the ECHO 4 flight at ground level (see e.g. Monson and Kellogg, 1978). An artificial beam sent up through the ionosphere into space terrestrially could possibly duplicate the passage of an impulsive electron stream passing out through the solar corona from a flare disturbance. The passage of Type III streams through interplanetary space has now been directly observed, establishing the correctness of Wild's original hypothesis. This is a beam-plasma phenomena truly on a cosmic scale, and its duplication in the laboratory is impossible. The role of electron beams as generators of cosmic radio emission is widely recognized (Carr and Gulkis, 1969, Gurnett, 1974).

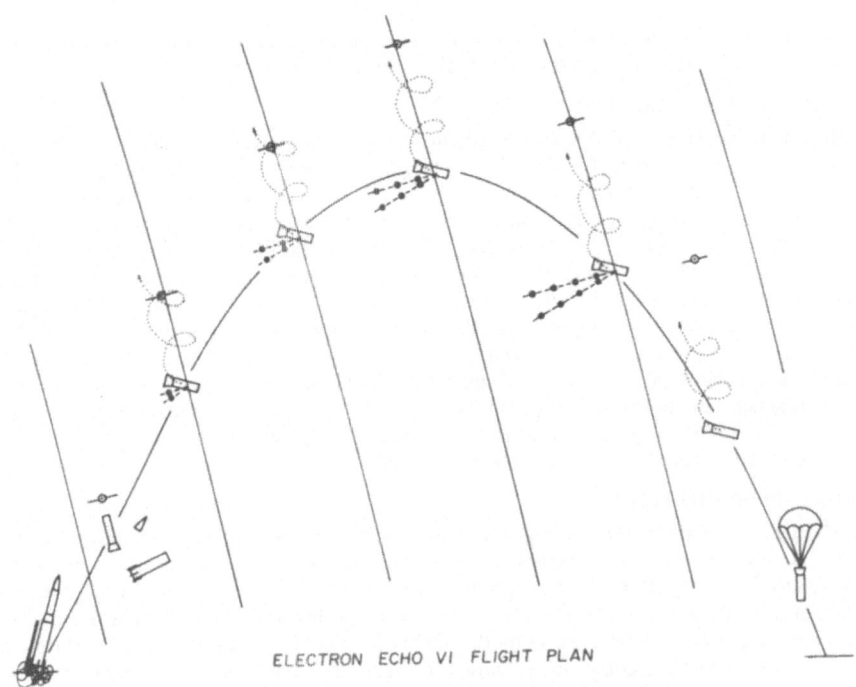

ELECTRON ECHO VI FLIGHT PLAN

Figure 31.   Launch, TAD deployment and recovery plan for ECHO 6.

VIEW IN MAGNETIC INJECTION PLANE AND PAYLOAD REFERENCE
FRAME OF POSSIBLE ECHO LOCUS AND TADS TRAJECTORIES

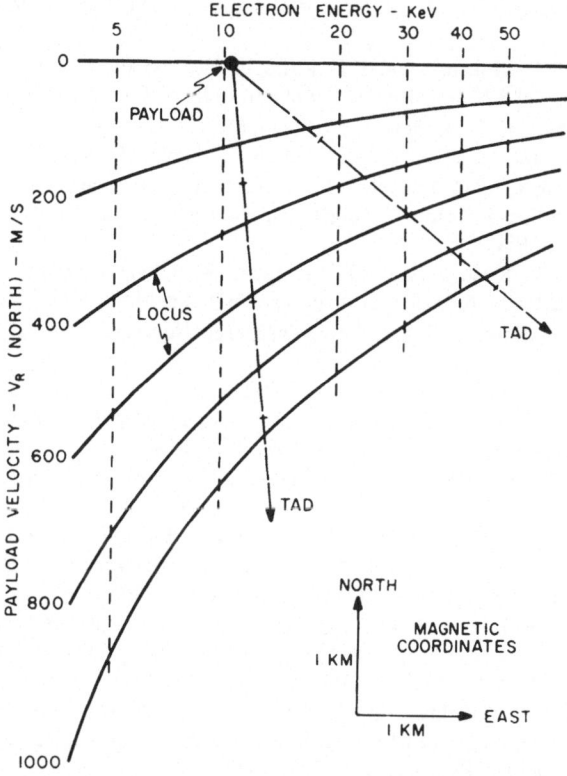

Figure 32.   TAD trajectories are shown in the rocket reference frame intersecting echo loci at various times in the flight corresponding to various north magnetic horizontal velocities.

5. ACKNOWLEDGEMENTS

The preparation of this manuscript was greatly aided by Rick Swanson with recent ECHO 5 results, by Perry Malcolm with new work on Echo Loci, and by James Steffen in many ways including the computer formating of the paper. This program is supported at Minnesota by the National Aeronautics and Space Administration, under grants NSG-07005 and NSG-5088.

6. LITERATURE REFERENCES

Arnoldy, R.L., and J.R. Winckler, The Hot Plasma Environment and Floating Potentials of an Electron-Beam-Emitting Rocket in the Ionosphere, J. Geophys. Res., 86, 575, 1981.

Ashour-Abdalla, M., and C.F. Kennel, Diffuse Auroral Precipitation, J. Geomag. Geoelectr., 30, 239, 1978.

Bernstein, W., H. Leinbach, P.J. Kellogg, S.J. Monson, and T. Hallinan, Further Laboratory Measurements of the Beam-Plasma Discharge, J. Geophys. Res., 84, 7271, 1979.

Cannara, A.B., and F.W. Crawford, Electron-Beam-Probing Studies of Beam- Plasma Interactions, J. Appl. Phys., 38, 583, 1967.

Cambou, F., U.S. Dokoukine, V.N. Ivchenko, G.G. Managadze, V.V. Migulin, O.K. Nazarenko, A.T. Nesmyanovitch, A.Kh. Pyatsi, R.Z. Sagdeev, and I.A. Zhulin, The Zarnitza Rocket Experiment of Electron Injection, Space Res., 15, 491, 1975.

Cambou, F., V.S. Dokoukine, J. Lavergnat, R. Pellat, H. Reme, A. Saint- Marc, R.A. Sagdeev, I.A. Zhuline, General Description of the Araks Experiments, Ann. Geophys., 36, 271, 1980.

Carr, T.D., and S. Gulkis, The Magnetosphere of Jupiter, Annu. Rev. Astron. Astrophys., 7, 577, 1969.

Cartwright, D.G., S.J. Monson, and P.J. Kellogg, Heating of the Ambient Ionosphere by an Artificially Injected Electron Beam, J. Geophys. Res., 83, 16, 1978.

Davis, T.N., T.J. Hallinan, G.O. Mead, J.M. Mead, M.C. Trichel, and W.H. Hess, Artificial Aurora Experiment: Ground-based optical observation, J. Geophys. Res., 76, 6082, 1971.

Davis, T.N., W.N. Hess, M.C. Trickel, E.M. Wescott, T.J. Hallinan, H.C. Stenbaek-Nielsen, and E.J.R. Maier, Artificial Aurora Conjugate to a Rocket-Borne Electron Accelerator, J. Geophys. Res., 85, 1722, 1980.

Dechambre, M., Yu.V. Kushnerevsky, J. Lavergnat, R. Pellat, S.A. Pulinets, and V.V. Selegei, Waves Observed by the Araks Experiments: The Whistler Mode, Ann. Geophys., 36, 341, 1980.

Getty, W.D., and L.D. Smullin, Beam-Plasma Discharge: Buildup of Oscillations, J. Appl. Phys. 34, 3421, 1963.

Gringauz, K.I., N.M. Shutte, L.P. Smirnova, H. Reme, A. Saint-Marc, and J.M. Vigo, Natural Precipitation of Electrons and Effects Observed During the Operation of the Electron Gun During the Araks Experiments, Ann. Geophys., 36, 363, 1980.

Gurnett, D.A., The Earth as a Radio Source: Terrestrial Kilometric Radiation, J. Geophys. Res., 79, 4227, 1974.

Hallinan, T.J., H.C. Stenbaek-Nielsen, and J.R. Winckler, The Echo IV Electron Beam Experiment: Television Observation of Artificial Auroral Streaks Indicating Strong Beam Interactions in the High-Latitude Magnetosphere, J. Geophys. Res., 83, 3263, 1978.

Hendrickson, R. A., The Electron Echo Experiment--Observations of the Charge Neutralization of the Rocket and Analysis of Echoes from Electrons Artificially Injected into the Magnetosphere, Ph.D. Thesis, University of Minnesota, Minneapolis, Mn., 1972.

Hendrickson, R.A., R.L. Arnoldy, and J.R. Winckler, Echo III: The Study of Electric and Magnetic Fields with Conjugate Echoes from Artificial Electron Beams Injected into the Auroral Zone Ionosphere, Geophys. Res. Lett., 3, 409, 1976.

Hess, W.N., M.G. Trichel, T.N. Davis, W.C. Beggs, G.E. Kraft, E.Stassinopoulos and E.J.R. Maier, Artificial Auroral Experiment: Experiment and Principal Results, J. Geophys. Res., 76, 6067, 1971.

Israelson, G.A., and J.R. Winckler, Effect of a Neutral N (2) Cloud on the Electrical Charging of an Electron Beam-Emitting Rocket in the Ionosphere: Echo IV, J. Geophys. Res., 84, 1442, 1979.

Jacobsen, T.A., and N.C. Maynard, Polar 5 - An Electron Accelerator Experiment Within an Aurora. 3. Evidence for Significant Spacecraft Charging by an Electron Accelerator at Ionospheric Altitudes, Planet. Space Sci., 28, 291, 1980.

Jost, R.J., H.R. Anderson, and J.O. McGarity, Measured Electron Energy Distributions During Electron Beam/Plasma Interactions, submitted to Geophys. Res. Lett., 1980.

Kennel, C.F. Consequences of a Magnetospheric Plasma, Rev. Geophys. 7, 379, 1969.

Leadabrand, R.L., M.J. Baron, J. Petricks, and H.F. Bates, Chatanika, Alaska, Auroral Zone Incoherent Scatter Facility, Radio Sci., 7, 747, 1973.

Lyons, L.R., Electron Diffusion Driven by Magnetospheric Electrostatic Waves, J. Geophys. Res., 79, 575, 1974.

Maehlum, B.N., K. Maseide, K. Aarsnes, A. Egeland, B. Grandal, J. Holtet, T.A. Jacobsen, N.C. Maynard, F. Soraas, J. Stadsnes, E.V. Thrane, and J. Troim, Polar 5 - An Electron Accelerator Experiment Within an Aurora. 1. Instrumentation and Geophysical Conditions, Planet. Space Sci., 28, 259, 1980.

McEntire, R.W., The Electron Echo Experiment: A Comparison Between the Observed and Predicted Trajectories of Electrons Artificially Injected into the Magnetosphere, Ph.D. Thesis, Sch. of Phys. and Astron., Univ. of Minn., Minneapolis, 1972.

Mills, D.M., E.E. Abraham, and F.W. Crawford, Beam-Plasma Interactions with Transverse Modulation, J. Appl. Phys., 38, 4767, 1967.

Mishin, E.V., and Yu.Ya. Ruzhin, Beam-Plasma Discharge During Electron Beam Injection in Ionosphere: Dynamics of the Region in Rocket Environment in ARAKS and Zarnitza 2 Experiments, Rep. 21a, b, Inst. of Terr. Magn., Ionos., and Radio Wave Propagat., Acad. of Sci. USSR, Moscow, 1978.

Mishin, E.V., and Yu.Ya. Ruzhin, The Model of Beam-Plasma Discharge in the Rocket Environment During an Electron Beam Injection in the Ionosphere, Ann. Geophys., 36, 423, 1980.

Monson, S.J., and P.J. Kellogg, Ground Observations of Waves at 2.96 MHz Generated by an 8- to 40-keV Beam in the Ionosphere, J. Geophys. Res., 83, 121, 1978.

Morgan, B.G., and R.L. Arnoldy, A Determination of F Region Convective Electric Fields From Rocket Measurements of Ionospheric Thermal Ion Spectra, J. Geophys. Res., 83, 1055, 1978.

Olson, W.P. and K.A. Pfitzer, Magnetospheric Magnetic Field Modeling, Annual Scientific Report, Contract F44620-75-C-0033, McDonnel Douglas Astronautics Co., 5301 Bolsa Ave., Huntington Beach, California, 92647, 1977.

ONeil, R.R., F. Bien, D. Bunt, J.A. Sandock, and A.T. Stair, Jr., Summarized Results of the Artificial Auroral Experiment Precede, J. Geophys. Res., 83, 3273, 1978.

O'Neil, R.R., E.T.P. Lee, and E.R. Huppi, Auroral O (S) Production and Loss Processes: Ground-Based Measurements of the Artificial Auroral Experiment Precede, J. Geophys. Res., 84, 823, 1979.

Paton, B.E., D.A. Dudko, V.N. Bernadsky, G.B. Asoyantz, Yu.N. Lankin, O.K. Nazarenko, V.D. Shelyagin, V.V. Pekker, V.V. Stesin, V.I. Kirienko, E.N. Bajshtruk, V.K. Mokhnach, Yu.V. Neporozhny, Yu.I. Drabovich, G.F. Pazeev, "A Powerful Electron Accelerator for Active Space Experi- ments", Space Science Instrumentation, 4, (2-3), 131, 1978.

Pierce, J.R., Theory and Design of Electron Guns, D. vanNostrand Company, Inc., 1949.

Pyatsi, A.Kh., and Yu.F. Zarnitsky Electron Precipitation in Magnetically Conjugated Region in the First Araks Experiment from Radar Data, Ann. Geophys., 36, 297, 1980.

Roeder, J.L., W.R. Sheldon, J.R. Benbrook, E.A. Bering, and H. Leverenz, X-ray Measurements During the Araks Experiment, Ann. Geophys., 36, 401, 1980.

Roederer, J. G., Dynamics of Geomagnetically Trapped Radiation, Springer Verlag, Berlin-Heidelberg (1970).

Sellen, J.M. Jr., "AMPS Particle Accelerator Definition Study", prepared for NASA Marshall Space Flight Center, TRW Systems Group, One Space Park, Redondo Beach, California, 1975.

Swanson, R. L., Electron Intensity and Magnetic Field Changes at synchronous orbit for the auroral electrojet, M.S. thesis, University of Minnesota, Minneapolis, Mn., 55455, 1978.

Uspensky, M.V., E.E. Timopheev, and Yu.L. Sverdlov, "Araks" Doppler Radar Measurements of the Ionospheric Effects of Artificial Electron Beam in the North Hemisphere, Ann. Geophys., 36, 303, 1980.

Walker, R.J., Quantitative Modeling of planetary fields, in Quantitative Modeling of Magnetospheric Processes, Geophys. Monogr. Ser., Vol. 21, edited by W.P. Olson, AGU, 1909 K Street, N.W., Washiongton, D.C., 20006, 1979.

Wild, J.P., and S.F. Smerd, Radio Bursts from the Solar Corona, Annu. Rev. Astron. Astrophys. 10, 1599, 1972.

Wilhelm, K., W. Bernstein, and B.A. Whalen, Study of Electric Fields Parallel to the Magnetic Lines of Force Using Artificially Injected Energetic Electrons, Geophys. Res. Lett., 7, 415, 1980.

Winckler, J.R., R.L. Arnoldy, and R.A. Hendrickson, Echo II: A Study of Electron Beams Injected into the High-latitude Ionosphere from a Large Sounding Rocket, J. Geophys. Res., 80, 2316, 1975.

Winckler, J.R., The Application of Artificial Electron Beams to Magnetospheric Research, Rev. Geophys. Space Phys., 18, 659, 1980.

Winckler, J.R., Probing the Magnetosphere with Artificial Electron Beams, Adv. Space Res., 1, 17, 1981.

RECENT OBSERVATIONS OF BEAM PLASMA INTERACTIONS IN THE IONOSPHERE AND A COMPARISON WITH LABORATORY STUDIES OF THE BEAM PLASMA DISCHARGE

W. Bernstein,[1] P. J. Kellogg,[2] S. J. Monson,[2]
R. H. Holzworth,[3] and B. A. Whalen[4]

[1] Dept. of Space Physics and Astronomy
Rice University
Houston, Texas 77001
[2] School of Physics
University of Minnesota
Minneapolis, Minnesota 55455
[3] The Aerospace Corp.
El Segundo, California 90245
[4] Herzberg Institute of Astrophysics, NRC Canada
Ottawa, Ontario K1A OR6
CANADA

Abstract

NASA Rocket 27:010 AE (EllB) launched April 9, 1978 from the Churchill Research Range, carried a modest accelerator which injected programmed electron beams of <100 ma at 2 and 4 kV into the ionospheric plasma over the altitude range 120 - 240 km. A major objective of this experiment was the study of beam-plasma interactions and the possible identification of the ignition of the Beam-Plasma Discharge which has been intensively studied in laboratory configurations.

The evidence for the BPD in the following flight data will be presented:
1. the dependence of the 3914 Å light intensity on the spatial configuration, altitude, and beam current and voltage;
2. the energy spectrum of the electron flux returning to the payload during injection;
3. characteristics of the energetic electron flux and spatial distribution in the disturbed region surrounding the payload;
4. VLF and RF wave spectrums.

These data will be compared to those obtained in the laboratory experiments for similar operating conditions. Many features
are clearly consistent with BPD ignition; however, other features
are ambiguous. No correlations of the BPD features with intense
auroral precipitation were apparent.

Introduction

The objectives of the first rocket borne electron beam injections experiments were primarily centered on the use of the beams
as passive probes of the geomagnetic and geoelectric field configurations (Hess et al., 1971; McEntire, 1974). In such experiments
it is tacitly assumed that the beams obey single particle dynamics
throughout their passage through the magnetospheric and ionospheric
plasmas. Confusing results from these early flights indicated that
this assumption was not strictly valid and that significant interaction with the ambient medium did occur particularly at the injection point. Thus subsequent experiments have concentrated on the
plasma physics aspects of the beam-plasma interactions (Maehlum et
al., 1979; Kaneko et al., 1978; Cambou et al., 1980; Hallinan et
al., 1978). The occurence of strong beam-plasma interactions is
now well established; however, this has not totally obviated the
use of electron beams in geophysical applications (Winckler, 1980).

This paper summarizes these experimental results which are
particularly revelant to collective beam-plasma interactions from a
recent electron beam injection rocket flight (27:010 AE) launched
into an active aurora.

DESCRIPTION OF THE PAYLOADS

The 27:010 AE payload was extremely complex; for a variety of
reasons it was subdivided in flight into six discrete sections,
each with its own telemetry to ground and aspect magnetometers.
The essential objectives and components of each section were as
follows:
(a) The forward section was intended to provide measurements
of ambient auroral conditions and to provide a remote platform for
wave diagnostics free of possible EMI associated with accelerator
operation. The diagnostics included two sets of particle detection
systems, and VLF and RF wave diagnostics.
(b) The aft section carried the electron accelerator system
and instrumentation for study of the perturbed region at the beam
injection point. The diagnostics included a particle detection
array and a pair of 3914 Å photometers. The conducting rocket
engine casing remained attached to the aft section to increase the
return current collection area.

(c) Two pairs of TADS (throw away detection systems) instrumented with electrostatic analyzers were deployed.

Table I summarizes the deployment/"turn on" times (and corresponding altitudes) of the various payload components together with the separation velocities of the various sections.

ACCELERATOR OPERATION

Essential features of the experiment required pulsed emission of selected beam currents at constant energy together with periods of 3 kHz modulation of the beam current. These requirements implied the use of a triode rather than diode electron gun configuration; in the triode configuration the magnitude of the emitted current is grid controlled. All commercially available triode configurations use oxide rather than refractory metal cathodes. Although their lower operating temperatures, and consequently lower power requirements, are advantageous applications, activation and contamination of the cathodes present problems in flight. It is necessary to launch these devices in a sealed, activated condition and to subsequently open them, in flight, at sufficiently high altitude (> 125 km) so that contamination by the ambient atmosphere is slow. For this flight, we employed two adjacently located Machlett EE-65 electron guns operated in parallel and mounted at 45° to the vehicle spin axis. The EE-65 uses a low voltage-high current (supplied by batteries) thermal shock breakseal arrangement. Both guns were opened at the desired altitude and operated satisfactorily throughout the flight. The 45° orientation of the guns coupled with the vehicle coning allowed injection over the range 180° to 45° during the flight.

The beam current-voltage programming system malfunctioned throughout the flight; in addition severe arcing occurred at voltages $\geqslant$ 4 kV. Figure 1a shows the current-voltage program in effect until approximately 30 sec after "turn-on"; following several short duration intermediate formats, the program shown in figure 1b continued throughout the remainder of the flight. Surprisingly after reentry, when the accelerator had ceased functioning, the programmer returned to its prescribed mode of operation.

The current-voltage program consists of 8 cycles of sequence A followed by 1 cycle of sequence B which then repeats throughout the flight. Although the current program applied to the grid continues to operate during the "0" accelerating voltage period, return current is not detectable; thus emission during the "0" V period is negligibly small. The I_m pulse currents were ~ 25 ma at 2 kV and ~ 70 ma at 4 kV. If interpreted as a space charge limited

Table 1. Operational Sequence for 27:010 AE

Step	Time After Launch (sec)	Separation Velocity	Altitude (km)
Despin	74		98
Nose cone eject	76		100
Boom extension	82		110
Forward Payload Separation	93	~10 m sec^{-1}	126
Aft tumble gun operation*	97		132
Gun break seals (sequentially)	105		144
TAD 1 + 2 release†	110	~ 1 m sec^{-1}	150
Aft diagnostics "on"^{X}	115		157
Accelerator HV "on"	125		169
TAD 3 + release°	161	~ 1 m sec^{-1}	206

*The tumble gun was employed to induce a large (± 45°) coning motion in the aft section. This coning allowed a wide range of beam injection pitch angles with the fixed position electron gun. The coning half angle of the forward section was ~ 10°.

†The deployment of TADS 1 + 2 prior to accelerator "turn on" precluded measurements of the disturbed region surrounding the rocket with the TAD diagnostics.

^{X}Turn on of the aft section diagnostics 10 sec prior to accelerator turn on allowed this brief period for observation of the ambient plasma without any perturbing effects possibly associated with beam injection.

°The late deployment of TADS 3 + 4 allowed measurements of the near region during gun operation. It had been expected that TAD deployment would occur with the payload spin axis parallel to B. Thus the separation velocity vector would be in the horizontal plane approximately perpendicular to B. Unfortunately, deployment occurred when the payload spin axis was oriented perpendicular to B (because of the coning); thus we have not been able to specify the orientation of the separation velocity vector relative to B in even a qualitative fashion.

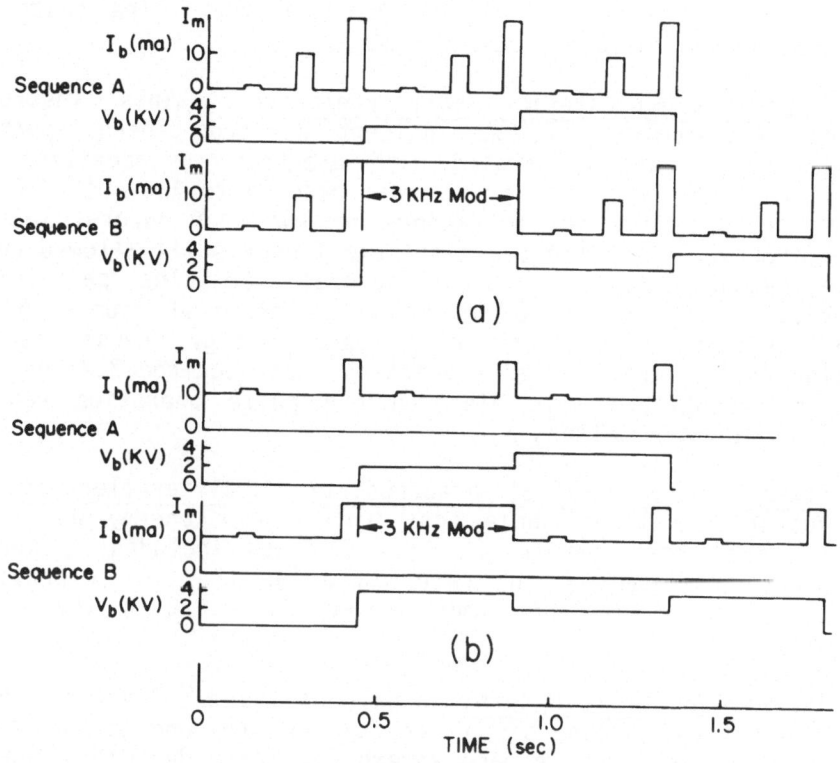

Figure 1. Beam current and voltage program.
a) program in effect until 30 sec from gun "turn on."
b) program in effect after 45 sec from tun "turn on."

current the gun perveance is 2.8×10^{-7} amp volt$^{-3/2}$; this perveance is somewhat less than observed with similar guns in the laboratory.

DIAGNOSTICS

The following diagnostics yielded useful data during the flight.

Aft Section

Combined particle detection array. This instrument package was mounted at the front end of the aft section; all entrance apertures (with one exception) were aligned at 45° to the spin axis. A similar particle detection array was included in the forward payload for measurements of the unperturbed ambient. The package was located 180° in azimuth from the electron gun. The instrument consisted of the following components:

Stepped electrostatic analyzer-electrons. This instrument provided measurements of the incident electron energy spectrum. Because high fluxes associated with electron gun operation were anticipated, the geometric factor was reduced to 1.6×10^{-7} E(keV) cm^2 str keV from the larger factors employed for auroral measurements. However, this reduced geometric factor still allowed qualitative measurements of natural auroral particles particularly > 10 keV. The instrument energy range extended from ~ 160 eV-20 keV in sixteen logarithmically spaced energy steps; the step duration was 200 msec. A single energy scan required 3.2 sec; thus spectrums associated with electron gun pulse operation required data from many gun pulses.

Fixed energy (slaved) electrostatic analyzer-electrons. An electrostatic analyzer whose pass band center energy was slaved to the actual beam energy (0, 2, 4 kV) was included. Thus the instrument only sampled the return flux at the beam energy; its geometric factor was ~ 36 times larger than that of the stepped analyzer.

Thermal Ion Drift Meters. A pair of thermal ion drift detectors (Whalen and Green, 1974), one at 45° and one at 90° to the spin axis was included in each assembly. These detectors can provide evidence for vehicle charging during and immediately after beam emission, indications of ion heating, and of course, measurements of the E x B drift of the ambient plasma. The repetition rate of the ion energy sweep was ~ 5 Hz.

3914 Å Photometers

3914 Å filter photometers (H. A. Cohen, AFGL) were installed in the aft section to provide measurements of the optical emission produced by the injected beams. A pair of photometers, each with 15° full field of view, were positioned to view in opposite directions perpendicular to the spin axis in the same azimuthal plane as the electron gun. A logarithmic current amplifier was employed to provide a dynamic range of ~ 10^4. Only relative rather than absolute calibration of the two photometers response was performed.

Aft payload geometric configuration

As can be seen in the schematic representations shown in figure 2, the particular geometric relationship employed in the aft section for the detector array and the electron gun proved somewhat limited. The combined variation in detection-injection pitch

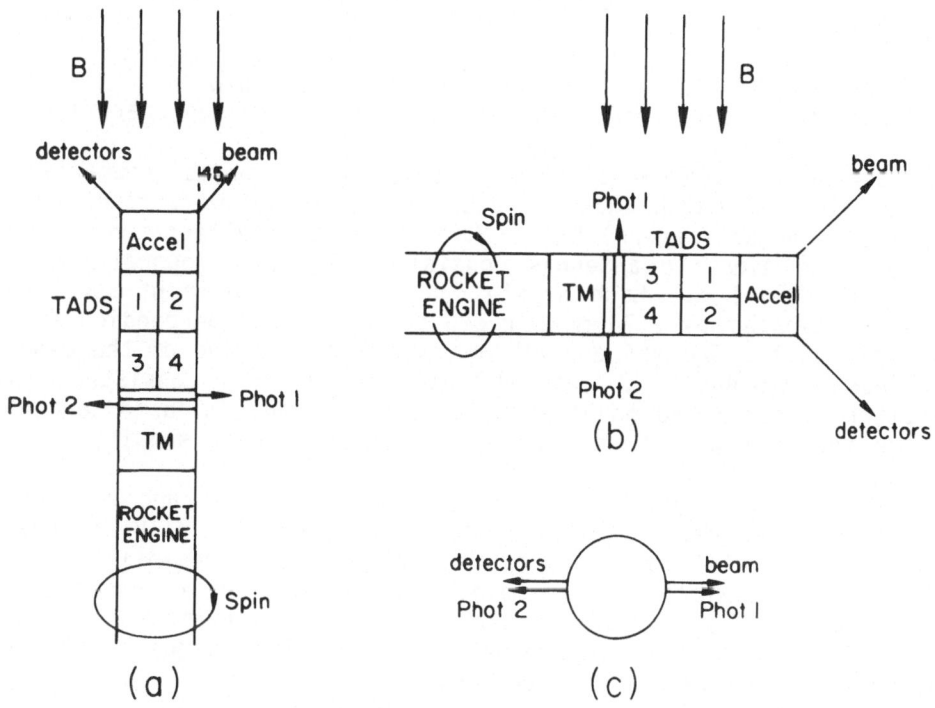

Figure 2. Schematic representation of the payload configuration after separation but before TAD deployment.

angles produced by the large amplitude (± 45°) coning allowed the following observations: injection into (90°-180°) and detection from (0°-90°) the upper hemisphere, injection into the upper hemisphere (90° to 180°) and detection from the lower hemisphere (90° to 180°), and injection into the lower hemisphere (0° to 90°) and detection from the upper hemisphere (0° to 90°). The combination of injection into (0-90°) and detection from the lower hemisphere (90° to 180°) was precluded by the geometry. This prevented observations of "near echos" (Winckler, 1980) produced by atmospheric scattering or magnetic mirroring of downward directed beams at altitudes below the rocket.

Similarly the photometer configuration was not optimum for viewing beam effects which could only usually be observed when either photometer pointed near the zenith. The presence of the aurora, ground reflections, and air glow precluded beam observations in the downward hemisphere or perpendicular to B except toward the end of the flight when the increased atmospheric density and decreased auroral luminosity improved the signal to background ratio greatly. The occurrence of unknown intermittent malfunctions limited useful data to the downleg of the flight.

Wave Diagnostics

For this flight, all wave diagnostics were placed in the forward section to avoid EMI problems associated with gun operation.

Separate instrumentation was used to provide measurements of
the DC field strength and VLF wave amplitude (Holzworth and Koons,
1980) and the RF (several KHz-13 MHz) spectrum (Kellogg and Monson,
1980). The electric antennas consisted of two orthogonal dipoles
separated by ~ 1 m. Each dipole consisted of a pair of spherical
probes separated by 2.75 m on rigid booms mounted perpendicular to
the spin axis. The VLF and RF systems each used one of the dipole
antennas independently. The VLF magnetic antenna consisted of a
multiturn ferrite rod mounted inside the payload; EMI unfortunately
caused the magnetic channel to operate in its least sensitive mode
throughout the flight.
For the VLF system, following preamplification and AGC, both
electric and magnetic signals up to 16 KHz were directly transmitted to ground via wide bandwidth telemetry. The preamplifier also
fed fixed frequency channel amplifiers with digital telemetry outputs. One of these fixed frequency channels was centered at 3 KHz
which allowed accurate determination of absolute signal amplitude
during the 3 KHz modulation periods.

A sweeping receiver spectrum analyzer was used for the RF measurements. The frequency sweep was non uniform in the sense that
the linear frequency scan rate was ~ doubled above 3.5 MHz. A complete frequency scan was obtained every 21 msecs; i.e. at least two
sweeps per 50 m sec gun pulse. In order to extend the instrument
response dynamic range, periodic order of magnitude changes in gain
were included.

TAD Instrumentation

TAD #2 contained four electrostatic analyzers operating at 3
fixed energies of 1.9, 4.0, 8.0 keV. The 1.9 keV and 8 keV channels viewed in opposite directions; each was accompanied by a 4.0
channel (Wilhelm et al., 1980).

TAD #3 was instrumented with two pairs of electrostatic analyzers (D. S. Evans, NOAA) viewing in opposite directions. In each
pair, one ESA covered the energy range from ~ 3 keV-14 keV; a
second parallel ESA provided a more expanded coverage of the low
energy spectrum (a few eV to 4 keV). These analyzers were operated
in a swept mode at a 20 Hz repetition rate. The 50 msec sweeptime
allowed a complete energy spectrum to be obtained during each
50 msec Im pulse.

LAUNCH CONDITION

The Nike-Black Brant V rocket was launched 4 hr 51:10.55 sec UT, 9 April 1978, from the Churchill Research Range. The launch azimuth was ~ 152°; apogee altitude of ~ 246 km was reached 255 sec after launch. A brief break-up event had occurred about 1/2 hr. prior to launch. An arc pattern then reformed in the north and moved southward. The first two arcs faded rapidly as they approached the zenith; however, the third arc passed to the south of the zenith. Ground based photometric measurements indicated an intensity of 70-100 kR (5577 Å) at launch which decreased slowly throughout the flight. With the launch azimuth of 152°, the rocket entered about midway through the bright arc, penetrated the equatorward border near apogee and then remained in a quiet environment for the entire downleg. Thus measurements in the presence and absence of intense auroral precipitation were accomplished.

D. S. Evans has provided a brief summary of the features of the auroral precipitation measured with his instruments on the forward payload. Prior to 4:55:10.0 UT, the precipitated flux was intense and energetic; the characteristic energy peak lay between 14 and 20 keV during this period. Subsequently, the intensity and characteristic energy both decrease until at ~4:57:10.0 UT, the characteristic energy feature has about disappeared. For the remainder of the flight, only a low intensity, low energy precipitating flux was observed. No meaningful association of the beam injection results with auroral conditions was seen.

EXPERIMENTAL RESULTS

161 eV-20 keV Electrons--Aft Section

Figure 3 is a complex presentation of the following simultaneously recorded data: (a) particle flux during one complete energy scan (3.2 sec duration) of the stepped analyzer, (b) particle counting rates recorded by the slaved analyzer, (c) the accelerator current-voltage program, and (d) the pitch angles of the injected and detected particles. As can be seen from the pitch angle data, these data samples have been selected for the configuration where the vehicle spin axis was aligned nearly perpendicular to the geomagnetic field allowing a range of injection-detection pitch angles. As noted previously, an important limitation of the configuration is that only injection into and detection from opposite hemisphere occurs except that both injection and detection pitch angles simultaneously pass through 90° twice each spin cycle.

The following features are evident in the data:
• The differential fluxes are relatively constant up to about

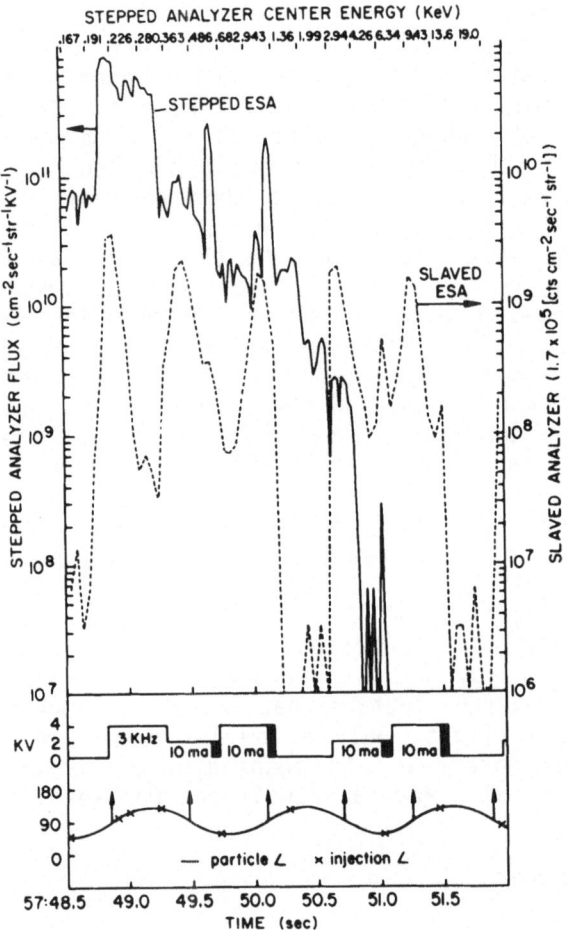

Figure 3. One complete energy scan of the "stepped" ESA together
with simultaneous counting rates recorded by the
"slaved" ESA. Also shown are the stepped ESA energy
steps, the injection-detection pitch angle configurations (indicate 90°) and the gun current-voltage program
(shading indicates Im pulses).

1 keV, then decrease rapidly with increasing energy and merge into
the background above 4 keV. Flux increases ranging from factors of
3-10 are associated with each 2 and 4 kV, Im pulse.

    • Above 4 keV, the fluxes are in good agreement (within the
uncertainties in geometric factor, etc.) with auroral measurements
provided by ESA's included in the forward section. Flux increases
associated with the Im pulses are not apparent in this energy
range.

    • Despite the wide variation in the pitch angles of the injec-
ted and detected particles, the measured fluxes do not show any

correlated variations indicating an isotropic distribution of the return flux. Of course, the configuration of downward injection coupled with the detection of upward moving particles was not accessible.

• No significant decrease in the continuous flux is observed during the 450 msec period when both the accelerator voltage and return current were measured to be zero.

• With this instrument, it was impossible to derive an energy spectrum from any single Im pulse. It was therefore necessary to assemble the pulse spectrum from many pulses distributed over all the energy steps of the analyzer. Figure 4 shows the average pulse energy spectrum for all 4 kV, Im and 2 kV Im pulses obtained during the downleg of the flight, together with a spectrum during the "0" V period. The same spectrum is obtained if only pulses restricted to a smaller range of detected particle pitch angles (0-90°) are

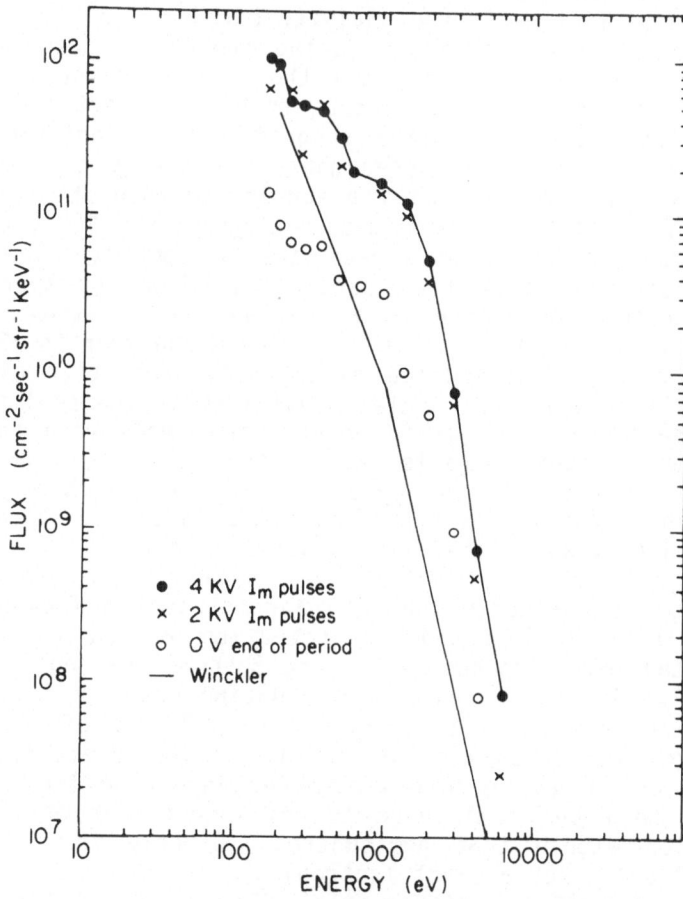

Figure 4. Energy spectrums recorded by the "stepped" ESA during Im pulse emission and the "0" V step.

included again indicating isotropy. Figure 4 also shows the energy spectrum of the return flux during emission derived in a similar averaging fashion by Winckler (1980). Note that not only the spectral shape but the absolute magnitudes of the fluxes are consistent.

• For all Im pulses, the flux magnitudes fall from the pulse level to that of the continuum within the 20 msec resolution set by the acumulation time. No long decay times are observed.

The slaved detector results show significant differences. These results include:

• Counts are observed throughout the 450 msec duration of each accelerator 2 and 4 kV voltage step. This is probably consistent with the ~ 10 ma beam current during this period arising from the malfunction of the programs. However, early in the flight, when the current program was operating as planned, continuous counts were also observed. The appearance of Im associated counts was not as well defined as in the stepped analyzer data.

• The count rates observed during the "0" V step are some 2 orders less than those observed for the 2 and 4 kV steps. This of course does not necessarily indicate that the detected fluxes were low; but rather that the slave program set the analyzer applied voltage to near "0" V where presumably its energy band pass would be small. Because of some uncertainty in the identification of the magnitude of the analyzer band pass energy in the "0" V condition, counting rates rather than fluxes are presented. The observed residual counts may have been particle produced or spurious, but provide no indication of the fluxes during the "0" step.

• The particles detected by the slaved analyzer for 2 and 4 kV beams show a severe dependence on injection-detection pitch angle with the maximum when both angles equal ~ 90°. The reality of this angular dependence is shown in observations made when the vehicle spin axis was aligned parallel to B; for this case the injection pitch angle of 135° and the detected particle pitch angle of 45° remained approximately constant during a spin cycle. No modulation of the 2 and 4 kV associated fluxes were observed.

It might be expected that the largest return fluxes would be associated with 90° injection and detection because the beam would return to the injection point. Figure 5 shows the magnitude of the 2 kV and 4 kV Im associated return counting rates on both injected and detected particle pitch angle for ~ 37 sec of flight time. Although counting rates for the 90°-90° condition are the highest, high rates are always observed whenever injection into and detection from the upper hemisphere occurs independent of the angular configuration within that hemisphere. Clearly then, the return fluxes at 2 and 4 keV are not isotropic.

• Estimates of the differential flux at 2 and 4 keV are not grossly inconsistent with the 2 and 4 keV fluxes recorded by the

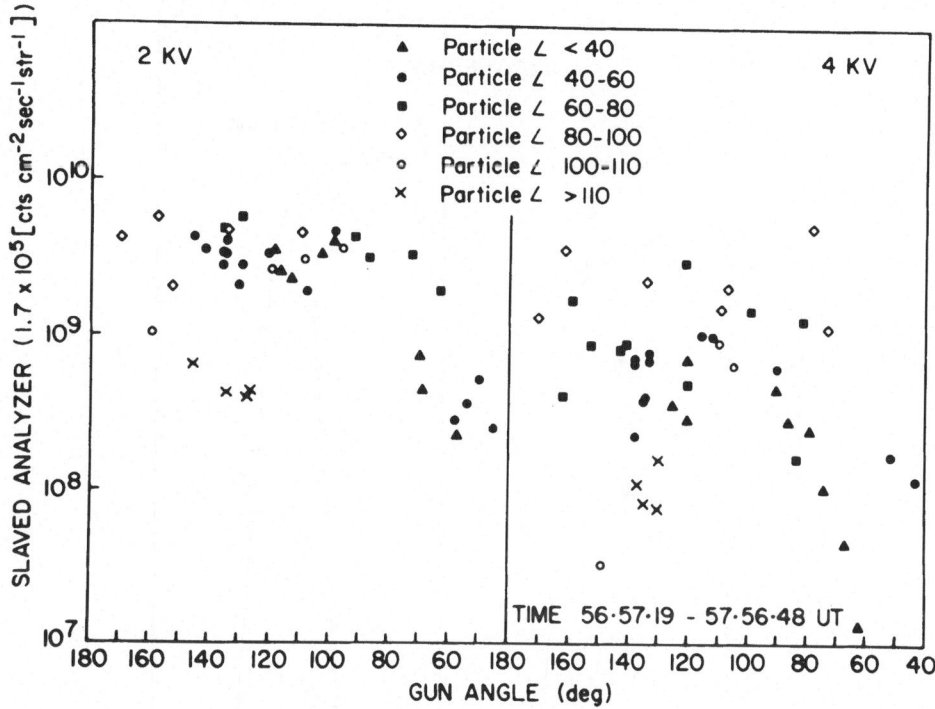

Figure 5. The dependence of the counting rates observed by the "slaved" detector on the injection-detection pitch angle configuration.

stepped analyzer given the uncertainty in relative geometric factors and sampled energy.

Thermal Ions

The thermal ion detectors operated satisfactorily only until apogee. Figure 6 shows typical thermal ion scans obtained for the 45° instrument during (a) the "O" V step and, (b) the 10 sec period prior to gun turn on. Both curves show the typical energy spectrum peaked near 2 eV; the energy above 2 eV together with the directional dependence of the flux allow estimates of the gross plasma drift. During emission all features of the peaked distribution disappear and much lower particle fluxes are detected throughout the sweep range.

In the early part of the flight, the gun current program operated satisfactorily with a 100 msec interval between the 50 msec gun pulses. In this case, 2 kV beams showed abrupt decreases in thermal ion flux during the gun pulses. When the energy range

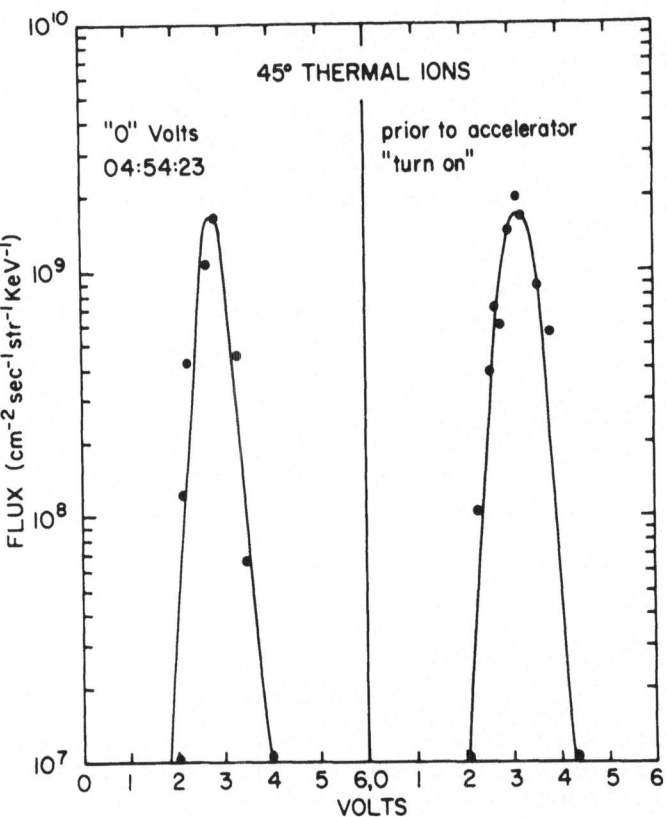

Figure 6. Thermal ion energy scans during a "O" V period and prior
         to gun "turn on" (b).

2-3 eV was sampled during the 100 msec "off" interval, the normal
peaked pattern was observed. The response during the 4 kV accelerator voltage steps was grossly different but could be attributed to
spurious counts associated with the severe breakdown at this
energy.

3914 Å Emission Patterns

     The pair of 3914 Å photometers malfunctioned during the upleg
of the flight, but provided useful data throughout most of the
downleg. As noted earlier, because of the intense aurora, ground
reflection and night glow, beam pulses were only observed when
either photometer viewed near to zenith at high altitude. Late in
the flight, when the rocket had fallen to low altitudes and the

aurora had faded, beam pulses could be observed when the photometers viewed the downward hemisphere.

Two features are evident in the data:
(a) Large amplitude light intensities are only observed over
a portion of the spin cycle; that is when either photometer scans
the angular range 180° to 0° rather than 0° to 180°. In fact
simultaneous large amplitude signals are only observed when the two
photometers are aligned approximately parallel and antiparallel to
the geomagnetic field. this suggests that the payload is not
immersed in a symmetric light emitting region but rather that it is
located near the radial boundary of the emission region. This one-
sidedness can be clearly seen in figure 7 which shows the dependence of the 3914 Å intensity on photometer look angle during one
of the 450 msec 3 kHz modulation periods; the angular dependence of
the background intensity determined during "0" voltage periods is
also shown.

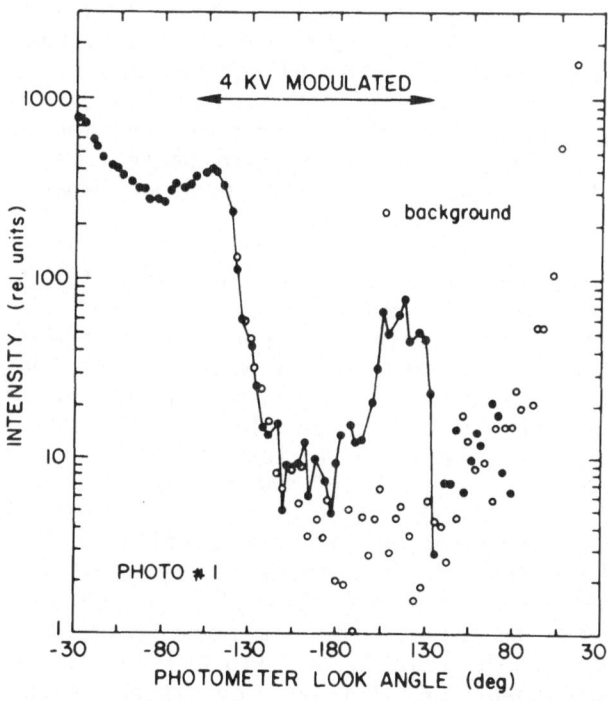

Figure 7. 3914 Å intensity pattern observed during one complete
angular scan of photometer 1; a 3 kHz modulation period
occurred from ~ -90° to +130°. Also shown is the background intensity observed during neighboring "0" V
periods.

(b) As noted earlier, photometer 2 is oriented antiparallel to the electron gun. Yet it clearly detects all gun pulses (when operating properly) although the beam is injected into the opposite hemisphere. This suggests that the illuminated region extends significantly ($\gtrsim$ 10 m) in the direction opposite to the beam injection direction. This effect is also observed during the long 10 ma injection periods.

These two observations suggest that the light emission region is best described as a field aligned cylinder extending below and above the payload with the payload located at or near the boundary of the cylinder. It should be noted that the maximum variation in injection azimuth during a spin cycle is ± 45° for this configuration.

It is far more difficult to characterize the emission intensity patterns associated with the ~ 10 ma beams. As noted earlier, the beam current program during the downleg was characterized by a 450 msec duration, ~ 10 ma current. The low intensity level coupled with the absence of pulse temporal features made identification of the 10 ma associated light emission far more uncertain than for the Im pulses. Qualitatively the same one-sidedness and extension of the beam emission region in the direction opposite to injection, appear in the 10 ma data for both 2 and 4 kV. The background intensity is always too large to permit observation of 10 ma associated emissions in the downward hemisphere; thus simultaneous observations of emissions in both the upper and lower hemisphere is not possible.

A very large variation in light intensity for identical beam energies and currents is observed over a small altitude range. Presumably much of the variation results from the simultaneously varying injection and viewing angles which do not reproduce in the limited number of beam pulses. Figure 8 shows the measured light intensity for 2 and 4 keV 10 ma and Im beam currents as a function of altitude over the downleg (240-120 km); the lack of reproducibility is evident. The ratio of intensities for 10 ma/Im is qualitatively consistent with the beam current ratio. Also shown in figure 8 is a representative altitude profile of the N_2 abundance taken from CIRA (1972) for moderately active conditions and a 00.00 local time atmosphere. At low altitudes (< 150 kms), the 3914 Å intensity for the Im pulses and 10 ma beam currents tracks the N_2 abundance well. Above 150 km, the light intensities appear to be relatively independent of altitude, consistent with the observations of Grandal et al. (1979). However, the intermittent nature of the observations, arising primarily from the vehicle coning and resultant inability of the photometers to view the zenith, can introduce doubt as to the validity of this conclusion.

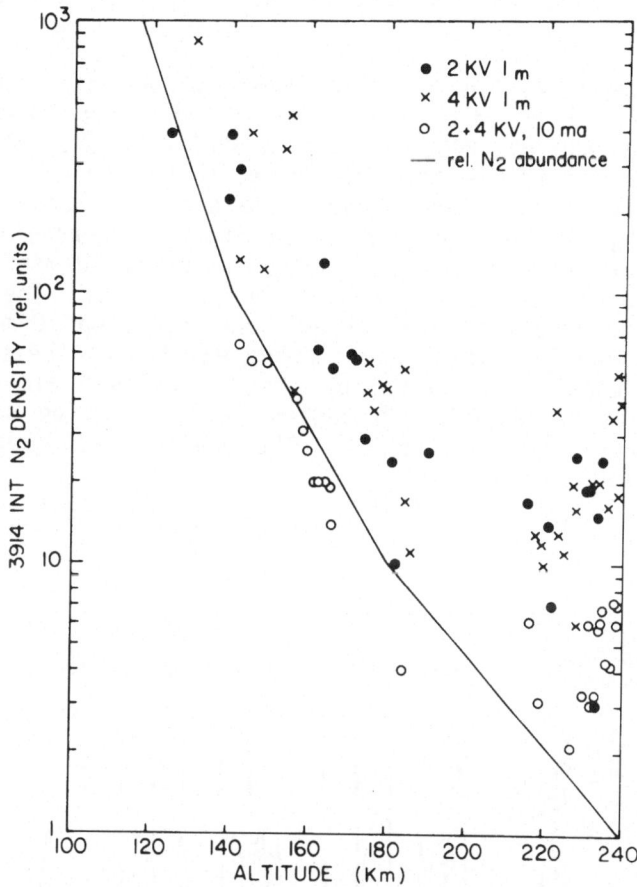

Figure 8. Summary plot of the altitude dependence of the altitude
dependence of the 3914 Å intensity for 10 ma and Im beam
currents at 2 and 4 kV. Only data in the favorable look
angle range is presented. Also shown is a crude altitude profile of the N₂ density.

TAD Data

The energetic particle detectors included in TAD-3 provided
important measurements of the spatial distribution of energetic
electrons surrounding the vehicle. Because the maximum accelerator
energy was 4 kV, we limit our attention to the 0-4 keV pair of
detectors.   As noted earlier, the TAD separation velocity was
determined entirely by the vehicle spin and was ~ 1 m sec⁻¹. The
azimuth of the separation velocity vector was unknown; it is
assumed that a significant component was aligned perpendicular to B
so that an approximately radial traversal of the disturbed region

was achieved. Release occurred ~ 161 sec after launch (4:53:51 UT).

For convenience in studying temporal effects, the 0-4 keV energy range was divided into 5 energy channels 0-800 eV, 800-1600 eV, 1600-2400 eV, 2400-32 eV, and 3200-4000 eV. Although this procedure underestimates the low energy contribution to the total flux in each of these energy bands, the error is large only in the 0-800 eV range. The analyzer sweep repetition rate of 20 sec^{-1} allows one data sample in each energy bin every 50 msec and a complete spectrum during each 50 msec Im gun pulse. The completely modified beam current program of 450 msec duration 10 ma beam with 50 msec 1 ma and Im pulses superimposed existed at the time of TAD-3 release. Lastly the motion of the TAD allowed almost the complete range of particle pitch angles 0-180°-0 to be sampled at a repetition rate of ~ 0.3 sec^{-1}. The two 0-4 keV detector look angles were 180° apart.

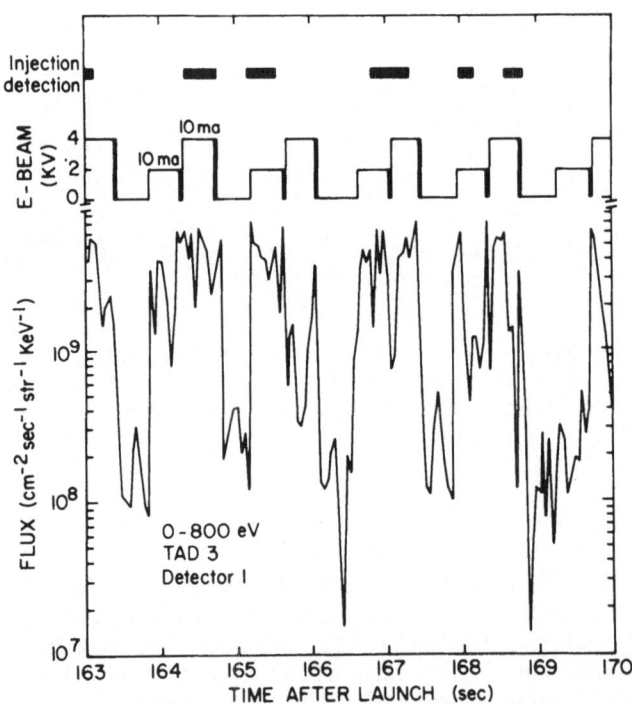

Figure 9. Fluxes observed in the 0-800 eV bin of detector 1 TAD 3 in the time interval 163-170 sec after launch (2-8 sec after release). Also shown are the gun current-voltage program and the periods of injection into and detection from the same hemisphere.

Figure 9 shows the flux measured in the 0-800 eV bin detector 1 as a function of time for a few seconds after release (163-170 sec after launch); also shown are the time intervals when particles were injected into and received from the same hemisphere together with the gun current-voltage program. Several features are evident:

(a) Large fluxes are associated with the 2 and 4 kV accelerator voltage steps but the fluxes fall to background during the "0" V periods.

(b) No clear cut association with the injection-detection patterns is discernible suggesting an isotropic flux.

(c) The response of the detector appears independent of elapsed time.

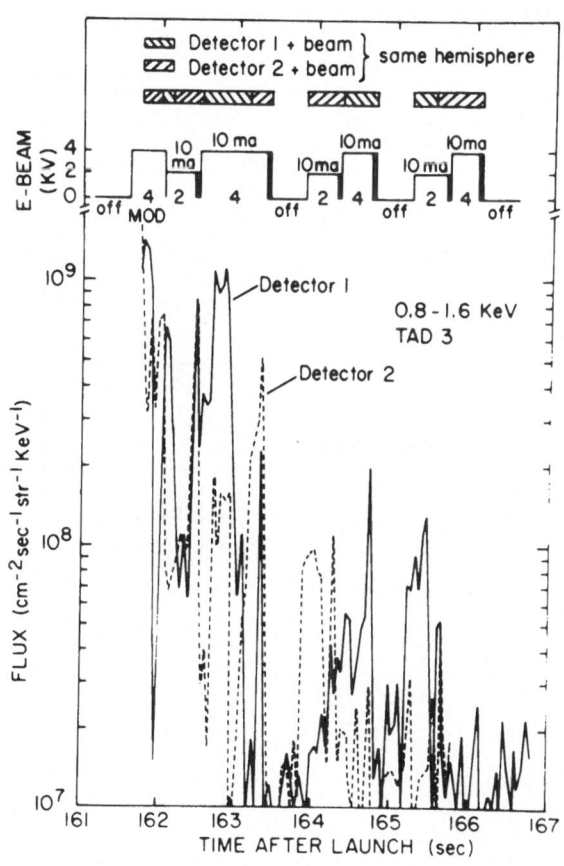

Figure 10.   Fluxes observed in the 800-1600 eV bins of detectors 1 and 2 of TAD 3 in the time interval 161.75-168.0 sec after launch (0.75-7.0 sec after release). Also shown are the gun current-voltage program and the periods of injection into and detection from the same hemisphere for each detector.

Figure 10 shows the temporal behavior of the 800-1600 eV bins of detectors 1 and 2 after release; also shown are the periods when each detector is aligned so that injection into and detection from the same hemisphere occurs together with the gun program. The important features include:

(a) These fluxes decay to background levels after ~ 5 secs (166 sec after launch) while the 0-800 eV have not shown any decrease. Higher energy fluxes disappear even more rapidly.

(b) An obvious dependence on injection-detection pitch angle is apparent with maximum fluxes observed in each detector for the configuration of injection into and detection from the same hemisphere. This pattern appears similar to that given by the "slaved" detector in the aft payload.

(c) The measured fluxes decrease to background in both detectors during the "0" volt periods.

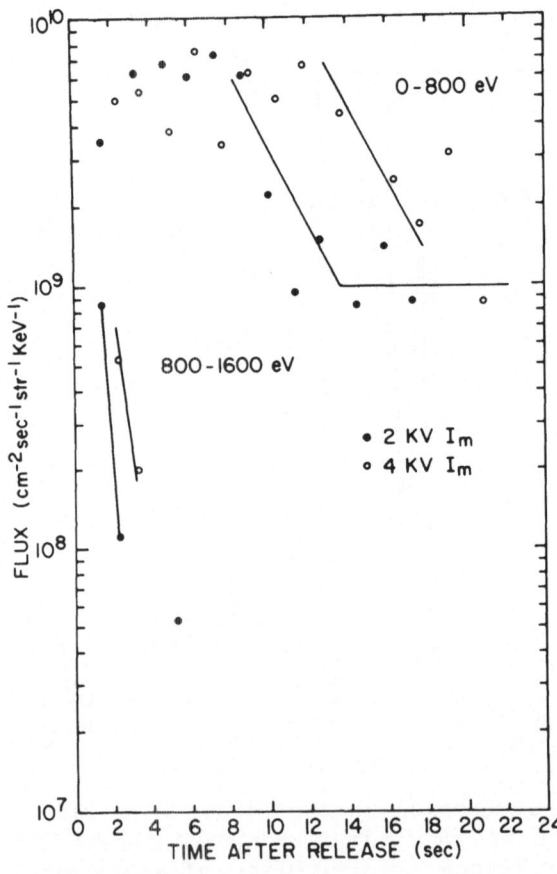

Figure 11. The longer duration temporal (radial) dependence of the fluxes recorded in the 0-800 eV and 800-1600 eV bins of detector 1, TAD 3.

Figure 11 shows the temporal behavior of the 0-800 and 800-1600 eV fluxes on a longer time scale for 2 and 4 keV beams. The 0-800 eV fluxes associated with the 2 kV injections appear to fall to auroral background in ~ 13 sec whereas those associated with 4 kV injections fall to background in ~ 19 sec. Assuming that the separation velocity perpendicular to B is ~ 1 m sec^{-1}, then the 2 and 4 keV beams create disturbed regions of 12 and 19 m in radial extent from the vehicle assuming release in the plane perpendicular to B. These distances are a factor of 2 larger than the corresponding radial widths based on the gyradii of the injected beams. No apparent flux enhancements are evident in the data at 2 R_c where $R_c = v_\perp/\omega_c$. The relatively rapid fall off of the 800-1600 eV is obvious.

The TAD measurements indicate that
1.   The region surrounding the payload is populated with energetic electrons during periods of beam emission. These fluxes disappear during the 450 msec duration "0" V periods.
2.   The energy spectrum is soft and softens still more with increasing radial distance from the payload.
3.   The radial extent of the perturbed region is greater for 4 kV than 2 kV beams.
4.   The 0-800 eV fluxes appear to be isotropic; the > 800 eV fluxes are anisotropic with maximum fluxes observed for injection into and detection from the same hemisphere.

Wave Measurements

VLF. A 20 second sequence of spectral data from the low frequency electric receivers on board the forward section is shown in figure 12. The 0 to 16 KHz electric spectrum shows discrete spikes at the times when the electron accelerator located on the aft section was operated at 2 and 4 kV with currents of 35 and 70 ma respectively. The longer pulse of 450 msec duration is where the beam current was modulated at 3 Khz is seen near 04:53:30 secs. Figure 13 shows the pulse associated spectrums on an expanded scale for comparison with examples of pre-BPD and BPD spectrums observed in the SESL vacuum chamber shown in figure 14. The clear spectral peaks in the range 5-6 KHz seen in the flight data are consistent with similar features in the laboratory BPD. The propagation of 3 KHz waves from the aft to forward payloads during the current modulation periods has been described by Holzworth and Koons (1981) and will not be discussed here.

RF Emissions. Figure 15 presents samples of flight spectrums during 1m pulses together with a typical laboratory BPD spectrum. Both spectrums show nearly frequency independent, large amplitude signals which cut off abruptly at the electron cyclotron frequency (the laboratory magnetic field strength was ~ 0.9 gauss). Also obvious in the laboratory spectrum is a higher frequency band believed to be the plasma frequency corresponding to the enhanced

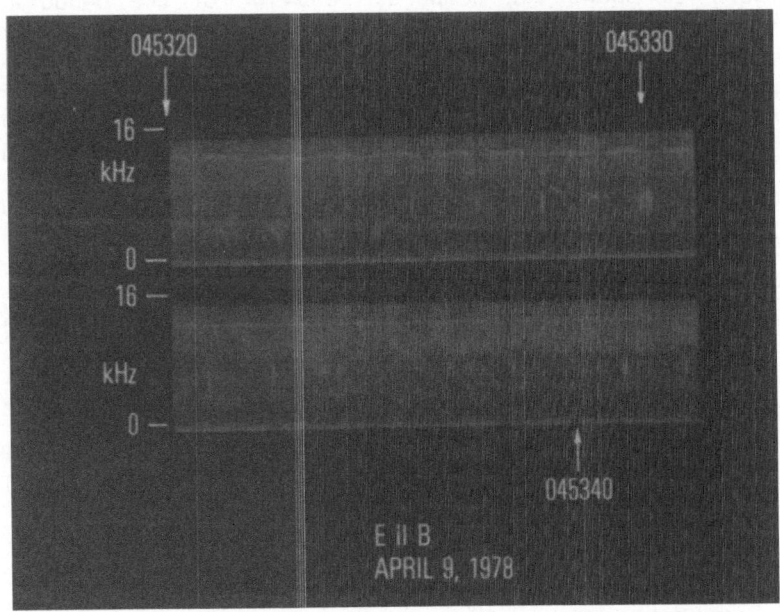

Figure 12.   20 second sequence of spectral data in 0-16 kHz range.

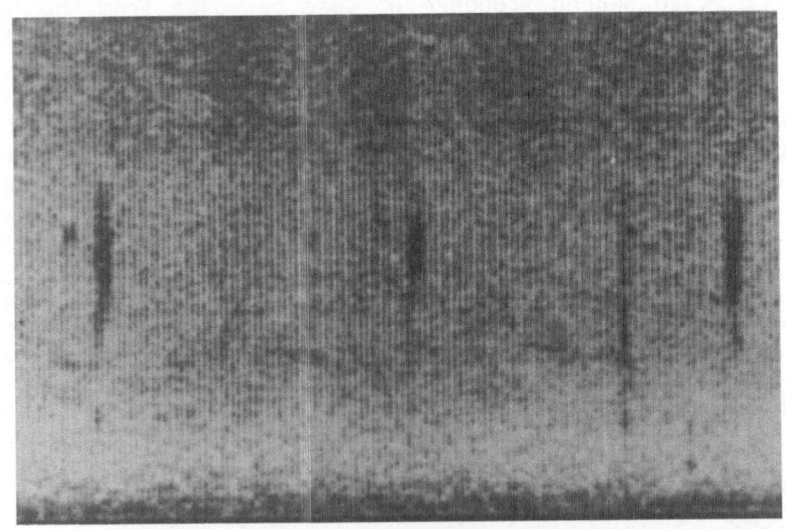

Figure 13. Expanded Im pulse spectrum observed in the flight
during the periods 0453:25-28.

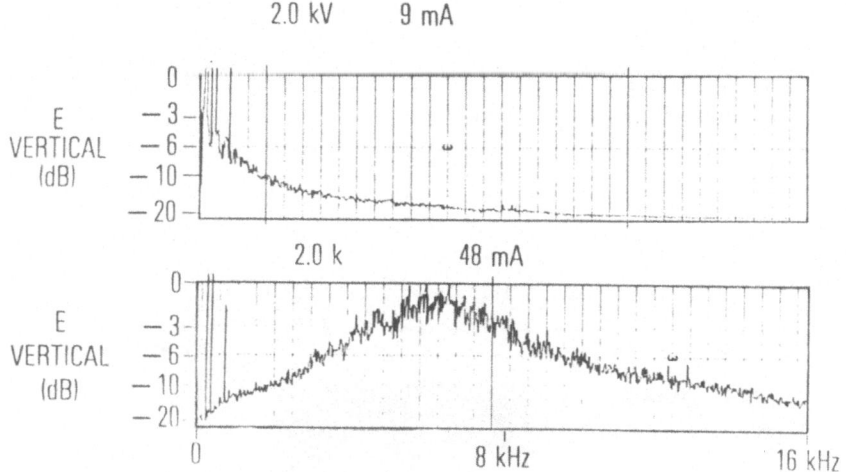

Figure 14. Lab VLF spectrums: upper-single particle trajectory and lower-BPD.

density characteristic of the BPD. This higher frequency component was not observed in the flight data, although beam associated emissions were sometimes observed at the local ionospheric plasma frequency (~ 3 MHz). The very great similarity in the detected wave spectrum below f_{ce} in the laboratory and flight data is clear.

A marked feature of these emissions below f_{ce} is their amplitude dependence on beam injection pitch angle. Without any exceptions, these emissions are not detected when the beam is injected downward and are almost always detected for upward injection. The forward section was ejected upward at separation and consequently reached higher altitudes throughout the flight. The direction of the separation velocity relative to B has not been determined however.

For the 10 ma beams, wave emissions in the RF and VLF frequency range have not been clearly identified above background. Similarly emissions above background are not observed during the 450 msec periods of 0 accelerating voltage and 0 beam current. Because of this, it is difficult to specifically identify these periods as "inactive" or undisturbed from the wave data alone.

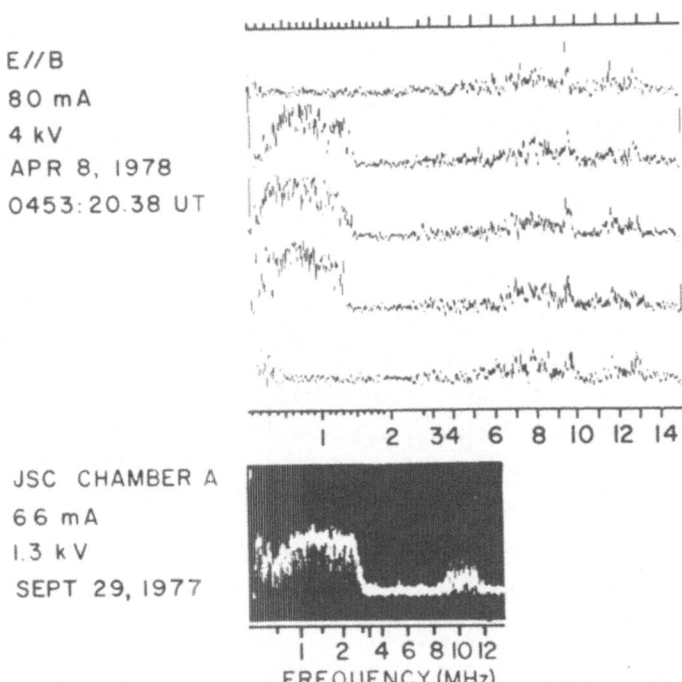

Figure 15. A typical RF spectrum observed in flight during an Im
          pulse compared with a typical BPD spectrum observed in
          the laboratory.

LABORATORY RESULTS

Before attempting a comparison of laboratory and flight
results, we present a brief summary of the pertinent optical and
energetic particle laboratory results. Laboratory wave measurements have already been presented together with the flight data.

Energetic particles

Jost et al. (1980) have presented extensive measurements of
the forward component of the energetic particle thus characteristic
of the BPD. The heating of the primary beam coupled with the
appearance of large fluxes of degraded (or accelerated) particles
has been amply demonstrated. However, these measurements have been
limited to only the forward component at energies > 200 eV.

Attempts have been made to measure the backscatter enhancements possibly associated with the BPD. However, because a beam

catcher was not employed, wall backscatter appeared to completely dominate any antiparallel fluxes with E > 200 eV associated with the BPD. No significant differences could be identified in the antiparallel intensities for pre-BPD and BPD conditions after normalization for differences in primary beam current.

To date, the energy range between a few eV, accessible to Langmuir probes, and the > 200 eV range, described by Jost et al. (1980) has not been studied. Furthermore the Jost et al. (1980) meaurements utilized a grounded electron gun; vehicle charging even to small potentials would significantly modify the return flux characteristics in the energy range < 200 eV. Thus this existing laboratory data is not directly comparable to the flight data.

Photometric observations

Detailed laboratory studies of the optical emissions associated with the BPD have been carried out by Hallinan et al. (1981) using TV and 3914 Å photometric techniques. The important features are:

(a) The total 3914 Å volume emission rate associated with the BPD is ~ a factor of 25 times greater than that corresponding to classical collisions by the primary beam itself with the ambient gas. Because the transition from single particle behavior to BPD is associated with significant changes in the radial configuration of the emission region, the measured increase in light intensity may be significantly reduced but a discontinuous increase in emission intensity is expected to characterize the transition from single particle behavior to the BPD.

(b) For parallel injection, the radial extent of the BPD is ~ 3 times greater than the corresponding single particle configuration [neglecting refocus nodes]. With increasing injection pitch angle, the radial extent of the BPD appears to conform more closely to the gyroradius of the single particle helix. For beam currents moderately greater than the BPD threshold at high pitch angles, the BPD configuration resembles a somewhat hollow cylindrical shell with significant boundary brightening. Increased beam current produces a more uniform radial emission profile.

(c) In some of the experiments the electron gun was mounted ~ 2 m above the chamber floor. For single particle trajectory conditions, no illumination was observed in the region between the gun and the floor. For BPD conditions the illumination extended from the ceiling to the floor, indicating significant extension of the illuminated region in the antiparallel direction.

(d) For beam currents above BPD threshold, the 3914 Å total volume emission rate showed an approximately linear dependence on the quantity (beam current minus threshold current) with perhaps a tendency toward saturation at the highest beam currents. Radial broadening of the BPD emission region with increasing beam current was not observed. Therefore, the 3914 Å intensity for beam currents reasonably above threshold should be linearly dependent on beam current similar to the single particle trajectory case.

(e) Only very qualitative measurements of the dependence of the 3914 Å volume emission rate on N_2 density at fixed current and beam energy during BPD have been carried out to date. These measurements indicate that the beam width is independent of pressure and that the emission rate increases with increasing pressure. However, the base pressure was ~ 7×10^{-6} torr during this run and only the very limited range up to 2×10^{-5} torr could be studied. As shown by Bernstein et al. (1979) this pressure range is approximately centered about the minimum in the threshold current-pressure relationship; thus the threshold current is relatively independent of pressure in this range. It is probably invalid to extrapolate this volume emission rate-pressure dependence into the region of ~ 1×10^{-6} torr where the threshold current appears to be linearly dependent on pressure.

DISCUSSION

The preponderant evidence appears to indicate that BPD ignition occurred for Im current pulses at 2 and 4 kV and during the 3 kHz modulation periods. This specific evidence includes:
1.   The asymmetrically located, field aligned 3914 Å emission region extending above and below the payload.
2.   The similarities seen in the VLF wave spectrums.
3.   The similarity in the "whistler" ($< f_{ce}$) wave spectrum. It is not expected that the higher frequency Langmuir waves, characteristic of the higher density BPD, would propagate to the remotely located wave detection system through the lower density ionospheric plasma.
4.   The particle detection systems on the aft payload and TAD indicate electron heating or more likely the production of suprathermal tails in the injection region. The increased ionization rates characteristic of ignition indicate the presence of such particles (not directly studied to date) in the laboratory BPD.

Both the TAD and the 3914 Å data indicate that if BPD ignition occurred for the Im pulses, it must also have occurred during periods of 10 ma emission. Subsequent laboratory studies with a very similar accelerator system operated with emission limited Im currents of ~ 50-60 ma indicated that 2 kV Im pulses produced BPD, 2 kV, 10 ma pulse produced a marginal BPD, 4 kV Im pulses produced marginal BPD, and 4 kV, 10 ma, 8 kV, Im and 8 kV, 10 ma pulses showed single particle behavior. Although laboratory experiments indicated an inverse dependence of the threshold current on system length, the extrapolation of this relationship to unbounded space conditions is uncertain; it is not unreasonable that 10 ma beams could produce the BPD in the flight environment.

The laboratory experiments provide little insight into the apparent altitude independence of the 3914 Å emission intensity at altitudes > 150 km. This feature appears characteristic of many rocket borne accelerator flights (Kaneko et al., 1979; Israelson and Winckler, 1979; Grandal et al., 1979). Israelson and Winckler (1979) suggested that this effect resulted because of N_2 leakage from the altitude stabilization system which produced a local N_2 density enhancement around the vehicle; in an earlier flight they reported consistency between the altitude profiles of the N_2 density given by several model atmospheres and the 3914 Å intensity (Israelson and Winckler, 1975). The significance of leakage and outgassing remains unclear.

Major difficulties exist in the interpretation of the energy measurements of the return flux to the aft section. The Im pulse energy spectrum appears very similar in both absolute flux and spectrum shape to the equivalent data given by Winckler (1980). However, Winckler indicates that the intense return flux decreases rapidly after termination of injection and disappears completely after ~30 msec presumably because the payload has left the perturbed region and also because of the escape of the energetic electrons along the lines of force. In the present experiment, the return flux also decreases rapidly after termination of injection, but then remains at a very high level throughout the 450 msec duration "0" volt period. As shown in figure 4, the form of the spectrum observed during the pulses and during the "0" volt periods are similar. On the other hand, all TAD energy bins clearly show decreases to auroral background during the "0" volt period. The thermal ion data, which represent a sensitive measure of vehicle charging either because of beam emission (positive) or the presence of ambient hot electrons (negative) indicate an unperturbed environment (neither beam emission or the presence of large fluxes of heated ambient electrons) during the "0" volt period. The presence of the continuous background flux in the "stepped" ESA data during the "0" V period obviously throws doubt upon the validity of its measurements during emission.

CONCLUSION

Although a variety of failures and malfunctions obviate a quantitative interpretation of the data, a very qualitative assessment indicates that BPD ignition was produced by both 10 ma and Im beams at 2 and 4 kV. Many of the observed characteristics are similar to if not almost identical to the BPD characteristics observed in the laboratory.

We acknowledge the major contributions provided by our collaborators in the flight program including: K. Wilhelm, H. A. Cohen,
and D. S. Evans. Similarly our thanks to collaborators in the
laboratory program including: F. H. Leinbach, T. Hallinan, H. R.
Anderson, R. J. Jost, and E. P. Szuszczewicz. We also thank D. A.
Burnside, M. Beghetto, B. Baker, and J. Stevenson for their
engineering support during the entire program. This work was
supported by NASA Grant NAGW-69.

REFERENCES

Bernstein, W., Leinbach, H., Kellogg, P. J., Monson, S. J., and
    Hallinan, T., 1979, Further laboratory measurements of the
    beam-plasma discharge, J. Geophys. Res., 84:7271.
Cambou, F., Dokoukine, V. S., Laveignat, J., Pellat, R., Reine, H.,
    Saint-Marc, A., and Zhuilin, I. A., 1980, General Description
    of the ARAKS Experiments, Ann. De. Geophys., 36:271.
Grandal, B., Throne, E. V., and Troim, J., 1980, Polar 5--An elec-
    tron accelerator experiment within an aurora, 4. Measurements
    of the 391.4 nm light produced by an artificial electron beam
    in the upper atmosphere, Planet. Space Sci., 28:309.
Green, D. W., and Whalen, B. A., 1974, Ionospheric Ion flow velo-
    cities from measurements of the ion flow distribution function
    technique, J. Geophys. Res., 79:2829.
Hallinan, T. J., Stenback-Nielsen, H. C., and Winckler, J. R.,
    1978, The Echo 4 electron beam experiment: Television obser-
    vations of artificial streaks indicative strong beam plasma
    interactions in the high latitude magnetosphere, J. Geophys.
    Res., 83:3263.
Hallinan, T. J., Leinbach, F. H., and Bernstein, W., 1981, Studies
    of the beam--plasma discharge optical emissions, in prepara-
    tion.
Hess, W. N., Trichel, M., Davis, T. N., Beggs, W. C., Kraft, G. E.,
    Stassinopoulos, E., and Maier, E. J., 1971, Artificial Aurora
    experiments: Experiment and principal results, J. Geophys.
    Res., 76:6067.
Holzworth, R. H., and Koons, H. C., 1981, VLF emission from a
    modulated electron beam in the auroral ionosphere, J. Geophys.
    Res., 86:853.
Israelson, G., and Winckler, J. R., 1975, Measurements of 3914 Å
    light production and electron scattering from electron beams
    artificially injected into the ionosphere, J. Geophys. Res.,
    80:3709.
Jost, R. J., Anderson, H. R., and McGarity, J. O., 1980, Electron
    energy distributions measured during electron beam/plasma
    interactions, Geophys. Res. Lett., 7:509.
Kaneko, O., Sasaki, S., and Kawashima, N., 1979, Active experiment
    in space by an electron beam, report, Inst. of Space and
    Aerosp. Sci., Univ. of Tokyo, Komaba, Megura-ku, Tokyo.

Kellogg, P. J., and Monson, S. J., 1980, Rocket Borne Electron Accelerator results pertaining to the beam plasma discharge, presented at the COSPAR Symposium on Active Experiments, Budapest, Hungary.

Maehlum, B. N., Måseide, K., Arsnes,K. A., Egeland, A., Grandal, B., Holtet, J., Jacobsen, T. A., Maynard, N. C., Soras, Γ., Stadsnes, J., Thrane E. V., and Troim, J., 1980, Polar 5--An electron accelerator experiment within an aurora, 1. Instrumentation and Geophysical Conditions, Planet Space Sci., 28:259.

McEntire, R. W., Hendrickson, R. A., and Winckler, J. R., 1974, Electron echo experiment, 1. Companion of observed and theoretical motion of artificially injected electron in the magnetosphere, J. Geophys. Res., 79:2343.

Wilhelm, K., Bernstein, W., and Whalen, B. A., 1980, Study of electric fields parallel to the magnetic lines of force using artificially injected electron beams, Geophys. Res Lett., 7:117.

Winckler, J. R., 1980, The application of artificial electron beams to magnetospheric research, Revs. Geophys. and Space Physics, 18:659.

DISCUSSION

Lavergnat: Could you comment on the value of 8 kHz as compared with the lower hybrid frequency either in the E-parallel-B rocket experiment or the tank experiment?

Bernstein: The observed frequency of 8 kHz is larger than the LHF in both lab and space experiment.

Kintner: Why should an 8 kHz frequency have the same physical meaning in the chamber as in the ionosphere?

Bernstein: If we are looking at small scale phenomena compared with experiment dimension, the simulation should be valid, at least in the gross picture although there would be no reason to expect identical frequencies but rather scaled frequencies.

LeQueau: What is the temporal dynamics of the whistler emission (below the electron gyro frequency) observed during your experiment?

Bernstein: Emission appears to be continuous during 50 msec pulses and during the 3 kHz modulation periods. Modulation can be observed in wave amplitude display.

Winckler: In regard to observation of light at one side of the rocket, as if BPD: the same observation was made by Echo 4, but in that case occurred when rotating gun injected into a region fixed in the rocket system, as if a wake phenomena.

Bernstein: This certainly represents an alternative explanation for one-sided configuration.

CHARGED PARTICLE MEASUREMENTS FROM A ROCKET-BORNE ELECTRON ACCELERATOR EXPERIMENT

G.R.J. Duprat, A.G. McNamara and B.A. Whalen

Herzberg Institute of Astrophysics
National Research Council of Canada
Ottawa, Ontario K1A 0R6 Canada

ABSTRACT

A Nike Black Brant rocket (NVB-06) was launched from Churchill Rocket Range (Manitoba) on December 3, 1979 into a bright auroral display. The primary objective of the flight was to use an electron beam to probe the auroral field lines for electric fields parallel to the magnetic field. The secondary objectives were to study electron beam interactions in the ionosphere and spacecraft charging effects. In this report results relating to the secondary objectives are described. In particular, charged particle observations are presented which relate to (a) the spatial distribution of energetic (keV) charged particles surrounding the accelerator during gun firings, (b) the energy distribution of energetic electrons produced in the plasma by the electron beam, and (c) the dependence of these characteristics on the beam energy, current, and injection angle. It is shown that certain similarities exist between these observations and those made in laboratory simulations of the experiment where a specific beam-plasma instability was stimulated.

INTRODUCTION

The results presented in the paper were derived from a sounding rocket experiment, which was the third in a series of four, designed to use an electron gun as a magnetospheric probe as well as to study electron beam-plasma interactions. The rocket payload consisted of two separable sections, that is, a mother and daughter, each section containing its own telemetry system. A set of charged particle detectors covering the energy range from the thermal ionospheric plasma up to several hundred keV was carried on both the

65

mother and daughter payloads. The mother, or aft, section was attached to the rocket motor and also contained a programmable electron accelerator and a bank of photometers (F. Harris, HIA, NRC)
to measure optical emissions induced by beam firings. The daughter
payload which was ejected forward of the mother carried a plasma
wave sensor (P. Kellogg) along with charged particle sensors.

In the next section a brief description of the electron gun
system and charged particle sensors on the daughter payload is presented. Results from these sensors are discussed in the following
section and a summary of these results and their interpretation
appears in the final section.

INSTRUMENTATION

The electron accelerator consisted of a power converter and
two Machlett EE-65 triode guns operated in parallel and was programmed to produce bursts of electrons at varying currents and voltages. Two different sequences were employed. The primary sequence
(A) called for three different voltage steps (at 1.9, 4.0 and 8.0
keV) and three different current steps (1, 10 and 100 ma) at each
voltage. A schematic diagram of three A sequences is shown in the
bottom two panels of Figure 1.

The second sequence (B) called for the emission of a 100 ma
square wave modulated (at 3 kHz) beam current at 8 and 4 keV. No
data during this sequence are presented here.

The energetic particle detectors consisted of collimators,
cylindrical plate electrostatic analysers and channel electron multiplier sensors. Four geometrically identical detectors were carried
on the daughter payload. The deflection plate voltages on three of
these devices were fixed to select electrons at the nominal beam
energies. The fourth instrument had its deflection plate voltage
stepped logarithmically through sixteen energy steps from 160 eV to
20 keV every 2.9 seconds.

Two thermal electron probes were also included in the daughter
payload, one a spherical langmuir probe and the other a hemispherical
retarding potential analyser (R.P.A.). The langmuir probe, consisting of a rhodium plated sphere 0.635 cm in diameter and mounted on
a 50 cm long non-conducting boom, was deployed perpendicular to the
payload spin axis. The voltage on the probe was programmed to remain fixed at -2.96 V bias for ∿0.6 sec, then to sweep up to +6.18
V and to remain fixed at this voltage for 1.4 seconds. This probe
was used to measure the ambient electron density and temperature.

The R.P.A. was constructed in the form of a hemisphere and
consisted of an inner electrode biased at +22 V and an outer grid

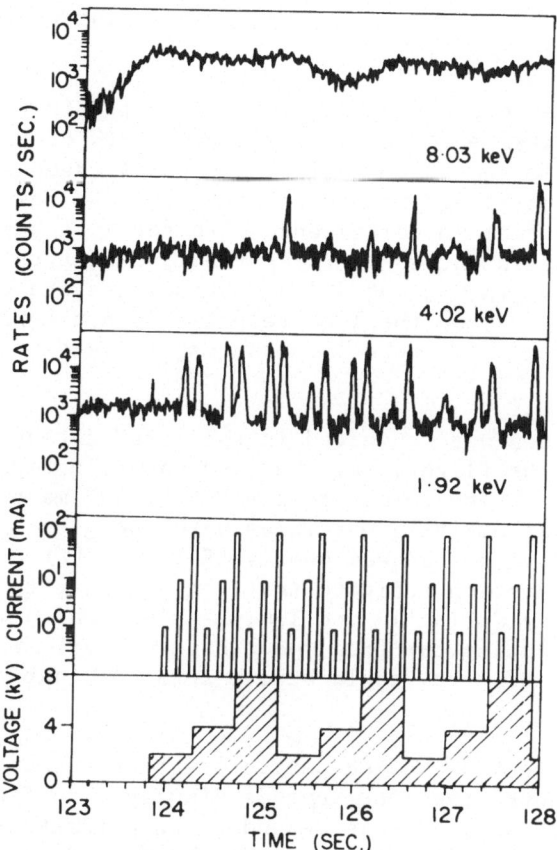

Fig. 1. Energetic particle detector count rates and gun program for time immediately after payload separation.

whose bias voltage tracked the langmuir probe voltage but had a voltage range from -3.47 V up to +5.81 V. When biased at -3.47 V, the probe collected electrons with energies $\gtrsim 4$ eV. This probe was also mounted on a non-conducting boom 50 cm long and deployed perpendicular to the spin axis.

RESULTS

Rocket NVB-06 was launched on December 3, 1979 at 05 h 32 m 47.0 s UT (23 h 32 m 47 s LT) into a bright auroral display, on an azimuth of 10° east of north and reached an apogee of 338 km at T + 299 seconds. The daughter payload was ejected forward of the mother at T + 119.6 s at a velocity of 1.62 m/sec and at 37° to the local magnetic field line. The first gun pulses occurred at T + 123.85 s, at an altitude of 203 km when the mother-daugter separation

distance d was 6.9 m which corresponded to a distance across magnetic field lines $d_\perp$ of 4.2 m.

The first sequence of pulses is represented in Figure 1 along
with the count rates observed in the three fixed energy electron
detectors. The 10 and 100 ma pulses are easily seen above the background due to auroral electron precipitation. No response is observed in any channel in coincidence with the 1 ma pulses. Note
that the response in the energetic particle detectors was not linear
with beam current, the 10 and 100 ma intensities are comparable for
the first sequence A and the 1 ma response is not apparent, which
it would have been if it were 1/10 the magnitude of the 10 ma response.

No beam pulses were observed in the 8 keV detector channel at
any time during the flight. At the time shown in Figure 1 the converter system was being loaded by the 8 keV, 100 ma pulses and as a
result the gun was emitting particles with energies slightly (∿10%)
less than the anticipated beam energy (E_B) of 8 keV. Also data
from the mother section indicate that, at high altitudes, the mother
charged positively during 10 and 100 ma pulses which would have the
net effect of reducing the free space beam energy. Thus the absence
of a response at 8.02 keV could be due to a combination of these
two effects.

Variations in the charged particle sensor responses which are
clearly visible over the four seconds displayed in Figure 1, indicate
that the response amplitude is dependent on parameters other than
simply the beam intensity and energy. The dependence on beam injection pitch angle (Θ_B) and particle detection angle is not investigated in any detail here.

Table 1 lists $d_\perp$ (max), the distance $d_\perp$ at which the last beam
pulse was detected in the energetic particle detector (EPD) array,
for 10 and 100 ma pulses at three values of E_B. Also listed is the
value of Θ_B for that pulse, and the value of $d_\perp$ (max) normalized
to the beam gyroradius (ρ_B).

Table 1. Last Observable Responses in EPD Array

E_B (keV)	I_B (mA)	Last Pulse $d_\perp$ (m)	$d_\perp/\rho_B$	Θ_B
2	10	5.7	2.26	69°
2	100	5.9	2.33	57.1°
4	10	13.22	3.60	77.5°
4	100	22.6	6.16	73.3°
8	10	5.3	0.89	84.6°
8	100	30.2	5.79	86.9°

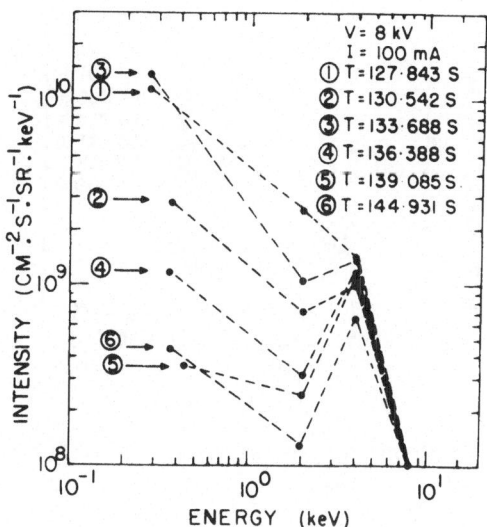

Fig. 2.   Induced electron energy spectra as a function of time for
          8 keV, 100 ma pulses.

     Although a considerable amount of scatter is evident in the
table, probably due to beam injection and particle detection angle
effects, it is seen that $d_\perp$ (max) increases with increasing beam
energy (E_B) and current (I_B).   Generalizing somewhat, it appears
that $d_\perp$ (max) is 5 or 6 ρ_B for 100 ma and approximately 3 ρ_B for the
10 ma beam.   Of course no responses due to the 1 ma beams were detected, therefore $d_\perp$ (max) for 1 ma was less than or of the order
of 2 ρ_B.   Note that energetic particles are observed at a maximum
distance perpendicular to the magnetic field for beam injections
near 90° pitch angle.

     Four point energy spectra derived from the three fixed energy
analysers and the stepped analyser are plotted in Figures 2, 3 and
4 for the 100 ma pulses at 1.9, 4.0 and 8.0 keV.   The times at which
samples were taken at 8 keV are shown in Figure 2.   Near the gun,
sample (1), the spectrum is characterized by a monotonically decreasing intensity with increasing energy out to the approximate
beam energy where the intensity drops rapidly to background levels.
As mentioned previously E_B was slightly less than 8 keV at this
time and particles at this energy would be below the bandpass of
the 8.03 keV detector.

     At increasing distances from the mother (increasing time) the
intensity near E_B remains relatively constant while the lower energy
electrons rapidly decrease in flux resulting in an energy distribution peaked near the beam energy.   All beam induced electron fluxes
drop rapidly to background at $d_\perp$ (max) (see Table 1).

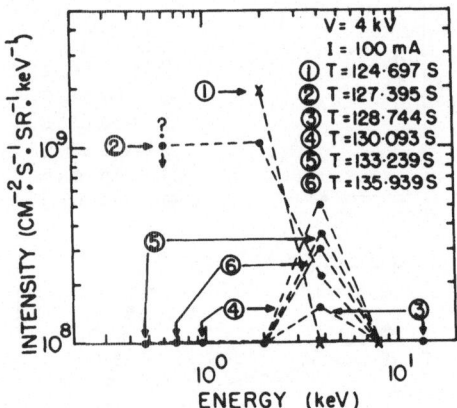

Fig. 3.   Induced electron energy spectra as a function of time for
         4 keV, 100 ma pulses.

In Figure 3 the corresponding results for 4 keV, 100 ma pulses
are presented. Similar, though less systematic, responses are observed here. Near the mother intense fluxes of low energy electrons
are observed while at larger distances the beam induced electron
distributions take on a mono-energetic appearance with a peak energy
near or at E_B.

Because of the poor sampling statistics (only the first two
pulses produced measurable responses), very little is known of the
spectrum during 1.9 keV 100 ma firings. It is clear from Figure 4,
however, that no significant fluxes of electrons above E_B were observed and a large flux was present near E_B.

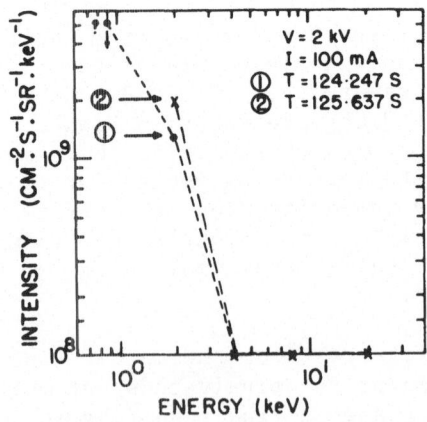

Fig. 4.   Induced electron energy spectra as a function of time for
         2 keV, 100 ma pulses.

Measurements of the supra-thermal electron flux (STEF) and the thermal electron flux (TEF) along with the electron gun program for times immediately after payload separation are displayed in Figure 5. The sampling period of this sensor was 50 ms which produced the delay apparent in these data. As can be seen in the top panel, when the R.P.A. was biased to measure the STEF (bias voltage = -3.47 V) a logarithmically increasing flux was observed for the 1, 10 and 100 ma, 4 keV beam pulses. Thus the STEF increased approximately linearly with beam current. On the next series of gun pulses with E_B = 8 keV, no response is observed at 1 ma whereas large fluxes are seen for the 10 and 100 ma pulses. No further clearly identifiable responses to 1 ma pulses were observed thereafter.

Profiles of the STEF response as a function of $d_\perp$ were similar to those of the EPD array, that is, relatively flat out to $d_\perp$ (max) and then falling rapidly to zero. Table 2 is a listing of $d_\perp$ (max) in the STEF channel along with the corresponding gun injection angle. The 4 and 8 keV channels are identical to those of Table 1; only the 2 keV channel differs, with the 2 keV responses being observed out to a larger $d_\perp$. This difference is presumably due to the more omnidirectional response of the STEF sensor.

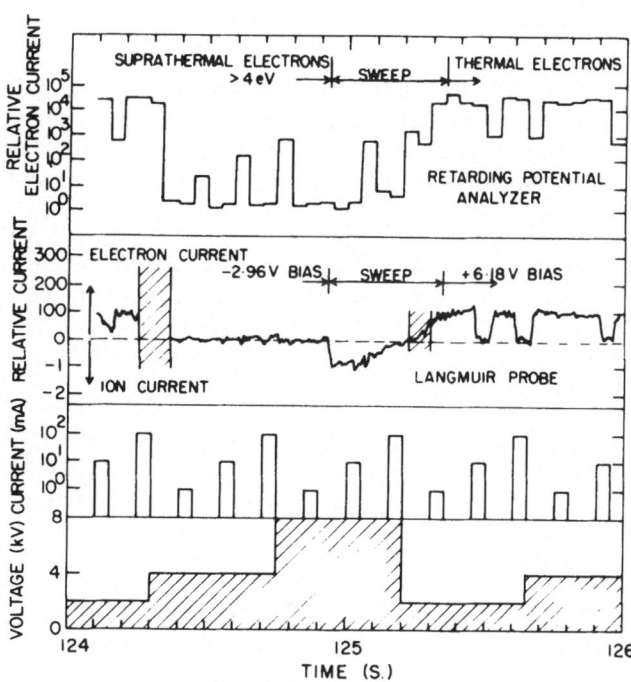

Fig. 5. Suprathermal electron and thermal electron fluxes observed immediately after payload separation. Also shown is the corresponding gun program.

Table 2. Last Observable Responses in STEF

E_B (keV)	I_B (mA)	Last Pulse $d_\perp$ (m)	$d_\perp / \rho_B$	Θ_B
2	10	9.7	3.83	23.7^o
2	100	9.85	3.89	36.1^o
4	10	13.22	3.60	77.5^o
4	100	22.6	6.16	73.3^o
8	10	5.3	0.89	84.6^o
8	100	30.2	5.79	86.9^o

It should also be noted in Figure 5 that during 10 and 100 ma beam firings the thermal electron flux to the sensor decreases drastically as indicated by both sensors when in the thermal electron collection mode (T+ $\geq$ 125.35). This response is most easily interpreted as being due to a negative charging of the daughter section, a large increase in electron temperature, or more likely a combination of both. A detailed study of this response during these periods is required to define the relative contributions of the two effects.

SUMMARY AND DISCUSSION

The data presented here clearly show that, during high current ($I_B \geq$ 10 ma) electron beam firings, an intense suprathermal as well as energetic electron population is created on flux tubes near the beam. Some of the characteristics of this population generated by beam are:

(i) Close to the gun, that is, at distances across magnetic field lines $d_\perp \leq 2 \rho_B$, the induced energetic electron distribution has a monotonically decreasing intensity with increasing energy out to the beam energy E_B where the intensity drops sharply below the limit of detectability (Figure 2, curve (1)).

(ii) At increasing $d_\perp$ the intensity near E_B remains relatively constant while the flux of electrons with $E \leq E_B$ decreases rapidly with $d_\perp$. The intensities detected at all energies then falls rapidly below the limit of detectability at $d_\perp = d_\perp$ (max) where $d_\perp$ (max) $\cong$ 5 ρ_B for 100 ma beams.

(iii) The dimensions of the perturbed volumes $d_\perp$ (max) are dependent on beam current. The responses to the 1 ma beams were restricted to $d_\perp \leq 2 \rho_B$ whereas $d_\perp$ (max) $\cong$ 3 ρ_B for 10 ma and $d_\perp$ (max) $\cong$ 5 ρ_B for 100 ma. This result is not dependent on instrument sensitivity (see results section).

The scatter in the measured $d_\perp$ (max), shown in Tables 1 and 2, is probably due to the dependence of this parameter on the beam injection and particle detection angles. One trend which is clearly evident in the data is for $d_\perp$ (max) to occur for beam injection angles $\Theta_B \cong 90°$.

The most obvious scenario suggested by these observations is the following:

For low beam currents (~ 1 ma), the electron beam is at least quasi-stable, that is, the growth rates for instabilities are low enough that no beam scattering or diffusion occurs locally and the beam electrons travel away from the gun in single particle orbits and are confined to within one gyrodiameter of the gun.

For 10 and 100 ma pulses a local instability occurs which diffuses the primary beam in pitch angle, energy, and across the local magnetic field, simultaneously heating the ambient electron population. The radial distance across magnetic field lines populated by this hot electron plasma is dependent on beam current (increasing with increasing current) and scales according to the primary electron beam gyroradius.

Certain similarities exist between these measurements and corresponding ones made in the Houston vacuum tank (see Bernstein et al. and Kellogg et al., this volume) suggesting that the same instability observed in the laboratory is occurring at high altitudes in the ionosphere.

ON THE USE OF ARTIFICIALLY INJECTED ENERGETIC ELECTRONS AS INDICATORS OF MAGNETOSPHERIC ELECTRIC FIELDS PARALLEL TO THE MAGNETIC LINES OF FORCE

Klaus Wilhelm

Max-Planck-Institut für Aeronomie
D-3411 Katlenburg-Lindau 3
Federal Republic of Germany

INTRODUCTION

The observation of magnetospheric electric fields parallel to the magnetic field direction could conveniently be performed by measuring their effects on the propagation of artificially injected energetic charged particles. Unfortunately, the experimental difficulties are tremendous. Nevertheless, initial results in this context have been obtained and have been presented by Haerendel et al. (1976) for ions and Wilhelm et al. (1980) and Wilhelm (1980) for electrons. With a view to formulating an adequate procedure for interpreting these and future measurements, an inversion procedure will be devised that would allow the computation of the electric field distribution as a function of height from the particle transit time measured as a function of pitch angle.

In general, the presence of a sufficiently strong electric field with appropriate direction above the magnetospheric injection point of the energetic charged particles will lead to their reflection and provide the basis for observing the particle transit time T. It normally will be a function of the particle energy and its magnetic moment

$$T = T(W, \mu) \tag{1}$$

where $\mu = W(s_0)B^{-1}(s_0)\sin^2\alpha(s_0)$ is the magnetic moment
s_0 the injection point measured from the earth's surface along the magnetic field
$W(s_0)$ the injection energy
$B(s_0)$ the magnetic field induction at s_0
and $\alpha(s_0)$ the particle injection pitch angle.

Particles will be considered as moving adiabatically with constant μ allowing a description of their motion in the guiding centre approximation (cf. e.g. Roederer, 1970). A conversion procedure yielding the electric field distribution from observations of the particle transit time as a function of the injection energy was formulated by Wilhelm (1977) for particle injection with constant magnetic moment. A constant magnetic moment μ at s_0 leads to the requirement

$$\alpha(s_0) = \text{arc sin } \sqrt{\mu B(s_0)/W(s_0)} \tag{2}$$

which is experimentally difficult to fulfill for varying $W(s_0)$. It would be much easier to vary the injection pitch angle at constant energy and determine the transit time as a function of μ. It therefore seems desirable to extend the inversion procedure to the case $W = \text{const}$ and $T = T(\mu)$.

CONVERSION OF $T(\mu)$

When an electric field $E_{\shortparallel}(s)$ parallel to the magnetic field direction is present, the influence of the potential

$$\phi(s) = - \int_{s_0}^{s} E_{\shortparallel}(s)\, ds \tag{3}$$

with $\phi(s_0) = 0$ on the motion of a charged particle can be described by a generalization of the magnetic moment in Eq. 1

$$\mu = [W(s_0) - q\phi(s)]B^{-1}(s)\, \sin^2 \alpha(s) \tag{4}$$

where q is the electric charge of the particle. By noting that the non-relativistic particle kinetic energy can be written as

$$W(s_0) - q\phi(s) = \frac{m}{2}(v_\perp^2(s) + v_{\shortparallel}^2(s)) \tag{5}$$

in terms of the transverse and parallel velocities with respect to $\underline{B}$ and that

$$v_\perp^2(s) = \frac{2}{m}[W(s_0) - q\phi(s)]\sin^2 \alpha(s) \tag{6}$$

the parallel velocity can be obtained as

$$v_{\shortparallel}(s) = \pm\sqrt{\frac{2}{m}[W(s_0) - q\phi(s) - \mu B(s)]} \tag{7}$$

Reflections below the peak altitude of a magnetic field line can only occur if

$$\text{sign } [q\phi(s)] > 0 \tag{8}$$

which is a necessary but not sufficient condition for the occurrence of fast particle echoes. The echo or transit time is

$$T = 2 \int_{s_o}^{s_m} \frac{ds}{v_{,,}(s)} \tag{9}$$

where the reflection, defined by $v_{,,} = 0$, occurs at $s = s_m$. Note that the integral in Eq. 9 in general is finite even if $v_{,,}$ approaches zero. Following the arguments presented in the introduction, the energy $W(s_o)$ should be kept constant during injection while varying the magnetic moment. By inserting the velocity $v_{,,}$ of Eq. 7 into Eq. 9, the transit time $T(\mu)$ can be written in a form that is amenable to further deductions. To simplify the situation, only magnetic dipole configurations should be considered thus allowing the substitution

$$B(s) = B_o R_E^3 (R_E + s)^{-3} \tag{10}$$

with R_E the earth's radius and $B_o = B(o)$ the magnetic field on the surface of the earth.

Under the assumption that the transit time $T(\mu)$ in Eq. 9 is known from observations, the integral equation has to be solved in order to find $\phi(s)$. Analogously to the treatment by Wilhelm (1977), the integral equation will be transformed into an equation of Abel's type. Here a rectified magnetic moment

$$M(s) = [W(s_o) - q\phi(s)] B^{-1}(s) \tag{11}$$

will be defined as a first step. From Eqs. 4 and 11 it follows that

$$M(s) = \mu \sin^{-2} \alpha(s) \tag{12}$$

which may be helpful in interpreting some graphical presentations of M(s) curves in Fig. 1. It should be noted in this figure that $M_{min} > 0$ does only exist when the particle energy is sufficiently high to permit penetration of the electric potential barrier. At s_o only particles with

$$\mu \leq W(s_o) B^{-1}(s_o) = M(s_o) \tag{13}$$

can be injected. They belong to a magnetic moment range specified as regime I. If they travel up the field line, they will be reflected at s_m and be lost in the atmospheric loss cone should $\mu < M_A$. Otherwise, they will be magnetically reflected between s_o and s_A and consequently be trapped between two mirror points. A particle with $\mu' > M(s_o)$ (regime II) injected above s_o will similarly be trapped between s_m and s_o'. In order to be able to substitute M in Eq. 9, a solution of M(s) in terms of s has to be found. The type 3 curve in Fig. 1, which we will consider as the general case with one upward-directed electric field region above the injection point, shows that there is no unique solution.

However, unique solutions exist in certain ranges as follows

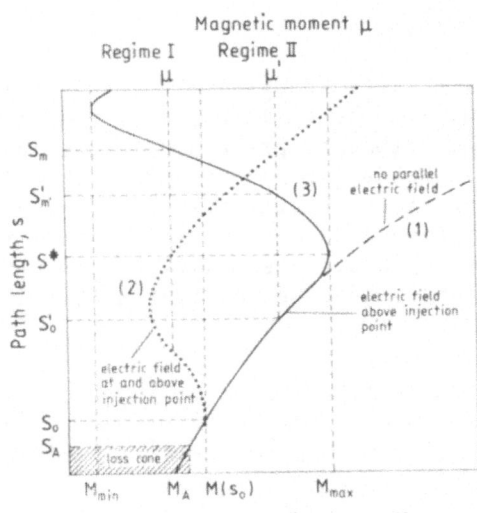

Fig. 1. Three typical patterns of rectified magnetic moment functions. Type 1 describes configurations without parallel electric field. Type 2 is representative for a strong electric
field at and above the injection point, whereas Type 3 results from an electric field region above the injection
point. Two test particles with magnetic moments μ and μ' are
considered in configuration 3. A particle injected at s_0
with μ will be reflected at s_m and then be lost in the atmosphere. A particle with μ', however, will be trapped
between s'_m and s'_0.

$$s = \begin{cases} g_1(M) \ ; & s_0 \le s \le s^* \\ g_2(M) \ ; & s^* < s \le s_m \end{cases} \tag{14}$$

Differentiation yields

$$ds = \begin{cases} g'_1(M)dM \ ; & s_0 \le s \le s^* \\ g'_2(M)dM \ ; & s^* < s \le s_m \end{cases} \tag{15}$$

From the discussion of Fig. 1 it follows that Eq. 9 can be written
in the form

$$T(\mu) = 2 \begin{cases} \int_{s_0}^{s^*} \dfrac{ds}{v_{||}} + \int_{s^*}^{s_m} \dfrac{ds}{v_{||}} & \text{for } \mu \text{ in regime I} \\[2em] \int_{s'_0}^{s^*} \dfrac{ds}{v_{||}} + \int_{s^*}^{s'_m} \dfrac{ds}{v_{||}} & \text{for } \mu \text{ in regime II} \end{cases} \tag{16}$$

thereby extending the range of definition beyond $\mu \leq M(s_0)$ of formula 13 to values of $\mu \leq M_{max}$. It should be noted, however, that $T(\mu)$ cannot be observed from s_0 in the range $M(s_0) < \mu \leq M_{max}$. In principle, the measurements could be made on a sounding rocket payload travelling upwards to at least $s*$. In view of the experimental difficulties involved, it is rather fortunate that reasonable estimates of the unaccessible portion of $T(\mu)$ can be obtained without measurements as will be explained later.

Combining now Eqs. 7, 10, 11, 14, 15 and 16 gives

$$T(\mu) = \sqrt{\frac{2m}{B_0 R_E^3}} \int_\mu^{M_{max}} \{\sqrt{[R_E + g_1(M)]^3}\, g_1'(M) - \sqrt{[R_E + g_2(M)]^3}\, g_2'(M)\} \frac{dM}{\sqrt{M-\mu}} \qquad (17)$$

if we define $g_1'(M) \equiv 0$ for all M in regime I. This definition is appropriate as it means that the transit time is reckoned from and to the injection point. Abel's integral equation can be obtained by the final substitution

$$G_{1,2}'(M) = \sqrt{[R_E + g_{1,2}(M)]^3}\, g_{1,2}'(M) \qquad (18)$$

resulting in

$$T(\mu) = \sqrt{\frac{2m}{B_0 R_E^3}} \int_{M_{max}}^\mu [G_2'(M) - G_1'(M)] \frac{dM}{\sqrt{M-\mu}} \qquad (19)$$

A solution of this equation can be found in textbooks (e.g. Schlögl, 1956) under the assumption that $T(\mu)$ is a continuously differentiable function vanishing at $\mu = M_{max}$. The fact $T(M_{max})=0$ immediately follows from physical considerations as such a particle is sitting at $s*$ without $v_{\shortparallel}$ and, therefore, will give a delay time of the order of one cyclotron period which is insignificant in our guiding centre approximation. $T(\mu)$ may however by a discontinuous function at $\mu = M_{max}$. A simple example would be a particle in a field distribution corresponding to a harmonic oscillator as discussed by Roederer (1970). In this particular case, the transit time would not depend on μ. More generally, it can be shown that the transit time will only be continuous near $\mu = M_{max}$ if $v_{\shortparallel} \propto ds^n$ with $n < 2$. It will be demonstrated later that this difficulty will not lead to serious implications in the context of our problem.

Another singularity at $\mu = M(s_0)$ deserves greater attention. Injection at s_0 with $\mu = M(s_0)$ gives a finite transit time T for case No. 3. A decrease in μ will lead to decreasing T as particles with greater parallel velocity travel faster than those with $\sin \alpha(s_0) = 1$. Only when $\sin \alpha(s_0) \ll 1$, the greater penetration depth into the electric field region might eventually cause an enhancement of T with decreasing μ. On the other hand, the transit time T is normally increasing with decreasing μ in the interval $\langle M(s_0), M_{max}\rangle$. A typical transit time function is given in Fig. 2. The values in regime I are observable from s_0 whereas in regime II either calculations or interpolations are required. It follows that

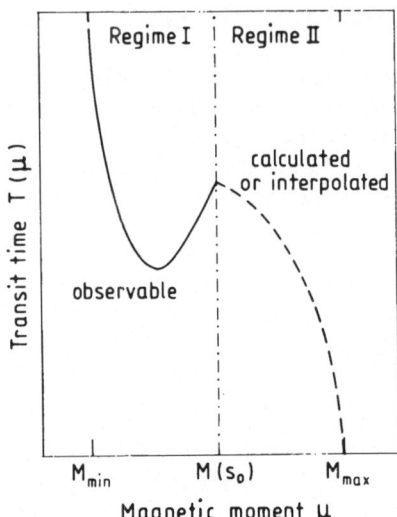

Fig. 2.   Transit time T as function of the magnetic moment of the
test particle. The curve represents the characteristics
of case No. 3 in Fig. 1. The function is observable from
a spacecraft located at s_o in regime I. The function $T(\mu)$
obviously is not continuously differentiable at $\mu = M(s_o)$.

the function $T(\mu)$ cannot be assumed as continuously differentiable
at $\mu = M(s_o)$. This behaviour is clearly related to the definition
$g_1'(M) = 0$ for all M in regime I made in order to suppress the influence of the lower reflection. If we refine this definition by

$$g_1^*{}'(M) \;=\; f(M,\delta)\; g_1'(M) \tag{20}$$

$$\text{with}\quad f(M,\delta) = \begin{cases} 0 \\ [M-M(s_o)+\delta]/2\delta \;\text{for all M in} \\ 1 \end{cases} \begin{cases} \langle M_{min},\; M(s_o)-\delta \rangle \\ \langle M(s_o)-\delta,\; M(s_o)+\delta \rangle \\ \langle M(s_o)+\delta,\; M_{max} \rangle \end{cases}$$

and modify the substitution (18) by writing $g_1^*{}'(M)$ instead of $g_1'(M)$,
the integrand in Eq. 19 will become continuous and $T(\mu)$ therefore
continuously differentiable. The usefulness of this defintion follows
from the fact that the contributions of the integral

$$\int_{M(s_o)-\delta}^{M(s_o)+\delta} [\, G_2'(M) - G_1'(M)\, \frac{M-M(s_o)+\delta}{2\delta} \,]\; \frac{dM}{\sqrt{M-\mu}} \tag{21}$$

to the transit time vanishes with diminishing δ. It is easy to give
a physical interpretation of this mathematical formalism. In Fig. 3
showing an enlarged portion of Fig. 1, a modified rectified magnetic
moment function has been drawn near $M = M(s_o)$. The modification

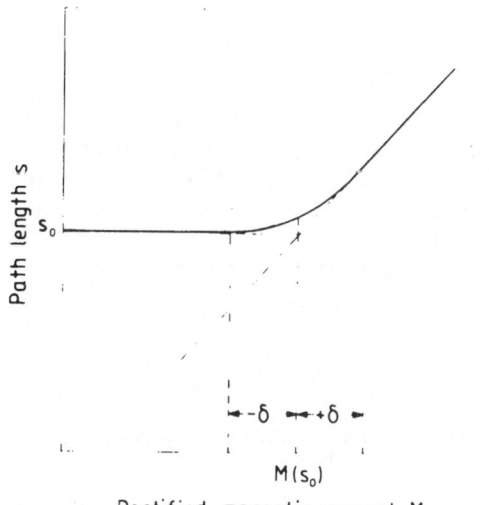

Fig. 3. Schematic presentation of the rectified magnetic moment function and its modification in the interval $\langle M(s_0)-\delta,\ M(s_0)+\delta\rangle$

obviously amounts to adding a strong electric field directed downwards along the magnetic field direction near s_0. In calculating
the total transit time, the second (lower) reflection will thus
not be suppressed. Its contribution to the transit time, however,
can be neglected.

A solution of Abel's integral equation can now be written in
the form

$$G_2(M) - G_1'(M) = -\frac{1}{\pi}\sqrt{\frac{B_oR_E^3}{2m}}\ \frac{d}{dM}\int_{M_{max}}^{M}\frac{T(\mu)d\mu}{\sqrt{\mu-M}} \qquad (22)$$

Before Eq. 22 can be integrated, the effect of M_{max} in definition 20
has to be considered as far as Eq. 15 is concerned. By using geometric arguments derived from Fig. 3, we conclude that the impact
on s also becomes insignificant with diminishing δ. Taking into
account Eq. 18, we can thus write

$$\sqrt{[R_E+g_2(M)]^5}\ \Big|_{M_{max}}^{M} - \sqrt{[R_E+g_1(M)]^5}\ \Big|_{M_{max}}^{M} = \frac{-5}{2\pi}\sqrt{\frac{B_oR_E^3}{2m}}\int_{M_{max}}^{M}\frac{T(\mu)d\mu}{\sqrt{\mu-M}} \qquad (23)$$

and $\quad \sqrt{(R_E + s)^5} - \sqrt{(R_E + s_o)^5} = AI(M) \qquad (24)$

with $\quad A = \frac{5}{2\pi}\sqrt{\frac{B_oR_E^3}{2m}}$

and $\quad I(M) = \int_{M}^{M_{max}} \dfrac{T(\mu)d\mu}{\sqrt{\mu - M}}$

Eqs. 23 and 24 are valid for all M in regime I. In regime II, the point s_0 has to be replaced by the lower reflection point s_0'. Since s_0' is not known, the contributions of s and s_0' cannot be separated in this instance. As in any case, we only get transit time observations if there is a significant electric field effect in regime I, the restriction of Eq. 24 to that regime does not seriously impact its applicability. Eq. 24 can be solved to give

$$ s = \sqrt[5]{[AI(M) + \sqrt{(R_E + s_0)^5}]^2} - R_E \qquad (25) $$

as a function of M in regime I. Applying Eqs. 10 and 11, finally, provides the electric potential

$$ \phi[s(M)] = -\frac{1}{q} \left\{ \frac{MB_o R_E^3}{[R_E + s(M)]^3} - W(s_0) \right\} \qquad (26) $$

as a function of M or s. In calculating the integral I(M), the complete $T(\mu)$ curve has to be known. Yet the portion in the interval $\langle M(s_0), M_{max} \rangle$ is not observable from s_0. Realizing that the function $T(\mu)$ is bounded in the unknown regime and, for small M, contributes relatively little to the integral there provides the possibility of using suitable interpolations without seriously compromising the accuracy of the results. It is clear that such an interpolation should fulfill the requirements on $T(\mu)$ for the existence of a solution of Abel's integral equation. Rather than theoretically discussing the effects of an interpolation in more detail, a numerical example will be treated in the next section.

MODEL CALCULATIONS

Some applications of the conversion procedure derived in the previous section will be performed by assuming certain electric field configurations, calculating their effects on the transit time functions, executing the inversions and comparing the resulting electric fields with the original model assumptions. The model electric field was assumed to be either constant or following a Gaussian distribution along the field line of the form

$$ E_{\shortparallel}(s) = U(2\pi\sigma^2)^{-1/2} \exp{-(s-\bar{s})^2/2\sigma^2} \qquad (27) $$

with $\quad \bar{s} \quad$ mean value of s (representative of the location of the electric field maximum)

$\quad\quad\quad \sigma^2 \quad$ variance of s (representative of the width of the electric field region)

$\quad\quad\quad U \quad$ potential difference between $s = \pm \infty$

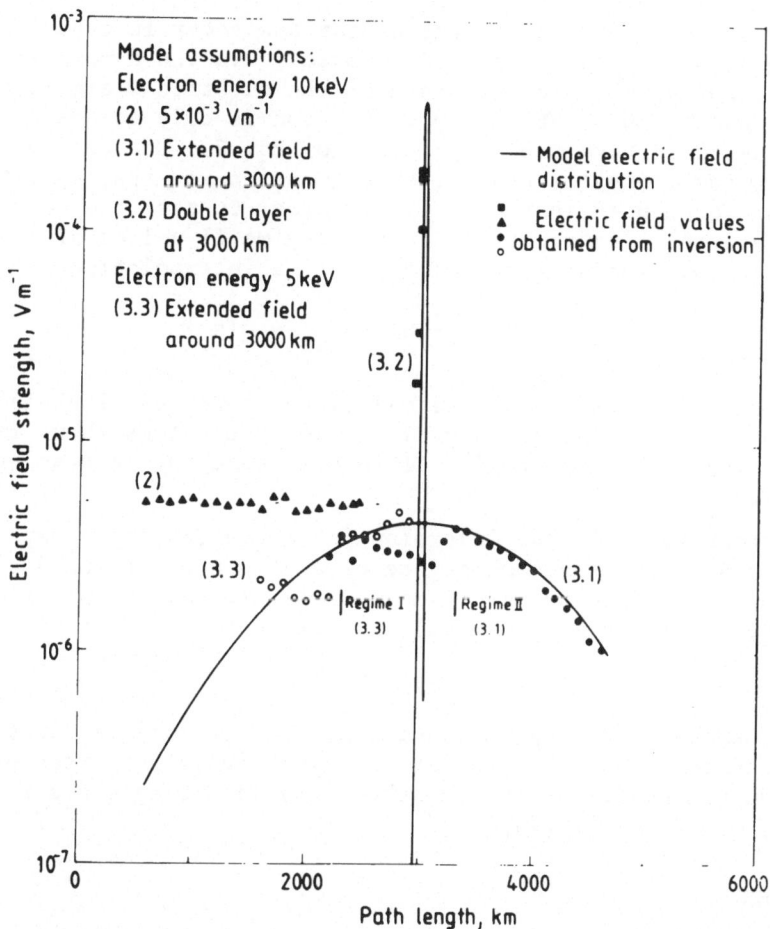

Fig. 4. Comparison of the model electric field distributions with the electric field values determined with the help of the inversion procedure.

Fig. 4 demonstrates the results obtained for electrons with 10 keV of energy and an electric field of 5 mv/V extending from s_0 = 500 km upwards (case 2) and for Gaussian distributions with $\bar{s}$ = 3000 km, σ = 1000 km, U = 10 kV and s_0 = 500 km (case 3.1) and $\bar{s}$ = 3000 km, σ = 10 km, U = 10 kV and s_0 = 500 km (case 3.2). In addition, the electric field with s = 3000 km, σ = 1000 km, U = 10 kV is probed with 5 keV electrons in case 3.3. The models with σ = 1000 km or 10 km may be representative for an extended or an localized electric field configuration, respectively. The comparison of the model electric field with the electric field obtained in the inversion process can be considered to be quite favourably for all cases. Small discrepancies mainly result from a finite step width during the numerical integrations that can introduce considerable errors near the reflection points. The most significant deviations occur in case 3.2 where the integration step width of 0.5 km used 11 in the numerical calculations is not matched to the width of the electric field layer. It also becomes obvious that the approximation is only acceptable in regime I as could have been expected from Eq. 24. One of the difficulties in inverting the transit time function was that an interpolation had to be used in regime II. In order to test the sensitivity of the resulting electric field on the regime II values of $T(\mu)$, the case 3.1 calculation were repeated with the interpolation

$$T(\mu) = T [M(s_o)] \sqrt{(M_{max} - \mu)/(M_{max} - M(s_o))} \qquad (29)$$

for $M_o < \mu \leq M_{max}$ instead of the calculated function. The electric field thus obtained deviated by -15 % from the values shown in Fig.4 near the transition from regime II into I where the largest errors should occur.

In order to apply the inversion procedure developed here to observational data with some degree of confidence, it would be required to measure the transit time as a function of μ in a wide range, preferably from 0 to $M(s_o)$, a task that should be undertaken in future projects.

ACKNOWLEDGEMENTS: W. Degenhardt and H. Lindner assisted in the numerical computations. The project was financially supported by the German Bundesministerium für Forschung und Technologie under grant No. 01 OM 046-ZK-A-WRK 274/3.

REFERENCES

Haerendel, G., Rieger, E., Valenzuela, A., Föpple, H., Stenbaek-Nielsen, H.C., Wescott, E.M., 1976, First observation of electro-static acceleration of Barium ions into the magnetosphere, Proceedings of the symposium on European programmes on sounding-rocket and balloon research in the auroral zone, ESA SP 115:203.
Roederer, J.G., 1970, Dynamics of geomagnetically trapped radiation. Physics and Chemestry in Space, Vol. 2 (ed. J.G. Roederer and J. Zähringer), Springer-Verlag, Berlin-Heidelberg-New York.
Schlögl, F., 1956, Randwertprobleme, in: Handb. Phys., Bd. 1 (S. Flügge, ed.), Springer-Verlag, Berlin-Göttingen-Heidelberg.
Wilhelm, K., 1977, Remote sensing experiment for magnetospheric electric fields parallel to the magnetic field, J. Geophys. 43:731.
Wilhelm, K., 1980, Natural and artificially injected electron fluxes near discrete auroral arcs. Proceedings of the Vth ESA-PAC Symposium on European Rocket & Balloon Programmes & Related Research, Bournemouth, ESA SP-152:407.
Wilhelm, K., Bernstein, W., Whalen, B.A., 1980, Study of electric fields parallel to the magnetic lines of force using artificially injected energetic electrons. Geophys. Res. Lett. 7:117.

DISCUSSION

<u>Koons</u>: Is it possible that you are seeing "reflection" from strong diffusion in a region of anomalous resistivity rather than from a parallel electric field?

<u>Wilhelm</u>: The experiment concept depends on measuring the transit time of energetic electrons. The detector determines the energy and pitch angle of the particles and thus the calculated travel time on the downleg can be compared with half the measured transit time in order to arrive at an estimate of the reflection height. The reflection process itself cannot be deduced without further considerations. Should the assumption not be valid that the observed particles behaved adiabatically (except for the injection process), even the height estimate may be in error.

THE FRENCH-SOVIET EXPERIMENTS ARAKS : MAIN RESULTS

J. Lavergnat

L.G.E.
4 Avenue de Neptune
94100 Saint-Maur, France

INTRODUCTION

The French-Soviet Araks experiments (Artificial Radiation and Aurora between Kerguelen and the Soviet Union) were designed to study the injection of an electron beam into the ionospheric plasma. To this end two rockets were launched on 26 January and 15 February 1975 from the Kerguelen Islands who have the special property to be magnetically conjugated with a land-point close to Karpagor. Many ground-based measurement facilities were set up in conjunction with these experiments, with emphasis on radar measurements in the Northern Hemisphere as well as on the VLF and VHF measurements at both hemispheres.

The scientific objectives of the Araks experiment are mainly twofold (i) to study the motion of an electron beam inside the magnetosphere with a particular attention focused on the large scale phenomena (e.g. location of the conjugate point) (ii) to study the interaction of the beam with the surrounding plasma.

Both of these purposes are common to most of the active experiments, moreover as extended results were published elsewhere (special issue of Ann. de Geophys. 1980), so this paper is concentrated on three points where the Araks experiments brought fundamental insights in the process involved by particle injection.

GEOMETRY AND EXPERIMENTAL SET-UP

The final stage of each rocket included two complementary experiment systems : an electron gun, indirect potential measurement devices, particle flux detectors, and a cone ejected at a speed of

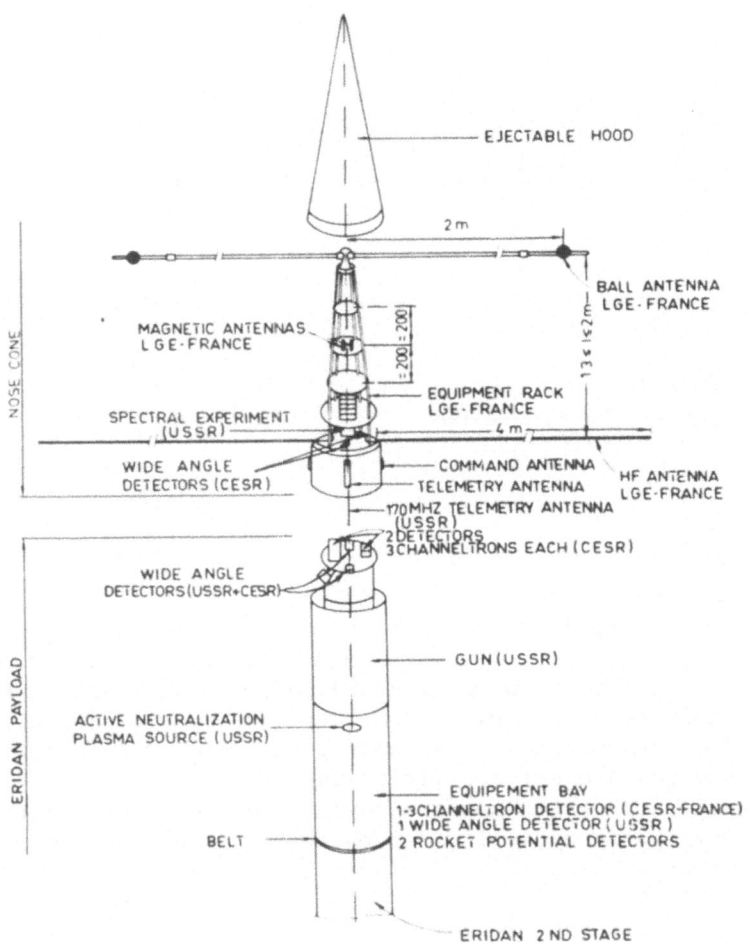

ANGLE BETWEEN THE BEAM AND THE ROCKET AXIS	ENERGY keV	CURRENT (AMPERES)	PROGRAM
30° FOR THE EASTWARD LAUNCH 0° FOR THE NORTHWARD LAUNCH	27 keV THEN 15 keV	0,5	
70°	27 keV THEN 15 keV	0,5	
140°	27 keV THEN 15 keV	0,5	
TOTAL DURATION OF THE SEQUENCE : 12.80 SECONDS			

Fig. 1. Scientific equipment in the payload and the nose cone together with the gun sequence.

40 ms^{-1} from the main payload (Fig. 1). This cone carried antenna to detect the waves created by the electron beam during its penetration into the ionosphere. Details can be found in Cambou (1980). Here it is sufficient to mention the cesium plasma source mounted in principle to ease the neutralisation but which has important "secondary" effects. Two rockets were launched : (i) one was fired towards magnetic east in order to compensate for the curvature and gradient drift of reflected electrons in the magnetic field. (ii) the other was fired towards the magnetic north, first, in order to reduce the size of the impact area of the beam at the conjugate point and thus to facilitate its observation, and second to obtain more information from the observation of the beam emission. The culmination in both cases was around 190 km and the distance between the nose cone and the electron gun reached 10 km but was less than 2 km from the magnetic field line passing through the main payload during the North launch.

The electron beam had the following characteristics (Fig. 1) : intensity 0.5 A ; energy 27 keV or 15 keV ; 3 emission angles 0° for the Northward launch or 30° for the Eastward one, 70° and 140° with the respect to the axis of rotation of the rocket. A variable pulse duration (20 ms, 1.28 s, 2.56 s) was adopted in order to have either an accurate definition of the injection angle (20 ms) or a large amount of energy deposited in the atmosphere (2.56 s).

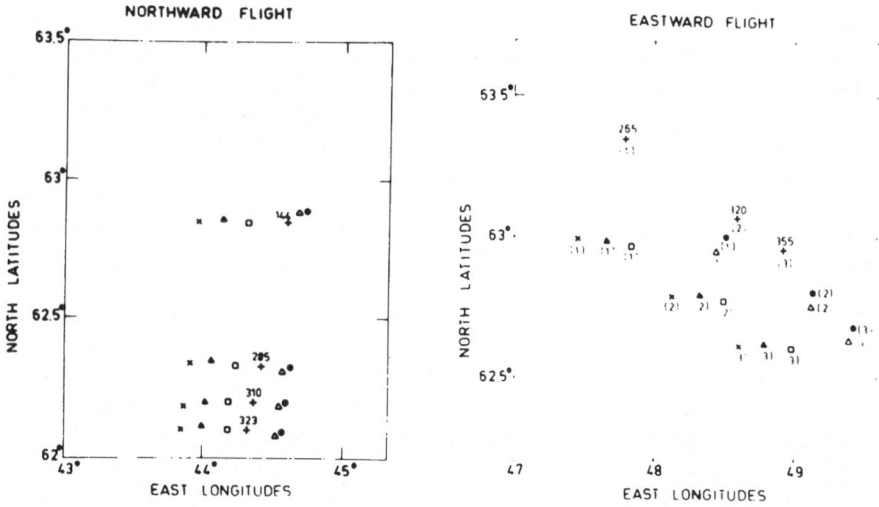

Fig. 2. Geographical position of radar echoes and calculated Araks rocket conjugate points for different models during the flights. The numbers linked to the radar echoes give the corresponding flight time of the rocket (in seconds).

CONJUGATED OBSERVATIONS

    As mentioned above the first scientific objective was to investigate conjugate effects. Observations of visible artificial aurora were however prevented by the bad weather in the Arckhangelsk region during all the campaign. Nevertheless Araks has demonstrated that it is possible to detect the ionisation trails created by the beam penetrating in the conjugate atmosphere by ground based radar observation. This fact is, by itself, of importance when considering the use of particles injection for the mapping of the magnetospheric field lines because the radar technic is free from climatic conditions. It must be noted that the equivalent power of radar π was around 220 dB for a frequency close to 40 MHz. ($\pi = P_t\,G_t\,G_r\,P_r^{-1}$ where P_t is the radiated power, $G_{t,r}$ the antenna gains and P_r the noise power of the receiver).

    Besides this general consideration two main phenomena appear : the location of the radar echoes and their temporal characteristics.

    When interpretating the data the first question which arises is : what accuracy is achievable ? For the Northward flight we are only able to say that the total error in the distance determination is less than ± 2 km, the accuracy in azimuthal estimates being limited by the width of the radar beam at ± 6°. For the Eastward flight a biangulation method allows to reduce the total uncertainty to ± 10 km.

    The Figure 2 shows the more significant results together with the computed values provided by different models (a combination of internal and external models was used) for both flights. For the Northward experiment all the models give the good latitude but the experimental inaccuracy in longitude prevents us to draw any conclusion. For the Eastward experiment some discrepancy appears in latitude determination, this can be attributed to a long period of moderate magnetic activity. Nevertheless these results show the value of this kind of technic for checking the different magnetic field models.

    The temporal characteristics of the echoes exhibit behaviour which has not been forecast. The variation of the slant distance is caused by the lateral movement of the rocket during electron injection. The interval of time, in which this distance is reduced (for the Northward flight), is fairly equal to the time when the pitch angle of injection is close to 0°. In each sequence the echoes pursue more than 10 sec after the full injection time but without change in the mean distance. Fig. 3 summarizes the observations during the clearer period of gun working. A spectral analysis of the echoe amplitude shows two periods : 4.16 s and 1.72 s. The latter is close to the bounce period of 15 keV electrons. This fact

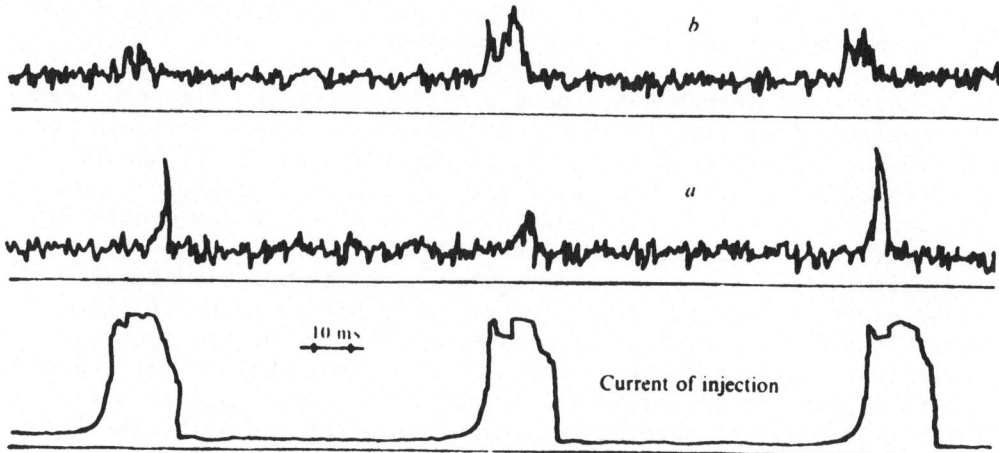

Fig. 3. Correlation between the evolution of radar echoes during the Northward flight and the sequence of electron gun.

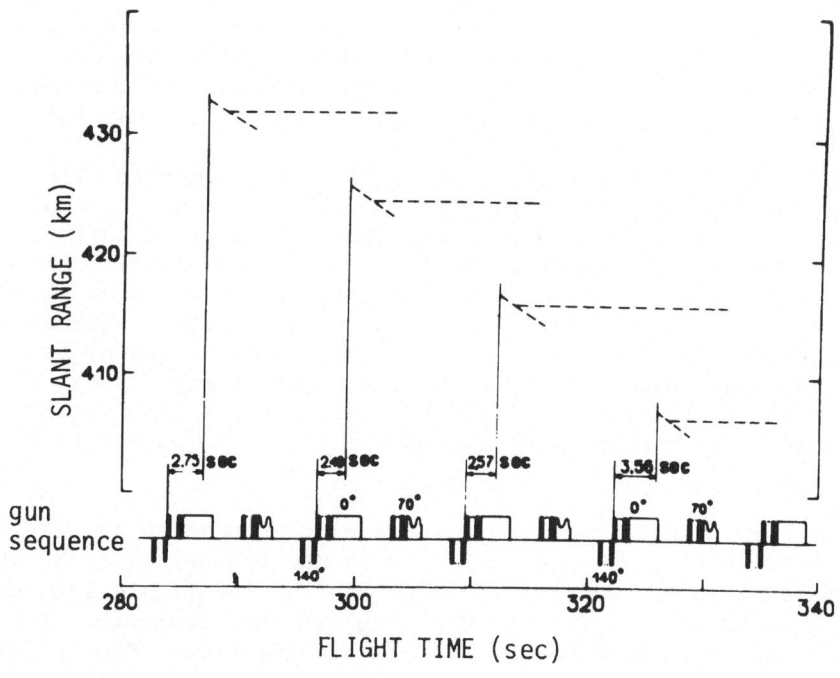

Fig. 4. Radio noise at Kerguelen : a 50 and b), 75 MHz. The lowest curve represents the current of the injected electron beam.

suggests that a great part of the beam particles are trapped inside the magnetosphere and are released gradually.

Before the launches all people expected for the electron beam a travelling time from the Southern hemisphere to the Northern one of 0.6 and 0.8 s depending on the energy (27 or 15 keV). The time difference between the onsets of the injection of electrons into the ionosphere at Kerguelen and the onsets of the radar echoes is larger (2÷3 s for 15 keV) than the 1/2 bounce period. The time of formation of the ionospheric irregularities responsible of the radar echoes is of about the time of their disappearance (20 m sec). Thus one can exclude this time as a reason of the abnormally large delay observed. These delays in the occurrence of the echoes are thus probably caused by delays of particle precipitation in the magnetically conjugated region. Braking of the particles in unlikely because of the bounce period found in the amplitude of the echoes, already mentioned. Whatever this delay is not yet understood and is certainly the most in intringuing observation made during Araks experiment.

NEUTRALIZATION OF THE ELECTRON GUN

Direct measurement of the vehicle potential during the beam injection is a difficult task and the results, more exactly their interpretation, are depending on the technic (RFA, suprathermal detectors ...) which has been utilized. Nevertheless as for the other experiments (Echo, Polar, etc.) the potential of the electron gun is low as to compare with the acceleration of the initial beam.

This implies the existence of an efficient neutralisation process. Although this process is very sensitive to ambient conditions (neutral atmosphere pressure, plasma density) Araks experiments bear out the observations made in Zarnitza (Cambou 1975) : HF radio waves are received at the ground. The frequencies of these emissions (up to 75 MHz) are well above the ionospheric plasma frequency. This fact indicates that very likely the local plasma frequency is increased by a considerable factor (two order of magnitude). This observation should coincide with the presence of a halo around the electro gun, visible from the ground when it is a low altitude as in Zarnitza.

The Fig. 4 shows the signal received at Kerguelen by two narrow band receivers tuned on 50 and 75 MHz (bandwidth = 250 kHz) as compared with the current of injection. In the figure 5 a summary of the observations is presented a long with the evolution of different parameters (altitude, pitch angle of injection). Four points of importance have to be underlined : (i) observation of these HF emissions is possible only when the angle between the line of sight and the magnetic field is large (electromagnetic conversion) (ii) these emissions are largely modulated by the rotation of the rocket (iii) the amplitude of emissions is very dependent on the altitude

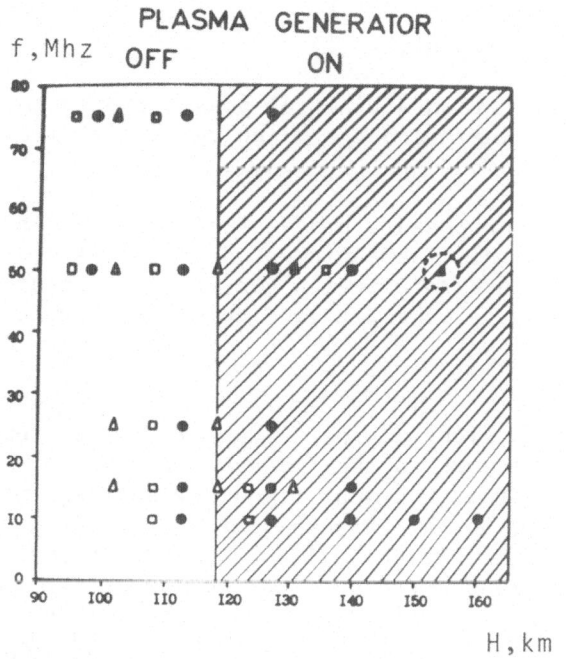

Fig. 5. Emission spectrum versus altitude for different injection series during the descent part of the Eastward flight.

▼ $(f_p^2 + f_b^2)^{\frac{1}{2}}$ UPPER HYBRID FREQUENCY

▲ fp PLASMA FREQUENCY

+ fb ELECTRONIC GYROFREQUENCY

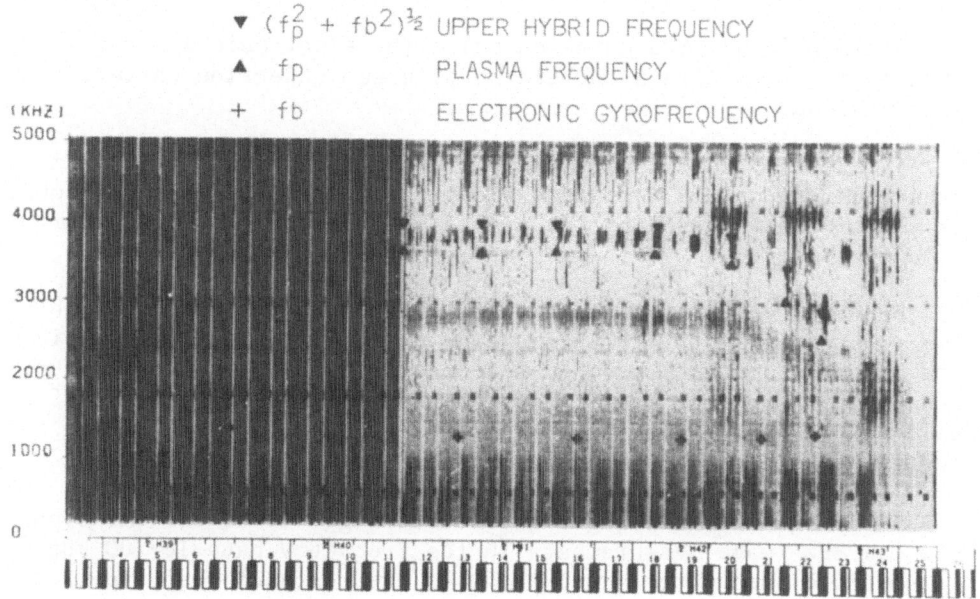

Fig. 6. High frequency waves observed during the Northward flight displayed like a sonagram (see text for details).

(maximum for h ≃ 105 km) (iiii) no HF waves were detected for small
pitch angle of emission (< 20°).

The presence of an artificial plasma around the rocket may be
caused by a ionizing discharge which certainly allows the neutralization to be achieved.

It has been proposed a process called the beam plasma discharge
(BPD) to explain this increasing of density (Galeev et al 1975).
Investigations close to electron gun, in the perturbated region, are
needed if we want to make detailed comparisons with the theory.

WAVES EMISSION

The figure 6 gives a general overview of the emissions observed
on board the nose eone in the frequency range 0.1 - 5 MHz. This
sonogram arises from the swept frequency analyser which is calibra-
ted periodically, so that black squares, regularly spaced, ranged
on four harmonic levels (0.6, 1.8, 3,0, 4.2 MHz) are apparent. The
vertical white lines correspond to the use of SFA as an other use.
The change in the intensity of the background grey between the lift
and the right parts of this figure is a consequence of the 20 dB
commutation of the gain of the receiver. Theoretical scheme of the
gun sequence is drawn at the bottom.

Associated signals may be seen in two main frequency ranges :

- A large bandwidth emission below the electronic gyrofrequency
(≃ 1.2 MHz). It is always correlated with gun injection whatever the
pulse duration.

- A narrow band emission whose frequency is close to 3.8 MHz
at the culmination and decreases along the descent of the nose cone.
This emission is bounded by the plasma frequency and by the upper
hybrid frequency.

The first emission ranges within the whistler mode. It exhibits
a one to one correlation with the beam current (Fig. 7). Its variations with the distance between the nose cone and the electron gun
shows no specific dependance (Fig. 8). Its spectrum is a very difficult point because the maximum of amplitude is the most often
situated near the lower boundary of the receiver (Fig. 9).

We can equally talk about a continuous growth of the emission
below 0.1 MHz (wave measurements in the frequency band 10-100 kHz
make possible such conclusion) or about the resonant shape (peaked
spectrum). Although some difficulties appear in the Eastward launch
the results seem to favour the second assumption. By comparison
with the Echo observations we must note that the front effect has
disappeared and that the level of emission is equivalent in both

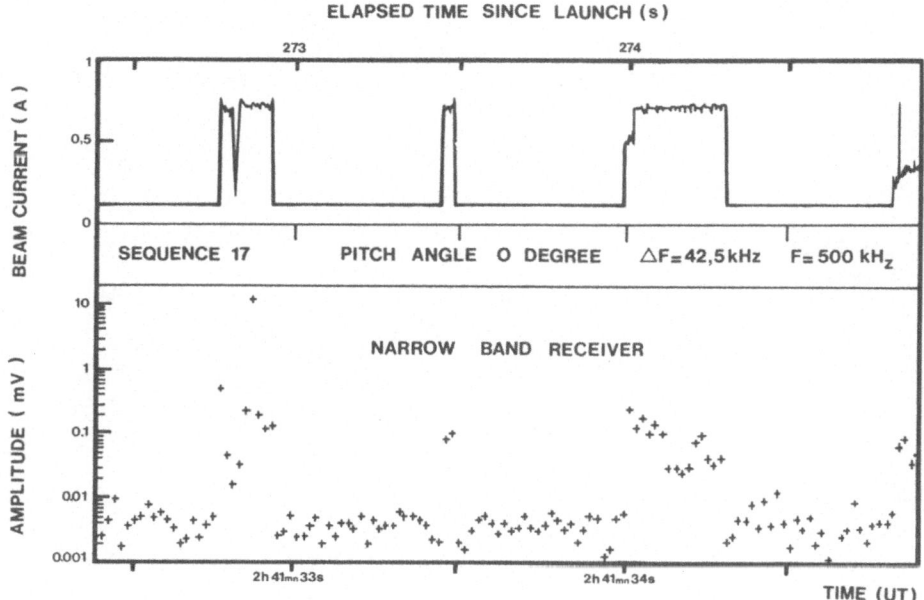

Fig. 7. Envelope of the signal obtained from the narrow band receiver around frequency of 500 kHz.

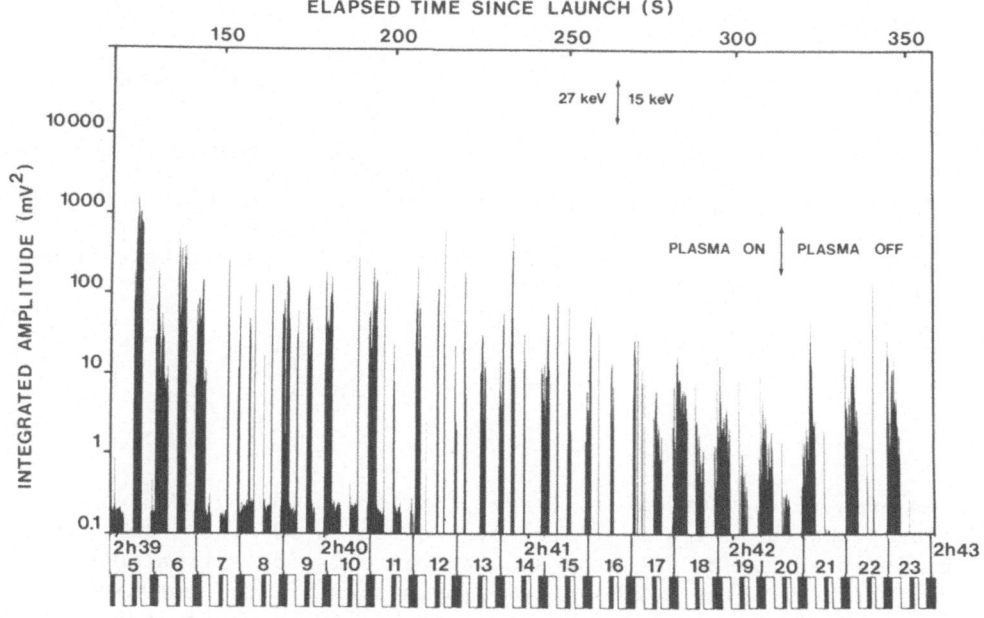

Fig. 8. Evolution of the integrated squared amplitude of the waves versus time. The theoretical scheme of sequences of the electron gun is drawn at the bottom.

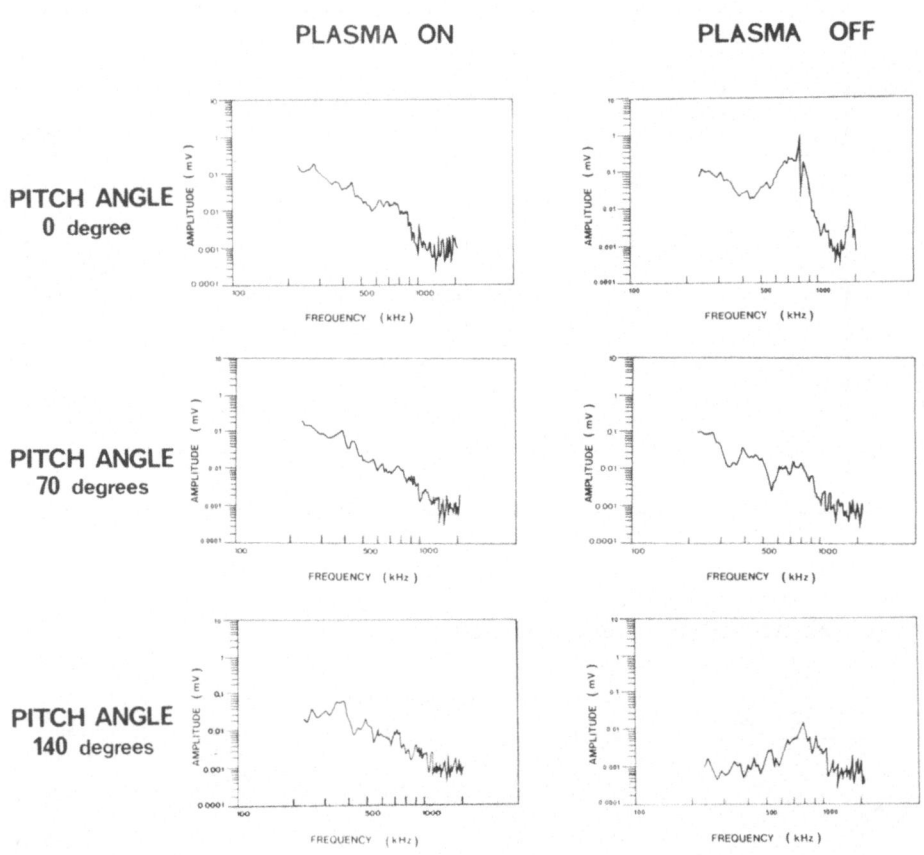

<u>Fig. 9</u>. Examples of spectra encountered where the plasma source works (plasma ON) and after if has been cut off (plasma OFF) for the 3 injection pitch angles. These spectra have been obtained during the Northward flight.

cases even if the current is largely different (80 mA, 0.5 A). As during Araks the pitch angle is never close to 90° the former statement is in agreement with the spontaneous emission estimated by Lavergnat and Pellat (1979). Beam plasma instability, non linear beating of HF waves and diochroton instability in the neutralization cloud have been proposed as mechanisms of whistler emission which is the most efficient as regarding the energy transfer from the electron beam to electromagnetic waves.

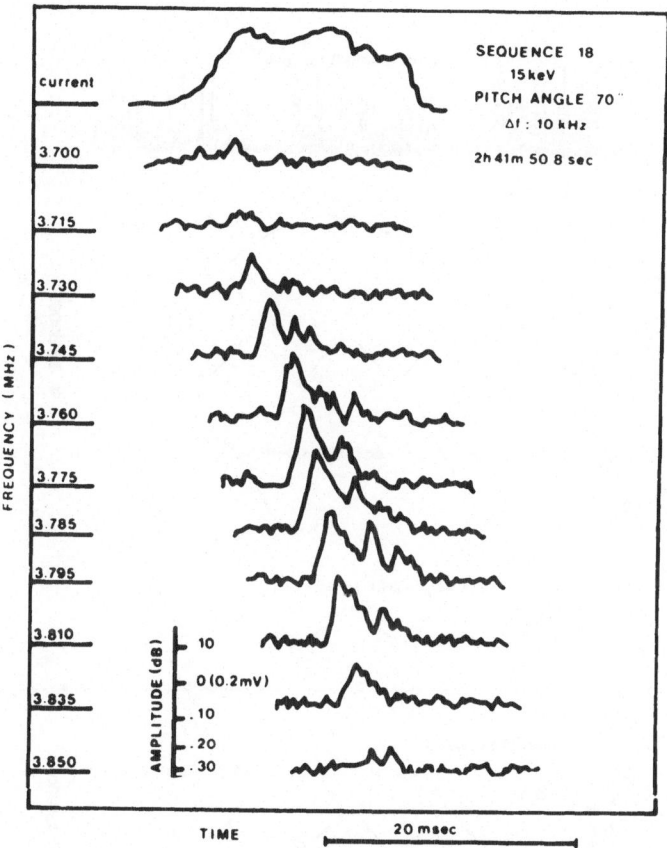

Fig. 10. Example of the dynamics of the plasma mode emission during one short pulse of 20 ms.

The emission near the plasma has a lower cutoff which is in excellent agreement with the theoretical value $\omega_p (1 + \dfrac{\omega_b{}^2}{8\omega_p{}^2})$

(Lavergnat and Pellat 1979) and exhibits the same temporal evolution as in Echo I (Cartwright and Kellog 1974). The amplitude of emission reaches a maximum in about 1-2 ms after the beginning of an electron impulse and then decreases continuously. Fig. 10 gives an example of this temporal evolution. The moderate values of emission (< 200 µV) as compared to those recorded during Echo I (3 mV) suggest that during Araks it is the incoherent Cerenkov radiation which has been recorded. Nevertheless if we can explain by this process the delay it is not yet understood the resume of background value at the end of the pulses.

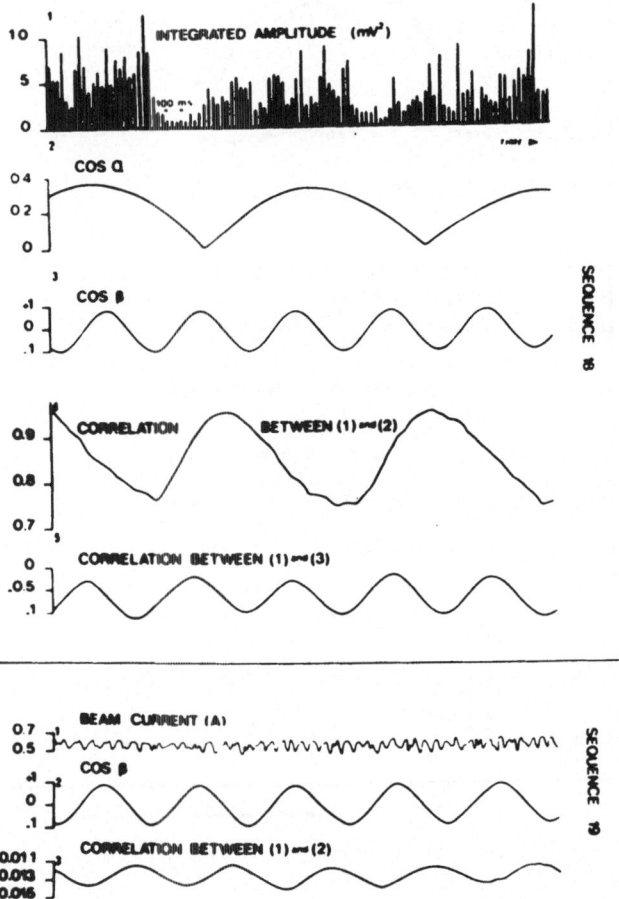

Fig. 11. Example of correlation between the integrated
amplitude of waves or beam current and cos α (α is the angle
between dipole and Bo) and cos β (β is the angle between the
velocity of the rocket and Bo).

For both frequency a very important result is the deep modulation of the radiated waves by the rocket spinning. This fact is
attested by Fig. 11 for the whistler range even when the electron
beam is injected strickly along the rocket vertical axis so that its
spinning could not tell on the injection pitch angle. We must conclude that the electron injection does not take place with the
forecast direction as concerned with the magnetic field. In other
words the electrons are deflected from their trajectory after their
coming out of the electron gun.

CONCLUSION

Several other results which are interesting but difficult to understand - even partially - have not been described in this short paper. The present selection, as any one, may be critisized but leads to the following conclusion : active experiment is promising for field line mapping in the auroral zone and precise investigation of beam-plasma interaction. However this aim will be never reached if we do not understand (i) the process of injection of a particle beam and first of all the neutralization aspects (ii) the own dynamic of the beam : what is the evolution of the distribution of the particles along it ? What is its spatial distribution ? What is the level of the waves trapped inside the beam ? Further experiments devoted to these aspects must be undertaken to answer to these questions.

ACKNOWLEDGEMENTS

The work exposed in this paper is due to the joined efforts of many people. Among them it is a pleasure for me to name : F. Cambou, R. Pellat, A.Kh. Pyatsi, H. Reme, R.Z. Sagdeev, I.A. Zhulin and Yu Ya Ruzhin.

REFERENCES

Cambou F., V.S. Dokoukine, V.N. Ivchenko, G.G. Managadze, V.V. Migulin, O.K. Nazarenko, A.T. Nesmyanovitch, A. Kh Pyatsi, R.Z. Sagdeev, I.A. Zhulin, 1975, "The Zarnitza rocket experiment of electron injection". Space Res. 15, 491-500.

Cambou F., V.S. Dokoukine, J. Lavergnat, R. Pellat, H. Reme, A. Saint-Marc, R.Z. Sagdeev, I.A. Zhulin 1980, "General description of the Araks experiments", Annales de Geophys., 36, 3, 271-283.

Cartwright D.G., P.J. Kellog, 1974, "Observations on Radiation from an electron beam artificially injected into the ionosphere". J. Geophys. Res., 79, 1439-1457.

Galeev A.A., E.V. Mishin, R.Z. Sagdeev, V.D. Shapiro, V.I. Shevchenko, 1976, "Discharge in the region around a rocket following injection of electron beams into the ionosphere". Doklady Akad. Nauk SSR, 231, 71.

Lavergnat J. and R. Pellat, 1979, "High Frequency spontaneous emission of an electron beam injected into ionospheric plasma". J. Geophys. Res., 84, 7223-7238.

"Special issue on the results of the active French-Soviet Araks experiments". Ed. by H. Reme, 1980. Ann. de Geophys., 36, 3.

DISCUSSION

Banks: Would you please clarify the nature of the radar
echo observed in the northern hemisphere conjugate
region? Does it result from plasma irregularities or an
overdense plasma with specular reflection?

Lavergnat: Aspect angle sensitivity and Doppler shift
measurements of the echoes show that the echoes most
likely are obtained from an underdense volume.

Kellogg: I noticed that several of the echo delay
times seem to be $(n+\frac{1}{2})$xbounce period. Is this a common
feature?

Lavergnat: It seems that there is some confusion because
the $\frac{1}{2}$ bounce period is around 0.8 seconds and for
example, the most common delay, 2.7 seconds, is not of
the kind $(n+\frac{1}{2})$x1.6 s.

Bernstein: Does the amplitudes of the whistler and high
frequency waves depend upon the injection pitch angle,
particularly, the up-down configurations?

Lavergnat: The amplitude of the waves depends on the
pitch angle of injection whatever the frequency range,
in fact, decreases when the frequency grows.

WAVE EXCITATION IN ELECTRON BEAM EXPERIMENT ON

JAPANESE SATELLITE " JIKIKEN (EXOS-B) "

Nobuki Kawashima and the JIKIKEN
(EXOS-B) CBE Project Team

Institute of Space and Astronautical Science
Komaba, Meguro-ku, Tokyo, JAPAN

ABSTRACT

Beam-plasma interaction experiment has been made in the magnetosphere by emitting an electron beam (100 ~ 200 eV, 0.25 ~ 1.0 mA) from the satellite " JIKIKEN " (EXOS-B). Various types of wave emission are detected by LF and HF wave detectors. Waves near at upper-hybrid frequency and at electron cyclotron frequency are detected in a low L-value region, which will be useful diagnostic means for plasma density and magnetic field. Vehicle charging up to the beam energy is also observed outside the plasmapause.

1. Introduction

A scientific satellite " JIKIKEN " (EXOS-B) was launched from Kagoshima Space Center, University of Tokyo, at 14:00 (JST) on September 16, 1978, which has an eccentric orbit with an apogee of 6 earth radii. JIKIKEN is now operating successfully, transmitting very interesting observation data in the magnetosphere. Purposes of the Controlled Beam Experiment (CBE) are to control the satellite potential by an electron beam emission, and to study the wave excitation (linear and non-linear wave phenomena due to the beam-plasma interaction).

2. Instrumentation

An electron gun on board " JIKIKEN " is a Pierce type axi-symmetric parallel beam electron gun with Wehnert electrode. The beam diameter at the exit is 5 mm and the beam divergence is about 3 rad. Both the beam current and beam voltage can be changed in 4 steps as in the following:

$$I = 0.25 \text{ mA, } 0.5 \text{ mA, } 0.75 \text{ mA, } 1.0 \text{ mA}$$

and

$$V = -100 \text{ V, } -125 \text{ V, } -150 \text{ V, } -200 \text{ V.}$$

The cathode heater used in this electron gun is a direct-type tungsten heater, and is heated by a 16 kHz rectangular AC current with 3.4 V (1 W) and supplied from a DC-AC Converter. A DC voltage of $-100 \sim -200$ V from a high voltage DC-DC Converter is applied between the cathode and anode to accelerate the electron beam. Diagnostic instruments are

   i) Wave detectors (NPW)
      VLF range (750 Hz ~ 10 kHz)
      (NPW (V))
      HF range
         (10 kHz ~ 3 MHz)
      (NPW (A) or NPW (S))
  ii) Particle detector (EPS)
      (5 eV ~ 11 keV)

3. Experimental Results

3-1 Wave Excitation

Typical frequency spectra of the waves received in NPW receiver, when the CBE system is operated in the maximum beam voltage and beam current mode (V = - 200 V, I = 1.0 mA), is shown in Fig. 1. A significant interaction between the surrounding plasma is clearly seen in this figure. It is found that several types of the waves appear making interactions with CBE electrons through the magnetospheric plasma.

The first type of the wave (we call it Type-A wave hereafter) is excited in a relatively low L-value region and its frequency increases as the L-value decreases. The time variation of the Frequencies of the Type-A wave is shown in Fig. 2 with Type-B, C waves (which will be discussed later) for the case of Rev. 654 (May 4, 1979).

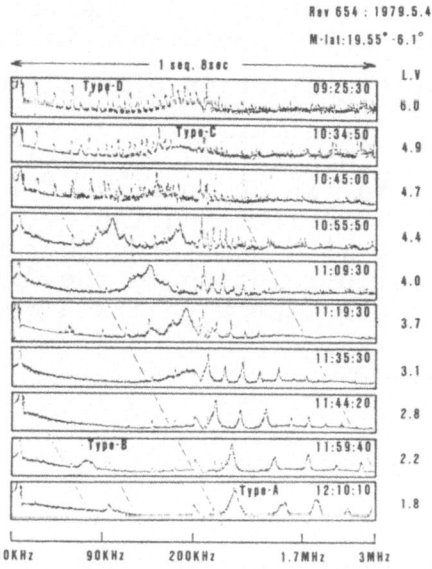

Fig. 1 Typical frequency spectra of waves
excited by the beam - plasma interaction.

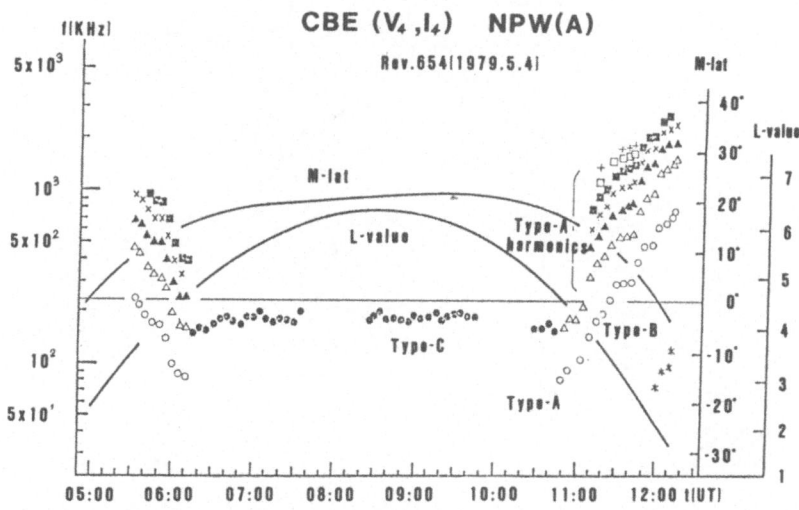

Fig. 2 The time variation of the frequency of
the Type-A, B and C waves (Rev. 654,
May 4, 1979).

The several harmonic waves are accompanied, which are
supposed to be generated in the instruments due to the
saturation of the signal level. The frequency of the
wave (fundamental) is the upper-hybrid frequency f_{UHR} =
$\sqrt{f_{pe}^2 + f_{pe}^2}$ or the electron plasma frequency f_{pe}.
Figure 3 shows the time variation of the Type-A wave
spectra. The wave changes its amplitude in the satellite
spin period; that is, in the period of T ~ 90 sec. Such
a spin dependency is reflecting the evidence that the
emission correlates with something that has one directional characteristic relating to the direction of sun
or the magnetic field line with its direction taken into
consideration.

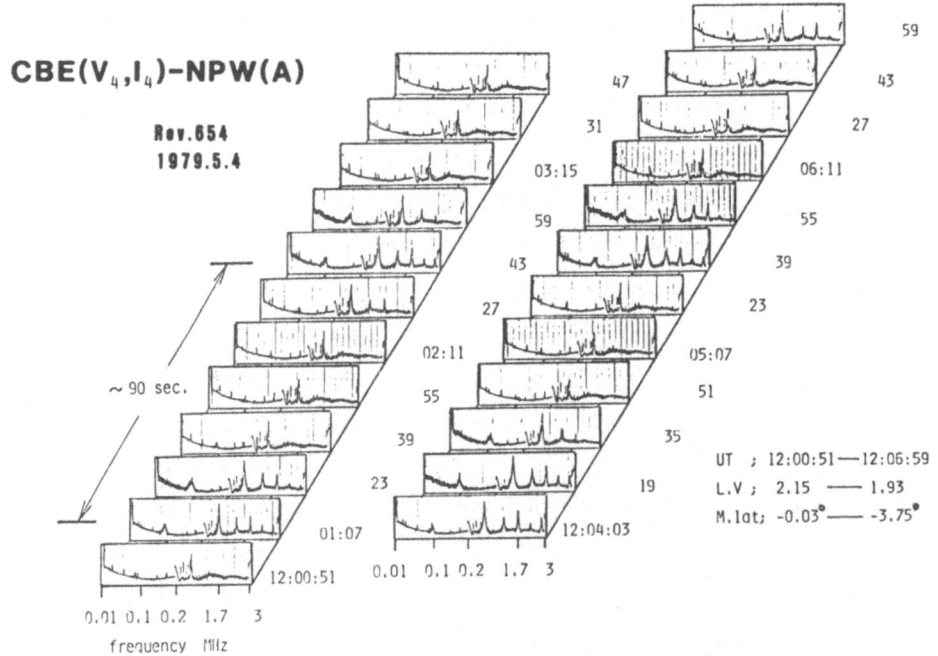

Fig. 3   The time variation of the spectrum
         pattern of the Type-A, B waves; These
         waves change their amplitudes in the
         satellite spin period, T ~ 90 sec.

The wave intensity increases as the beam current and/
or beam voltage is increased.

This emission is a very good indication of plasma
structure in the magnetosphere and the density variation
can clearly be seen in Fig. 4.

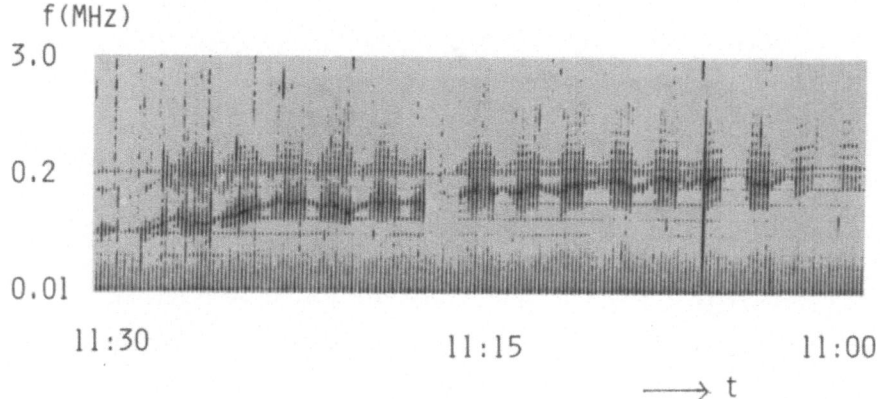

Fig. 4   f-t diagram of Type-A emission showing
         the plasma structure in the magnetosphere.

Some interesting non-linear phenomena have been
obtained. One example is shown in Fig. 5 and satellites
(Side band) formation around a strong Type A emission
can be seen.

The second type of the wave (Type-B wave) is excited
in a region closer to the earth (a lower L-value) than
that of the Type-A wave and as in the case of the Type-A
wave emission, its frequency increases when the L-Value
of the satellite position decreases (see Figs, 1 and 2).
The observation result along an orbit of the JIKIKEN
indicates that the emission reduces its amplitude and
finally disappears as the L-value increases as seen in
Fig. 1. The surrounding plasma density N or the ratio
N/N_b of the density of the emitted beam density N_b is
possibly controling the excitation of this wave. The
frequency of Type B emission ranges from half of the
electron cyclotron frequency $f_{ce}/2$ to f_{ce}. The spin
dependence is the same as Type-A emission.

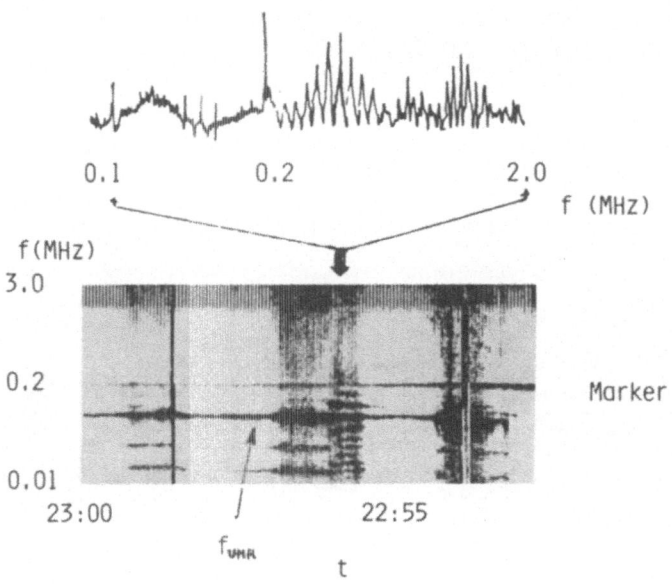

Fig. 5   Strong type A emission accompanied
         with side bands.

The third type of the wave (we call here Type-C wave)
shown in Figs. 1 and 2 is excited in a relatively high
L-value region and is observed for a long period during
each orbit. The maximum of the spectra of the emission
frequency f_c is usually in a range from 150 to 200 kHz,
independent of the L-value (so the local plasma density
and also local magnetic field intensity) and the shape
of the spectra is rather broad. Since the frequency is
not dependent on the local plasma parameters, the wave
is not possibly generated in situ process but that may
come from a source at some distant place.

The fourth type of the wave (we call here Type-D wave)
is the harmonics of 16 kHz signals mentioned earlier (see
Figs. 1 and 6). As already decribed, the heater of the
cathode is direct-heating type using the AC power supply
of 16 kHz. The Type-D wave is attributed to the 16 kHz
AC power supply.

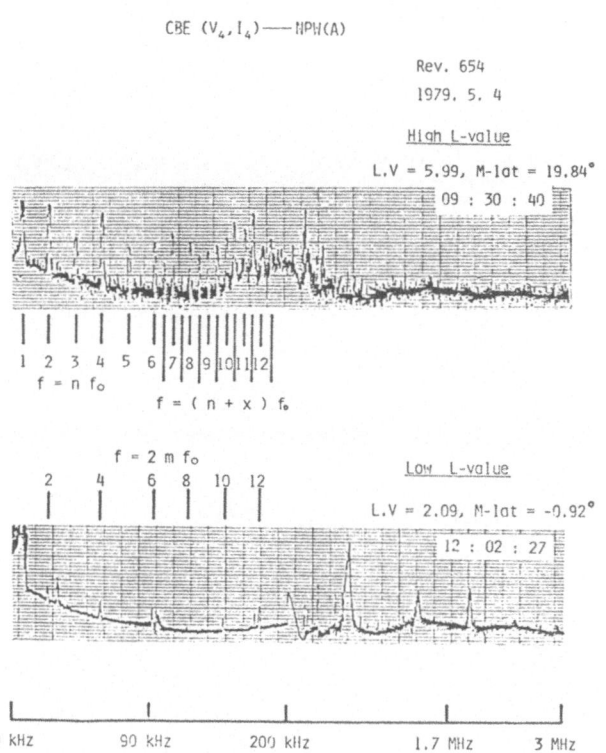

Fig. 6 Typical frequency spectrum of the Type-D, E waves.
Upper: In the case of a high L-value; The harmonics $f_D = nf_0$ ($f_0 = 16$ kHz; $n = 1, 2, ...$) and $f_E = (n + x)F_0$ ($0 < x < 1$) are excited. Lower; In the case of a low L-value; the even harmonics $f_D = 2mf_0$ ($m = 1, 2, ...$) are excited.

The fifth type of the wave (we call here Type-E wave) appears in a high L-value region, in which its frequency is modulated by the satellite spin phase angle. Its frequency is $f = (n + x)f_0$ ($n = 1, 2, ..., f_0 = 16$ kHz).

Various kind of emission are observed in VLF range. One example is shown in Fig. 7. In general, the signal intensity increases in the low frequency region. Sometimes a strong discrete emission is excited by the beam emission, but other time, natural discrete emission such as whistler is suppressed by the beam emission.

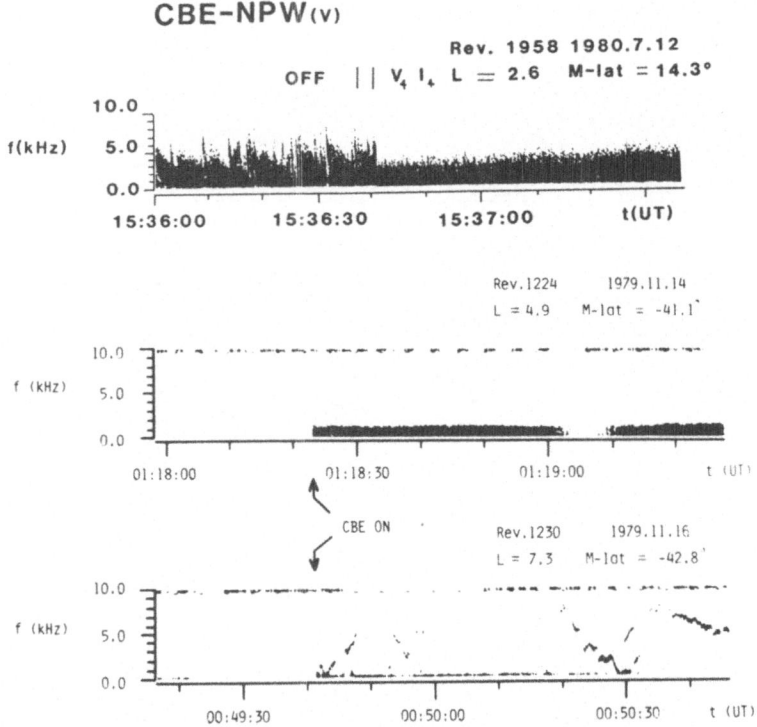

Fig. 7   Effect of the beam emission
in VLF range

3-2   Charging of Satellite

The satellite charging is observed by detection of
returning electrons measured by an electron energy
analyser. Fig. 6 shows the spatial dependence of the
energy spectrum of electrons and it is clear that above
a certain L-value, a peak is observed at the energy
corresponding to the beam energy. The L-value corresponds to the plasmapause. It indicates that the
satellite potential rises to the value correspoinding
to the beam energy.   As the density increases within
the plasmasphere, the peak in the energy spectrum
disappears.

4.   Summary and Discussion

The electron beam emission experiment on satellite
JIKIKEN have produced a large amount of interesting

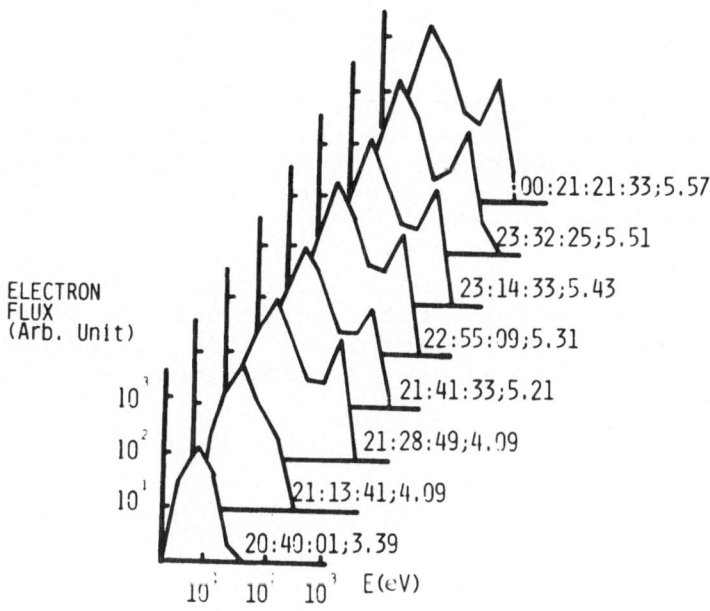

Fig. 8 Energy spectrum of electrons
during the beam emission for
various L-value.

scientific data on the electron beam - space plasma
interaction. Among them, HF wave excitation near the
upper hybrid frequency offers a useful tool for
determining the plasma structure in the magnetosphere.
Non-linear behavior of this wave is also very interesting.
Waves excited in the beam - plasma interaction is
strongly dependent on plasma and other parameters in the
magnetosphere so that it will provide important knowledge
on the magnetospheric plasma processes.

REFERENCES

Kawashima, N., Ushikoshi, A., Murasato, Y., Morioka, A.,
    Oya, H., Ejiri, M., Miyatake, S. and Matsumoto, H.,
    1981, J Geomag Geoelectr, 33:145-159.

DISCUSSION

Kellogg: What is perigee? Is there any indication of
BPD, as your perveance is large enough if the density
is large enough?

<u>Kawashima</u>:   The perigee is 250 km.   No evidence of BPD near the perigee is observed.

<u>Jost</u>:   A comment on your observed harmonics of the upper hybrid frequency: During certain cases in the beam electric field measurements at NASA JSC, harmonics of the upper hybrid are observed that are believed to be real and not instrumental.

<u>Kawashima</u>:   At present, we believe the harmonic generation is instrumental, but I agree that we cannot completely reject the possibility that it is coming from real non-linear beam plasma interaction.

PLASMA WAVES AND ELECTRICAL DISCHARGES

STIMULATED BY BEAM OPERATIONS ON A HIGH ALTITUDE SATELLITE

H. C. Koons[*] and H. A. Cohen[+]

[*]Space Sciences Laboratory
THE AEROSPACE CORPORATION
P. O. Box 92957
Los Angeles, California 90009

[+]Air Force Geophysics Laboratory
Hanscom Air Force Base, MA 01731

INTRODUCTION

The P78-2 (SCATHA) satellite was launched on 30 January 1979 to measure the characteristics of the spacecraft charging process near synchronous orbit. A particle beam system was included in the payload to investigate the phenomenon of spacecraft charging by modifying the spacecraft environment. The beam system has both an electron gun and an ion gun that can be used to alter the spacecraft ground with respect to the plasma either positively or negatively. A charging electrical effects analyzer (CEEA) was provided for the payload to verify that electrical discharges are occurring when other instruments measure large differential potentials between surface materials on the vehicle. The CEEA consists of three instruments, a Pulse Shape Analyzer, a VLF Analyzer, and an RF Analyzer. The Pulse Shape Analyzer measures the number of pulses, their amplitudes and shapes on four sensors. The VLF Analyzer measures the electric and magnetic field spectra of waves in the frequency range from ≈100 Hz to 300 kHz. The RF Analyzer measured the electric field intensity on a 1.8-m monopole antenna in the frequency range from 2 to 30 MHz. The instruments on the SCATHA satellite are described by Stevens and Vampola (1978).

In this paper we describe plasma waves and electrical discharges measured by the CEEA instruments during electron and ion beam operations.

ELECTRON BEAM OPERATIONS

Plasma Waves

     The VLF Analyzer has detected plasma waves stimulated by the
beams during their operation. In order to better understand the
features of the spectra that are related to the beams we first
show in Fig. 1 spectrograms from a quiet time period when essentially no plasma wave emissions are present. Fig. 1a shows the
normal receiver background. The receiver noise peaking at 1.3 kHz
is apparent in the magnetic antenna (B) data. At the level used
to produce these spectrograms there is no background visible in
the electric antenna (E) data. Fig. 1b shows a similar quiet time
spectrogram showing the only significant source of electromagnetic
interference.   A 700 Hz tuning-fork driver circuit in another
instrument produces monochromatic interference at 700 Hz and 2100
Hz and weak interference at 1400 Hz. Beating among the oscil-
lators produces a slight modulation of the receiver AGC circuit.
These signals are seen only in the data from the magnetic antenna. The instrument producing this interference is often turned
off during the transmission of the broadband VLF data.

     Fig. 2a shows a spectrogram during electron beam operations
on April 24, 1979 at 100µA, 150 V. On that date the tuning forks
were off. The electron gyrofrequency at the satellite was 2200
Hz. Electromagnetic chorus emissions with a well defined gap at

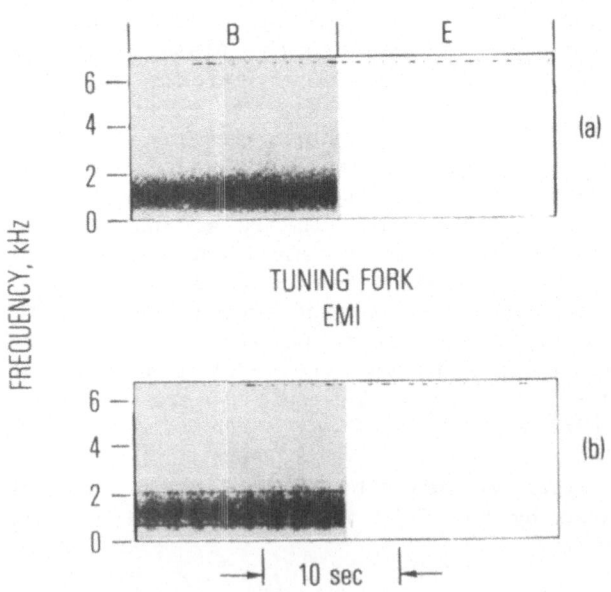

Fig. 1. VLF receiver background noise.

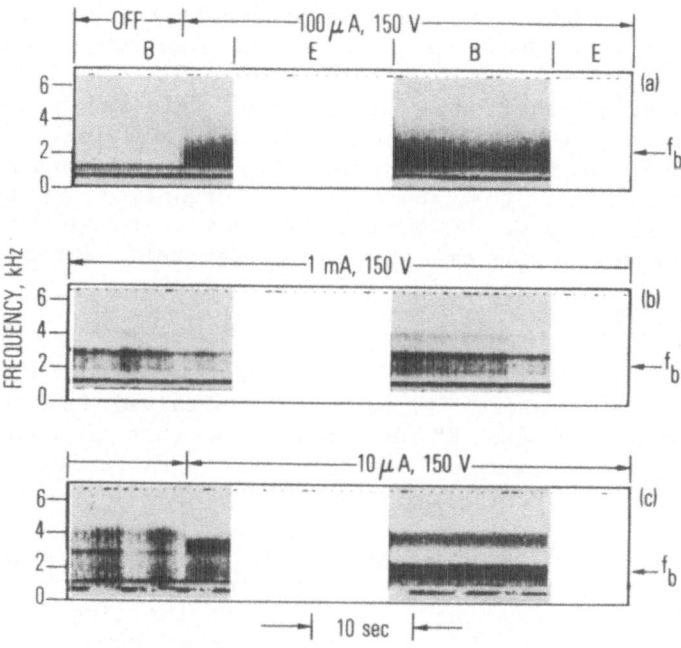

Fig. 2. VLF spectrograms of emissions detected during electron beam operations.

one-half of the gyrofrequency were present below the gyrofrequency. Above the gyrofrequency electrostatic emissions commonly referred to as $(n + 1/2)$ f_b emissions were detected on the electric antenna. The gun was turned on in the 100μA, 150 V mode during the time period shown in the spectrogram. Before the gun was turned on only the chorus was detected on the magnetic antenna. After the gun was turned on there was a wideband white-noise background extending from about one-half of the gyrofrequency to twice the gyrofrequency. The electron gun operated in this mode for 95 s. Just before it was commanded to a 100μA, 50 V level the wideband noise decreased in amplitude and an emission about 500 Hz wide appeared near $f \sim 1.5$ f_b. Fig. 2b shows a spectrogram during operations at 1 mA, 150 V. A wide band of emissions extended from f_b to 2.3 f_b. A well defined narrower emission can be seen at 1.5 f_b. Data from the final mode in this sequence is shown in Fig. 2c. When the current was reduced from 1 mA to 10μA at 150 V very well defined emission bands appeared. At first a single band was located between 1.5 f_b and 2 f_b. Shortly thereafter two bands can be seen centered near f_b and $2f_b$ respectively. The narrowband channels responded at frequencies as high as 10 kHz, the fifth harmonic of the local electron gyrofrequency.

To summarize then the VLF Analyzer detected plasma waves stimulated by the electron beam during its operation. The spectrum changed significantly when the beam parameters were changed. Generally the spectrum was organized by the local electron gyrofrequency. A puzzeling aspect of these observations is the occurrence of the emissions in the magnetic antenna data.

Electrical Discharges

During the electron beam operations on 30 March 1979 two plasma potential sensors failed. Data obtained from the Pulse Shape Analyzer and the RF Analyzer show that discharges were occurring on the vehicle at the time of the failure. Two of the largest pulses occurred at the times of the plasma probe failures.

Table 1 shows the order of events during the electron gun experiments on 30 March. Pulses exceeding the 0.165 V analysis threshold abruptly onset at 15:12:08 UT at the time the electron gun was commanded to a -3 kV potential at 6 mA current.

Table 1.   Electron Gun Operations on March 30, 1979
            at the time of the SC2-1, -2 plasma potential
            sensor failures

15:12:08	Operation of electron gun at -1.5 kV and 6 mA: No observed deleterious effects
15:12:08	Electron gun commanded to -3 kV TPM records 8.4 V negative pulse on Lo-Z sensor Pulse analyzer begins detecting pulses above 0.165 V threshold
15:12:09	TPM records 7.0 V negative pulse on Lo-Z sensor
15:12:10	SC2-1 plasma potential sensor fails
15:12:16	Data system begins scrambled operation
15:12:29	Data system corrects itself (at main frame 0)
15:12:40	SC2-2 plasma potential sensor fails
15:12:08- 15:13:29	TPM records numerous pulses in the 6 V to 15 V range, pulse characteristics measured by pulse analyzer
15:13:29	Electron gun commanded to -1.5 kV. Pulses drop to 1 V to 3 V range
15:17:36	Electron gun current lowered to 0.01 MA. Pulses cease on pulse analyzer

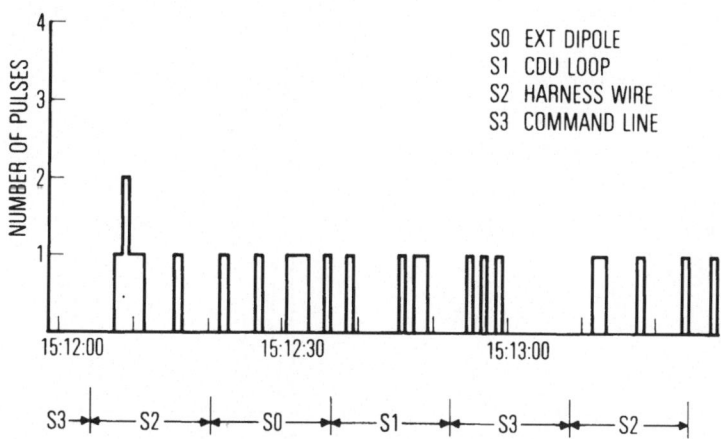

Fig. 3.   Number of pulses above a 7.18 V threshold.

Pulses of comparable magnitude occurred on each of the four
Pulse Analyzer sensors. The largest count rate threshold was 7.18
V. The number of pulses above this threshold is shown in Fig. 3
as a function of time.   Only once did two pulses occur in one
second. The number of pulses above a threshold of 0.47 V is shown
in Fig. 4.   Typical pulse shapes on the harness wire sensor are
shown in Fig. 5.   The pulses on the same sensor tend to have the
same shape. This suggests that the larger discharges are occurring at the same point on the vehicle. Pulses of differing shapes
are seen sufficiently often to rule out an instrumental effect in
the shapes. The pulse that occurred at the time of one of the
probe failures is identified on Fig. 5.

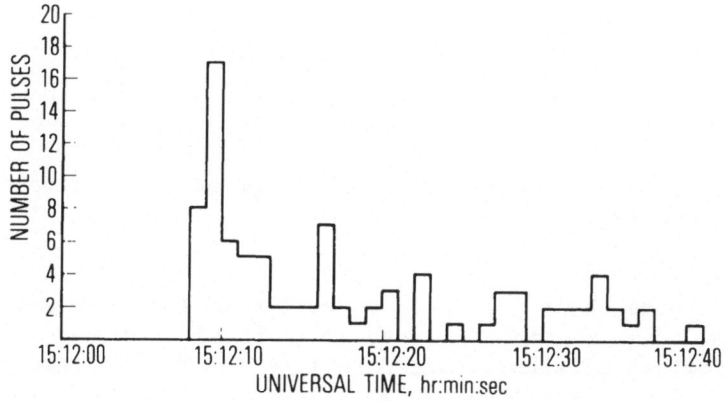

Fig. 4.   Number of pulses above a 0.47 V threshold.

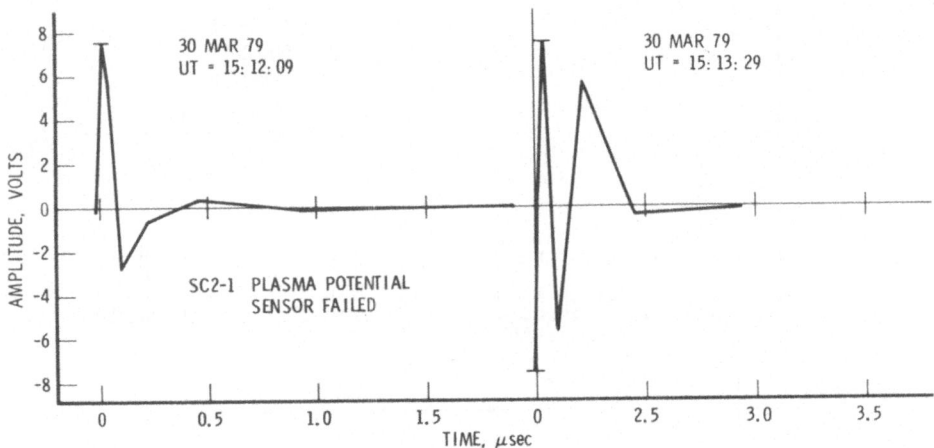

Fig. 5.   Pulse shapes measured in the low resolution mode.

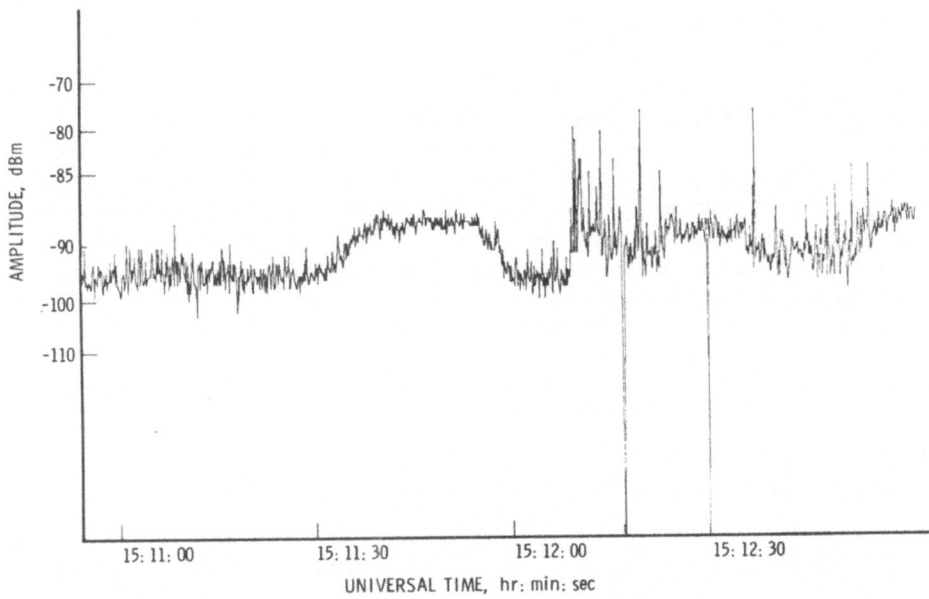

Fig. 6. RF analyzer data during electron beam operations on 30
         March 1979.

     At the time these pulses occurred the RF Analyzer was operat-
ing on the 1.8-m monopole antenna at a fixed frequency of 20 MHz
with a bandwidth of 4 kHz. The data from the RF Analyzer is shown
in Fig. 6. The pulses began at 15:12:08 UT.

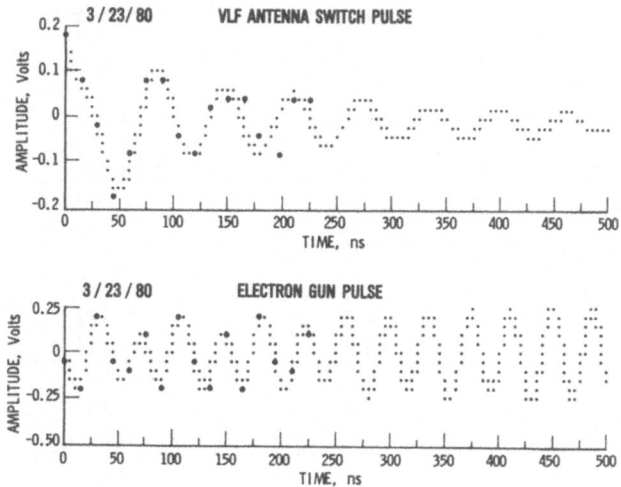

Fig. 7. Pulse shapes measured in the high resolution mode. The heavy dots are the measured points.

On March 23, 1980, the Pulse Shape Analyzer detected a pulse during electron beam operations at 1.5 kV, 1 mA. At that time the analyzer was in its high-resolution time-analysis mode. The pulse shape is shown in Fig. 7. A computer fit of the functional form

$$V = V_o + \Sigma V_i \, e^{-k_i t} \cos (2 \pi f_i t + \Phi_i)$$

was made to the sixteen sample points. For the pulse during the electron gun operations that best fit was obtained for two frequencies. The fitting parameters are shown in Table 2. One of the two frequencies showed a slight growth rate while the other was slightly damped. Apparently the damping rate is very small and the data set is too short to determine the damping coefficent.

A pulse that occurred on one of the four sensors when the VLF antenna switches is also shown in Fig. 7. The best fit was again obtained for two frequencies. The frequencies differ significantly from those obtained for the electron beam pulse and for the natural discharges also shown in Table 2.

ION BEAM OPERATIONS

Both plasma waves and electrical discharges have been stimulated by ion beam operations. Fig. 8a shows the turn on of the ion beam on October 12, 1979. These data were taken during attempts to detect the Xenon ions from this experiment by instruments on the GEOS-2 satellite. The vertical lines across the spectrogram represent discharges.

Table 2.   Discharge fitting parameters

Date	Sensor	i	Frequency MHz	Amplitude V	Damping ns^{-1}
Electron Gun Pulse					
3/23/80	CMD	0		-0.003	
	Line	1	14.1	0.089	0.004
		2	26.4	0.140	-0.001
VLF Antenna Pulse					
3/23/80	CMD	0		-0.007	
	Line	1	9.0	0.397	0.037
		2	16.0	0.135	0.004
1/24/80	CMD	0		0.01	
	Line	1	5.0	0.38	0.000
		2	25.7	0.13	0.015
4/16/80	Dipole	0		0.00	
		1	11.1	0.89	0.022
		2	25.0	0.68	0.008
6/13/80	Harness	0		0.08	0.009
	Wire	1	19.5	0.25	0.006
		2	31.8	0.80	0.021
6/14/80	Dipole	0		0.00	
		1	13.3	0.06	0.016
		2	26.1	0.08	0.025
6/20/80	Dipole	0		0.24	0.069
		1	21.7	0.12	0.050

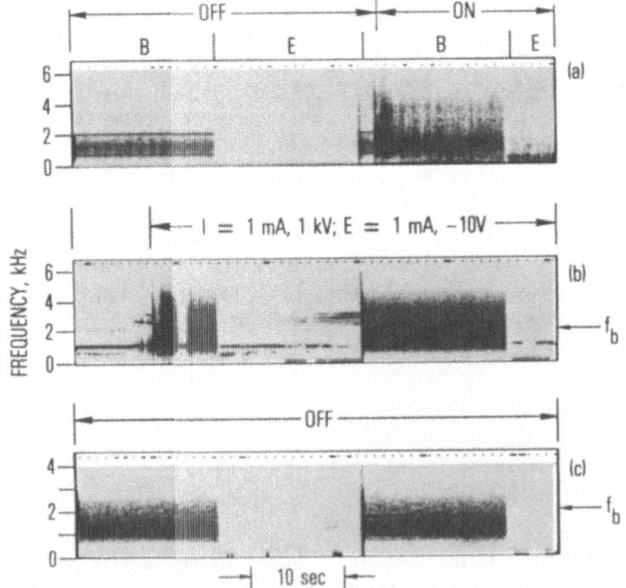

Fig. 8.   VLF spectrograms during ion beam operations.

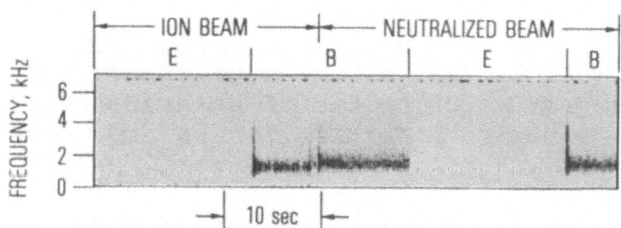

Fig. 9. VLF spectrogram during Xe$^+$ transport experiment.

Downstream of the decelerator grid is an electron emission filament that can be biased either positive or negative with respect to spacecraft ground. When operating, a quasi-neutral beam of ions and electrons will flow into the ambient plasma. The potential of the vehicle with respect to the ambient plasma shortly after the ion beam alone was turned on October 12 was -800 to -960 V. During the time period shown in the spectrogram in Fig. 9 the electron emission filament was turned on. The discharges abruptly ceased and the vehicle potential dropped to -40 to -80 V. After that time 1 mA of electrons at 100 V and 1 mA of ions at 1 keV would be leaving the SCATHA vehicle into the ambient plasma.

The spectrogram in Fig. 8 shows broadband emissions, again on the magnetic antenna, during ion beam operations on April 24, 1979. The most intense emissions occurred between 0.4 and 1.0 times the local electron gyrofrequency. At this time the ion beam was operating in the 1 mA 1 kV mode with 1 mA of electrons from the filament biased at -10 V. The broadband emission is similar to that shown in Fig. 2a when the electron gun alone was on.

Several minutes after the ion gun was turned off on another day the unusual striated emissions shown in Fig. 8c were detected on the magnetic antenna. The striations occurred between 0.3 and 0.9 times the local electron gyrofrequency.

Summarizing the ion beam observations we find discharges when the ion beam was on alone. When the neutralizing filament was turned on the discharges ceased and plasma wave emissions were detected. The most intense emissions are below the local electron gyrofrequency, typically 0.3 to 0.9 times f_b. These emissions lasted for several minutes after the ion beam system was turned off.

REFERENCES

Stevens, J. R., and Vampola, A. L., 1978, Description of the Space Test Program P78-2 Spacecraft and Payloads, SAMSO TR-78-24, U. S. Air Force Space Division, El Segundo, California.

DISCUSSION

Lavergnat: What do you think about the electrostatic or
the electromagnetic character of an emission when the
observation is made very close to the emission production?

Koons: I do not think that you can properly characterize
a wave in the near field as either electrostatic or
electromagnetic. A measure of the wavelength will
identify the mode. We do not make this measurement in
our experiments.

Kintner: I am confused by your report that the particle
beams are primarily producing magnetic fluctuations
instead of electric fluctuations. Is it possible that
the magnetic sensor is acting as a short electric sensor
while the electric booms are sensitive only to long
wavelength emissions?

Koons: We are also very puzzled by the response of our
magnetic antenna to electrostatic waves. The magnetic
antenna responds properly to electromagnetic waves such
as whistler mode chorus. The index of refraction computed from the ratio of B to E is reasonable and the
magnetic component is perpendicular to the geomagnetic
field. However, presumably electrostatic emissions
near $3/2\ f_c$ show up strongly on the magnetic antenna.

PLASMA DIAGNOSTICS BY ELECTRON GUNS AND ELECTRIC

FIELD PROBES ON ISEE-1

Arne Pedersen

Space Science Department of ESA, ESTEC
Noordwijk
The Netherlands

INTRODUCTION

The ISEE-1 satellite, launched in October 1977, carried two small electron guns for control of the satellite potential. They formed part of a spherical double probe instrument for measurements of electric fields,(Mozer et al., 1978.) The intention was to reduce the asymmetrical photo-electron cloud emitted sunward and the possible influences of its negative space charge on the electric field measurements. Early diagnostic experiments with the electron guns lead to the conclusion that the photo-electron space charge problem is a relatively small source of interference for the electric field measurements, and the guns were therefore not used in connection with routine electric field measurements, but only operated during special diagnostic periods in order to understand better spacecraft charging and potential control. It is important to notice that it is only meaningful to speak about satellite potential control for a satellite with surfaces which are sufficiently conductive to allow currents to flow so that charge distributions are smoothed to the extent that voltage differences do not exceed typically 1.0 Volt. This can be achieved in practice with a layer of indium oxide ($\sim 10^5$ /$\square$). ISEE-1 is one of the few magnetospheric satellites with such conductive surfaces, the only other satellites are EXOS-B, GEOS-1 and GEOS-2.

The potential of the conductive ISEE-1 satellite can be determined, with or without electron gun operation, by measuring the potential difference between the satellite and one electric field probe (36m from the satellite) which is biased with a high impedance current source to be very near the plasma potential. The electric field probe is therefore the reference for measuring the

satellite potential and a necessary tool for observing the effects
of the electron guns. The probe biasing, which will be explained
later, is based on probe photo-emission and does only work in sunlight. In eclipse does this method not work, and the satellite ·
potential, which tends to go negative in this case, can only be
measured by observing the threshold due to ambient cold ions accelerated to the satellite. The usefulness of this latter method to
measure large negative satellite potentials has been demonstrated
by DeForest and McIlwain (1971) and DeForest (1972). These authors
were also the first to publish data and explain spacecraft charging
of the geostationary ATS satellites in the magnetosphere. Charging
to large negative potentials were most frequently observed during
eclipses, however several cases were also observed during sunlight.
This is in contrast to ISEE-1 and the GEOS satellites (GEOS-2 in
geostationary orbit like the ATS satellites) where no charging
effects to large negative potentials have been observed in sunlight.
A simple model for respectively non-conductive and conductive
satellites will be developed and will show that it is possible to
explain these observations.

   The electron guns have produced unexpected results in another
field, namely stimulation of plasma waves, (Mozer et al., 1979).
Emitted electron currents of typically 0.1 mA at a few times 10 eV
are obviously sufficient to trigger waves. This will not be pursued
further in this paper, but will be the subject of a separate paper
at this meeting.

NON-CONDUCTIVE AND CONDUCTIVE SATELLITES

   Most satellites with solar cells have cover glasses on the
solar cells which are perfect insulators. Fig. 1 shows how a sphere
with insulated surfaces will charge in a hot plasma in sunlight.
The sunlit surfaces will tend to be charged positive to maintain a
balance between escaping photo-electrons and ambient electrons. Ion
currents are negligible in comparison. The shadowed surfaces will
go to a negative potential where a sufficient number of ambient
electrons are repelled to balance the attracted ions. This results
in a large negative surface charge on the shadowed side, which for
a hot magnetospheric plasma will totally dominate over the small
positive surface charge on the sunlit side. The result is the
potential distribution sketched in Fig. 1, where equipotential surfaces negative relative to the ambient plasma can be observed all
around the satellite. Photo-emission is controlled by a local
potential barrier on the front side. Such potential distributions
have been discussed by Fahleson (1973) and more elaborate models
have been discussed by Rubin et al.(1979). A satellite will in
reality have some conductive surface elements which in the case
of a conductive path between sunlit and shadowed surfaces, will
modify the simple picture in Fig 1, i.e.reduce the differential
charging. The purpose of this explanation is to make it clear that

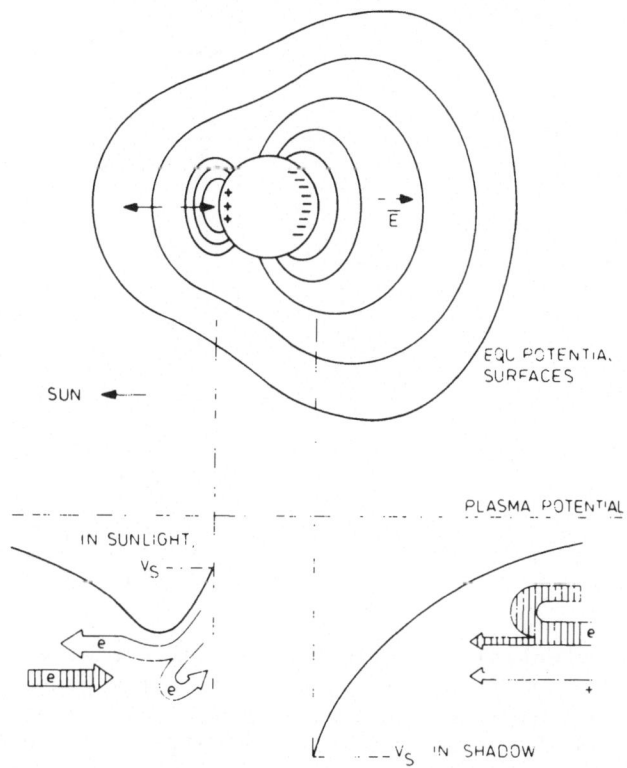

Fig. 1. Potential distribution near a sphere with insulating surface material during sunlight and in a hot plasma.

it is not very meaningful to attempt active control of the potential of a satellite with non-conductive surfaces.

A conductive sphere in a plasma in sunlight will acquire a
potential as explained in Fig. 2 (left). The potential of the sphere
is positive and determined by the ambient electron current to the
whole sphere balancing the fraction of photo-electrons escaping
from the sunlit side. Lower energy photo-electrons will orbit and
return. An electron gun current emitted from a conductive satellite
will be added to the photo-electrons and a new more positive potential equilibrium will be achieved, Thus an electron gun can only
drive the satellite more positive.

A conductive satellite in eclipse will charge negative like a
non-conductive satellite. The only difference is that for a conductive satellite in eclipse, with an electron gun, it is possible to
emit electrons and in a fashion identical to Fig. 2 (left) bring
the satellite positive. This positive potential can be a comfortable

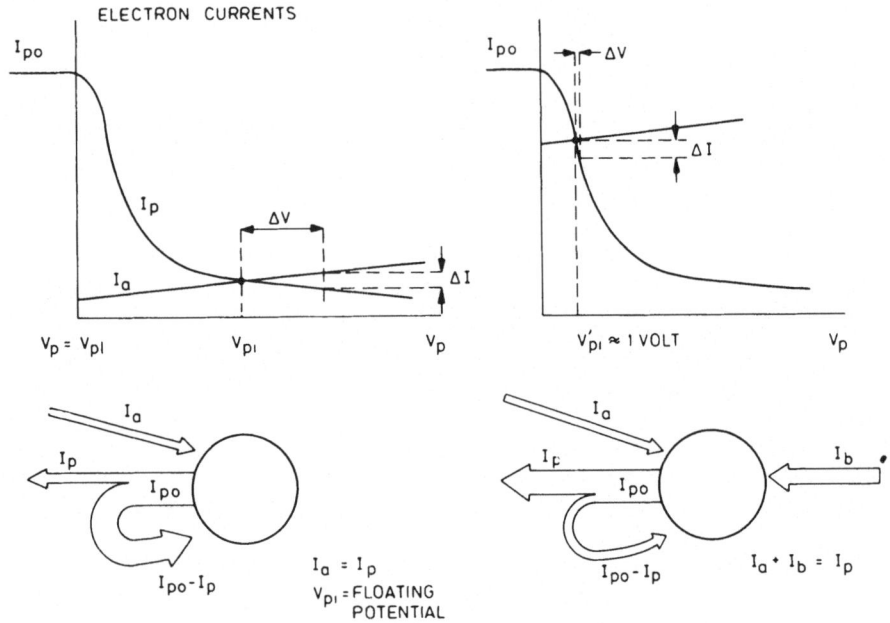

Fig. 2: Probe potential is determined by balance of electron currents to and from a probe. I_p = escaping photo-electrons
with maximum value I_{po}. I_a = ambient electrons to the
probe. With a negative bias current, I_b, the probe is
forced from its floating potential, V_{pl}, to a potential,
V'_{pl}, closer to the plasma potential.

+10 or +20 Volt instead of thousands of Volt negative.

MEASUREMENTS OF SATELLITE POTENTIAL

The negative charging of a satellite can be measured by an ion
particle detector because ions will not be observed below a certain
cut-off in energy corresponding to e V_s, where V_s is the satellite
potential (DeForest and McIlwain, 1971 and DeForest, 1972). A
typical uncertainty in this method is 10 Volt.

We will in the following explain how electric field probes on
ISEE-1 can be used to determine the positive potential of this
conductive satellite in sunlight. Fig.2 (left) shows the floating
potential for a conductive sphere under given plasma conditions.
The sphere can represent, either the satellite in an approximate
way, or one electric field sensor which actually has a spherical
shape and has a diameter of 8 cm. Fig. 2 (right) illustrates that

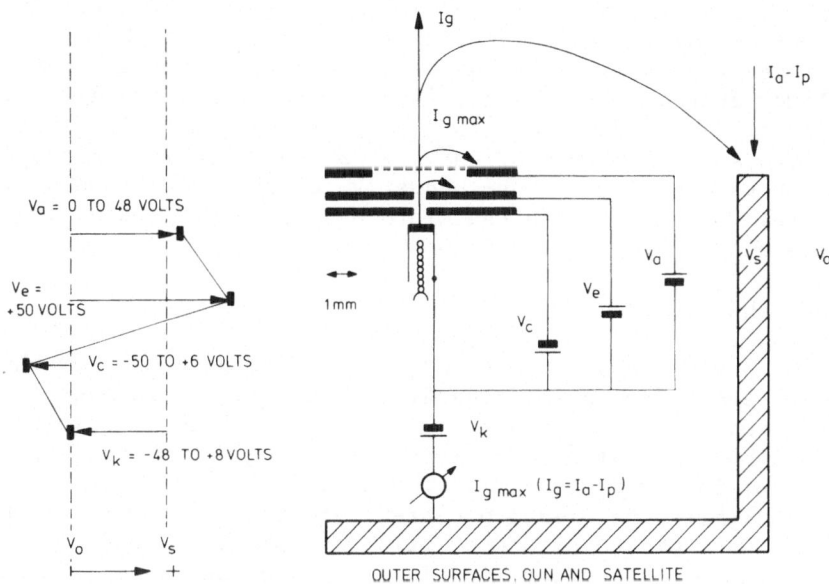

Fig. 3: Electron gun electrode arrangement.

if we drive a negative bias current to the electric field sensor, its potential will be close to and slightly positive relative to the plasma potential. The bias current is driven by a high impedance current source, and the satellite being larger than the probes is driven only slightly positive by the current source. The probe/plasma impedance is smaller than the current source impedance and the satellite floating potential, V_s, can be written

$$V_s \approx V_s - V_p + 1.0 \text{ Volt}$$

where $V_s - V_p$ is the measured potential between the satellite and one probe located 36m away from the satellite as a reference. The GEOS satellites also have conductive surfaces and nearly identical electric field experiments. A low energy electron experiment on GEOS-1 was capable of detecting a peak of accelerated electrons to the positive satellite in sunlight. The satellite floating potential determined in this way was in good agreement with the electric field probe technique, (Decreau et al., 1978).

It is interesting to notice that neither ISEE-1 nor the GEOS satellites have been charged to a large negative potential in sunlight as far as it has been possible to survey all data collected. This is in agreeement with the following simple numerical example; the photo-emission over a surface πr_s^2 (r_s = satellite radius) of 40 μA/m^2 will dominate over an ambient electron current over a

surface $4\pi r_s^2$ at the rate $1.5.10^9$ $s^{-1}cm^{-2}$ $ster^{-1}$. The latter figure
is a typical maximum electron flux in a substorm. A photo-emission
current density of 40 μ A/m^2 has been found to be typical for average
satellite materials. In reality is the effective surface for photo-
emission closer to $2\pi r_s^2$, and this makes it even more likely that a
conductive satellite in sunlight will be positive in a substorm.
The negative charging of non-conductive or mostly non-conductive
satellites in sunlight can therefore only be explained by the
effect shown in Fig. 1.

ELECTRON GUN DESIGN

Fig. 3 shows how the electron gun is designed and how the
different electrodes are arranged. The cathode (k) is made from
Barium-fused material and has a diameter of 0.8 mm. The voltages
of the cathode and all other electrodes are given on the left of
the figure. The grid closest to the cathode is the control grid
(c) which limits the electron current provided for extraction by
the positively biased extractor grid (e). The last grid is an
anode (a) normally kept at, or close, to the satellite potential.
The electrons which pass the control grid will be accelerated to
an energy e (V_a-V_k) before leaving the gun and they will loose this
energy by passing from the gun to the ambient plasma i.e. the
satellite will be forced to a positive potential (V_a-V_k) relative
to the plasma. It is possible to measure the electron current (I_{gmax})
which leaves the electron gun, irrespective of smaller losses to
the different electrodes. A certain fraction of the gun electrons
(I_g) will together with a smaller current of escaping photo-electrons
(I_p) balance the current from arriving ambient electrons (I_a), so
that $I_g + I_p = I_a$ or $I_g = I_a - I_p$. We can therefore monitor directly the capability of the gun to emit electrons by measuring I_{gmax},
which is independent of ambient conditions, as expressed by I_a.
However we cannot see what fraction of gun electrons escape (I_g)
and what fraction returns to the satellite. The electron guns, their
design and early diagnostics have been described in more detail by
Gonfalone et al (1979).

Fig. 4 is a picture showing the size and shape of one electron
gun and the mounting of two guns on the edges of the solar panels
on ISEE-1. The emitted currents from two guns is nearly 1 mA at
energies between 10 and 48 eV.

The guns were baked out and sealed before launch and a special
opening system was designed to open them,(Arends and Gonfalone, 1976).
The lifetime of the cathodes has been much longer than expected,
because they can easily be poisoned by small quantities of oxygen
and other gases. The indirectly heated filament of the cathode is
burning continuously to reduce pickup of contaminants whereas the
other grids are operated to emit electrons only during short diagnostic runs. At present a lifetime of 3.5 years has been reached.

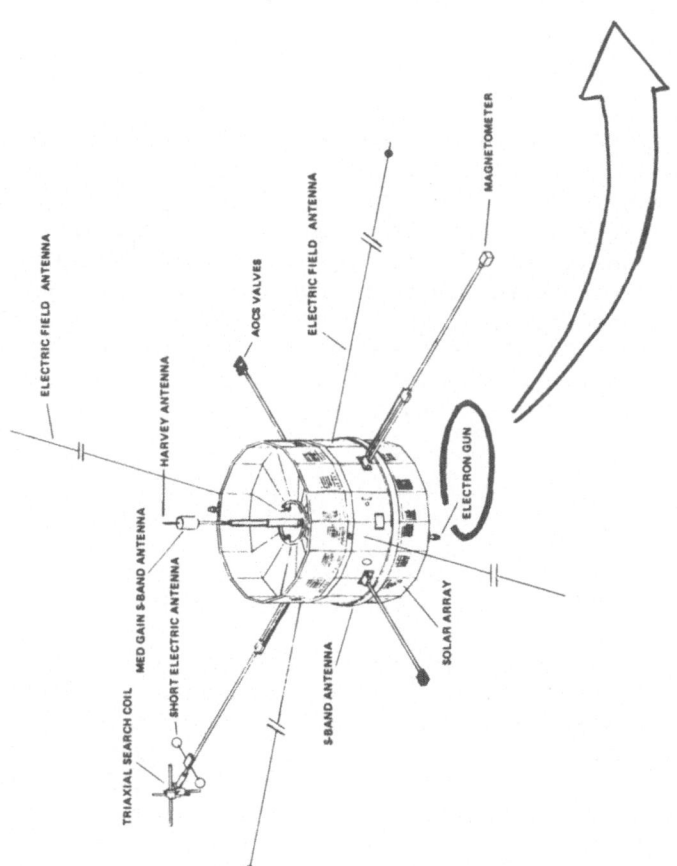

Fig. 4: Mounting of the two electron guns on ISEE-1 and picture of one gun (height ∼ 12 cm).

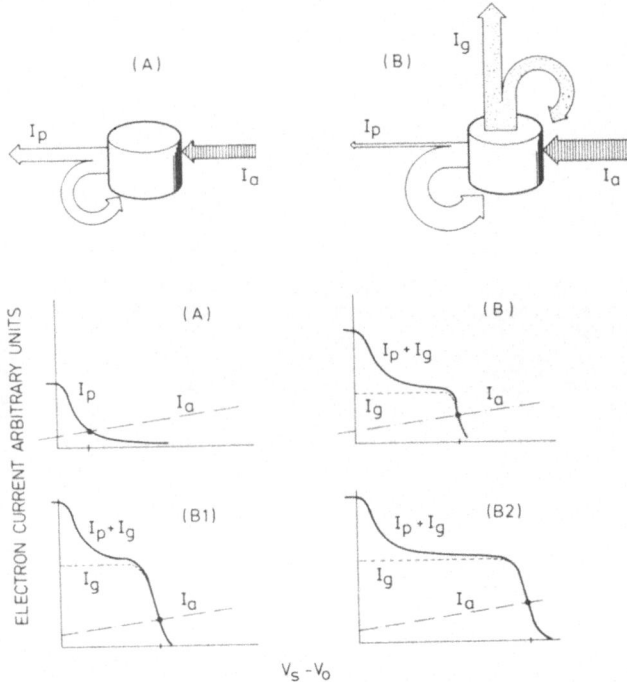

Fig. 5 : A; satellite comes to a positive floating potential as
consequence of escaping photo-electrons (I_p) and ambient
electrons attracted by the satellite (I_p). B; By electron
gun operation the equilibrium of electron currents are at
a more positive satellite potential. B1; The maximum
available electron current from the gun is increased with
the consequence that more gun electrons return to the
satellite. B2; The energy of gun electrons is increased
and the satellite is forced to a potential more positive
than the previous case.

ELECTRON GUN OPERATION

Fig. 5(A) shows in a way similar to Fig. 2 (left) how a conductive satellite in sunlight will come to a positive floating
potential due to a balance of escaping photo-electrons (I_p) and
ambient electrons coming to the satellite (I_a). When the electron
gun is operated with a sufficient current of emitted electrons at
energy $e(V_a-V_k)$, the satellite will move to a more positive potential provided $(V_a-V_k) > V_s$ (floating), see Fig. 5(B). Practically
all photo-electrons will return to the satellite. If $e(V_a-V_k)$

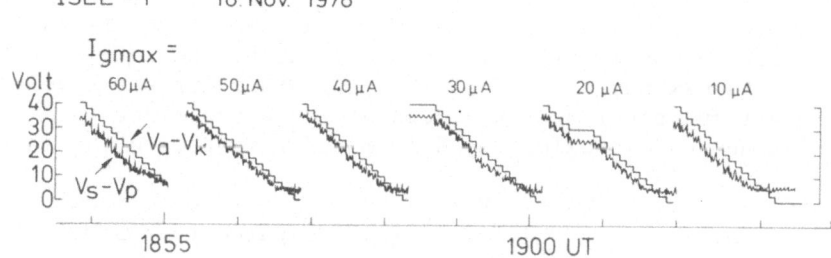

Fig. 6: Systematic variation of the ISEE-1 potential, V_s, by step-
         ping (V_a-V_k) in 16 steps of 2.5 Volt starting at 40 Volt.
         Six such sequences are run with I_{gmax} set at 60, 50, 40,
         30, 20 and 10 μA. V_s is measured relative to an electric
         field at potential, V_p, which is approximately 1.0 Volt
         positive relative to the plasma.

is set to values much larger than typical photo-electron energies
(1-10 eV). For the gun to keep the satellite positive at a potential (V_a-V_k) must I_{gmax} be larger than I_a. However, setting I_{gmax}
too large will result in a large fraction of the gun electrons
returning to the satellite as illustrated in Fig. 5(B1). In fact,
such an operation of the gun during early operation caused some
concern amongst experimenters with electron detectors. Gun electrons
returned to the positive satellite with energies up to 48 eV and
gave high count rates in channeltron detectors. This incident caused
a limitation of gun operations to short diagnostics intervals. It
is easy, however, with the experience gained, to adjust the current
to a level where the return current will be small and harmless.
Fig. 5 (B2) shows that by increasing (V_a-V_k), the satellite potential
will follow if the gun current is sufficiently large.

    Fig. 6 shows an example of potential control on ISEE-1 when
the satellite was in the magnetosphere at a radial distance of
approximately 10 Earth radii; the local time was 0700 and the geomagnetic latitude was 25°N. The electron guns were operated so that
I_{gmax} was at six current levels; 60, 50, 40, 30, 20, 10 μA, and
(V_a-V_k) was stepped at each current level 16 times in intervals of
2.5 Volts starting at 40 Volt. The variation of I_{gmax} and (V_a-V_k)
is illustrated in Fig. 6, and the corresponding change of the
satellite-probe potential (V_s-V_p) is also given. (V_s-V_p) is nearly
identical to the satellite-plasma potential as explained in paragraph 3.

The variation of the satellite potential looks nearly the same for all current levels of I_{gmax}, which means that 10 μA is sufficient to balance the ambient electron current to the satellite for (V_a-V_k) at its maximum of 40 Volt. For all other current levels of I_{gmax}, did all gun electrons in excess of 10 μA orbit out to a radial distance of approximately 5-10 m and returned to the satellite.

A closer examination of (V_s-V_p) for I_{gmax} = 10 μA shows that the maximum value of (V_s-V_p) is $\sim$32 Volt compared to $\sim$35 Volt for other current levels. This indicates that 10 μA is very close to the ambient electron current for the satellite at (32 + 1 = 33 Volt). By approximating the satellite with a sphere of diameter 1.0 m, and assuming that ambient electrons are freely attracted (long Debye length), it is possible to determine electron density, N_e, or temperature, T_e, if one of these parameters is known. N_e is determined by active wave techniques on ISEE-1 to be of the order 1.0 cm^{-3} (C. Harvey, private communication), which leads to a value for kT_e/e of the order 1 eV. This means that as far as electron flux is concerned, low energy electrons dominate. It is worth noting that this is the only instrument on ISEE-1 which can determine electron energies below approximately 10 eV. In other cases can both N_e and T_e be determined if I_{gmax} and (V_a-V_k) is stepped in a suitable way relative to the ambient electron current to the satellite.

CONCLUSIONS

The use of electron guns to control the potential of a satellite with conductive surfaces has been clearly demonstrated on ISEE-1. The electron guns used in this case can emit electrons with energies up to 48 eV, and the emitted electron current can be controlled, and has a maximum value of 0.5-1.0 μA for two guns. The potential of the ISEE-1 satellite, with or without gun operation, can be measured with reference to one or two spherical electric field probes positioned on booms at a distance of 36 m from the satellite. The electric field probes are biased with a negative current from a high impedance source to be slightly positive (0.5-1.0 Volt) relative to the plasma, and the spacecraft is normally floating several Volts more positive or can be pushed further positive by operating the electron guns.

From a technical point of view it is interesting that the guns at the time of this conference have been operational for 3.5 years without any sign of deterioration.

The electron guns could be used to prevent a satellite from charging to large negative voltages in eclipse in the magnetosphere. Instead of being at several kV negative, it is possible to keep a satellite (provided it has conductives surfaces) at more comfortable voltages of a few Volts positive. This way of controlling satellite potential may be of use for technological as well as scientific reasons. This procedure has not been tested in space because ISEE is very rarely in eclipse in the outer magnetosphere.

Plasma diagnostics can be carried out by appropriate sweeps of gun currents and energy of emitted electrons to obtain information about density and characteristic energy of ambient electrons. This is in fact obtained by using the whole satellite as a probe varied in potential. The electron density is determined on ISEE-1 by other techniques (particle and active wave experiments), however electron energies less than ~10 eV are not determined by these techniques.

A somewhat surprising observation was the stimulation of waves by the emitted gun electrons. This could be clearly demonstrated by correlating electric wave data with electron gun sweeps. This topic will be dealt with in a subsequent paper in these proceedings.

REFERENCES

Arends, H.J. and Gonfalone, A.A., Mechanical opening system for vacuum tubes in space environment, Rev. Sci. Instrum., Vol. 47, no.1, pp 153-155, 1976.

Decreau, P.M.E., Etcheto, J., Knott, K., Pedersen, A., Wrenn, G.L., and Young, D.T., Multiexperiment determination of plasma density and temperature. Space Science Reviews, 22, pp 633-645, 1978.

DeForest, S.E., McIlwain, C.E., Plasma clouds in the magnetosphere, J. Geophys. Res. 76, pp 3587-3611, 1971.

DeForest, S.E., Spacecraft Charging at Synchronous orbit, J. Geophys. Res., 77, pp 651-659, 1972.

Fahleson, U.V., Plasma-vehicle interactions in Space. Some aspects on present knowledge and future development, in: "Photon and Particle Interactions with Surfaces in Space", D. Reidel Publ. Co., pp 563-569, 1973.

Gonfalone, A.A., Pedersen, A., Fahleson, U.V., Fälthammar, C-G., Mozer, F.S., and Torbert, R.B., Spacecraft potential control on ISEE-1, in "Spacecraft Charging Technology - 1978", NASA Conf. Publication 207, pp 256-267, 1979.

Mozer, F.S., Torbert, R.B., Fahleson, U.V., Fälthammar, C-G., Gonfalone, A.A., and Pedersen, A., Measurements of quasistatic and low frequency fields with spherical double probes on the ISEE-1 spacecraft. IEEE Trans. Geosci. Electron.GE-16, (3) pp 258-259, 1978.

Mozer, F.S., Torbert, R.B., Anderson, R.R., Gonfalone, A.A., Etcheto, J., and Harvey,C.C., Observation of radiation from an electron beam on-board ISEE-1, in "Chapman Conference on Waves and Instabilities in Space Plasma", Denver Colorado, August 1979.

Rubin, A.G., Bhavnani, K.H., Tautz, M.F., Charging of Spinning Spacecraft. Air Force Geophysical Laboratory Report TR-79-0261, 1979.

DISCUSSION

<u>Koons</u>: Herb Cohen on SCATHA has not always been able to
reduce the vehicle ground potential using the electron
gun alone. He has routinely succeeded using the neutralized ion beam.

<u>Pedersen</u>: I understand that this is for eclipse condi-
tions. It is not clear to me how the potential of the
conductive surfaces on SCATHA can be measured unless you
look at cut-offs in electron and ion spectra. I expect
that some of the low energy electrons used to neutralize
the ion beam may leave the satellite at a higher rate
than the ions and thus bring a strongly negatively
charged payload to a smaller negative potential. This
is very similar to emitting low energy electrons from
ISEE-1; the difference is that in this latter case the
gun electrons have enough energy to bring the satellite
positive. This can hopefully be tested in a future
eclipse.

STIMULATION OF PLASMA WAVES BY ELECTRON GUNS ON THE ISEE-1 SATELLITE

Jean-Pierre Lebreton[1], Roy Torbert[2], Roger Anderson[3],
Christopher Harvey [4]
1) Space Science Department of ESA, ESTEC, Noordwijk
The Netherlands
2) Space Science Laboratory, University of California,
Berkeley, California 94720, U.S.A.
3) Department of Physics and Astronomy, The University
of Iowa, Iowa 52242, U.S.A.
4) Observatoire de Paris, 92190 Meudon, France

INTRODUCTION

The ISEE-1 satellite carries two electron guns for control of
the satellite potential. The characteristics of these guns and their
functioning in space are described in a parallel paper (Pedersen,
these proceedings). Here it is sufficient to repeat that both guns
were designed to inject electrons at energies up to 48 eV and with
a total current up to 500 μA into the ambient plasma. Significant
enhancements in the electric field wave spectrum have been observed
during the gun operations (Mozer et al., 1979).

The purpose of this paper is to describe typical observations
of the waves stimulated during the electron injections, when the
spacecraft is passing through the magnetosphere, the magnetosheath
and the solar wind. Phenomena induced by electron beams injected
in the ionospheric plasma have been studied in a number of experiments, using rocket-borne accelerators at high energies and currents
(Winckler, 1979 and references therein). Only a few experiments
have been carried out so far injecting electrons into the magnetosphere from a satellite; SCATHA and EXOS-B are the only ones beside
ISEE-1. Kawashima et al. (1979) have described an electron gun
experiment on EXOS-B with 100 to 200 eV in energy and 0.25 to 1 mA
in current. The gun was used to control the satellite potential and
to exite plasma waves by beam-plasma interactions. In this experiment we demonstrate that even a beam with lower energy and lower
current produces waves in a wide frequency range,which can extend
to a few times the local ambient plasma frequency.

ORBITS OF GUN ELECTRONS

When the guns are not operating, the satellite is floating at a positive potential. It has been demonstrated that, provided the current injected is sufficient, the potential of the satellite goes further positive to a value close to the accelerating voltage of the beam (Mozer et al., 1978). The electric field caused by the positive satellite and the negative space charge region around the spacecraft will decelerate the gun electrons, impeding them to escape out to the ambient plasma. A sufficient number of gun electrons will, after deceleration, escape to the ambient plasma to balance the ambient electron current to the satellite. The escaping electrons have energies of the order 1-2 eV, which is the typical energy spread of the gun electrons. The remaining electrons will return to the spacecraft, as observed by low energy electron detectors on ISEE-1. This shows that we have to consider two possible mechanisms for explaining the waves observed; either they are generated by orbiting and returning electrons inside the negative space charge region around the satellite, or they are generated by the interaction of the escaping electrons which go further into the ambient plasma. When the gun is not operating, the potential of the satellite (V_0), is determined by the equation;

$$I_p(V_o) + I_a(V_o) = 0 \qquad\qquad (1)$$

I_p and I_a being respectively the photo-electron current escaping and the ambient electron current collected, Fig. 1(A). Ion currents are negligible in comparison. Operating the gun, Fig. 1(B), at a current I_g introduces a new parameter in the equation which becomes

$$I_p(V_1) + I_a(V_1) + I_g(V_1) = 0 \qquad\qquad (2)$$

It is worthwhile to notice that, for the same ambient conditions, the potential rises up from V_0 to V_1, V_1 being the accelerating voltage of the beam. Equation (2) is, within certain limits, independent of the total beam current I_{go}, that is to say, increasing I_{go} will increase $(I_{go}-I_g)$ which is the part of the beam electrons orbiting inside the negative space charge region. This region, which is of the order a few times the Debye length, can extend from 10 m to 40 m as shown in Table 1 for typical parameters.

Table 1: Typical parameters along the orbit of ISEE-1.

	Density $N_e (cm^{-3})$	Temperature $T_e (^oK)$	Debye length λ_D (m)
Magnetosphere	1	$2\ 10^4$	9.7
Magnetosheath	50	$40\ 10^4$	6.2
Solar wind	20	$10\ 10^4$	4.9

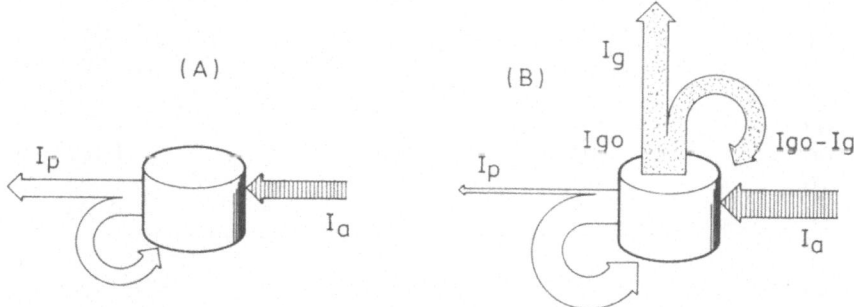

Fig. 1: Contribution of the different electron currents to achieve the equilibrium of the satellite potential: I_p photo-emission, I_a current from the ambient plasma, I_g gun current.

The density inside the negative space charge region which looks like an electron plume, can be estimated by using a simple model describing the potential profile along the axial direction of that plume, Fig. 2.

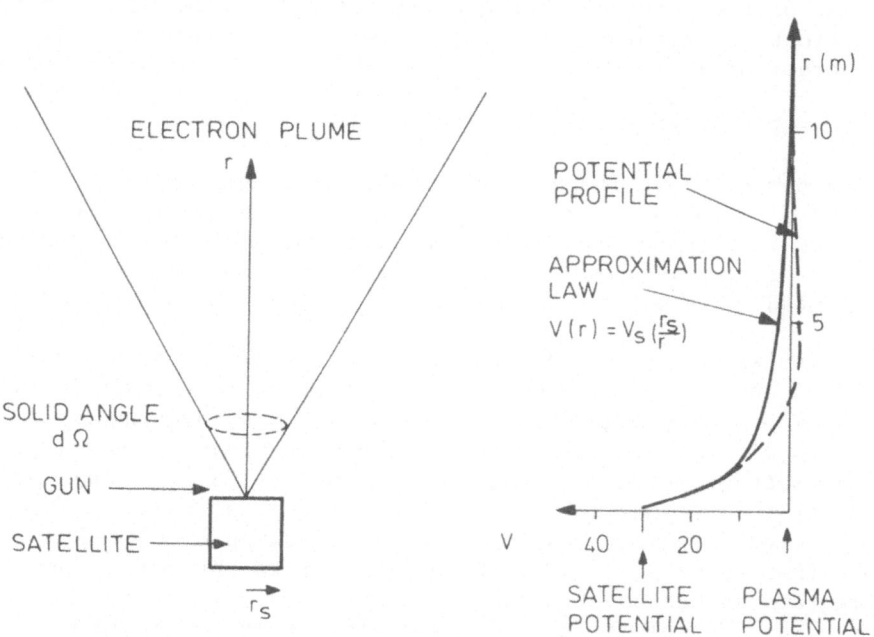

Fig. 2: Axial potential profile inside the plume.

Table 2: Electron density inside the plume.

V(volts)	I_g (A)	n_g $\frac{cm^{-3}}{}$ (10 m)	n_g $\frac{cm^{-3}}{}$ (40 m)
40	60	1,4	0,17
40	10	0,24	0,03
10	60	2,8	0,35
10	10	0,5	0,06

The potential distribution is approximated by $V(r) = Vs(r_s/r)$, V_s and r_s are the potential and the radius of the satellite. The velocity distribution function of the electrons is related to their initial velocity v_o at the output of the gun by the relation $v(r) = v_o(r_s/r)^{\frac{1}{2}}$. The gun current, I_g, through the surface $r^2 d\Omega$ is given by $I_g = n_g(r) e v(r) r^2 d\Omega$. Then the density follows the relation

$$n_g (r) = \frac{I_g r_s^{\frac{1}{2}} d\Omega}{ev_o r^{3/2}} \qquad (3)$$

The relation (3) has been evaluated assuming $d\Omega = \pi$ for 2 distances $r = 10$ m and 40 m for $r_s = 1$ m. The results are shown in table 2.

The above calculations give a good estimation of the plume density when most of the electrons are escaping into the ambient plasma. If electrons are returning the density will be increased by a factor of the order of 2.

ELECTRIC FIELD ANTENNA

On board ISEE-1, the electric field measurements can be performed by 3 antennas, of different lengths which can be connected to various instrumentation. Here, we will summarize the performances of the two long antennas, in the configuration they are used in the present study, and shown in fig. 3.

A fine wire electric dipole antenna, with a tip to tip length of 215 m connected to a 20 channel spectrum analyser (Gurnett et al., 1978) provides high time resolution spectra covering the range 5.62 Hz to 311 kHz. We will refer to the corresponding spectrograms as GUM spectrograms. A double sphere antenna of 73.5 m connected to a swept frequency analyzer covering the range 0-51 kHz in 128 steps, provides high frequency resolution during 1 sweep completed in 16 seconds (Harvey et al., 1979). We will refer to the corresponding spectrograms as HAM spectrograms. The potential of the satellite is measured with reference to one spherical electric field sensor at 36 m from the satellite. This sensor is biased to be near the plasma potential. In addition, the electric field up to a few

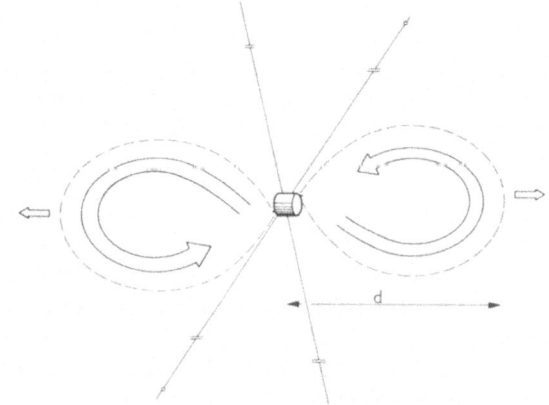

Fig. 3: Schematic representation
of ISEE-1 showing the long antenna
and the electron plume created
by the guns.

hundred Hz is measured deifferentially between two identical
spherical sensors, giving the low frequency wave power in three
bands centered at 4 Hz, 32 Hz and 256 Hz. When referring to these
data we will name the corresponding channel.

The 2 guns, aligned along the spinning axis of the satellite,
can be operated either separately, or together; gun 1 and gun 2 are
emitting in opposite directions. We have have not observed phenomena
which can be attributed to one gun or the other; therefore we will
not differentiate between them. In fact, in most of the operating
modes chosen in the study, they are operating together, set at the
same parameters, according to a sequence depicted in fig. 4. The
actual sequence can deviate a little from this typical sequence
due to delay in commands.

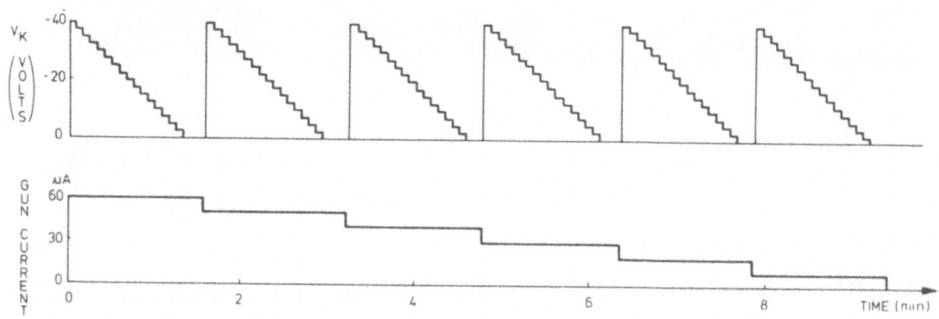

Fig. 4: Typical sequence applied to both guns.

OBSERVATIONS

We are presenting data obtained on the 18th and 19th of November, when ISEE-1 was inside the magnetosphere around 1900 UT, in the magnetosheath at 2325, crossing the bow shock around 2340 and in the solar wind afterwards.

Gun operation in the magnetosphere

Two gun sequences were started at 1853 and 1913. At that time the density was of the order of .75 cm^{-3} as determined by the propagation experiment (Harvey et al., 1979). Using that value, the electron temperature has been found to be of the order of 1 eV by Pedersen (1981, these proceedings).

The HAM spectrogram obtained before and during the first sequence of the guns operation is shown in fig. 5. The regular structure seen prior to 1852 is often seen when the shields of the double sphere antenna are driven negatively at -10V with respect to the spheres biased at the plasma potential. This configuration was

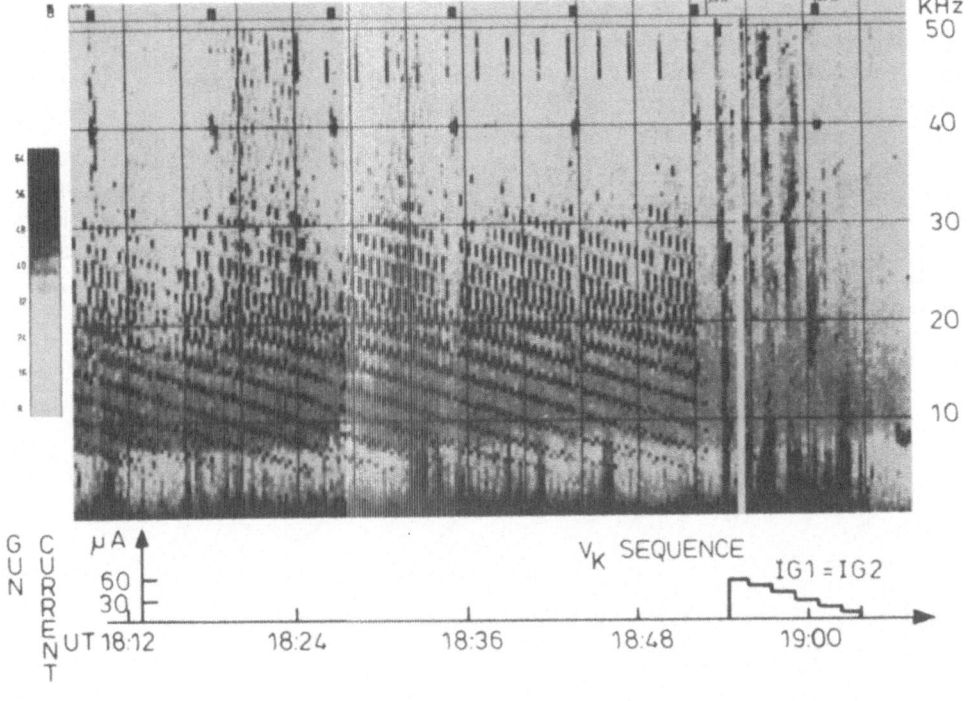

Fig. 5: HAM spectrogram in the magnetosphere. The timing of the gun operations is shown in the lower part of the figure. HAM data are not available for the second sequence starting at 1913.

not used during the gun operation, during which the booms were grounded to the satellite. The maximum current of photo-electrons escaping to the ambient plasma was 20 µA, assuming a typical current density of 40 µA/m^2 emitted by the booms. The noise spikes are thought to be due to the modulation of the photoemission from the shields by the spacecraft rotation, and the diagonal structure is due to the receiver sweep period of 16s not being an integer multiple of the spin period of 3.04 s. This brief description shows that there is some similarity between the process involved due to the photoemission of the negatively polarized booms and the controlled injection of electrons by the guns. This will be the subject of an other study. The lower cut-off of the thermal continuum at 7.5 kHz gives a good evaluation of the electron plasma frequency, f_{p-}, compatible with the value given by the propagation experiment. The electron gyrofrequency was of the order of 1 kHz. During the first gun sequence, the enhancement of the HAM spectrum over the full frequency range is correlated to the different steps applied during the sequence. It is possible to distinguish an oblique structure which is related to the 5 s duration of each energy step applied to the guns. A first type of wave is detected up to f_{p-}. Another type, above f_{p-} is detected up to 50 kHz; the frequency band for which the power is maximum has a tendency to decrease when the gun current is decreasing.

All these observations are generally confirmed by the GUM spectrograms obtained during the 2 sequences, which, in fact, look very similar, fig. 6. In addition, due to the good frequency resolution in the lower part of the frequency scale, we observe an enhancement of the mean power below 100 Hz, while the peak power, also enhanced, is sharply modulated. This modulation is correlated to the abrupt change in energy of the electron beam, which changes the potential of the satellite. This low frequency noise is also detected in the 3-channels at 4, 32 and 256 Hz.

Nevertheless differences are appearing when comparing the spectrum given by both antennas. For example there is no enhancement in the channel 17.8 kHz at 18.57 of GUM spectrograms, fig. 6, while we see a broad enhancement between 15 to 25 kHz in the HAM spectrogram, fig. 5. A detailed comparison of the power received by the two antennas, not done yet, could provide useful information on the wavelength as shown by Harvey et al. (1981). The structure is more peculiar in the GUM spectrogram during the 10 µA step, fig. 6. The enhancement appearing first in the channel 5.62 kHz, becomes apparent in upper channels with increasing time, which corresponds to decreasing energy of the beam. In the next chapter we will tentatively interpret the waves as a beam plasma instability due to a coupling beween the electron plasma mode and the electron beam. For higher gun currents, the density under the plume increases as shown by the results of our calculations in table 2. We should get the correct figures of the electron density distribution around the satellite to interpret the observations.

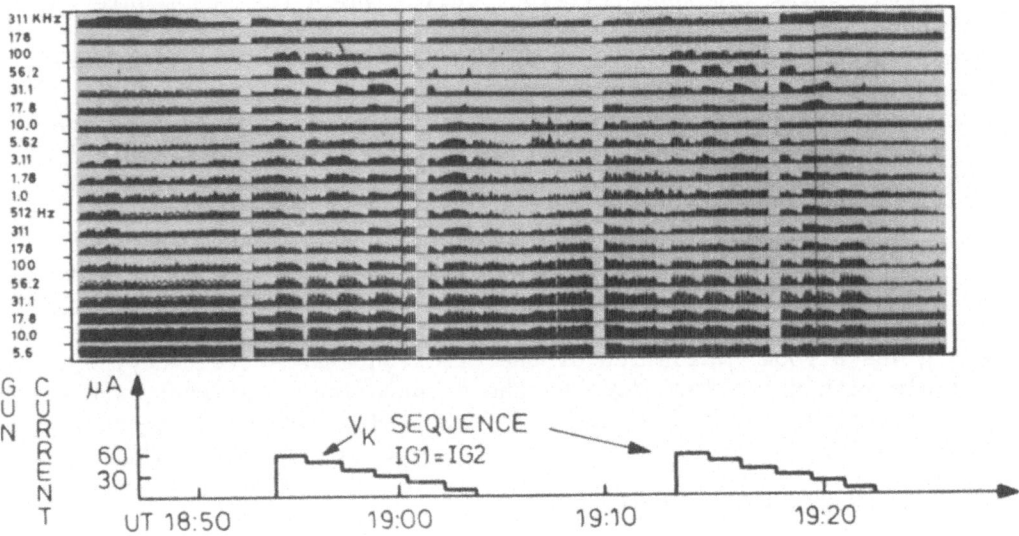

Fig. 6: GUM spectrogram in the magnetosphere. See caption of fig. 5 for the description of the timing of the gun.

Natural waves in the magnetosheath and the solar wind

Similar gun sequences were started at 2319, 234530 and 240530. In addition each gun is run separately at a fixed energy and current between 2354 and 2404 as shown in fig. 7 and 8. During the period under study, the spacecraft passed from the magnetosheath into the solar wind. The spectrum shown in the HAM spectrogram, fig. 7, is typical of the sheath until about 2340. During the bow shock crossing, between 2339 and 2342, the magnetic field fell from 21 γ (f_{c-} = 595 Hz) to 12 γ (f_{c-} = 340 Hz). The HAM spectrogram changes from typical sheath to typical solar wind in the presence of electrons backstreaming from the bow shock, as occurs when the field line through the spacecraft is nearly tangent to the shock. (Filbert and Kellogg, 1979); these electrons produce intense noise near to the local plasma frequency, which was about 47 kHz. At about 2354 to HAM spectrogram changes again; the magnetic field magnitude did not change, but its direction changed to become nearly perpendicular to the shock surface. During the period 2354 to 2405 the magnetic field direction and the HAM spectrum are consistent with the presence of upstream m.h.d. waves, as described by Harvey et al. (1981); the presence of large magnetic field fluctuation confirms this identification. At 2406 a further rotation of the interplanetary magnetic field brough the spacecraft onto a field line not connected to the Earth's bow shock, and the HAM spectrogram corresponds to the quiet solar wind.

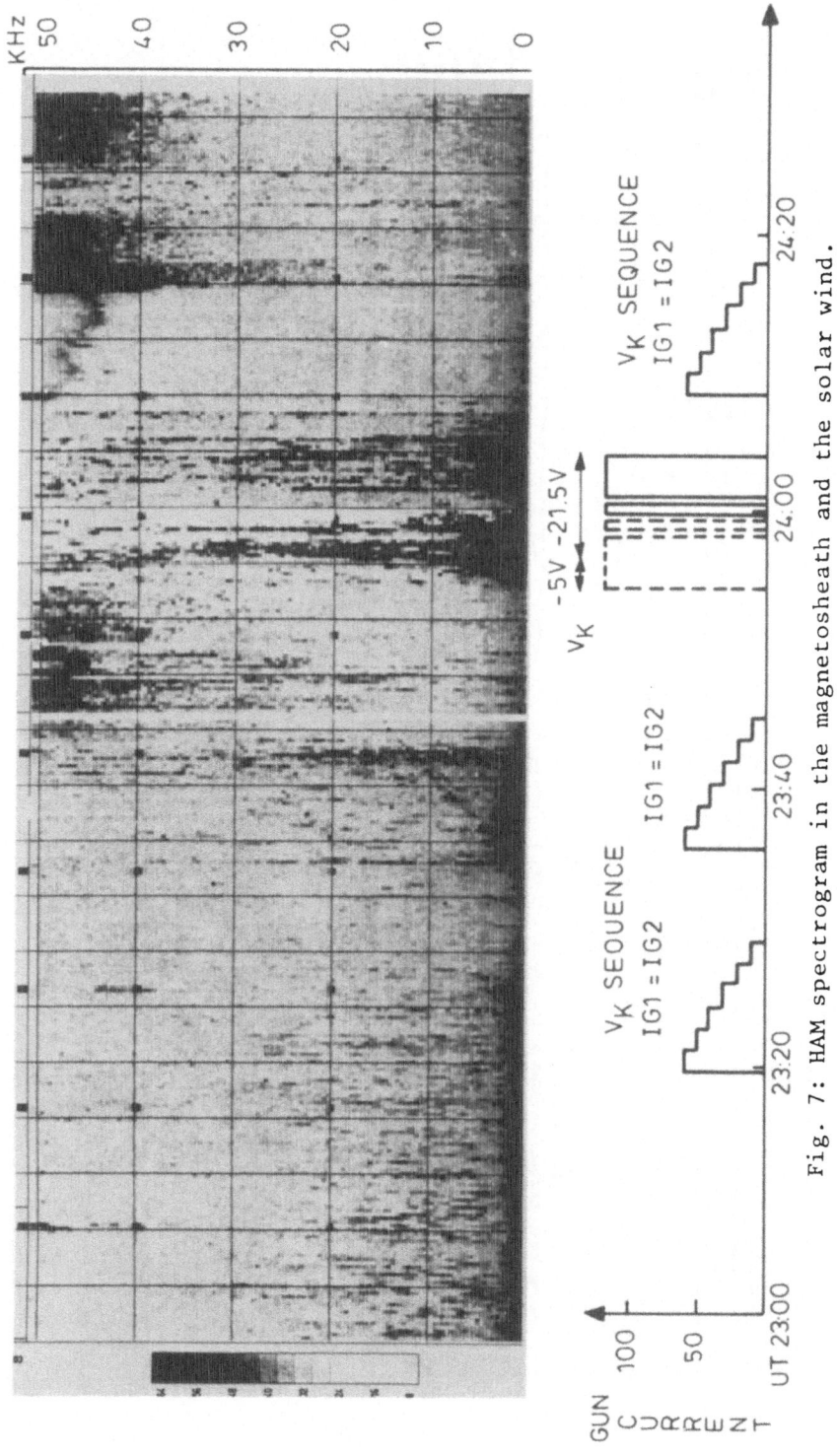

Fig. 7: HAM spectrogram in the magnetosheath and the solar wind.

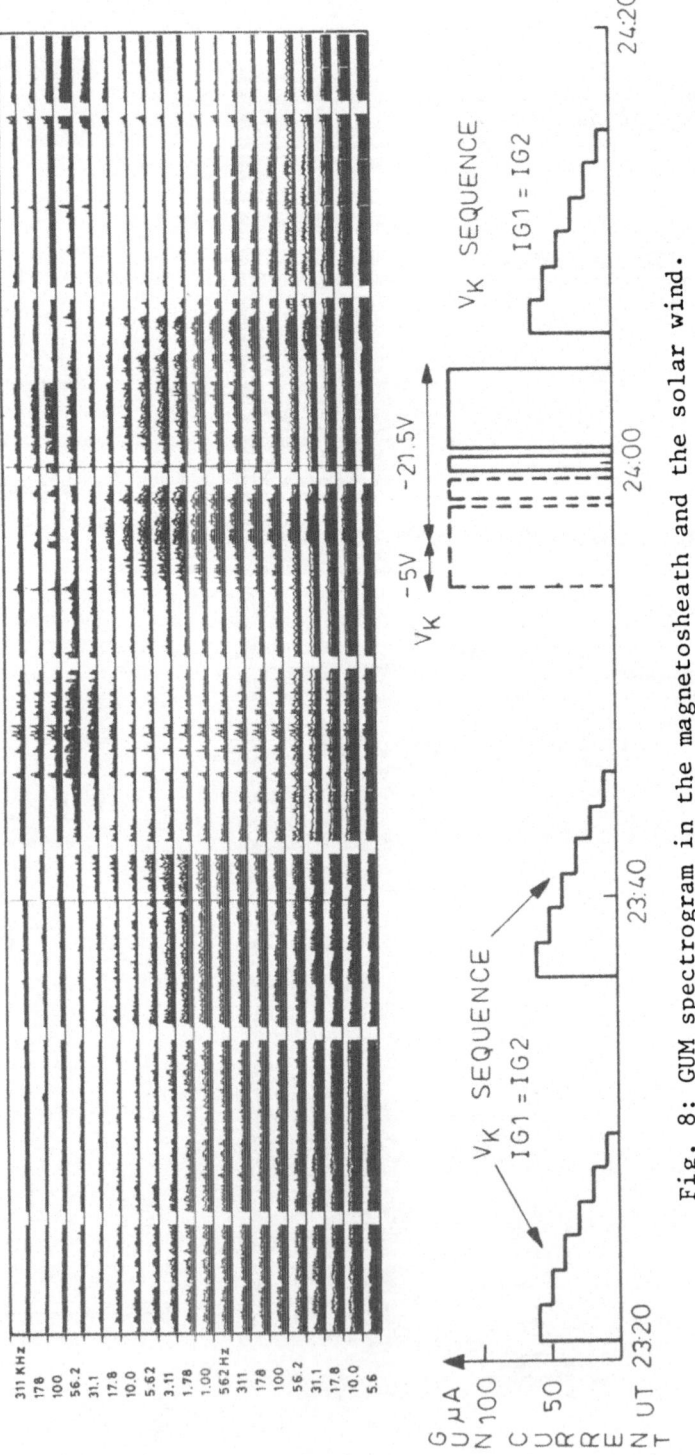

Fig. 8: GUM spectrogram in the magnetosheath and the solar wind.

At 001520 a further small rotation of the interplanetary field
brought the spacecraft field line once more nearly tangent to the
shock, and again HAM sees intense noise in the vicinity of the
electron plasma frequency.

Gun operation in the magnetosheath

Inside the magnetosheath, the satellite was floating around
2V positive at 2321 and 1.5 positive at 2327. During the gun sequence, the potential of the satellite was not increasing up to
the accelerating voltage of the beam, except for the lowest energy
step. Higher values of the gun current would have been necessary
to clamp the satellite potential as it happened in the magnetosphere.
This situation is typical when the ambient electron current collected
by the satellite is high due to a combination of high density and
high temperature. In these conditions, the gun electrons were escaping into the ambient medium. Only for the lowest values of the
beam energy, part of the beam current was orbiting, creating a
negative space charge region much reduced in dimension as compared
to the previous case, when ISEE-1 was in the magnetosphere. Correlated to the gun sequence, the lower part of the spectrum is slightly enhanced, up to 100 Hz as shown in the GUM spectrogram, fig. 8.
Noise is also observed in the two channels 32 and 256 Hz. There is
no wave stimulated in the vicinity,or above the plasma frequency,
as seen in magnetospheric conditions. The ratio between the beam
density and the electron ambient density is less than 1%, and too
small to produce a detectable instability in the high frequency range.

Gun operation in the solar wind

When ISEE-1 was in the solar wind, one gun was operated after
the other at 235430, first at a low energy 5V and later at 21,5 V
with a current of 120 μA. The GUM spectrogram, fig. 8, shows correlated emissions above 100 kHz, which is of the order of 2 f_{p-}. A
new gun sequence was started at 000430. Low frequency noise, below
1 kHz, is detected as shown by the GUM spectrogram, fig. 8,with a
maximum power in the channels 176, 311 and 562 Hz. This is confirmed
by the noise received on the double sphere antenna; the noise is
high only in the upper channel at 256 Hz.

DISCUSSION OF THE OBSERVATIONS

We have also looked at the magnetic wave spectrum available
in the range 5,6 Hz to 10 kHz for our study, but we never observed
any signifcant enhancement during the guns operation. That is why
we will interpret our obervations in terms of electrostatic mode
only.

We have seen a signal correlated in time with the change
of energy step of the guns on the peak power of the low frequency

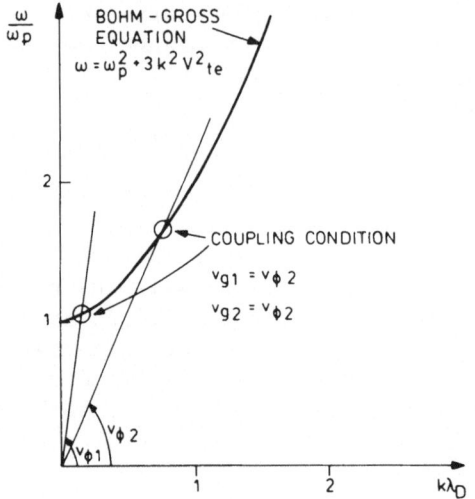

Fig. 9: Coupling mechanism f_{p-} between beam electrons and electron plasma wave mode.

part of the GUM spectrograms and in the HAM spectrograms which presents a serie of 3 parallel lines spaced in time by the duration of 1 energy step. We will not try to interpret these observations. We are not sure they are not instrumental.

Owing to the good frequency resolution of the GUM spectrograms and to the data available in the 3 channels of the filter bank connected to the two spherical sensors, we have observed waves below the ion plasma frequency which was of the order of 175 Hz in the magnetosphere, 1400 Hz and 1100 Hz respectively in the magneto-sheath and the solar wind during the period under study. We have tried to identify these waves as ion acoustic waves although we are very far from the existence condition $T_e/T_i \gg 1$ established for a plasma with a Maxwellian ion and electron population. As an example the hot ions had a typical energy of ~4 keV in the magentosphere at 1900 (V. Formisano, private communication); this gives a ratio $T_e/T_i \ll 1$. These arguments are not in favour of the ion acoustic mode unless another excitation mechanism is important.

Finally we also observed, especially in the magnetosphere, but also in the solar wind, waves above and up to several times f_{p-}. These observations can be explained invoking a coupling between the electron plasma mode and gun or ambient plasma electrons with a velocity greater than the thermal velocity of the ambient electrons. This is illustrated in Fig. 9, in which we used the Bohm-Gross dispersion relation for the simplicity of the demonstration. The wave phase velocity decreases with increasing frequency, and we obtain coupling with beam particles at lower energies for increasing frequency as observed around 1903 and 1923 (fig. 9). Although the damping, which is not described by the Bohm-Gross relation, is important above 2 f_{p-}, it is not surprising to detect a signal up to several times f_{p-}, the corresponding waves having a wave length of the order of 100 m.

CONCLUSIONS

It has been clearly shown that, on board ISEE-1 the injection of an electron beam current of the order of 10 to 60 μA with energies ranging from 0 to 40 eV was producing enhancements in the electric wave spectrum. We have tried to identify the low frequency electrostatic wave observed below the ion plasma frequency as an ion acoustic mode although the excitation mechanism has still to be understood. A coupling mechanism between the electron plasma mode and streaming electrons with energies higher than the thermal speed of the cold electron population (T_e = 1 eV) has been proposed to explain the observations above the electron plasma frequency. We have not taken into account neither the hot ion population (T_i = 4 keV) nor any Doppler shift which are thought to be able to modify the dispersion relation. These effects are under study.

We are convinced that performing such active experiments in controlled conditions may give a good contribution to the understanding of the properties of the waves naturally existing in the magnetosphere or the solar wind.

ACKNOWLEDGEMENTS

We wish to thank A. Pedersen for useful discussions and D.A. Gurnett for the use of the ISEE plasma wave experiment data. We also acknowledge many discussions with A. Gonfalone and V. Formisano. The research at the University of Iowa was supported by NASA through contact NAS5-20093 with Goddard Space Flight Center. The sounder and propagation experiment was financed by the CNES under contract no. CNES/214 and SL.180.01.E.

REFERENCES

Filbert, P.C., Kellogg, P.J., Electrostatic noise at the plasma frequency beyond the Earth's bow shock, J. Geophys. Res. 84, 1369, 1979.

Gurnett, D.A., Scarf, F.L., Fredricks, R.W, and Smith, E.J.,: The ISEE-1 and -2 plasma wave investigation, IEEE Trans. Geosci. Electron. GE-16, 225.

Harvey, C.C., Etcheto, J., and Mangeney, A., Early results from the ISEE electron density experiment, Space Sci. Rev., 23, 39-58, 1979.

Harvey, C.C., Bavassano-Cattaneo, M.B., Dobrowolny, M.,Orsini, S., Mangeney, A., and Russell, C.T., Correlated wave and particle

observations upstream of the Earth's bow shock. J. Geophys.
Res., 1981, accepted for publication.

Kawashima, N., Murasato, Y., Kubo, H., Mukai, T., Ejiri, M.,
Miyatake, S., Matsumoto, H., and Oya, H.,: Controlled beam
experiment (CBE) in "Jikiken"(Exos-B), Proceeding of the
International Workshop on Selected topics of Magnetospheric
Physics, Tokyo, 1979.

Mozer, F.S, Torbert, R.B., Fahleson, U.V., Fälthammar, C-G.,
Gonfalone, A., Pedersen, A.: Measurements of quasi-static and
low frequency electric fields with spherical double probes on
the ISEE-1 spacecraft, IEEE transactions on Geoscience Electronics, Vol. GE-16, no. 3, 1978.

Mozer, F.S., Torbert, R.B., Anderson, R.R., Gonfalone, A., Etcheto, J.
and Harvey, C.C.,: Observation of radiation from an electron
beam on-board ISEE-1. Proc. Chapman Conference for Waves and
Instabilities in Space Plasma, Denver Colorado, August 1979.

Pedersen, A.: Plasma diagnostics by electron guns and electric field
probes on ISEE-1. These proceedings.

Winkler, J.R.: The application of artificial electron beams to
magnetospheric research,Univ. of Minnesota, Cosmic Physics,
Technical Report ## 183, 1979, Minneapolis, Minnesota 55455.

EVIDENCE FOR BEAM-STIMULATED PRECIPITATION OF HIGH ENERGY ELECTRONS

E. Bering, J. Benbrook, J. Roeder, and W. Sheldon

Physics Department
University of Houston, Central Campus
Houston, Texas 77004, U.S.A.

INTRODUCTION

The interaction of artificial and natural electron beams with the ionosphere and magnetosphere has been observed to give rise to a rich variety of phenomena. Most of the work on this topic has concentrated on the interaction of a beam with the ambient thermal plasma, on plasma wave generation near a beam source, etc. Relatively little attention has been paid to the effects of a beam on the distant magnetosphere, and on high energy particles. This paper will discuss two observations in which an electron beam apparently stimulated the precipitation of high energy electrons from the deep magnetosphere. These observations were made by the UH Group in conjunction with two active magnetospheric experiments, the first Araks experiment on January 26, 1975 (Roeder et al., 1980 (paper 1)) and the Trigger experiment on February 11, 1977. (Bering, et al., 1980 (paper 2)). In both cases, the data were obtained from rocket-boosted, parachute-borne X ray bremsstrahlung detectors which were initially deployed of an altitude of 80 km near the foot of the flux tubes affected by each experiment. The X ray detectors provided the capability for remote sensing of energetic electron precipitation which might not be measured by the electron detectors on the F region payloads in the two experiments for a variety of reasons. The X ray results from these experiments have been published elsewhere (paper 1, paper 2). However, the similarities between the results have not been presented or discussed. This paper will compare the X ray results from the Araks and Trigger experiments and discuss this comparison in terms of beam physics.

147

INSTRUMENTATION

The instrument system used for the X ray measurements was described in paper 2. The payloads were boosted to a nominal apogee of 80 km by a Super Arcas sounding rocket. At that point they were deployed on a parachute with the detectors uncovered providing about 12 minutes of data above 30 km altitude. The primary detector was a NaI(Tl) scintillation detector with a thin (0.127 mm) beryllium window. The scintillator was 3.18 cm in diameter and 2 mm thick, with an acceptance cone of approximately 50° half-angle. The X ray data were telemetered in two formats: time to accumulate 2^n counts in four energy level discriminators (>5, >15, >30, and >50 keV) and 0.25 S count accumulations in a fifteen channel pulse height analyzer (PHA) covering the energy range 10 to 95 keV.

ARAKS EXPERIMENT

The Araks experiment was a Franco-Soviet rocket-borne electron gun experiment (Cambou, et al., 1980). The gun emitted a modulated 0.5 A beam at 27 keV for 140.8 seconds and at 15 keV for 102.4 seconds. Two UH X ray payloads were deployed in coordination with each Araks experiment. The first, launched 11 minutes before the Eridan Araks rocket, was intended to provide background data. A second X ray detector was launched 1 minute before the Eridan, and deployed on its parachute 20 seconds before operation of the electron gun began. In order to correlate the X ray data with operation of the electron gun, particular attention was given to the spatial relationships of the three payloads. The trajectory of the first Araks vehicle is drawn as a solid curve in Figure 1. Arrows mark significant events during the Araks experiment with the elapsed time from the Eridan launch. The solid straight lines connect these points to the 100 km altitude level along the geomagnetic field lines. A dotted line connects each of these bremsstrahlung production points in the upper atmosphere to the position of the second X ray payload (shown as #2 in the figure) at the respective times marked by the arrows. The positions of the lower altitude X ray detector are marked as #1 for two times in the figure. No enhancements in the X ray flux corresponding the operation of the electron gun were readily apparent in the data. Therefore, in order to detect any small periodicity in the X ray flux due to the Araks experiment the integral energy level X ray data were subjected to a superposed epoch analysis keyed to the electron gun sequence. The results of this procedure on the data from payload #2 are shown in Figure 2. In the bottom panel one cycle of the nominal gun current sequence is plotted. There was nineteen such 12.8 second epochs during the flight. The superposed X ray data are displayed as 0.4 second averages in the other four panels. The mean for each channel has been subtracted out and the result scaled in terms of standard deviations. The standard deviations used were computed as rms averages of the deviations from the means. These deviations are nearly equal to those predicted assuming Poisson statistics, an indication that the distribution of the data is roughly stationary in time.

The analysis shows a small enhancement at 3.6 seconds epoch time, and a hint of an increase at 8.8 seconds. They seem to be associated with the two wide accelerator pulses injecting electrons up the field line. To investigate the possibility that these peaks are mere coincidence, various checks were performed. A pseudo-random transition data set was created, and subjected to an identical epoch analysis. The results show approximately the same amount of statistical fluctuation as the real data, but no significant enhancements. Examination of the distribution of counting rates contributing to each point of the epoch analysis revealed that the enhancements were not due to a high intensity burst in any one epoch but resulted from many small increments over the nineteen epochs. To further investigate the validity of this result a power spectral analysis was performed on the >50 keV

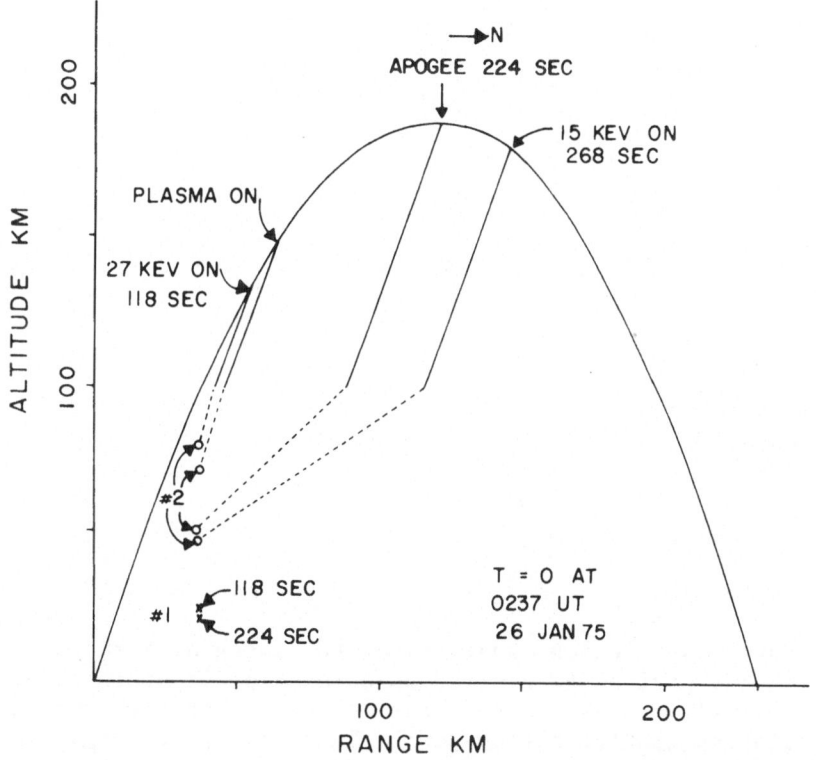

Fig. 1. The trajectory of the first Araks payload with the positions of the X ray payloads marked at various times.

data, the energy at which the strongest signal appears in the epoch analysis. Peaks of moderate significance were found at harmonic frequencies of the gun period in confirmation of the peak in the epoch analysis. Hence it is concluded that electron precipitation was stimulated by operation of the Araks electron gun.

An epoch analysis was also performed on the PHA data. The results are shown in Figure 3, plotted as an integral energy spectrum. The spectrum derived from the integral channel data during the peak is also shown, as are the corresponding background spectra. Since the accumulation period was not commensurate with the gun operation period, the PHA data frames from the times most nearly corresponding to the main peak in the integral level epoch analysis are still somewhat contaminated with background data.

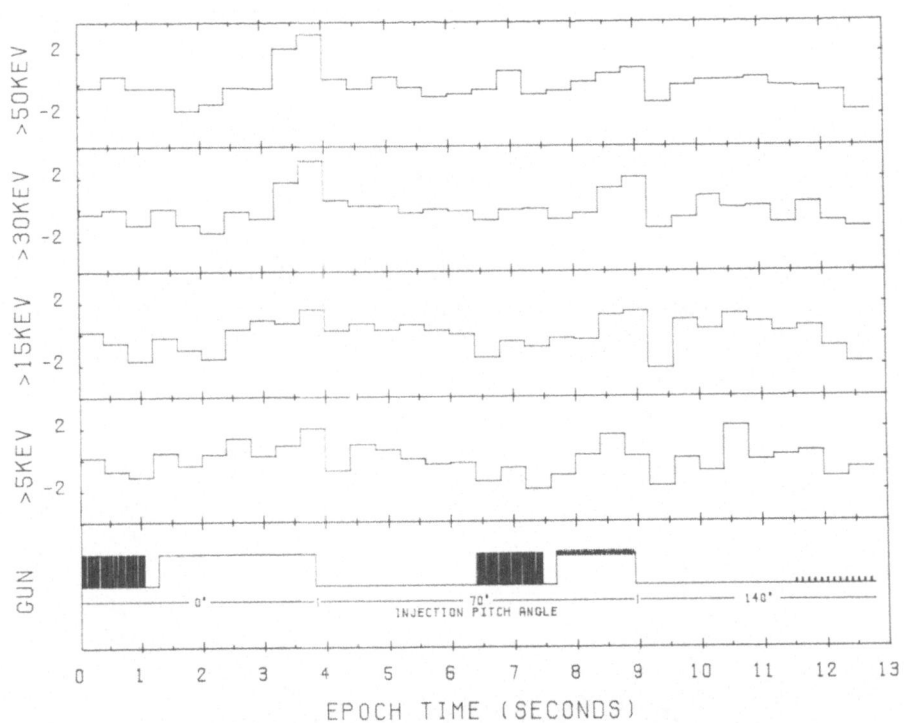

X-RAY EPOCH ANALYSIS, ARCAS 15.130UE
0.4SEC AVE,19 EPOCHS,STRT 2:39:10.335,JAN 26,1975

Fig. 2 Superposed epoch analyses of the integral X ray data from X ray payload #2 plotted as a function of epoch time. The nominal Araks electron gun current sequence is plotted in the bottom panel. Injection pitch angle is also indicated. The ordinate is in terms of standard deviations from the mean counting rate.

Hence the peak in the integral data is higher than in the PHA data. Nonetheless, Figure 3 shows clearly that the enhancement extends to photon energies of at least 95 keV. Under the assumption that the triggered precipitation during Araks was monoenergetic electrons at 100 keV, the peaks in the epoch analysis correspond to an excess of approximately 10^{19} electrons sec.$^{-1}$ The fact that the PHA peak spectrum shows little sign of sagging toward background even at 95 keV indicates that the energy of the precipitated electrons was probably in the 200- 300 keV range.

TRIGGER EXPERIMENT

The Trigger experiment was not intended to be an electron beam experiment, since the perturbation mechanism was a cesium cloud release (Holmgren, et al., 1980). In this experiment, a 2 km radius cloud containing some 10^{22} Cs^+ ions was created at 164 km altitude by the detonation of a 12 kg TNT-AlO-$CsNO_3$ bomb. However, one of the major effects of this release was the apparent acceleration of a field-aligned beam of 1.5 keV electrons with a peak flux of $\sim 2 \times 10^9$ electrons cm^{-2} sec^{-1} sr^{-1} keV^{-1} (Lundin and Holgren, 1980). One X ray detector was deployed in coordination with this experiment, 68 seconds prior to detonation of the cesium cannister, and remained above 25 km for about 12 minutes.

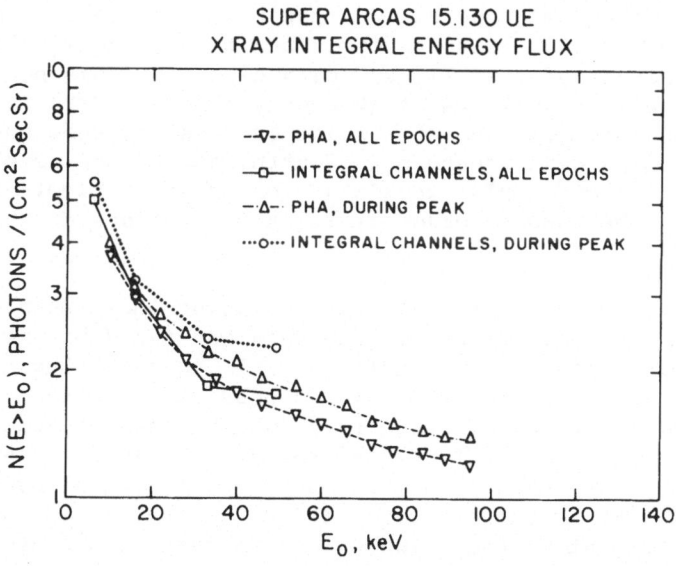

Fig. 3. Integral energy spectra of the peak found in the epoch analyses. Spectra from the integrated PHA data, the integral energy level data, and the corresponding backgrounds are shown. The statistical uncertainties are smaller than the symbols.

As in the Araks experiment, no enhancement in the X ray flux was observed in the few seconds immediately following the release. However, there are two lines of evidence indicating that delayed precipitation was produced by the interaction of the 1.5 keV beam with the magnetosphere.

The first such evidence was found by examining the power spectral density of the data. Power spectra were computed from 128-s blocks of 0.5-s averages of the >5 keV data. The window was stepped forward in 1-min intervals. The main feature of interest was a peak at ∿0.13 Hz(7.7 second period) in the spectra of the data blocks starting 1 and 2 minutes after the release. This peak exceeded the 99% confidence limit in these spectra, and exceeded the 95% level in all of the variety of confidence tests applied (paper 2). To view the temporal development of the event in the frequency domain, a dynamic set of spectra has been calculated. Since the total event was ∿10 periods long, a 10-period window length of 76.8 s was used, advancing one period per step starting 1 min 4s prior to the explosion. The result is shown in Figure 4. The 0.13-Hz event is indicated by the prominent extended feature in the middle of the figure.

There are a number of reasons for relating the 0.13-Hz signal in the data to the 1.5 keV beam. First, the effect was very local.No evidence of a similar signal was found in spectrograms of the ground-based micropulsation magnetometer data. Second, no obvious instrumental source for the signal could be found. Spectral analysis of the >15-keV data also show this peak. Third, spectral analysis of the high-energy electron data taken by the Tomahawk payload prior to the explosion does not show any evidence of a signal at this frequency. Fourth, the signal develops after the explosion and only persists for a short time. Fifth, 7.7 seconds is the bounce period of 1.5 keV electrons at the latitude of ESRANGE.

The second piece of evidence indicating the possible presence of a beam related effect in the data was found by examining the time development of the differential energy flux after the explosion, shown in Figure 5. It is possible that the increase in the 9.5-keV to 20-keV counting rate, which began at about 2056 UT, might have been a beam effect. As shown in paper 2, this increased low energy flux cannot be accounted for by the bremsstrahlung produced by background precipitation or by known cosmic ray sources. The peak in these low energy photons occurs at a depth of ∿7 gm/cm , which is too shallow to attribute this peak to the altitude profile of the secondary cosmic ray flux. Unfortunately, attributing this increase to the beam is also difficult because of the absence of quiet time high-latitude control data. However, the present data do not agree in altitude profile with slightly lower latitude (L=4) control data (paper 1). Therefore, it seems likely that the increase was a temporal variation related to the explosion generated beam.   Calculations based on the work of Seltzer et al.(1973) show that the peak counting rate increase could have been produced by a 30-keV monoenergetic beam of electrons with an intensity of 2×10^7 cm^{-2} s^{-1} over the viewing circle of the instrument or by an electron flux with a differential spectrum of 4×10^6 exp (-E/7 keV) electrons cm^{-2} sr^{-1} s^{-1} keV^{-1}.

DISCUSSION

The common feature of these two events is the observation that injection of electron beams into the magnetosphere produces delayed

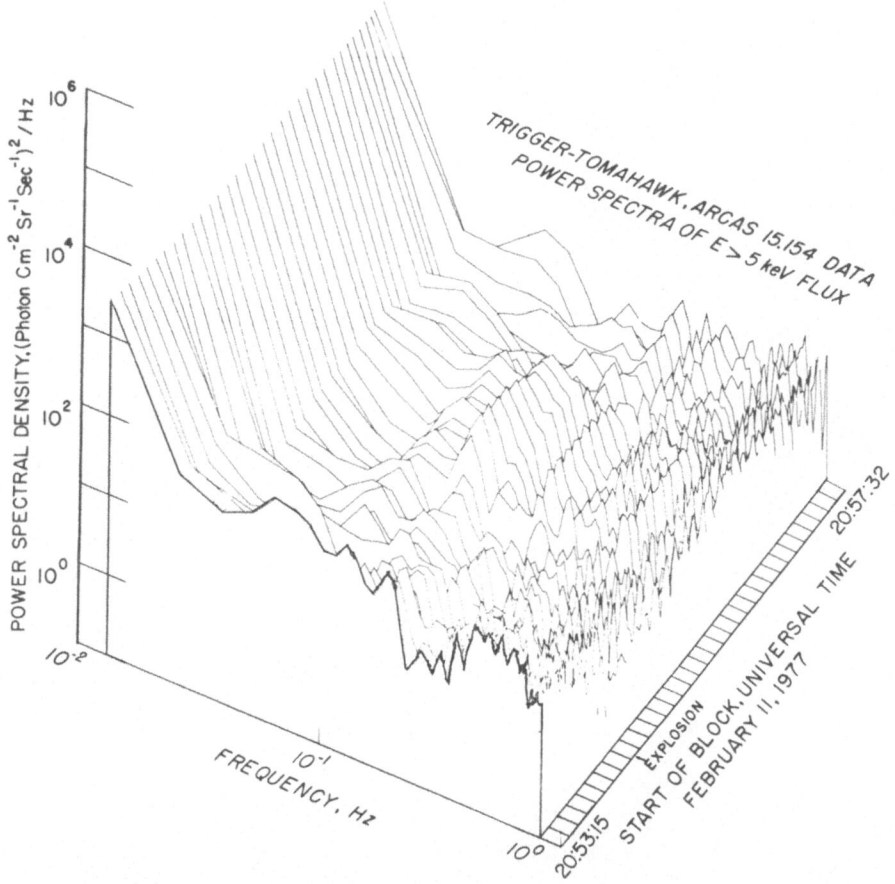

Fig. 4 A perspective view of the dynamic power spectral density of the 5-keV X ray observed during the Trigger experiment as a function of frequency and time. The spectra have been computed from 76.8 -S blocks of 0.6-S averages starting at the indicated times.

precipitation of electrons with roughly an order of magnitude more
energy than the injected electrons had. Because of the nature of
bremsstrahlung it is not possible to determine either the energy
spectrum or the pitch angle distribution of the precipitating
electrons, thus it is difficult to determine their source or the
precipitation mechanism.   However, some premliminary conclusions
can be reached. Thermodynamic arguments suggest that acceleration
of the beam itself is unlikely. The possibility also exists that
these electrons represent bursts of very hot plasma being created

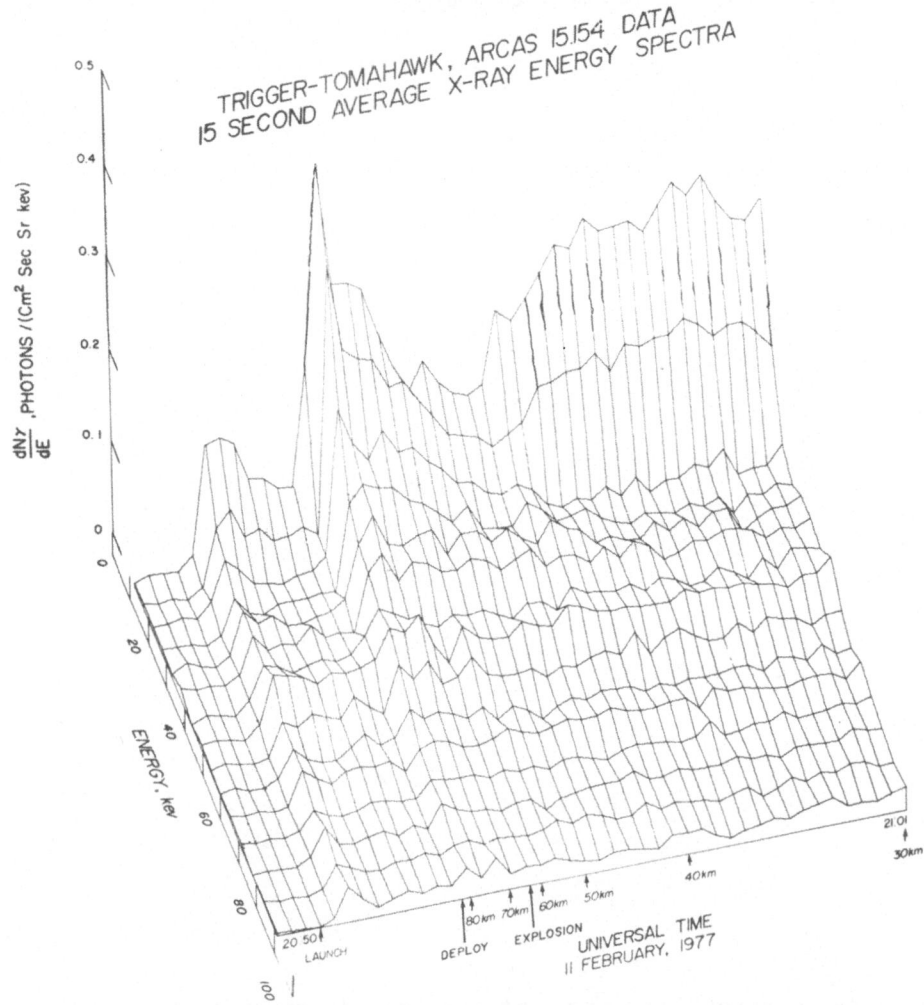

Fig. 5. Fifteen-second averages of the X ray differential energy
         flux observed during the Trigger experiment are plotted
         on a linear scale as a function of time and energy.

by the beam.   The time structure of the events, including the
relatively rapid rise and fall times of the pulses and the extreme
delay seen in the Trigger experiment, argue against this
explanation.

A model which would explain the observations is one in which
the precipitation was produced by pitch angle scattering of
radiation belt electrons via some form of particle-wave-particle
interaction.   We have attempted to ascertain some details of the
interactions by careful examination of the temporal structure of
the precipitation.   In the case of the Araks experiment this
objective proves to be unattainable due to the irregular nature of
the electron gun operation.   Because an overly sensitive arcing
protection circuit, only 3 of the planned 11 2.56 long pulses were
actually emitted at 27 keV.   It is therefore not clear if the peak
should be attributed to the long pulses, or the burst of 32 20ms
pulses which initiate the sequence.   It should be noted that the
bounce times involved are long enough to make the latter a more
likely possibility.

The Trigger data have provided better ground for development
of a detailed model. In order to precipitate short pulses of 15-30
keV electrons at the bounce period of 1.5 keV
electrons, the wave mode responsible must itself be in bounce
resonance with the 1.5 keV electron beam, but the beam must only be
unstable to these waves in a relatively small region of the
magnetosphere.   The wave mode must also be one which can pitch
angle scatter high energy electrons elsewhere in the magnetosphere.
In paper 2, we developed such a model. The model proposes that the
beam was unstable to the electrostatic two-stream instability.
Near the equator the frequency of maximum growth rate of this
instability, 1.7 kHz, was close to the frequency of a whistler mode
wave which was simultaneously in equatorial gyro resonance and
bounce resonance with 1.5 keV electrons. After $\sim$15 bounces the
wave field ultimately acquired enough energy to produce appreciable
pitch angle scattering. The model also proposes that as the wave
amplitude increased, the beam is also scattered in pitch angle.
Therefore, the beam and the interaction became dispersed due to
bounce time differences. The extended, structureless precipitation
which began to develop at the end of the periodic precipitation is
attributed to this phase of the interaction.

CONCLUSION

One of the processes which artificial electron beams injected
from the ionosphere can stimulate in the magnetosphere is the pitch
angle scattering and precipitation of higher energy electrons via a
particle-wave-particle interaction.   This mechanism appears to
operate over at least one order of magnitude of injection energy.

ACKNOWLEDGEMENTS

    This work was supported by the U.S. National Science
Foundation under grant ATM-76-82653, by the U.S. National
Aeronautics and Space Administration under grant NSG-6008, and by
the U.S. Air Force Office of Scientific Research. Figures 4 and 5
were reproduced with permission from Bering et al. (1980),
copyrighted by the American Geophysical Union.

REFERENCES

Bering, E.A., Benbrook, J.R., Stansbery, E.G., Sheldon, W.R., and
        Roeder,   J.L.,   1980,   The   results   from   the   X ray
        bremsstrahlung experiment of Project Trigger, J. Geophys.
        Res., 85: 5079.
Cambou,   F., Doukoukine, V.S. Lavergnat, J., Pellat, R., Reme, H.,
        Saint-Marc, A., Sagdeev, R.Z., and Zhulin, I.A., 1980,
        General description of the Araks experiments, Ann.
        Geophys., 36: 271.
Holmgren, G., Bostrom, R., Kelley, M.C., Kintner, P.M., Lundin, R.,
        Fahleson, U.V., Bering, E.A., and Sheldon, W.R., 1980,
        Trigger, an active release experiment that stimulated
        auroral particle precipitation and wave emissions, J.
        Geophys. Res., 85: 5043.
Lundin,   R.,   and   Holmgren,   G.,   1980, Rocket observations of
        stimulated electron acceleration associated with the
        Trigger experiment, J. Geophys. Res., 85: 5061.
Roeder, J.L., Sheldon, W.R., Benbrook, J.R., Bering, E.A., and
        Leverenz, H., 1980, X ray measurements during the Araks
        experiments, Ann. Geophys., 36: 401.
Seltzer, S.M., Berger, M.S., and Rosenberg, T.J., 1973, Auroral
        bremsstrahlung at balloon altitudes, Spec. Publ. 3081,
        Nat. Aeronaut. and Space Admin., Washington, D.C.

DISCUSSION

Gough: We have observed in the "mother-daughter" payload Electron 2 what we believe to be modulations at
1 MHz in precipitating auroral particles induced by the
presence of gun electrons (100 mA at 10 keV). A wave
experiment with electrostatic probes onboard the "mother"
measured a 1 MHz electrostatic whistler mode emission
that started at gun turn on and lasted to 260 seconds
into flight but only during gun pulses. At the same time
an identical frequency was observed superimposed upon

the 10 keV electrons collected at the "mother". However, this modulation continued in between gun pulses requiring that the natural auroral 10 keV electrons were bunched by a wave induced by the gun electrons (see Nature, 287, 15, 1980). Although a somewhat different scenario, this was another example of gun effects on natural auroral electrons.

Bering: The time scales are vastly different. The basic time scale of the phenomena I have presented is the bounce time, which is of the order of 1-10 seconds. In both the ARAKS radar observations and the Trigger observations, the interactions continue for tens of bounce periods. Furthermore, the best available model for the Trigger data indicates that the wave mode responsible for scattering the ambient particles had a frequency of 1.7 kHz.

HIGHLIGHTS OF THE OBSERVATIONS IN THE

POLAR 5 ELECTRON ACCELERATOR ROCKET EXPERIMENT

Bjørn Grandal

Norwegian Defence Research Establishment
Kjeller, Norway

INTRODUCTION

The POLAR 5 experiment consisted of a sounding rocket, carrying an electron accelerator, launched into the high latitude upper atmosphere in order to study the interaction between the artificially injected electron beam, the auroral electron precipitation, the ionospheric plasma and the ambient neutral atmosphere.

In this paper only a selection of the observed beam induced effects will be reviewed. They are the observations during beam injection of the scattered beam electrons, the low energy electron production, the luminescence at 391.4 nm and the observations after beam injection of the delayed stimulated wave emission.

A complete description of the POLAR 5 experiment is given in a series of papers: Mæhlum et al (1980a), Mæhlum et al (1980b), Jacobsen and Maynard (1980), Grandal et al (1980a), and Grandal et al (1980b).

The Polar 5 sounding rocket was launched from Andøya, Norway on February 1, 1976 at 1929 UT across two auroral regions. An apogee of 221 km was reached after 232 s of flight time as shown in Figure 1.

The Polar 5 sounding rocket was of the "mother"-"daughter" configuration. The "daughter" payload was separated from the "mother" payload at an altitude of 90 km. The axial separation speed relative to the

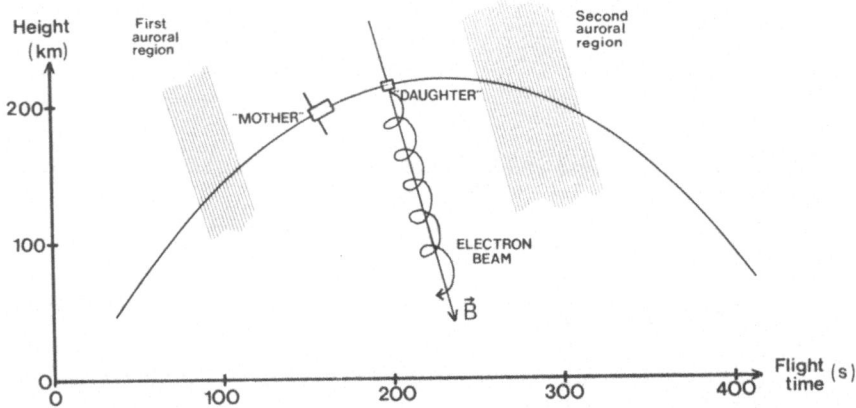

Figure 1.   Polar 5 trajectory relative to the auroral
            regions

"mother" was 0.34 m/s.   In addition, a velocity component perpendicular to the axial direction gave rise to
a separation between the "daughter" and the "mother"'s
plane of trajectory, which at 400 s, near the end of
the flight, was 40 m.   The distance between the payloads at that time, was 118 m and between the geomagnetic field lines intercepting the two payloads 80 m.

     The electron accelerator, mounted on the
"daughter", was capable of emitting a maximum current
of 130 mA of 10 keV electrons.   The beam was pulsed at
a repetition rate of about 2.5 Hz.   In addition, a
signature was attached to the beam by letting each main
pulse consist of five 2 ms sub-pulses separated from
each other by 2 ms as shown in Figure 2.

     The "daughter" carried, in addition to the accele-
rator, the 391.4 nm photometer and a few other experiments, while the "mother" carried most of the diagnostic experiments including particle counters and wave
receivers.

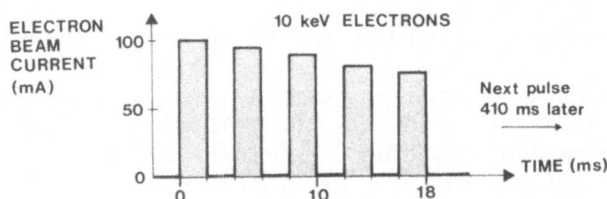

Figure 2.   The electron pulse signature

SCATTERING OF BEAM ELECTRONS

     By using an omnidirectional electron counter
mounted on the "mother", the scattering of the beam
electrons could be studied at various distances from
the center of the beam. The electron counter used here
had a fixed energy range of 8-10 keV and was sampled
every 0.5 ms.

     The Larmor column traced out by the electron beam
is located at various positions relative to the orientation of the "daughter" and "mother". The positions
depend upon the beam injection angle relative to the
local geomagnetic field as shown in Figure 3.

     During the 2 ms electron pulses the following
observations were made:

     Early in the flight when the separation between
the "daughter" and the "mother" was small, the electron
counter experienced saturation when the beam was injected in the eastward direction but not for westward
injection. In particular, the saturation effects
started at a flight time of 111 s when the distance
from the mother to the magnetic field line intercepting
the daughter was 11 m and ended at a flight time of
123.5 s when this distance was 14 m.

     We note that the gyro diameter of 10 keV electrons
is 13.9 m. From this we conclude that the energy of
the beam electrons did not differ significantly from
the nominal value of 10 keV.

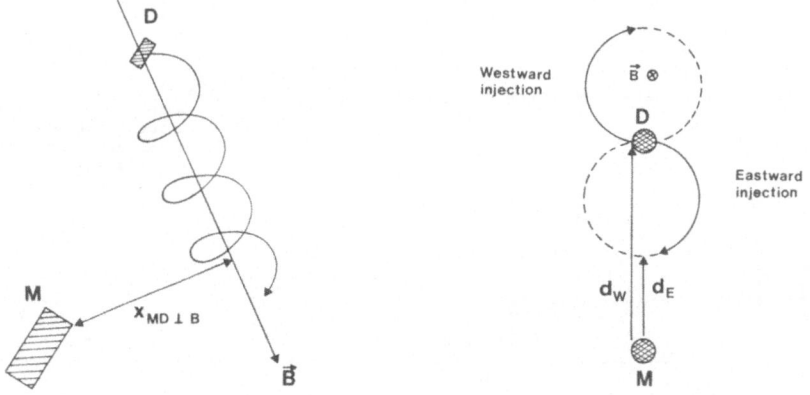

Figure 3.   The geometry of the beam relative to the
            "mother" (M) and the "daughter" (D)

If the above positions for entering and leaving saturation is interpreted as a measure of the size of the beam core, then the angular spread of the beam core after half a Larmor orbit is $35^{O}-40^{O}$ (FWHM).

As the separation between the two payloads increased further, the electron counters on the "mother" were able to study the "halo" of scattered electrons around the beam. The characteristic parameter was here the distance, d̲, from the "mother" to the nearest point on the Larmor column traced out by the beam electrons which defines the outer boundary of the beam core.

The observed fluxes of 8-10 keV beam electrons in the halo as a function of the distance d̲ is given in Figure 4.

In the altitude range 150-180 km the fluxes varied according to:

$$I = 3x10^{7} e^{-0.07 d (m)} \quad (electrons/cm^{2} s sr keV)$$

In the altitude range 215-220 km the fluxes were about one order of magnitude lower than predicted by this equation. Furthermore, below 145 km the flux of scattered electrons is higher than expected from the above equation. Thus a strong altitude effect is observed for the halo surrounding the beam.

Our results have been compared with the calculations of beam scattering made by Berger et al (1970) even though our observations are made less than 100 m

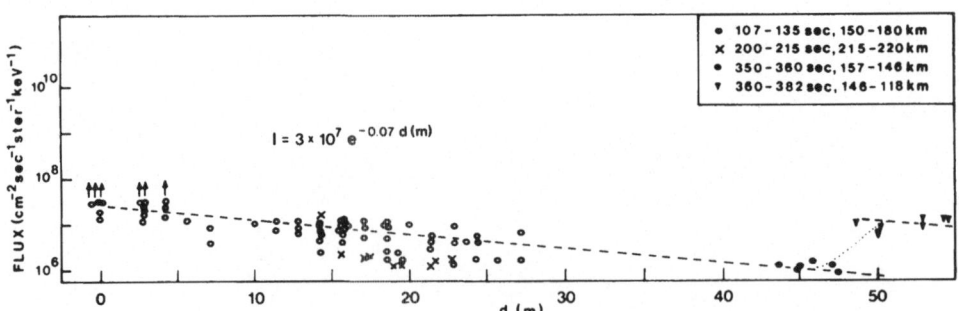

Figure 4.   Observed fluxes of 8-10 keV electrons
            inside the "halo"

from the place of beam injection while Berger et al considered the scattering of a beam far from the injection point. In spite of this difference, the comparison clearly shows that the strong scattering of the beam electrons observed here cannot be explained by diffusion due to Coulomb scattering.

LOW ENERGY ELECTRON PRODUCTION BY THE BEAM

An electron spectrometer mounted on the "mother" was employed to study the flux of low energy electrons in the vicinity of the beam. The spectrometer covered the energy range 0.13-12.8 keV and used 34 ms per scan. The sampling period was 1 ms.

The following observations of the low energy electrons in the "halo" were made during the 2 ms electron pulses at a flight time of 108-125 s when the distance d, between the "mother" and the nearest point on the Larmor column traced out by the beam electrons, was in the range 10-15 m.

The electron flux was strongly pitch angle dependent with counter saturation close to 90°.

Outside this region, i e pitch angles less than 70° and larger than 110° the electron fluxes were fairly independent of the orientation of the counter. The average energy spectrum of the electrons in the "halo" is given in Figure 5.

The theoretical spectrum of secondary electrons produced through ionization processes by the beam electrons, derived from the calculation by Banks et al (1974) is shown as a dashed line in Figure 5. The observed low energy spectrum is significantly different from the theoretical electron spectrum both in shape and intensity (by a factor 250).

Furthermore, when energy absorption is neglected, the ratio between the fluxes of secondary and primary electrons should decrease with increasing altitude. However, using electron counters, fixed in energy to 100 eV, we find in the "halo" that the ratio between the fluxes of 100 eV electrons and the 10 keV electrons is higher at an altitude of 165-185 km than at 125-145 km.

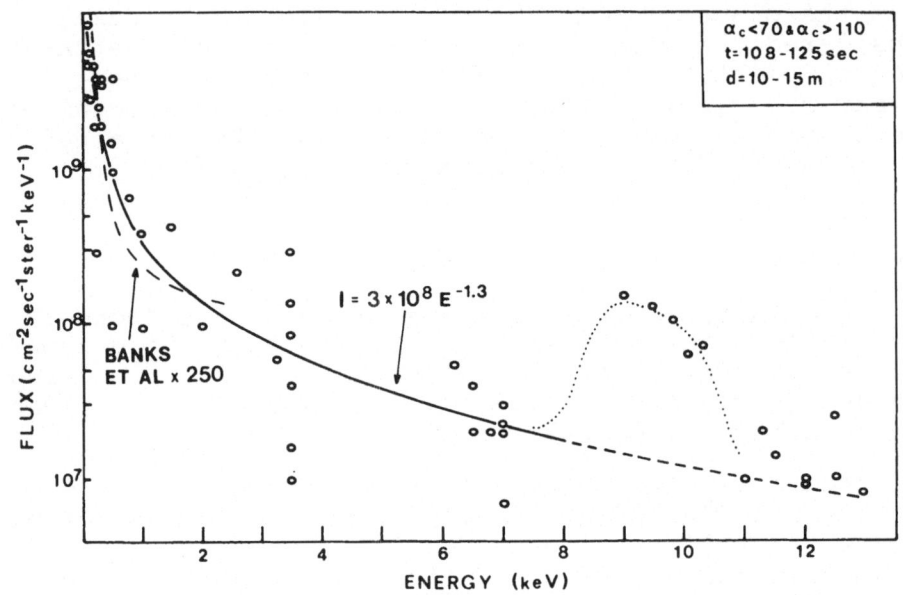

Figure 5.   Observed average energy spectrum of elec-
            trons in the "halo". The dashed line refers
            to the theoretical spectrum of secondary
            electrons produced by ionization processes
            but multiplied by a factor of 250

      Thus we conclude that the observed low energy
electron population in the "halo" is not produced
directly through ionization processes near the "mother".

      Let us consider other sources of low energy elec-
trons in the halo.

      Secondary electrons produced in the core of the
beam may diffuse perpendicular to the geomagnetic field
but not in sufficient number within the 2 ms electron
subpulse.

      A non-neutralized "daughter" could accelerate
thermal electrons, but this seems to be an unlikely
source at a distance of 10-15 m for potentials on the
daughter of less than 1 kV.

      Finally, wave-particle interactions may produce an
enhanced low energy electron population. The mechanism
consist of beam produced plasma waves which decay into
waves whose phase velocity enable them to closely

interact with ambient thermal electrons and accelerate these.   This mechanism would be altitude dependent. Above 165 km the collective, non-collisional processes would dominate as the period for wave growth would be smaller than the typical time between the collisions experienced by the electrons. Below 145 km, however, the collisional processes would dominate. Thus this energy transfer from the beam to ambient electrons via wave-particle interactions will break down between 165 and 145 km thus yielding the observed decrease in the flux of 100 eV electrons with decreasing altitude.

THE LUMINESCENCE AT 391.4 nm PRODUCED BY THE ELECTRON BEAM

The 391.4 nm photometer was mounted on the "daughter" and had an opening angle of 9.6° and a bandwidth of 5 nm. The sampling rate of the photometer was 2 kHz. The photometer's field-of-view intersected the electron beam about 1 m from the accelerator as shown in Figure 6.

The observed light at 391.4 nm during the 2 ms electron subpulses is given in Figure 7. The most striking features are the nearly constant light level from 150 km up to apogee and down to 150 km again and the large value of this level. Below 130 km the intensity of the observed light increases exponentially with decreasing altitude.

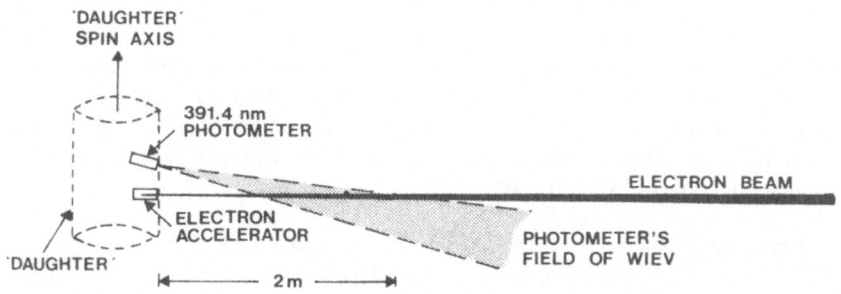

Figure 6.   The geometry of the photometer and the
            electron accelerator on the "daughter"

The radiation at 391.4 nm is due to the excitation of the (0,0) first negative band of N_2^+ by electron impact through the interaction:

$$e+N_2 \rightarrow N_2^{+*}+e+e \rightarrow N_2^{+}+h\nu_{391.4}+2e$$

whose cross-section $\sigma_{391.4}(E)$ is given by Stair and Gauvin (1967) and Borst and Zipf (1970) as a function of electron energy, E, above a threshold of 18.8 eV. The number of photons emitted per second is given by:

$$C(ph/s) \propto n_{N_2}(h) \sum_E N_e(E) \; \sigma_{391.4}(E) \; v_e(E)$$

where $n_{N_2}(h)$ is the molecular nitrogen number density at the altitude h, the sum is over the different electron populations of energy E, $N_e(E)$ and $v_e(E)$ are the number and velocity of electrons of energy E, respectively.

Assuming an ambient N_2 density profile equal to that of CIRA 1972 and a narrow electron beam relative to the photometers field-of-view, the expected count rate is given as a dashed line in Figure 7. If the electron beam is assumed to be broad, then the resulting count rate is the dotted line in Figure 7.

Thus below 130 km the luminescence at 391.4 nm can be explained by a broad electron beam.

Above 130 km additional sources of ambient N_2 or energetic electrons are needed to explain the observed results. The possibility of increasing the ambient N_2 density by outgassing from the rocket motor seems to be very small in this case as a "mother-daughter" configuration was employed. Here the photometer is mounted on the "daughter" whose distance from the "mother" with the motor increases throughout the flight, but an equivalent decrease in the light level with time is not observed. Let us therefore assume an ambient N_2 density profile equal that of CIRA 1972, and consider possible sources for the energetic electrons giving rise to the observed luminescence.

The beam electrons will not suffice as shown above. Locally produced secondary electrons, which decrease in number with increasing altitude, are not candidates either. The return current electrons will not do as their energy seem to be below 4 eV according

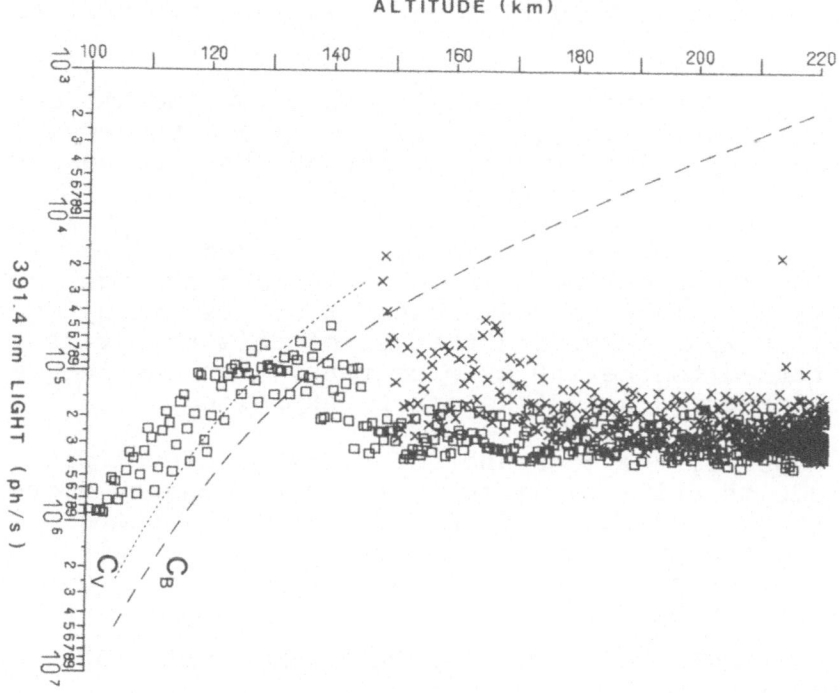

Figure 7.   Observed beam induced luminescence as a
            function of altitude

to Jacobsens RPA measurements, Jacobsen and Maynard
(1980).

      Let us consider the possibility of a beam produced
low energy electron population different from the
above.   In order to arrive at a lower bound on their
number, these electrons are assumed to have an energy
of 100 eV which is the energy where electrons are most
efficient at producing luminescence at 391.4 nm.

      By using the observed luminescence at 220 km we
may estimate the minimum density of these low energy
electrons needed to explain the observations:

$$n_{necess}(220 \text{ km}) \sim 6 \times 10^6 \text{ cm}^{-3}$$

where we have assumed that the light producing region fills the photometers field of view to a depth of about 10 m.

The above electron density is below the ambient electron density (maximum: 10^6 cm^{-3}) and the equivalent density of low energy electrons observed near the "mother" (about 10 cm^{-3}).

Thus additional ionization is needed and we will consider the presence of a beam plasma discharge (BPD). Suppose the electron beam loses 5% of its energy during the first 20 m (BPD) and that 1% of this energy is transferred to the 100 eV electrons, then the luminescence observed at 220 km can be explained.

However, the constancy of the light level with altitude is still a puzzle. If the N_2 density decreases with altitude as in CIRA 1972, then the number of exciting electrons must increase correspondingly with altitude. A possible mechanism has now been suggested by Papadopoulos (1981).

THE STIMULATED WAVE EMISSION OBSERVED AFTER BEAM INJECTION

The wave receivers were mounted on the "mother" payload. The HF (high frequency) receivers had four broad channels: 0.1-0.25 MHz, 0.25-0.75 MHz, 0.75-2.0 MHz and 2.0-5.0 MHz and one narrow channel at 2.8 MHz with a bandwidth of 0.1 MHz. These channels were sampled every 0.5 ms. The VLF (very low frequency) receiver were broadbanded covering the range 0.1-100 kHz. The receivers employed a single dipole antenna, 9.3 m tip-to-tip.

During most of the flight the received signal in most of the frequency channels reflected the signature of the current of the ejected electron beam as shown in Figure 8a. However, at certain flight periods, for certain frequencies the received signal amplitude peaked after the end of the beam injection as shown in Figure 8b. It should be emphasized that these after-effects were only observed in conjunction with the injection of the electron beam and in the VLF range in several periods throughout the flight while in the HF range only inside the second auroral region.

In particular, the received signal observed in the 2-5 MHz channel is found to be closely correlated with

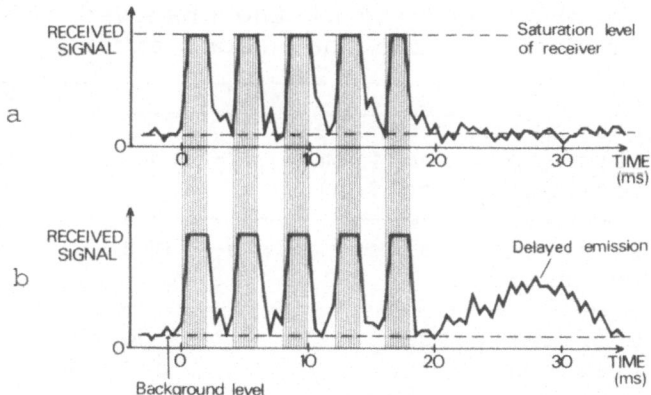

Figure 8.   a)   Typical signal received in most of the
                 receiver channels during most of the
                 flight
            b)   Signal received on some frequencies
                 inside the second auroral region

the flux of auroral electrons in the range 4-5 keV, as
shown in Figure 9.   The characteristics of the observed after-effects inside this second auroral region
are given in Table 1.

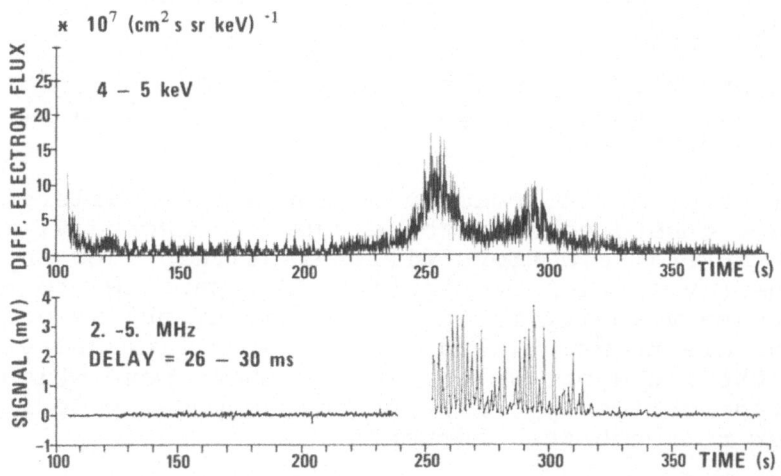

Figure 9.   Comparison between the occurence of auroral
            electron precipitation and the beam in-
            duced wave after-effects

Table 1.   Characteristics of the observed after-
           effects inside the second auroral region

	HF	VLF
Time delay to maximum of the after-effects (measured from the start of beam injection)	22-30 ms	30-40 ms
Frequencies at which after-effects are observed	0.75-2 MHz 2-5 MHz	5.6 kHz 2 kHz
Dependence of the pitch angle of the beam	No clear dependence	
Polarization of received signal	Signal mini-mum for an-tenna perpen-dicular to $\vec{B}$	No such depen-dence
Duration	Same order as injec-tion period $\sim$ 18 ms	

The distance between the "daughter" and the "mother" was of the order of 75 m inside the second auroral arc as seen in Figure 10. The dashed cylinder is the surface bounding the Larmor columns as they move around the "daughter" for beam injection at various azimuths.

In order to explain the observed after-effects in the HF range, the following selective amplification scheme is proposed:

Firstly, during beam injection a wide variety of waves are generated near the daughter. Secondly, the 4-5 keV auroral electrons will selectively feed energy into the waves which propagate nearly parallel to $\bar{B}$ and whose phase velocity is nearly equal to the velocity of these electrons. Thirdly, the large time delays are due to the low group velocities of those wave which are travelling obliquely, relative to the phase velocities, in order to reach the "mother".

CONCLUSION

The Polar 5 observations of the scattering of beam electrons, the low energy electron spectrum, the lumin-

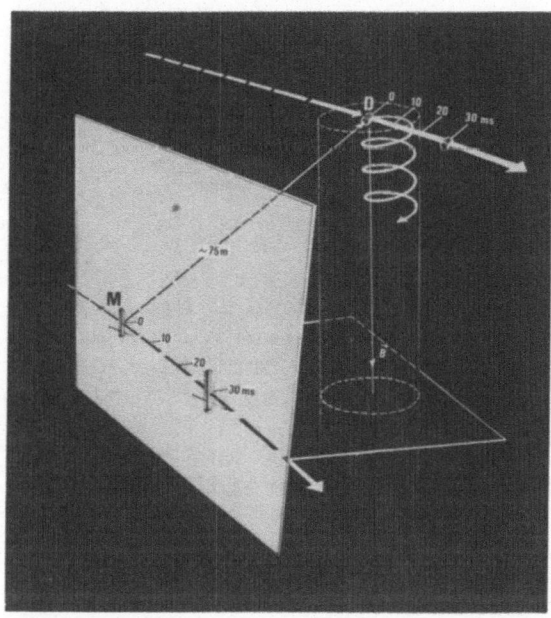

Figure 10.   The relative geometry of the "mother" (M)
             and the "daughter" (D) payloads inside the
             second auroral region

escence at 391.4 nm and the stimulated wave after-
effects indicate the dominant role played by the following processes at the various altitudes:

     The collective beam-plasma interactions dominate
at high altitudes (above 160 km). At intermediate
altitudes (130-150 km) there is competition between the
collective and collisional processes while at low altitudes (below 130 km) the collisional processes
dominate.

ACKNOWLEDGEMENTS

The author would like to thank B N Mæhlum, the project
scientist of the POLAR 5 experiment, for his inspiring
support. Furthermore, discussions with T A Jacobsen,

E V Thrane, J Trøim and J A Holtet are gratefully
acknowledged.

REFERENCES

Banks, P. M., Chappel, C. R. and Nagy, A. F., 1974, A
    new model for the interaction of auroral electrons
    with the atmosphere: Spectral degradation, back-
    scatter, optical emission and ionization, J Geo-
    phys Res, 79:1459.
Berger, M. J., Seltzer, S. M. and Maeda, K., 1970,
    Energy deposition by auroral electrons in the
    atmosphere, J atmos terr Phys, 32:1015.
Borst, W. L., and Zipf, E. C., 1970, Cross-section for
    electron impact excitation of the (0,0) first
    negative band of N_2^+ from threshold to 3 keV,
    Phys Rev A, 1:834.
CIRA 1972 Cospar International Refence Atmosphere, 1972,
    Akademie-Verlag, Berlin.
Grandal, B., Thrane, E. V., and Trøim, J., 1980a, POLAR
    5: An electron accelerator experiment within an
    aurora: 4. Measurements of the 391.4 nm light
    produced by an artificial electron beam in the
    upper atmosphere, Planet Space Sci, 28:309.
Grandal, B., Holtet, J. A., Trøim, J., Mæhlum, B. N.,
    and Pran, B., 1980b, Observations of waves arti-
    ficially stimulated by an electron beam inside a
    region with auroral precipitation, Planet Space
    Sci, 28:1131.
Jacobsen, T. A., and Maynard, N. C., 1980, POLAR 5 -
    An electron accelerator experiment within an
    aurora: 3. Evidence for significant spacecraft
    charging by an electron accelerator at ionospheric
    altitudes, Planet Space Sci, 28, 291.
Mæhlum, B. N., Måseide, K., Aarsnes, K., Egeland, A.,
    Grandal, B., Holtet, J., Jacobsen, T. A., Maynard,
    N. C., Søraas, F., Stadsnes, J., Thrane, E. V.,
    and Trøim, J., 1980a, POLAR 5 - An electron ac-
    celerator experiment within an aurora: 1. Instru-
    mentation and geophysical conditions, Planet Space
    Sci, 28, 259.
Mæhlum, B. N., Grandal, B., Jacobsen, T. A., and Trøim,
    J., 1980b, POLAR 5 - An electron accelerator ex-
    periment within an aurora: 2. Scattering of an

        artificially produced electron beam in the atmos-
        phere, Planet Space Sci, 28, 279.
Papadopoulos, K., 1981, Theory of beam plasma discharge,
        in: Artificial particle beams in space plasma
        studies, B. Grandal, ed., Plenum, New York.
Stair, A. T., and Gauvin, H. P., 1967, Research on opti-
        cal infrared characteristics of aurora and airglow
        (artificial and natural), in: "Aurora and Airglow",
        B. M. McCormac, ed., Reinhold, New York.

DISCUSSION

Kawashima: How large is the bandwidth of the 3914Å
detector? Can you reject the possibility that the 3914Å
signal you obtained may be contaminated by other emission
lines or bands?

Grandal: The bandwidth is 50Å. Yes, the N_2^+ first
negative band is the only candidate at this wavelength.

Folkestad: To explain your measurements of the 3914Å
emissions you postulated a density of 6×10^6 cm^{-3} of
electrons at 100 eV. What do your observations indicate
about the lifetime of this ionization?

Grandal: The enhanced luminosity at 3914Å was only
observed during beam injection with no after-effects.
The sampling rate was 2 kHz, thus the lifetime was much
less than 0.5 ms.

OBSERVATIONS OF PLASMA HEATING EFFECTS IN THE

IONOSPHERE BY A ROCKET BORNE ELECTRON ACCELERATOR

Tore A. Jacobsen

Norwegian Defence Research Establishment

Kjeller, Norway

INTRODUCTION

In most of the electron beam experiments carried out in space so far it seems necessary to assume that some sort of a plasma discharge takes place that can provide a source for the high return currents needed to explain the neutralization of the accelerator vehicle.

The discharge processes may be of different types depending on the beam parameters, the characteristics of the surrounding plasma and the neutral gas density.

If the discharge is created by a high vehicle potential through an acceleration of the ambient ions and electrons, the discharge region should be fixed to the magnetic fieldline associated with the accelerator vehicle at the time of injection. In the case of a beam-related discharge, however, the position of the discharge region should be related to the position of the gyro center of the beam. If the accelerator is mounted on a spinning vehicle one should therefore observe that the center of the discharge region rotates around the vehicle in the same sense as the gyro center of the beam.

A possible way of discriminating between these two situations is to fly a mother-daughter configuration with the accelerator mounted on a spinning daughter that flies ahead of the mother payload. If the mother and the daughter have approximately the same trajectory

175

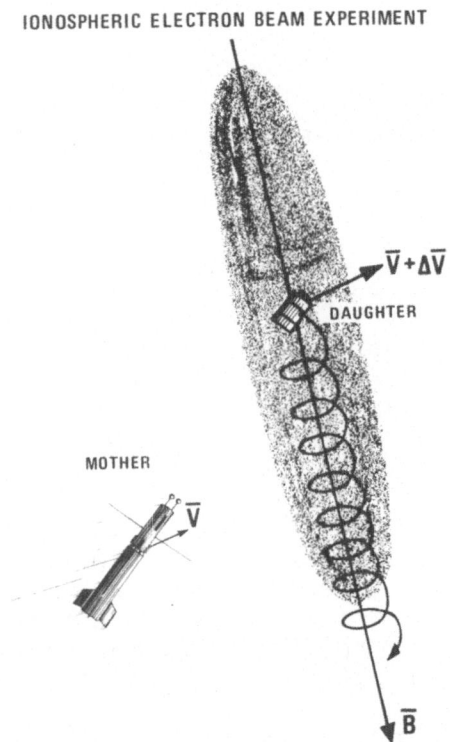

IONOSPHERIC ELECTRON BEAM EXPERIMENT

Figure 1.   Survey of the ELECTRON 2 mother-daughter
            experiment

plane, the instrumented mother will fly through the
discharge region after a certain time delay.

     Such a configuration has many advantages. In
addition to providing information on the position of
the discharge region the diagnostic instruments on the
mother can also be used to study the extent of the
discharge, the characteristics of the plasma inside it,
as well as the temporal changes. It should also be
pointed out that the measurements on the mother are not
contaminated by the possible interference produced by
the pulse emission on the vehicle carrying the electron
accelerator.

     In the Norwegian/US rocket borne electron beam
experiment POLAR 5 launched in 1976 the configuration
discussed above was applied. The outcome of this
experiment has been described in a series of papers

(Mæhlum et al 1980a, Mæhlum et al 1980b, Grandal et al 1980a, Grandal et al 1980b, Jacobsen and Maynard 1980).

A more recent rocket experiment ELECTRON 2, was launched in 1978 with a similar mother-daughter flight configuration (Jacobsen et al 1981) (Figure 1). The diagnostic instruments on the ELECTRON 2 mother payload were better suited for studies of the plasma in the discharge region than those on POLAR 5. In this paper results from two plasma probes on the ELECTRON 2 mother payload will be presented. These measurements made it possible to determind that the position of the discharge region was related to the position of the beam. The probe data also provided information on the plasma heating in the discharge region and on the extent and life time of the region. The observed altitude - and beam current dependence will also be discussed.

EXPERIMENTAL CHARACTERISTICS

The ELECTRON 2 rocket was one in a pair of rockets launched simultaneously from Andøya rocket range, Norway on November 27th, 1978 at 1855UT. The campaign was a joint project with participation from Austrian, British, Norwegian and US-groups. A Nike Tomahawk rocket carried the mother-daughter payload to an apogee of 192 km during quiet night time ionospheric conditions.

An electron accelerator emitting short pulses of 10 keV electrons was mounted on the daughter together with a series of return current monitors. The pulse current was varied between 100 mA and 10 mA and the pulse sequence was a combination of 1,2 and 10 ms pulses according to the scheme presented in Figure 2. The mother payload included a variety of wave receivers, particle detectors, photometers and two plasma probes.

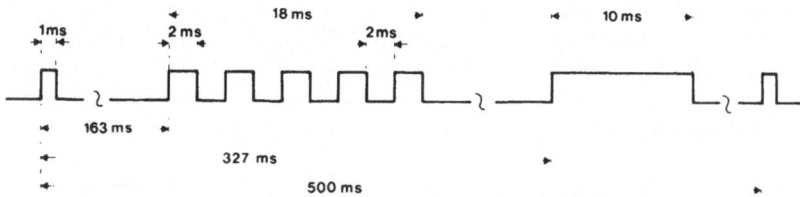

Figure 2. The ELECTRON 2 pulse sequence

The daughter was released in the trajectory plane in the northward direction 70 s after take-off with a separation speed of 0.4 m/s. At apogee the separation distance was 60 m increasing to about 120 m near the end of the flight.

The two plasma probes were mounted on short booms extending along the rocket spin axis on the front end of the mother (Figure 1). One of the probes, supplied by the NDRE, and in the following referred to as the electron temperature probe (ETP), was a spherical electrostatic analyzer of the Sagalyn type (Sagalyn et al, 1963). In order to achieve a high time resolution in the measurements the sweep voltage was restricted to the range from -1.4 V to +2.7 V. The probe current was sampled 8 times in each sweep period which had a duration of 9.6 ms. Once each 0.5 s the current was sampled at a rate of $5 \cdot 10^3$ s^{-1} corresponding to 50 samples pr sweep. The data was fed into a memory and read out at a lower rate between the two subsequent pulse-emissions. This procedure made it possible to obtain an accurate determination of the background electron temperature.

The other probe, provided by the Technical University of Graz, Austria, was a spherical capacitance probe (CP) similar to the one used on POLAR 5 (Mæhlum et al, 1980). This instrument is basically a free running LC-oscillator in which the spherical probe constitutes part of the impedance of the oscillator circuit. The impedance of the sphere depends mainly on the ambient electron density and the thickness of the ionosheath around the probe. Changes in those parameters will cause a shift in the oscillator frequency. The probe frequency, which had a free space value of 1.2 MHz, was sampled at a rate of 833 s^{-1}.

DETERMINATION OF THE SIZE OF THE DISCHARGE REGION

The electron accelerator was turned on at 93 s after take-off at an altitude of 122 km. In the subsequent 163 s, that covered the flight past apogee down to an altitude of 180 km more than 1000 crossings of the discharge region created by the beam were observed by the plasma probes on the mother.

A typical example of the data obtained during one of the discharge crossings is shown in Figure 3. The upper curve represents the LSB TM-word of the CP-data.

On the lower section on the figure data from seven
succeeding sweeps of the ETP are presented. The time,
running along the horizontal axis, is measured relative
to the start of the 10 ms pulse number 920 at t = 243 s.
The interpretation of the probe responses in the discharge region will be presented below. At this stage
it is of interest to note that each discharge crossing
can be characterized by the times when the mother enters
and leaves the disturbed region. These times are
marked in Figure 3 by t_B and T_E, respectively.

The amplitude or strength of the disturbance as
measured by the probes is defined as the deviation of
the CP-TM signal from the background level marked by
A on Figure 3. This quantity is proportional to the
period of the probe oscillations. A decrease of the
TM-signal corresponds to an increase in the probe
frequency or equivalently a decrease in the effective
probe capacitance. The frequency shift does normally
reflect changes in the ambient concentration of thermal
electrons or in the thickness of the ion sheath surrounding the probe.

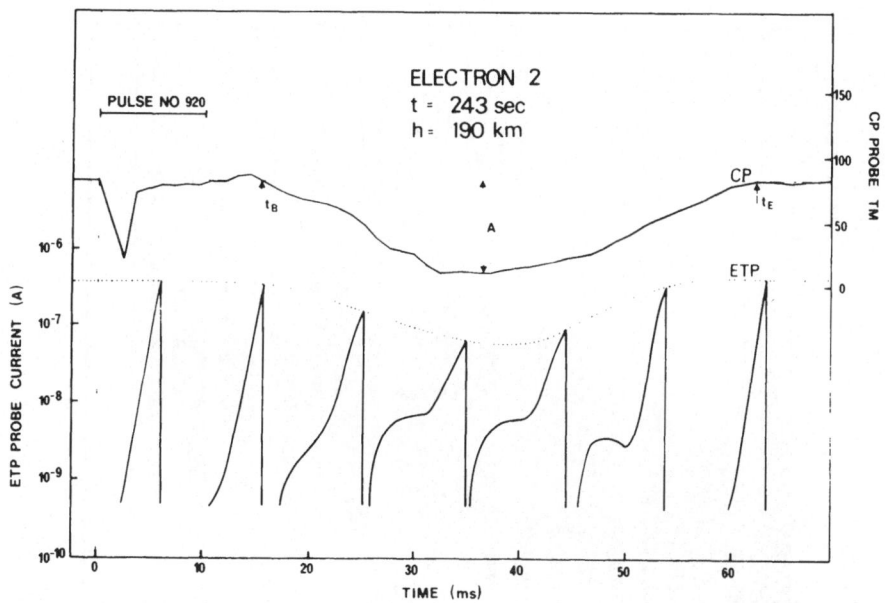

Figure 3.   A typical example of the effects observed by
            the capacitance probe, CP and the electron
            temperature probe, ETP in the discharge region
            during the last part of the flight

On Figure 4 the three quantities t_B, t_E (shaded area) and A (upper curve) are plotted versus pulse number. Only data from the 10 ms pulses are included on this display. The reason for selecting only pulses with the same pulse length is that the extention of the discharge in the direction along the trajectory depends on the pulse length (cf Figure 7). Plots based on the 1 ms pulses or the "train pulses" consisting of five 2 ms pulses are similar to Figure 4.

The lower border of the shaded area corresponds to the time, t_B when the mother enters the discharge region and the upper border represents the time, t_E when the probe signal returns to the background level. The six equally spaced gaps in the data are the periods when the accelerator is in the low current mode. During these periods no significant effects were observed by the probes except for the first period that took place at an altitude of about 130 km.

For the first 100 pulses the probe effects were observed to start immediately after the start of the

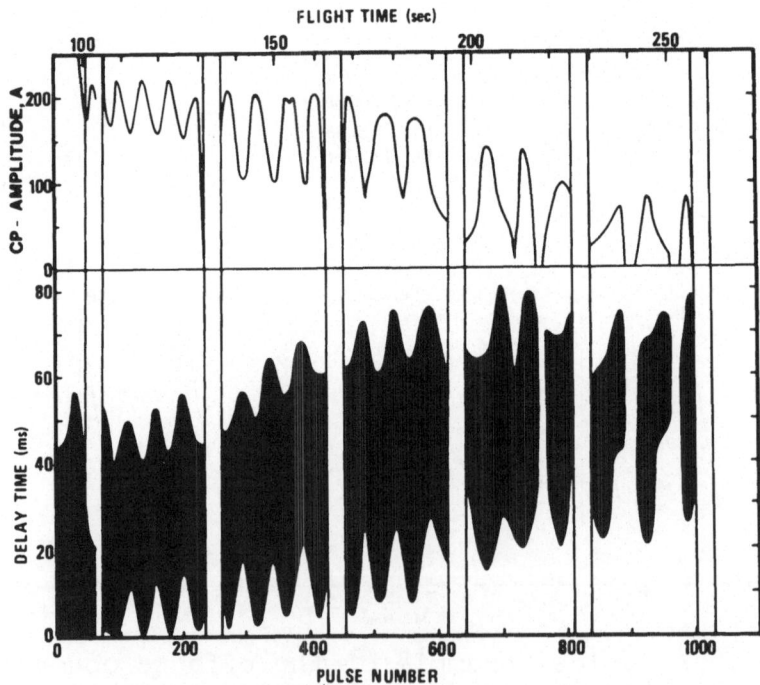

Figure 4.   Delay and amplitude of the observed probe effects for the 10 ms pulses

pulses. Thus, during this part of the flight, when the distance between the mother and daughter was small, the mother was already inside the plasma region modified by the beam at the time of pulse emission. When the separation distance increased the time delay from the start of the pulse until the time, t_B when the mother payload entered the discharge region increased from zero to more than 40 ms later in the flight. During the last part of the period with data beam effects were observed up to 80 ms after the start of the pulses.

The amplitude, A and the delaytimes, t_B and t_E reveal a clear, periodic modulation. These modulations are not related to the spin of the mother. The spin period of the mother stayed constant at about 5.6 s throughout the flight while the spin period of the daughter increased from 5.6 s at the time of separation to about 12 s at the flight time 256 s when pulse number 1000 was emitted. This pronounced change in the spin was the result of a $6 \cdot 10^{-7}$ W dissipation of the rotational energy of the daughter. The modulation period of the probe data presented on Figure 4 increases exactly in the same way as the daughter spin period, indicating that the modulations are related to the azimuth direction of the beam injection.

A detailed model calculation of how the delay and the amplitude, A should change with time has been carried out. The model is based on the assumption that the center of the discharge region coincides with the gyro center of the beam. Further it includes the separation velocity components parallel and perpendicular to the trajectory plane, the changing injection angle of the beam and a model of how the amplitude of the observed effects decreases with increasing distance from the center of the region. This model reproduced most of the details shown on Figure 4. The agreement between the model and the observations could, however, only be established if the daughter rotation was opposite to the spin direction acquired during the spin up phase of the rocket flight. In fact this was proved to be right as the aspect gyro dato from the mother showed that the rocket was dispun through zero at t = 68 s due to a mechanical failure and obtained a spin in the opposite direction with a period of 5.6 s (Maynard, 1979).

Based on the study discussed above it can be concluded that the position of the region of modified plasma, observed by the probes and interpreted as the

discharge region, was closely related to the gyro
center of the beam. The cross-section of the region
perpendicular to the geomagnetic field had a diameter
of 20-25 m, corresponding to 1.5-2 times the gyro
diameter of the beam electrons.

The study made it possible to discriminate between
temporal and spatial variations of the amplitude, A.
It was found that no significant changes in the strength
of the observed effects took place within a time period
of 60 ms. The observed amplitude, A did, however, dis-
close a strong dependence on the distance, S_{11B} measured
along the geomagnetic field from the position of the
daughter at the time of beam injection.

A presentation of the mother-daughter-discharge
geometry is given in Figure 5. The reference system
is placed at the daughter, D. In this system the
mother moves downwards along the separation line. The
positions of the mother at four different times are
indicated by open circles and the directions of the
velocity vector of the mother are shown as dotted
arrows. The shaded area represents the discharge region
for southward or northward beam injections.

The amplitude, A measured at the field line $\overline{B}$ is
plotted versus the distance S_{11B} on Figure 6. The
amplitude decreases almost linearly as S_{11B} increases.
An extrapolation indicates that A should be zero for
$S_{11B} \approx 100$ m. At this time (t ~ 260 s) and later in
the flight only very weak pulse effects were found in
the probe data. When the pulse effects disappeared the
distance between the trajectory planes of the mother and
the daughter was less than 10 m. The effects should
therefore still be observable for a southward beam
injection if the region was extended below 100 m from
the daughter. After t = 217 s, corresponding to pulse
number 760 in Figure 4, the mother fails to hit the
discharge region for the northward beam injections in
agreement with the geometrical model.

Based on the observations discussed above it can
be concluded that the extention of the discharge region
was about 100 m along the B-field measured downwards
from the daughter. As no significant differences in the
observed effects for downgoing and upgoing beams were
noted, it seems reasonable to assume that the discharge
region had an equal extension of 100 m above the
daughter. The total length of the discharge region
along the B-field should then be ~ 200 m.

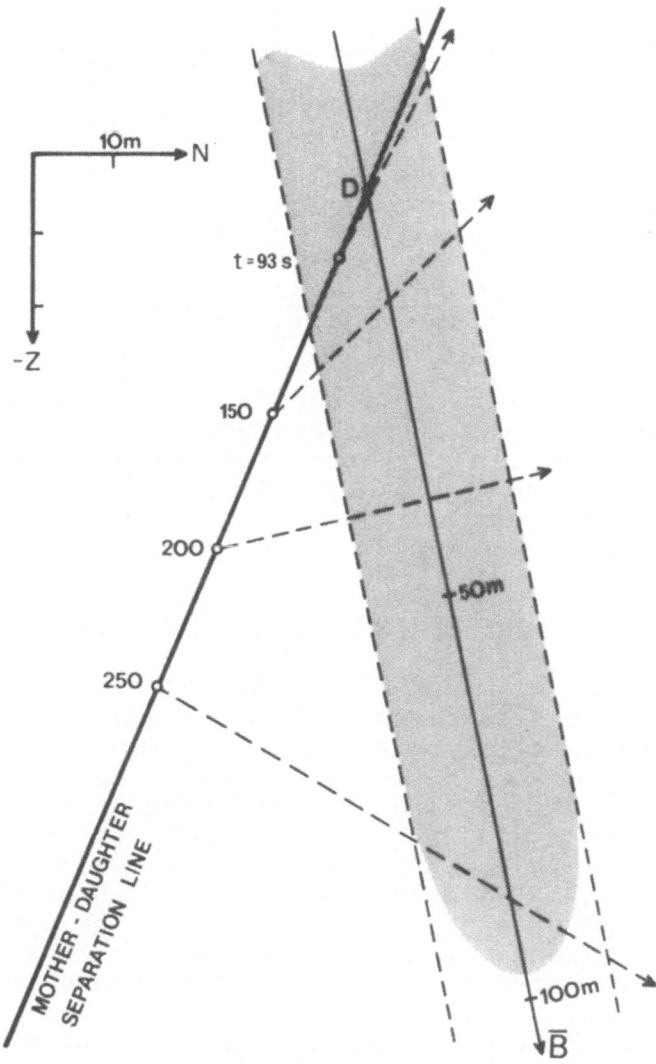

Figure 5.   The mother-daughter-discharge geometry

        The region of intensive luminosity observed close
to the ZARNITSA-2 rocket had approximately the same
characteristic size (Mishin and Ruzhin, 1978). This
was found to be in agreement with rough theoretical
estimates of the length of the discharge region (Mishin
and Ruzhin, 1978).

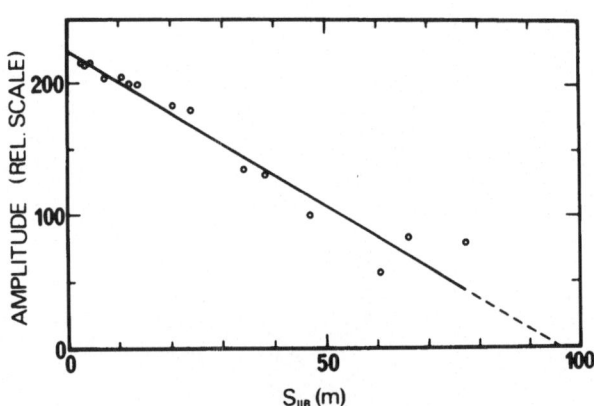

Figure 6.   The maximum amplitude of the discharge effects
            observed by the capacitance probe as a
            function of distance from the daughter. S_{11B}
            is measured along the B-field

INTERPRETATION OF THE OBSERVED PROBE EFFECTS

        Each of the 7 sweeps of the ETP experiment dis-
played on Figure 4 is based in 8 samples of the probe
current. The time period between the samples is 1.2 ms
and each sweep covers a voltage range of 4.1 V (from
-1.4 V to +2.7 V). During the last part of the flight
the probe current did not reach the saturation level at
the last sample in the standard sampling mode. However
in a fast sampling mode (4 sweeps every 0.5 s) when the
last samples were taken close to +2.7 V, the typical
knee in the current characteristics were always seen.
It normally occurred at a current level close to the
value of the last sample in the standard mode. The
dotted line in Figure 4 does therefore fairly well
represent the saturation current of the probe except
for the sweeps inside the disturbed region where the
knee is normally not observable.

        The most striking feature of the ETP-data is the
high concentration of suprathermal electrons observed
inside the discharge region. The ETP data presented in
Figure 4 is typical for the last half of the period
during which pulse effects were observed by the probes.
The maximum fluxes of the suprathermal electrons had
energies of about 3 eV. The peak densities varied somewhat in the range 0.1 to 1 percent of the thermal
electron density.

The fraction of the current characteristics laying above the contribution from the suprathermal electrons have been used to derive the temperature of the thermal electrons inside the discharge region. Values in the range 3000-4000 K are found to be typical for this part of the flight. This is 3-4 times higher than the background electron temperature that varied between 1000 K and 1500 K in the altitude range from 160 km to apogee at 192 km.

Pronounced heating of the thermal electrons in the discharge region has previously been reported both in the ionosphere (Cartwright et al, 1978) and in the laboratory (Szuszczewicz, 1979).

An example of the probe data from the first half of the period with pulse effects is presented in Figure 7. The ETP data is in this case difficult to analyze in a quantitative way. Therefore only a qualitative interpretation will be presented.

When the mother payload enters the discharge region it encounters an increasing flux of suprathermal

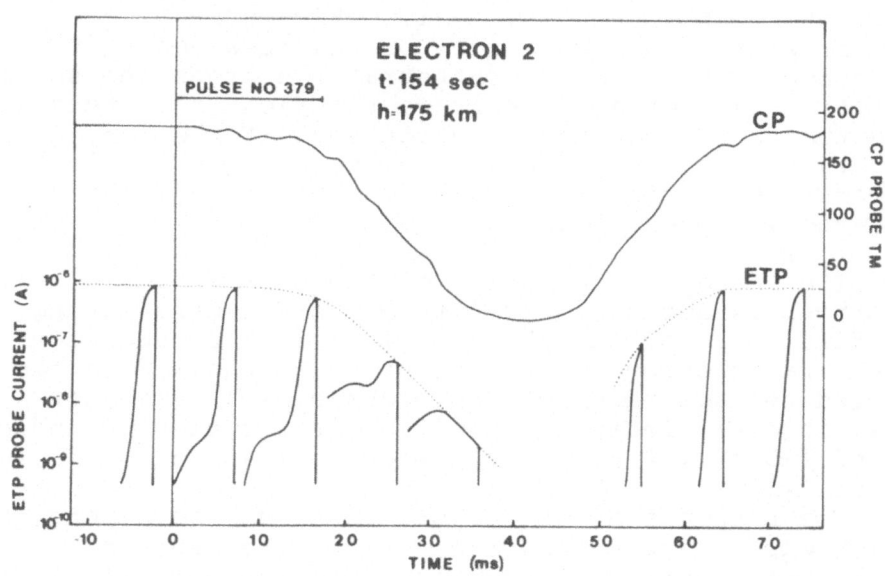

Figure 7.   Typical discharge effects observed by the plasma probes during the early part of the flight

electrons. Those electrons drive the floating potential of the payload negative. When the absolute value
of the negative payload potential becomes greater than
2.7 V, which is the upper limit of the probe sweep,
the thermal electrons can not reach the probe. This
occures during the fifth sweep on Figure 7, when only
suprathermal electrons are observed. During the sixth
sweep the electron current to the probe is below the
threshold level of the electrometer at $5 \cdot 10^{-10}$ A. Since
the suprathermal electrons are not observed the relative
consentration of those hot electrons must be so high
that the thermal electrons have a negligible influence
on the floating potential.

Theoretical estimates indicates that suprathermal
electron concentrations of a few percent of the ambient
plasma density is needed to explain the measurements.
Those estimates are, however, not quite reliable because
it is not known to which extent the ions are heated.
Further, the calculations does not include the emission
of secondary electrons from the payload due to the
bombardment of the hot electrons. If these two factors
are underestimated, the deduced consentration of
suprathermal electrons will be too low.

The effect discussed above can also explain the
apparent reduction of the maximum ETP-current in the
data presented in Figure 3. For the observed energy
and concentrations of suprathermal electrons one should
expect an even greater shift in the floating potential
than the current characteristics show if the ion
temperature is assumed equal to the background electron
temperature. This discrepancy may be an indication of
a heated ion population.

The energy distribution of the suprathermal
electrons cannot be derived from the ETP data during
the first part of the flight. This is a consequence
of the rapid change in the floating potential and the
limited voltage range of the probe sweep. The data
does, however, indicate that the distribution extended
beyond 4-5 eV (cf Figure 7, sweep no 4 and 5).

The response of the CP-probe inside the discharge
region was always an increase in the frequency or
equivalently a reduced oscillation period as illustrated
in Figures 3 and 7. In the given altitude range an
increased probe frequency is consistent with a decrease
in the ambient electron concentration or an increase in
the sheath thickness.

The thickness, S of the ion sheath is assumed to be proportional to the Debye length and will increase as the square root of the electron temperature, T_e. A change in the floating potential, V_f towards a more negative value must also be compensated by a thicker ion sheath. Based on the ETP-data it was concluded that both T_e and V_f increased inside the discharge region. Computations of the probe impedance as a function of N_e, T_e and V_f have been carried out based on an expression developed by Balmain (1966) for a cold magnetoplasma with collisions and a model for the ion sheath. For those parts of the flight where the ETP provided information on the electron temperature and the floating potential the observed increase in T_e and V_f were not sufficient to explain the pronounced change of the CP-frequency.

An explanation of the CP-effects in terms of a change in the ambient electron density would imply that N_e in some cases were reduced by up to two orders of magnitude inside the discharge region. This seems to be a very unlikely physical situation.

An alternative explanation is that the probe signal is subject to some sort of a loss process equivalent to an increase in the electron collision frequency, ν_e in the cold plasma approximation. A high value of ν_e reduces the probe capasitance to the free space value as illustrated on Figure 8 and is the reason why such probes cannot be used below 85-90 km.

No adequate theory exists for the probe behaviour in a warm, highly turbulent magnetoplasma typical for the discharge region. We will therefore use the cold plasma approximation with collisions to illustrate qualitatively the damping effect. It is seen in Figure 8 that the probe capacitance is most sensitive to changes in ν_e close to the resonance feature. The resonance is generally found near the electron gyro frequency, ω_c and depends only slightly on N_e and S (Jacobsen, 1972). The CP frequency on ELECTRON 2 was 1.2 MHz which is just below $f_c = 1.4$. A similar probe with a lower frequency, 0.6 MHz, flown on POLAR 5 (Mæhlum et al, 1980) did not show the characteristic CP-effects displayed in Figures 3 and 7. The probe effects seen on the POLAR 5 flight were generally observed during the pulse emissions and were probably caused by the beam and discharge generated wave phenomena.

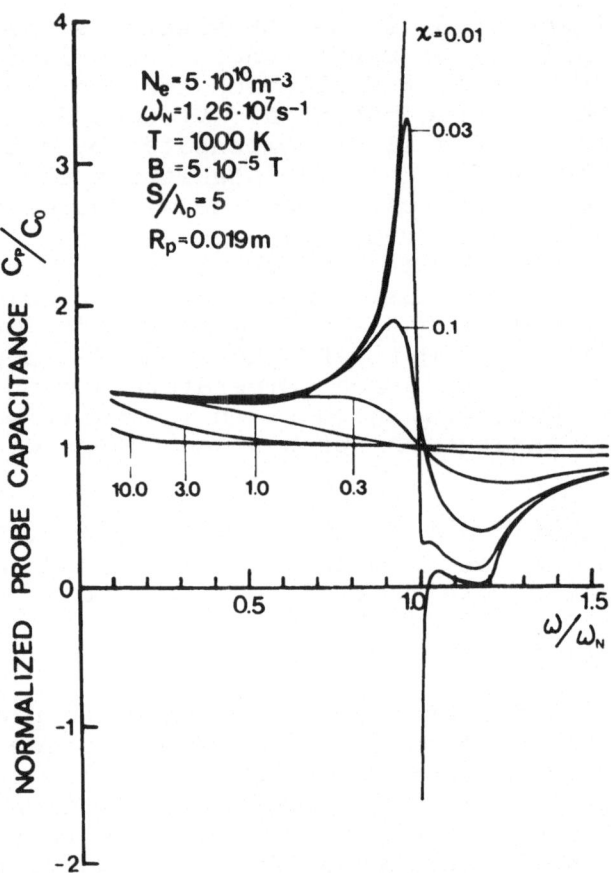

Figure 8.   Calculated probe capacitance, C_p as a
            function of frequency, ω for different values
            of the collision parameter, $\chi = \nu_e/\omega_N$. C_0 is
            the free space capacitance of the spherical
            probe that have a radius, $R_p = 1.9$ cm

Based on the cold plasma approximation the probe
capacitance has been calculated as a function of the
effective collision frequency, ν_{eff} for both the
ELECTRON 2 and the POLAR 5 probes. The results presented in Figure 9 demonstrate how a collision frequency
around 10^7 s^{-1} reduces the capacitance of the E2 probe
while the P5 probe remains unchanged. Within the frame
of this simple approach an effective collision frequency,
$\nu_{eff} \approx 10^3-10^4 \cdot \nu_e$ could explain the CP effects observed
on ELECTRON 2 and the lack of such effects on the POLAR
5 flight. The numerical results of this exercise must,
however, be treated with great care and should mainly be
considered as a qualitative support to a possible inter-

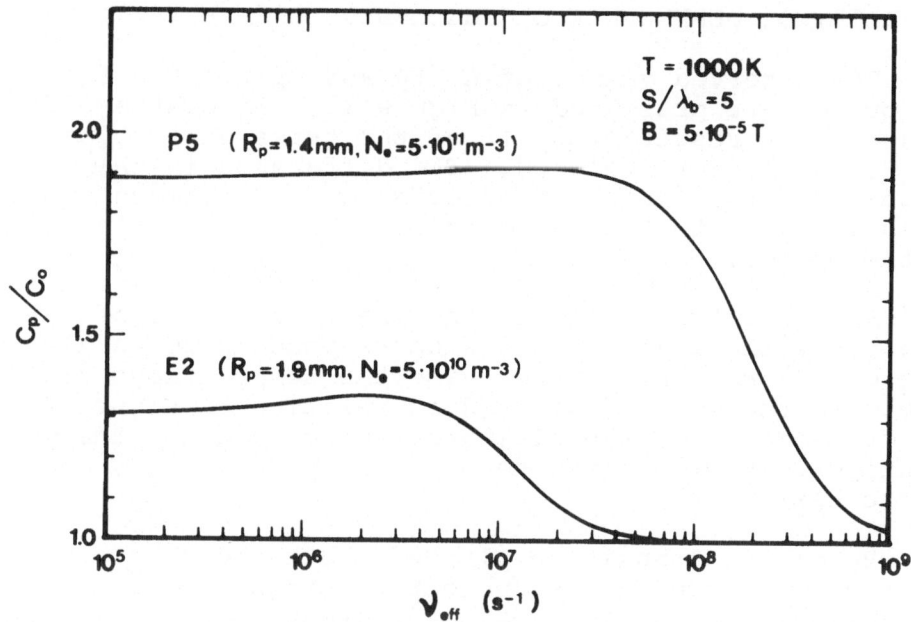

Figure 9.   Calculated probe capacitance as a function
             of collision frequency. The two curves are
             representative for the POLAR 5 (P5) and the
             ELECTRON 2 (E2) flights, respectively

pretation that has to be studied in more detail both
experimentally and theoretically.

Pronounced increases of N_e in the discharge region
have been reported in space (Cambou et al, 1978) and in
the laboratory (Szuszczewicz et al, 1979, Walker et al,
1981).

Neither the CP-data nor the ETP-data on ELECTRON 2
provided evidence of an enhanced electron density. For
the ETP-data this can be explained by the high negative
potential of the payload causing the thermal electron
population to drift out of the limited sweep range of
the probe. In the case of the CP-probe the possible increase of ν_{eff}, together with the heating of the electrons may compensate for the capacitance change due to
an enhanced electron concentration. It may also be of
importance that the probe measurements in the discharge
region were carried out after the beam had left and the
return current to the daughter had been drawn from the
plasma.

THE ALTITUDE DEPENDENCE OF THE OBSERVED EFFECTS

The observed values of A, t_B and t_E for the first 170 pulses are presented in Figure 10. In contrast to Figure 4 the results for all of the three pulse types (cf Figure 2) are included. The difference in the duration of the discharge effects for the various pulse types is equal to the 8-9 ms difference in the length of the pulses.

A marked transition in the characteristics of the data takes place at t = 106 s corresponding to an altitude of 137 km. After that time the observations fit nicely to the model discussed in relation to Figure 4. The crossection of the discharge region did not change significantly with altitude or with the injection angle of the beam.

Below the transition altitude both the cross-section and the amplitude, A are subject to strong variations with peak values up to a factor 2 greater than observed after pulse no 100. The possible altitude and injection angle dependence in the pronounced fluctuations are difficult to separate due to the limited amount of data and the 4 s period when the accelerator was in the low current mode.

Another interesting aspect of the data shown in Figure 10 is the relation between the pulse type and the amplitude, A. In the lower altitude regime the amplitude seems to be independent of the pulse type. However, after t = 106 s the effects produced by the 1 ms pulses are 35-40% weeker than those generated by the 10 ms and 18 ms pulses.

The altitude change in the characteristics of the discharge region, as observed by the probes, resembles the altitude dependence of the light emissions at 391.4 nm observed during the electron beam injections on POLAR 5 (Grandal et al, 1980a). A marked transition appeared in the altitude range 130-150 km between a region with a high light level independent of altitude and an exponential increase below 130 km. Simultaneous measurements by a retarding potential analyzer showed that the flux of suprathermal electrons decreased below the sensitivity level of the instrument at the lower border of the transition region (Jacobsen and Maynard, 1980).

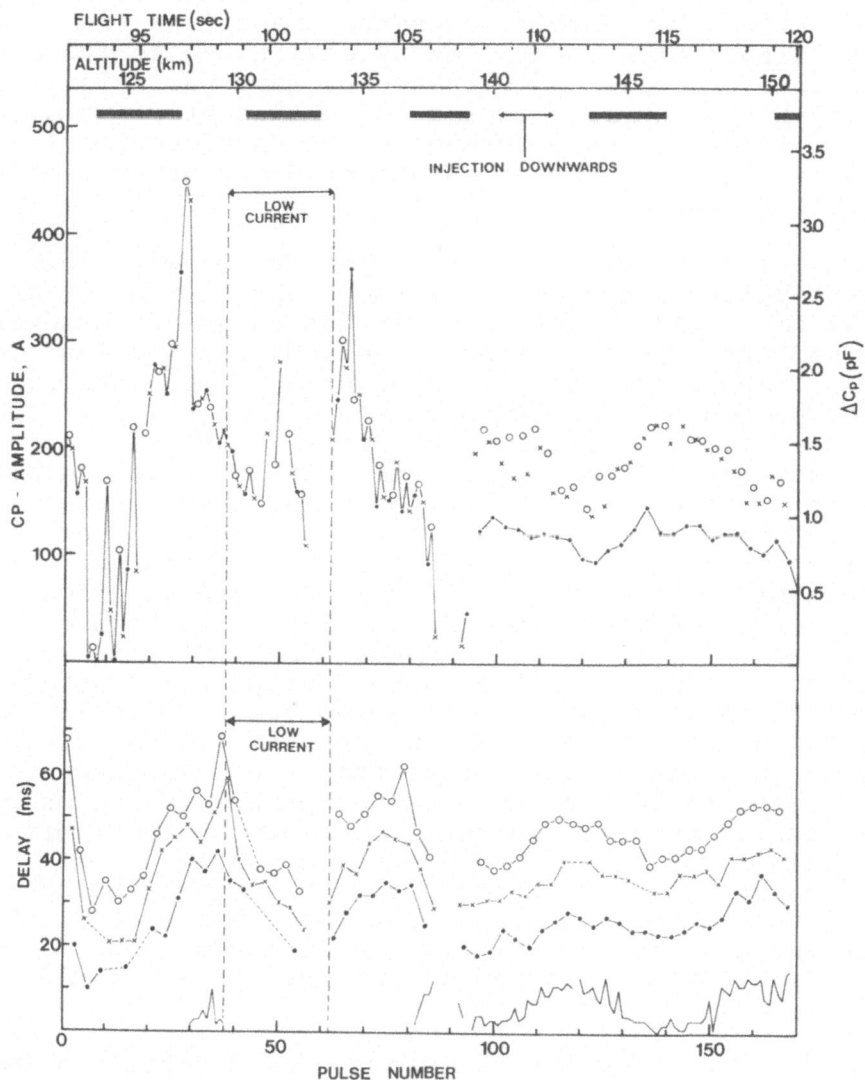

Figure 10. The observed amplitude of the discharge
            effects (upper panel) and the times for
            entering (t_B) and leaving (t_E) the discharge
            region plotted as a function of pulse number
            for all of the three different pulse types.
            The lower curve on the lower panel repre-
            sents t_B for all pulses. The amplitude, A
            and t_E are marked by 0, X and ● for the 18
            ms, 10 ms and 1 ms pulses, respectively

Based on the observed scattering of energetic electrons it was proposed by Mæhlum et al (1980b) that this altitude range represents the transition from beam energy dissipation controlled by collisions at low altitudes to a dissipation dominated by wave-particle interactions higher up. A comprehensive theoretical treatment of these processes presented by Papadopoulos (1981) supports this interpretation and predicts several of the observed features.

As already mentioned the first period with low current pulses occurred below the transition altitude of 137 km (cf Figure 10). Clear discharge effects were observed throughout that period except for the last few pulses. Above the transition no effects were observed by the probes for the 10 mA pulses (cf Figure 3). Due to a 1 s time constant in the change of the current level several pulses were emitted with an intermediate current. The observed CP-response has been plotted versus time in Figure 11 for three low current periods above the transition altitude. The zero reference time corresponds to the start of each low current period. The envelope of the pulse current level is shown on the lower section of the figure. Each vertical line represents one pulse. When the current drops the discharge effects disappear at a level somewhere between 10 mA and 100 mA and reappear again when the current rises. These data clairly demonstrate that the phenomenon being responsible for the observed probe effects only developes if the beam current is greater than a certain threshold level. The data are not suitable for a detailed study of the threshold. It can only be concluded that the critical current level was greater than 10 mA and less than 100 mA in the altitude range between 160 km and 190 km. Around 130 km the threshold was below 10 mA.

Due to a telemetry failure on the daughter payload and the absence of observable beam effects on the mother for the 10 mA pulses above 160 km, it is not possible to confirm whether the beam left the daughter during the low current modes. A charging of the daughter up to the 10 kV acceleration potential can therefore not be excluded for these events.

One of the important results of the comprehensive electron beam experiments carried out in the plasma chamber at NASA Johnson Space Center was that the beam current had to exceed a critical level, I_c in order to produce the beam-plasma-discharge (BPD) (Bernstein et al

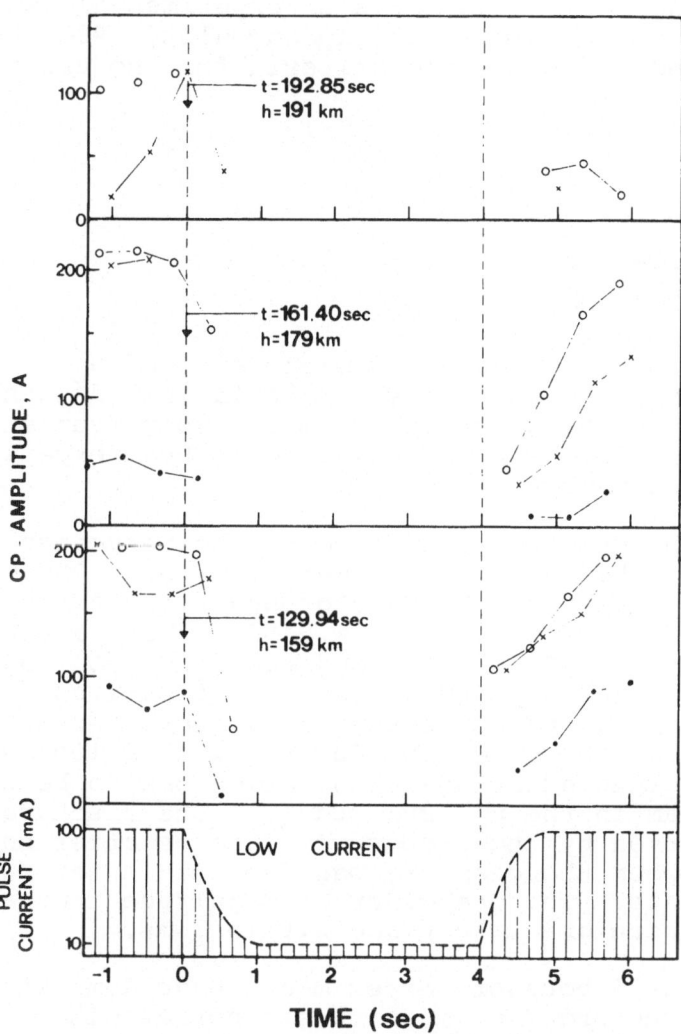

Figure 11.   The amplitude, A of the discharge effects
observed by the capacitance probe during
transitions between high and low beam currents. The zero reference time represents
the start of each of the three low current
periods shown. Each vertical line in the
lower panel represents one pulse.

o———o   18 ms pulses
x———x   10 ms pulses
·———·   1 ms pulses

1978, 1979). An empirical expression for I_c was
derived and later studies, both experimental (Walker et
al, 1981) and theoretical (Papadopoulos, 1981) have
established the following criteria for the BPD to take
place:

$$\frac{\omega_N}{\omega_C} \geq 5$$

$$K \geq 10^{-2} \frac{N_b}{N_e}$$

where K is the perveance in microperv and N_b is the
density of the beam. These criteria includes the
boundary conditions of the chamber experiment and are
not necessarily representative for a beam experiment
in the ionosphere.

In the ELECTRON 2 experiment the perveance for the
100 mA and the 10 mA pulses were 10^{-1} and 10^{-2} microperv, respectively. The background electron density was
typically $5 \cdot 10^{10}$ m^{-3} and $\omega_C \approx 8$ MHz. With a beam
density of 10^9 m^{-3} (Linson and Papadopoulos, 1980) the
above criteria, applied without modifications, require
that the N_e has to be increased by one order of magnitude to meet the condition for BPD with a 100 mA pulse
current. This additional ionization has to be created
by the beam in the pre-BPD stage. The discharge effects
were observed by the probes within 1 ms after pulse
emission when the geometry was favourable for such
measurements. The one order of magnitude increase in
N_e must therefore take place within 1 ms.

Recent laboratory experiments have demonstrated
that a three order of magnitudes increase in N_e can
take place within 5 ms with a 34 mA, 1.9 keV beam at a
comparable pressure level (Szuszczewicz and Lin, 1981).
The ELECTRON 2 payload had a velocity component perpendicular to the B-field of less than 1 m/ms and the spin
period was greater than 6 s. It therefore seems reasonable to assume that the 10 ms and 18 ms pulses produced
sufficiently high electron concentrations for the BPD
to occur. The fact that the strength of the discharge
effects produced by the 1 ms pulses were significantly
lower than for the other pulses probably indicates that
the BPD did not have sufficient time to develop fully
for those short pulses.

In the altitude range below the transition at 137 km strong currents were collected by the ETP-probe throughout the whole sweep range. The low sampling rate of the electrometer range bits, imposed by limitations in the TM-system, made it impossible to decide whether this was due to a pronounced general heating of the plasma or the presence of a high density suprathermal electron population.

Finally, the upper limit of about 3 eV in the observed suprathermal electrons during the last part of the flight does not necessarily mean that electrons are not heated to higher energies. Since the observations are generally, made several tens of ms after the beam injection the high energy part of the population is probably reduced due to ionization and excitation processes. Some evidence of a decrease in the density of suprathermal electrons below 3 eV are possibly caused by inelastic scattering of electrons by vibration-rotation excitation of N_2 as observed on the AE-E satellite by Doering et al (1976).

ACKNOWLEDGEMENTS

The author would like to thank Mr Friedrich, Technical University of Graz who kindly supplied the capacitance probe data and J Trøim who designed and tested the electron temperature probe. I am also grateful for valuable discussions with my colleagues B N Mæhlum, J Trøim, B Grandal and E V Thrane at the Norwegian Defence Research Establishment.

REFERENCES

Balmain, K.G., 1966, Impedance of a spherical probe in a Magnetoplasma, IEEE Trans. on Antennas and Propagation, AP-14: 402.

Bernstein, W., Leinbach, H., Kellogg, P., Monson, S.J., Hallinan, T., Garriott, O.K., Konradi, A., McCoy, J., Daley, P., Baker, B., and Anderson, H.R., 1978, Electron beam experiments: The beam-plasma discharge at low pressures and magnetic field strengths, Geophys. Res. Lett., 5:127.

Bernstein, W., Leinbach, H., Kellogg, P.J., Monson, S.J., and Hallinan, T., 1979, Further laboratory measurements of the beam-plasma discharge, J Geophys Res, 84:7271.

Cambou, F., Lavergnat, J., Migulin, V.V., Morozov, A.I.,
      Paton, B.E., Pellat, R., Pyatsi, A.Kh, Reme, H.,
      Sagdeev, R.Z., Sheldon, W.R., and Zhulin, I.A.,
      1978, ARAKS-Controlled or Puzzling Experiment?,
      Nature, 271:723.

Cartwright, D.G., Kellogg, P.J., and Monson, S.J., 1978,
      Heating of the Ambient Ionosphere by an Artifici-
      ally Injected Electron Beam, J. Geophys. Res,
      83:16.

Doering, J.P., Peterson, W.K., Bostrom, C.O., and Potemra,
      T.A., 1976, High resolution daytime photoelectron
      energy spectra from AE-E, Geophys. Res. Lett., 3:129.

Grandal, B., Thrane, E.V., and Trøim, J., 1980a, POLAR
      5: An electron accelerator experiment within an
      aurora: 4. Measurements of the 391.4 nm light
      produced by an artificial electron beam in the
      upper atmosphere, Planet Space Sci., 28:309.

Grandal, B., Holtet, J.A., Trøim, J., Mæhlum, B.N., and
      Pran, B., 1980b, Observations of waves artifici-
      ally stimulated by an electron beam inside a region
      with auroral precipitation, Planet. Space Sci.,
      28:1131.

Jacobsen, T.A., 1972, High latitude ionospheric studies
      by rocket measurements with emphasis on RF-capaci-
      tance probes, Intern Rap. E-204, Norw. Def. Res.
      Estab., Kjeller, Norway.

Jacobsen, T.A., and Maynard, N.C., 1980, POLAR 5 - An
      electron accelerator experiment within an aurora:
      3. Evidence for significant spacecraft charging
      by an electron accelerator at ionospheric altitudes,
      Planet. Space Sci., 28:291.

Jacobsen, T.A., Trøim, J., Maehlum, B.N., and Friedrich, M.
      1981, Ionospheric electron heating by a rocket borne
      electron accelerator, Adv. Space Res., 1:123.

Linson, L.M. and Papadopoulos, K., 1980, Review of the
      status of theory and experiment for injection of
      energetic electron beam in space, 1980, SAI Tech.
      Report No. LAPS 69.

Maynard, N.C., 1979, Private communication.

Mishin, E.V., Ruzhin, Yu. Ya., 1978, Beam-plasma discharge during electron beam injection in ionosphere: dynamics of the region in rocket environment in ARAKS and Zarnitza-2 experiments, Preprint No. 21, (a & b), Academy of Sci. USSR, Institute of Terrestrial Magnetism, Ionosphere and Radio Wave Propagation, Moscow.

Mæhlum, B.N., Måseide, K., Aarsnes, K., Egeland, A., Grandal, B., Holtet, J., Jacobsen, T.A., Maynard, N.C., Søraas, F., Stadsnes, J., Thrane, E.V., and Trøim, J., 1980a, POLAR 5 - An electron accelerator experiment within an aurora: 1.   Instrumentation and geophysical conditions, Planet. Space Sci., 28:259.

Mæhlum, B.N., Grandal, B., Jacobsen, T.A., and Trøim, J., 1980b, POLAR 5 - An electron accelerator experiment within an aurora: 2.   Scattering of an artificially produced electron beam in the atmosphere, Planet. Space Sci., 28:279.

Papadopoulos, K., 1981, Theory of beam plasma discharge, in: Artificial particle beams in space plasma studies, B. Grandal, ed., Plenum, New York.

Sagalyn, R.C., Smiddy, M., and Wisnia, J., 1963, Measuremets and interpretation of ion density distributions in the daytime F-region, J. Geophys. Res., 68:199.

Szuszczewicz, E.P., Walker, D.N., and Holmes, J.C., 1979, Plasma diffusion in a space-simulation beam-plasma-discharge, Geophys. Res. Lett., 6:201.

Szuszczewicz, E. and Lin, C.S., 1981, Time-dependent plasma behavior triggered by a pulsed electron gun under conditions of beam-plasma-discharge, in: Artificial particle beams in space plasma studies, B. Grandal, e.d., Plenum, New York.

Walker, D.N., Szuszczewicz, E.P., and Lin, C.S., 1981, Ignition of the beam-plasma-discharge and its dependence on electron density, in: Artificial particle beams in space plasma studies, B. Grandal, e.d., Plenum, New York.

PLASMA WAVES PRODUCED BY THE XENON ION BEAM EXPERIMENT

ON THE PORCUPINE SOUNDING ROCKET

Paul M. Kintner and Michael C. Kelley

Dept. of Electrical Engineering
Cornell University
Ithaca, N.Y. 14853

ABSTRACT

The production of electrostatic ion cyclotron waves by a perpendicular ion beam in the F-region ionosphere is described. The ion beam experiment was part of the Porcupine program and produced electrostatic hydrogen cyclotron waves just above harmonics of the hydrogen cyclotron frequency. The plasma process may be thought of as a magnetized background ionosphere through which an unmagnetized beam is flowing. The dispersion equation for this hypothesis is constructed and solved. Preliminary solutions agree well with the observed plasma waves.

INTRODUCTION

Substantial progress in the last several years has been made in understanding both the physics of ion beams and their significance in space plasmas. Laboratory studies of collisionless ion beams propagating through magnetized plasmas have emphasized the importance of collective effects produced by fluctuating electric fields. In space it has become increasingly apparent that natural ion beams are a frequent and even common component of ionospheric and magnetospheric plasmas (Sharp, et al., 1977; Klumpar, 1979; Gorney, et al., 1980). The production and propagation of these beams is a current topic of intense interest (Lysak et al., 1978; Dusenbery and Lyons, 1981). At the same time it has become possible with sounding rocket technology to produce ion beams in ionospheric plasmas and to study the ion beams as an active, controlled experiment. The controlled experiment is already producing results pertinent to natural ion beams.

The propagation of ion beams may be thought of as occurring in two limiting extremes; propagation parallel or perpendicular to the ambient magnetic field. The subject of this paper is the production of electrostatic ion cyclotron (EIC) waves by the perpendicular ion beam experiment on the Porcupine sounding rockets. However, the parallel case both occurs in space and has been studied in the laboratory. It has also been produced in space actively with barium shaped charges. Since this paper will limit itself to the discussion of perpendicular propagation we refer the reader interested in parallel propagation to Bohmer et al. (1976), Hendel et al. (1976), Kintner et al. (1979) and references therein.

Several laboratory experiments employing perpendicular beams have been performed in Q machines which are relevant to space plasmas. Generally a plasma is produced which is partially confined by an axial magnetic field and then a perpendicular ion beam is introduced. The beam drift velocity typically has an effective temperature of 10 to 300 times the temperature of the background plasma and the beam is usually less dense than the ambient plasma. The principal result of these experiments is that the perpendicular beams are unstable to the production of electrostatic ion cyclotron waves. That is fluctuating electric fields are produced at frequencies between multiples of the ion gyrofrequency (Bohmer, 1976) and the most intense emissions occur near the lower hybrid frequency (Seiler et al., 1976). Usually the lowest order emissions are missing. The instability process may be modeled with a dispersion relation that assumes perpendicular wave propagation and an unmagnetized beam. The assumption of an unmagnetized beam is justified if the beam ion gyroradius is large compared to the ambient plasma gyroradius. The assumption of perpendicular wave propagation implies that Landau damping by electrons with velocities in the parallel direction is nil and hence large growth rates are possible ($\omega \simeq .3\omega_{ci}$). We will apply a similar theory to understand electrostatic ion cyclotron waves produced in the ionosphere although the dispersion relation is solved in a parameter regime different from the laboratory.

In the next section we describe the Porcupine ion beam experiment and the plasma wave receiver results. In the third section we discuss the linear plasma wave dispersion relation for an unmagnetized beam and illustate its application to the Porcupine experiment.

ION BEAM GENERATED PLASMA WAVES

The Porcupine sounding rockets were composed of a central instrumented payload and four subpayloads which were ejected in the radial direction during the ascending leg of the flight. One of the subpayloads contained an xenon gun which directed a 200eV Xe$^+$ beam roughly perpendicular to the magnetic field (Galeev et al.,

1977). The xenon beam was exercised for periods of several seconds
with intervals of tens of seconds between each ion beam exercise as
the subpayload moved radially away from the central payload. The
ion beam subpayload was spinning at 3Hz so that the beam
illuminated the central instrumented payload every .33 seconds. In
this way a radial profile of the xenon beam interaction with the
ambient ionospheric plasma could be constructed using instruments
on the central spacecraft. We report here measurements made by the
$\delta n/n$ experiment on the central payload during the flight of F4.
Further wave measurements are described in Kintner and Kelley
(1981) and Jones (1981).

The $\delta n/n$ experiment on Porcupine measured plasma waves between
30Hz and 24kHz by sensing changes in plasma density. Since density
fluctuations are characteristic of electrostatic waves in warm
plasmas, we anticipated that ion acoustic and/or ion cyclotron
waves would be primarily observed. During the first exercise of
the xenon plasma beam electrostatic ion cyclotron waves were
observed.

Figure 1 is a frequency-time-gray scale plot of plasma waves
observed during the first xenon beam exercise. The xenon beam is
turned on at 118.5 sec and turned off at 136.2 sec which
corresponded to radial separations of 9 m and 88 m between the ion
beam and the main payload. Beginning at about 123 sec several line
emissions occurred which were separated in frequency by about
700Hz. The frequency spacing was the hydrogen cyclotron frequency
(f_{H^+}). The 3Hz amplitude modulation feature was produced by the
spinning xenon beam subpayload. The line emissions spaced at
frequency intervals of f_{H^+} are EIC waves and they ceased with the
xenon beam turn off at 136 sec. Additional ion cyclotron features
can be seen in the line "splitting" at higher frequencies, above
5kHz. In this case the frequency separation of the two closely
spaced lines corresponds to the helium gyrofrequency.

The observation of hydrogen cyclotron waves was surprising
since the payload altitude at 130 sec was 240km. In this part of
the F-region the dominant ion is O^+ and H^+ is only a minor
constituient. Figure 1 also indicates that the fastest growing
modes occcur near 4kHz which corresponds to the H^+ plasma frequency
or to the lower hybrid frequency if ions other than H^+ are ignored.
The latter correspondence agrees with the Seiler et al. (1976)
laboratory result.

Figure 2 shows the $\delta n/n$ power spectrum on an expanded
frequency scale at 135 sec. The emission peak near 2200Hz
corresponds to the lowest frequency emission line of Figure 1. The
emission peak occurred just above the third multiple of f_{H^+} as
expected for the EIC mode. Additionally the width of the spectral
line appears to be given by oxygen cyclotron frequency (f_{O^+}).

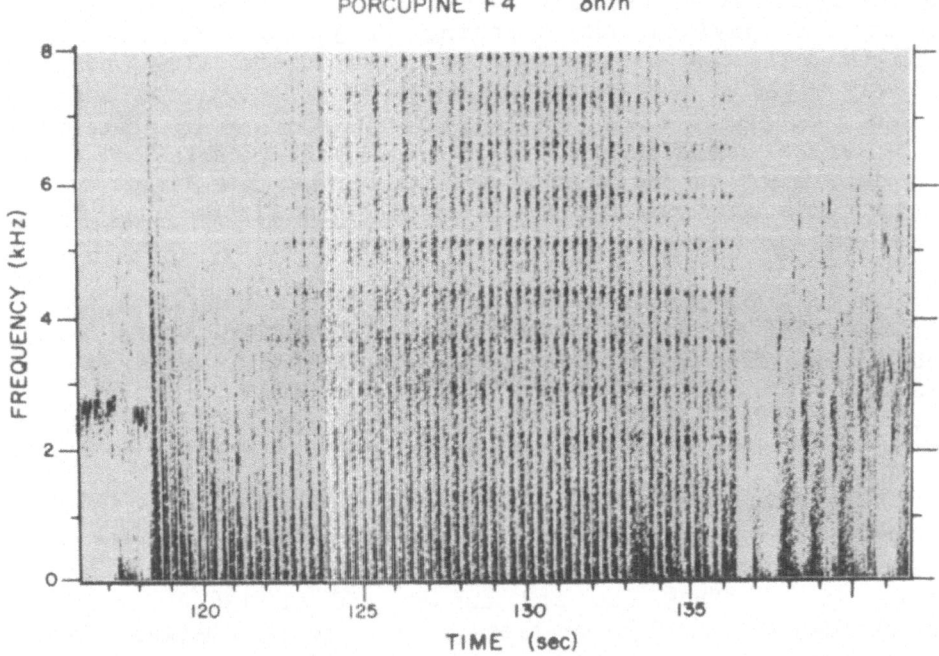

Fig. 1. Frequency-time-gray scale plot of density fluctuations
during the first xenon ion beam exercise on Porcupine
flight F4. The spacing of emission lines at intervals of
700 Hz correspond to the hydrogen cyclotron frequency.

Whether the correspondence of the line width to f_{0^+} is accidental
or a consequence of O^+ being the dominant ion remains to be seen.

Since the EIC waves corresponded to periods of time when the
xenon beam illuminated the payload it seems reasonable to conclude
that the beam provides the source of free energy for the plasma
waves. In this case the Xe^+ with a large gyroradius appears as an
unmagnetized perpendicular beam to the background ionosphere. EIC
waves whose perpendicular phase velocity corresponded to the xenon
beam velocity coupled to the beam. Since there is no requirement
on the parallel phase velocity electron Landau damping can be
neglected. For a single species plasma growth rates of up to $.3f_{ci}$
are predicted (Seiler et al. (1976). The growth rates for a
multispecies plasma have not yet been calculated although Hamelin
and Beghin (1976) have estimated damping rates for a propagating
EIC wave. Their estimates predict that the lightest ion
establishes the frequency spacing and heavier ions establish the
line width which is consistent with the results presented here.

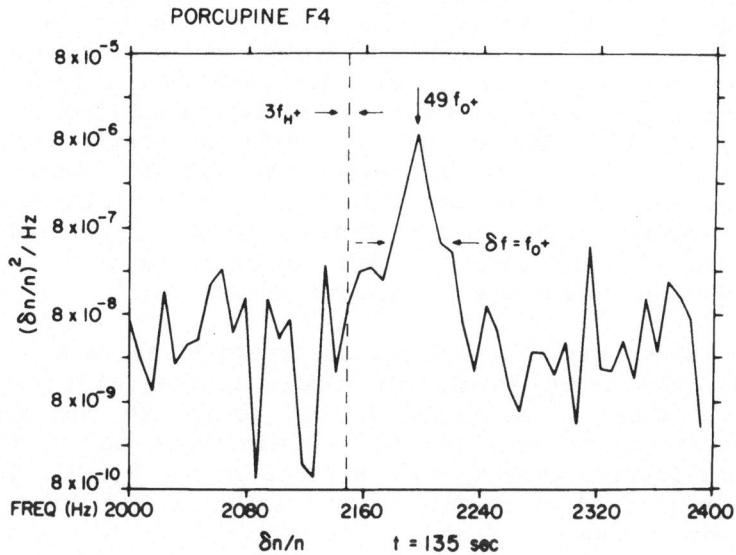

Fig. 2.   The power spectrum of the lowest frequency emission line
of Fig. 1. on an expanded frequency scale.

THE ION CYCLOTRON WAVE-ION BEAM DISPERSION RELATION

      If we assume that the frequency is much less than either the
electron gyro or plasma frequency and that $k_{11}=0$, we can use
Seiler's approach to write the dispersion relation for a multi-ion
plasma with an unmagnetized perpendicular beam as,

$$1 - \sum_j \left\{ \frac{\omega_{pj}^2/\omega_{cj}^2}{\lambda_j} \sum_{n=1}^{\infty} \frac{2n^2 \, e^{-\lambda_j}}{\omega^2/\omega_{cj}^2 - n^2} \, I_n(\lambda_j) \right\}$$

$$- \frac{1}{2} \frac{\omega_{pb}^2/\omega_{cb}^2}{\lambda_b} \, Z' \, (\frac{\omega - ku_\perp}{kv_b}) = 0$$

where $\lambda_j = k^2 \rho_j^2/2$

           $u_\perp$ = beam perpendicular drift velocity

and v_b = beam thermal velocity

The last term represents the perpendicular beam contribution while
the proceeding term represents the contribution for each ambient
magnetized ion. The ion plasma and cyclotron frequencies are given
by ω_{pj} and ω_{cj} respectively. We have developed a technique for
solving the dispersion equation numerically and applied it first to
laboratory results (Seiler et al., 1976) to verify its correct
application. In the case of Seiler the ambient plasma density
greatly exceeded the beam density. For the Porcupine experiment
the beam density was the order of the O^+ ion density and greatly
exceeded the H^+ ion density. Thus far we have only solved the
dispersion relation for one background ambient ion.

The assumptions of $k_{11}=0$ and an unmagnetized beam are easily
justified. The $k_{11}=0$ assumption simply implies that the plasma
will select those modes which do not couple to the electrons.
Since the effect of coupling to the electrons is to damp wave
growth, the largest growth rates will occur for $k_{11}=0$. The ratio
of the beam gyroradius to the H^+ gyroradius is 3×10^2 and hence
the beam may by thought of as being unmagnetized.

Figure 3 shows the dispersion relation near the third multiple
of f_{H^+} for a single ion H^+ plasma and a ratio of beam density to
ambient H^+ density of 200. The other parameters have been chosen
for typical ionospheric conditions and the published ion beam

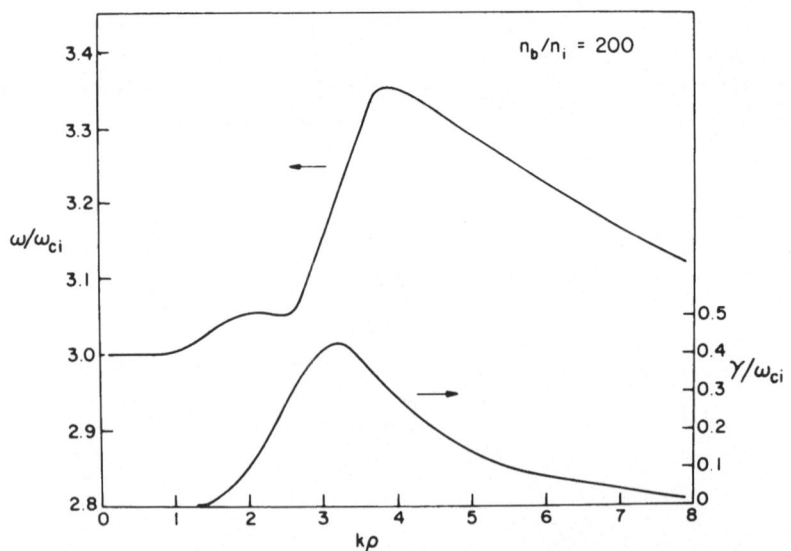

Fig. 3. The real and imaginary parts of the dispersion relation
for the F-region ionosphere and a perpendicular Xe^+ beam.
The ionosphere is assumed to be composed only of diffuse
H^+ and the O^+ contribution is neglected.

characteristics (Galeev et al., 1977). The frequency range was chosen so that it could be compared to Figure 2. The dispersion relation in Figure 3 illustrates several features which agree with the observed plasma waves. For example in the frequency range where waves were observed, about $\omega/\omega_{ci} = 3.1$, there is a large growth rate ($\gamma = .4\omega_{ci}$). The largest growth rate corresponds to a frequency of $\omega = 3.2\omega_{ci}$ which is somewhat higher than the observed value. The frequencies corresponding to large growth rates are band limited to between $3.05\omega_{ci}$ and $3.35\omega_{ci}$. This frequency range is somewhat larger than the observed frequency range but it does include the observed emissions. The inclusion of O^+ as an ambient magnetized ion in the dispersion relation may narrow the frequency band of growth as suggested by Hamelin and Beghin (1976).

CONCLUSIONS

A Xe^+ beam directed perpendicular to the earth's magnetic field produced electrostatic waves in the earth's ionosphere. The plasma waves received by $\delta n/n$ receiver indicate the wave energy occurs just above multiples of the H^+ gyrofrequency. Since the ionosphere at the altitude of the experiment is primarily O^+ and H^+ is only a minor constituient, the observed frequency structure is somewhat of a surprise. However the dispersion relation assuming an unmagnetized beam whose density is much greater than the ambient plasma yields solutions which have large growth rates at the observed frequencies.

ACKNOWLEDGEMENTS

We thank the Porcupine experimenters for their cooperation and interest and Carl Siefring for his programming assistance. This research was supported by ONR Contract N00014-81-K-0018 and NASA Contracts NGR-33-010-161 and NAGW-90.

REFERENCES

Bohmer, H., 1976, Excitation of ion cyclotron harmonic waves with an ion beam of high perpendicular energy. Physics of Fluids,19:1371.
Bohmer, H., Hauck, J. P., and Rynn, N., 1976, Ion-beam excitation of electrostatic ion-cyclotron waves, Physics of Fluids,19:450.
Dusenbery, P. B. and Lyons, L. R., Generation of ion-conic distributions by downward auroral currents, J. Geophys. Res.,submitted for publication.
Galeev, A. A., Dokukin, V. S., Zhulin, I. A., Kapitanov, V. Ja., Kozubski, K. N., Morozov, A. I., Mishin, E. V., Ruzhin, Ju. Ja., Sagdeev, R. Z., Haerendel, G., Shubin, A. P. and Snarski, R. N., 1977, Porcupine, Project, experiment no. 12. Dynamics of a plasma beam, in: "Issledovanija po Problemam

Solnechno-Zemnoi Fiziki", Moscow, p. 152.

Gorney, D. J., Clarke, A., Croley, D., Fennell, J., Luhmann, J. and Mizera, P., 1981, The distribution of ion beams and conics below 8000 km, J. Geophys. Res.,86:83.

Hamelin, M. and Beghin, C., 1976, Electromagnetic and electrostatic waves in a multi-component plasma near the lower hybrid frequency, J. Plasma Physics,15:115.

Hendel, H. W., Yamada, M., Seiler, S. W. and Ikezi, H., 1976, Ion-beam-driven resonant ion-cyclotron instability, Phys. Rev. Letts,36:319.

Jones, D., 1981, Xe^+-induced ion-cyclotron harmonic waves, in: "(COSPAR) Space Research", Vol. XX, M. J. Rycroft, ed., Pergamon Press, Oxford-New York.

Kintner, P. M. and Kelley, M. C., 1981, Ion beam produced plasma waves observed by the $\delta n/n$ plasma wave receiver during the Porcupine experiment, in: "(COSPAR) Space Research", Vol. XX, M. J. Rycroft, ed., Pergamon Press, Oxford-New York.

Kintner, P. M., Kelley, M. C., Sharp, R. D., Ghielmetti, A. G., Temerin, M., Cattell, C., Mizera, P. F. and Fennell, J. F., 1979, Simultaneous observations of energetic (keV) upstreaming ions and electrostatic hydrogen cyclotron waves, J. Geophys. Res.,84:7201.

Klumpar, D. M., 1979, Transversely accelerated ions: An ionospheric source of hot magnetospheric ions, J. Geophys. Res.,84:4229.

Lysak, R. L., Hudson, M. K. and Temerin, M., 1978, Enhanced ion heating by coherent electrostatic ion cyclotron waves, Trans. American Geophysical Union,59:1155.

Seiler, S., Yamada, M. and Ikezi, H., 1976, Lower hybrid instability driven by a spiraling ion beam, Phys. Rev. Letts.,37:700.

Sharp, R. D., Johnson, R. G. and Shelley, E. G., 1977, Observation of an ionospheric acceleration mechanism producing energetic (keV) ions primarily normal to the geomagnetic field direction, J. Geophys. Res.,82:3324.

DISCUSSION

Koons: What was the fractional concentration of H^+ in the ambient plasma when the H^+ gyroharmonic emissions were observed?

Kintner: The ratio of beam density to H^+ density used in the model was 200. The H^+ density was not measured and we are not aware of any H^+ density measurements in the night time auroral zone at this altitude. H^+ densities measured elsewhere at a similar altitude imply that the H^+ is a fraction of a percent of the total ion density.

THE EXCEDE SPECTRAL ARTIFICIAL AURORAL EXPERIMENT: AN OVERVIEW

R. R. O'Neil and A. T. Stair, Jr.

Radiation Effects Branch
Optical Physics Division
Air Force Geophysics Laboratory
Hanscom AFB, Massachusetts 01731, U.S.A.

W. R. Pendleton, Jr.[§] and D. A. Burt

Center for Atmospheric and Space Sciences
Utah State University
Logan, Utah 84322, U.S.A.

INTRODUCTION

EXCEDE is a Defense Nuclear Agency and Air Force Geophysics
Laboratory program designed to study atmospheric radiative
processes resulting from the controlled deposition of energetic
electrons from rocketborne electron accelerators. On October 19,
1979, the 2600-kilogram EXCEDE SPECTRAL payload was successfully
lauched from Poker Flat, Alaska, into a dark, clear and aurorally
inactive night atmosphere. The stabilized payload contained: a
60 kilowatt (3 kV) electron accelerator, an array of ultraviolet,
visible, and cryogenic infrared spectrometers, photometers, and
both photographic film and video cameras. Atomic and molecular
emissions induced in the atmosphere by the pulsed, rocketborne
electron accelerator and radiating in the 0.15 to 22 micron wavelength region were recorded at altitudes from 70 to 128 km.
Observed emissions included: the N_2 Lyman Birge Hopfield system,
the N_2 Herman Kaplan system, the N_2 first and second positive

---

[§]Present address: Department of Meteorology
University of Stockholm
Arrhenius Laboratory
S-106 91 STOCKHOLM, Sweden

systems, the N_2 Wu Benesch infrared system and the N_2^+ first negative and Meinel systems. In addition, the beam-induced emissions recorded by the cryogenic infrared instrumentation included CO_2 at 4.3 microns, NO at 5.4 microns, and a feature at 4.5 microns tentatively identified as NO^+. The comprehensive set of spectral measurements are column emission rates and will be analyzed in terms of production and loss mechanisms.

The EXCEDE SPECTRAL experiment also used remote measurement platforms in an attempt to measure visible and infrared time dependent pulse shapes and to record the spatial extent of the 3 kilovolt electron beam with imaging systems. The AFGL KC-135 aircraft was instrumented with a 2.7 micron radiometer, a 3914A photometer and a low-light-level television system. The Hilltop Optics Site at Poker Flat was instrumented with an array of film and video cameras and a dual-channel telephotometer which monitored optical emissions at 3914A and 5577A.

The primary scientific interest in this experiment is investigation of the detailed production and loss processes of various excited electronic and vibrational states that result in optical and infrared emissions as energetic primary electrons and their secondaries and all subsequent generation electrons are stopped in the atmosphere. In this artificial auroral experiment, the dosing conditions (primary electron energy, beam power, deposition volume, deposition altitude, and dose duration) are parameters which are reasonably well controlled and monitored. In natural aurora, these excitation conditions must be inferred, and the observed atmospheric emissions typically are effects integrated over a range of conditions (electron energy, electron flux density, altitude, and dosing time). These integral effects complicate the interpretation of auroral optical/infrared emissions in terms of basic production and loss processes.

Results from a number of previous launches in the EXCEDE series of artificial auroral experiments have been reported.[1-3] The major innovations in the EXCEDE SPECTRAL experiment included: a significant increase in the number and size of the payload instruments, increased beam power, and the addition of optical- and infrared-spectral instruments to record detailed band profiles. The previous experiments had utilized photometric and radiometric instruments which merely sampled all emissions within the instrumental band passes. The experimental approach is described by O'Neil et al.[4]

The dose level realized in the EXCEDE SPECTRAL experiment was on the order of 10^{13} ion prs cm^{-3} sec^{-1} at 75 km on rocket descent. The high dose level significantly enhanced the concentration of species (metastable atoms and molecules, nitric oxide, various positive ions, and electrons) which are important to the production and loss processes of the excited states resulting in many of the optical and infrared emission. The beam-induced concentration of electrons at low altitudes was sufficient to make collisional deactivation of $N(^2D)$ by electrons and dissociative recombination of vibrationally excited NO^+ competitive with other loss mechanisms for these species.

EXPERIMENTAL APPROACH

The EXCEDE SPECTRAL flight profile is presented in Figure 1. As indicated, the payload was despun and oriented such that the long dimension was elevated at an angle of approximately 43 degrees after nosecone and door ejection and payload separation. The proposed nominal electron-accelerator power was 100 kW (3 kV, 32A) using 4 electron-gun modules each providing an 8-A beam. The electron accelerators initiated a pulse sequence at approximately 115 km on payload ascent which continued through apogee (128 km) and continued to approximately 70 km on payload descent, providing a total experiment duration of 180 sec.

The payload orientation during the experiment positioned the electron accelerator and instruments so that: (1) the electron-beam injection angle was canted 30 degrees from the normal to the payload; (2) the fields of view of the optical sensors were normal to the payload, intersected the magnetic field (and the electron-beam axis) at 30 degrees, a few meters from the accelerator port, and were aligned with the plane of the vehicle trajectory to

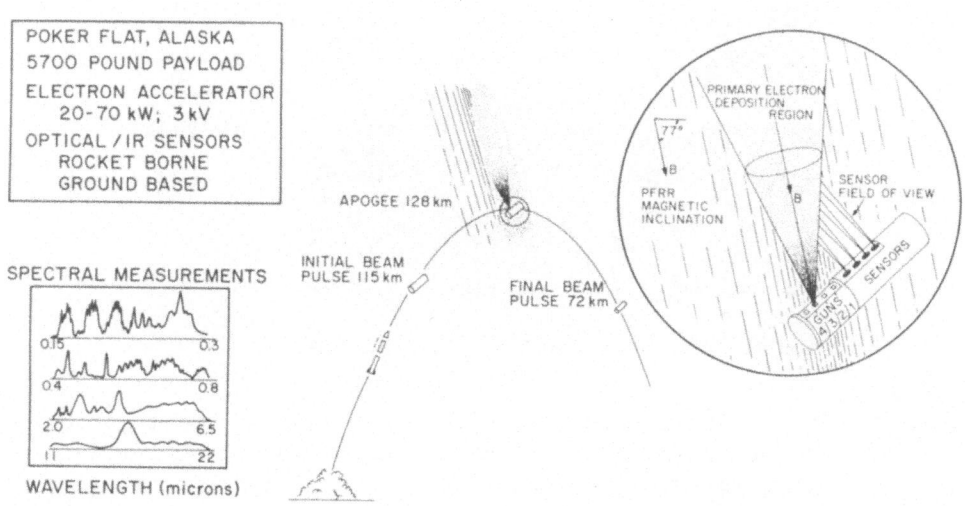

Figure 1.   EXCEDE SPECTRAL flight profile.

observe both the prompt emissions in the primary electron deposition region and slower emissions in the electron-beam afterglow.

In the EXCEDE SPECTRAL experiment, the accelerator was periodically activated for approximately four seconds in which the spectral
instruments completed one or more wavelength scans. After accelerator turn-on, the radiance of each emitting species, as observed by
the payload-based sensors, was determined in part by: the characteristic production and/or emission time constant of a given excited
state, payload velocity, and the instrumentation viewing aspect relative to the payload velocity. Since the horizontal component of
payload velocity was on the order of 300 m/sec, production and/or
emission processes having characteristic times on the order of
0.001 sec or longer were anticipated to show an afterglow emission
displacement that is significant in terms of the primary electron
depostion volume near the payload. Preflight estimates of N_2^+
first negative 3914A band radiance indicated a single electron
accelerator module should produce approximately 100 megarayleighs
at lower altitudes. The time dependent emission profiles to be
observed by the remote sensors, aircraft and ground based, were
intended to record the total volume of the electron-excited
atmosphere and to monitor the rise and decay of the emission signature of a complete 4-second accelerator pulse at selected wave-
lengths.The remote sensors supplement and assist the analysis and
interpretation of the rocketborne measurements.

INSTRUMENTATION AND FLIGHT RESULTS

Figure 2 is a photograph of the optical and infrared instrumentation in the payload sensor module. Payload instruments, in

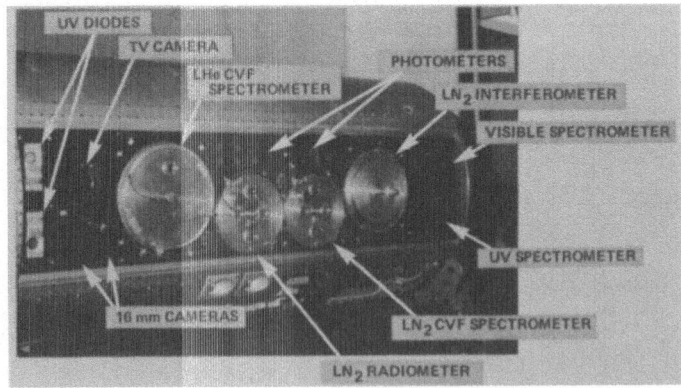

Figure 2.   The EXCEDE SPECTRAL sensor module showing the location
            of the optical instruments. The payload diameter was
            0.8 meter and the sensor module was 1.8 meters long.
            The accelerator was of comparable size and was attached
            near the ultraviolet and visible spectrometers.

addition to the electron accelerator, included ultraviolet and
visible grating spectrometers, liquid-nitrogen- and liquid-helium-
cooled circular variable filter (CVF) spectrometers and a newly
developed liquid-nitrogen-cooled Michelson interferometer. Electron-
impact-induced atomic and molecular emissions in the wavelength
range from 0.15 to 22 microns were monitored by the rocketborne
spectrometers. Selected measurements will be presented in this
review of the EXCEDE SPECTRAL experiment. Data from the rocket-
borne film cameras are presented in a separate contribution.[5]

The electron accelerator modules were designed to be synchron-
oulsy pulsed for a period of four seconds, followed by a two-second
off interval, and every third pulse was to be produced solely by
the master gun module (the unit closest to the sensor module). The
operation of either one or four accelerator modules provided a
measure of atmospheric emissions as a function of dose magnitude.
High-voltage breakdown of the electron accelerator occurred during
the flight, different in time and duration for each module, which
reduced the duty cycle and beam power from the design levels.
Single modules operated at power levels of 15 to 30 kilowatts during
the experiment, with the smaller values occurring during the final
portion of the experiment. A maximum accelerator power level of
approximately 60 kilowatts was achieved at times when three of the
four modules contributed to a stable beam pulse.

Figure 3 presents the ultraviolet emissions recorded by the
grating spectrometer during a 2.8-second period when the payload
was at an altitude of approximately 125 km on ascent and the accelerator was operating at 20 kilowatts. The spectrometer was a 1/4-

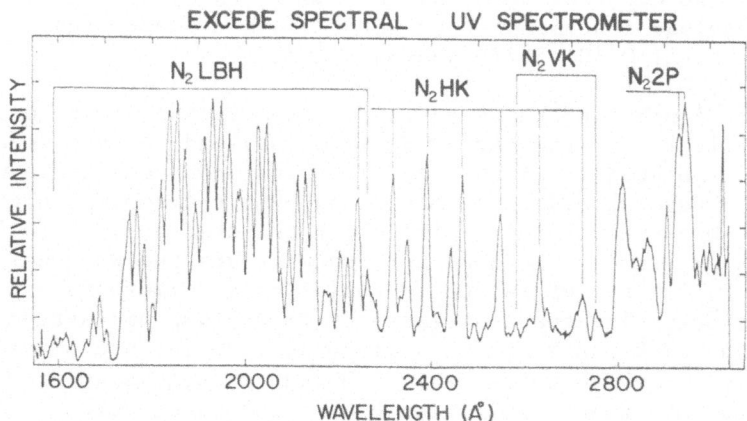

Figure 3.   Beam-induced ultraviolet emissions at 125 km on rocket
            ascent

meter Ebert-Fastie scanning instrument and utilized a photon
counting detection system. The Lyman Birge Hopfield (LBH) system
($a^1\pi_g \to X^1\Sigma_g^+$), Herman Kaplan (HK) system ($E^3\Sigma_g^+ \to A^3\Sigma_u^+$) and Vegard
Kaplan (VK) system ($A^3\Sigma_u^+ \to X^1\Sigma_g^+$) of N_2 are the dominant band systems
recorded in this wavelength region. The data, recorded at high
count rates, were measured with excellent precision and provide the
initial definitive observation of the N_2 Herman Kaplan system in an
electron excited atmosphere.

Data comparable to Figure 3 were recorded by a grating
spectrometer scanning from 3500 to 8000A. The principal emissions
are the first negative system ($B^2\Sigma_u^+ \to X^2\Sigma_g^+$) of N_2^+, the first
positive system ($B^3\pi_g \to A^3\Sigma_u^+$) of N_2, the second positive system
($C^3\pi_u \to B^3\pi_g$) of N_2, the first negative system ($b^4\Sigma_g^- \to a^4\pi_u$) of O_2^+
and several atomic transitions. These features are anticipated to
serve as monitors of parameters essential to understanding and
interpreting the infrared measurements. The emissions provide a
measure of the relative abundance of N_2 and O_2 and possibly O in
the electron excited atmosphere as well as a measure of the electron
energy deposited in the field of view of the rocketborne sensors.
The $O(^1S)$ emission at 5577A was not readily observed by the rocket-
borne visible spectrometer even at higher altitudes where collisional deactivation by ambient atmospheric species (O and O_2) is
negligible. The suppression of this feature is attributed to
collisional deactivation of $O(^1S)$ by H_2O which was outgassing from
the vehicle and contaminating the volume near the payload. The rate
coefficient for the collisional quenching of $O(^1S)$ by H_2O is
6×10^{-10} cm^3 s^{-1} (6). This interpretation is qualitatively consistent with the infrared sensors which revealed H_2O emissions at
2.7 and 6.3 microns. The ground based dual channel telephotometer
observed the $O(^1S)$ 5577A emission with photon emission rates similar
to the N_2^+ 1N 3914A band. This photometer observed the total volume
excited by the electron beam and the small fraction of this volume
near the payload contaminated by water vapor did not significantly
affect the total $O(^1S)$ 5577A photon yield.

Figure 4 presents the spectra recorded by the low resolution
circular filter spectrometer at 74 km on payload descent when the
accelerator emitted 15 kilowatts. The beam induced spectra were
enhanced by several orders of magnitude over the ambient night
atmosphere measurement (beam off) which detected only the 4.3
micron CO_2 emission. The dominant features are the Wu Benesch
($W^3\Delta_u \to B^3\pi_g$) system of N_2, CO_2 emission at 4.3 microns, and NO
at 5.3 microns. The spectral range of this cryogenic system
operating at 77°K was 2.0 to 5.4 microns. The intense emission
at 2.7 microns is H_2O which is attributed to outgassing from the
payload and subsequent excitation by the pulsed electron accelerator. The CO_2 emission at 4.3 microns and NO at 5.3 microns
saturate the data channel presented in Figure 4. The spectrum
is also available for analysis on a lower-gain channel.

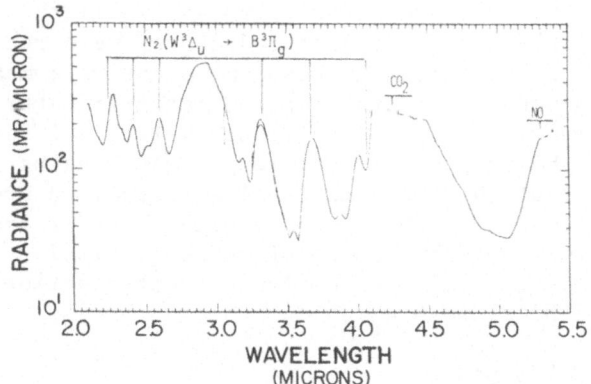

Figure 4. Liquid nitrogen CVF spectrometer measurement of beam
           induced emission at 74 km.

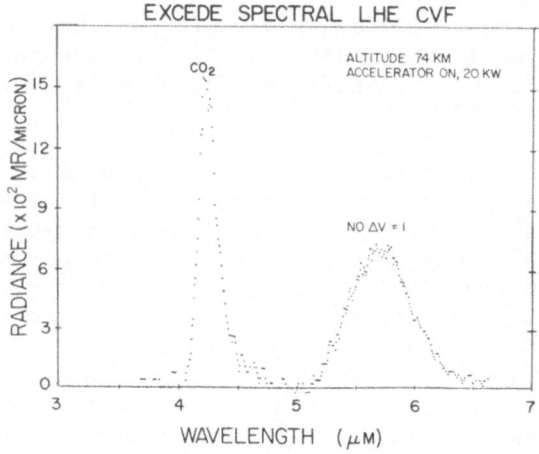

Figure 5. Liquid helium cooled CVF spectrometer measurement of
           beam induced emission at 74 km.

A 3.6-to-6.8-micron spectrum recorded by the liquid-helium-cooled CVF spectrometer, also at 74 km on descent, is presented in Figure 5. The CO_2 (001-000) transition at 4.3 microns and the NO fundamental emission, the dominant radiators, were recorded with large signal-to-noise ratios. These emissions were measured by the liquid-nitrogen-cooled spectrometer and interferometer as well as by this instrument. However, the liquid helium CVF spectrometer is unique in that the responsivity at longer wavelengths fully defines the band profile of the NO fundamental emission. A preliminary analysis indicates that vibrational energy transfer from electron-excited N_2 is an important source of the CO_2 emission and that the NO emission results from the chemiluminescent reaction of electron-impact-produced $N(^2D)$ with O_2.

SUMMARY AND CONCLUSIONS

The extensive set of spectra measured in this experiment will be analyzed to determine production mechanisms for each excited state, to determine electron-induced luminous efficiencies and to determine collisional deactivation rate coefficients in the 72 to 128 km altitude range. The spectra also serve as a diagnostic measure of the electron energy distribution within the plasma produced by the electron beam in favorable cases where the emitting species has well-established population and depopulation properties.

The N_2 Wu Benesch system was the dominant electronic transition measured at infrared wavelengths greater than 2 microns. The electron-induced luminous efficiencies for the N_2 Wu Benesch (2,0) and (3,1) transitions, at 3.3 and 3.6 microns, have been assigned a preliminary value in the range $2-5 \times 10^{-4}$. The data indicate this band system is collisionally deactivated at lower descent altitudes (90 km to 70 km).

Excited states with Einstein coefficients of 10^5 sec^{-1} or less are vulnerable to collisional deactivation in the lower-altitude range (70 km to 80 km) of the EXCEDE SPECTRAL measurements. Electronic states in this category included the parent states of the Wu Benesch, the Herman Kaplan, the Lyman Birge Hopfield, the Vegard Kaplan and the first positive systems of N_2 and the Meinel system of N_2^+. The electron-induced luminous efficiencies for these and other band systems will be determined in the complete interpretation of the EXCEDE SPECTRAL data. In addition, these systems will be analyzed to determine either a definitive value or an upper bound for the collisional deactivation rate coefficient of an air-like mixture for each observed vibrational level.

REFERENCES

1. R. R. O'Neil, F. Bien, D. Burt, J. A. Sandock, and A. T.
    Stair, Jr., Summarized results of the artificial auroral

experiment, PRECEDE, J. Geophys. Res., 83 (No. A7): 3273 (1978).

2. R. R. O'Neil, O. Shepherd, W. P. Reidy, J. W. Carpenter, T. N. Davis, D. Newell, J. C. Ulwick, and A. T. Stair, Jr., EXCEDE II Test, An artificial auroral experiment: Ground based optical measurements, J. Geophys. Res., 83 (No. A7): 3281 (1978).

3. R. R. O'Neil, E. T. P. Lee, and E. R. Huppi, Auroral $O(^1S)$ production and loss processes: Ground based measurements of the artificial auroral experiment PRECEDE, J. Geophys. Res., 84 (No. A3): 823 (1979).

4. R. R. O'Neil, E. T. P. Lee, A. T. Stair, Jr., and J. C. Ulwick, EXCEDE II, AFGL-TR-76-0308, Environmental Research Papers, No. 586 (1976).

5. I. L. Kofsky, R. B. Sluder, and D. P. Villanucci, On-board radiometric photography of EXCEDE SPECTRAL's ejected-electron beam, These proceedings (1981).

6. T. G. Slanger and G. Black, $O(^1S)$ Interactions--The product channels, J. Chem. Phys., 68 (No. 3): 989 (1978).

DISCUSSION

Schmerling: To maintain a beam current of 7 amps, 7 amps of return current must flow to the rocket, and there is a tendency for the rocket to charge to a positive potential. How does the return current flow, and to what potential is the rocket charged? I would expect the beam (or beam-plasma-discharge) to spread out so that its projected area perpendicular to the magnetic field can collect enough return electrons from the ionosphere.

Pendleton: The details of the charge-production mechanisms and charge-return paths involved in maintaining the high-power EXCEDE accelerators at relatively low "steady-state" potentials (less than +200 V) have not been established with certainty. Each primary (~3 keV) electron stopped in the atmosphere produces ~100 secondary electrons. Hence, this mechanism probably contributes significantly to the source of negative charges for vehicle neutralization at lower altitudes (≲110 km). However, at high altitudes, where a significant portion of the beam may "escape" to the conjugate hemisphere, this mechanism probably becomes unimportant. Discharge phenomena may contribute, especially at higher altitudes, but this has not been established. The spatial extent of the beam in the EXCEDE SPECTRAL experiments was reasonably well determined by ground-based and on-board cameras and is not grossly different from the expected dimensions.

ONBOARD RADIOMETRIC PHOTOGRAPHY

OF EXCEDE SPECTRAL'S EJECTED-ELECTRON BEAM[*]

I.L. Kofsky, R.B. Sluder, and D.P. Villanucci

PhotoMetrics, Inc.
4 Arrow Drive
Woburn, Mass. 01801 U.S.A.

INTRODUCTION

A radiometrically-calibrated wide angle camera system was installed on EXCEDE SPECTRAL (O'Neil et al., 1981) to determine spatial distributions of energy deposition within and near the fields of view of its UV-visible-IR-sensitive instruments. High signal/noise, on-scale photographic images were obtained over the full range of electron injection altitudes. We describe here the cameras and film data reduction procedure, and present measured column emission rate distributions within a few 10's m from the rocket between 123 and 85 km. These are found to differ considerably from predictions of an independent particle transport model.

The luminosity has sufficient cylindrical symmetry to permit unfolding of volume excitation rates. Spectroscopic data taken from onboard EXCEDE contain further diagnostic information about plasma conditions in the beam region.

CAMERA SYSTEM AND CALIBRATION

Specifications of the two coaligned 16 mm cameras are in Table 1. The monochrome camera is short-pass filtered to measure the electron impact-excited fluorescence from N_2 molecules at wavelengths between 3800 and 4600 Å (FWHM). Exposure is due principally to the 3914 Å (0,0) and 4278 Å (0,1) bands of the N_2^+ First Negative ($B^2\Sigma_u^+ \rightarrow X^2\Sigma_g^+$) system (known to remain closely proportional to column rate of N_2^+

---

[*]*Work supported by the U.S. Defense Nuclear Agency.*

Table 1. Specifications of EXCEDE SPECTRAL'S Photographic Cameras

Cameras:	Two coaligned -- Photo-Sonics 1VN, 16 mm full frame, pin-registered with magazine load.
Film:	Double perforated, 0.010 cm Estar base, 38 m length (125 ft), spooled to Kodak Specification 430.
Lenses:	5.7 mm fℓ, f/1.8 (on optic axis). T/2.2 at 4000 Å on axis, T/2.4 at 20°, T/2.6 at 35°, T/4.5 at 47° from axis.
Fields of view:	62° x 94° on 16 mm film frame (7.6 x 12.0 mm image). Refer to Figure 1 for viewing geometry.
Emulsion types:	Kodak 2475 (monochrome), Kodak EF 7241 (color).
Film processing:	Standard dynamic range 2.7 log units on monochrome film.
Optical filters:	For 2475: -- Schott BG-12, FWHM 3450 Å - 4650 Å For 7241: -- none (quartz window).
Exposure cycle:	3 x 0.008 sec, 4 x 0.078 sec, 1 x 0.80 sec; total duration 1.75 sec nominal. Individually programmed intervalometers for redundancy in case of failure.
Frame Identification:	Camera shutter opening signals on telemetry channel.
Start time:	84 sec after rocket launch, 78 km altitude.

ionizations down to ~ 40 eV incident electron energy), and secondarily to N_2 Second Positive ($C^3\pi_u \rightarrow B^3\pi_g$) bands. The blue filter attenuates delayed, potentially smearing radiation from chemiluminescence and metastable species, in particular (by a factor 10^3) the $O^1S \rightarrow {}^1D$ line at 5577 Å. The fast film was developed to a moderate contrast to provide a useful exposure range of about 500 in individual images.

The second camera records on three-color film, principally to assess the contribution to the glow radiances from Coulomb-scattered (and possibly other) soft electrons and also, since it is about a factor 10 less sensitive, to increase the photographic system's dynamic range. Secondary electrons selectively excite the triplet-manifold transitions of N_2: First Positive ($B^3\pi_g \rightarrow A^3\Sigma_u^+$) bands mainly in the red and near IR, Second Positive bands in the blue-violet-near UV, and other features outside the camera's wavelength sensitivity range whose intensities were measured from EXCEDE SPECTRAL. The color positives above ~ 80 km indeed show the expected magenta (blue + red) glow extending outward from a bluish-white (which is overexposed blue) "core" along the expected beam trajectory.

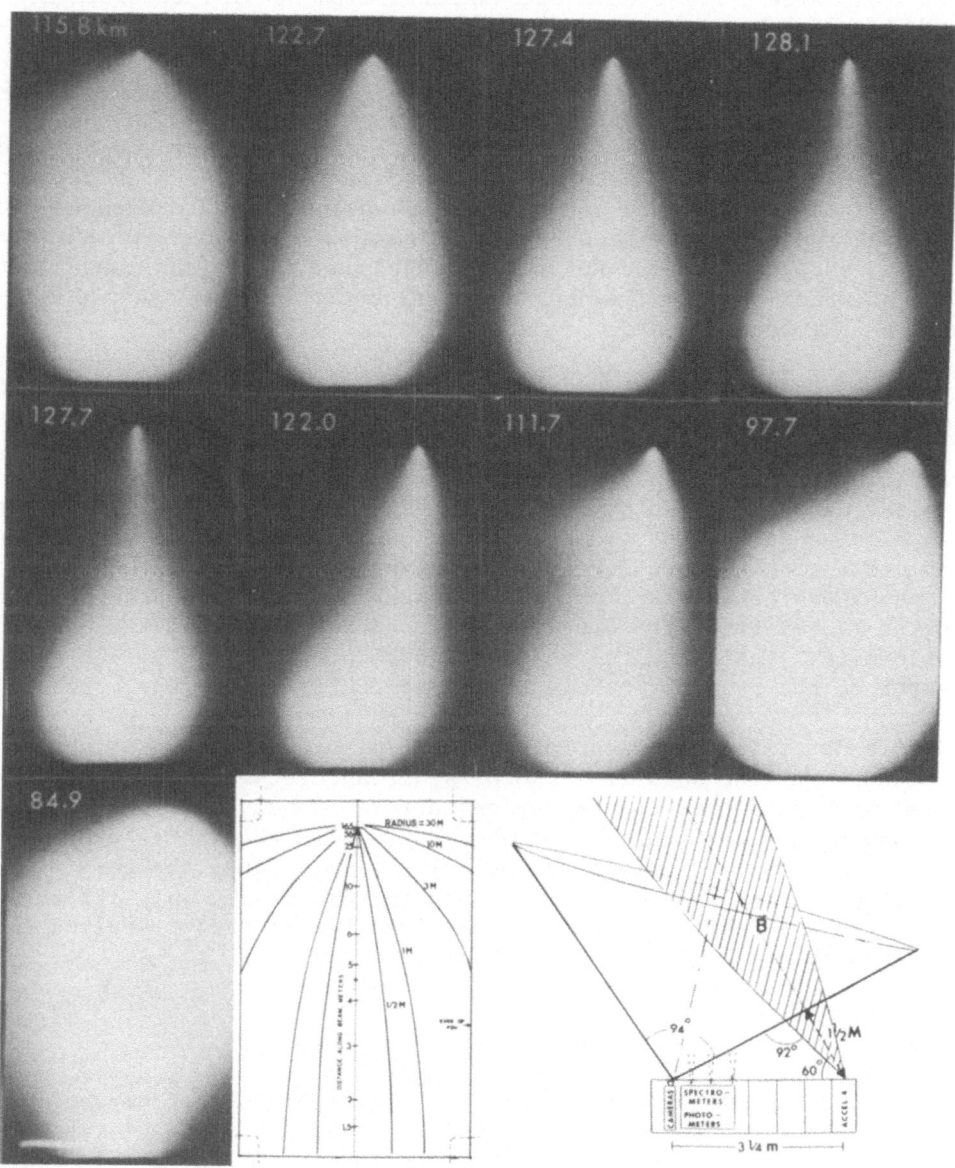

Fig. 1. Typical 3800–4600Å 0.08 sec photographs from onboard
EXCEDE SPECTRAL. The inserts show the projection geometry
(refer to text) and the position of the electron accelerator
relative to the cameras and other optical instruments.

Both wide angular coverage (which limits spatial resolution) and high light gathering power are needed to image the airglows effect-ively. All fast, moderate-distortion lenses have decreasing aperture ratio away from the optic axis (vignetting), which makes the original film density distributions qualitatively different from the scene radiance distributions. A correction was also made for geometric dis-tortion of the lens, a second-order effect. A program of alternate short, intermediate, and long exposures was designed to keep on-scale the expected extremely broad range of maximum brightness of the air fluorescence between 128-km rocket apogee and 72 km (the lowest alti-tude at which the accelerator operated) and the large dynamic range within individual scenes.

The procedures for data access (Kofsky, 1967a) and radiometric calibration of the black-and-white film and lenses have been commonly applied to photometry of other upper-atmospheric glows, such as those resulting from chemical releases (Golomb and MacLeod, 1966) and nuclear explosions (Kofsky, 1967b). (Absolute radiance distributions can also be determined from the three-color film.) Lens vignetting and distortion were measured, film calibration strips were prepared at sensitometer wavelengths and exposure times close to those on the rocket (to minimize transfer errors), and the film was developed for uniformity. The estimated radiometric accuracy is +60%, -40% in the center of the frames, increasing to ± a factor 2 near the corners.

To determine the exposing spectral distribution, we weighted altitude-averaged data from EXCEDE's ultraviolet-visible spectro-meter with data from the (E-region) aurora (Jones, 1971). The effect-ive wavelength was found to be about 4000 Å in the "core" and 3900 Å in the magenta-toned areas of the color film. Since the camera's sensitivity to monochromatic illumination differs by only 12% between these two wavelengths (Eastman Kodak specification), we applied the same absolute calibration to both image areas.

VIEWING PERSPECTIVE, DERIVATION OF ISOPHOTE CONTOUR PLOTS

The data presented here refer to a 7 ampere beam from accelerator section #4; refer to the insert at lower right of Figure 1, which also illustrates EXCEDE's spectroradiometers' pointing. 3 keV electrons were emitted in a 30° full angle cone (as measured in a laboratory tank) with axis maintained parallel to the geomagnetic field (upward) on upleg and about 4° southwest of the field direction on downleg, and 60° from the rocket's long axis. Each camera viewed toward the beam from a point in (on upleg) or 21° east of (on downleg) the meri-dian plane 3¼ m north of the exit port. The injection field line lies within the image area down to 1½ m from this anode.

The representative images in Figure 1 (each has identical ex-posure and printing) are not the familiar far-field, "stand-off"

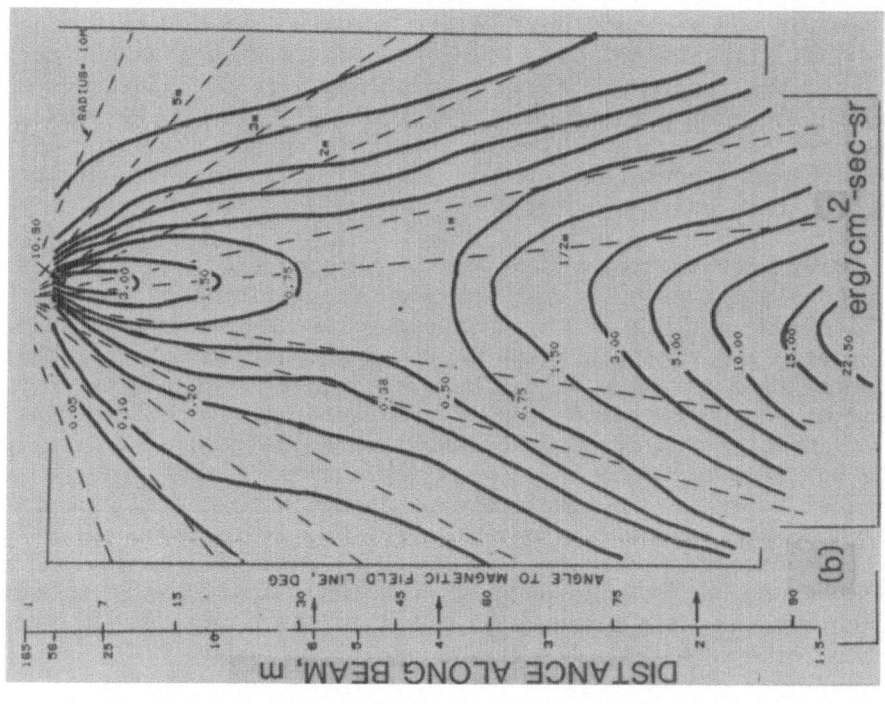

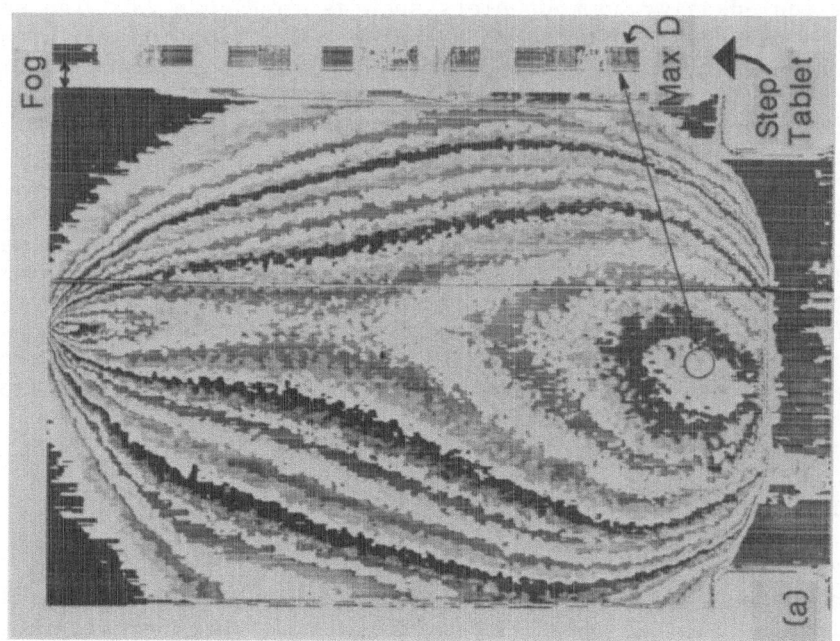

Fig. 2. Contour plots of film density (a) and glow radiance (b) from the 0.08 sec exposure at 122.7km on upleg of EXCEDE SPECTRAL. (b) appears larger because of the correction for lens distortion.

views, and furthermore are not corrected for vignetting. Irradiances
at the film plane are as usual two-dimensional projections to the
camera lens of three-dimensional volume emission rate distributions
in these optically-thin glows. Figures 1 and 2b show distance from
the anode to the intercept on the beam's long axis of columns extending from the lens as a function of vertical position on the image,
the distance scale perpendicular to this axis in the plane normal to
these intercepts (curved lines), and angles to the magnetic field
line through this axis.

The rebrightening toward the top of the images is interpreted as
being due the sight path through the glow increasing more rapidly
(due to decrease of this angle) than the volume emission rate is decreasing; note its narrowing with increasing rocket altitude. Maximum expected Larmor radius of the beam electrons is just under 0.9 m,
and their crossover distance is 21 m. No evidence of this refocusing
node is evident in the frames microdensitometrically analyzed to date,
with the possible exception of a weak inflection in the radiance in
the frame at 97.7 km. No discontinuous change with altitude in the
appearance of the images is apparent. The beam's increased slope to
the right on downleg is due to a change in azimuth of the payload's
long axis, which was rotated under ground control to line it up with
the trajectory and thus increase EXCEDE's infrared spectroradiometers'
sight path through previously-irradiated air.

Both the film and calibration tablet were scanned with an Iso-
densitracer® (Kofsky, 1967b; Miller, Parsons, and Kofsky, 1964) to
produce equi-density contour plots such as in Figure 2a. The set of
exposures was then converted to a radiance distribution using the
position-dependent lens aperture ratios. The outer region of image
where distortion is present was then re-contoured to make horizontal
displacements from the image center proportional to angles from the
optic axis in object space.

Figure 2b is the contour plot of 3800-4600 Å radiance projected
to the onboard camera at 123 km upleg altitude. The essentially-
cylindrical symmetry of the glows allows 3-dimensional volume emission rate distributions -- the more fundamental physical quantity --
to be unambiguously unfolded from the 2-dimensional projections.
Data from a 4°-field 3914 Å photoelectric photometer pointing toward
the beam from ½ m closer to the exit port (see Figure 1) are closely
consistent with the mid-field photographic radiances.

DIMENSIONS AND RADIANCE DISTRIBUTIONS

The equi-radiance contour lines are open in the bottom half of
the image area (Figure 2b is an example); they would close outside
the camera's field of view. In the negatives, in contrast, they
appear to close near the edge of the frame. Brightness of the 123 km

glow increases toward the rocket body not only along its field-aligned axis, but at all positions in the image within ∿ 4 m from the anode.

Figure 3 plots radiance along the long axis of the beam volume at four altitudes spanning an ambient air density range of 500 (from contour maps such as Figure 2b), and Figure 4 shows transverse cuts at 123 km. At the two higher altitudes the radiance viewing perpendicular to this axis (i.e., in an orthographic projection) appears nearly constant between 4 and ∿ 40 m, beyond which foreshortening severely degrades the spatial resolution. The "radius" of the emitting volume at 123 km, determined by extrapolating to zero the nearly straight sections of transverse radiance plots (such as in Figure 4) becomes ∿ 2¼ m beyond 4 m. The effective full angle of the ejected-beam region is thus nearer 60° than 30°, and the glow's width shortly grows to over twice the expected Larmor radius.

The large brightness increase within ∿ 3 rocket diameters is qualitatively similar at all four ejection altitudes. The onboard photometer data (Figure 3) also show the small change of midfield radiance over a wide range of ambient air densities, as well as a

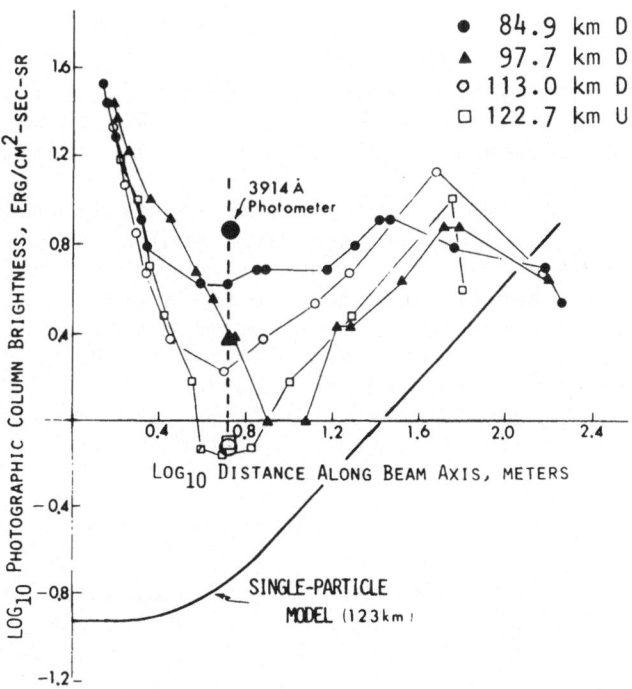

Fig. 3. Radiance projected to the black-and-white camera along the injected electron beam's long axis at four rocket altitudes. Data from the narrow-field 3914Å photometer and individual-particle model calculations for 123 km are also shown.

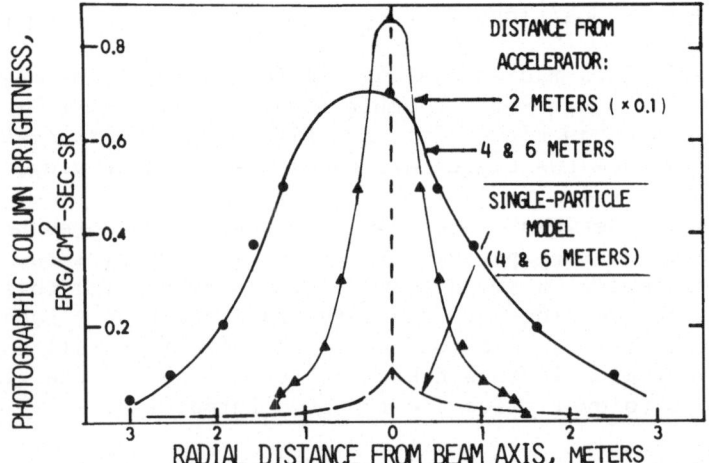

Fig. 4. Radiance projected to EXCEDE SPECTRAL'S camera along lines
         perpendicular to the injected electron beam's long axis
         (see arrows on Figure 2b). Rocket altitude 122.7 km upleg.

factor 2-10 higher signal on upleg than on downleg. This is indicative of outgassing from the rocket, which is also shown by strong
H_2O emission and quenching of O^1S (O'Neil et al., 1981).

     Radiances predicted by an independent-particle transport model
(Archer, 1980), which considers the usual Coulomb scattering, geomagnetic confinement, and impact ionization-excitation (of secondaries as well as energy-degraded primary electrons), are included
in Figures 3 and 4. The predictions for injection at 123 km would
scale with ambient density down to $\sim$ 94 km altitude. The model
underestimates the measured radiances beyond about $\sim$ 4 m at the two
higher altitudes, overestimates them below $\sim$ 113 km, and in particular
does not predict the observed rapid increase close-in to the rocket.

DISCUSSION, RELATION TO OTHER OPTICAL DIAGNOSTIC DATA

     Many other physical phenomena inconsistent with simple independ-
ent-particle transport in the ambient ionosphere, most of them associated with suprathermal electrons thought to play a part in neutralizing the vehicle and beam, have been reported (as reviewed by
Winckler, 1980 and Linson and Papadopoulos, 1980). These first radiometric images from onboard show a largely altitude-independent close-
in glow, whose volume emission rates permit the fluxes of $\gtrsim$ 20 eV
electrons to be estimated (refer to Figure 5). The radiance plots

thus represent a major source of information on the space-charge spreading, return-current, instability-generation, and other processes relating to initial injection and charge neutralization. Further analysis of the photographic data from a series of altitudes will clarify the roles played in exciting the air fluorescence and determining particle flux and energy distributions by outgassing from the rocket, generation of secondaries by impact of backflowing electrons on its surface and air molecules, and beam-plasma interactions. (We note that extrapolation of the laboratory tank discharge data of Bernstein et al. (1979) with energy absorption length parameter taken as 100 m would indicate that ignition threshold is ∿ 90 km altitude for EXCEDE SPECTRAL's current-voltage conditions.)

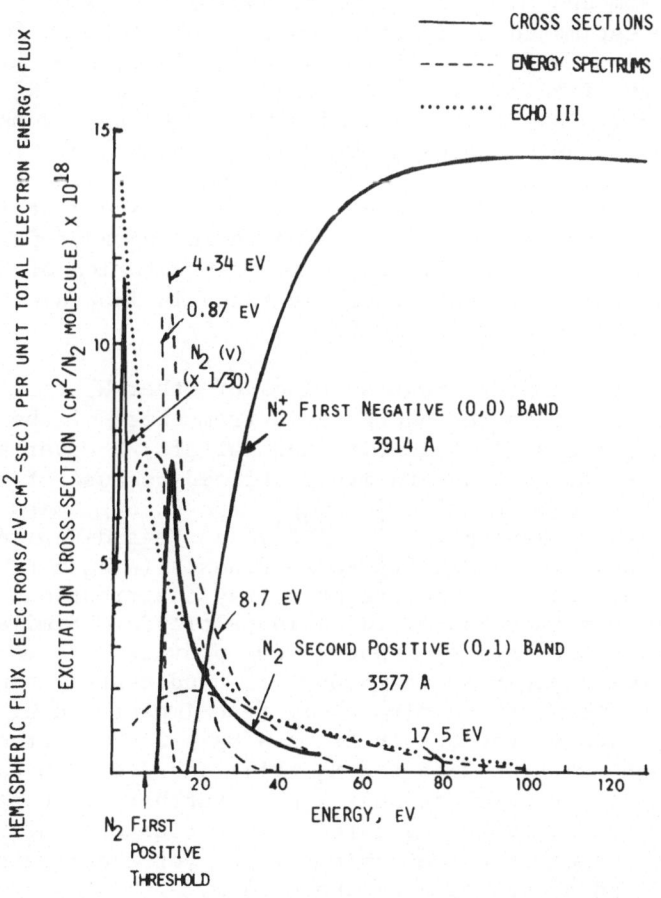

Fig. 5.   Excitation cross-sections of N_2, and electron energy flux spectrums at four temperatures and as measured from Echo III. The optical cross-sections for the other features of the N_2 triplet sequence have a steep rise and sharp falloff similar to that shown for the Second Positive (0,1) band.

The projections to long-focus photographic and video cameras at ground stations, which provide information at a $\sim$ 30 m structure scale, complement the onboard-camera data. A very preliminary analysis of these images shows no obvious anomalies in the beam's long-distance propagation, and a small (and possibly unresolved) high-brightness region surrounding the vehicle, qualitatively similar to that seen at Zarnitsa I (as reported in Winckler, 1980).

The intensities of molecular emissions measured from EXCEDE provide further critical information about plasma conditions in and near the ejected beam. From the relative and absolute radiances in features whose excitation cross-sections and radiative-collisional quenching parameters are known, the energy distribution and (to the extent that atmospheric densities are known) fluxes of $\leq$ 100 eV electrons can be estimated. Figure 5 shows cross-sections for impact excitation of two well spectrally isolated, intense — i.e., measurable with high signal/noise — N_2 Second Positive and N_2^+ First Negative bands, along with electron number flux spectrums at four temperatures and as measured on Echo III (Arnoldy and Winckler, 1981). The predicted ratio N_2 3577 Å/N_2^+ 3914 Å is 2.8 at 4.34 eV temperature, falls to 0.19 at 20 eV, and is 0.22 for Echo III's distribution. The ratio within a region traversed by energetic degraded-primary electrons would of course be lowered by their excitation of N_2^+ 3914 Å. EXCEDE'S spectrums give this ratio averaged in instrument fields of view through the beam-excited volume.

The electron impact cross-sections of other N_2 triplet-manifold features are offset on the energy axis from those of the Second Positive, and show the same rapid rise and falloff; however actual excitation rates are not as quantitatively defined because of complexities introduced by cascading and quenching. The strong emission in the ultraviolet Herman-Kaplan ($E^3\Sigma_g^+ \to A^3\Sigma_u^+$) and Vegard-Kaplan ($A^3\Sigma_u^+ \to X^1\Sigma_g^+$), First Positive, and infrared Wu-Benesch ($W^3\Delta_u \to B^3\pi_g$) sequences will provide an improved measure of energy distributions of $\sim$ 10-100 eV electrons as a function of injection altitude. Similarly, the infrared radiations from atmospheric CO_2 provide information on fluxes at lower energies. Its 4.3μm (ν_3 fundamental) band's upper state is populated both by direct electron impact and by collisions with N_2 molecules vibrationally excited by $\sim$ 2 eV electrons (Figure 5 shows the high cross-section for N_2 (v)); and its 15μm band is enhanced by a process resonant near 4 eV. Further diagnostic information may be present in the measured intensities of other EXCEDE features whose excitation mechanism is not yet understood, for example those of H_2O and at wavelengths above 16μm.

Spatial distributions of emission in these band systems can be assessed with the aid of unfolds from EXCEDE'S radiometrically-calibrated photographs. The camera data thus serve both to normalize the infrared emission intensities to the energy input by electrons with sufficient energy to excite visible air fluorescence and to one another (as each sensor's pointing is somewhat different from the others'), and in deriving volume excitation rates by suprathermal electrons.

The authors express their thanks to the staff of the Defense Nuclear Agency (RAAE Division) and Air Force Geophysics Laboratories (OPR Branch) for their support, and to Dr. M. Chamberlain for emission band intensity ratio calculations.

REFERENCES

Archer, D.H., 1980, "EXCEDE Energy Deposition: Theory and Experiment Compared", Final report on U.S. DNA contract 001-79-C-0016, Mission Research Corporation, Santa Barbara, California.

Arnoldy, R.L., and Winckler, J.R., 1981, The hot plasma environment and floating potentials of an electron beam-emitting rocket in the ionosphere, J. Geophys. Res., 86:575.

Bernstein, W., Leinbach, H., Kellogg, P.J., Monson, S.J., and Hallinan, T., 1979, Further laboratory measurements of the beam-plasma discharge, J. Geophys. Res., 84:7271.

Golomb, D., and MacLeod, M.A., 1966, Diffusion coefficients in the upper atmosphere from chemiluminous trails, J. Geophys. Res., 71:2299.

Jones, A.V., 1971, Auroral spectroscopy, Space Sci. Rev's. 11:776.

Kofsky, I.L., 1967a, Reduction of pictorial data by microdensitometry, in: "Aerospace Measurement Techniques", G.G. Mannella, ed., NASA SP-132, Washington, D.C.

Kofsky, I.L., 1967b, Clarification of airglow processes by nuclear excitation, in: "Aurora and Airglow", B.M. McCormac, ed., Reinhold, New York.

Linson, L.M., and Papadopoulos, K., 1980, "Review of the Status of Theory and Experiment for Injection of Energetic Electron Beams in Space", LAPS 65, Science Applications, Inc., LaJolla, California.

Miller, C.S., Parsons, F.G., and Kofsky, I.L., 1964, Simplified two-dimensional microdensitometry, Nature, 202:1196.

O'Neil, R.R., Stair, A.T. Jr., Pendleton, W.R. Jr., and Burt, D., 1981, The EXCEDE SPECTRAL artificial auroral experiment: an overview, these Proceedings.

Winckler, J.R., 1980, The application of artificial electrons to magnetospheric research, Rev's. Geophys. Space Phys., 18:659.

DISCUSSION

<u>Hallinan</u>:   There are three observations which imply that something similar to BPD did occur in the EXCEDE SPECTRAL flight.

1) Weak pulses, produced when the accelerator failed to operate at rated current, were seen with our television camera aboard the payload. Some, but not all of these pulses had the noded configuration indicative of single-particle trajectories. Kofsky reported that photographic images of the full-power pulses did not show the nodes. This suggests that the beam was heated within a short distance from the accelerator.

2) We have determined from ground-based television cameras that the total luminosity within 100 m of the payload (our limiting resolution) was at least four times the value expected on the basis of a smooth extrapolation of the luminosity further from the rocket.

3) Telemetry AGC (automatic gain control) records show that there were abrupt changes in the power received on the ground corresponding to some of the accelerator pulses. In some cases the signal was enhanced. In other cases it was attenuated. When two guns had partially overlapping pulses, the combined effect of the two guns was not the same as the sum of the effects of the individual guns. The most plausible interpretation is that the antenna pattern of the rocket was altered by local reflections from plasma produced by the pulses. To reflect a 2.2 GHz signal, a plasma density of at least 5×10^{11} cm^{-3} is required. One of the most obvious examples of the AGC level changes occurred when the rocket was at an altitude of 87 km on the downleg. The output of the accelerator was 21 amps at 3 kV. Over a 3 meter diameter, this should produce approximately 10^{13} ions cm^{-3} s^{-1} by direct collisions. If the only loss is recombination, the equilibrium density is 6×10^{9} cm^{-3}. This is two orders of magnitude below the level required for reflection of the telemetry.

GENERAL DISCUSSION ON

ACCELERATOR EXPERIMENTS IN SPACE

Chairman: J R Winckler

There is evidence of the beam plasma discharge (BPD) in a number of rocket experiments: E-parallel-B, Polar 5, Zarnitza, Electron 2 and ARAKS, but the BPD is not observed in the Echo experiments. By ordering the experiments according to the perveance of the electron accelerator employed (see Winckler, Table II) a threshold perveance for BPD in space of 1.2×10^{-8} amps/volts $^{3/2}$ is obtained.

Are the BPD-like processes observed in space the same as the BPD observed in the laboratory?

First, the determination of the presence of the BPD depends critically upon the diagnostic evidence one uses as the signature of the BPD.

Second, the condition for BPD may be expressed as follows:

$$K > 10^{-4} \frac{n_{beam}}{n\ plasma} A\ V^{-3/2}$$

where K is the perveance of the electron gun, n_{beam} and n_{plasma} are the electron number densities in the beam and the plasma, respectively. Furthermore, the electron density in the plasma may depend upon beam induced ionization. In space the determination of the beam electron density is difficult. There are few observations of the beam width, but the Larmor diameter of the beam electrons is typically taken as a measure of the beam width.

It was pointed out that in the laboratory multiple reflections may be important for the occurrence of the BPD.

There are a number of curious observations in the space experiments:

The drop-out of the telemetry signal observed in the Zarnitza experiment during part of the spin cycle.

Higher luminosity than expected in the vicinity of the rocket as observed in the E-parallel-B and Polar 5 experiments.

In Echo V there was no evidence for reflections of the electron beam from the magnetic conjugate point. However, in the ARAKS experiment for beam injection nearly parallel to the geomagnetic field, the electron beam never seemed to have entered the loss cone at the conjugate point. Thus the beam must have suffered large pitch angle scattering.

How do we explain these observations? What happened to the ARAKS beam?

The loss mechanisms may differ and each case must be studied separately. While the magnetosphere will have affected the ARAKS beam, it seems more likely that the major changes took place in the ionosphere as the observed radiation during beam injection indicated the presence of large electron density enhancements. Furthermore, the BPD, if present, could cause serve pitch angle scattering.

There are, however, accelerator experiments in space where no anomaly was observed. In the second experiment of Hess the beam arrived at the conjugate point and both arrival time and brightness seemed to fit with theoretical estimates.

The discharge region around the accelerator rockets may be characterized by a size and a lifetime. Using the observations in the Electron 2 experiment, the lifetime of the disturbed region was estimated to much larger than 50 ms for beam injections lasting up to 18 ms. It was suggested that after the end of beam injection, the thermal electrons created by the beam will return to the initial hot spot, thus prolonging the duration of the disturbance.

Finally, the question of whether the enhanced secondary electron production can be explained by non-collisional processes was addressed.

The proposed mechanism consists of a selective acceleration of thermal electrons which may yield a population of superthermal electrons which in turn give rise to enhanced secondary electron production. The acceleration takes place as follows: Very large plasma waves, created by the beam, decay in localized regions. The associated electric fields will in these regions give rise to "transit time" acceleration of ambient charged particles. These particles will thus receive a kick one way or the other.

The observed density of the secondary electrons, possibly produced by the above mechanism, is much higher than that of the primary beam electrons, thus these secondary electrons cannot be thermalized beam electrons.

Chapter 2: NATURAL BEAM PLASMA INTERACTIONS IN
           SPACE

OBSERVATIONS OF NON-LINEAR PROCESSES IN THE IONOSPHERE

A. D. Johnstone

Mullard Space Science Laboratory
University College London
Department of Physics and Astronomy
Holmbury St. Mary
Dorking, Surrey
England

1. INTRODUCTION

The purpose of this paper is to review observations of auroral electron distributions in the light of the physics of beam-plasma interactions. I will do this by relating the discussion to a particular problem in auroral physics which I believe can only be solved with reference to such processes. The question is - how are electrons accelerated to form discrete auroral arcs? It is usually accepted that they are accelerated by an electric field parallel to the magnetic field. Some of the earliest evidence for this view came from measurements of electron spectra which contained sharp peaks at energies in the kiloelectron volt range and pitch-angle distributions that were strongly field-aligned. (Albert 1967, Evans 1968, Westerlund 1969). An example is shown figure 1 (Arnoldy et al 1974). These observations raised a number of problems. The first problem to receive attention was to understand how electric potential differences of kilovolts could be maintained along magnetic field lines. It is now known that several processes might contribute, e.g. the magnetic mirror force, anomalous resistivity, formation of double layers, electrostatic shocks, although it is not known which is the most important. The second problem, currently being studied, is to find out how the potential differences are created and what the potential distribution is above an auroral form (Lyons 1980 etc.). There is now a third problem which concerns those who have been making measurements of auroral electron velocity distributions. Despite the fact that much of the initial

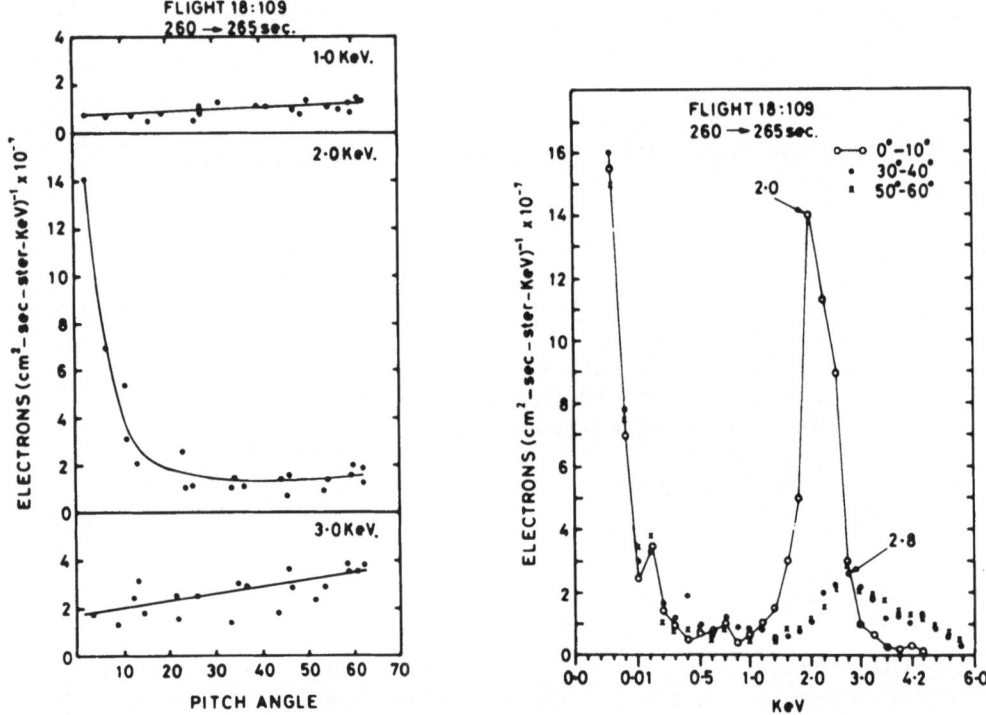

Fig. 1. Energy spectra and pitch angle distributions measured by a sounding rocket in a discrete auroral arc showing an intense, field-aligned energy peak, and a weaker, isotropic energy peak (Arnoldy et al., 1974).

impetus for the idea of a parallel electric fields came from measurements of auroral electrons it is now clear that the observations do not agree with what is calculated by modelling the acceleration of plasma sheet electrons by a parallel electric field. One response to this question has been to argue that wave-particle interactions will resolve the differences. If that is so, then it ought to be critically and quantitatively evaluated. This is an attempt to make that evaluation.

2. MODEL DISTRIBUTION

To begin with, it is worth briefly reviewing the capabilities of models of accelerated distributions. The calculation involves tracing trajectories backwards in time through the combined electric and magnetic fields to a source distribution, and then using Liouville's theorem to obtain the density in the model distribution from the source density (Kaufmann et al. 1976,

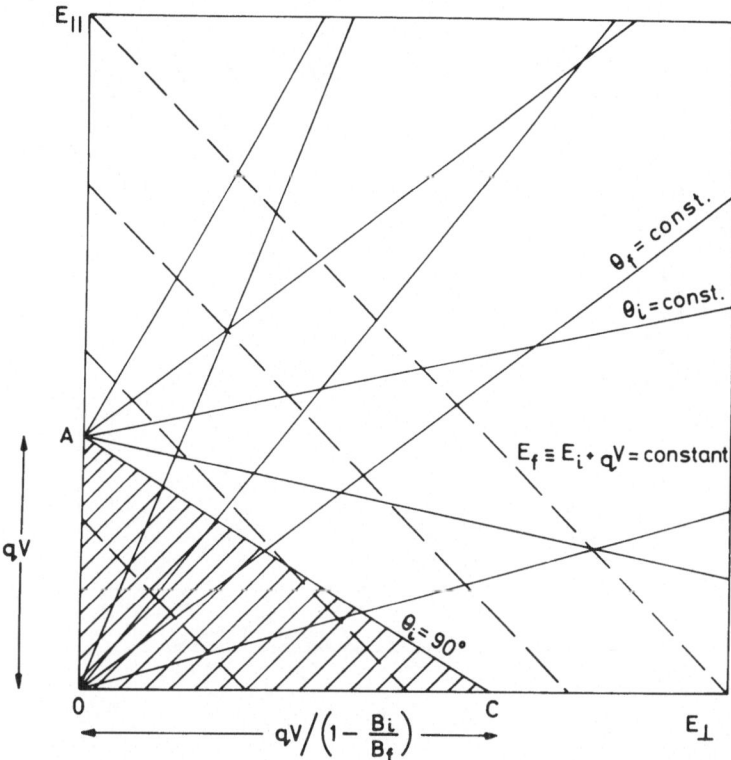

Fig. 2. The parallel electric field divides the distribution of
precipitated electrons into two parts. Electrons in the
low energy region shown shaded can be traced back to the
atmosphere. Those in the higher energy region come from
the plasma sheet. Lines marked θ_i = constant are lines
of constant pitch angle in the original distribution.
Cold electrons E_i << qV in the original distribution are
found near point A in the precipitated distribution.

Whipple 1977, Johnstone 1978a). The only new feature which the
electric field introduces into this pattern is to divide the
velocity space of the model distribution into two source regions
(Figure 2). Electrons on the high energy side of the boundary
have come from the plasma sheet and been accelerated through the
full potential drop. Those on the low energy side originate in
the atmosphere and have travelled up the field line and have then
been reflected downwards by the electric field. They have not
been accelerated. In general, the density in the two
distributions will not be identical at the boundary and therefore
there will be a discontinuity. Since the density of accelerated
electrons on the high energy side is usually greatest, and the

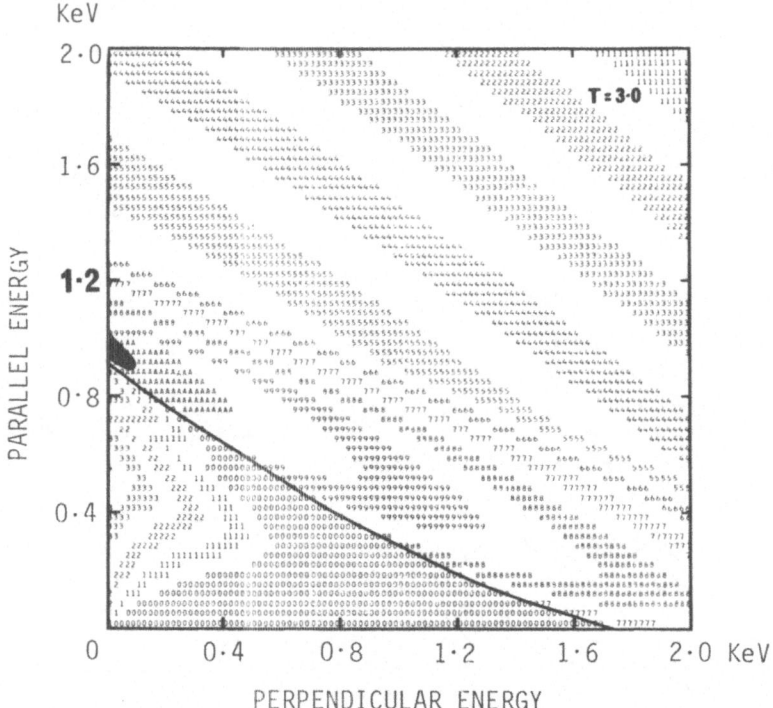

Fig. 3. A contour plot of a model of an auroral distribution accelerated by a parallel electric field. The characters in the contour indicate the electron density in velocity space ranging from 0 to 9 then A to J. A black line is drawn along the boundary between the two source regions where there is a discontinuity in density. The black shaded region near E parallel = 1 keV is produced by the acceleration of electrons from a low temperature distribution in the plasma sheet.

boundary itself does not lie along a line of constant pitch angle, then an energy peak and a field-aligned distribution will result. For any quantitative model some source distributions must be put in. Most models include

(a) a hot Maxwellian distribution representative of the plasma sheet which is found on the high energy, or accelerated side of the boundary

(b) a distribution backscattered by the atmosphere from the precipitation of the energetic particles found on the low energy side.

The model shown in Figure 3 includes a third distribution which is

(c) a cool distribution on the plasma sheet side presumed to be of ionospheric origin but which has drifted outward along the field line.

This third distribution is necessary to give the intense, field-aligned energy peak in Figure 1, in addition to the broad, near isotropic energy peak which comes from the acceleration of the plasma sheet distribution.

The important point to notice here is that it is possible to construct different distributions by making suitable assumptions about the source characteristics. Many discrepancies between observed electron distributions and model distributions have been itemised by various authors (O'Brien 1970, Whalen and Daly 1979, Bryant 1980) and some of these could perhaps be explained by variations in the source distributions or by time variations in the electric field. The only specific test for a parallel electric field (Kaufmann et al 1976) is to identify the existence of a discontinuity. Of course, none has been found because it is obviously unlikely to exist as such in any real physical situation. The discontinuity provides the free energy to drive the beam-plasma interactions which then modify its shape and possibly makes other changes in the distribution. It is these changes which are the subject of this paper.

3. APPROACH

I am not attempting a theoretical review of beam-plasma interactions, nor their application to the aurora. I intend to survey the phenomenology of the observed distributions to identify those features which could be the result of beam-plasma interactions.

Where might wave-particle interactions be important in shaping the distribution? First of all they will play a major role in determining the shape of the positive slope where the discontinuity might originally have been found. Second, they are likely to create suprathermal electrons from the ambient plasma which then act as a source for the distribution on the low energy side of the boundary. Thirdly, some electrons may be accelerated to energies beyond the peak.

Before considering wave particle interactions it is necessary to take account of the effects of atmospheric collisions which

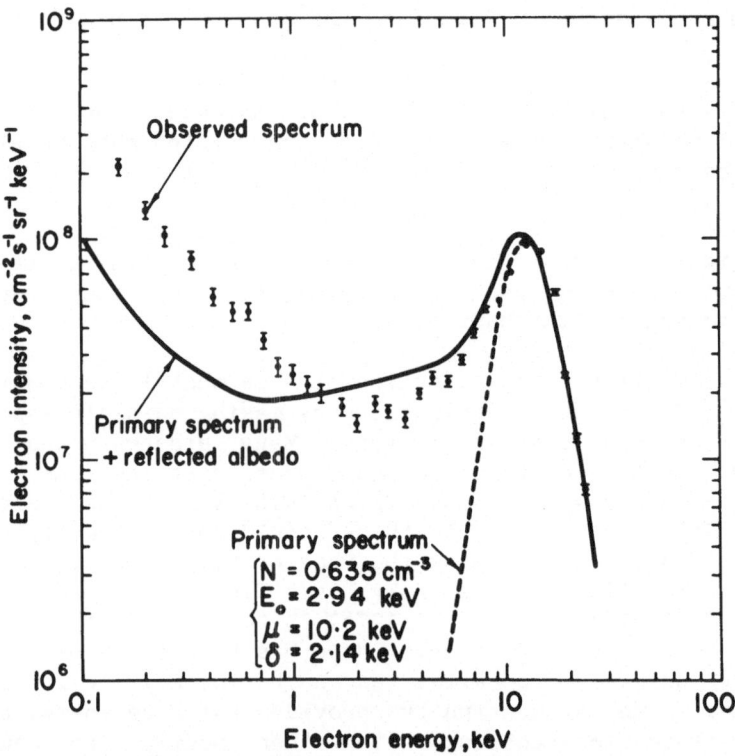

Fig. 4. The primary spectrum is modelled by a Maxwellian
          distribution, density N, temperature Eo, accelerated by
          an electric potential with mean value μ and fluctuations
          δ. The thick black line includes the calculated effect
          of backscattered atmospheric secondaries reflected
          downward by the electric field. The observed spectrum
          exceeds the calculated atmospheric values by a factor of
          3 at a low energies (Bryant et al., 1978).

contribute to the source distribution through backscattering and
modify the precipitating distribution as it descends the field
line.

4.   ATMOSPHERIC EFFECTS

The first question to answer is to determine whether
atmospheric backscattering could be the only source for electrons
in the low energy region of Figure 2. There are a number of
observational tests one can apply.

(a) The low energy electron fluxes below the energy peak which identifies the accelerating potential should be equal to the backscattered flux from the entire precipitating distribution. The backscattered intensity is given by

$$I_b \, (E,\alpha) \sin\alpha = \int_E^\infty \int_0^{\frac{\pi}{2}} 2\pi \sin\alpha' \, I \, (E',\alpha') \, K \, (E,\alpha,E',\alpha') \, d\alpha' dE'$$

where $K(E,a,E',a')$ is the probability that an electron with primary energy E', and pitch angle a' will be backscattered with E,a. (Evans 1974, Johnstone 1978a). This backscatter probability must include both the backscatter of the primary electron with degraded energy and the backscatter of secondary electrons created in ionising collisions (Banks et al 1974). Backscatter probabilities have been derived from Monte Carlo calculations of electron precipitation into the atmosphere and can be used to calculate the backscattered fluxes to be expected from any observed precipitating electron distribution. Such calculations have been performed several times. Sometimes a reasonable fit is obtained (Evans 1974) but in general it is found that the low energy intensities are higher than can be accounted for by simple backscattering (Bryant et al 1978, Matthews et al 1976). The results of one of these calculations is shown in Fig.4. Furthermore, the spectrum at low energies varies as $I(E) \propto E^{-r}$ where r <1.5 while the backscatter spectrum, also in power law form, should be steeper, with r = 2 to 3.

A further indication that the suprathermal electron distribution is not atmospheric is demonstrated by diagrams which show that the intensity at low energies does not correlate well with the intensity measured near the peak (Whalen & Daly 1979, Hall 1980). In fact, this lack of correlation may be the first positive evidence for a parallel electric field since it suggests that the low and high energy parts of the distribution are coming from separate sources.

(b) The second test is a test of the reflection hypothesis and is that the downgoing intensity should equal the upgoing intensity at the corresponding pitch angle

i.e. $I(E,a) = I(E, 180-a)$

This should be true for all E below the peak energy and all pitch angles. Figs. 5 and 6 show some typical auroral spectra obtained in discrete aurora. Each shows a peak in the energy spectrum. At energies below the minimum in the intensity i.e. 20 eV < E < 500 eV the angular distribution decreases monotonically from 0° to 180° (Fig.7) although the overall decrease is only a factor of 1.5. Between the minimum and the peak the upgoing intensity is significantly lower than the downcoming. The upgoing

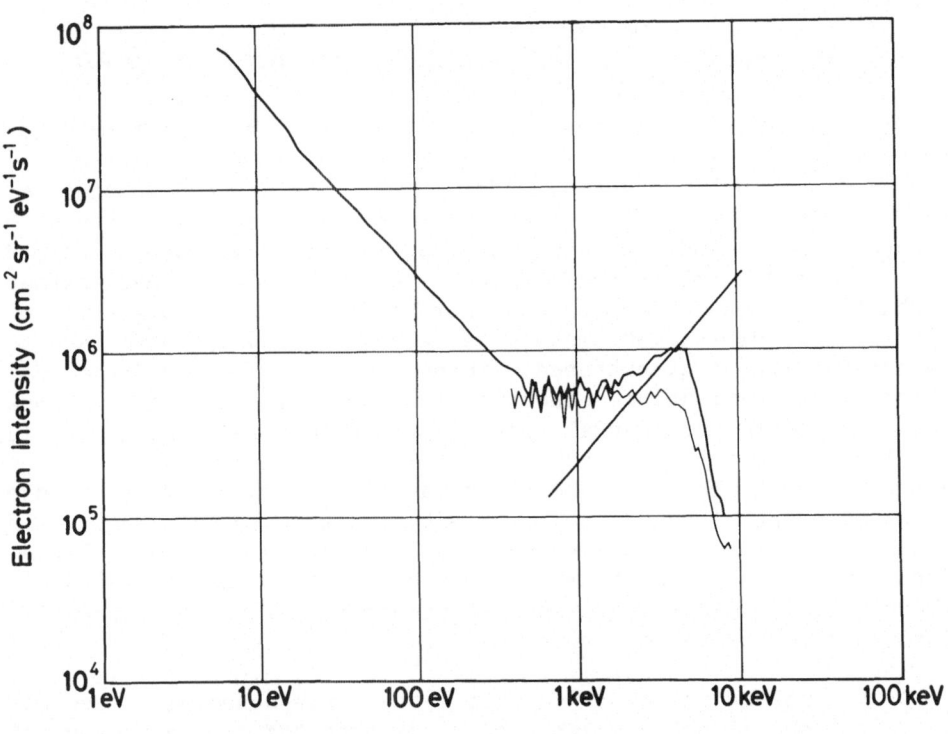

Electron Energy

Fig. 5.   The electron spectrum measured by Terrier Malemute F38
on Nov 27, 1976 launched from Andoya Norway into a
discrete auroral arc.   Data for E > 480 eV are obtained
from   sensors   provided   by the Appleton Laboratory
(courtesy D. A. Bryant, D.S.   Hall).   Data from a
sensor directed at 30° to the spin axis of the rocket
are plotted as a thick line; those from a sensor at
135° (backscattered) as a thin line. The fluctuations
between 400 eV and 2 keV are statistical. Below 500 eV
data come from a sensor provided by Mullard Space
Science Laboratory directed at 15° to the spin axis.
This sensor had a much greater geometric factor and
hence statistical fluctuations are negligible.   The line
at an angle of 45° to the axes through the peak
indicates the slope that would correspond to a plateau
in the distribution function.   This spectrum gives a
monotonically decreasing distribution function.   The
altitude of the rocket was 510 kms.

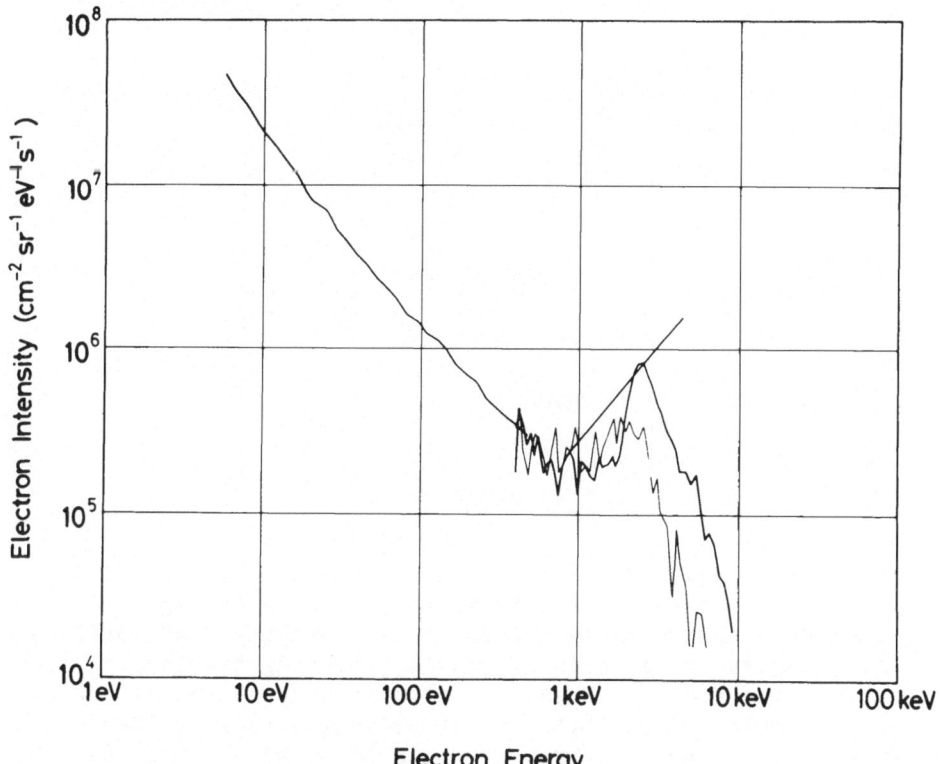

Electron Energy

Fig. 6. This electron spectrum was measured on the same flight
at figure 6, 100 secs later at an altitude of 420 kms.
Data from the same sensors as fig 5 are shown here in
the same way. This spectrum gives a positive slope in
the distribution function.

and downcoming spectra above 500 eV have been drawn separately in
Figs. 5 and 6 to demonstrate this.

(c) It is important to apply the third test before attempting
an interpretation of the results described above. If there is no
electric field to reflect the backscattered electrons then one
should not expect to observe the same distribution. The same low
energy distribution as found in discrete aurora is found in
diffuse aurora where there is no evidence for a parallel electric
field. Below approximately 500 eV the spectrum has the same power
law form with the same exponent i.e. $r < 1.5$ with a nearly

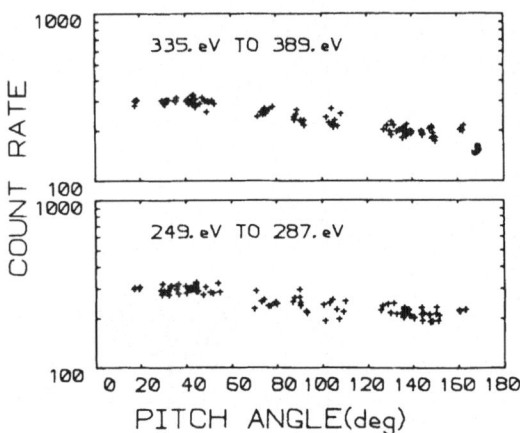

Fig. 7. Pitch angle distributions compiled over a period of 3
         secs at the same time as the spectrum in Figure 5. The
         data are assembled from four detectors at different
         angles on a spinning rocket, covering overlapping pitch
         angle ranges. These distributions are typical of the
         angular distributions at all energies from 20eV to 450
         eV in the spectra of both figures 5 and 6. The
         distribution is flat from 10° to 40°, decreases by a
         factor of 1.5 between 40° and 140° and is flat from 140°
         to 170°.

isotropic angular distribution. Above 500 eV the distribution is
approximately Maxwellian with an atmospheric loss-cone. Evans &
Moore (1979) have suggested that the shape of the distribution is
the result of precipitation from a primeval plasma sheet
distribution combined with electrons backscattered from the
atmosphere which have travelled out along field lines and then
been scattered in pitch angle to form part of a new plasma sheet
distribution. While this explanation is successful in explaining
the shape of the distribution above 500 eV, it cannot account for
the power law distribution at low energies.

The atmospheric source must be important and on some occasions may be sufficient to account for all the low energy electrons, but there must also be another stronger and more variable source which does not rely on reflection by the electric field.

The atmosphere also modifies the shape of the distribution as it descends the field-line. The only way to remove the effect on the observations is to make them at a sufficiently high altitude, in practice, above 400 kms.

5. EVIDENCE FOR BEAM-PLASMA INTERACTIONS IN THE VELOCITY DISTRIBUTION OF PRECIPITATING ELECTRONS

5.1. Positive Slope

Any discontinuity with a positive increase formed at the boundary would very quickly be eroded to a region of finite width with a finite positive slope. As long as there is a positive slope in the velocity distribution wave growth can occur and the interaction of those waves with the electrons then leads to further erosion of the slope. The shape of the distribution in the neighbourhood of the peak in the energy spectrum is therefore a sensitive indicator of the regime of beam-plasma interactions the beam has experienced. Figures 5 and 6 show two electron energy spectra, obtained by a sounding rocket in the same discrete auroral form 100 secs apart in time. The ordinate is intensity $I(E)$ (cm^{-2} sr^{-1} eV^{-1} s^{-1}) rather than density in velocity space $f(v)$ (cm^{-6} s^3) so that to indicate how steep the slope must be to give a real positive slope in the velocity distribution a dashed line proportional to E has been drawn since

$$f(v) \propto I(E)/E$$

The two cases illustrated by Figures 5 and 6 give some idea of the range of slopes observed. The earlier one, Fig. 5, does not have a real positive slope in velocity distribution; the later one, Fig. 6, has one of the steepest slopes ever observed. The pitch angle of the measurement averaged 30° but the distribution from 20° to 70° was nearly isotropic in both cases. There were no dramatic changes in auroral conditions between the two observations which were both made at high altitude, well above any atmospheric degradation. The distribution gradually evolved from the flattened distribution of Figure 5 into the peaked distribution of Figure 6 with a decrease in the energy of the peak by approximately 2 KeV and a decrease in the total number flux associated with the beam. This suggests that, with the decay of

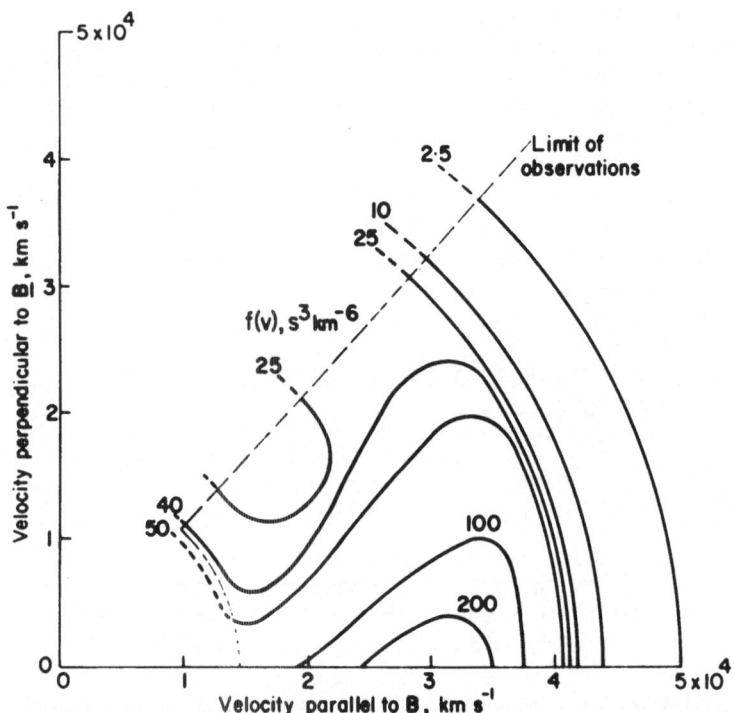

Electron Distribution Contours in Velocity-Space, Measured on Fulmar F3 at 220 km Altitude

Fig. 8. The distribution shows strong field alignment over a factor of 2 in velocity (4 in energy) suggesting modification of the parallel velocity component by electrostatic waves after acceleration of a cold distribution (D.A. Bryant, private communication).

the beam density and velocity, the distribution is moving from the quasilinear regime towards the linear regime (Maggs 1981).

Figure 8 shows an example of a strongly field-aligned energy peak superimposed on an isotropic peak in intensity at the same energy. This looks very much like the model distribution of fig 2 where the cold ionospheric-type particles are combined with the hot plasma sheet electrons in the high-altitude source except that the field alignment extends over a wider range of energy. It appears that waves have grown on the strong feature with $df/dv_{\parallel} > 0$, and these waves have then modified the parallel component of velocity. As a result the field-alignment is maintained but extended in energy. The field-aligned peak still has a positive slope in the velocity space density at the point of observation.

A proper analysis of the stability of the precipitating electrons requires consideration of the complete distribution function. An examination of the slope at the particular pitch angle only gives qualitative information. Kaufmann and co-workers (Kaufmann et al 1978a, 1978b, Kaufmann 1980, Dusenbery and Kaufmann 1980) have made a quantitative evaluation of the stability of a measured auroral electron distribution function to the growth of electrostatic waves. Although the distribution contained a parallel peak i.e. $df/dv_\parallel > 0$, and a perpendicular peak $df/dv_\perp > 0$, it was only weakly unstable and the growth rates were much too small to produce any observable effects.

Since the measurements were made at an altitude of 250 km by a sounding rocket the distribution observed was probably the result of the evolution of the plasma towards a marginally stable state. The rocket only observes the distribution after the strong interactions have taken place. Whatever the initial state was, this analysis shows that it evolves to a state which is only weakly unstable, presumably as a result of beam-plasma interactions.

5.2 Power Law Spectrum below 500 eV

At first sight the spectrum below 500 eV appears featureless and uninteresting. This is probably why it has received little attention although there are a number of problems in interpreting the data. It was argued in section 4 that this distribution could not be the result of atmospheric backscattering alone because it is too intense and the spectrum is too hard

The spectrum usually has the power law form

$$I(E) \propto E^{-r}; \qquad 0.5 < r < 1.5$$

over an energy range that spans 10 eV to 500 eV on occasions. The power law form has been reported in many observations (Reasoner and Chappell, 1973; Arnoldy and Choy, 1973; Feldman and Doering, 1975; Matthews et al., 1976; Doering et al., 1976; Peterson et al., 1977; Johnstone, 1978b) from low-altitude satellites and sounding rockets. In the spectra shown in Figures 5 and 6 the power law exponent has values of $r = 1.05$, 1.04 respectively.

The angular distribution decreases by a factor of 1.5 from 0° to 180° (figure 7); i.e. the upgoing intensity is slightly less than the downcoming. The power law energy spectrum and this angular distribution is found both in discrete auroral forms, where there may be a downward reflecting electric potential, and in diffuse aurora (Evans & Moore 1979) where there is almost certainly not. Since in the latter case there is nothing to

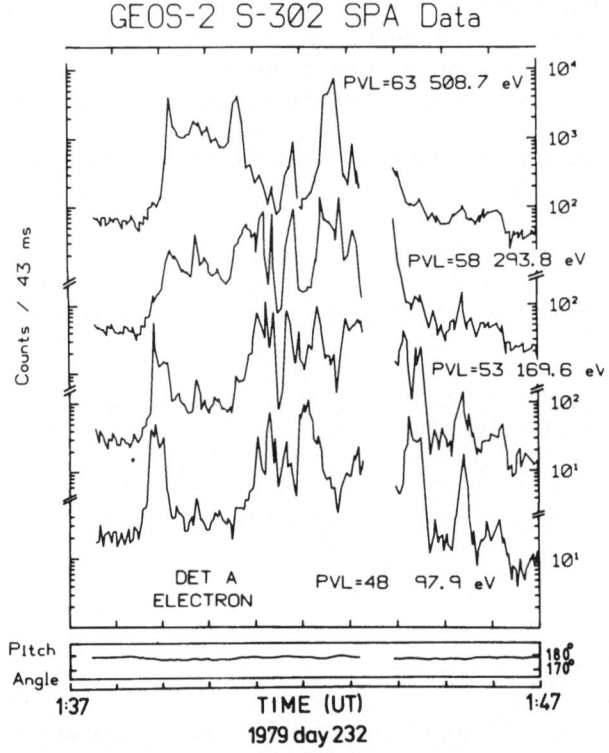

Fig. 9. Suprathermal electron fluxes in the atmospheric loss-cone observed by GEOS-2 at the geomagnetic equator at four fixed energy levels. Notice that there are significant differences between the four energies showing that the spectrum is highly variable on a short time scale (Wrenn et al., 1981).

prevent these suprathermal electrons from travelling up the field line to the equator it might be expected that they could be found there by high-altitude satellites.

The two suprathermal particle analysers on the GEOS satellites should be able to detect this distribution. One analyser is directed along the spin axis of the satellite (detector A) while the other (detector B) views at 80° to this direction. When the satellite is near the geomagnetic equator the spin axis is nearly aligned with the magnetic field during undisturbed conditions, although during the course of a substorm this direction may fluctuate considerably. The satellite was stationed near the geomagnetic equator during August 1979 at a longitude of 15° E where it crosses the geographic equator.

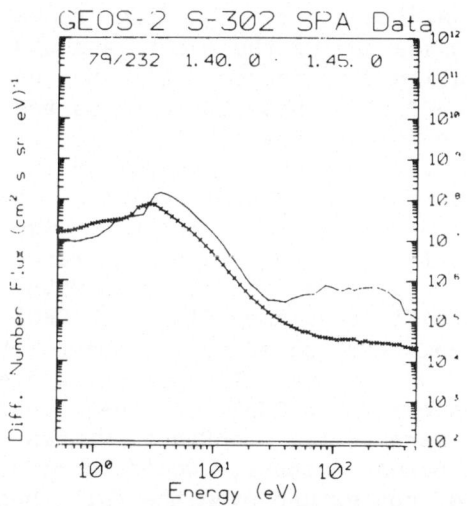

Fig. 10. The spectrum of field-aligned detector A (dots) and
           perpendicular detector B (crosses) averaged over 5
           minutes during the event of figure 9. The broad peak
           extending up to 20 eV is caused by electrons in the
           spacecraft sheath. Before the event both detectors
           measured the same as detector B does during the event.
           The averaged spectrum of the field aligned electrons
           from 20 eV to 500 eV is given by the difference between
           the   two   spectra   (A.   J.   Norris,   private
           communication).

During this period detector A was able to detect fluxes in the
atmospheric loss-cone. Field-aligned fluxes of suprathermal
electrons were often observed at the time of substorm activity
(Wrenn et al. 1981). Figure 9 shows an example of the time
variations in four energy channels through one of these events.

If the power law distribution could reach the equator then
this is perhaps what would be expected i.e. strongly fluctuating
suprathermal electron fluxes in the atmospheric loss-cone. A more
stringent test is to compare the spectrum and density of the
electrons observed by GEOS with the suprathermal electrons
measured at low altitudes by sounding rockets. The spectrum
averaged over 5 minutes during the event is shown in figure 10 for
both the field-aligned detector (A) and detector B nearly
perpendicular to the field. Detector B shows the same spectrum as
it did before the event began; it is essentially flat in the
energy range 50 eV to 500 eV. The average spectrum of the
field-aligned electrons is therefore given by the difference
between the spectra of detectors A and B. Like the background
spectrum it is nearly flat over the energy range 50 eV to 500 eV.
It is nothing like a power law distribution with r ~ 1.0. The
intensities of the field-aligned fluxes at GEOS are in the range
10^8 - 10^9 electrons/(cm^2.sr.eV.s), as are the power law
distributions at low altitude. The power law distributions ought
therefore to be observable at GEOS if they reached there even
though, if they were still confined within the atmospheric
loss-cone, their apparent intensity would be reduced because the
distribution would not extend over the full opening angle of the
detector. It is possible that the probability of the electrons
reaching the equator is energy-dependent, so that the power-law
source distribution is modified by processes occurring along the
field line. All that can be concluded now is that the power law
distribution does not originate in the plasma sheet, or come from
the conjugate point. The field-aligned fluxes observed at GEOS
may be the remnants of the upgoing suprathermal fluxes produced at
low altitudes.

Power law distributions of suprathermal electrons are
produced in beam-plasma experiments (Christiansen, 1981) and so it
is quite possible that the power law distribution is produced by
the interaction between the precipitating auroral electrons and
the ionosphere. This has been suggested on theoretical grounds by
Papadopoulos and co-workers (Papadopoulos and Coffey 1974a, b,
Matthews et al. 1976, Rowland and Papadopoulos 1977, Papadopoulos
and Rowland 1978). They have shown that a spectral exponent r of
the order of 1.0 is to be expected theoretically from the
operation of the oscillating-two-stream instability excited by
unstable, beam-like velocity distributions of electrons. They
found a similar result with a computer simulation. Matthews et
al. (1976) found a roughly linear relationship between

suprathermal density and beam density but no other dependence of the suprathermal distribution on beam parameters. The numerical values of their spectra are consistent with the theory. The natural power-law distributions are extremely wide spread and occur even when the beam-like qualities of the distribution are rather weak. The mechanism of Papadopoulos and co-workers may be important but it does not seem possible that it can account for all the cases. Further studies of the quantitative relationship, which may be non-linear, are still needed before a proper understanding can be achieved.

5.3 Suprathermal Halo

The power law distribution is an almost permanent feature of the auroral electron spectrum. Another secondary distribution has been identified by its correlation with time variations in the precipitating intensity. This distribution is the quasi-Maxwellian halo which accompanies suprathermal electron bursts. An energy/time spectrogram of a burst, a common occurrence outside discrete auroral forms, is shown in Fig. 11. The data were obtained by a sounding rocket at an altitude of 780 kms. The burst consists of an increase in the downward field-aligned intensity of electrons in the suprathermal energy range by more than an order of magnitude. The burst is usually confined within 20° of the magnetic field direction. The field-aligned portion of the distribution alone is shown in the spectrogram in fig 12 where the difference between the upgoing and downgoing flux is plotted. It is accompanied by a nearly isotropic increase in the intensity at lower energies. The isotropic distribution can be deduced by subtracting the field-aligned portion in fig 12 from the total distribution in fig 11. Often this isotropic distribution is nearly Maxwellian in form (Johnstone and Sojka 1980). When the field-aligned burst distribution has been subtracted from the total a temperature kT (eV) and a density n (cm^{-3}) can be derived for the ´halo´ electrons. The variation of n and T through a typical burst is given in figure 13 which also compares the variation of the energy density E in this secondary distribution where $E = (3/2)nkT$, with the energy flux in the field-aligned primary beam.

Temperatures range up to 10 eV (116,000 $^{\circ}$ K), or more than 20 times the ambient electron temperature, while the densities may be as much as 100 cm^{-3} which is approximately 1% of the ambient ionospheric electron density. The energy density in the secondary electrons is, at most, 20% of the ionospheric electron energy density. These events have also been found by the ISIS-2 satellite at an altitude of 1400 kms and a selection of spectra are shown in fig 14.

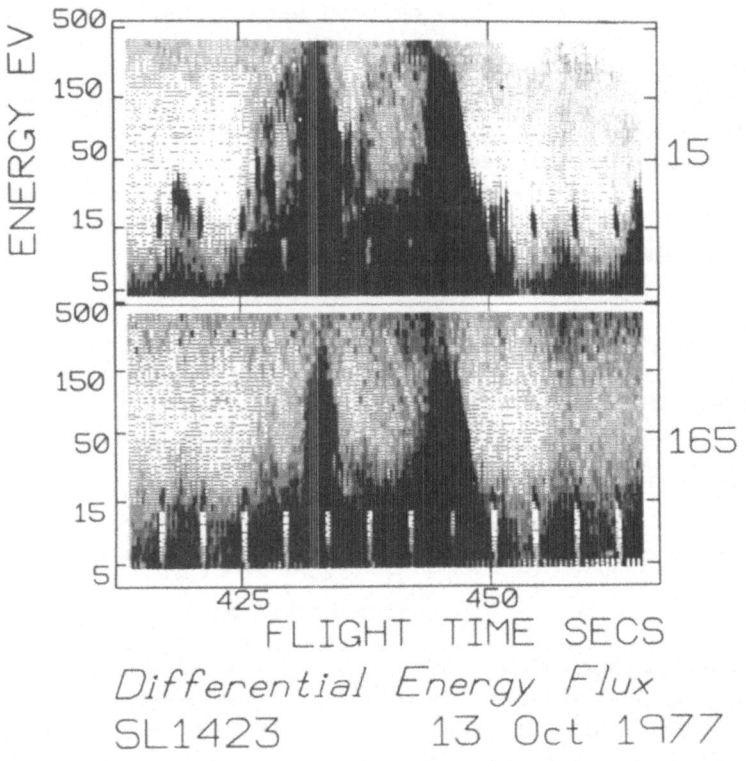

Fig. 11. A spectrogram of two intense suprathermal electron
         bursts observed at 780 kms altitude from a sounding
         rocket. The upper panel was measured by a detector at
         15° to the rocket spin axis (i.e. near field-aligned,
         downcoming electrons) while the lower panel shows data
         from a detector at 165° (field-aligned, upgoing). The
         shading from white to black is scaled logarithmically
         over a factor of 40. The structure occurring regularly
         at 4 sec intervals is caused by the switching-on of a
         retarding potential of 18 volts in the detector
         collimator (Johnstone et al., 1981).

    Fig 13 shows that the secondary energy density is
proportional to the beam energy flux over a wide dynamic range in
the event. Similar values of the ratio are obtained in different
events.

    The ratio

Primary Energy Flux/Secondary Energy Density = L/t

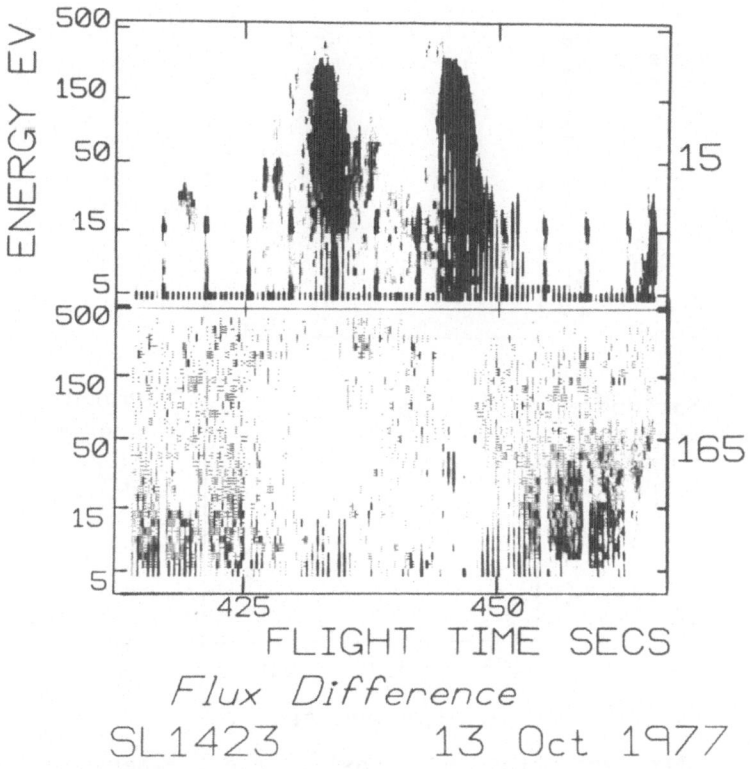

Fig. 12. The data of figure 11 are replotted here as the
difference between the two detectors to the same
intensity scale. If the 15° detector has the greater
flux the difference is plotted in the upper panel and
the corresponding element in the lower panel is left
blank and vice versa. This spectrogram only shows the
downcoming primary beam in the burst; the isotropic
halo has been removed. Some idea of the contribution
from the isotropic halo is gained by comparing with
figure 11.

where t is the lifetime of the secondary electrons. A
maximum value for t from the data is 0.34s giving L < 600 km for t
< 0.34s where L is the beam penetration depth.

Two aspects of the data suggest that the secondary electrons
are locally produced by a beam-plasma interaction:

(a) the close proportionality between beam and secondary
distribution;

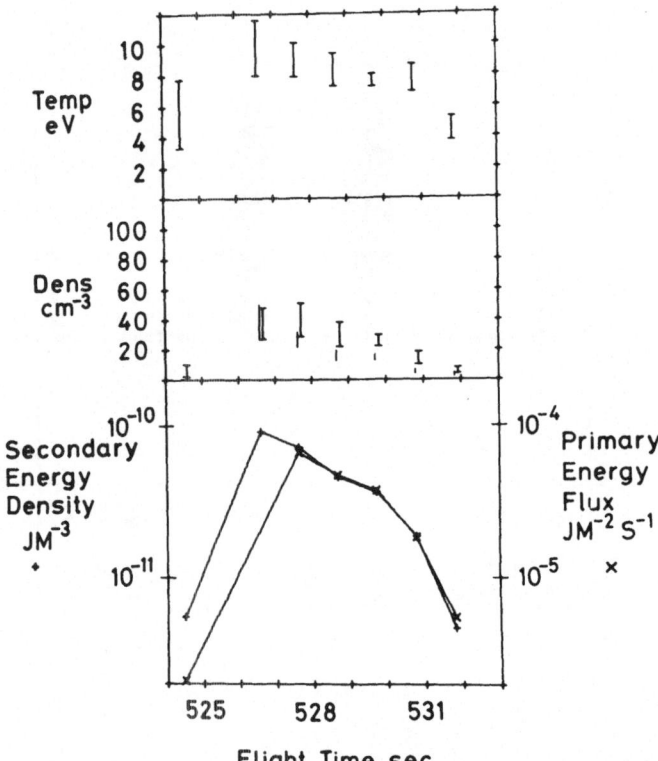

Fig. 13. The variation of density and temperature of the Maxwellian halo in a suprathermal burst (not the event in fig 11). The distribution was measured by four detectors at angles 15°, 65°, 115° and 165° to the spin axis of the rocket. The length of the temperature bar indicates the spread of the four values. The density has been plotted separately for the 'parallel' detectors (15°, 165° straight line) and the perpendicular detectors (65°, 115° barred line) (Johnstone and Sojka, 1980).

(b) the close temporal correlation between the beam and both up and downgoing electrons in the secondary distribution.

This interpretation is supported by measurements of high frequency (200 KHz - 5 MHz) electrostatic waves on the same vehicle (P. Christiansen private communication). There is a burst of broad-band electrostatic noise simultaneously with the

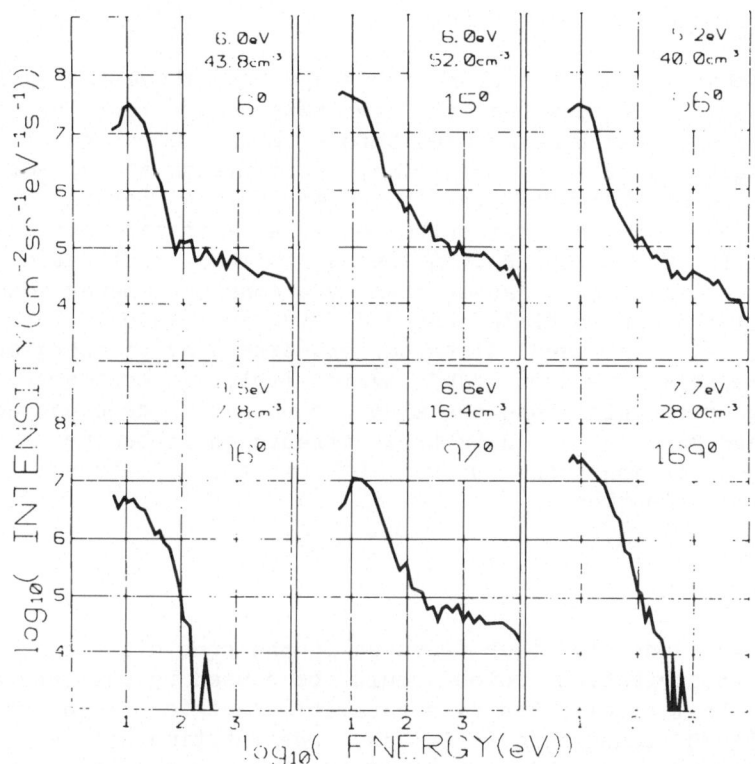

Fig. 14. Six examples of Maxwellian haloes measured by the ISIS-2 satellite at an altitude of 1400 kms. The temperature (eV) and density (cm^{-3}) of the best Maxwellian fit to each spectrum is shown in the upper right panel on each plot, together with the pitch angle of the measurement. Each event was only detected during one energy sweep (approx 1 sec duration) (Johnstone and Winningham, 1981).

electron bursts which implies that the noise is also being locally generated.

There is therefore direct evidence that the Maxwellian 'halo' distribution is caused by a beam-plasma interaction. The only direct link between the power-law distribution and a beam is the linear dependence of suprathermal density on beam density found by Matthews et al. (1976), but the power law distribution is also found when it is hard to identify a beam at all.

5.4 Acceleration beyond the peak

The shape of the electron energy spectrum above the peak is often used as a measure of the temperature of the source distribution in the plasma sheet before it has been accelerated. If the static models of parallel electric fields are correct then the temperature measured in this way should remain constant however much the whole distribution is accelerated. In the high energy tails the distribution seems to be accelerated by a constant energy factor rather than by a constant energy amount as the parallel electric field should do (Bryant 1980). On other occasions field-aligned distributions extending to energies well beyond the peak have been found again implying acceleration to much higher energies than the electric potential drop can achieve (Whalen and Daly 1979). Both these effects could be the result of beam-plasma interactions driven by unstable features in the original distribution.

6. CONCLUSIONS

I have identified four features in the precipitating auroral electron distribution which could be caused by the interaction between a beam accelerated by a parallel electric field and the high-altitude ionosphere. At the present time little is known about the quantitative relationships between the modified distributions and the original accelerated beam. The problem is that by the time the distribution is usually observed it has already evolved to the point where it is only weakly unstable, and the quantitative relationships cannot be deduced without knowing what the unmodified distribution was like. It will therefore be difficult to confirm whether the acceleration is produced by a quasistatic electric field without observing the way the distribution evolves as it travels down the field line. The features which have been observed correspond to those expected from laboratory observations and theoretical studies of beam-plasma interactions. This is itself evidence for the existence of strong beam-like characteristics in the precipitating electrons at some point along the field line but it is not yet possible to deduce from the low-altitude distributions what acceleration mechanism is responsible.

ACKNOWLEDGEMENTS

It is a pleasure to acknowledge stimulating discussions with D. A. Bryant, D.S. Hall, G.L. Wrenn, A.J. Norris and J.E. Maggs during the preparation of this work.

REFERENCES

Albert, R.D., 1967, Energy and flux variations of nearly
        monoenergetic auroral electrons, J. Geophys. Res., 72:5811.
Arnoldy, R.L., and Choy, L.W., 1973, Auroral electrons of energy
        less than 1 keV observed at rocket altitudes,
        J. Geophys. Res., 78:2187.
Arnoldy, R.L., Lewis, P.B., and Isaacson, P.O., 1974, Field
        aligned auroral electron fluxes, J. Geophys. Res., 79:4208.
Banks, P.M., Chappell, C.R., and Nagy, A.F., 1974, A new model for
        the interaction of auroral electrons with the atmosphere;
        spectral degradation, backscatter, optical emission,
        ionisation, J. Geophys. Res., 79:1459.
Bryant, D.A., 1980, Rocket studies of particle structure
        associated with auroral arcs, in: "Proc. Chapman
        Conference on the formation of auroral arcs," (In press).
Bryant, D.A., Hall, D.S., and Lepine, D.R., 1978, Electron
        acceleration in an array of auroral arcs,
        Planet. Space Sci., 26:81.
Christiansen, P.J., Jain, V.K., and Bond, J.W., 1981, Laboratory
        studies of beam-plasma interactions linear and non-linear,
        in: "Proc. NATO Adv. Res. Inst. on Artificial particle
        beams utilised in space plasma physics".
Doering, J.P., Potemra, T.A., Peterson, W.K., and Bostrom, C.O.,
        1976, Characteristic energy spectra of 1 to 500 eV
        electrons in the high latitude ionosphere from Atmospheric
        Explorer C, J. Geophys. Res., 81:5507.
Dusenbery, P.B., and Kaufmann, R.L., 1980, Properties of the
        longitudinal dielectric function: an application to the
        auroral plasma, J. Geophys. Res., 85:5969.
Evans, D.S., 1968, The observations of a near monoenergetic flux
        of auroral electrons, J. Geophys. Res., 73:2315.
Evans, D.S., 1974, Precipitating electron fluxes formed by a
        magnetic field aligned potential difference,
        J. Geophys. Res., 79:2853.
Evans, D.S., and Moore, T.E., 1979, Precipitating electrons
        associated with the diffuse aurora: evidence of electrons
        of atmospheric origin in the plasma sheet,
        J. Geophys. Res., 84:6451.
Feldman, P.D., and Doering, J.P., 1975, Auroral Electrons and the
        optical emissions of nitrogen, J. Geophys. Res., 80:2808.
Hall, D.S., 1980, Unpublished manuscript.
Johnstone, A.D., 1978a, Precipitation of charged particles by a
        parallel electric field, Planet. Space Sci., 26:581:
Johnstone, A.D., 1978b, The spectrum of auroral suprathermal
        electrons, in "Proc. Esrange Symposium, Ajaccio", ESA
        SP-135:57.
Johnstone, A.D., and Sojka, J.J., 1980, A beam/plasma interaction
        in the high altitude auroral ionosphere,
        Planet. Space Sci., 28:467.

Johnstone, A.D., Sojka, J.J., Gibbons, W., Madahar, B.K., and
    Woolliscroft, L.J.C., 1981, An intense wave/particle event
    in the auroral ionosphere, Geophys. Res. Letters, 8. (In
    press).
Johnstone, A.D., and Winningham, J.D., 1981, Satellite
    observations of suprathermal electron bursts, to be
    submitted to J. Geophys. Res..
Kaufmann, R.L., Walker, D.N., and Arnoldy, R.L., 1976,
    Acceleration of auroral electrons by a parallel electric
    field, J. Geophys. Res., 81:1673.
Kaufmann, R.L., Dusenbery, P.B., Thomas, B.J., and Arnoldy, R.L.,
    1978a, Auroral electron distribution function,
    J. Geophys. Res., 83:586.
Kaufmann, R.L., Dusenbery, P.B., and Thomas, B.J., 1978b,
    Stability of the auroral plasma: Parallel and
    perpendicular propagation of electrostatic waves,
    J. Geophys. Res., 83:5663.
Kaufmann, R.L., 1980, Electrostatic wave growth: Secondary peaks
    in a measured auroral electron distribution function,
    J. Geophys. Res., 85:1713.
Lyons, L.R., 1980, Generation of large scale regions of auroral
    currents, electric potentials and precipitation by the
    divergence of the convection electric field,
    J. Geophys. Res., 85:17.
Maggs, J.E., 1981, Interaction between natural particle beams and
    the magnetospheric plasma, in, "Proc. NATO Adv. Res.
    Inst. on Artificial particle beams utilised in space
    plasma physics.
Matthews, D.L., Pongratz, M., and Papadopoulos, K., 1976, Non
    linear production of suprathermal tails in auroral
    electrons, J. Geophys. Res., 81:123.
O´Brien, B.J., 1970, Considerations that the source of auroral
    energetic particles is not a parallel electrostatic field,
    Planet. Space Sci., 18:1821.
Peterson, W.K., Doering, J.P., Potemra, T.A., McEntire, R.W.,
    Bostrom, D.O., Hoffman, R.A., Janetske, R.W., and Burch,
    J.L., 1977, Observations of 10 eV to 25 keV electrons in
    steady diffuse aurora from Atmosphere Explorer C and D,
    J. Geophys. Res., 82:43.
Papadopoulos, K., and Coffey, T., 1974a, Non-thermal features of
    the auroral plasma due to precipitating electrons,
    J. Geophys. Res., 79:674.
Papadopoulos, K., and Coffey, T., 1974b, Anomalous resistivity in
    the auroral plasma, J. Geophys. Res., 79:1558.
Papadopoulos, K., and Rowland, H.L., 1978, Collisionless effects
    on the spectrum of secondary auroral electrons at low
    altitudes, J. Geophys. Res., 83:5768.
Reasoner, D.L., and Chappell, C.R., 1973, Twin payload
    observations of incident and backscattered auroral
    electrons, J. Geophys. Res., 78:2176.

Rowland, H.L., and Papadopoulos, K., 1977, Simulations of non linearly stabilised beam-plasma interactions, Phys. Rev. Letters, 39:1276.

Westerlund, L.H., 1969, The auroral electron energy spectrum extended to 45 eV, J. Geophys. Res., 74:351.

Whalen, B.A., and Daly, P.W., 1979, Do field aligned auroral particle distributions imply acceleration by quasistatic parallel electric fields? J. Geophys. Res., 84:4175.

Whipple, E.C. Jr., 1977, The signature of parallel electric fields in a collisionless plasma, J. Geophys. Res., 82:1525.

Wrenn, G.L., Johnson, J.F.E., and Sojka, 1981, The suprathermal plasma analysers on the ESA GEOS satellites, Space Science Instrumentation, (in press).

DISCUSSION

Anderson: We have electron spectral data in the discrete aurora from rocket 29.007 UE which can be accounted for by upward atmospheric scattering with the electrons multiply reflected from a region of parallel electric fields above. A primary spectrum consisting of an accelerated Maxwellian was inferred. No wave-particle effects need be invoked to account for most of this data.

Johnstone: Although it is possible to account for the lower energy spectrum by atmospheric backscattering on some occations (e g Evans 1974) it is not always true. On perhaps the majority of occasions some other process must also be operating. The variety in the observations emphasizes the complextiy of the factors controlling the relative importance of these processes and also how little we know about them.

Koons: Ion beam conics demonstrate that ions are accelerated perpendicular to the magnetic field at relatively high altitudes presumably by ion cyclotron waves. Satellites measure strong electrostatic waves above 1500 km i e above the altitude accessible to rockets. These waves are intense enough to cause strong diffusion in the auroral electrons.

Johnstone: Your comment points to one of the principle difficulties in trying to observe the effect of wave-particle interactions on the electron distribution, namely that the interaction probably takes place in a region remote from the observation point making it impossible to correlate wave and particle observations.

<u>Haerendel</u>: Did you ever undertake an interpretation of the observed electrostatic noise in terms of anomalous resistivity and calculate the parallel electric field it could maintain in the presence of the deduced current density? Is this value of the parallel electric field sufficient for local generation of the burst electrons below 1 keV?

<u>Johnstone</u>: The observed electrostatic noise is generated by the interaction between the accelerated burst electrons and the local ionospheric plasma. It is this interaction which I believe heats the isotropic halo of secondaries. The acceleration of the suprathermal electron bursts is an entirely different matter.

INTERACTION BETWEEN NATURAL PARTICLE BEAMS AND SPACE PLASMAS

James E. Maggs

Institute of Geophysics and Planetary Physics
University of California, Los Angeles
Los Angeles, California 90024

INTRODUCTION

In discussing the interaction of naturally occurring particle beams with space plasmas it is convenient to divide the plasma processes involved into two classes - weakly turbulent and strongly turbulent. Naturally occurring particle beams generate a variety of wave modes, and the beam dynamics depend upon several channels of both weakly and strongly turbulent wave-particle interactions. Wave modes generated by the particle beams are divided into two very distinct groups - electrostatic and electromagnetic. The main problem of theoretical interest is to identify the plasma processes that play the major role in the spatial and temporal evolution of the particle beam. The main topic of this paper is a discussion of the role of electrostatic waves in determining the evolution of the auroral electron beam through weakly turbulent processes. This discussion allows the identification of the parameter regime in which strongly turbulent processes are likely to occur. A brief discussion of some wave spectra observed in conjunction with Jupiter's bow shock is included as an illustration of nonlinear plasma processes.

While the emphasis of this paper is electrostatic waves, it should be noted that large power fluxes of electromagnetic waves can also be generated by the auroral beam. The phenomenon of auroral kilometric radiation (AKR) storms involves the generation of bursts of electromagnetic radiation over a broad range of frequencies (Gurnett, 1974). The plasma processes leading to AKR are clearly complex as, while the range of frequencies generated is broad, it consists of many individual bursts of narrow banded rising and falling tones (Gurnett et al., 1979). Several processes, both linear and nonlinear, have been suggested as causes of AKR (Grabbe, 1981).

Among the linear variety is cyclotron resonance amplification of right hand polarized electromagnetic waves (Melrose, 1976; Wu and Lee, 1979). The cyclotron maser instability which depends upon electron phase bunching in velocity space has been investigated in the linear and quasilinear regime by Wu and Lee (1979, 1980). This instability depends not upon the existence of a particle beam but rather the occurrence of an enhanced loss cone in the associated backscattered particle fluxes. The nonlinear processes suggested as sources of AKR commonly involve electrostatic wave fluxes as a crucial part of the generating mechanism. Large amplitude fluxes of electrostatic upper hybrid waves are needed in processes suggested by Barbosa (1976) and Roux and Pellat (1979). On the other hand the presence of large amplitude low frequency electrostatic ion cyclotron waves are needed in the process suggested by Grabbe et al. (1980). While some of the analyses of the various processes suggested as sources of AKR have been carried to the point of considering saturation mechanisms they have all, to date, involved only local effects. No detailed study of the effects of convective processes in the inhomogeneous auroral ionosphere have yet been reported.

On the other hand, the process of electrostatic wave generation discussed here is treated as an inherently convective process in which changing ionospheric parameters and auroral arc geometry play dominant roles. This approach necessarily embodies the fundamental concepts of weak turbulence - the WKB approximation and quasilinear diffusion. The WKB approximation leads to the wave kinetic equation while the concepts of quasilinear theory lead to a diffusion equation for the spatial and temporal evolution of the particle velocity distribution. The range of validity for the application of weak turbulence theory to auroral beam dynamics is determined by demanding that the time scale for nonlinear processes is much longer than the time scale associated with wave packet propagation. The nonlinear time scale is associated with mechanisms that modify the high frequency dispersion relation in processes such as the oscillating two stream instability (Papadopoulos and Coffey, 1974) or spiky turbulence (Morales and Lee, 1976) and is, typically $\tau_{NL} = (2\pi/\omega_{pe}\bar{W})$ where $\bar{W}$ is the ratio of electric field energy density to plasma thermal energy density, $\bar{W} = |E|^2/8\pi nT$. The time scale associated with a wave packet is $\tau_L = 2\pi/\Delta\omega$, where $\Delta\omega$ is the frequency band width of the packet. In a one dimensional plasma where electrostatic plasmons are involved, $\Delta\omega = V_g\Delta k \simeq a_T^2(k/\omega)\Delta k \simeq \omega_{pe}(k\lambda_D)^2$ (a_T is the thermal velocity). In this case the criterion for applying weak turbulence theory becomes $\bar{W} \ll (k\lambda_D)^2$. However, the auroral beam cannot generally be treated as one dimensional. Typically a broad spectrum of waves with band widths on the order of ω_{pe} is generated for a wide range of beam densities so that weak turbulence theory should be applicable when $\bar{W} \ll 1$. Weakly turbulent processes subsequently play a major role in auroral beam dynamics over a fairly broad range of beam and ionospheric parameters.

WEAK TURBULENCE THEORY APPLIED TO THE AURORAL BEAM

The theoretical approach used to study the auroral electron beam dynamics is to first find the power flux of electrostatic waves generated by the beam in a given spatial region and then to determine the changes in the beam velocity distribution caused by quasilinear diffusion. The problem will be treated as one of steady state. That is, all purely temporal changes will be considered zero ($\partial_t = 0$). In this section the methods employed in calculating the electrostatic differential power flux are reviewed. The changes in the beam velocity distribution due to quasilinear diffusion are discussed in a subsequent section.

The power flux spectra of auroral beam generated electrostatic waves is estimated by using equations based on the WKB theory. For a given frequency the wave power levels at a point in space are calculated by summing the noise from all ray paths of the wave mode of interest passing through the point. Along a given ray path the power level is found by integrating the wave kinetic equation (Maggs, 1976)

$$\frac{dP}{d\omega} (\omega, \underline{k}, s_1) = \int_{s_1}^{\infty} ds \ \varepsilon F \ \exp 2\left(\int_{s_1}^{s} \frac{\gamma}{V_g} ds' \right) \tag{1}$$

In (1) the point of observation is s_1 while s denotes distance along the ray path. Each unit volume of plasma along the ray path emits an incoherent level, ε, of electrostatic noise. F is a geometric factor representing the spreading of ray paths. The incoherent noise levels convectively amplify or decay along the ray at a local rate γ/V_g. The temporal growth rate, γ, is not just determined by the auroral beam but by the total electron distribution. If the wave amplitude is large, γ is found by including quasilinear diffusion of the beam. The group velocity of the waves, V_g, depends upon the dispersion characteristics of the mode of interest which are primarily determined by the relatively cold dense ionospheric plasma.

The auroral beam exists only over a limited spatial region. Over this region it represents a source of nonthermal noise levels which can be calculated using the concepts of convective amplification. The amount of amplification achieved depends greatly upon changing ionospheric and beam parameters along the ray path. The determination of the wave power flux depends upon establishing a detailed model for the auroral beam and ionosphere (Maggs and Lotko, 1981).

PROPERTIES OF THE AURORAL BEAM

A detailed model of the auroral electron distribution is developed based upon rocket and satellite observations. The auroral

electron precipitation typically shows a peak in the directional
differential intensity between one and ten keV (Meng, 1976; Liu and
Hoffman, 1979). The peak in the intensity has been observed, at
times, to rise fast enough to produce a region of positive slope in
velocity space and thus may be expected to be unstable to waves with
phase velocities in this region. The peak in the differential intensity of auroral precipitation is produced in the model by using a
beam of electrons with a drifting Maxwellian velocity distribution. In addition a hot Maxwellian plasma of temperature about 1.5
keV is used to give better agreement with observation. This hot
Maxwellian component plays an important role in establishing thresholds for wave growth. The electron fluxes below the peak intensity
region increase with decreasing energy, usually in a power law fashion. These electrons of moderate energies are called warm electrons.
For ease of mathematical manipulation the warm electron population
is modeled by using a sum of six Maxwellians to reproduce their power
law distribution. The model distribution has been chosen to be representative of measured distributions as illustrated in Figure 1.

The electron distribution is observed to vary with altitude.
At altitudes of a few hundred kilometers the peak in electron intensity is observed to occur almost uniformly with pitch angle in the
downward hemisphere (Kaufman et al., 1978). At higher altitudes the

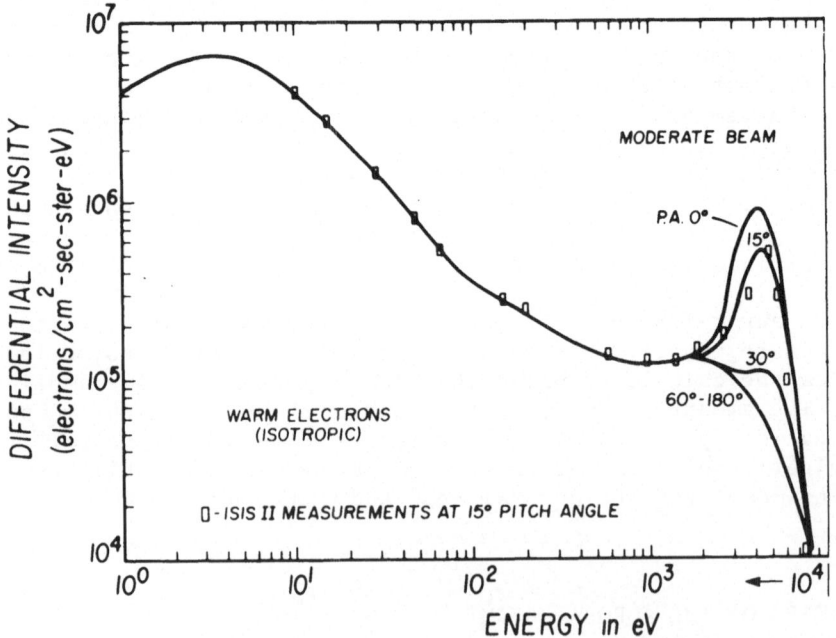

Fig. 1.   The model electron distribution contains a warm population
that closely resembles measured distributions.

peak intensity is more nearly field aligned (Frank and Gurnett, 1971; Eliasson et al., 1979). The changes in the beam pitch angle distribution are modeled by assuming the beam evolves adiabatically with altitude. The model beam is pictured as starting at some source altitude with a drifting Maxwellian distribution of equal temperature, T_b, perpendicular and parallel to the magnetic field. The initial beam electron flux is highly field aligned, however, so that T_b is much smaller than the beam energy, E_b. As the beam moves down the field line away from the source its pitch angle distribution evolves in a fashion such that total energy and first adiabatic invariant, $\mu = V_\perp^2/2B$, are conserved. The region below the source is assumed to be free of parallel electric field. The resulting particle motion produces, at low altitudes, a hemispherical shell distribution of beam electrons. On the other hand, the hot Maxwellian component and warm electrons are assumed to be isotropic in pitch angle so that their density and pitch angle distributions are uniform in altitude.

The electron velocity space distribution is shown at various altitudes in Figure 2. The parameter α is the ratio of source magnetic field strength B_s to local magnetic field strength B and therefore decreases away from the source. The smallest value of α achieved by the beam before entering the collisional regime of the ionosphere depends upon the source altitude. Small α values are important primarily for generating upper hybrid electrostatic noise. Values of α below .05, and thus nearly isotropic pitch angle distributions, are achieved for source altitudes beyond ten thousand kilometers.

THE MODEL IONOSPHERE AND AURORAL ARC GEOMETRY

The plasma number density in the topside auroral ionosphere decreases steadily with increasing altitude. Data from the S3-3 satellite (Mozer et al., 1980) indicate that the plasma density drops so rapidly the electron plasma frequency, ω_{pe}, is less than the electron cyclotron frequency, ω_{ce}, at altitudes over a few hundred kilometers. The model of the ionosphere adopted for use here is based on a model presented by Maeda (1975) which has an altitude variation similar to the S3-3 data, but the plasma density is about four times larger everywhere. The important feature of these models is that a large extent of the auroral field line has densities low enough that $\omega_{pe} < \omega_{ce}$. The important parameters in the model ionosphere in regards to convective amplification are the scale sizes associated with changes in magnetic field strength and plasma density. These parameters determine the shape of the ray path and the rate of change in wave number vector along the path (Maggs, 1978). The scale sizes for changes in the number density along the magnetic field are readily obtainable from the vertical density profile of the model. Scale sizes perpendicular to the

field are another matter. The model ionosphere is assumed to vary
in density in the magnetic meridian plane, i.e. in the direction
across the auroral arc. The scale sizes for this variation at a
given altitude are taken to be equal to the vertical scale size at
that altitude. This completely ad hoc assumption is made in lieu of

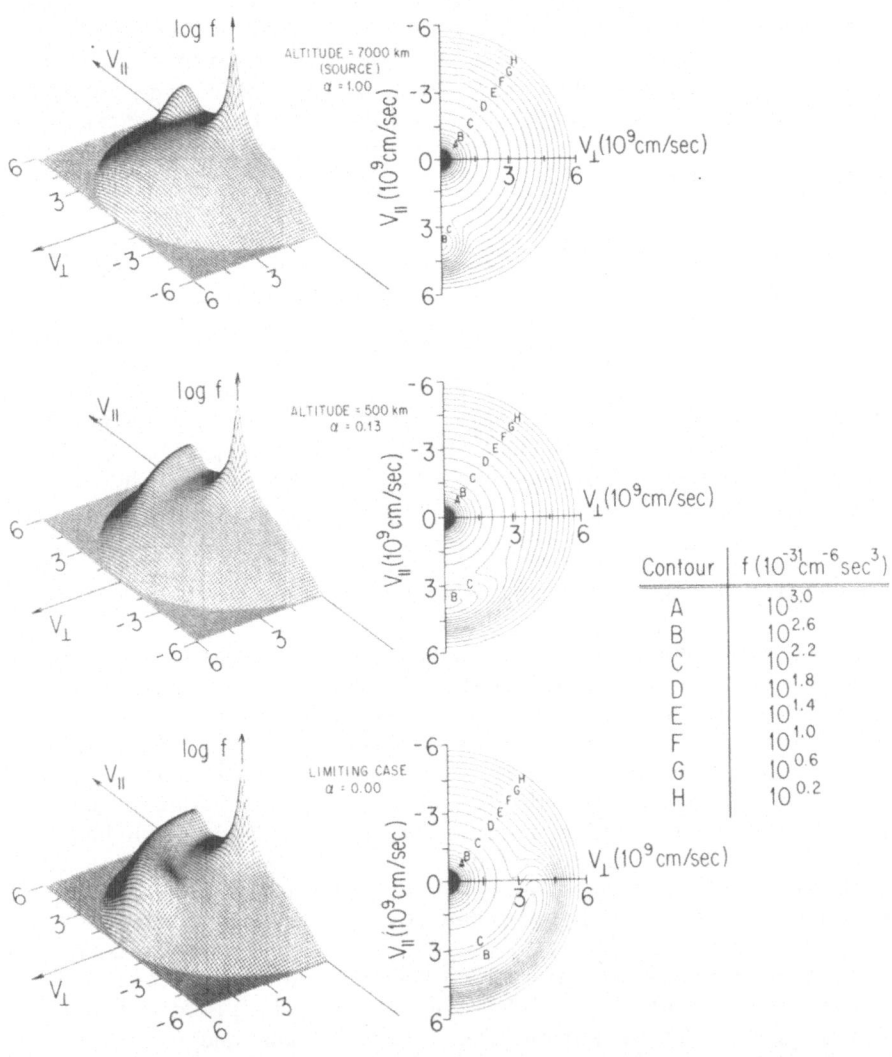

ELECTRON DISTRIBUTION FUNCTION   ISODENSITY CONTOURS

Contour	f (10^{-31} cm^{-6} sec^3)
A	$10^{3.0}$
B	$10^{2.6}$
C	$10^{2.2}$
D	$10^{1.8}$
E	$10^{1.4}$
F	$10^{1.0}$
G	$10^{0.6}$
H	$10^{0.2}$

Fig. 2. The model electron beam starts out as a drifting Maxwellian
but evolves towards a hemispherical shell distribution as
it moves down into increasing magnetic field strength.

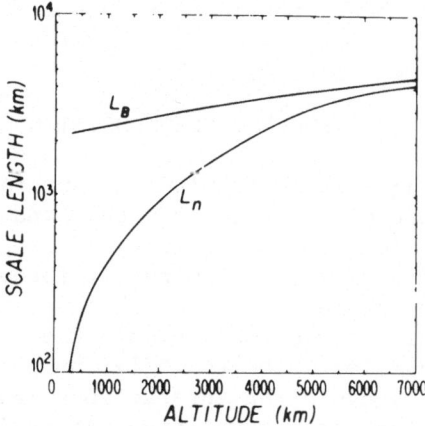

Fig. 3. The model ionosphere density scale length L_n largely determines the rate of ray refraction because it is smaller at
all altitudes than the magnetic scale length L_B.

observational data. The perpendicular scale sizes of the plasma
density are important for determining the range of beam number densities over which linear growth occurs as will be discussed below.

The magnetic field lines are assumed to be uniformly directed
at an angle of 6° to the density gradient, but the field strength
is taken to vary as in a dipole field (viz, r^{-3}). The scale lengths
for variation in the magnetic field at any altitude are readily obtained from this model. The vertical scale sizes for magnetic field
strength and plasma density associated with the model ionosphere are
illustrated in Figure 3. Clearly the number density varies more
rapidly than the field, and as a consequence the number density profile almost completely determines the ray path shape and rate of
change of wave number vector.

The auroral arc is modeled as being field aligned with a uniform velocity distribution at constant altitude across the arc in
the magnetic meridian plane (north-south direction). The arc is
modeled as extending indefinitely in the direction perpendicular to
the magnetic meridian plane (east-west direction). Ray paths do not
traverse along the arc in the east-west direction indefinitely, however, because of wave refraction. Arc curvature along the east-west
direction can also lead to limitation of the distance a ray traverses
the arc. This effect can be simulated by introducing a limited
extent to the arc in the east-west direction. The arc width varies
with altitude in proportion to the radius of a flux tube (i.e., as
$B^{1/2}$). The width of the model arc at the source is taken as 20 km

because arc widths of tens of kilometers are commonly observed by
satellite (see, for example, Meng, 1976). At an altitude of 300 km
the model arc width is only 7 km.

CHARACTERISTICS OF HIGH FREQUENCY ELECTROSTATIC MODES

Waves interacting resonantly with the auroral electron distribution either interact directly through the Landau resonance (the
wave phase velocity along the magnetic field $\omega/k_{\shortparallel}$ equals the particle
velocity along the field, $v_{\shortparallel}$) or indirectly through the doppler
shifted cyclotron resonances, $\omega - n\omega_{ce} = k_{\shortparallel}v_{\shortparallel}$. In the weakly turbulent process of quasilinear diffusion only the Landau resonant inter-
action alters the parallel velocity while the perpendicular velocity
is altered through the cyclotron interactions (Barbosa, 1976).
Electrostatic waves typically have large refractive indexes and thus
slow phase velocities. Since the energy of auroral beams is only a
few keV the parallel velocity is much less than light speed. Thus
only slow phase velocity waves can interact with the beam through
the Landau resonance and subsequently alter the parallel beam velocity distribution. For this reason, as well as their slow group
velocities which lead to rapid convective amplification, electrostatic modes play the dominant role in weakly turbulent auroral beam
dynamics.

The dispersion characteristics of electrostatic waves arise
primarily from the thermal properties of the plasma. Terms proportional to the plasma temperature introduce an infinite number of
wave modes called Bernstein modes. A separate electron Bernstein
mode exists in the frequency band between each of the harmonics of
the electron cyclotron frequency. Two of these modes are distinctive, however, in that they couple at wave lengths on the order of
the gyroradius (or Debye length if it is larger) to two cold plasma
modes. For a given frequency the two cold plasma modes propagate
for all values of wave number at a single angle called the resonance
cone angle. One mode referred to here as the electrostatic whistler,
propagates in the frequency band between the lower hybrid frequency,
ω_{LH}, and the smaller of ω_{pe} or ω_{ce}. The other, referred to as the
upper hybrid mode, propagates in the band between the larger of ω_{pe}
or ω_{ce} and the upper hybrid frequency, ω_{UH}. At long wavelengths
the nature of both of these modes changes. The index of refraction
of the modes approaches unity, and they acquire a magnetic component.
This feature gives the index of refraction surfaces for these modes
the appearance of a truncated cone. It is also at these wavelengths
that the whistler mode acquires the dispersive properties more commonly associated with it. In the electrostatic regime the group
velocity of both modes is perpendicular to their wave number vector.
Electrostatic whistler waves in Landau resonance with the auroral
beam propagate downwards while Landau resonant upper hybrid waves
propagate upwards. The electrostatic modes most readily amplified
by the model are the two cold modes.

POWER FLUXES IN THE AURORAL ARC

Power flux spectra of beam generated noise can be found at any
altitude in the model auroral arc and ionosphere by integrating the
wave kinetic equation, (1). In this section a method is described
for approximately evaluating (1) in the power level regime where the
convective growth rate is not affected by quasilinear diffusion.
The next section deals with the modifications introduced when quasilinear diffusion alters the beam velocity distribution.

Ray refraction in the inhomogeneous ionosphere leads to ray
paths that remain in the arc for only a limited distance. Since wave
amplification occurs only in the arc where the beam-like velocity
distribution occurs the refraction process limits the amount of power
produced by convective amplification. Figure 4 depicts the behavior
of a typical electrostatic whistler ray path in the model ionosphere.
The changing number density in the direction across the arc causes a
steady change in the perpendicular wave number leading to nearly
parobolic shapes for paths projected down to the observers altitude.
The path arriving at the observer with a given angle in the veritcal
plane (and thus a single frequency for the cold electrostatic modes)
that produces the highest power flux is the one which remains in the
arc for the longest distance. This ray is displayed as the maximally
amplified ray in Figure 4. The shapes of the ray paths are determined by the scale lengths of the model ionosphere. The amount of
amplification along the ray path is determined, on the other hand,
by the beam velocity distribution. In the method used to approximately evaluate (1) the amplification factor for the maximally amplified ray,

$$M = \int_{s_1}^{s} \frac{\gamma}{V_g} ds', \tag{2}$$

is evaluated by solving for M the equation (Maggs, 1978)

$$W = \sum_{n=1}^{M/2} [V_{g\perp}(n-1)\dot{k}_x/\gamma^2]\{k_y^2 + [(n-1)\dot{k}_x/\gamma]^2\}^{-\frac{1}{2}} \tag{3}$$

where W is the arc width. The properties of the ray geometry leading to (3) are illustrated in Figure 4. The perpendicular wave
vector in the east-west direction is denoted by k_y, while k_x denotes
the component in the north-south direction. Once the amplification
factor M is determined using (3) the power flux is evaluated by
using the approximation to (1)

$$\frac{dP}{d\omega} \simeq W \frac{\Lambda k_\perp}{k_\parallel} \frac{m_-\omega_{pe}\omega_{ce}^2 n_b}{n_o a_b 8\pi^{7/2}} \frac{\epsilon e^{2M}}{M \cos\theta} \tag{4}$$

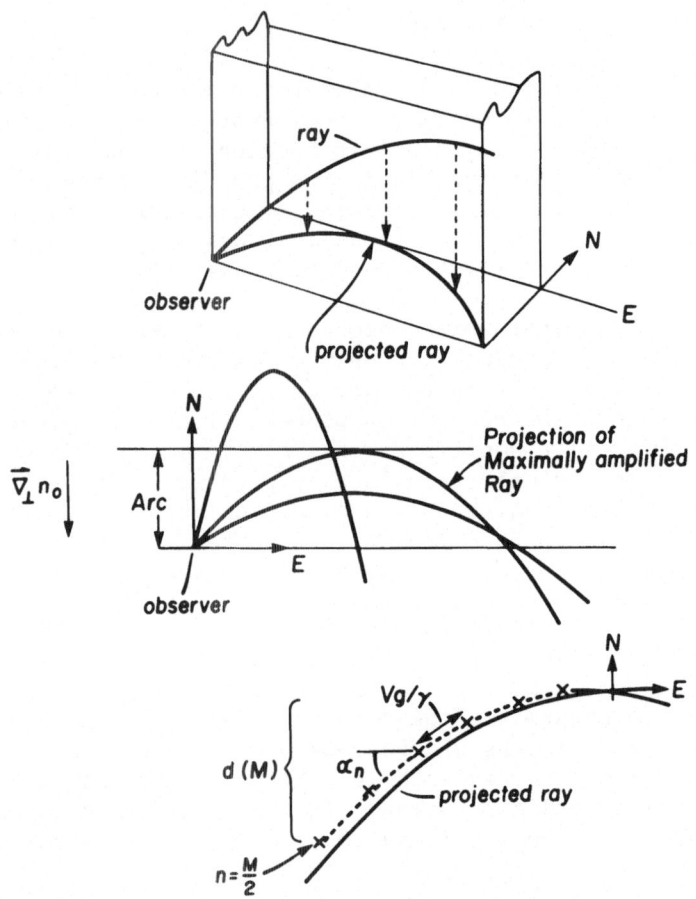

Fig. 4. Rays are refracted out of the model arc because of the
         north-south density gradient. The ray path can be approxi-
         mated by a path consisting of segments of V_g/γ in length.

In (4), $\Delta k_{||}$ is the range of parallel wave numbers over which waves
are amplified, and the ratio $\Delta k_{||}/k_{||}$ is approximately the same as the
ratio of beam thermal velocity to drift velocity. In using (3) and
(4) all parameters (e.g. growth rates, group velocity, plasma frequency, etc.) are evaluated at the altitude of observation. Despite
the somewhat crude approximations used to arrive at (3) and (4),
they can be used to obtain order of magnitude estimates of power
flux levels and relatively accurate power flux spectral shapes
(Maggs, 1978).

     Predictions of power flux levels in the model auroral arc made
by using the theory of convective amplification of incoherent levels
of Cerenkov radiation at the linear growth rate with the amplitude
limited only by ray refraction out of the beam are shown in Figure 5.

Power flux levels are evaluated using (3) and (4). The regions of expected linear growth for the electrostatic whistler and upper hybrid modes lie between the lines marked "stable" and "nonlinear." The linear growth regions are functions of altitude and beam source density. The thresholds for wave growth occur along the lines marked "stable." The whistler threshold decreases nearly as B^{-1} because the beam density increases almost in proportion to a flux tube cross section. The upper hybrid mode grows only at relatively low altitudes where $\omega_{pe} > \omega_{ce}$ because at higher altitudes the entire wave band is very near ω_{ce} and cyclotron damping by warm electrons is strong.

    The upper range of the linear regime is bounded by the line marked "nonlinear." This line indicates the point at which beam generated power fluxes are comparable to the beam energy. It is in the vicinity of this line that quasilinear diffusion will begin to modify the beam velocity distribution and power fluxes will be limited by quasilinear processes rather than by ray refraction. The lines marked "5," "10" and "20" denote the location of the "non-linear" line if propagation is limited to an east-west distance of 5, 10 or 20 times the arc width. These lines illustrate the effects arc curvature may have on the extent of the linear growth regime. The effects of changing $L_{n\perp}$ can be illustrated by noting that if the perpendicular scale lengths were half the values used to obtain

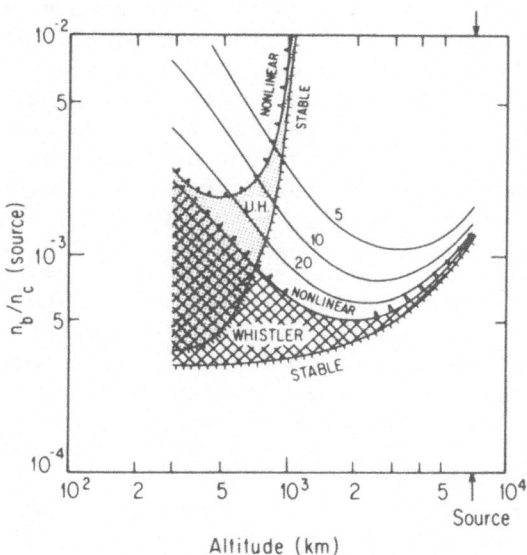

Fig. 5. The model beam generates electrostatic noise at the linear growth rate only in the regions bounded by the lines marked stable and nonlinear. The numbered lines indicate the extent of the linear growth regime when the arc is limited in east-west extent.

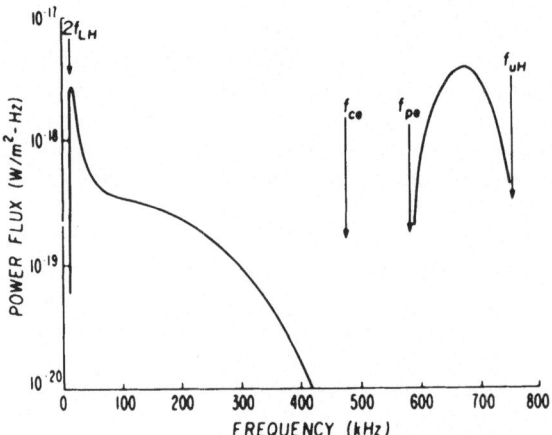

Fig. 6. The model beam produces broadbanded whistler noise that
         tends to peak near the lower hybrid frequency. Upper hybrid
         noise peaks in the middle of the propagation band.

Figure 5 the extent of the linear growth region would be to about
the line marked "10."

     Representative whistler and upper hybrid spectra are shown in
Figure 6. The whistler power flux tends to peak near the lower
hybrid frequency, especially at higher altitudes. The spectra is
broadband and may be relatively constant in frequency for certain
choices of ionospheric parameters. Effects of propagation outside
the arc tend to accentuate the peak at ω_{LH} (Yamamoto 1979) because
downward propagating whistlers reflect at the local value of ω_{LH}
resulting in a sharp cutoff at this frequency. The features of the
predicted whistler power flux agree well with the observed properties
of VLF hiss. Amplified upward propagating upper hybrid noise produces spectra peaked in the middle of the propagation band. Observed
properties of upper hybrid noise are not well documented so that predicted results are not readily compared to observation. Regions of
growth for the electrostatic Bernstein modes are not shown because
growth of these modes is severely limited by refractive effects, and
they are not amplified by the model beam parameter range shown in
Figure 5.

     Noise near the electron plasma frequency is not strongly ampli-
fied because of refractive effects. The group velocity of strictly
parallel propagating plasma frequency noise in resonance with the
beam is along the magnetic field, and its magnitude is much less
than the thermal velocity. However, wave refraction in the presence
of a magnetic field causes initially parallel propagating noise to
propagate at angles for which the magnitude of the group velocity is
nearly the thermal speed and its direction is across the magnetic

field. Thus refraction plays a key role in producing a broad power flux spectrum rather than an essentially one dimensional spectrum strongly peaked at the plasma frequency. Wave refraction then extends the range of beam source densities over which weak turbulence theory is applicable from the one dimensional criterion [$E^2/8\pi nT \ll (k\lambda_D)^2$] to the broadband criterion ($E^2/8\pi nT \ll 1$).

The changing beam pitch angle distribution results in a velocity distribution at low altitudes with a positive slope in perpendicular velocity ($\partial_{v_\perp} f > 0$). This feature of the velocity distribution provides a channel for growth of electrostatic waves interacting with the beam through the cyclotron resonances. This channel proves effective for amplification of upper hybrid noise but not the Bernstein or whistler modes (Lotko and Maggs, 1981). Because of strong refraction effective growth of electrostatic upper hybrid modes is achieved only at small values of α (= B_S/B), or large beam densities. The effect of wave refraction on the growth of cyclotron resonant upper hybrid noise is shown in Figure 7. Here the ratio of the linear growth time to the wave resonance time is plotted as a func-

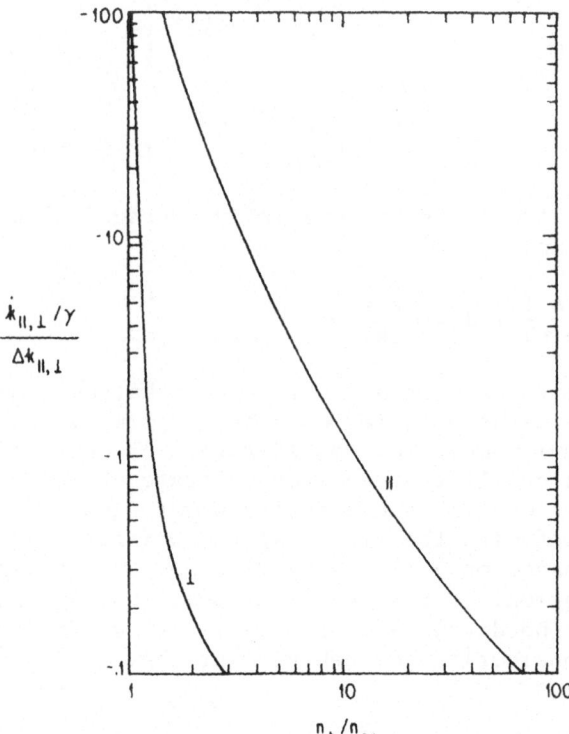

Fig. 7. The changes in wave number during one growth period for upper hybrid waves in cyclotron resonance with the beam decrease rapidly with increasing beam number density.

tion of beam source density. The beam density at marginal stability is indicated by n_{cr}. Increasing the growth rate by increasing the density helps produce wave growth rapid enough to overcome the effects of wave refraction. A similar effect is achieved by decreasing α because the magnitude of $\partial_{v_\perp} f$ is then increased. Decreasing α at a given altitude corresponds to increasing the source altitude. Strong power fluxes of upper hybrid waves growing through cyclotron interactions with the model auroral beam are not produced by the beam densities and source altitude of 7000 km shown in Figure 3. Electrostatic noise growing from a beam with a source altitude of 7000 km interacts with the beam exclusively through the Landau resonance.

QUASILINEAR DIFFUSION IN THE AURORAL ARC MODEL

In weak turbulence theory the wave kinetic equation is used to determine wave amplitudes while the quasilinear diffusion equation determines the behavior of particles in the presence of the waves. For electrostatic waves in a magnetized plasma the quasilinear diffusion equation is (Kennel and Engelman, 1966)

$$\partial_t f_o + \underline{v} \cdot \partial_{\underline{x}} f_o = -i \frac{e^2}{m^2} 8\pi \int \frac{d^3k}{(2\pi)^3} \frac{\xi(k)}{k^2} \left(\partial v_\perp \frac{n\omega_{ce}}{v_\perp} + k_{\shortparallel}\partial v_{\shortparallel} \right)$$

$$\frac{J_n^2(k_\perp v_\perp/\omega_{ce})}{\omega - n\omega_{ce} - k_{\shortparallel}v_{\shortparallel}} \left(\frac{n\omega_{ce}}{v_\perp} \partial v_\perp + k_{\shortparallel}\partial v_{\shortparallel} \right) f_o \qquad (5)$$

where f_o is the slowly varying spatially averaged electron distribution function and

$$\xi(k) = \lim_{V\to\infty} \frac{1}{V} \int d^3x \frac{|E(k)|^2}{8\pi} \qquad (6)$$

From the previous section it is clear that over most of the auroral field line the whistler mode is of primary importance in terms of wave amplitude. The consideration of quasilinear diffusion in the auroral arc will therefore concentrate solely on the beam electron interaction with the whistler power flux. Since the beam electrons interact with the electrostatic whistlers solely through the Landau resonance only the parallel component of velocity changes. The cyclotron harmonic terms in (1) proportional to $n\omega_{ce}$ need not be considered. The discussion of quasilinear diffusion is further simplified by considering the behavior of the reduced distribution function

$$g = 2\pi \int v_\perp dv_\perp f_o \qquad (7)$$

Using (7) and considering only resonant diffusion gives

$$\partial_t g + \underline{v} \cdot \partial_{\underline{x}} g = 8\pi^2 \frac{e^2}{m^2} \partial_{v_{\shortparallel}} \int \frac{d^3 k \xi(k)}{(2\pi)^3 k^2} k_{\shortparallel}^2 \delta(\omega - k_{\shortparallel} v_{\shortparallel}) \partial_{v_{\shortparallel}} g \tag{8}$$

Equation 8 is a convenient form to use for studying the quasilinear behavior of the auroral beam.

Temporal Behavior

In order to obtain an initial understanding of the quasilinear diffusion of the auroral beam it is useful to consider the temporal evolution of a spatially homogeneous beam. The analysis presented in this section is patterned after a similar analysis by Melrose and White (1980). Assuming that the reduced auroral electron distribution can be modeled as a drifting Maxwellian

$$g = \frac{n_b}{(2\pi)^{\frac{1}{2}} a_b} e^{-(v_{\shortparallel} - v_b)^2 / a_b^2} \tag{9}$$

where n_b is the beam number density and a_b the beam thermal velocity. Equations governing the temporal evolution of the average parallel velocity and beam temperature can be found by averaging (9) over parallel velocity,

$$\partial_t v_b = \partial_t \langle v_{\shortparallel} \rangle = \frac{4\sqrt{\pi}\, \omega_{pe}^2}{mn_o a_b^2} \int \frac{d^3 k \xi(k)}{(2\pi)^3 k^2} k_{\shortparallel} \zeta \, \exp{-\zeta^2} \tag{10a}$$

and

$$\tfrac{1}{2}\partial_t a_b^2 = \tfrac{1}{2}\partial_t \langle (v_{\shortparallel} - v_b)^2 \rangle = \frac{4\sqrt{\pi}\, \omega_{pe}^2}{mn_o a_b^2} \int \frac{d^3 k}{(2\pi)^2 k^2} k_{\shortparallel}^2 \zeta^2 \, \exp{-\zeta^2} \tag{10b}$$

where $\zeta = (\omega - k_{\shortparallel} v_b)/a_b$. Using the group velocity to relate the power flux to the energy density gives, for the rate of change of beam speed and temperature

$$\partial_t v_b \simeq -2\sqrt{\pi}\, \omega_{pe}^2 \, Pk_{\shortparallel}/mn_o V_g k^2 a_b^2 \tag{11a}$$

$$\partial_t a_b \simeq \sqrt{\pi}\, \omega_{pe}^2 \, k_{\shortparallel} P/mn_o V_g k^2 a_b^2 \tag{11b}$$

where $P = \int d\omega (dP/d\omega)$ is the total band-integrated power flux. Using the approximation that the whistler noise is propagating at the resonance cone angle where $V_g \simeq v_b$, equations 11a and 11b become

$$\partial_t v_b \simeq -2\sqrt{\pi}\, \omega_p P/mn_o a_b^2 \tag{12}$$

and

$$\partial_t a_b \simeq \sqrt{\pi} \, \omega_p \, P/mn_o a_b{}^2 \tag{13}$$

The subscript "p" refers to the value of frequency for which the peak flux occurs. Since $a_b \ll v_b$ for the auroral beam, equations 12 and 13 indicate the time scale for a change in v_b is v_b/a_b larger than the time scale for changes in a_b. The primary effect of quasilinear diffusion is to change the beam temperature! It is important to note however that the ultimate energy source for the waves and random beam motion is the coherent or average motion of beam particles along the magnetic field. The beam is heated at the expense of beam drift energy.

The time scale for reaching the quasilinear regime can be estimated following Melrose and White (1980) as

$$\tau_{QL} = a_b/\partial_t a_b = mn_o a_b{}^3/\sqrt{\pi} \, P\omega_p \tag{14}$$

For times before τ_{QL} the power flux is exponentiating at a constant rate. For later times the growth rate $\gamma(t)$ is time dependent and for a drifting Maxwellian is inversely proportional to the beam temperature. Thus

$$\partial_t a_b \simeq \partial_t P \simeq P/a_b{}^2 \tag{15}$$

The long time behavior ($t > \tau_{QL}$) of a_b and P is a slow growth proportional to $t^{\frac{1}{2}}$. Figure 8 illustrates the temporal behavior of P.

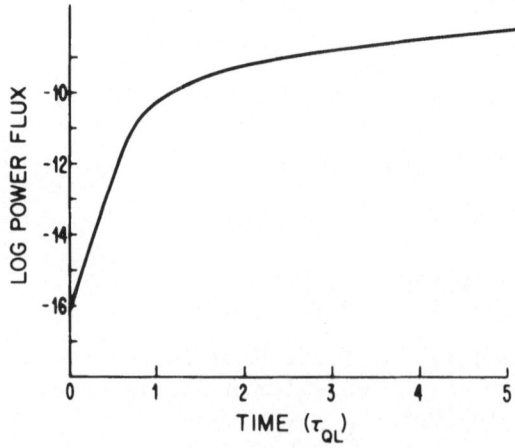

Fig. 8.   The power flux grows exponentially until the quasilinear regime is reached where it grows as $t^{\frac{1}{2}}$.

Taking the time rate of change of a_b at τ_{QL} to be proportional to $\gamma(t)a_b$ and using (15), the power flux at the quasilinear transition time can be estimated as

$$P_{QL} = m\, n_b\, v_b^2\, a_b \tag{16}$$

where γ has been taken to be $n_b/n_0 (v_b/a_b)^2 \omega_p$.

In the auroral beam model the hot Maxwellian component makes a significant contribution to damping in the vicinity of the beam. The growth rate is given by

$$\gamma = \gamma_b - \gamma_h \tag{17}$$

where $\gamma_b \simeq \gamma_h$ i.e. $(\gamma \ll \gamma_b, \gamma_h)$. In the analysis of the temporal behavior of the beam, quasilinear diffusion was important when the growth rate had changed by a factor of two. Given the relation (17) the growth rate in the auroral beam model will change by a factor of two when the beam temperature changes by a factor

$$\Delta a = (\gamma_b - \gamma_h)/4\gamma_b \equiv \delta/4 \tag{18}$$

Thus the time needed to reach the quasilinear stage in the auroral problem is $\delta\tau_{QL}$, and the power flux at τ_{QL} is δP_{QL}. This corresponds to a wave energy density of

$$\varepsilon_{QL} = \delta P_{QL}/V_g = 2\delta a \varepsilon_b/v_b \ll \varepsilon_b \tag{19}$$

where $\varepsilon_b = m n_b v_b^2/2$ is the initial beam energy density.

Convective Effects

In the steady state model the evolution of the power flux in the auroral beam is governed by convective rather than temporal growth. The most important aspect to consider in the convective case is the finite extent of ray paths within the auroral arcs. To assess the effect of convective quasilinear diffusion on the beam, the changes in beam parameters caused by increasing magnetic field strength are first ignored. Denote the distance from the source that the beam traverses along the field line before it produces power flux levels above P_{QL}, by L_0. Then, if the ray path length of the maximally growing wave is denoted by L_r, the power flux behavior depends upon the ratio of L_r to L_0 as illustrated in Figure 9. If $L_r \gg L_0$ the beam power flux evolves as in the temporal case because quasilinear diffusion occurs well before the finite length raypath comes into play. If $L_r \ll L_0$ the amplitude is path limited and quasilinear diffusion amplitudes are never

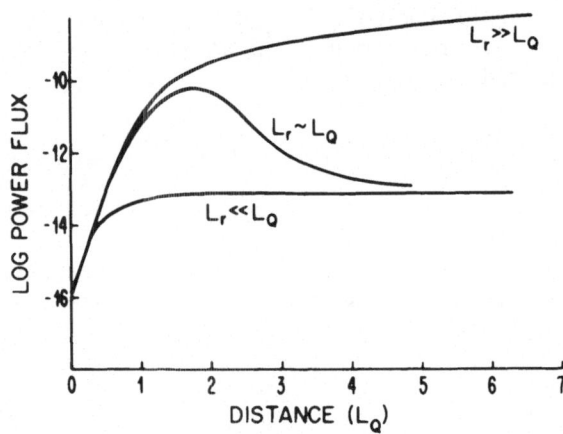

Fig. 9.   In convective amplification the growth of the power flux
away from the source depends upon the ratio of L_r to L_Q.

reached.   If $L_r \simeq L_Q$ the power flux reaches quasilinear levels but
then decreases again because waves growing near and beyond the peak
flux are growing in regions where quasilinear diffusion has reduced
the growth rate.   Power fluxes beyond the peak are much lower than
the peak flux because growth of waves is exponential and quasilinear
diffusion has reduced the growth rate by a factor of about two.   In
the lower power regions beyond the peak quasilinear diffusion is
negligible.

As illustrated by Fig. 5 the auroral arc usually starts out in
the regime $L_r \ll L_Q$ and approaches the case $L_r \simeq L_Q$ as the beam
travels down the field line.   This occurs because the number density
of the beam increases with decreasing altitude due to the converging
magnetic field.   The density continues to increase until $B/B_s = 1/\alpha$
$\simeq v_b/a$ at which point it decreases with altitude (Barbosa, 1976).
The power flux does not behave as discussed above, however, because
the increasing magnetic field plays a crucial role in determining
the beam dynamics.

Once the beam traverses the distance L_Q and thus produces
enough power to drive quasilinear diffusion it then evolves at a
rate dependent upon the magnetic field gradient.   This is clear from
the following argument.   If the power flux drops below levels around
P_{QL} the beam temperature remains constant because quasilinear diffusion is negligible over distances on the order of L_r.   In order for
power levels to drop below P_{QL} it is necessary that the convective
growth rate be less than the value at L_Q.   This will not be the case,
however.   While the above argument indicates the beam temperature
remains constant, the beam density, and thus the growth rate,
increases due to the converging magnetic field.   Thus the power

levels must be comparable to or higher than P_{QL} because the growth rate must have increased. On the other hand, if the power levels are much higher than P_{QL} quasilinear diffusion over the distance L_r is large enough to heat the beam faster than the number density is increasing. Since the growth rate is inversely proportional to beam temperature the growth rate must decrease. However, if the growth rate is less than the value at L_Q, the convective amplification is not rapid enough to produce power levels as large as P_{QL}. These arguments imply that the power levels must not be much higher than P_{QL}. We conclude then, that the power levels produced are sufficient to increase beam temperature through quasilinear diffusion at the same rate the beam density is increased by converging magnetic field; and that these power levels are comparable to P_{QL}. The above reasoning is valid until the beam temperature increases to the point that $E_b/T_b \sim (1/\alpha)^2$. At this point the number density begins to decrease with increasing magnetic field strength. While the growth rate drops at this point and quasilinear diffusion becomes negligible the positive slope in the distribution does not persist. This is not an effect comparable to quasilinear plateauing but occurs because the beam number density is now decreasing with altitude.

The height evolution of the beam can be found from the above arguments if it is assumed the beam remains a drifting Maxwellian. If power levels generated over the distance L_r are to remain

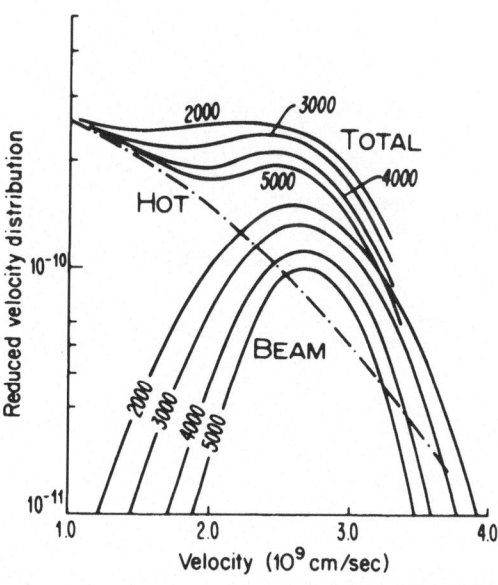

Fig. 10. The model beam is heated by quasilinear diffusion but increases in density due to increasing B. The total model distribution (beam plus hot electrons) is shown at various altitudes as marked on the curves.

comparable to P_{QL} the total convective amplification ($\simeq \exp 2\gamma L_r/V_g$) must remain nearly constant. This requirement gives the relation between beam temperature and local magnetic field strength,

$$\frac{T_b}{T_{bQ}} = \frac{\alpha_Q}{\alpha}\left[1 + \delta_Q\left(\frac{L_{rQ}V_g}{L_r V_{gQ}} - 1\right)\right]^{-1} \tag{20}$$

where the subscript "Q" indicates the value of the subscripted quantity at the altitude where quasilinear power levels first occurred. Using (20) and the fact that the beam speed evolves at a rate a_b/v_b slower than the beam thermal velocity the beam evolution in altitude can be determined in the region of increasing beam number density. If the actual ratio of beam number density as a function of altitude is used instead of α_Q/α an accurate picture of beam evolution can be obtained as illustrated in Figure 10 for a beam with a source density of $8 \times 10^{-4} n_s$. The altitude at which the positive slope disappears can be estimated as the point where $\alpha^2 = T_b/E_b$. Using (20) gives the value of α at which this occurs.

$$\alpha_d = \left(\frac{\alpha_Q T_{bQ}}{E_{bQ}}\right)^{1/3}\left[1 + \delta_Q\left(\frac{L_{rQ}V_g}{L_r V_{gQ}} - 1\right)\right]^{-1} \tag{21}$$

The behavior of the whistler power flux as a function of altitude including the effects of quasilinear diffusion are illustrated in

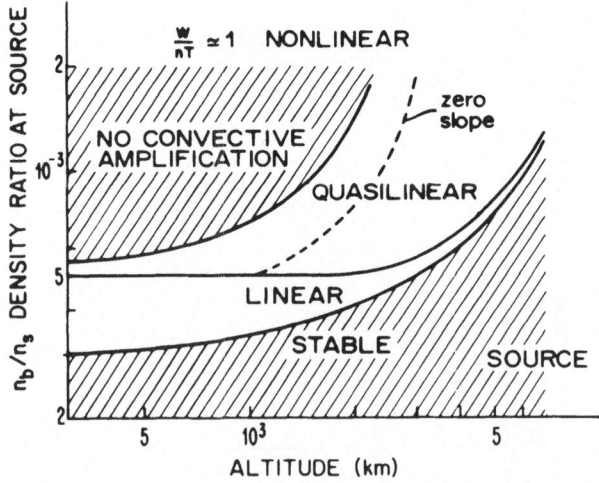

Fig. 11. Quasilinear diffusion limits power fluxes and produces a region of no growth at low altitudes below the altitude at which the velocity distribution has zero slope as marked by the dashed line. The strongly turbulent regime may start at beam densities around $2 \times 10^{-3} n_s$.

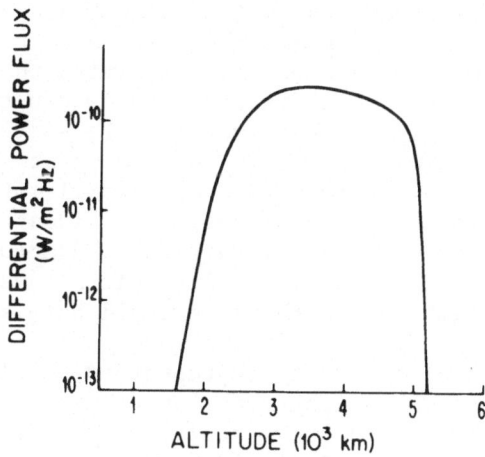

Fig. 12. The peak power flux as a function of altitude for a model beam with $n_b/n_s = 8 \times 10^{-4}$.

Figure 11. Note that there are regions at low altitude where growth does not occur because of the effects of quasilinear diffusion. The behavior of the peak power flux for the beam shown in Figure 10 is illustrated by Figure 12. The power flux reaches levels above 10^{-10} W/m^2-Hz and remains relatively constant over an altitude range of several thousand kilometers.

STRONGLY TURBULENT WAVE SPECTRA

An example of a strongly turbulent wave spectra produced by an electron beam has been observed by Voyager 1 near Jupiter's bow shock (Gurnett et al., 1981). An electron beam of estimated energy about 10 keV is generated at the bow shock. The beam travels away from the bow shock into the solar wind plasma along the inner plane-tary magnetic field. The solar wind plasma can be considered as essentially unmagnetized for investigating the beam plasma inter-action. Plasmons grow up with k-vectors in the direction of the beam and are swept back toward the shock by the streaming solar wind which flows much faster than the plasmon group velocity. The direc-tionality of the plasmon spectrum makes the problem essentially one dimensional, and the strong turbulence regime can be expected to occur at wave energy densities $E^2/8\pi nT \gtrsim (k\lambda_D)^2$.

As the spacecraft approaches Jupiter upstream from the bow shock a narrow band of wave noise is observed at the plasma frequency with wave numbers $k\lambda_D \simeq .015$. The field intensity increases for a time until a peak amplitude around 1 mV/m is reached after which the field intensity remains relatively constant around 100 μV/m. The bandwidth of the noise remains narrow until the region of

constant intensity is reached at which point sideband emissions begin
to appear. The sideband emissions therefore appear to result from a
saturation mechanism (i.e. a process that limits the amplitude of
the beam produced noise). The frequency bandwidth covered by the
sidebands continues to grow steadily until the spacecraft enters
the bow shock where it disappears. The event lasts about 20 seconds
in the solar wind frame. Figure 13 shows details of the wave spectrum at a point where the sideband spread is widest, just before the
bow shock. The frequency spectrum in Figure 13 actually corresponds
to a k-spectrum because of large doppler shifts caused by the solar
wind flow. The largest wave numbers observed have $k\lambda_D \simeq .15$.

This wave spectra appears to represent an excellent example of
the processes expected in strong turbulence, namely, rapid transfer
of long wavelength energy to short scale lengths with the resultant
formation of localized electric fields and density cavities (Morales
and Lee, 1975; Nicholson et al., 1978; Smith et al., 1979). The
onset of the transition to the nonlinear regime can be represented
by the oscillating two streams (OTS) instability, a parametric process (Papadopoulos et al., 1974). The onset of the nonlinear regime
and subsequent field saturation is predicted from the OTS instability
to occur when $\bar{W} = E^2/8\pi nT > (k\lambda_D)^2$, where k is the pump wave number.
This threshold is never exceeded by the observed field strength.
The highest field observed has $\bar{W}$ over an order of magnitude below
threshold while the typically observed $\bar{W}$ is over three orders of
magnitude below threshold. Furthermore, the transition to localized
fields (Zakharov, 1972) once threshold is exceeded should occur with
time scales on the order of the ion acoustic travel time across a
typical ion wavelength, $2\pi/c_s k \simeq .25$ sec. Rather than making this
rapid transition, the wave number spectrum is observed to steadily
broaden at a rate about two orders of magnitude slower.

The transition to short wavelengths appears to occur, not from
a parametric instability process, but rather as a result of scattering from a nonthermal flux of ion acoustic waves always present in

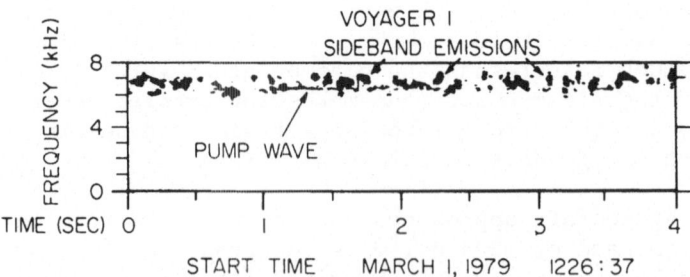

Fig. 13. The frequency spectrum of waves produced by an electron
          beam near Jupiter's bow shock shows the production of
          short wavelength plasmons from the long wavelength pump.

the solar wind. The observed steadily broadening frequency structure results from multiple scattering in which the wave number increases by the maximum allowable amount $\Delta k \lambda_D = 2(m_-/m_+)^{1/2}/3 \simeq .015$ at each scattering. The observed ion wave electric field strengths and spectra are calculated to correspond to a scattering rate rapid enough to account for the observed rate of broadening. This scattering process does not result in field saturation because the beam growth rate must exceed the scattering rate for plasmons to grow up out of the noise in the first place. If scattering is indeed occurring the noise must saturate as a result of some other process. Perhaps convective effects are as important at Jupiter's bow shock as they are in the auroral ionosphere.

CONCLUDING REMARKS

Ray refraction in the inhomogeneous auroral ionosphere plays an important role in limiting wave noise not only over the entire frequency band of propagating electrostatic modes but especially near the electron plasma frequency. The resulting broad band wave spectra permits the application of weak turbulence theory over a considerable range of auroral beam densities. The application of the concepts of convective amplification and quasilinear diffusion to the problem of electrostatic wave generation by the auroral beam results in predicted power flux spectra and beam evolution in general agreement with observation. From a microscopic viewpoint the dominant interaction of electrostatic modes with the electron beam is through the Landau resonance resulting directly in parallel heating of the beam. From a global point of view the beam evolution is controlled primarily by the converging magnetic field and the principal mode generated by the auroral beam is the electrostatic whistler. While the concepts of weak turbulence are evidently widely applicable to the auroral beam problem they are not exclusively so. The strong turbulence regime can certainly be achieved by reasonable values of beam fluxes and ionospheric parameters and undoubtedly is important some, if not much of the time. Perhaps auroral kilometric storms are an indirect manifestation of the strong turbulence regime. On the other hand complex and dynamic wave spectra appear to occur even in the weak turbulence regime as evidenced by observations near Jupiter's bow shock.

ACKNOWLEDGEMENTS

This work was supported by the Office of Naval Research, contract N00014-76-C-0307.

REFERENCES

Barbosa, D.D., 1976, Electrostatic mode coupling at $2 \omega_{UH}$: A gener-
    ation mechanism for auroral kilometric radiation, Ph.D. thesis,
    Univ. of Calif., Los Angeles.
Eliasson, L., Holmgren, L.-A. and Rönnmark, K., 1979. Pitch angle
    and energy distributions of auroral electrons measured by the
    ESRO 4 satellite, Planet. and Space Sci., 27:87.
Frank, L.A. and Gurnett, D.A., 1971, Distributions of plasma and
    electric fields over the auroral zones and polar caps, J.
    Geophys. Res., 76:6829.
Grabbe, C., Palmadesso, P. and Papadopoulos, K., 1980, A coherent
    nonlinear theory of auroral kilometric radiation: I. Steady
    state model, J. Geophys. Res., 85:659.
Grabbe, C.L., 1981, Auroral kilometric radiation: A theoretical
    review, Revs. of Geophys. and Space Phys., in press.
Gurnett, D.A., 1972, The earth as a radio source: Terrestrial kilo-
    metric radiation, J. Geophys. Res., 77:640.
Gurnett, D.A., Anderson, R.R., Scarf, F.L., Fredericks, R.W. and
    Smith, E.J., 1979, Initial results from the ISEE-1 and -2 plasma
    wave investigation, Space Sci. Reviews, 23:103.
Gurnett, D.A., Maggs, J.E., Gallagher, D.L., Kurth, W.S. and Scarf,
    F.L., 1981, Parametric interaction and spatial collapse of beam-
    driven Langmuir waves in the solar wind, J. Geophys. Res., in
    press.
Kaufmann, R.L., Dusenbury, P.B., Thomas, B.J. and Arnoldy, R.L.,
    1978, Auroral electron distribution, J. Geophys. Res., 83:586.
Kennel, C.F. and Engelman, F., 1966, Velocity space diffusion from
    weak plasma turbulence in a magnetic field, Phys. Fluids,
    9:2377.
Lee, L.C. and Wu, C.S., 1980, Amplification of radiation near cyclo-
    tron frequency due to electron population inversion, Phys.
    Fluids, 23:1348.
Lin, C.S. and Hoffman, R.A., 1979, Characteristics of the inverted-V
    event, J. Geophys. Res., 84:1514.
Lotko, W. and Maggs, J.E., May 1981, Amplification of electrostatic
    noise in cyclotron resonance with an adiabatic auroral beam,
    J. Geophys. Res., 86.
Maeda, K., 1975, A calculation of auroral hiss with improved models
    for geoplasma and magnetic field, Planet. and Space Sci.,
    23:843.
Maggs, J.E., 1976, Coherent generation of VLF hiss, J. Geophys. Res.
    81:1707.
Maggs, J.E., 1978, Electrostatic noise generated by the auroral
    electron beam, J. Geophys. Res., 83:3173.
Maggs, J.E. and Lotko, W., May 1981, Altitude dependent model of the
    auroral beam and beam generated electrostatic noise, J. Geophys.
    Res., 86.

Melrose, D.B., 1976, An interpretation of Jupiter's decametric radiation and the terrestrial kilometric radiation as direct amplified gyroemission, Astrophys. J., 207:651.

Melrose, D.B., and White, S.M., 1980, Amplified Cerenkov emission of auroral hiss: Limitations implied by quasilinear theory, J. Geophys. Res., 85:3442.

Meng, C.-I., 1976, Simultaneous observations of low energy electron precipitation and optical auroral arcs in the evening sector by the DMSP 32 satellite, J. Geophys. Res., 81:2771

Morales, G.J. and Lee, Y.C., 1976, Spiky turbulence generated by a propagating electrostatic wave of finite spatial extent, Phys. Fluids, 19:690.

Mozer, F.S., Cattel, C.A., Hudson, M.K., Lysak, R.L., Temerin, M. and Torbert, R.B., 1980, Satellite measurements and theories of low altitude auroral particle acceleration, Space Sci. Reviews, 27:155.

Nicholson, D.R., Goldman, M.V., Hoyng, P. and Weatherall, J.C., 1978, Nonlinear Langmuir waves during Type III solar radio bursts, Astrophys. J., 223:605.

Papadopoulos, K. and Coffey, T., 1974, Anomalous resistivity in the auroral plasma, J. Geophys. Res., 79:1558.

Papadopoulos, K., Goldstein, M.L. and Smith, R.A., 1974, Stabilization of electron streams in Type III solar radio bursts, Astrophys. J., 190:175.

Roux, A. and Pellat, R., 1979, Coherent generation of auroral kilometric radiation by nonlinear beatings between electrostatic waves, J. Geophys. Res., 84:5189.

Smith, R.A., Goldstein, M.L. and Papadopoulos, K., 1979, Nonlinear stability of Type III solar radio bursts. I. Theory, Astrophys. J., 234:348.

Wu, C.S. and Lee, L.C., 1979, A theory of the terrestrial kilometric radiation, Astrophys. J., 230:621.

Yamamoto, T., 1979, On the amplification of VLF hiss, Planet. and Space Sci., 27:273.

Zakharov, V.E., 1972, Collapse of Langmuir waves, Sov. Phys. JETP, 35:908.

DISCUSSION

Papadopoulos: First a comment: One has to be careful in interpreting local measurements of electric fields, E, as to whether they satisfy

$$\frac{E^2}{8\pi nT} > k^2 \lambda_D^2$$

or not, due to the inadequacy of measuring techniques to

resolve the peak value of the electric fields.   In
addition when

$$\frac{E^2}{8\pi nT} \approx (\frac{\omega_{ce}}{\omega_{pe}})^2$$

one can get quasi-stable solitons even in three dimensions.

Maggs: The pump and side band wave forms were completely
resolved by the wave experiment and were very coherent
so amplitude measurements are thought to be very accurate.   At Jupiter

$$(\omega_{ce}/\omega_{pe})^2 \triangleq 10^{-4} \approx (k\lambda_D)^2$$

so the latter condition is not satisfied either.

Papadopoulos: The spatial model of the beam (i e n_b)
depends strongly on the adiabatic invariance.   How is
this affected by the diffusion due to low frequency
instabilities?

Maggs: Diffusion of beam electrons by low frequency
noise was not considered.

Chapter 3:   ACCELERATOR EXPERIMENTS IN THE
             LABORATORY

LABORATORY SIMULATION OF INJECTION PARTICLE BEAMS

IN THE IONOSPHERE

P.J. Kellogg[1], H.R. Anderson, W. Bernstein[2],
T.J. Hallinan[3], R.H. Holzworth[4], R.J Jost[5],
H. Leinbach[6], and E.P. Szuszczewicz[7]

[1]University of Minnesota, Mpls., MN
[2]Rice Univer., Houston, TX 77001
[3]University of Alaska, Fairbanks, AK 99701
[4]Aerospace Corp., Los Angeles, CA 90009
[5]Johnson Space Center, Houston, TX 77058
[6]NOAA, Boulder, Co 80302
[7]Naval Research Laboratories, Washington, D.C. 20375

INTRODUCTION

When Hess (1971) and Winckler (Hendrickson et al., 1971) carried
out the first rocket-borne, electron beam injection experiments,
the results appeared to be consistent in some ways with the concept
that the beam electrons demonstrated single particle behavior over
their entire trajectory from injection to detection. These results
were somewhat surprising; it was expected that beam plasma instabilities would occur and significantly modify the beam characteristics.
Nevertheless, the success of the flight experiments led to the design
of several experiments in which electron beams would be used as
non-perturbing remote probes in magnetospheric and auroral studies.
A fairly complete listing and description is given in USRA (1976).
Generally, these experiments assumed single particle motion at even
greater power levels than had been used in the rocket experiments.

The earlier experiments, however, did present a number of
puzzles, both in the particle behavior and more particularly in the
behavior of the plasma waves generated by these beams (Cartwright
and Kellogg, 1974; Monson et al., 1976; Monson et al., 1978).
Further, continuing experiments demonstrated even more sharply these
anomalies (Cambou et al., 1975, 1978; Monson and Kellogg, 1978;
Maehlum et al., 1980, Pyatsi and Zarnitsky, 1980). It was clear
that some of the puzzles were so sharply at varience with

289

expectations that there must be some fundamental processes occurring
which were not understood. William Bernstein (Bernstein et al.,
1975, 1977, 1978, 1979) then organized a series of experiments in
large vacuum chambers, test chambers of NASA, in order to try to
understand the physics of the beam injection experiments. The first
experiments took place at Plumbrook near Cleveland, Ohio. When that
chamber was closed, the experiments were moved to Chamber A at
Johnson Space Center, where the addition of some features specifically
for plasma work, particularly a stronger magnetic field, led to
considerable progress.

   Chamber A is a cylinder with hemispherical top and bottom ends.
It is about 20 m in diameter (horizontal) and the usable vertical
space is about 30 m high. The experimental configuration is shown
in Fig. 1. The experimental conditions were as follows:

1) A tungsten cathode, Pierce, electron gun was used to generate
    the beam. The maximum beam currents and voltages employed were
    100 mA and 2 kV, respectively. For voltages somewhat less than
    2 kV, the gun current was space-charge limited at a perveance
    of $\sim 1.4 \; 10^{-6}$ $A/V^{3/2}$. The gun was operated both in DC and in
    a pulsed mode. Although the gun could be isolated electrically,
    so that it was forced to collect its return current from the
    plasma, it usually was grounded to the chamber walls.

2) A set of three coils have been added around the chamber peri-
    phery to create a magnetic field; the total variation in field
    strength along the beam path was $\sim 15\%$. Most measurements were
    made at mean field strengths ranging from 1.0-1.45 G. Typical
    beam injection pitch angles were < 20°. The path length between
    gun and collector was ~ 20 m.

3) The base pressure in the system was 1×10^{-6} torr, consisting
    primarily of water vapor (30%) and N_2. Increases in pressure
    to 2×10^{-5} torr were accomplished with the addition of dry N_2.
    This pressure range corresponds to the altitude range 120-180
    km; although rocket-borne accelerators have been flown at higher
    altitudes, it is probable that the rockets are always surrounded
    by gas clouds of similar density which are produced by outgassing
    and residual motor exhaust.

4) As shown in Fig. 1, the primary diagnostics consisted of:

    (a) Several low light level television systems (total light)
         to provide a picture of the overall beam configuration and
         rough measurements of emission intensity. These were
         provided and operated by Tom Hallinan, University of Alaska,
         and by Johnson Space Center.
    (b) A scanning 3914A photometer to provide more quantitative
         measurements of the beam profile and the rate of ionization

of the neutral gas. The photometer scanned perpendicularly
to the beam about midway along the beam path. This was
provided by H. Leinbach, NOAA, Boulder.

(c) A circularly segmented collector to provide both measure-
     ments of the beam current profile and a known termination
     of the beam.

(d) A high resolution curved plate electrostatic energy analyzer,
     mounted in front of the collector. This was provided by
     H.R. Anderson, Rice University.

(e) Three different dipole antenna-spectrum analyzer systems
     distributed in the chamber to provide measurements of the
     RF wave spectrum. These were provided by the University
     of Minnesota, by W. Bernstein, and by R.H. Holzworth,
     Aerospace Corporation.

(f) An array of Langmuir probes, some fixed, some movable, to
     provide measurements of the radial and longitudinal density
     and temperature distributions. These were provided by the
     University of Minnesota, by E.P. Szuszczewicz, NRL, and by
     A. Konradi, Johnson Space Center.

(g) A plasma generator (a Kaufman ion engine) to generate a
     background plasma. This was operated by A. Konradi and
     B. McIntyre of Johnson Space Center for the runs presented
     here.

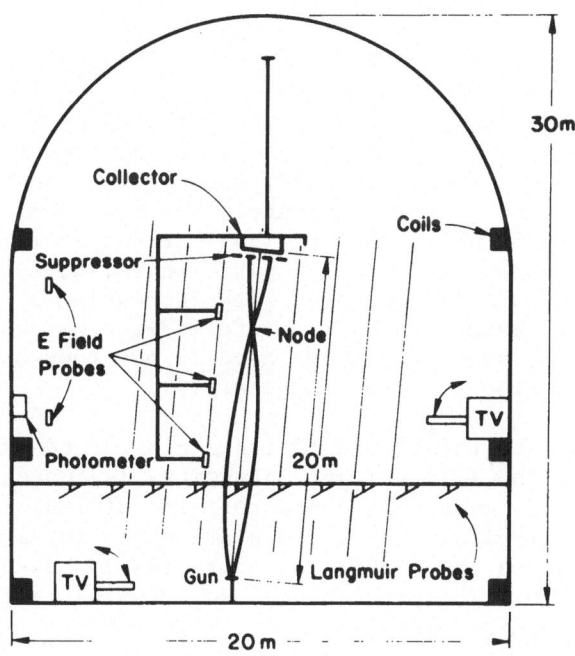

Fig. 1.   Schematic representation of the experimental arrangement.

IGNITION CONDITIONS FOR BEAM PLASMA DISCHARGE

In the first series of experiments, the electron beam was injected into an initially neutral gas. A plasma then accumulated as a result of ionization of the gas by the electron beam. The long interaction length and the longitudinal magnetic field resulted in the accumulation of plasma densities considerably greater than the beam density.

The most significant experimental result is that electron beams in the low keV energy range can ignite the beam plasma discharge when their perveance is $> 5 \times 10^{-7}$ $AV^{-3/2}$, for pressures typical of the ionosphere. This is a typical perveance for flight experiments. Despite the very different conditions from earlier laboratory BPD experiments (Getty and Smullin, 1963; Karchenko et al., 1962; Alexeff et al., 1964), all qualitative features of the ignition and of the BPD are duplicated and the entire process scales quantitatively between the earlier and present experiments.

The characteristic features of the BPD include:

1)   Ignition occurs at and above a specific beam current I_c for a given set of experiment parameters. The dependence of I_c on these parameters is summarized by the empirical experimentally determined relationship (Bernstein et al., 1979)

$$I_c \sim \frac{E_b^{3/2}}{BL} \, f(p) \qquad (1)$$

where E_b is the beam energy, B is field strength and L is interaction length. p is the pressure and $f(p)$ is a function with a minimum at $1-2 \ 10^{-5}$ torr, and varying roughly as $p^{\pm.5}$ to $p^{\pm 1}$ above and below this pressure. For L = 20 m, B = 1 G, E = 1 keV and p = $5 \ 10^{-6}$T, the critical current is 20 mA. I_c appears to be independent of injection pitch angle, α, for angles $< 60°$; for $60 < \alpha < 80°$, I_c decreases with increasing α. We have not observed any maximum in I_c for intermediate values of α. The exponent of B, namely 1, given here is now thought to be more accurate than the original value of .7.

2)   At ignition, the total rate of ionization of the neutral N_2, derived from the λ 3914 light intensity, is increased by about a factor of 100 over that characteristic of ionization by the beam alone. Typical observations are shown in Fig. 2. The left panel shows the λ 3914 intensity (A) in sub-BPD conditions and (B) in BPD. The right hand panel shows a hysteresis effect which will be discussed later.

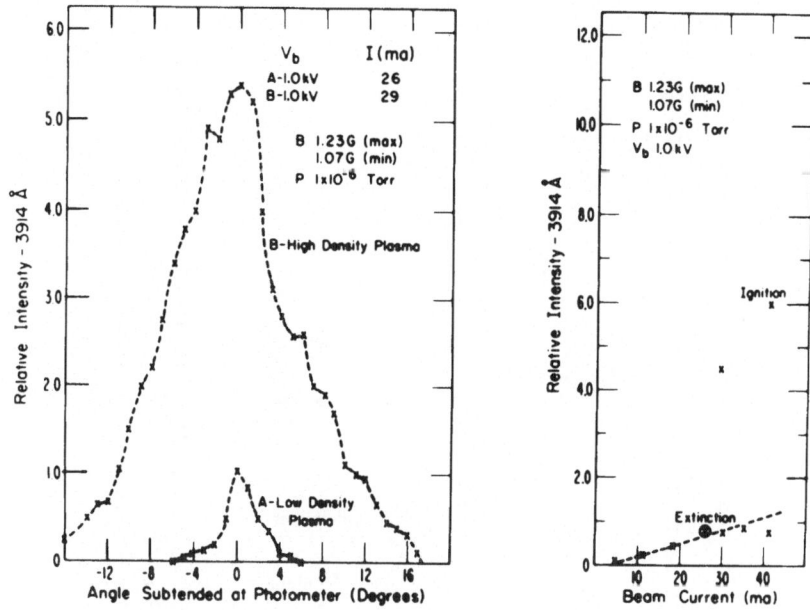

Fig. 2.   Illustration of the 3914 scanning photometer data: a)
comparison of the intensity profiles in the low and high
density states, b) the dependence of peak 3914 intensity
on beam current showing the abrupt transition for increasing
beam current at 42 mA and subsequent hysteresis as the beam
current is reduced to 26 mA, the high density state is
extinguished and the beam returns to the low density
condition.

3) The increased rate of ionization leads to an increased plasma
    density. Radial diffusion is also enhanced so that the density
    at any point is increased by less than the total chamber content.

4) The primary beam is severely perturbed with some particles
    accelerated to energies 30% above the nominal beam energy and
    also a large increase in the flux of degraded particles. An
    example of beam heating is shown in Fig. 3. These observations
    were made with a high resolution curved plate analyzer mounted
    on the collector. Two pairs of examples of beam energy distri-
    bution are shown, one each in the sub-BPD regime (50 mA beam
    current in each case) and the other in BPD (80 and 86 mA beam
    current).

5) Intense broad band HF waves are observed; the frequency appears
    to correspond approximately to the ambient plasma frequency.
    This band shows strong structure at harmonics of the electron
    cyclotron frequency. In addition, waves with a flat frequency

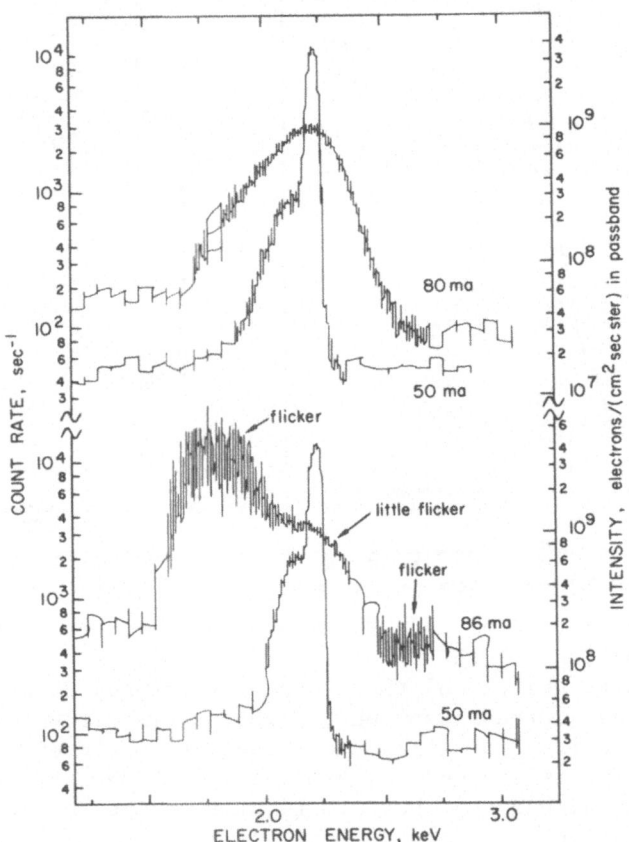

Fig. 3.   Energy distributions of beam electrons for pre-BPD ($I<I_c$),
          and BPD ($I>I_c$). Note that intensity in the passband
          rather than differential flux is plotted. The passband is
          proportional to the energy.

spectrum extending from about 100 kHz to the electron cyclotron
frequency are observed; i.e., whistler mode waves.

6) Severe pitch angle scattering and radial diffusion of the primary
    beam occur. Coupled with the energy spreading, these effects
    obliterate the single particle trajectory features of the beam
    within a few meters of the injection point.

7) Once ignited, the strength of the beam plasma discharge increases
    with increasing beam current.

8) The addition of an auxiliary plasma with density comparable to
    that at ignition (10^6 cm^{-3}) has no effect on I_c or actually
    slightly increases I_c.

Ignition of the BPD may provide an explanation for many of the puzzling results observed in flight injection experiments (Galeev et al., 1976; Cambou et al., 1978; Winckler, 1979; Grandal et al., 1980). It will have very serious implications for those experiments planned for Spacelab which assume complete single particle behavior of the beam. Although the ambient neutral density in the upper ionosphere and magnetosphere may be insufficient for ignition of the BPD, vehicles such as rockets and Spacelab may be surrounded by their own neutral clouds arising from a variety of out-gassing and off-gassing processes.

EFFECT OF AN AMBIENT PLASMA

The vacuum chamber experiments have been undertaken with an aim to understanding what happens in similar rocket experiments. A major difference between the vacuum chamber results and the rocket experiments is the presence of an initial background plasma in the latter. Indeed, one might have believed that the scaling laws, eq. (1), reflect merely a critical current necessary to generate enough plasma for the instabilities to have sufficiently fast growth rates. We therefore have undertaken a series of experiments using an ion thruster to generate a background plasma in the chamber. This thruster was a plasma source of the Kaufman type and generated a very broad fan of plasma consisting of argon ions of a few ten's of eV energy and sufficient electrons to neutralize them. The ion thruster was capable of making a plasma density of the order of .5 - 1 x 10^6 cm^{-3}. The plasma was reasonably uniform.

We were very surprised to find that this background plasma had only a very slight effect on the threshold conditions expressed in eq. (1) and, in fact, made it slightly more difficult to generate beam plasma discharge, rather than easier as we had expected.

This slight increase of the threshold current for beam plasma discharge can be partly understood on the following basis. For gun currents which are below beam plasma threshold, the plasma in the chamber is very unstable. There are density fluctuations of more than an order of magnitude and large low frequency fluctuations of the electric field. When the beam plasma discharge is ignited, the plasma density is greatly enhanced, the growth rates of instabilities are increased, and the system remains in BPD even when the gun current is reduced well below the threshold current.

An example of this hysteresis is shown in the right hand panel of Fig. 2, where for some currents, the system may be either in BPD (high light intensity) or not. Note that there is almost a factor of two difference between the current necessary to ignite the BPD, and the current at which it extinguishes, once ignited. It is clear from our observation that the fluctuations may occasionally lead to

BPD conditions. The beam plasma discharge then is excited, and the system remains in BPD. The actual threshold current expressed in eq. (1) then may be an underestimate. However, when the ion thruster is used to provide a background plasma in the chamber, the sub-BPD plasma is more stable and fluctuations are less pronounced. It therefore appears that the slightly higher threshold current obtained when the thruster is operating is a truer measure of the critical current for BPD.

While the hysteresis effect may provide some partial under-standing, we believe it is still a puzzle as to why the background plasma has so little effect, for as we shall show in a later section the plasma densities at BPD ignition can be changed by a factor of the order of 2 when the thruster is used, without affecting the critical current I_c.

THE LANGMUIR PROBE SYSTEMS, AND ANALYSIS

Two Langmuir probe systems were operated in the vacuum chamber for the determination of plasma density, temperature and plasma potential. The probe systems of the University of Minnesota group consisted of 28 long cylindrical probes strung in two strings across the center of the chamber. Usually one string was about six meters from the floor of the chamber and the other about three meters. Most of the probes were 45 cms long and .32 cms in diameter but the two central ones in each string were only half as long. The probes were gold-plated. Each probe was connected to the electronics box, located at the side of the chamber through a long wire. The necessity of driving the capacitance of this wire dictated probes of fairly large area.

The Naval Research Laboratory group (E.P. Szuszczewicz and co-workers) operated a pulsed plasma probe, a design which they have frequently flown on rockets and satellites. Their probe sits at a constant potential most of the time, the probe characteristic being traced out by raising the potential of the probe to the desired voltage for a very brief time, making a quick measurement of the current and then returning to its constant potential. Because of the necessity of rapid changes in the probe potential, the probe must be very close to its electronics and so the whole system of probe plus electronics box was suspended from a cable and physically moved across the chamber in order to make density and temperature profiles.

The data from fixed Langmuir probes in the vacuum chamber are subject to three errors. First, while the Langmuir probes are being swept in voltage in order to measure the current-voltage character-istic, they change the plasma distribution around them. (We refer to this as the draining problem.) Second, at the floating potential

the probes apparently acquire a bi-polar layer of atoms which
results in a potential at the outside of the layer which is several
volts different from the probe voltage itself. In the course of
the sweep, this layer is partially depleted, so that the floating
potential appears to undergo a change of a few volts between sweeps.
We minimize this effect by sweeping both up and down in voltage and
averaging the sweeps. Also, it is somewhat reduced by first raising
the probe to a positive voltage for a short length of time before
beginning the sweep, in order that the collected electrons can knock
off some of the atom layer.

The draining problem, i.e., the draining of all the electrons
available to a probe, may be illustrated as follows. We assume that
the electrons available for collection by the probe are those
contained in a volume whose dimensions are the length, L, of the
chamber (20 meters), the length of the probe ℓ and a width equal to
the diameter of the probe plus one Larmor diameter on each side.
The number of electrons in this volume is then:

$$N_e = n\ell L(d + 4r_L) \tag{2}$$

where n is the electron density and r_L is a typical Larmor radius.

For electron saturation current to the area of the probe

$$I = J_e \pi d \ell$$

where J_e is the electron saturation current

$$J_e = \frac{n\bar{v}e}{4} = ne\sqrt{\frac{kT_e}{2\pi m}} \tag{3}$$

and the Larmor radius is taken as

$$r_L = \frac{mvc}{eB} = \frac{mc}{eB}\sqrt{\frac{2kT_e}{m}} \tag{4}$$

the time to drain out the electrons in this volume is

$$T = \frac{n\, Le(d + 4r_L)}{J_e \pi d} \sim \frac{8}{\sqrt{\pi}} \frac{L}{d} \frac{1}{\omega_{ce}} \tag{5}$$

in the limit $4r_L \gg d$. Here $\omega_{ce} = eB/mc$.

For typical operating values L = 20 m, d = .32 cm and B = 1
gauss, this gives

$$T = 1.6 \text{ msec}$$

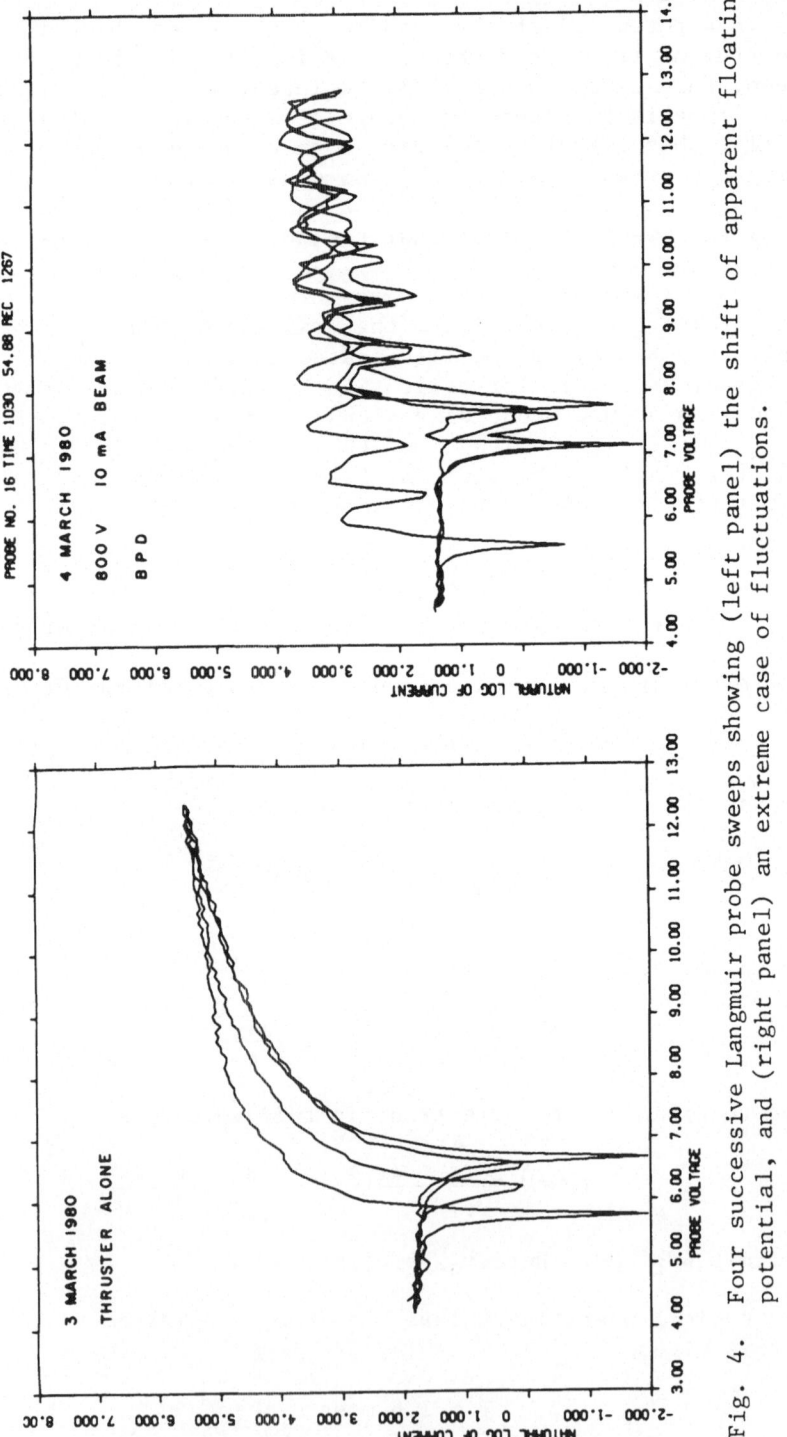

Fig. 4. Four successive Langmuir probe sweeps showing (left panel) the shift of apparent floating potential, and (right panel) an extreme case of fluctuations.

The time to sweep one of the University of Minnesota probes from its lowest to its highest voltage was of the order of a second so that the drainage problem is severe and the saturation current being measured is limited by electron diffusion onto the field lines containing the probe. Comparison with plasma frequency measurements shows that the error is less than a factor of two, however. In more recent experiments, this problem has been reduced by pulsing the probe, but cable capacity prevents pulsing at a rate which is very fast compared to 1 msec.

In the left panel of Fig. 4 we illustrate the second problem, the apparent change of the potential of the probe with respect to the plasma as it is swept up and down. Shown at the left in this figure are two sweeps toward positive voltage and two sweeps toward negative voltage of a probe. The left or most negative sweep is the first and the successive sweeps move toward more positive floating potentials.

Before making these sweeps, the probe has already been "cooked" at a potential of about 4 volts positive for about 300 msec in order to reduce the change of potential during the sweep. In Fig. 5 we show the floating potential of the string of probes which has been sitting without applied voltage for some length of time together with floating potentials obtained during sweeps. It will be seen that the sitting probe has a floating potential which is even more negative, presumably due to the cooking, and that the total change of the floating potential from the sitting probe to the last of the four sweeps can be as much as four volts. Nevertheless, the potentials obtained under various conditions seem to track each other quite well, so we guardedly accept relative potential differences as being accurate.

Both problems can be alleviated as far as temperature and density determinations go by the use of the pulsed plasma probe of the Naval Research Laboratory group. This probe generally sits at some fixed potential but is stepped to another potential for about 50 µsec, the current measured, and then is returned to the base potential. After a certain time, another step is made to another voltage and the entire I-V characteristic is traced out in this way. Since the probe is actually only away from its base potential for 50 µsec, the draining problem is smaller for the step and depends mainly on the draining at the base potential. The change of probe potential due to removal of the layer of atoms is also negligible; however, the measured floating potential is again not the true floating potential of the plasma because presumably the layer of atoms is undisturbed during the entire measurement.

The third problem is that the Langmuir probe characteristics that are obtained in the chamber are not very "clean" because of

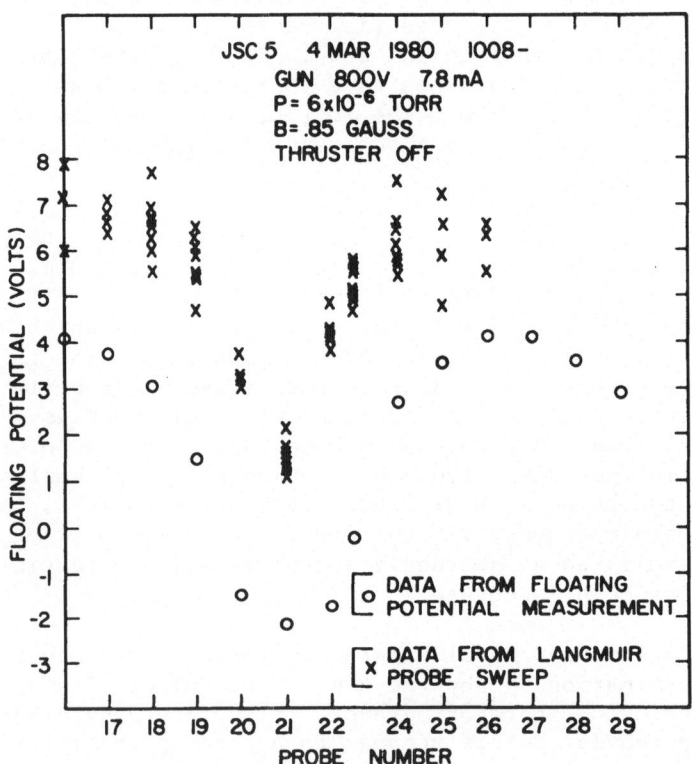

Fig. 5.   Floating potentials of probes obtained by direct measurement
at equilibrium (0) and during successive sweeps (x).

fluctuations of plasma density and possibly of floating potential, which are almost always present. These fluctuations, in fact, become most extreme in just the situations of most interest. An example is shown in the right panel of Fig. 4. It is therefore rather difficult to see the knee of the Langmuir probe characteristic, i.e., the point at which the electron saturation current is being collected and at which the probe is at the same potential as the plasma. Identification of this point is crucial to the normal Langmuir probe analysis. We have therefore reduced our Langmuir probe data using a method which depends on the following property. When a long cylindrical probe is sufficiently positive with respect to the plasma, the collected current is linearly related to the square root of the potential. As is shown already by Langmuir and Mott-Smith (1926), when there are no trapped orbits the current is

$$I = n \sqrt{\frac{m}{2\pi kT_e}} \, A_p \int_o^\infty \sqrt{v^2 + \frac{2e\phi_p}{m}} \, \exp\left(-\frac{mv^2}{2kT_e}\right) dv$$

$$\approx nA_p \sqrt{\frac{e\phi_p}{2m}} \left[1 + \frac{1}{4}\frac{kT}{e\phi_p} + \dots \right]$$

(6)

where ϕ_p and A_p are the potential and area of the probe. This result is only valid provided that there are no trapped orbits through the probe, which implies that the "collection distance"

$$s^2 = r_p^2 \left(1 + \frac{2e\phi_p}{mv_o^2}\right) \approx r_p^2 \left(1 + \frac{e\phi_p}{kTe}\right)$$

(7)

be smaller than a Debye length.

The temperature and density are evaluated by a computer and the algorithm is as follows: First, the floating potential is found, i.e., the point at which the probe current changes sign. Then the plasma potential is estimated at 2 volts above the floating potential and the data fitted to a curve of the form

$$\ln I = A_1 + B_1 V$$

(8)

for probe potentials in the electron retarding region and to a curve of the form

$$I^2 = A_2 + B_2 V$$

(9)

for probe potentials above the plasma potential. The temperature is then taken from the coefficient $B_1 = e/kT_e$ in the electron retarding region and the plasma density is obtained from the coefficient B_2

$$n = \sqrt{\frac{2mB_2}{e}} \, \frac{1}{A_p}$$

(10)

The electron saturation current, I_{sat}, can now be evaluated. These numbers allow us to calculate a new estimate of the plasma potential V_0 from eq. (8) above; i.e.,

$$\ln I_{sat} = A_1 + B_1 V_0 \tag{11}$$

and the whole process is repeated. This iteration is performed until the change in plasma potential is less than 1% or until five iterations have been performed. Each of four sweeps, two up and two down, is fitted separately and the final results averaged.

A second method of analysis was used to investigate fluctuations of plasma density. In this the probes were successively biased positive with respect to the chamber walls by two different potentials and a number of samples of current taken (36 successive samples with an adjustable time period between). From each run the maximum, average, and minimum current were obtained and then the two runs determine the slope of I^2 vs. voltage (the coefficient B_2 above) and hence the maximum, average, and minimum densities. This method, of course, does not give a temperature, or any details of the electron distribution.

PLASMA PROPERTIES EXPECTED FROM THE SINGLE PARTICLE PICTURE

We consider the conditions in the system immediately after the beam has turned on. We suppose that the gun has been aligned along the magnetic field, our most common operating condition. Since the divergence angle of the gun is only about 5° (half width), all electrons have very nearly the same velocity along B and so all will return to the field line through the gun nozzle at the same distance. This point, the "node", is generally easy to see on the TV even for weak beams. It will be seen that the envelope of the beam has a radius of twice the beam electron Larmor radius, or, for the aligned beam

$$r_b = 18.6 \left(\frac{E_b}{1 keV} \right)^{1/2} \left(\frac{1 \text{ Gauss}}{B} \right) \text{ cm.} \tag{12}$$

Initially, the beam itself represents a negative charge distribution in the chamber.

The beam immediately begins to ionize the background gas. The electrons from these ionizations will have energies of a few electron volts and will escape from the chamber along the magnetic field lines in a time of about 10 µsec. The ions receive negligible energy from the ionization so they are expected to be at room temperature. If they are collisionless, their time to escape from the chamber along the field lines is given by:

$$T = \frac{10 \text{ meters}}{\sqrt{\frac{kT_i}{M}}} \tag{13}$$

which is about 35 msec for N_2 ions at 300°K. The effective collision cross section for N_2^+ on N_2 is about 75 A^2, obtained from measurements of mobility (Loeb 1960, p. 58). The cross sections for other diatomic ion species with their neutrals is smaller but of the same order. The mean free path of an N_2 ion is then (Loeb loc. cit.)

$$\lambda_i = \frac{1}{\sigma n} \sqrt{\frac{M_i}{M_i + M_a}} = 5.4 m \times \frac{5 \ 10^{-6} \text{ Torr}}{P} \tag{14}$$

so that N_2^+ ions are marginally collisional. If the ions are collisional, their loss rate is governed by ambipolar diffusion rather than simple time of flight, but in the marginally collisional case, as here, the loss rate is not greatly different. Furthermore, we shall present reasons to believe that the ions are hotter than room temperature, and that neither ambipolar diffusion nor thermal escape time gives the lifetime, but that it is determined by instabilities.

The cross·section for ionization of electrons in the energy range under consideration is given to good accuracy by:

$$\sigma = .91 \ 10^{-6} \left(\frac{E_b}{1 kev}\right)^{-.63} cm^2 \tag{15}$$

(Brown, 1966). The mean free path of the beam electrons for ionization is then

$$\lambda = \frac{1}{n\sigma} = .62 \text{ km} \times \left(\frac{5 \ 10^{-6} \text{ Torr}}{P}\right) \left(\frac{E_b}{1 kev}\right)^{.63} \tag{16}$$

so that the mean free path is longer than the 20 meter path length used in the chamber for all energies and pressures which we use. However, it is only a few times longer than the chamber at the lowest energies and highest pressures. The probability that a beam electron makes an ion pair is then λ/L, and the number of ions made in the chamber per second is:

$$\text{ionizations/sec} = \frac{I_b}{e} \frac{L}{\lambda}$$

$$= 2.0 \ 10^{15} \left(\frac{I_b}{10 mA}\right) \left(\frac{P}{5 \ 10^{-6} \text{Torr}}\right) \left(\frac{1 \text{ kev}}{E_b}\right)^{.63} \tag{17}$$

The number of beam electrons in the chamber at a given time

$$N_b = \frac{I_b}{e} \frac{L}{\sqrt{\frac{2eE_b}{m}}} = 4.71 \ 10^{10} \left(\frac{I_b}{10mA}\right)\left(\frac{1kev}{E_b}\right)^{1/2} \tag{18}$$

At a time of the order of tens of microseconds, therefore, a sufficient number of positive ions have been made in the chamber to neutralize the beam electrons. Beyond this point, the potential of the center of the beam begins to become positive. Presumably then a sheath forms at the chamber ends and the secondary electrons from the ionizations begin to be trapped in the potential trough to maintain an approximately neutral system. We therefore expect that the center of the beam should be positive with respect to the end walls of the chamber in order to confine the secondary electrons, to maintain the necessary neutrality.

The transverse structure of the beam plasma system adds a complication. Even at room temperature, the Larmor radius of a typical ion, say N_2^+, is given by

$$r_L = \frac{1}{\omega_{ce}} \sqrt{\frac{2kT_i}{M}} = 125 \text{ cm x } \left(\frac{1 \text{ Gauss}}{B}\right) \tag{19}$$

and is therefore considerably larger than the radius of the beam. We expect these ions, then, to form a postive charge layer around the beam and secondary electron system, and that this charge layer forms a potential which confines the ions. We find then a system in which the ions are confined to the central column by electrostatic forces, but in which the center of the beam must be negative with respect to the plasma radially out from the beam, whereas the electrons must be confined by a potential in which the center of the beam is positive with respect to the end plates. The confining potentials in each case would be expected to be a few times kT for the confined species, which would only be a fraction of a volt for the ions. We presume that the plasma would resolve the sign conflict by some more complicated distribution in the corners of the chamber.

These considerations lead us to a number of expectations which we will now show are not borne out. These include 1) the plasma should be confined in a small region right around the beam, and the density should be very small, essentially zero, outside this region; 2) the density within the beam should be proportional to the beam current; 3) the ions should be confined by a small radial potential of a fraction of a volt; 4) this small potential should be a few times the ion temperature, and independent of beam current; 5) the electrons should be confined by a longitudinal potential of the order of a few volts.

COMPARISON WITH LANGMUIR PROBE MEASUREMENTS

When we confront these considerations with the results of
measurements using the University of Minnesota Langmuir probe system,

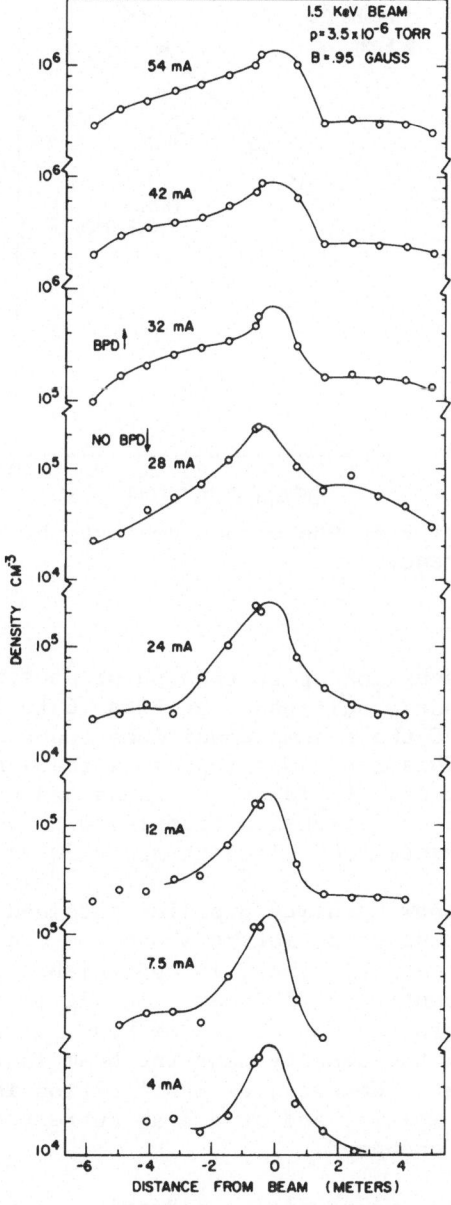

Fig. 6.   Plasma density profiles in the chamber for various gun
          currents.

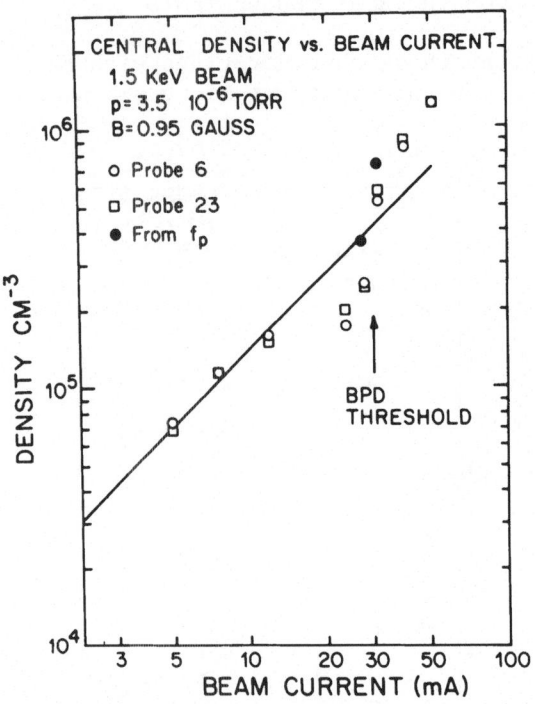

Fig. 7. Plasma density at the probes nearest the beam, as a function
        of gun current.

we find that the discussion up to this point does not give a correct
account of the chamber conditions. In Figs. 6 to 11 are presented
results of a study of the plasma conditions under approach to beam
plasma discharge. This study was made at a chamber pressure of
3.5 10^{-6} torr, a magnetic field of .95 gauss and beam voltage of
1.5 kV. Under these conditions, a transition to beam plasma discharge
(for increasing current) took place at about 30 mA beam current.

In Fig. 6 are shown density profiles obtained from the lower
string of probes, about three meters above the chamber floor. Even
for the lowest beam current, there is appreciable electron density
quite far from the center of the beam. In Fig. 7 we show the density
as measured by two probes closest to the beam, as a function of beam
current. Initially, the density near the beam increase linearly
with the beam current. However, as BPD ignition is approached, the
density rises less rapidly, and at a beam current of 12 mA the
decrease is quite pronounced.

The straight line representing a density proportional to beam
current can be used to calculate the ion lifetime in the pre-BPD
region. If we take merely the central part of the beam, then the

total electron content out to a radius of .7 meters is approximately (at, say 10 mA)

$$N_e = \pi r_b^2 LM$$

$$= \pi(70cm)^2 \ (2000 \ cm) \ (1.2 \ 10^5 \ cm^{-3}) \qquad (20)$$

$$= 3.7 \ 10^{12} \ \text{electrons}$$

Using eq. (17), the number of ionizations per second for a 10 mA beam is

$$\frac{dN}{dt} = \frac{I_b L}{e\lambda} = 1.0 \ 10^{15} \ \text{electrons/sec}$$

This results in a "central" lifetime of 3.7 msec, or considerably shorter than we estimated above due to the thermal motion of ions. On the other hand, if we include all of the plasma in the chamber, then we estimate that the total electron content of the chamber is about ten times higher than the content of this center of the beam, and so a lifetime for ions to escape from the chamber is given fairly well by the thermal lifetime.

Both of these results then can be interpreted to show that diffusion is quite important in the chamber. First, as the beam current is increased, we should expect the level of instabilities, which presumably are driving the diffusion, to increase and so the lifetime should decrease. This is a particularly strong effect when we have almost reached the beam plasma discharge threshold as can be seen from Fig. 7. Also, from the chamber profiles it can be seen that as the beam current increases, the central peak of density at the beam becomes less and less prominent relative to the density in the rest of the chamber and that just before beam plasma discharge the central peak amounts only to about a factor of two. This shows that radial diffusion increases with beam current. In BPD the central peak amounts to relatively little and the chamber is filled with an electron density which decreases rather smoothly to the walls.

In the next figure, Fig. 8, we showed direct observations of what we believe to be a cause or part of the cause of the diffusion. In this figure, we plot, as a function of time, the probe current in the successive probes which have been biased at five volts positive with respect to the chamber wall. We also show both the upper and lower strings. One of the upper probes, probe number 7, did not operate properly during this run. As we proceed from the bottom, that is, inward from the chamber wall, and at the lowest current, the outer probes show no measurable current. The upper string of probes begins to show some current above threshold first, indicating that there is also a certain amount of diffusion going

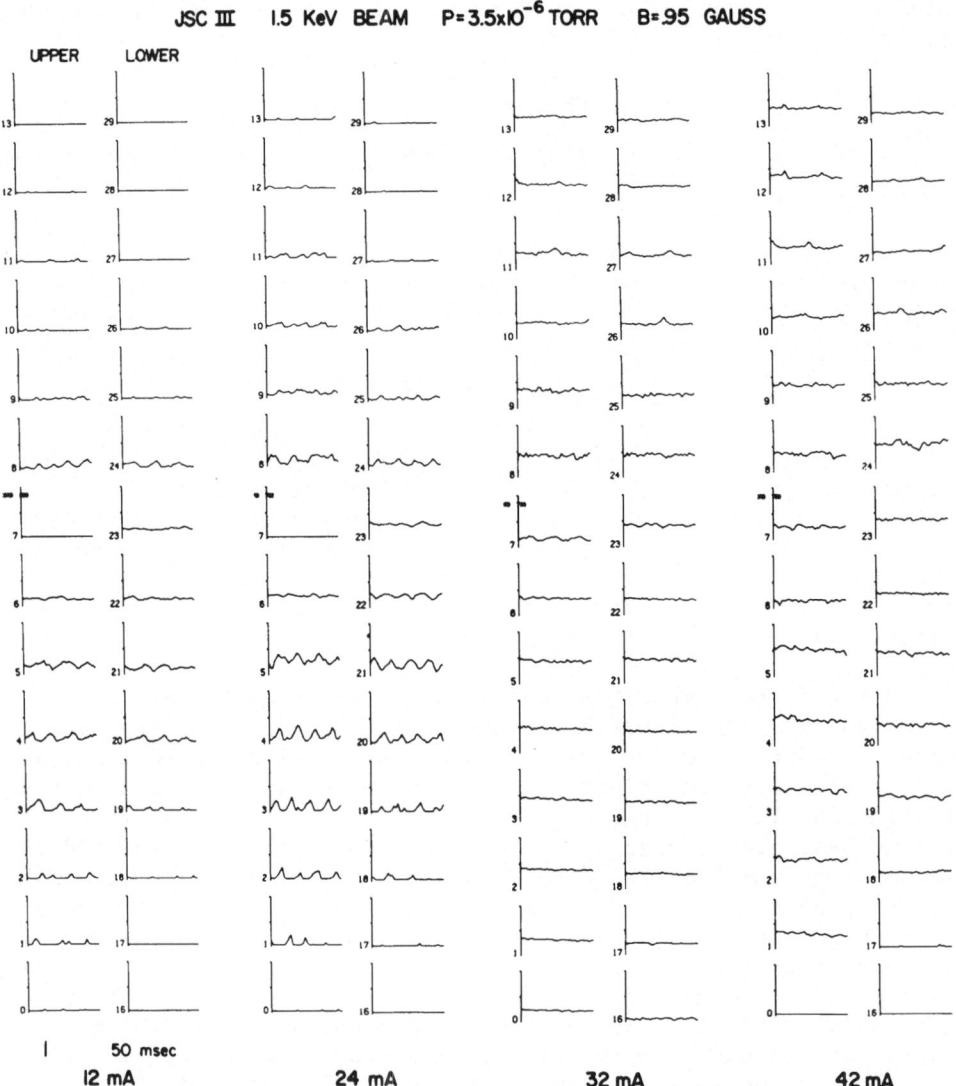

Fig. 8.   Current drawn to probes biased five volts positive.   The
          current is proportional to plasma density.

on along field lines as well as across, so that the density near the
ends of the chamber is reduced relative to that higher up. As we
come to probes number 3 to 5, we begin to see considerable fluctuations in the plasma density even at 12 mA. At 24 mA, i.e., a little
below beam plasma discharge, these oscillations are more than a
factor of two in density and become quite regular. (Inspite of
their appearance, the measurements are not simultaneous and so the
fact that some of the oscillations seem to line up in a vertical

direction is purely an accident.) At the center of the beam (probes 7 and 21) the density is higher than it is on the sides, but the oscillations are not very prominent. The other side of the beam we again have oscillations and then the density again decreases as we go toward the wall.

The probe system has a response time of about a millisecond, so does not respond well to fluctuations faster than about 200 Hz. This means that it can respond to waves of the order of the N_2^+ cyclotron frequency of 50 Hz, like those shown, but could not respond to the lower hybrid or even H^+ cyclotron frequencies. RF probe measurements show that there are also strong waves in the kHz range which also may drive diffusion.

Once the threshold for beam plasma discharge has been reached, the density oscillations at these low frequencies are very much

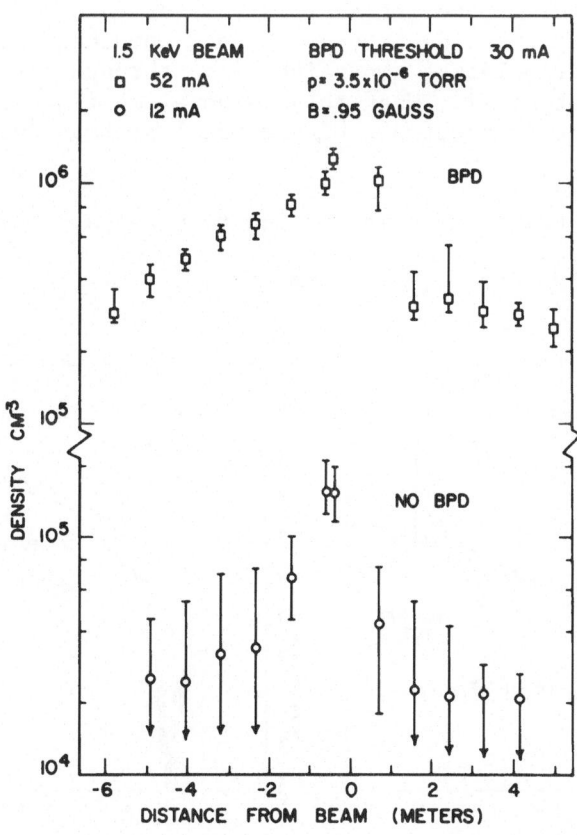

Fig. 9. Effect of BPD on density fluctuations. The bars indicate maximum and minimum densities found in 36 measurements.

reduced, although they become somewhat more prominent again at the
highest beam current shown, namely 42 mA. This effect, the
reduction of low frequency fluctuations by the BPD, is shown more
clearly in Fig. 9. In the lower panel is shown the average density
together with bars indicating the density obtained from maximum and
minimum currents as described above. In the upper panel, we show
the same thing for beam plasma discharge conditions. Not only is
it obvious that the relative density fluctuations are much reduced
under beam plasma discharge conditions, but this figure also shows
more clearly how the greatest density fluctuations take place in an
intermediate region between the beam and the wall.

We note that the density profiles are not symmetrical with
respect to the center of the beam. This is probably because the
beam was striking near one edge of the target, so that the available
length to the plasma was greater on the left side of Figs. 6 through
9.

When the BPD is ignited, there is a hot electron component near
the beam. Langmuir probe characteristics for this case are shown
in Fig. 10 (left). The density of the hot component must be
estimated by extrapolating a straight line through its characteristic
to the plasma potential. When this is done, we find the density of
the hot component is about 30% of the total plasma, and its temperature to be 3 eV.

Since this temperature is too low to cause appreciable ioniza-

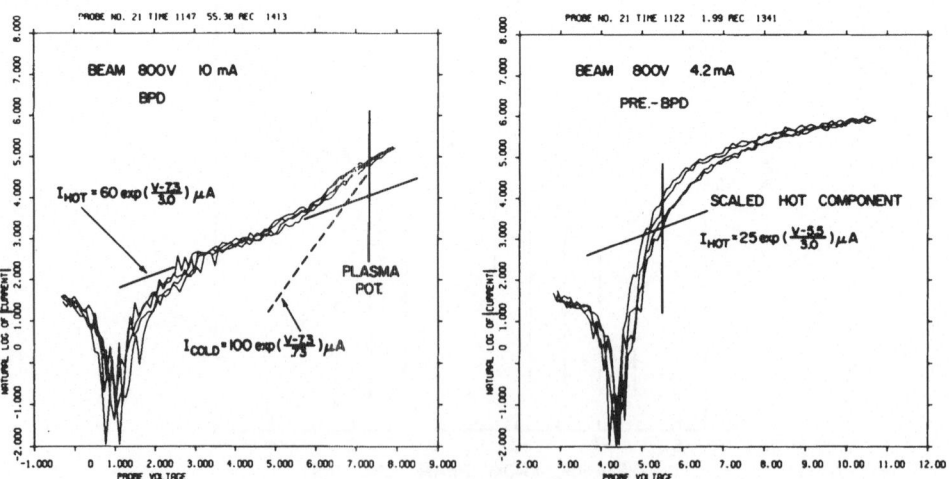

Fig. 10. Langmuir probe characteristics showing (left) hot electrons
          in BPD and their absence for low currents.

tion, we cannot immediately say that this hot component is the
expected group of electrons which, driven strongly by the electric
fields, are responsible for the BPD. However, it can be shown that
these electrons are not simply beam secondaries, but are apparently
accelerated by instabilities. If they were secondaries, their
number ought to be proportional to beam current. In the right panel,
the characteristic of the same probe is shown at a pre-BPD current
of 4.2 mA, and the line shows the expected current due to the hot
electrons of the left panel, but reduced by the ratio of the beam
currents. Clearly these hot electrons are not there in the pre-BPD
case. It is likely then, that these are a fringe of the BPD-driving
electrons, but that no probe was near enough to the core to see the
hottest component.

We now discuss the Langmuir probe measurements when the thruster
is used to generate a background plasma. In the lower right panel
of Fig. 11, we show the plasma density with a 4 mA beam which is
indistinguishable from that generated by the thruster operating by
itself, and see that the density is approximately 10^6 per cm^3 and
nearly uniform across the chamber. The floating potential of the
plasma when the thruster is operating becomes a few volts more
positive than it is in the absence of the thruster. The electron
temperature of this plasma is quite low, of the order of a fraction
of an electron volt. This temperature is colder than that expected
to be generated by the thruster and so is presumably the result of
a selective trapping effect as is probably also the case for
electrons without thruster operation. The ion temperature can, of
course, not be measured with cylindrical probes.

The remainder of Fig. 11 shows the densities produced as we
approach BPD ignition with and without the thruster in operation
(ignition occurred at 7.8 mA). We have discussed previously our
observation that the background plasma has only a very small effect
on the critical current for BPD ignition and that its effect is to
raise the critical current slightly instead of lowering it. In
Fig. 11, we see that at BPD ignition, the density when the thruster
is operating is nearly three times higher than the density without
the thruster and yet ignition occurs at the same current. It is
likely that no probe is exactly at the center of the beam but in
this configuration the plasma frequency line shows that the probe
density cannot exceed the measured density by a factor of as much
as two. It seems unlikely then that the hysteresis effect can
fully explain why the thruster has no effect, and so we have to
conclude that BPD ignition does not simply depend on plasma density.

Our expectation that a potential of a fraction of a volt would
be sufficient to confine the ions, is seriously in error. The
floating potential structure has already been shown in Fig. 5. We
show some further floating potential measurements in Fig. 12. Of
course, the measurement that is really wanted is the plasma potential,

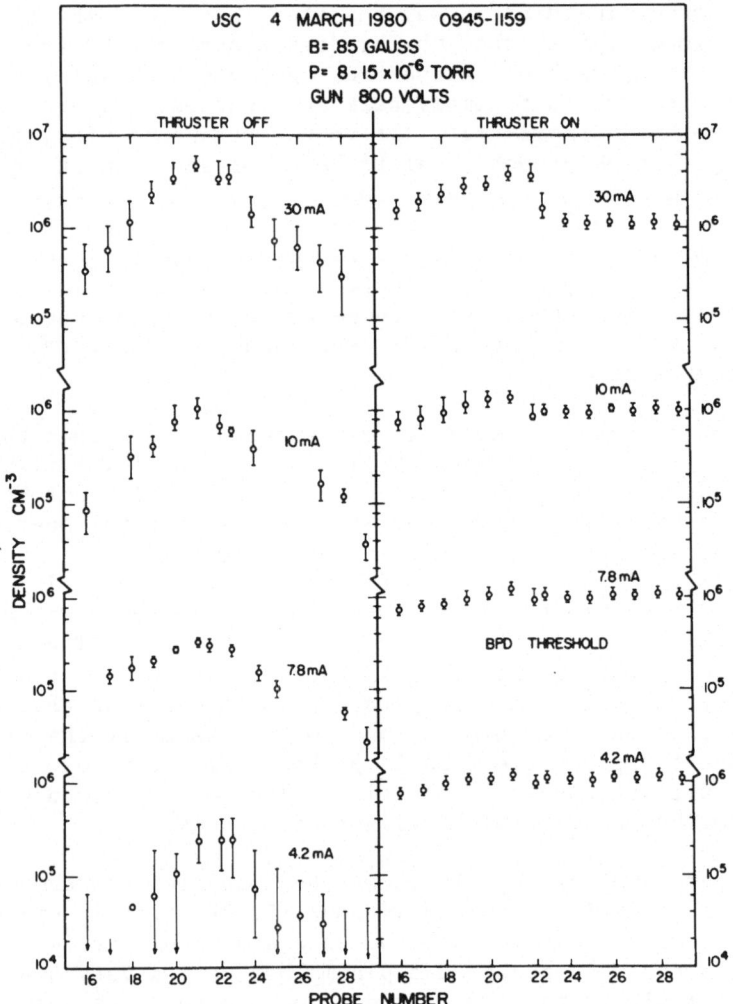

Fig. 11.   Plasma densities generated by the electron beam together
           with and without a background plasma generated by a
           Kaufman engine.

the potential at which the probe does not excert any electric forces
on the plasma particles, rather than the floating potential, which
is the potential at which the electron and positive ion currents
cancel. In pre-BPD conditions, however, the electron temperature
is small, and varies only by a factor of two across the chamber
(from about .6 eV at the center to .3 at the edge of the chamber,
typically), and so the floating potential should track the plasma
potential well enough for our purposes.

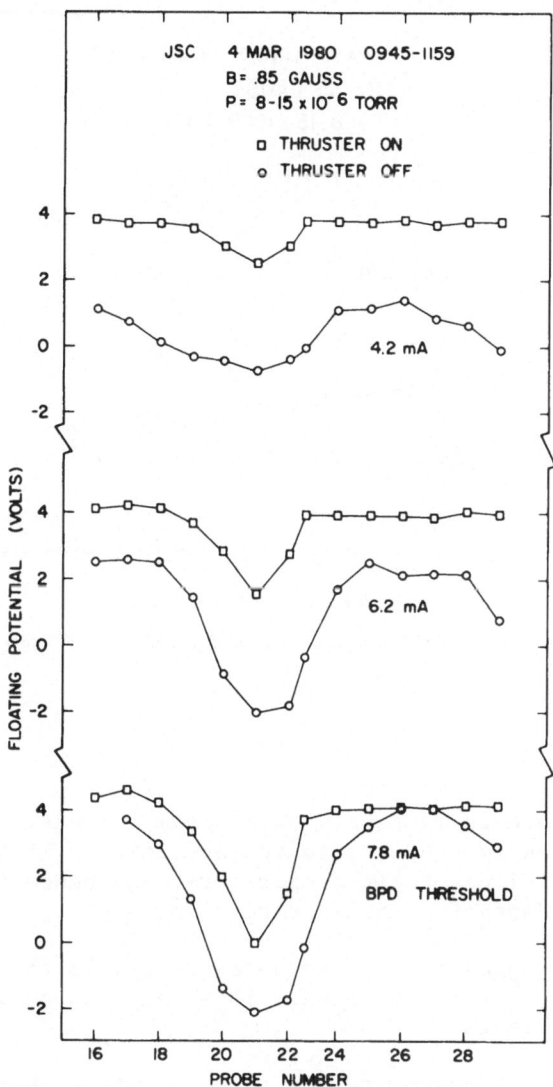

Fig. 12.   Floating potential of the probes as a function of beam
current and of position in the chamber.

We see that not only are the floating potential variations
larger than a fraction of a volt, but also the variations increase
with beam current. This is shown more clearly in the next figure,
Fig. 13, where we have plotted the difference between the probe most
central and one of the exterior probes as a function of beam current.
The conclusion is inescapable that the ion energy is considerably
above room temperature, and amounts in fact to a few volts. Further,

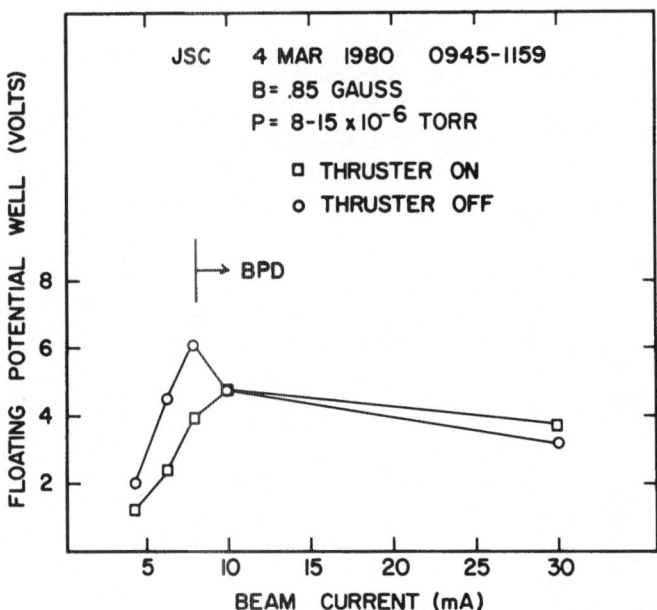

Fig. 13. Potential well depth as a function of beam current.

the ion temperature is increasing with beam current. This ion
heating is apparently also due to instabilities. If the hot ions
were due to collisions, their temperature, and hence the well depth,
would be independent of beam current.

There is a suggestion in the data of Fig. 13 that there is a
threshold current for ion heating. This needs to be investigated
in further experiments.

Unfortunately, the ion temperature cannot be measured with a
long cylindrical Langmuir probe. The reason is that in order for
the ion current to dominate the electron current, it is necessary
that the probe potential already be quite far away from the plasma
potential. The ions are then being strongly attracted and so the
current to a cylindrical probe is independent of temperature
according to eq. (6).

The unexpectedly strong electric field at the sides of the beam
gives a source of free energy. Electrons in this region will drift
around the beam at a speed determined by the electric field. For
values shown we may estimate that the electric field is about 2.5
volts per meter and that at a radius of 1.5 meters, an electron
drifts around the beam with the frequency:

$$f = \frac{c}{2\pi r} \ \frac{E}{B} = 3 \ \text{kHz} \tag{21}$$

This is in rough agreement with the main feature of the spectrum of low frequency waves measured by Holzworth (Holzworth et al., 1981). The actual mechanism for conversion of this free energy into fluctuations is presumably some gradient drift instability, and so the frequency estimate will give only an order of magnitude. This low frequency mode is a prime candidate for the primary cause of the BPD.

We should expect that the background plasma would short out the potential, which confines the ions to the vicinity of the beam. This is only partly true. In Fig. 12, we also show floating potential measurements with the thruster on. It will be seen that the effect of the thruster is to reduce the well potential but not to eliminate it. Further, it can be seen that the diameter of the well is also reduced, and the sides are steeper, so that the electric field is not reduced as much as the potential, and that the rotation frequency will not be changed by very much at all.

This suggests the following picture for the interaction of an electron beam with a plasma. At very low beam currents, true single particle interactions probably dominate, but this current is lower than any we have investigated. As the current is increased, instabilities, with a major component of the order of the ion cyclotron frequency, heat the ions. This hot ion distribution is confined to the central column by the potential of a double layer, and the field in this double layer causes strong electron drift around the column. This drift drives a second regime of instability, and builds to the point where electrons are accelerated to cause further ionization, i.e, the Beam Plasma Discharge.

THE "SHARP FRONT EDGE" EFFECT

In the earliest rocket experiment to measure the waves produced by these electron beams, the Electron Echo I experiment, the waves showed a puzzling feature in that they were strongly generated in the first few milliseconds after the electron gun was turned on, but disappeared long before the end of the gun pulse. This was a characteristic of most of the wave modes observed. In the next flight, Electron Echo II, to the contrary the waves had the same duration as the beam. Similar effects were observed in ARAKS (Dechambre et al., 1980). This "front edge effect" in Echo I has been attributed to coherent radiation by the sharp leading edge of the electron beam (Alekhim and Karpman, 1973), although we (Kellogg et al., 1978) have found it very hard to believe that the front edge could be sufficiently sharp, owing to the time necessary for the gun to reach full voltage.

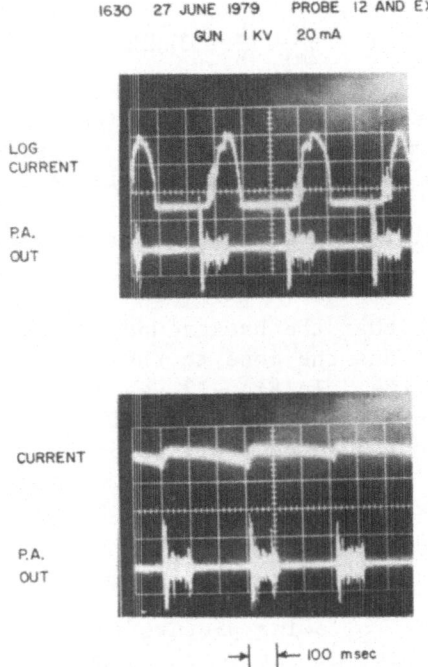

Fig. 14.   Changes in the broad band emission intensity as the plasma
density builds toward equilibrium during pulsed gun operation.

        We have been able to produce an effect in the chamber which
looks identical to this "front edge effect", but under conditions
which show that it cannot be a front edge effect. In Fig. 14 we
show, in the lower part of each oscilloscope photograph, the broad
band output of an electric antenna, during pulsed gun operation.
By using filters, we have confirmed that the major wave modes show
the same time structure separately. It will be seen that there is
an initial burst of radiation which is more than a factor of two
more intense (in amplitude), and which lasts for a few milliseconds.
This time cannot be associated with the front edge. The beam takes
only a few μsec. to travel the length of the chamber, and the propagation times for at least the plasma oscillation and whistler modes
are also only some microseconds. The initial radiation, and its
subsequent sharp decrease must be associated with the approach of
the plasma density to equilibrium. It is obvious that the plasma
density must decay between pulses, and then rebuild, but to confirm
the point, we also show the electrometer output from one of the
Langmuir probes which has been biased to a positive voltage. In
the lower panel is shown the lower part of the current range on a
linear scale, but which saturates at a current of 10 uA, and in the

upper panel is shown the logarithm of the current. The sharp increase
in the current at the end of the gun pulse is thought to be due to
an abrupt change in plasma potential, leading to a different
effective bias voltage on the probe, and so not a genuine increase
in plasma density.

THE PRIMARY INSTABILITY OF THE BEAM PLASMA DISCHARGE

Identification of the primary instability which drives the beam
plasma discharge has been a primary consideration in our laboratory
work. Previous work on the BPD has concentrated on Langmuir, cyclotron and upper hybrid waves as being the most important (Cabral et
al., 1969; Hopman et al., 1968; Stix, 1964). The situation we find
in the vacuum chamber is quite complex, with prominent modes both
at low frequency and at high.

At our present level of understanding, we think that there are
three frequency ranges which are important in energy transfer and
diffusion, i.e., (1) a usually featureless band near the ion
cyclotron frequency, of order 100 Hz; (2) a higher frequency band,
from a few to 10 kHz; (3) the "plasma line", usually 10-30 MHz.

The role of (1) and (2) have already been discussed at the end
of the section "Comparison with Langmuir Probe Measurements". In
this section we will present evidence that (3) is responsible for
energy diffusion of the primary beam. This will be followed by an
inconclusive discussion as to whether (2) or (3) are the primary
instability of the BPD.

The wave measurements were made with several antennas. The
University of Minnesota group used a crossed pair of electric dipole
antennas consisting of short aluminum cylinders of diameter 4 cm and
length 5 cm separated by 44 cms. Each cylinder was connected to a
short (22 cm) co-axial cable through a small capacitor, usually 1.5
to 2 pf. This capacitor raised the effective input impedance to a
value larger than the antenna impedance which varies with plasma
density, thus obviating correction factors for impedance. The short
cable led to the preamplifier.

A very short (1.5 cm) electric antenna was also used, but did
not give significantly different results.

A third antenna was a magnetic search coil consisting of 65
turns on a high frequency ferrite rod, and with its maximum response
at 2-3 MHz.

Other antennas were used by R.H. Holzworth, and by J. Jost.
Their results will be presented in other papers at these sessions.

These antennas together with preamplifiers were movable in a certain plane within the chamber by remote-controlled motors.

Wave amplitudes were measured in two different ways in two frequency ranges. For frequencies below 100 kHz, the preamplifier output was passed through an adjustable bandpass filter, and the filter output was measured with a RMS voltmeter. The filter band-width was usually half a decade in frequency. These measurements, after correction for preamp response, are shown for a typical BPD as the "low frequency" measurements in Fig. 15. Even with the long integration time of the RMS voltmeter (of the order of 1 s), the signals fluctuated strongly (factor of 2). The measurements shown are a visual estimate of the average amplitude. The peak amplitude is certainly higher. The preamp also fed a sweeping receiver which

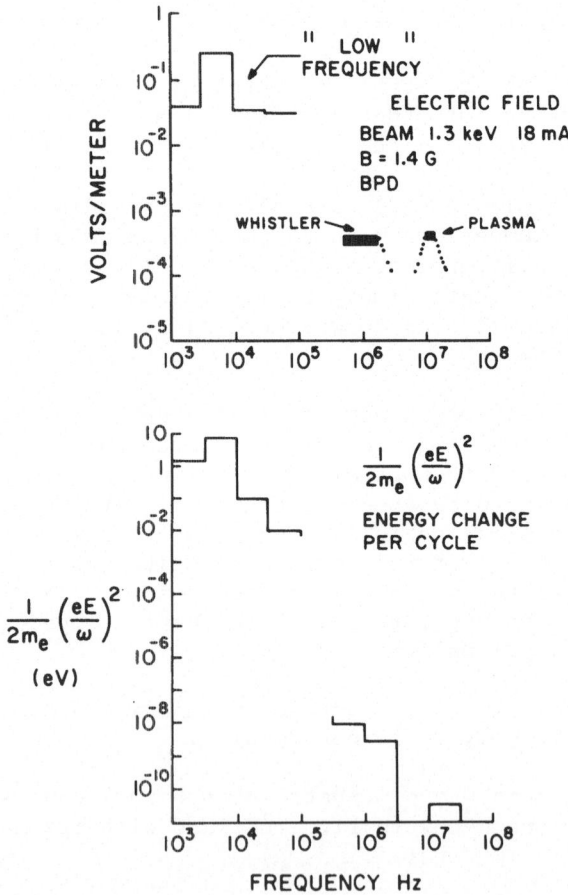

Fig. 15. Rough RF field strengths 3 meters from the beam together with the energization they would produce.

swept from 100 kHz to 14 MHz in 21 ms. Above 300 kHz, measurements
were made by estimating the amplitude from an oscilloscope display
of the sweeping receiver output which was logarithmic. We also show
the sweeping receiver results in Fig. 15. The two measurements did
not give entirely consistent results, probably owing to the different
way that these highly fluctuating signals are averaged. The sweeping
receiver gives lower amplitudes than the filter system. However,
the conclusions presented below are not affected by the maximum
observed discrepancy.

Recently, Jost (1981) has shown that, for the measurements of
the plasma frequency waves of Fig. 15, the probe was not positioned
properly. When the probe is carefully placed within the beam, field
strengths higher by a factor of many tens of dB are observed. Jost
will report these observations at this conference.

In a recent run in Chamber A, more careful evaluation of the
wave amplitude was made to confirm Jost's results. The probe was
placed in the center of the beam, and the signal passed through a
high pass filter to remove the always strong kHz component. Then
the envelope of the remaining high frequency waves was measured on
an oscilloscope. The peak field strength observed during a 200 msec
sweep was generally of the order of ten volts per meter, with 25 V/m
being the peak value observed for a strong BPD. More detailed peak
and estimated rms field strengths as a function of current have not
yet been reduced, and also these results will be underestimates if
the wavelength is shorter than our antenna.

The fields measured have roughly equal longitudinal and transverse electric fields. For low currents, the transverse field is
about twice the longitudinal field, while for high currents they
become more equal.

Jost, Anderson and McGarrity (1980) have measured the energy
spread of the primary beam due to its wave interactions, and have
found accelerations of about 30% in beam energy, as well as energy
degradation essentially to zero energy (see Fig. 3). In order to
gain an energy of 300 eV, a keV electron must see an average field
of 15 V/m over its entire trajectory from gun to target. We assume
that the electrons which do gain this energy are those electrons
which have caught the "perfect wave", so the field strength of this
wave must be about at least 15 V/m. We see that these upper hybrid
waves must be the waves responsible for beam spreading.

In recent experiment, Bernstein and co-workers have investigated
the structure of this "plasma frequency" line and found it to be
complex. A typical spectrum is shown in Fig. 16. The plasma "line"
is seen actually to consist of a number of lines. Their spacing is
quite uniform, is proportional to B, and is roughly consistent with
the electron cyclotron frequency (which is not accurately determined

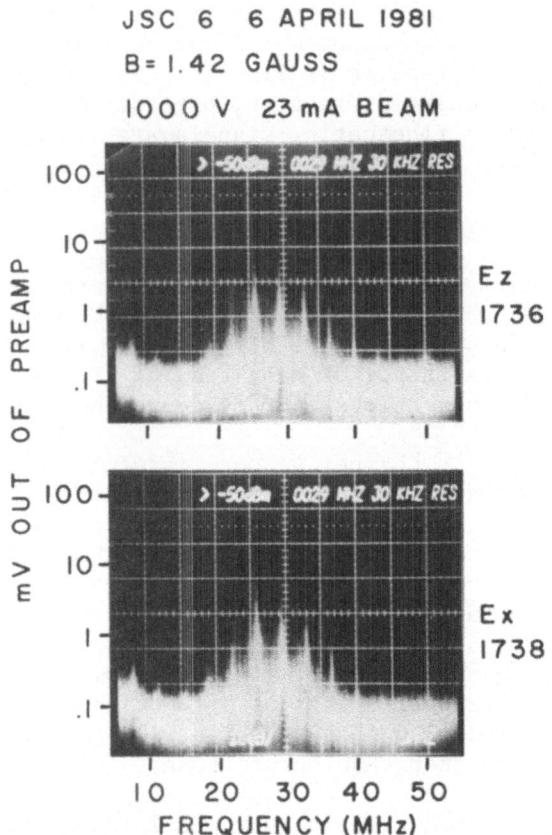

Fig. 16. Detailed spectrum of the "plasma line". This spectrum is
wildly fluctuating and strong averaging is necessary to
show this detail.

because of the variations of B in the chamber). The width of the
lines seems to be less than one would expect from the known
variations of B in the chamber, which is perhaps an indication that
the modes are localized.

If a particle interacts with a short wave train of the form
$E_m e^{i\omega't}$ ($E_0 = E_m/\sqrt{2}$ is RMS field strength), its velocity, taken as
v_0 initially, will be given by $v_0 + eE_m/i\omega't$, and its energy
therefore consists of three terms (the last term is time-averaged).

$$\frac{m}{2} \left(v_0{}^2 + 2 \frac{v_0 e \ E_m}{\omega'm} e^{i\omega't} + \left(\frac{eE_0}{\omega'm}\right)^2 \right) \qquad (22)$$

Two basic processes are then possible, either the particle enters the wave packet with a significant initial energy which is then randomly changed by a rms amount $\Delta E = eE_0 v_0/\omega'$, or its energy is raised on the average by an amount $(eE_0/\omega')^2/2m$. The parameter $1/2m$ (eE_0/ω') then has a two-fold importance; (1) it measures the energy increase when the last term dominates, and (2) the relation $(eE_0/\omega')^2/2m = 2 mv_0^2$ marks the transition from stochastic acceleration (middle term in eq. (22) dominates) to secular increase (last term dominates).

It is therefore of interest to consider the quantity $1/2m$ $(eE_0/\omega')^2$. For the set of measurements under discussion, a plot of $1/2m$ $(eE_0/\omega')^2$ expressed in electron volts is shown in the bottom panel of Fig. 15.

If an electron interacts with several, say N, successive wave packets, then the first will give it energy $(eE_0/\omega')^2/2m$, in rough approximation, which raises it to the stochastic acceleration threshold. Hence, subsequent interactions are stochastic, and the rms final energy will be approximately

$$E_{final} = \sqrt{N} \ (eE_0/\omega')^2/2m$$

For a corresponding treatment of the cyclotron harmonics which constitute the "plasma line", the ω' should be replaced by $\omega' - n\omega_c$ for the nearest n. Unfortunately, we have not yet had the time to measure the structure of the sub-lines in sufficient detail. At our present level of understanding, these waves also seem to be capable of driving the BPD.

Evidence that the low frequency instability is fundamental is provided by the pressure dependence of the critical current relation, eq. (1). It is clear that our picture of the beam plasma discharge is that of an RF discharge in which the RF driver is an instability field rather than externally imposed. The critical field strength for striking an RF discharge typically shows a minimum at a pressure such that the electron collision frequency is approximately the RF drive frequency, ω, in radians per sec (Brown, 1966). We can use the collision frequency at optimum pressure then to estimate the effective RF frequency.

In the chamber (Bernstein et al., 1979), the optimum pressure for BPD is about $1-2 \ 10^{-5}$ torr. The ionization cross section for a 5-25 eV electron on N_2 is about $1.1 \ 10^{-15}$ cm^2, giving a collision frequency (Brown, 1967) of $8 \ 10^4$ s^{-1} in rough agreement with the low frequency peak at $f = 3 \ 10^3 - 10^4$ Hz or $\omega = 2$ to $6 \ 10^4$ and in disagreement with the upper hybrid frequency of $\omega_p = 5 \ 10^7$ rad/s.

It is of some interest to consider also the excursion of an electron interacting with these waves. Integrating once again, we obtain

$$x = x_O + v_O t - \frac{eE_O}{m(\omega')^2} e^{i\omega' t} = x_O + v_O t + \Delta x \qquad (23)$$

For the low frequency waves (in the three to ten kHz range) we obtain $\Delta x = 3 \cdot 10^3$ cm, i.e., an excursion which is of the order of the length of the vacuum chamber. At still lower frequencies, fields which are large enough to make ionizing electrons would make excursions larger than the length of the chamber.

When a low frequency instability is dominant, then it must meet two conditions. It must be sufficiently intense to produce ionizing electrons:

$$\frac{1}{2m} (eE_O/\omega')^2 > eI_O$$

without driving all the electrons out of the chamber:

$$\frac{eE_O}{m\omega'^2} < L$$

Here L is the length of the chamber and I_O is the ionization potential of the neutral gas. These conditions define a region in E_O, ω space whose lowest corner is given by

$$E_O = \frac{2I_O}{L}$$

$$\omega^2 = \frac{2eI_O}{mL^2} \qquad (24)$$

The length of the chamber thus enters critically into the frequency range of important instabilities. In a shorter system, not only will the required field strength and frequency be increased, but also a higher beam current will be required to give the necessary plasma density to support these waves. At some value of L then, it may happen that Langmuir waves or other waves become the dominant driving mechanism, as was claimed in earlier work (Hopman et al., 1968).

To investigate further the structure of the low frequency emission, we made a series of measurements under sub-BPD conditions, as we found that certain features, which we now describe, were more easily isolated in this state.

The predominance of the low frequency emissions in the sub-BPD state is so strong that it was necessary to put a high pass filter, a single R-C filter with a corner frequency of 10 kHz, in the pre-amplifier. This should be kept in mind but does not have a major

effect on the interpretation of Fig. 17, to follow. The low
frequency emissions exhibit structure on a number of different time
scales. In Fig. 17, we show observations taken in the sub-BPD
state, 3 meters from the beam in a magnetic field of 1.45 gauss,
$p = 8 \ 10^{-6}$ torr. The top panel shows the signal on a time scale
of 50 ms per division. It will be seen that the emissions are very
bursty on this time scale and at a certain time the emissions almost
disappear. The next panel shows the wave emissons during what
corresponds to a fairly bursty period of the top picture, on a time
scale of .5 ms/div. It will be seen that the emissions consist of
sharp pulses separated by periods of relative quiet. These pulses
always have the same form. Both the transverse (to B) and the longitudinal electric field pulses always start in the same direction.
The initial radial field is away from the beam. This demonstrates
that the pulses are essentially non-linear. In the bottom frame of
Fig. 17, we show several sweeps taken on a still faster time scale

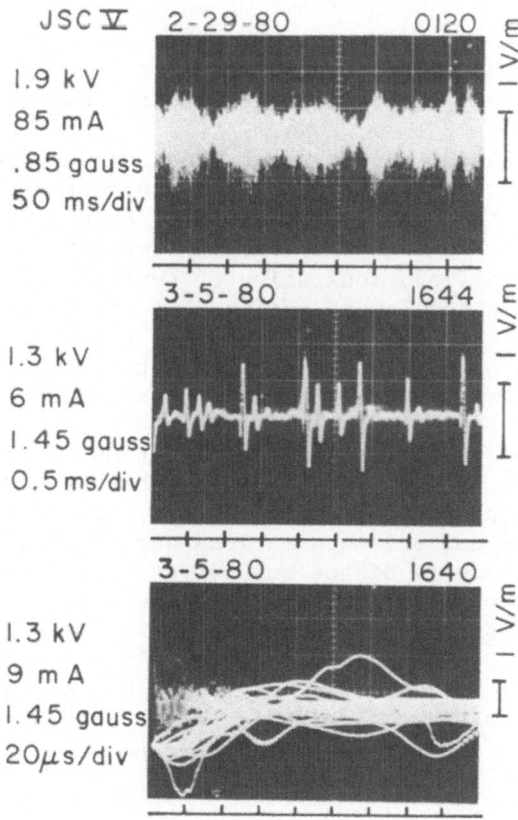

Fig. 17. Structure of the signals observed in pre-BPD. The waveforms are dominated by the low frequency signal.

(20 μs/div). The oscilloscope was triggering on the negative-going part of the strongest pulses and the majority of the pulses are seen to be a sharp rise and an exponential decay with a time constant of the order of tens of microseconds. However, two of the pulses, for which the envelope is otherwise similar to the others, are seen to consist of rapid oscillations at a frequency of about 1 MHz. These appear to be a sort of soliton, similar to whose which have been predicted to be formed by the collapse of plasma oscillations. However, the frequency of the rapid oscillations seems low.

The significance of the pulses which do not show the high frequency structure is uncertain to us. It may be that such pulses are only solitons which have passed nearby the antenna, but whose central ponderomotive core has never reached the antenna. Or it may be that these are different structures.

THE NEUTRALIZATION PROBLEM

Since the capacitance of a typical rocket is about 200 pf, an accelerator emitting a beam of 100 mA would cause a rocket in vacuum to charge at a rate of 500/volts per microsecond. If it is to continue, emitting a beam, therefore, the rocket must draw neutralization current which is very accurately equal to the emitted current.

In the ionosphere, this current tends to deplete the surrounding plasma. A number of processes have been suggested to replace it, including the BPD (Bernstein et al., 1979; Galeev et al., 1976), the cross field or Penning discharge (Cartwright et al., 1978; Galeev et al., 1976), and ionization of neutrals by the beam. If the beam current is too low to excite BPD, the voltage too low for the cross field discharge, and the neutral pressure too low for adequate production of secondaries, then the rocket should charge to a potential equal to the acceleration voltage.

In the vacuum chamber, the neutralization current is normally supplied through a ground wire to the chamber wall. However, it was possible to disconnect the gun and its power supply, and allow it to float. When suitable precautions were taken, i.e., metal surfaces in the vicinity of the gun but at chamber wall potential were covered with plastic sheet, and if there was no background plasma in the chamber, then when the gun was floated, it did rise to the accelerator potential. The beam voltages used were 500 and 750 volts. The TV then showed the gun to be surrounded by a luminous cloud with a rather sharp and regular outer boundary, with a diameter of about a meter. Clearly the electron beam was being stopped at this boundary.

On second thought, it was not clear why the beam should stop at one meter. If the field around the gun were that of a point charge, then even if the gun charged to a potential equal to the beam energy,

still the electron would escape to infinity, but arrive there with
zero velocity.

In their trajectories, a small fraction of the stopping electrons
must ionize a neutral atom. Then both the final electrons will fall
back to the gun, and the ion will be thrown outward. A small fraction
of the beam electrons must also escape to compensate for these ions.

It is reasonable to suppose that these escaping ions are slower
than the escaping beam electrons, and also they will certainly occupy
a cylinder of larger radius than the electrons, so the net charge
density will be positive at large distances. Hence, starting from
infinity, the potential must first become negative by a small amount,
and then rise to its large positive value at the rocket or gun.
There must, therefore, be a virtual cathode and a potential trough,
at some distance far from the gun, and this is what we saw.

This potential structure implies that electrons will be repelled
at large distances from the rocket, and therefore suggests that the
same virtual cathode structure with charging to the full beam energy
could exist also in the presence of plasma, provided that it was
sufficiently tenuous.

This experiment was tried also, with a tenuous plasma made with
an RF discharge. The resulting plasma was not very uniform and we
do not know what its density was in the vicinity of the gun. In
this case, the potential of the gun oscillated, staying at a value
which stopped the beam for a few seconds, then dropping almost to
the ground potential, then stopping the beam again, with a period
of a few seconds.

These experiments were rather qualitative, and no follow up
was made until W. Bernstein made a more systematic study in the
most recent experiment run. The results of this experiment will be
reported in due course.

Acknowledgements

Various aspects of this work have been carried out in
collaboration with J.O. McGarrity, B.A. Whalen, S.J. Monson, D.N.
Walker, O.K. Garriott, A. Konradi, J. McCoy, R. Mesli and G.
Mantzoukis.

We wish to thank the staff of Chamber A, Johnson Space Center
for their willing cooperation.

One of us (PJK) wishes to thank R.W. Boswell for several very
helpful discussions, and for his continued insistence on the
importance of the low frequency waves.

This work was supported by grants from the Southwest Research Institute and the National Oceanic and Atmospheric Administration (03-78-801-60).

REFERENCES

Alekhim, Ju.K., and Karpman, V.I., 1973, On Cerenkov Radiation by Electron Beam Injected Into the Ionosphere, Cosmic Electro 3: 406.

Alexeff, I., Neidigh, R.V., and Peed, W.F., 1964, Beam-Plasma Interaction Experiments and Diagnostics, Phys. Rev. 136:A689.

Bernstein, W., Leinbach, H., Cohen, H., Wilson, P.S., Davis, T.N., Hallinan, T., Baker, B., Martz, J., Zeimke, R. and Huber, W., 1975, Laboratory observations of RF emissions at ωpe and (n + 1/2)ω_{ce} in electron beam plasma and beam interactions, J. Geophys. Res., 80:4375-4379.

Bernstein, W., Leinbach, H., and Baker, B., 1977, Preliminary results from the electron beam-plasma experiments in the very large vacuum facilities at Plum Brook and Johnson Space Center, NOAA Technical Memorandum ERL-SEL 50.

Bernstein, W., Leinbach, H., Kellogg, P., Monson, S., Hallinan, T., Garriott, O.K., Konradi, A., McCoy, J., Daly, P., Baker, B., and Anderson, H.R., 1978, Electron beam experiments: The beam plasma discharge at low pressures and magnetic field strengths, Geophys. Res. Lett., 5, 127-130.

Bernstein, W., Leinbach, H., Kellogg, P.J., Monson, S.J., and Hallinan, T., 1979, Further Laboratory Measurements of the Beam Plasma Discharge, J. of Geophys. Res., 84, 7271.

Brown, S.C., 1966, "Introduction to Electrical Discharges in Gases", J. Wiley & Sons, New York.

Brown, S.C., 1967, "Basic Data of Plasma Physics", MIT Press, Cambridge, Mass.

Cabral, J.A., Hopman, H.J., Insinger, F.C., and Ott, W., 1969, Time resolved beam distribution functions in a beam plasma experiment, Plasma Physics and Controlled Nuclear Fusion Research, vol. II, International Atomic Energy Agency, Vienna.

Cambou, F., Dokoukine, V.S., Ivchenko, V.N., Managadze, G.G., Migulin, V.V., Nazarenko, O.K., Nexmyanovich, A.T., Pyatsi, A.Kh., Sagdeev, R.Z., and Zhulin, I.A., 1975, The Zarnitza Rocket Experiment on Electron Injection, Space Research XV.

Cambou, F., Lavergnat, J., Migulin, V.V., Morozov, A.I., Paton, B.E., Pellat, R., Pyatsi, A.Kh, Reme, H., Sagdeev, R.Z., Sheldon, W.R., and Zhulin, I.A., 1978, ARAKS-Controlled or Puzzling Experiment?, Nature, 271, 723.

Cartwright, D.G., and Kellogg, P.J., 1974, Observations of Radiation from an Electron Beam Artificially Injected into the Ionosphere, J. Geophys. Res., 79, 1439.

Cartwright, D.G., Kellogg, P.J., and Monson, S.J., 1978, Heating of the Ambient Ionosphere by an Artificially Injected Electron

Beam, J. Geophys. Res., 83, 16-24.

DeChambre, M., Gusev, G.A., Kushnerevsky, Yu.V., Lavergnat, J., Pellat, R., Pulinets, S.A., Selegel, V.V., and Zhulin, I.A., 1980, High Frequency Waves During the ARAKS Experiments, Annales de Geophys., 36, 333.

Galeev, A.A., Mishin, E.V., Sagdeev, R.Z., Shapiro, V.D., and Shevchenko, I.V., 1976, Discharge in the region around a rocket following injection of electron beams into the ionosphere, Sov. Phys. Dokl., Engl. Transl., 21, 641-643.

Getty, W.D., and Smullin, L.D., 1963, Beam-plasma discharge: Buildup of oscillations, J. Appl. Phys., 34, 3421-3429.

Grandal, B., Holtet, J.A., Troim, J., Maehlum, B., and Pran, B., 1980, Observations of Waves Artificially Stimulated by an Electron Beam Inside a Region with Auroral Precipitation, Planet. Space Sci., Vol. 28, 1131-1145.

Hess, W.N., Trichel, M.C., Davis, T.N., Beggs, W.C., Kraft, G.E., Stassinopoulos, E., and Maier, E.J.R., 1971, Artificial Aurora Experiment; Experiment and Principal Results, J. of Geophys. Res., 76, 6067.

Hendrickson, R.A., McEntire, R.W., and Winckler, J.R., 1971, Electron Echo Experiment, a New Magnetospheric Probe, Nature, 230, 564.

Holzworth, R.H., Koons, H.C., and Harbridge, W.B., 1981, Plasma Waves Stimulated by Electron Beams in the Lab and in the Auroral Ionosphere (this volume).

Hopman, H.J., Matitti, T., and Kistemaker, J., 1968, The Electron Cyclotron Instability and High Frequency Ionization in a Beam-Plasma Experiment, Plasma Phys., 10, 1051.

Jost, R.J., 1981, Radial Dependence of HF Field Strength in the BPD Column (this volume).

Jost, R.J., Anderson, H.R., and McGarrity, J.O., 1980, Electron Energy Distributions Measured during Electron Beam/Plasma Interactions, Geophys. Res. Lett., 7, 509.

Kellogg, P.J., Cartwright, D.G., Hendrickson, R.A., Monson, S.J., and Winckler, J.R., 1976, The University of Minnesota Electron Echo Experiments, Space Res. XVI.

Kharchenko, I.F., Fainberg, Ya.B., Nikoloev, R.M., Kornilov, E.A., Lutsenko, E.I., and Pedenko, N.S., 1962, The interaction of an electron beam with a plasma in a magnetic field, Sov. Phys. Tech. Phys., Engl. Transl. 6, 551-553.

Langmuir, I., and Mott-Smith, H.M., 1926, Studies of Electric Discharges in Gases at Low Pressures, in "Collected Works of Irving Langmuir" (G. Suits, ed.), Vol. 4, Macmillan (Pergamon) New York, 1961; originally published in Gen. Elec. Rev. 27, 449 ff (1924).

Loeb, L.B., 1960, "Basic Processes of Gaseous Electronics", Univ. of Calif. Press, Berkeley.

Maehlum, B.N., Maaseide, K., Aarsnes, K., Egeland, A., Grandal, B., Jacobsen, T.A., Maynard, N.C., Soraas, F., Stadsnes, J., Thrane, E.V., and Troim, J., 1980, Polar 5-An Electron Accelerator Experiment Within an Aurora, Planet. Space Sci., 28, 259.

Monson, S.J., Cartwright, D.G., and Kellogg, P.J., 1976, Whistler
    Mode Plasma Waves Observed on Electron Echo 2, J. Geophys. Res.,
    81.
Monson, S.J., and Kellogg, P.J., 1978, Ground Observations of Waves
    at 2.96 MHZ Generated by an 8- to 40-KEV Electron Beam in the
    Ionosphere, J. of Geophys. Res., 83, 121-131.
Pyatsi, A.Kh., and Zarnitsky, Yu.F., 1980, Electron precipitation
    in magnetically conjugated region in the first Araks experiment
    from radar data, Annales de Geophys., 36, 297.
Stix, T.H., 1964, Energetic Electrons from a Beam-Plasma Over-
    stability, Physics of Fluids, 7, 1960.
USRA, 1976, Final Report of the science definition panel for Atmo-
    sphere and Plasmas in Space Spacelab payload, Universities
    Space Research Association, Houston, Texas.

DISCUSSION

Haerendel: Could you clarify the respective roles
played by the high frequency noise at the "plasma line"
and the kHz noise created by the electron rotation around
the beam axis? What is heating what?

Kellogg: The "plasma line" waves, which we now know
have structure, are responsible for diffusing the energy
of the beam electrons. While these waves could possibly
heat the plasma electrons, I have given two pieces of
evidence that the kHz waves are responsible for this
heating and hence for driving the BPD.

Kofsky: Cannot the plasma outside the "core" arise from
secondary electrons emitted at the target and floor surfaces at large angles to the magnetic field? After all,
you cannot predict the energy deposition distribution
without a transport calculation that takes into account
all loss/production processes at initial beam boundaries.

Kellogg: We have estimated the transverse diffusion due
to scattering only in a crude way, by saying that an
electron can only move radially by one Larmor radius per
scattering, and that it makes about 200 lengthwise
traversals of the chamber before being lost. It can
diffuse radially by $\sqrt{200}$ Larmor radii which is only a
few tens of cm.

<u>Schmerling</u>: The accelerator takes in electrons at one
end, accelerates them and then pushes them out. In the
chamber, the circuit is completed through the cables and
power supply. I am, therefore, not surprised that the
Kaufman thrusters make no difference. In space, the
return current must come from the surrounding plasma.
When the gun was isolated, there was an effective capacitor from the gun cables to ground. There was, therefore, a relaxation oscillator which allowed the potential
of the gun to rise as it sucked out return electrons from
the space charge region, then choked off and then started
again when conditions were restored through leakage resistance.

<u>Kellogg</u>: Yes, I agree that it should not be too sur-
prising if the neutralization process is oscillatory
under some conditions.

THE NASA SPACE ENVIRONMENT SIMULATION LABORATORY

R. J. Jost

Space Environment Office
NASA Johnson Space Center
Houston, TX 77058

INTRODUCTION

The Space Environment Simulation Laboratory (SESL) at the
NASA Johnson Space Center operates a very large, low base pressure
vacuum chamber that has been modified to be utilized as a plasma
physics facility. Originally constructed as a thermal/vacuum test
chamber for manned spacecraft, the facility was initially used for
plasma physics research in 1977. Bernstein et al. (1975) developed a laboratory beam/plasma research program in the large vacuum
chamber at the NASA Plum Brook Station, Sandusky, Ohio; following
deactivation of that facility in 1975, the program was moved to the
SESL chamber in 1977. Since then a variety of plasma investigations have been conducted here with the emphasis to date placed
on programs studying specific space plasma physics phenomena of
the lower inonsphere. Approximate simulation of the lower ionosphere has been achieved in terms of the neutral pressure, magnetic field strength, plasma density, and electron temperature.
The chamber volume is sufficiently large that parameter scaling
is unnecessary.

The plasma programs undertaken have ranged from basic physics
to very applied investigations. Several beam/plasma studies, VLF
antenna impedance measurements, high voltage sheath investigations,
and several sounding rocket tests demonstrate the successful
achievement of a large scale, ionospheric modeling laboratory.

BASIC CHAMBER

As described by Pearson (1980) the test chamber is an up-
right, domed cylinder with a usable volume approximately 17 m in

331

diameter and 27 m high. The primary structural shell is constructed of non-magnetic stainless steel while most of the internal
construction material is aluminum. The non-magnetic nature of the
structure has been important for control of the internal field by
use of large current coils encircling the chamber.

Figure 1 is a cutaway view of the SESL test chamber showing
the principle access doors and the magnetic field current coils.
Most of the active hardware components for a given experimental
program are installed through the large 12 m diameter door at the
ground floor level. In addition, double man-lock entrances open
to the ground floor and a 10-meter level catwalk that encircles
most of the chamber near the inner wall. A single door access also
exists for a 19-meter level catwalk.

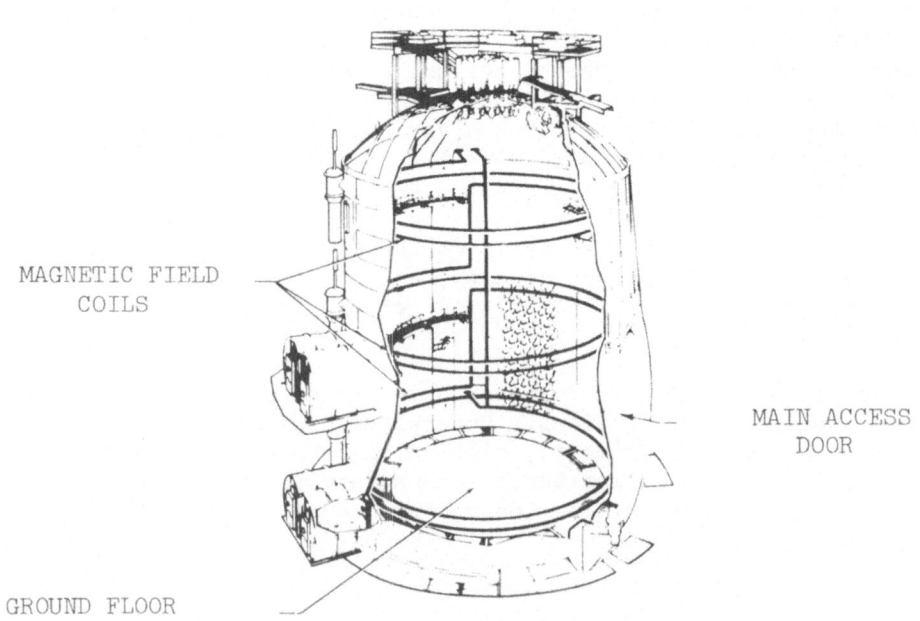

Fig. 1   Cutaway view of the large SESL plasma
chamber. The usable volume is 27 m
high and 17 m in diameter.

Chamber pressures are typically maintained in the 10^{-6} torr range. Dry gases can be leaked into the chamber to increase the pressure to higher values. The pumping system consists of 20°K helium, large-surface cryopanels for heavy, condensible gases and trapped oil diffusion pumps for hydrogen, helium, and neon. In the 10^{-6} torr range the pumping speed is $2(10^{4})m^{3}/s$ for condensibles and $3(10^{2})m^{3}/s$ for noncondensibles with a total chamber leak rate of less than $8(10^{3})m^{3}/s$ of air. Starting from atmospheric pressure with reasonable "clean" hardware, test conditions can normally be achieved within 6 hours; the pressure then gradually decreases throughout the duration of the test period.

Equipment positioning can be remotely controlled from outside the chamber during test conditions. The facility maintains several high torque, variable speed, DC motors which are usually directly coupled or indirectly coupled through nylon or steel cables with pulley arrangements to the hardware to be moved. A closed circuit television system, also incorporated inside, provides views of the hardware to facilitate proper alignment and position. This system enables an experimenter to achieve a linear resolution of position to generally within 10 cm and an angular resolution of approximately 5 degrees. For most large scale investigations this is adequate; improved position and alignment resolution has been a responsibility of the experimenter.

IONOSPHERIC SIMULATION

Since the ionospheric plasma experiment programs in 1977, several modifications to the basic thermal/vacuum test chamber have taken place. Primarily, plasma generation is provided by a 30 cm Kaufman thruster exhausting into the chamber volume. Stable densities in the 10^{5}-10^{6} cm^{-3} range with a uniformity of within a factor of three (3) over the usuable volume are readily achieved with this system in the normal configuration. Argon is typically used for the ionization gas but other ion species are available. A large fractional ionization is achieved with this technique, therefore the neutral density is not severely altered when high plasma densities are produced. The ions have a bulk flow energy of 20-50 eV and due to charge exchange with the ambient gas a cold isotropic ion component is also present. Other techniques of plasma generation have been used (i.e., RF discharge, electron beam ionization, and a hollow cathode source) but the Kaufman thruster has received the most usage. Important in controlling plasma distributions and for conducting magnetized plasma experiments is the configuration of the embedded magnetic field. A major modification to the SESL chamber was the addition of a large, chamber encircling solenoid consisting of three double loops of 2.5 cm copper tubing. The loops are located inside the chamber near the top, center, and bottom of the usable cylindrical volume. A zero to 1000 A DC power supply system is connected to

the six coils for up to 6000 A turns in the solenoid. The corresponding vertical steady state magnetic field in the central region
of the chamber ranges from zero to about 2.5 gauss. The ambient
geomagnetic field has a field strength of approximately 0.3 gauss
with an average dip angle of 60°, therefore the final magnetic
field vector resulting from the applied field and the geomagnetic
field is dependent on the coil current for both magnitude and dip
angle. At the highest current setting (2.5 gauss) the field is
very nearly vertical. Figure 2 shows the measured uniformity of
the field in the central region as a function of height above the
chamber floor for a 200 A coil current.

A limited set of plasma diagnostic equipment has been established in the facility. Five low light, remotely pointed television cameras are mounted in the chamber with controls and
monitors available to the users. An array of cylindrical Langmuir
probes are computer controlled and provide electron temperature
and density at specific probe locations. A mass spectrometer and
ion gauges are also available to monitor the ambient gas composition and pressure. Beyond these basic systems, other plasma
environment diagnostics must be provided by the investigator.

The SESL plasma chamber characteristics described above are
summarized in Table 1.

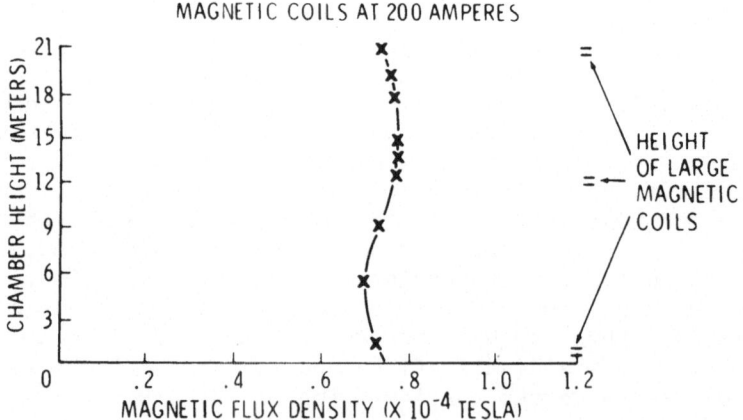

Fig. 2   Magnetic field variation along central vertical
         axis of the plasma chamber (Pearson, 1978).
         Several meters off-axis the variation is more
         pronounced (in angle and magnitude).

TABLE 1. CHARACTERISTICS OF SESL PLASMA CHAMBER

Working Dimensions	16.8m diameter
	27.4m height
Vacuum	1×10^{-6} torr
	7-hr pumpdown time
Plasma Density	10^4-10^6 cm^{-3}
Plasma Species	N_2^+, A^+, H^+
Magnetic Field	zero to 2 gauss
Electron Temperature	0.1 to several eV
Access	12.2m diam. door. Dual manlocks at floor and 9.5m level. Access door at 19m level.

REPRESENTATIVE EXPERIMENT PROGRAMS

The following representative experiment programs, among several programs successfully conducted in the plasma chamber, indicate the type of research that has been undertaken:

1) Electron beam-plasma interactions (Bernstein et al., 1979)

2) VLF antenna impedance measurements (Konradi et al., 1978)

3) High voltage plasma sheaths (McCoy et al., 1980)

NASA and Air Force sounding rocket payloads have also been tested in this plasma environment prior to launch.

The beam/plasma work, summarized in Bernstein et al. (1979) and Banks (private communication), has made significant impact to the present understanding of keV electron beams injected into ionospheric plasmas. In particular, results of studies of the beam plasma discharge (BPD) with its effect on the particle energy distribution, light emission, spatial dependence, and increased ionization rate has guided much of the recent development of ionospheric beam/plasma work. The BPD may be used to improve the

understanding of solar and auroral beam interactions. Future
Shuttle electron beam projects are being designed with major input
coming from the results of the chamber work.

Determination of large, VLF dipole antenna impedance characteristics in an ionospheric plasma has been undertaken by Konradi
et al. (1978), and Garriott (private communication). Results from
these efforts were used to determine the driving point impedance
as a function of input power levels, frequency, and magnetic field
orientation for the Dipole Antenna Subsystem of the Waves in Space
Plasmas (WISP) experiment to be flown as a spacelab project.

High voltage sheath experiments, such as those conducted by
McCoy et al. (1980), are particularly relevant to solar power
satellite technology. Because of the large volume, low magnetic
field, and cool, dilute plasma the sheath thicknesses of simulated
high voltage solar panels were of sufficient size (1-5m) that
impirical determination of their properties could be made. For
example, results indicate that equilibrium high-voltage leakage
currents to the plasma should be much less than some early predictions, particularly for very large solar arrays.

The large volume has also provided a unique plasma test bed
for several sounding rocket and Shuttle payloads. In some cases,
operation of the flight instruments and detectors in "flight"
conditions has uncovered problems that were rectified prior to
the actual flight. Additionally, the operational tests have
contributed or will contribute greatly to the understanding of
the flight data. The ability to determine or accurately anti-
cipate scientific spacecraft performance prior to the flight has
been of great value to a few investigators.

CONCLUSION

During the short operational period of the large SESL plasma
chamber many important investigations have been undertaken and
completed to at least a first generation level. The iterative
process between theory and experiment in understanding a physical
phenomenon has been undeniably aided by this chamber for certain
space science efforts. The balance between theory and experiment
for any space plasma program is of vital importance for its success.
Hopefully, the SESL plasma chamber can continue to provide this
link.

REFERENCES

Bernstein, W., H. Leinbach, Herbert Cohen, P. S. Wilson, T. N. Davis, T. Hallinan, B. Baker, J. Martz, R. Zeimke, and W. Huber, Laboratory observations of RF emissions at ωpe and (n + 1/2) ωce in electron beam-plasma and beam-beam interactions, J. Geophys. Res., 80, 4375, 1975.

Bernstein, W., H. Leinbach, P.J. Kellogg, S.J. Monson, and T. Hallinan, Further laboratory measurements of the beam plasma discharge, Jour. Geophys. Res., 84, 7271, 1979.

Konradi, A., W. Bernstein, and O.K. Garriott, Space plasma laboratory experiment in simulated ionospheric plasma, A Collection of Technical Papers, AIAA/IES/ASTM 10th Space Simulation Conference, 114, 1978.

McCoy, J.E., A. Konradi, and O.K. Garriott, Current leakage for low altitude satellites, in Space Systems and Their Interactions with the Earth's Space Environment, eds. Garrett and Pike, Progress in Astronautics and Aeronautics, 71, 523, 1980.

Pearson, O.L., Modification of a very large test chamber for plasmasphere simulation, Jour. Spacecraft and Rockets, AIAA, 17, 323, 1978.

VISIBLE SIGNATURES OF THE MULTI-STEP TRANSITION TO A BEAM-PLASMA-DISCHARGE

T. J. Hallinan, H. Leinbach[+], and W. Bernstein[*]

Geophysical Institute, University of Alaska, Fairbanks, Alaska 99701
[+]Space Environment Laboratory, National Oceanic and
[*]Atmospheric Administration, Boulder, Colorado, 80303
Space Plasma Center, Rice University, Houston, Texas 77001

ABSTRACT

The transition of an electron beam to a beam-plasma-discharge (BPD) is characterized in part by substantial changes in both the distribution and the magnitude of the visible emissions from the ambient gas. At low ambient pressure ($<4 \times 10^{-6}$ Torr) there are abrupt transitions between four visually recognizable states. These include the basic beam with its noded configuration (A_1), the noded beam surrounded by a weak halo (A_2), and two states (B and C) of the BPD characterized by very bright halos and significant alteration of the beam itself. The latter states are differentiated by the differing intensities and diameters of their associated halos. When the beam current is reduced, these transitions occur in reverse and with substantial hysteresis.

Analysis of the optical intensities (total light and $\lambda 3914\text{Å}$ (N_2^+)) suggests that, at the critical current, the BPD may be twenty to one hundred times as effective as direct collisions in ionizing the ambient gas. Even the weak halo of state A_2 (tentatively associated with emissions near the electron cyclotron frequency and its harmonics) may indicate an ionization rate several times that due to direct collisions of the beam electrons.

At higher pressures, the transition to BPD is less abrupt and may not include these separate intermediate states. There is little or no observable hysteresis; nor are the cyclotron emissions observed. The ratio of the ion production rate in BPD to that of the beam alone is lower than it is at low ambient pressure.

INTRODUCTION

One of the principal characteristics of a beam-plasma-discharge (BPD) is a bright optical glow surrounding the path of the electron beam. The glow is indicative of the ionization and excitation of the ambient gas by locally accelerated electrons. The purpose of this paper is to describe the characteristics of the glow at the transition from a stable beam to a BPD. The description is based on television recordings and 3914Å photometric measurements from BPD experiments in the large vacuum facility at Johnson Space Center (Bernstein et al., 1978).

Instrumentation

The overall experimental configuration is described by Bernstein et al. (1978). An electron accelerator on the floor of the chamber produced a field-aligned DC electron beam that was terminated at a distance of 20m by an aluminum target. The beam current was controlled by energizing a motorized variac that increased or decreased the voltage applied to the cathode heater in the accelerator. The speed of the variac motor precluded instantaneous changes in the beam current.

The optical diagnostic equipment consisted of a scanning photometer to measure the emission at 3914Å (directly proportional to the ionization rate of N_2) and two low-light level television cameras. The photometer was equipped with an interference filter having a width of 13Å centered at 3914Å. A rotating mirror was used to scan the 0.9° field of view horizontally to provide a profile of intensity approximately perpendicular to the beam. The photometer profiles intersect the beam at a fixed distance from the accelerator -- approximately midway between the accelerator and the target.

The television cameras are image-orthicon systems with extended red (S-25) photocathodes sensitive to light in the range 3900Å - 8500Å. No optical filters were used. One camera, located on the floor of the chamber, approximately 10 m from the accelerator, had a field of view of 12° x 16°. The other camera, having a 30° x 40° field of view, was located near the photometer. Both cameras were on fully steerable mounts and could view any part of the beam.

The images were recorded on video tape for subsequent viewing and analysis. A recently developed digital image processing system was used in the analysis of the data. This system consists of a Quantex digital image processor and frame store, a Nova mini-computer and disc, and a Tektronix 4010 terminal and hard-copy unit. A special interface card was designed for the Nova to allow direct random access to the Quantex memory.

This system allows the user to obtain horizontal profiles of luminosity through the beam at any location within the TV picture. Where appropriate, successive TV frames (1/60 second exposures) can be added for an improved signal to noise ratio. Similarly, adjacent horizontal scans can be summed and scans from a "background" frame in which the beam is off can be subtracted from the data scans to remove TV shading patterns.

Observations

The most detailed information concerning the transitions of the beam between steady-states comes from the television recordings. The more time-consuming photometer scans are available for some selected steady-state conditions. When available, these provide better quantitative estimates of the ion production rate.

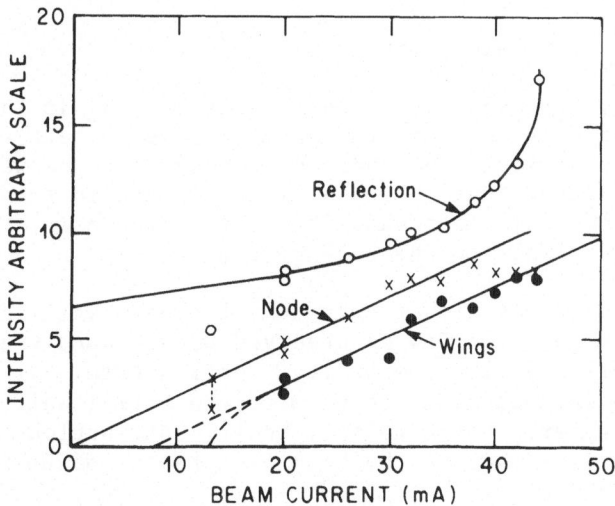

Figure 1. Intensity vs. beam current for a specular reflection from the accelerator's filament, the node of the beam, and the broad glow surrounding the beam. The beam energy is 1200 V, the pressure 2.5 x 10^{-6} Torr and the magnetic field strength 1.5 Gauss. BPD was not ignited.

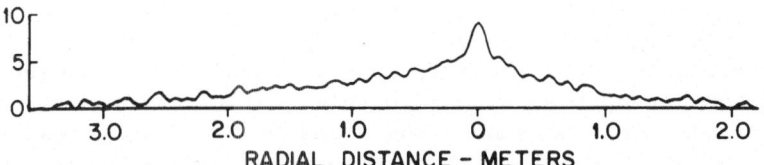

RADIAL DISTANCE – METERS

Figure 2. Horizontal scan of intensity through the beam node in the A_2 state. The central peak is the beam node. The broader wings are the A_2 glow. $V = 1200$ V, $P = 2.5 \times 10^{-6}$ Torr, $B = 1.5$ Gauss, $I = 44$ mA.

In general, we are concerned with the optical intensity as a function of beam current (I), distance (z) from the accelerator, and distance (r) from the axis. We are restricting ourselves to cases where the beam is injected parallel to the magnetic field and azimuthal symmetry is assumed. From the appearance of the visible emissions, four distinct states of the beam-plasma system can be identified. These are designated A_1, A_2, B and C. States A_1 and A_2 pertain to currents less than the critical current (I_c) required to ignite the BPD. States B and C are two states of the BPD.

A_1: At relatively low values of beam current, the luminosity is confined to the envelope of single-particle trajectories from the accelerator. This envelope is a noded configuration given by:

$$r = 2 \frac{v}{\omega_c} \sin \theta \sin \frac{\omega_c z}{v} \qquad (1)$$

where v is the velocity of the beam electrons, ω_c is the electron cyclotron frequency and θ is the angular spread (half angle) in the beam at injection. The nodes are observed to be well-defined, from which we infer that there is little spread in the parallel velocity of the beam electrons. The optical intensity (Fig. 1) is directly proportional to the beam current.

A_2: At somewhat higher currents, the beam still appears as in A_1, but a weak uniform glow is observed to surround the beam. This shows up as wings on the luminosity profiles (Fig. 2). The wings, although less intense than the glow of the beam itself, are considerably wider. Consequently, their contribution to the total luminosity exceeds that of the beam by a factor of up to six.

As seen in Fig. 1, the wings seem to require a threshold current, but are otherwise linear functions of current. Since the beam current is controlled by varying the filament temperature, the white-light background also correlates with beam current. That the "wings" are distinct from the white light is shown in Fig. 1 by the

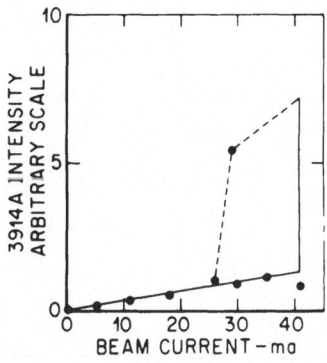 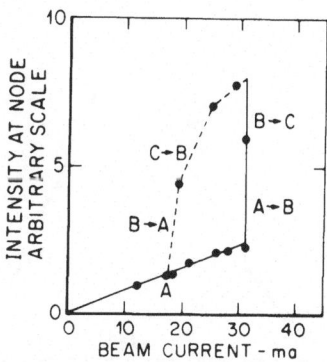

Figure 3.   Intensity vs. beam current for increasing and decreasing currents.   Left: 3914Å (N_2^+) intensity.   V = 1000 V, P = 1 x 10^{-6} Torr, B = 1.1 Gauss.   Measurements are not at node.   Right: Total light (TV).   V = 1000 V, P = 2.5 x 10^{-6} Torr, B = 1.5 Gauss. Measurements are at the beam node and the transitions between states are indicated by appropriate letters.

plot of the intensity of a specular reflection from the filament. As expected, this intensity is a nonlinear function of beam current.

In Fig. 1 the wings appear gradually.   In other cases the transition from A_1 to A_2 was more abrupt.   In at least some cases, this abrupt transition was observed to coincide with the sudden appearance of narrow-band fluctuations in the electric field at or slightly above the electron cyclotron frequency and its harmonics. (Bernstein et al., 1979).   Bernstein et al. also noted that the wings are evident in the scans from the 3914Å photometer.

B: The most dramatic transition is from A to B at the ignition of the beam-plasma-discharge. An infinitesimal increase in current above the critical value produces a large change in the peak luminosity (Fig. 3).   This increase occurs in the glow surrounding the beam rather than in the beam itself.   Fig. 4 shows a scan through the node of a 1500v beam in state B.   For comparison, there is also a scan through the node at a current slightly below the critical value.   A background scan (including the effect of A_2 wings) has been subtracted from both scans.   The dotted curve is a gaussian having the same area, center, and deviation as the actual data.

The total luminosity, proportional to the area under the curve, is 120 times that of the beam node alone.   If wings similar to those of Fig 2 are counted as part of the beam, the enhancement ratio would be reduced to approximately 17.

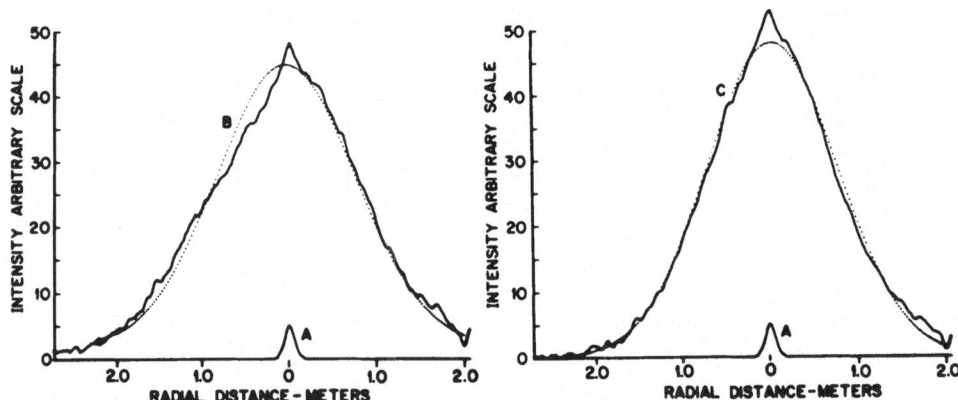

Figure 4. Scans through the first node (total light, TV) in the A, B and C states. The background, including the A_2 glow, has been subtracted from all scans. The beam current is approximately 45 mA for all scans. V = 1500 V, P = 2.5 x 10^{-6} Torr, B = 1.5 Gauss.

C: Within a second after the transition to BPD, there is a second abrupt transition to the C state. Possibly states B and C have different but similar threshold currents. Alternatively, state C simply follows B after some time delay. The experimental configuration did not allow a clear distinction between these possibilities.

The transition is marked by an increase in the peak luminosity accompanied by a reduction in the half-width of the glow (Fig. 4). In the case illustrated, there is little change in the total luminosity. In the case of 1000 v beam (Table 1) there was an approximately 20% enhancement in total luminosity over the B state.

TABLE I

V Volts	P Torr	B Gauss	Detector	R_B	R_B'	R_C	R_C'
1500	2.5x10^{-6}	1.5	TV	120	17	128	18
1000	2.5x10^{-6}	1.5	TV	75	11	92	13
1000	1x10^{-6}	1.1	3914Å photometer				20

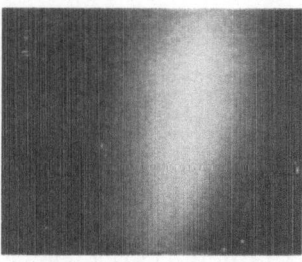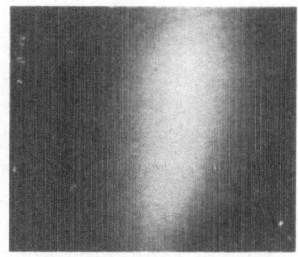

Figure 5. Television images of the first node in the A, B, and C states. The beam current is approximately 45 mA. V = 1500 V, P = 2.5 x 10⁻⁶ Torr, B = 1.5 Gauss.

As can be seen in Fig. 5, the first node remains visible in the B and C states. This is also evident in the luminosity profiles (Fig. 4). In fact, a careful comparison of the profiles shows no measureable difference in the shape and amplitude of the node between states, A, B, and C. For the 1000 v beam, the node is the same in states A and B, but is slightly attenuated in C.

Although the first node seemed largely unaffected by BPD, the second node disappeared entirely. It is not visible on the TV monitor; nor is it recognizable in the luminosity profiles. This pattern was typical for a variety of voltages. However, at very low voltages (~ 500 v) the first two nodes remained in BPD while subsequent nodes were lost.

Hysteresis

Once a BPD is ignited, the current may be reduced somewhat below the critical current without extinguishing the discharge. A typical hysteresis curve is shown in Fig. 3. The dropout current is approximately half the ignition current.

In addition to the obvious drop out of BPD, (B to A), there is also a transition from C to B that is identified by the abrupt change in the diameter of the glow. Thus the B and C states each exhibit hysteresis and each has its own dropout current.

Pressure Dependence

The behavior described here pertains to ambient pressures of less than 4 x 10⁻⁶ Torr. At this pressure, the mean free path approximately equals the 20 m pathlength of the beam.

It is beyond the scope of this paper to describe in detail the pressure dependence of the various features of BPD. Indeed, the data to do so is not available. Nonetheless, it is possible to make some general qualitative observations.

At higher ambient pressures, the A_2 state may not exist or may not be distinguishable from light produced by electrons scattered from the beam. The cyclotron-related emissions associated with the A_2 state do not occur.

The highest pressure at which the transition from B to C has been noted is 4.3×10^{-6} Torr. At higher pressures there seems to be a single BPD state. Also, the transition to BPD is less abrupt. There is less certainty in identifying the critical current. Finally, there is little or no hysteresis observed at higher pressures.

Discussion

Since the emission at 3914Å (N_2^+) is directly proportional to the ionization rate, the ratio of total luminosity in the BPD to that produced by the beam alone is an important parameter. For a 1000 volt beam at 1×10^{-6} Torr background pressure and a magnetic field strengh of 1.1 Gauss this ratio was 20:1 (Table I). Similar ratios determined from the television data (white light) suggest that the intensity as seen by the television is also proportional to the ionization rate.

The ratio (R') includes the A2 glow as part of the beam. If this glow is excluded from the definition of the beam, the value of R is of order one hundred. Thus it appears that, at pressures less than 4×10^{-6} Torr, the total production of ionization in the BPD is twenty to one hundred times that due solely to collisions between the beam and the ambient gas.

Although the plasma instabilities in the BPD are 20-100 times as effective as direct collisions in ionizing and exciting the ambient gas, this is still a small percentage of the total power of the beam. As an illustration, consider a 1000 V beam at an ambient pressure of 2.3×10^{-6} Torr. Using the ionization cross-section of N_2 for a 1000 V beam (Borst and Zipf, 1970), the total collisional ionization rate in a unit length is found to be:

$$Q_T = 4.2 \times 10^{10} \text{ ionization cm}^{-1} \text{ mA}^{-1} \text{ sec}^{-1}$$

Assuming an average energy loss of 32 electron volts per ion pair, the energy loss per cm for a 1 mA beam is 2.1×10^{-7} Watts. In a 20 m pathlength the loss is 4.2×10^{-4} Watts while the beam power is 1 Watt. For Q_c/Q_a ratios between 20 and 100, the power lost, through ionization alone (in a 20 m pathlength) is between 1% and 4% of the total beam power. (The redistribution of energy within the beam may of course be far more drastic than the power dissipation).

While the BPD is the most impressive beam-plasma interaction, the glow in the A2 state is also significant. The peak intensity

of the glow is approximately equal to that of the beam node, but
its width is several times that of the node. The total luminosity
is approximately six times that of the beam node. The cause of the
wings is uncertain. The possibility of scattered light from the
filament in the accelerator has been eliminated. The wings could
be caused by electrons that are scattered out of the main beam
within the accelerator or by secondary electrons ejected from the
aluminum target. Neither of these explanations accounts for the
abrupt transitions from the A_1 to the A_2 state.

Although not definite, a more plausible explanation is that
the glow is produced by ambient electrons that are energized by the
observed emissions near harmonics of the electron cyclotron frequency. This is consistent with the observation that these emissions appear at the same current as the A_2 glow and that neither
are observed at higher pressures. If this interpretation is correct, it follows that, even below the BPD threshold, collective
interactions are up to six times as effective as direct collisions
in ionizing the ambient gas. In short, the plasma density required
for BPD is produced primarily by indirect cyclotron interactions
rather than by direct collisions.

At pressures greater than 4×10^{-6} Torr, the mean free path is
comparable to or less than the 20 m pathlength. Evidently the
increased collisions suppress the cyclotron interaction, perhaps by
disordering the beam itself. (Deliberate collisional disruption of
the beam by placing a probe in the beam at a node also destroyed
the cyclotron emissions). At the same time, there is increased
collisional ionization so that the plasma density necessary for BPD
can be supplied by collisions.

The three abrupt transitions that are observed at low pressures
suggest that there are three important plasma instabilities (or
perhaps three combinations of two instabilities). The observed
hysteresis for the B and C states reinforces this conclusion. The
A_2 state seems to be associated with cyclotron-related emissions
while the C state seems to be dominated by plasma oscillations.
Except for the TV images, there are no available diagnostic data
for the B state.

While the total luminosity gives an indication of energy
dissipation, degradation of the nodes indicates redistribution of
the beam energy. More specifically, a spreading of the parallel
velocity results in a spreading of the downstream node. If there
is a substantial velocity spread near the accelerator, the first
node should be largely washed out and subsequent nodes nonexistent.

The observed coherence of the first node in BPD, combined with
the loss of downstream nodes, suggests that the beam remains
monoenergetic for the first few meters, but rapidly diffuses in

energy in the next few meters. This implies that the instability
responsible for spreading the beam energy is convective.

Summary

Observations of the beam-plasma-discharge at pressures below
4×10^{-6} Torr show that there are three abrupt transitions in the
beam-plasma interactions. The low-current A_1 state is dominated
by direct collisional ionization of the background gas. In the A_2
state, this ionization is supplemented by another mechanism, probably
involving cyclotron interactions. The B and C states are distinctly
separate forms of the BPD. Both involve a convective redistribution
of beam energy as indicated by the changes in the beam nodes. Both
also involve enhancements of between 20 and 100 in the power dissipation by ionization. In a 20m pathlength, approximately 1% to 4%
of the beam power is dissipated by the ionization associated with
the BPD.

ACKNOWLEDGEMENTS

This work was supported in part by contract number NA79RAC0056
from the National Oceanic and Atmospheric Administration and by
grant number ATM-8019445 from the National Science Foundation.
The participation of H. Leinbach and W. Bernstein was sponsored by
the National Aeronautics and Space Administration purchase order
GSFC/S-55762A.

REFERENCES

Bernstein, W., H. Leinbach, C. Kellogg, S. Monson, T. Hallinan,
    O. K. Garriott, A. Konradi, J. McCoy, C. Daly, B. Baker, and
    H. R. Anderson, Electron beam-injection experiments: The
    beam-plasma discharge at low pressures and magnetic field strength
    Geophys. Res. Lett. 5, 127-130, February 1978.

Bernstein, W., H. Leinbach, P. J. Kellogg, S. Monson, and T. J. Hallina
    Further laboratory measurements of the Beam Plasma Discharge,
    J. Geophys. Res., 84, Dec. 1979.

Borst, W. L., and E. C. Zipf, Cross-section for electron-impact excita-
    tion of the (0,0) first negative band of N_2^+ from the threshold
    to 3keV, Phys. Rev. A., 1, 834-840, 1970.

DISCUSSION

Stenzel: How much of the beam power goes into total line excitation and radiation?

Hallinan: Just above the BPD threshold for a 1.5 keV beam at 2.5×10^{-6} torr, the total ionization rate is of order 100 times that due to direct collisions between the beam and the ambient gas. Based on an assumed average energy loss of 30 eV per ion pair, this implies that 4% of the beam power is lost to ionizations.

Banks: Have you searched for plasma sheaths at the ends of the chamber? Can you describe their character?

Hallinan: We have looked at a sheath in BPD at the target. This shows up as a dark region, a few tens of cm thick next to the target.

Shawhan: Can you rule out the two step BPD being due to the ionization of two different ions with different ionization potentials?

Hallinan: This is possible and I have considered it. But it does not easily explain the change in the diameter of the glow. Also, if each ionization potential shows up as a distinct BPD transition, we might expect several BPD states rather than two.

Winckler: Rockets (Echo 5) show light during beam injection symmetric along and contrary to beam unlike BPD in the tank. Rocket photometers seem to be seeing return current glow discharge near rocket skin.

Hallinan: The BPD in the tank does show substantial luminosity below the accelerator although it is somewhat weaker than above. A complication in interpreting this is the fact that the floor is also a conducting boundary.

ELECTRON ENERGY DISTRIBUTION PRODUCED BY BEAM-PLASMA DISCHARGE

H. R. Anderson,[1] R. J. Jost,[2] and J. Gordeuk[1]

[1] Rice University
   Houston, Texas 77001
[2] NASA-Johnson Space Center
   Houston, Texas 77058

As an initially monoenergetic electron beam moves through a partially ionized gas with $kT \ll E_{beam}$, the beam electrons are scattered in energy and angle, the thermal electrons are similarly scattered, and new secondary electrons are created by ionization. In case only collisional scattering occurs, the energy of beam particles is lowered and a secondary distribution is built up, increasing along the beam path, and varying as $\sim E^{-3}$ according to various calculations. Applications to auroral studies are familiar (Vallance-Jones, 1974).

If the current density of the beam is raised sufficiently, a strong collective interaction known as the beam plasma discharge (BPD) occurs. The BPD onset is characterized by a large increase in the plasma production rate (ionization) and luminous emission, and the presence of large electric fields oscillating at frequencies up to the local plasma frequency (Bernstein et al., 1978, 1979). As observed at high density and magnetic field by Smullin and Getty (1962), the beam electrons are diffused radially and in energy to above and below the initial beam energy. A similar effect occurs under ionospheric conditions as approximated in the Bernstein experiments and described briefly by Jost et al. (1980).

Measurements were made in the Chamber A vacuum system at NASA-JSC using an electrostatic analyzer mounted 0.5 m in front of a 3m × 3m target ~ 20 meters above the electron accelerator at the floor of the chamber. The electrons were ejected upward anti-parallel to the 1.2 gauss magnetic field, passing through residual air at 3 to 7×10^{-6} torr pressure. The accelerator was mounted on a cart that could be driven about on the chamber floor carrying the

351

electron beam to various radial distances (perpendicular to B) from the fixed detector.

The detector employed a channel electron multiplier preceeded by a 63.5° cylindrical electrostatic deflection system with matched drift spaces to provide first order focusing, in the deflection plane, of the entrance on the exit slit. The detector was rotated about a fixed axis to accept particles with pitch angles from -90° through 180° (up coming field aligned) to +90°, the scan plane coinciding with the instrument's deflection plane. The pulse output fed a log-count-rate meter driving an x-y plotter so that graphs of count rate vs. energy or angle were produced. Other analyzer properties were:

1) Linear count rate response from < 1 to ~ 2×10^5/sec.
2) Energy resolution $\Delta E/E = 0.04$.
3) Analyzer constant 2.95 eV/volt. Balanced voltage on deflection plates.
4) Acceptance cone 9° × 3° (3° in deflection plane).
5) Geometric Factor 3×10^{-6} cm^2 ster (8.2×10^{-3} steradian x 3.6×10^{-4} cm^2).

RESULTS

Figure 1a shows a sample of the raw data produced in this manner, with scales for counting rate and intensity in the passband. (Numerically the intensity in passband when multiplied by $(.04)^{-1}$ gives energy flux (eV/cm^2 sec ster per keV energy interval).) Figure 1b shows the electron intensity per eV obtained by dividing the data in Fig. 1a by $\Delta E = 0.04$ E. A further division by E gives the velocity distribution f. Subsequent figures in this report will give intensity in passband as in Fig. 1a.

The energy dependence shown in Fig. 1a is typical. Beam currents below I_c, the critical current for BPD ($I_c \sim 30$ ma in this case) suffer only collisional scattering from the neutrals (mfp > 100 m) leaving most electrons at the accelerating potential (V = 1850 volts in Fig. 1). In BPD the beam is strongly heated, some electrons being accelerated and many being produced at lower energy. (The cutoff at 200 eV and the background of ~ 80 counts/sec in Fig. 1 are instrumental.)

The details of the spectra depend critically on radial offset from the beam and the pitch angle and azimuth sampled. The energy and pitch angle dependence of electron intensity was measured with the detector at radial distances of 30 cm, 102 cm, and 162 cm from the beam core and at azimuths of 0°, 45°, and 90° from a reference direction. Data in the following figures were obtained with the plane of pitch angle scan perpendicular to the radius from beam to detector. Particles were injected anti-parallel to B ($\alpha = 180°$)

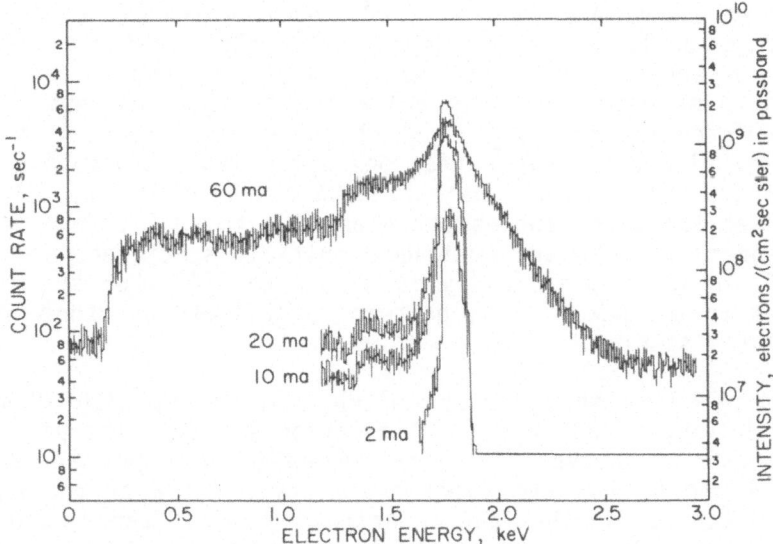

Fig. 1a Electron intensity 2m from the beam core with pitch
angle ~ 125°. Accelerating potential 1850 volts.

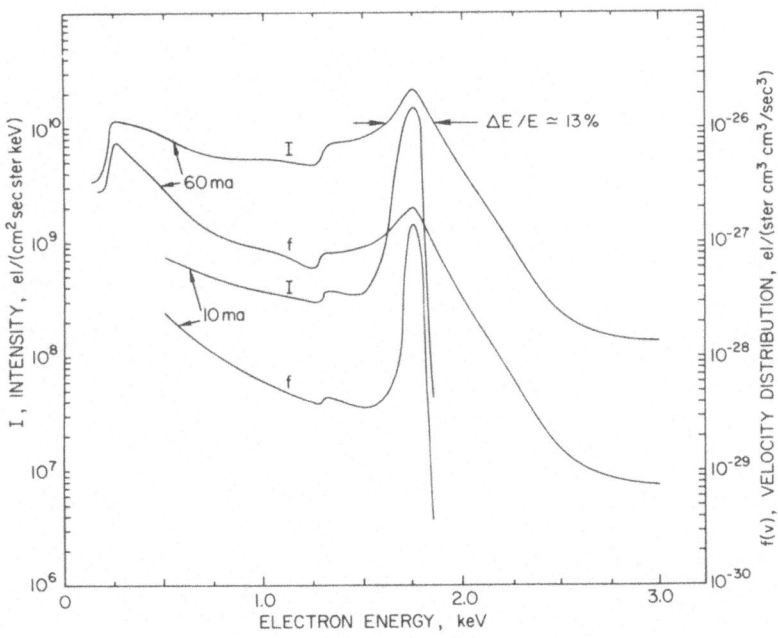

Fig. 1b. The differential energy spectra, I, derived from the data
in Fig. 1a, and the velocity distribution function f.

with 2 keV energy and 20 and 70 mA beam current. Figure 2 shows such pitch angle scans at 101cm and 162cm distance with the analyzer accepting 2 keV electrons (injection energy), and the gun emitting 20 mA beam, which is below the critical current for BPD. The peak intensity occurs at a pitch angle and azimuth that can be reached by electrons that suffer one pitch angle scattering out of the beam and then follow a helical path to the detector. Electrons observed at the pitch angles shown as 180° to +90° arrive from the complementary azimuth and must have multiply scattered. Note that the angle of maximum intensity is closer to 180° when the detector is closer to the beam. This pattern is followed at other loctions and azimuths sampled.

In Fig. 3 we show similar pitch angle scans with 70 ma beam current driving a BPD, again at injection energy. The pitch angle spectra 180 eV above or below injection energy are similar. Figures 4 and 5 show the electron energy spectra at the angle of maximum intensity in the favored azimuth (Figs. 2 and 3) and at the complementary azimuth. Beam heating by BPD is evident in Fig. 5 including the acceleration of some electrons to well above injection energy. The direction of arrival of highest flux at injection energy (Figs. 2 and 3) is favorable for all energies, but there is decreasing anisotropy at the lower energies.

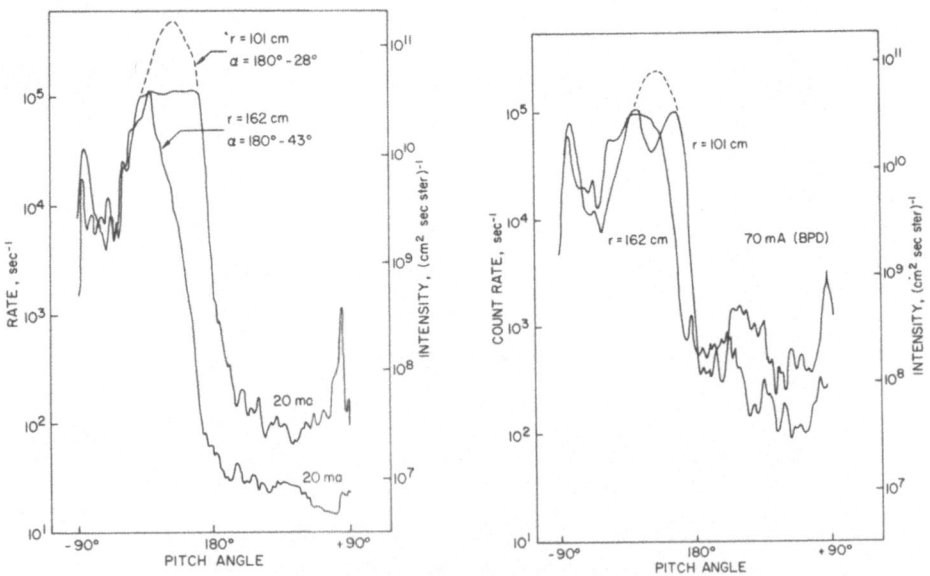

Figs. 2 and 3. Pitch angle scans with the energy analyzer accepting ~ 2000 eV electrons, the energy of injection. Fig. 2 $I < I_c$; Fig. 3, $I > I_c$.

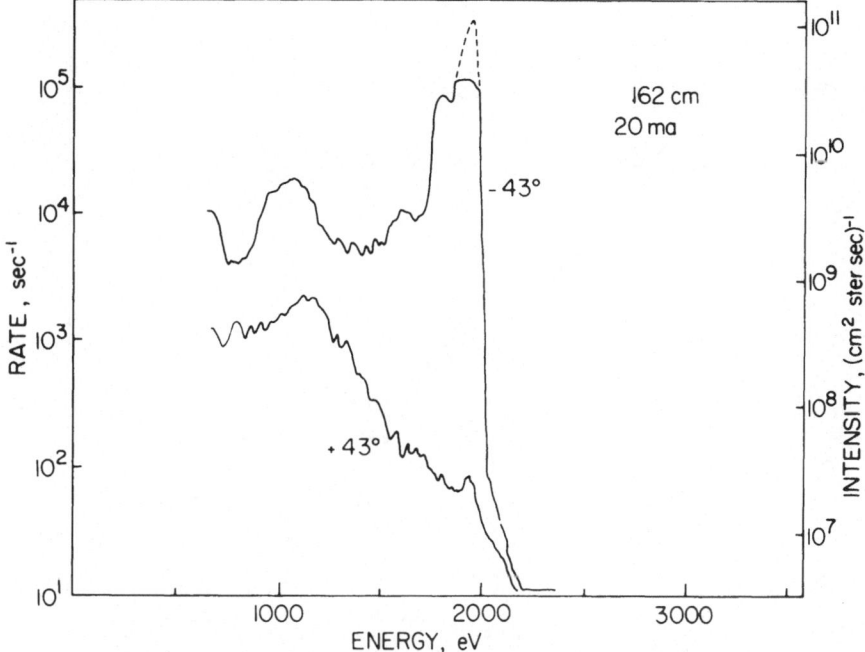

Fig. 4.   Energy scans of electron intensity at the pitch angle of
          maximum intensity at the favored azimuth ($-43° = \alpha$) and
          the complementary azimuth ($+43° = \alpha$). $I < I_c$.

It is possible to calculate the population of guiding centers
from these data, and this is shown by Fig. 6 in terms of counting
rate vs. guiding center distance from the original beam.   These
data are taken at two detector distances and with angles chosen for
each energy so that the closer guiding center defines a trajectory
that intersects the original beam.   The outer center is at the same
pitch angle and complementary (unfavorable) azimuth.   The general
result is that the population gradient steepens with increasing
energy except that it is less steep at accelerated energies than at
the initial injection energy.   There is a slight tendency for the
BPD to reduce gradients.   Data selected at constant pitch angle
show the same features.

In another observing period electrons were measured over a
more extended energy range down to 100 eV.   The data in Fig. 7 were
taken with particles injected anti-parallel to B at approximately
1500 eV energy and a variety of beam currents.   The critical current for BPD was ~ 19 mA.   The detector was ~ 25 cm from the beam
core and accepted particles with 180° pitch angle.   Simultaneous

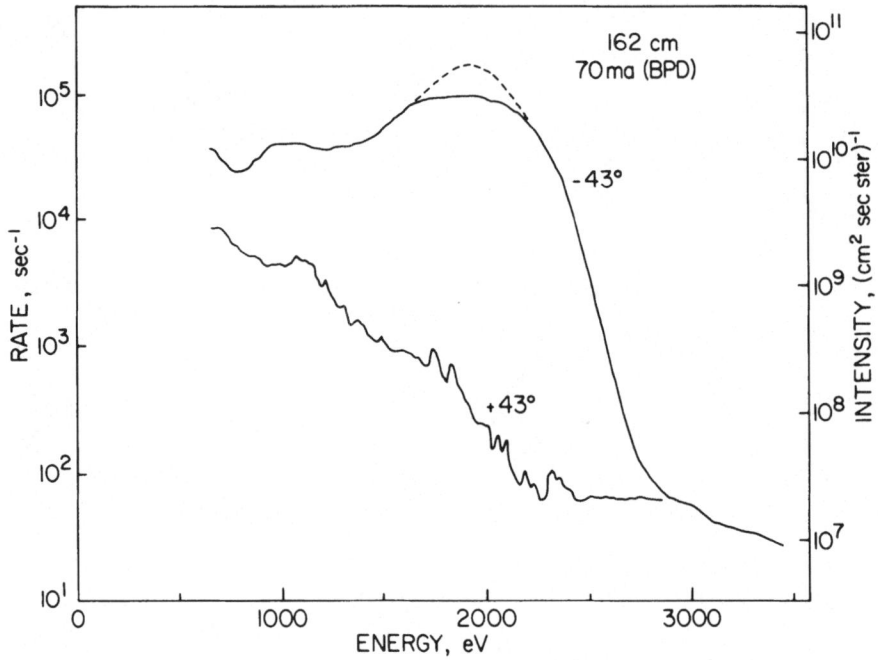

Fig. 5.   Same as Fig. 4 but $I > I_c$.

measurements were made at approximately the same distance by Wm. Sharp using a HARP analyzer measuring electrons from 1-500 eV. His data agree with those shown (private communication, 1981) and exhibit another increase in intensity from ~ 100 eV down to 5 eV.

The important feature of these spectra is the variation with electron energy of the relationship between intensity and beam current. Although the apparent intensity near 1500 eV may be distorted by saturation in counting rate, there appear to be at least three energy regimes: electrons below 600 eV; electrons from 600 eV to injection energy; and electrons above injection energy. Figure 8 shows how the intensity of each varies with current.

DISCUSSION AND CONCLUSIONS

From these data we conclude the following:

1)   The occurrence of BPD heats the initially cold electron beam from the accelerator.

2)   The directional intensity of electrons measured outside the beam core allows the inference that most particles suffer a single scattering in energy and pitch angle. At low currents this is expected as beam particles collide with the neutral atmosphere. In

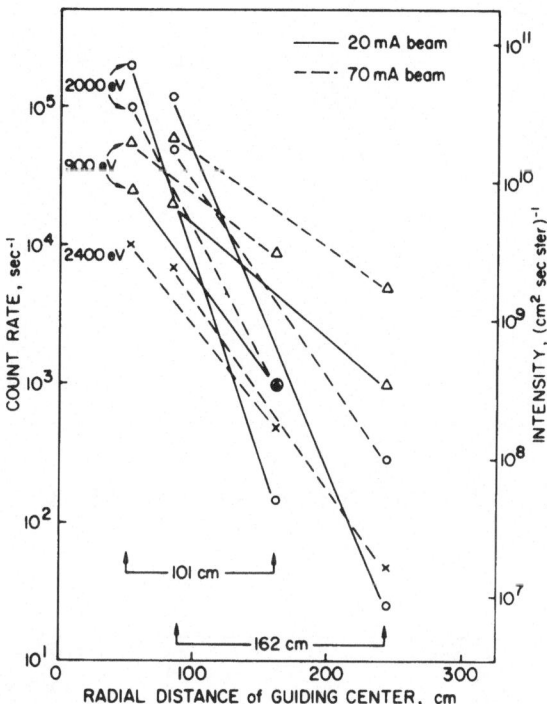

Fig. 6.   Population of guiding centers vs. radial distance from beam.

BPD the majority of particles still undergo a single scattering near the original beam core. This is consistent with the observation that the very strong radio-frequency electric fields are confined to near the beam core. Large angle scattering is more likely in BPD than in the lower current, non-BPD case. Also large angle scattering is more likely with the lower energy electrons.

3)   The extended energy spectra shown in Fig. 7 show two rather distinct plasma populations, one centered at the initial beam energy (~ 1500 eV) and the other at ~ 150 eV. The velocity distribution functions corresponding to these intensities are obtained by division by E^2. The magnitude of the peak at 150 eV is then ~ 1/2 that of the peak in the distribution at 1500 eV for a 32 mA beam current. Above 1500 eV the distribution function falls through three decades from the peak as $\exp[-E/E_0]$, the functional form of a Maxwellian. The equivalent temperature $E_0 \simeq 86$ eV. The slope in distribution function from 150 eV to 300 eV at 32 mA also has $E_0 \simeq 86$ eV, but this varies with beam current. The angular distribution of this low energy plasma has not been thoroughly measured, but preliminary data suggest that it is peaked towards 90° pitch angle.

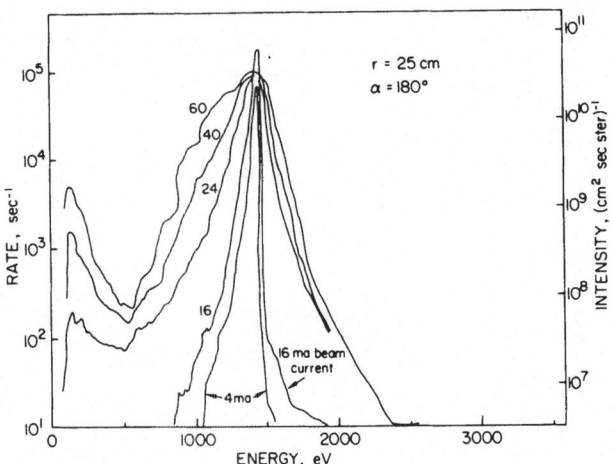

Fig. 7.   Energy scans at various beam currents

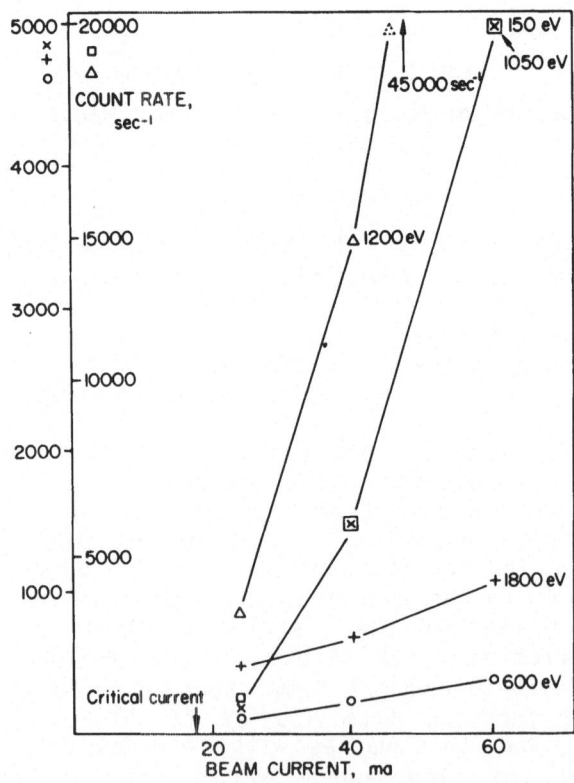

Fig. 8.   The variation with beam current of intensity at various
           energies, from Fig. 7.

ACKNOWLEDGEMENTS

It is a pleasure to thank William Bernstein for the loan of the electron accelerator and for many useful discussions of this work. The experiments were supported by NASA through availability of the JSC Chamber A and through Grant NGL44-006-012 to Rice University.

REFERENCES

Bernstein, W., H. Leinbach, P. Kellogg, S. Monson, T. Hallinan, O. K. Garriott, A. Konradi, J. McCoy, P. Daly, B. Baker, and H. R. Anderson, 1978, Electron beam injection experiments, Geophys. Res. Lett., 5:127.

Bernstein, W., H. Leinbach, P. J. Kellogg, S. J. Monson, and T. Hallinan, 1979, Further laboratory measurements of the beam-plasma discharge, J. Geophys. Res., 84:7271.

Jost, R. J., H. R. Anderson, and J. O. McGarity, 1980, Electron energy distribution measured during beam-plasma interactions, Geophys. Res. Lett., 7:509.

Smullin, L. D., and W. D. Getty, 1962, Generation of a hot dense plasma by a collective beam-plasma interaction, Phys. Rev. Lett., 9:3.

Vallance-Jones, A., 1974, Chapters 4 and 5, in "Aurora," Reidel Publ. Co., Dordrecht, Holland.

TIME-DEPENDENT PLASMA BEHAVIOR TRIGGERED BY A PULSED ELECTRON GUN

UNDER CONDITIONS OF BEAM-PLASMA-DISCHARGE

Edward P. Szuszczewicz

E.O. Hulbert Center for Space Research
Naval Research Laboratory
Washington, DC 20375

C.S. Lin

Bendix Field Engineering Corp.
9250 Route 108, Columbia, MD 21045

ABSTRACT

We have conducted a number of experiments intended to simulate spaceborne applications of energetic electron guns while exploring the "in situ" diagnostics of time-dependent beam-plasma behavior under pulsed electron gun conditions. The conditions include the beam-plasma-discharge (BPD) and the BPD afterglow that exists after gun-pulse termination. With electron gun characteristics set at $(I_g, V_g, T_{on}/T_{off})$ = (34 ma, 1.9 keV, 80 ms/270 ms) and a super-imposed magnetic field at 1.5 gauss, the results show that: (i) There is a three order-of-magnitude increase in plasma density within 5 ms of gun turn-on; (ii) During the pulse-ON time a quasi-steady-state BPD is maintained with characteristics identical to dc-BPD conditions; (iii) Plasma losses appear to be dominated by Bohm-like diffusion processes; and finally (iv) The afterglow can be characterized by an isodensity radial profile that decays with a 36 msec time constant and cools at a $3.8(10^3)$ °K/sec rate.

INTRODUCTION

An energetic electron beam, propagating through a neutral or charged-particle environment will, under various conditions, follow single particle trajectories or undergo collective effects that influence the energetic-particle orbits and render the beam-plasma system unstable to various plasma modes (Linson and Papadopoulos, 1980). From points of view focussed on single-particle

361

behavior, there are a number of valuable spaceborne applications,
among them being the mapping of geomagnetic field lines, detection
of geomagnetic conjugates by the generation of artificial aurora,
the study of beam spreading, atmospheric excitation and ionization
processes and the measurement of magnetic field-aligned potentials.
On the other hand the nonlinear processes that cannot be described
by classical single-particle behavior and result in varied unstable
states, represent an area of extreme interest not only to basic
plasma physics but also to a large number of space-plasma phenomena
that include anomalous spacecraft neutralization, enhanced ionization processes, wave-particle interactions and plasma turbulence,
to name a few. While collective phenomena can significantly limit
single-particle-trajectory experiments, they represent one of the
most unexplored areas of controlled beam experiments in space.

    One of the areas in space-related beam-plasma interactions to
receive considerable attention in recent years has been the
collective plasma process called the beam-plasma-discharge (Bernstein
et al., 1978; Bernstein et al., 1980; Szuszczewicz 1979; and Jost
et al., 1981). The beam-plasma-discharge (BPD) describes a beam-
plasma state that appears at a critical beam current I_{crit}. This
critical current level yields a marked increase in ion-pair-
production, a greatly enhanced 3914 Å emission, a modification of
the primary beam velocity distribution and the emission of intense
RF waves. The BPD has been the subject of a continuing series of
space-simulation experiments (Bernstein et al., 1978; Bernstein et
al., 1980, Szuszczewicz 1979; and Jost et al., 1981) that to large
measure have dealt with the steady-state beam-plasma behavior in
various stages ranging from pre-BPD (i.e., $I_{beam} < I_{crit}$), threshold
($I_{beam} \sim I_{crit}$) and solid BPD($I_{beam} > I_{crit}$). One such recent
work (Walker et al., 1981) has determined that the density-related
plasma condition for BPD threshold can be expressed as $\omega_p =
(5.8 ^{+1.3}_{-1.9}) \omega_c$, where ($\omega_p, \omega_c$) are the plasma- and electron-cyclotron
frequencies, respectively. Since this result and other steady-
state BPD signatures are finding what appear to be plausible
theoretical descriptions (Rowland et al., 1981 and Papadopoulos,
1981), it is fair to say that the steady-state space-simulated BPD
is approaching a reasonable level of accepted scientific understanding. The transfer of this understanding to spaceborne applications is accompanied by a number of conditions which to date
have not been adequately simulated or extensively studied. These
conditions include the existence of a uniform and quiescent pre-
beam plasma, the existence of a moving beam-plasma reference frame
(as would be the case for a Shuttle-borne accelerator), an unbounded
beam-length and temporal beam-plasma behavior. To close that gap
and develop another perspective on space-simulated beam-plasma
interactions we have conducted a number of experiments which
explore the time-dependent beam-plasma behavior under pulsed
electron gun conditions. The objectives included:

(i)   The determination of time-dependent electron density profiles under pulsed-BPD conditions, including the BPD state itself and the BPD afterglow that exists after gun-pulse termination. The BPD afterglow is of interest in itself and conceivably bears signatures relevant to the beam-plasma wake in space (Region IV described by Anderson et al., 1979).   In addition the BPD-afterglow could be used as the homogeneous pre-beam plasma for threshold studies more realistically simulating those to be investigated in space.

(ii)   The test of the pulsed-plasma-probe technique (Holmes and Szuszczewicz, 1975 and Holmes and Szuszczewicz 1981) and its ability to simultaneously determine electron density, temperature, space potential and density fluctuation power spectra under pulsed electron gun conditions, and finally

(iii)   The exploration and validation of previous observations of Bohm-like diffusion processes that appear to be active in a turbulent beam-plasma system (Szuszczewicz 1979).

These objectives were all accomplished and a selection of the results with experimental details are presented in the succeeding sections.

EXPERIMENT CONFIGURATION AND DIAGNOSTIC TECHNIQUE

The experiment was conducted in the large vacuum chamber (20 m diameter x 30 m high) facility at the NASA Johnson Space Flight Center in Houston, Texas.   The chamber, with base pressures in the range 5 (10^{-7}) to 1 (10^{-6}) torr, was equipped with large current-carrying coils to generate magnetic fields up to 2.1 gauss.   A steerable tungsten cathode gun was mounted near the chamber floor on a movable cart that allowed the beam to be injected upwards and parallel to the magnetic field $\bar{B}$, and terminated on a gridded 3x3 m collector suspended 20 m above the gun aperture.

The chamber was also equipped with a position-controlled cylindrical pulsed-plasma-probe P^3 (Holmes and Szuszczewicz, 1975 and 1981) that could be continuously varied in its radial separation from the beam core.   All radial traversals were along the local magnetic meridian at a height of 8 m above the gun aperture.   Care was taken to maintain the probe axis perpendicular to $\bar{B}$ in order to guarantee that radial profile information was not distorted by magnetic-aspect sensitivities (Takacs and Szuszczewicz 1979).   The pulsed probe itself is a specialized Langmuir probe technique which provides a high-time-resolution determination of relative electron density (1 msec resolution was utilized in this experiment) while simultaneously generating a "conventional" Langmuir probe character-istic for determination of absolute N_e, T_e and plasma potential V_∞.   The technique applies a chain of voltage pulses to the probe

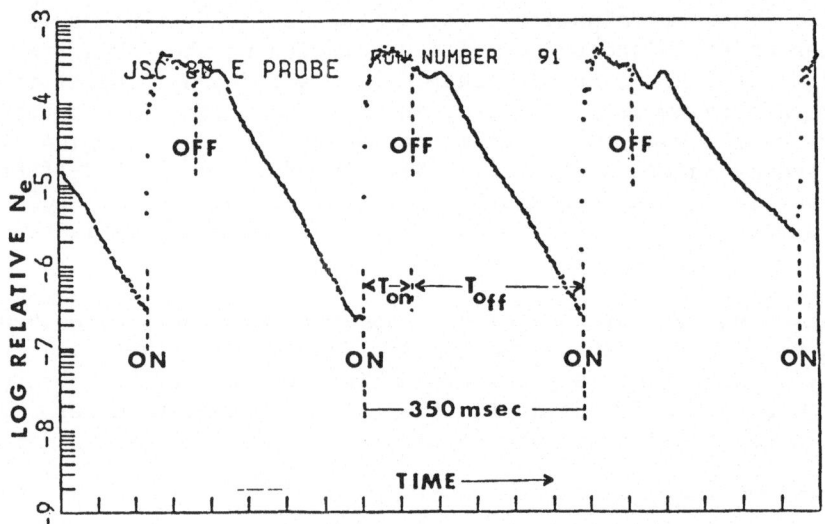

Fig. 1.   Time-dependent plasma response during three consecutive
gun-pulse periods. N_e^{max} = (1.3 ± 0.5) (10^7) cm^{-3}.

that follows a sawtooth envelope and generates the (I_{sweep},V_{sweep})
data pairs for the conventional Langmuir probe I-V characteristic.
During the interpulse period the probe is held at a fixed-baseline
level, V_B, in the electron-saturation portion of the characteristic.
The running measurements of I_B, during the baseline period V_B,
then provide a measure of relative N_e variations (assuming I_B =
I_e^{sat} ∝ N_e).

The P^3 procedure and its application to relative N_e measurements
during pulsed gun operation is illustrated with reference to
Figure 1 which shows a sample of time-dependent relative density
measurements (as indicated by msec baseline current samples)
during three contiguous pulsed gun cycles. (Simultaneously generated
sweep currents are not shown in the Figure.) The gun's current
and voltage were set at (I_g,V_g) = (34 ma, 1.9 keV) and the operation
cycle was at 80 msec ON and 270 ms OFF for a total 350 msec period.
The results in the figure can be characterized as follows:

(a) There is a rapid enhancement in plasma density as the
gun turns on (about 3 orders of magnitude increase in density in
approximately 5 msec);

(b) There is a "flat" beam-plasma state during the pulse-ON
time (in this case a beam-plasma discharge was achieved); and finally

(c) There is an exponential decay in plasma density once the
gun pulse is terminated. Electron decay time constants were found
to be 36 ± 8 msec.

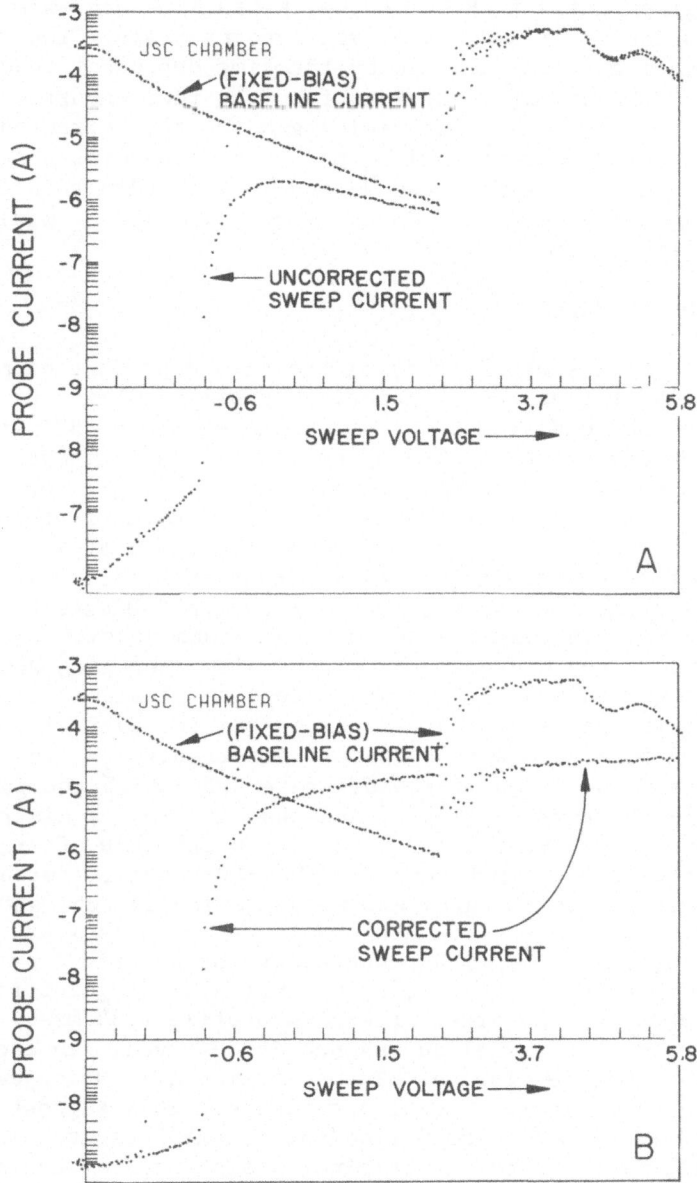

Fig. 2.   Raw P³ data (2A) showing the effects of density variations
          (baseline currents) of sweep current characteristic during
          gun-pulse period. Fig. (2B) shows the "corrected"
          characteristic after density variations have been unfolded.

Absolute electron densities and temperatures were determined by routine P^3 analysis procedures (Szuszczewicz and Holmes, 1977) summarized graphically in Figure 2, with baseline and sweep currents collected during their associated voltage intervals. The relative density variations (as indicated by the time-dependent behavior of the baseline current) were unfolded from the raw, uncorrected probe characteristic (Fig. 2a) yielding a smooth, corrected curve (Fig. 2b) to which conventional N_e analysis procedures were applied (Chen, 1965 and Szuszczewicz and Holmes 1977). With this procedure, the maximum observed electron density in Fig. 2 was $N_e(max)$ = (1.3 ± 0.5) (10^7) cm^{-3}.

THE PLASMA DECAY PROCESS

As an initial step in studying the time-dependent beam-plasma process and in testing the diagnostic capability of the P^3 technique under pulsed-gun conditions, focus was placed on the BPD afterglow... it's time-dependent density decay, possible relationships to diffusion processes and associated electron cooling. For beam conditions set at (I_g, V_g) = (34 ma, 1.9 keV), chamber pressure at $6.6(10^{-6})$ torr and with the superimposed field set at 1.5 gauss, a series of time-dependent beam-plasma density-profiles were generated with the gun cycle at 80 msec ON and 270 msec OFF. Because of any given position there were plasma density variations from pulse-to-pulse, a procedure of ten-pulse averaging was utilized to represent the pulsed BPD and BPD-afterglow plasmas. The relative density profiles that resulted from the averaging process are presented in Figure 3 by P^3 baseline currents I_B. The uppermost profile represents the quasi-steady-state BPD during the gun-ON period, while the four lower profiles show the radial distribution of plasma at successively later times during the BPD-afterglow, that is, during the gun-OFF period. The BPD profile (during the 10-80 msec period) is in agreement with previously published BPD conditions conducted under dc gun operation (Szuszczewicz, 1979), in that the plasma's radial dependence can be described by an exponential function, that is, $N_e = N_e^o \exp(-r)$. In going from the BPD (10-80 msec) to the first afterglow profile (133 msec) a marked difference in radial dependence is observed. In fact, for all times in the afterglow the plasma is within $\pm$ 10% of being an isodensity profile in its radial dependence. This suggests that in the immediate after-pulse period the plasma loss is dominated by radial diffusion, a result originally suggested by the Bohm-like diffusion coefficients found in the dc-BPD profiles (Szuszczewicz 1979) where the values for D_e were found to be orders of magnitude larger (e.g., $D_e(nom)$ = 2.2 $(10^6)cm^2/sec$) than would be expected for classical cross-field collisional diffusion in the presence of a superimposed magnetic field. If radial diffusion were not the dominant loss mechanism, the afterglow profiles would maintain qualitatively the exponential BPD (10-80 msec) profile while the plasma decayed axially. The profiles suggest just the opposite.

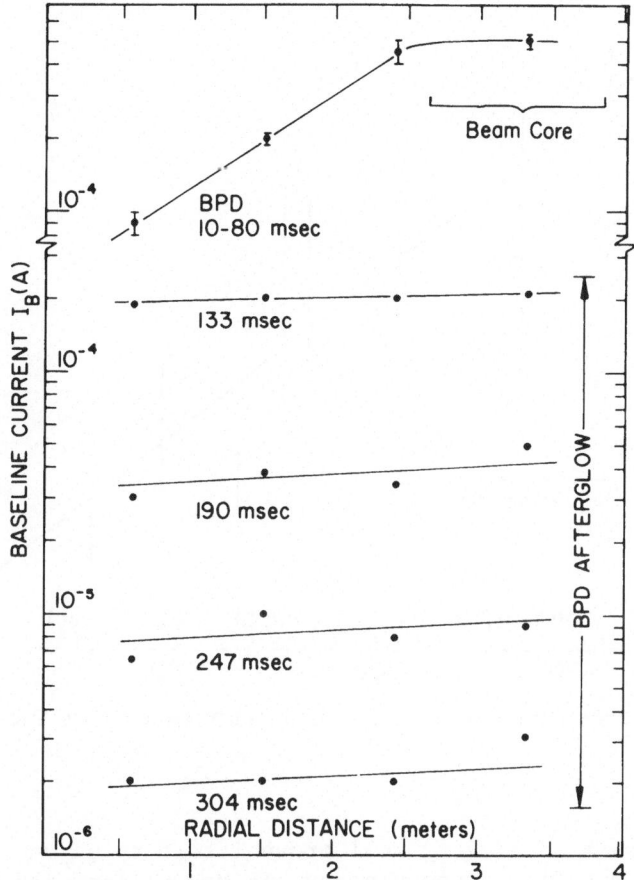

Fig. 3. Time-dependent radial profiles of relative plasma density covering BPD (gun ON) and afterglow (gun OFF) conditions. Absolute densities can be scaled within ± 20% by N_e/I_B = 0.39 (10^{11}) and 0.20 (10^{11}) for the BPD (10-80 msec) and afterglow profiles, respectively.

We interpret these findings as further support for the existence of an enhanced cross-field diffusion process in the turbulent beam-plasma system characterized by the BPD.

At present one further step has been taken in studying the BPD afterglow and associated applications of the P^3 technique. This step involved determination of the electron cooling rate, with P^3 measurements of T_e presented in Figure 4 as a function of time within the gun-pulse cycle. The results can be fit with a linear function of time that points to a constant electron cooling rate equal to 3.8 (10^3) $^{\circ}$K/sec. The composite (N_e,T_e) profile information provided in Figures 3 and 4 suggest that the BPD

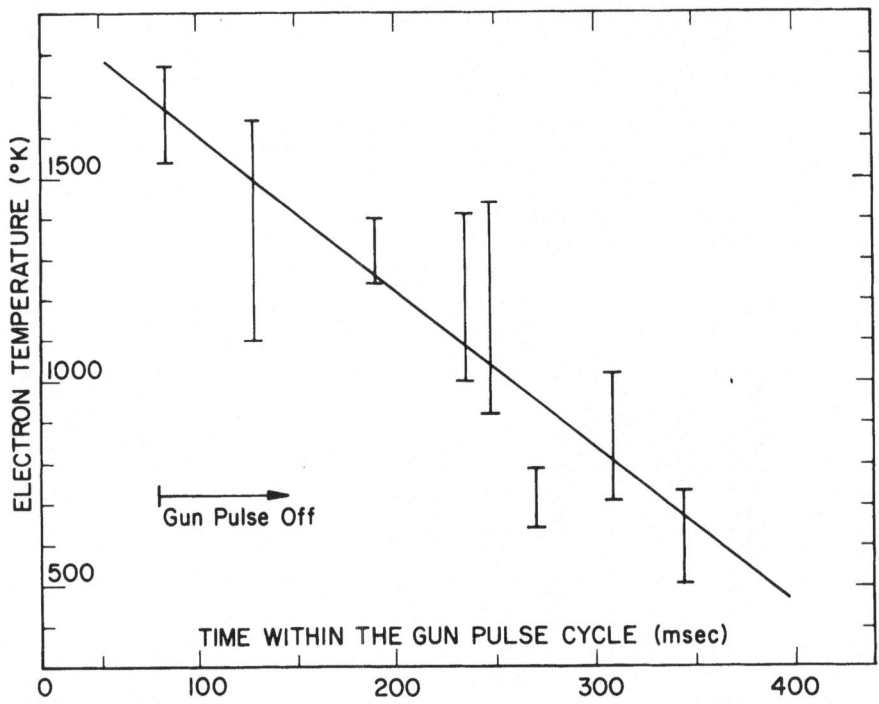

Fig. 4.   Time-dependence of electron temperature in the BPD afterglow.

afterglow plasma could very well prove itself to be an ideal pre-
beam environment (i.e., homogeneous and Maxwellian with temperatures
nearly equivalent to those found in the F-region ionosphere) for
space-simulation studies of beam-plasma interactions. Indeed, the
laboratory space-simulation afterglow has provided a valuable test
bed for potential spaceborne applications of P^3 to beam-plasma
investigations.

COMMENTS AND CONCLUSIONS

    We have conducted the first "in situ" measurements of electron
density profiles in a space-simulated pulsed-gun beam-plasma-
discharge covering the quasi-steady-state BPD condition, its onset
and its afterglow decay. While there are substantial variations
with gun settings and radial-position-sampling, the observations
can be characterized as follows:

    (i) There is a rapid enhancement in plasma density as the
gun turns on (about 3 orders of magnitude increase in approximately
5 msec for the case reported in this investigation); additional
results (not reported here) show that the total enhancement and
the BPD onset time are a function of gun current and energy, the

superimposed magnetic field and gun ON/OFF cycle time. Details will be provided in future publications.

(ii) During the pulse-ON time a quasi-steady-state BPD can be maintained with characteristics identical with its dc counterpart.

(iii) In the period immediately following gun-pulse termination the plasma loss process is dominated by cross-field radial diffusion in keeping with an earlier suggestion that the plasma turbulence in a BPD system results in an enhanced Bohm-like diffusion process.

(iv) The afterglow plasma is within $\pm$ 10% of being an isodensity contour in its radial extent and cools linearly with time at a 3.8 $(10^3)^{\circ}$K/sec rate with an average density decay time constant equal to 36 $\pm$ 8 msec. The BPD afterglow appears to provide an ideal pre-beam plasma environment (i.e., homogeneous and Maxwellian with $T_e \sim T_e$ (ionosphere)) for space-simulation studies of beam plasma interactions more likely to be encountered at F-region altitudes.

ACKNOWLEDGMENT

This work was supported in part by NASA/NOAA Contract No. NA79RAA04487. Support for analyses was supplemented by the Office of Naval Research under Program Element 61153N-33 in Task Area RR033-02. We wish to thank J.C. Holmes for his critical care in electronics design and L. Kegley for technical assistance in experiment execution.

REFERENCES

Anderson, H., Bernstein, W., Papadopoulos, K., Szuszczewicz, E.P., and Linson, L.M., 1979, "A Theoretical and Experimental Investigation of Beam-Plasma Physics", Proposal submitted to NASA.

Bernstein, W., Leinbach, H., Kellog, P., Monson, S., Hallinan, T., Garriott, O. K., Konradi, A., McCoy, J., Daly, P., Baker, B., and Anderson, H.R., 1978, "Electron Beam Injection Experiments: The Beam-Plasma Discharge at Low Pressures and Magnetic Field Strengths", Geophys. Res. Lett. 5:127.

Bernstein, W., Whalen, B.A., Harris, F.R., McNamara A.G., and Konradi, A., 1980, "Laboratory Studies of the Charge Neutralization of a Rocket Payload During Electron Beam Emission", Geophys. Res. Lett. 7:93.

Chen, F.F., 1965, Chapter 4, in "Plasma Diagnostics Techniques", R.H. Huddlestone and S.L. Leonard, ed., Academic, New York.

Holmes, J.C. and Szuszczewicz, E.P., 1975, "A Versatile Plasma Probe Rev. Sci. Instr. 46:592.

Holmes, J.C. and Szuszczewicz, E.P., 1981, "A Plasma Probe System with Automatic Sweep Adjustment", Rev. Sci. Instr.

Jost, R.J., Anderson, H.R. and McGarity, J.O., 1981, "Measured
    Electron Energy Distributions During Electron Beam-Plasma
    Interactions", Geophys. Res. Lett. 7:509.
Linson, L.M. and Papadopoulos, K., 1980, "Review of the Status of
    Theory and Experiment for Injection of Energetic Beams in
    Space", Report No. LAPS 69 SAI-023-80-459-LJ, Science
    Applications Inc., La Jolla, CA.
Papadopoulos, K., 1981, Private Communication.
Rowland, H.L., Chang, C.L., and Papadopoulos, K., 1981 (in press),
    "Sealing of the beam plasma discharge", J. Geophys. Res.
Szuszczewicz, E.P. and Holmes, J.C., 1977, "Observations of Electron
    Temperature Gradients in Mid-Latitude E$_s$-Layers", J. Geophys.
    Res. 82:5073.
Szuszczewicz, E.P., 1979, "Plasma Diffusion in a Space-Simulation
    Beam-Plasma-Discharge", Geophys Res. Lett. 6:201.
Takacs, P.Z. and Szuszczewicz, E.P., 1979, "Magnetosheath Effects
    on Cylindrical Langmuir Probes", Phys. Fluids, 22:2424.
Walker, D.N., Lin, C.S., and Szuszczewicz, E.P., 1981, "Ignition of
    the Beam-Plasma-Discharge and It's Dependence on Electron
    Density", in Beam Experiments in Space, Bjorn Grandal, Ed.,
    Plenum Publishing Corp., New York, NY.

DISCUSSION

Shawhan: Do you see the two temperature electron plasma
that was reported by Paul Kellogg?

Szuszczewicz: I have observed a transition from a
single-component Maxwellian plasma outside the beam core
(e g Geophys Res Lett, 6, 201, 1979) to a multi-component
(not necessarily two-temperature) plasma electron population inside the beam core. We have also had evidence
of non-Maxwellian electron populations as a function of
length along the beam. In that Kellogg reports the
existence of electron populations other than single-
Maxwellian, there is agreement.

IGNITION OF THE BEAM-PLASMA-DISCHARGE AND ITS DEPENDENCE ON ELECTRON DENSITY

D. N. Walker and E. P. Szuszczewicz

E.O. Hulburt Center for Space Research
Naval Research Laboratory
Washington, DC 20375

C.S. Lin

Bendix Field Engineering Corp.
9250 Route 108, Columbia, MD 21045

ABSTRACT

A cold electron beam, propagating through a weakly ionized plasma will, under proper conditions, produce a modified beam-plasma state known as the Beam-Plasma-Discharge (BPD). As the subject of a continuing series of experiments in a large facility chamber it was previously determined that the BPD had an abrupt ignition threshold as the beam current (I_B) was increased at fixed beam energy. While a specific empirical relationship was established among the controlling parameters of beam current, energy and length as well as ambient pressure and magnetic field, a dependence of the BPD on plasma density of the form $\omega_p \sim \omega_c$ was suggested. We have since conducted a survey of various beam-plasma conditions covering beam currents from 8 to 85 ma, beam energies from 0.8 to 2.0 keV and magnetic fields at 0.9 and 1.5 gauss. This survey includes full determinations of radial profiles of electron density for each of the selected conditions extending from a low-density pre-BPD state to a strong BPD condition. At BPD threshold N_e^{max} was determined and ω_p calculated with results that can be summarized by

$$\omega_p = (5.8 \, {+ 1.3 \atop - 1.9}) \, \omega_c$$

as the density dependent threshold condition for BPD. The experimental results are shown to compare favorably with a developing theoretical model that considers BPD to be triggered by electron plasma wave excitation of a beam-plasma instability.

INTRODUCTION

A cold electron beam, propagating through a weakly ionized plasma will, under proper conditions, produce a modified beam-plasma state known as the Beam-Plasma-Discharge (BPD). This discharge state has received considerable attention in recent years as a result of increased interest in mechanisms for vehicle neutralization during spaceborne accelerator experiments (Bernstein, et al., 1980; Cambou, et al., 1978), enhanced beam-plasma ionization processes (Bernstein, et al., 1978), and in general single-particle or collective phenomena initiated by beams injection into neutral gas and charged-particle environments (Hess et al., 1971; Winckler, et al., 1975; Hendrickson and Winckler 1976; Cambou, et al., 1975; Monson and Kellogg 1978a; Szuszczewicz 1979; Jost et al., 1980). As the subject of a continuing series of experiments in a large vacuum chamber facility (Bernstein et al., 1978) it was determined that the BPD appears at a critical energetic-electron-beam current I_B^c, following the relationship

$$I_B^c \propto \frac{V_B^{1.5}}{B^{0.7}PL},$$ (1)

where V_B, B, P and L are the beam energy (voltage), the superimposed magnetic field, the ambient pressure and the beam length (gun aperture-to-collector distance), respectively.

While the $I_B^c = I_B^c (V_B,B,P,L)$ relationship was established among the controlling system parameters, a dependence on plasma density was also expected, with early thoughts (Bernstein, et al., 1979) suggesting that $\omega_p = \omega_c$ satisfied ignition threshold criteria. We have conducted a survey of various beam-plasma conditions from 8 to 85 ma, beam energies from 0.8 to 2.0 keV and magnetic fields at 0.9 and 1.5 gauss. The survey included determination of radial profiles of electron density for each of the selected conditions extending from a low-density, pre-BPD state to a strong BPD condition. In summary, the results indicate that

$$\omega_p = (5.8 \, {}_{-1.9}^{+1.3}) \, \omega_c$$ (2)

is the density-dependent threshold condition for BPD. The experimental details and analysis procedures that led to this result are presented below and compared with the predictions of a theoretical model which assumes that the BPD is triggered by electron plasma wave excitation of a beam-plasma instability.

EXPERIMENT CONFIGURATION AND RESULTS

The experiment was conducted in a 20 m diameter by 30 m high vacuum chamber facility at the NASA Johnson Space Flight Center.

The configuration involved a pair of pulsed-plasma-probes mounted on a radial traversal mechanism positioned at approximately 8 m above the injection point of the beam. Each of the probes provided simultaneous measurements of electron density N_e, temperature T_e, plasma potential V_∞, and density fluctuation power spectra δN_e ($\rightarrow P_n(k)$) with capability for the associated diagnostics under dynamic plasma conditions and under environmental conditions that could contaminate electrode surfaces (Holmes and Szuszczewicz, 1975, 1981; Szuszczewicz and Holmes 1975, 1976). Both these conditions prevailed to various degrees.

A tungsten cathode gun was mounted near the chamber floor on a movable cart so that the beam could always be injected parallel to the magnetic field $\bar{B}$ and terminated on the 3 x 3 m target suspended about 20 m above the gun aperture. A combination of coil current and the Earth's magnetic field established the B-field at one of two levels, 0.9 and 1.5 gauss. The chamber was also equipped with a dipole-antenna/frequency-spectrum-analyzer system (Bernstein et al., 1979) which was used to determine BPD ignition from its characteristic plasma wave emissions. The dipole system was connected to a Tektronix spectrum analyzer with a frequency response from 200 kHz to 30 MHz. Because the high-frequency cut-off was not abrupt, frequencies up to 50 MHz could be detected readily.

In most cases the beam was injected into a neutral gas with no pre-beam plasma environment; however the experimental survey included two cases in which the chamber was filled with a pre-beam plasma created by a Kauffman-type argon ion thruster. In these cases the pre-beam plasma density was lower than the critical density at BPD ignition.

The survey included seven different conditions, each identified by pre-selected values for V_B, B, P and the existence or non-existence of a pre-beam plasma. For each condition a steady state value for I_B was set, a radial traversal was made and an electron density profile was recorded. A sample profile collected under pre-BPD conditions, is presented in Figure 1. The abscissa is time relative to the start of the radial traversal and the ordinate is relative electron density as determined by baseline electron-saturation currents collected by the E-probe. (The second in the two-probe configuration was defined as the I-probe because the associated baseline currents were collected in the ion-saturation portion of the probe's current-voltage characteristic (Holmes and Szuszczewicz, 1975, 1981).) At the start of each traversal the probe was at its outermost position relative to the center of the chamber. As time increased the probe was moved into and through the beam; at minimum radial distance from the chamber center, the traversal system was reversed, allowing a second measurement of the density profile as the probe moved back to its original outermost position. With this procedure the probe's minimum radial coordinate is identified by the symmetry point in the "double" profile.

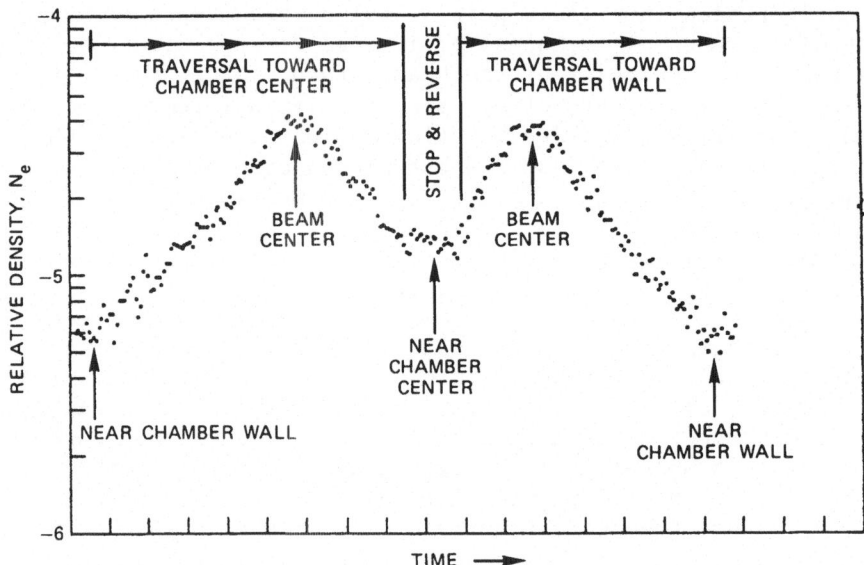

Fig. 1. Radial profile of relative electron density under pre-
         BPD conditions. Run #57, (I_B, V_B, B) = (7 ma, 1.3 keV,
         0.9G). The figure shows two cuts through the beam-plasma
         profile, as time increases from left-to-right the plasma
         density probe moves into and through the beam center, then
         reverses and passes through the beam a second time. The
         symmetry verifies that beam-plasma conditions were stable
         during the execution of the radial traversal.

Absolute electron densities were determined by standard P^3
analysis procedures summarized graphically in Figure 2. The technique
provides a determination of relative electron density through the
direct measurement of baseline electron-saturation-currents at a
sample rate of 1 kHz. Simultaneously, the technique generates a
"conventional" Langmuir probe characteristic. The relative density
fluctuations (as indicated by the variations in the baseline current)
are then unfolded from the raw, uncorrected probe characteristic
(Fig. 2A) yielding a smooth, corrected curve (Fig. 2B) to which
conventional N_e analysis procedures (Chen, 1965; Szuszczewicz and
Holmes, 1977) are applied. This procedure was utilized for all
beam-plasma conditions included in this investigation.

Relative electron density profile information and associated
plasma wave signatures are presented in Figure 3 for (V_B, B) = (1.3
keV, 0.9G) and for beam currents I_B stepped through a sequence
allowing for coverage of conditions which encompassed pre-, thresh-
old-, and solid-BPD. The conditions at threshold and under BPD are
summarized in Table 1 where the peak density N_e^{max}, associated plasma
frequency ω_p^{max}, and plasma-to-cyclotron frequency ratio ω_p^{max}/ω_c
are also listed. The results can be summarized by

$$\omega_p = (5.8 \, \begin{smallmatrix} +1.3 \\ -1.9 \end{smallmatrix}) \, \omega_c$$

as the density-dependent threshold condition for the BPD.

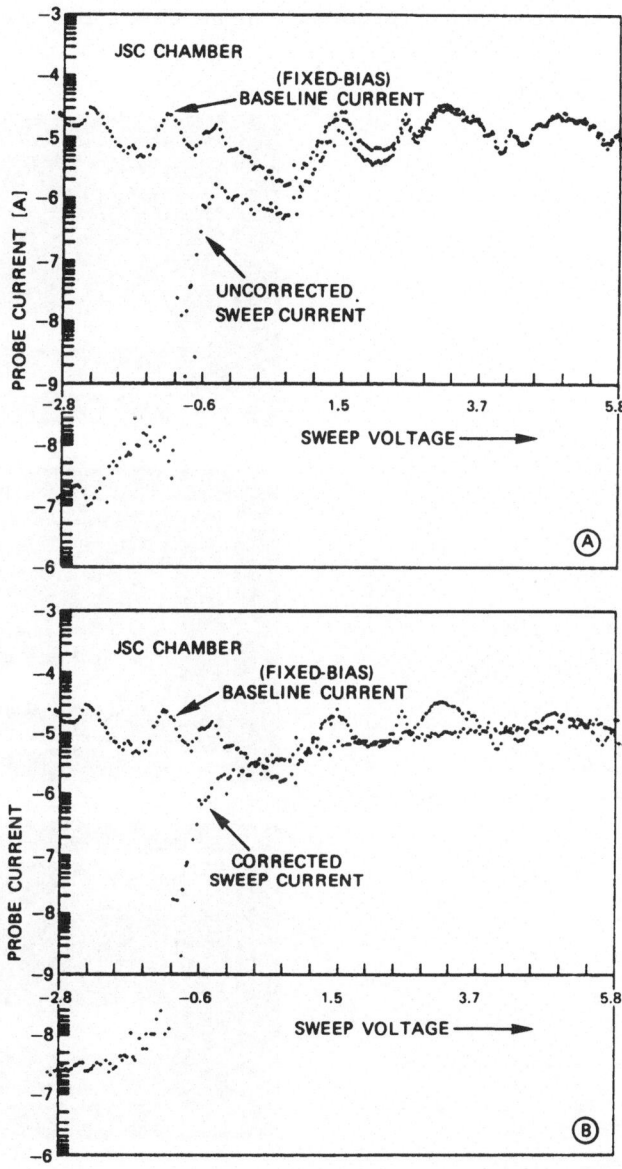

Fig. 2. Sample of raw probe data (2A) showing the effects on
density fluctuations (baseline electron-saturation-currents)
on the probe's current-voltage characteristics (sweep
currents). 2B shows the "corrected" characteristic.

DISCUSSION OF RESULTS

The experimentally derived threshold condition is reasonably consistent with the suggestion that BPD is triggered by the onset of a beam plasma instability excited by electron plasma waves (Rowland et al., 1981; Papadopoulos, private communication, 1981). Qualitatively the threshold process can be described as follows:

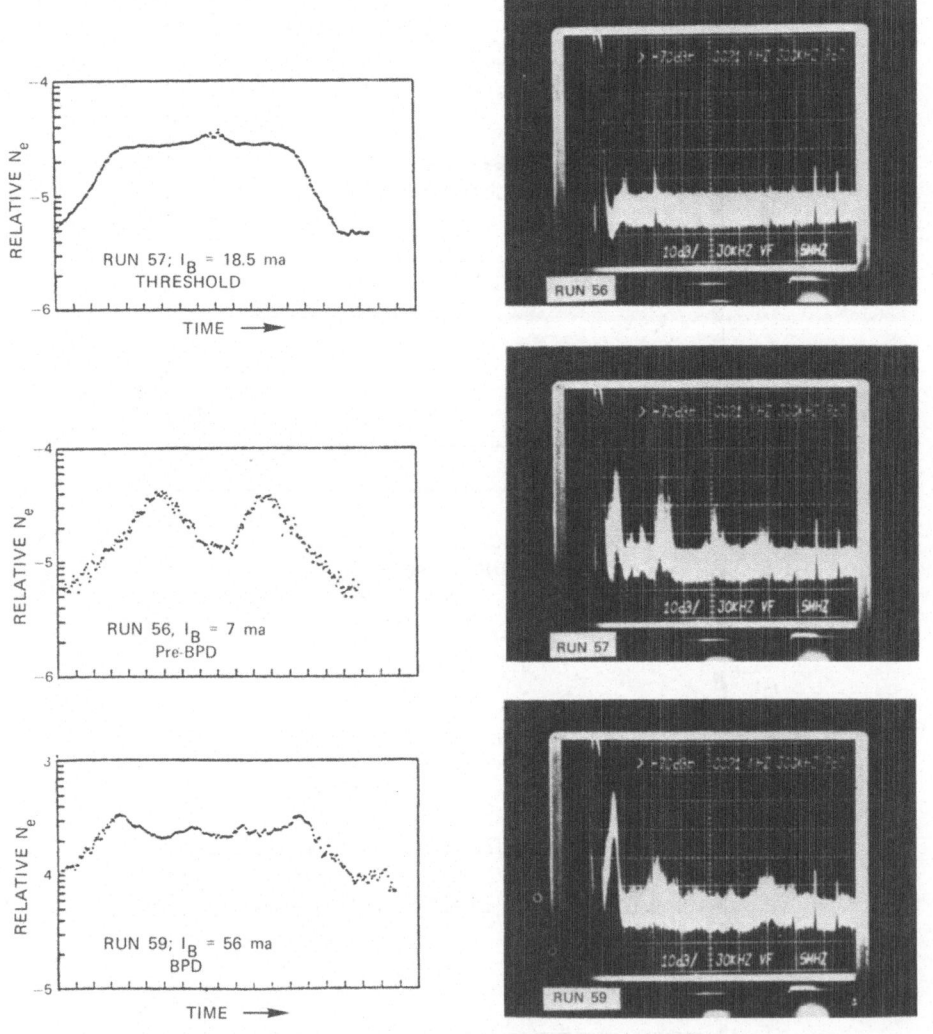

Fig. 3. Sequence of relative plasma density profiles and associated plasma wave signatures for increasing values of beam current I_B for a fixed condition (V_B, B) = 1.3 keV, 0.9G) encompassing runs 56 through 59 (pre-BPD through solid-BPD).

(i)   As an electron beam linearly interacts with a neutral gas, it collisionally produces a plasma with a density that varies directly with the magnitude of the beam current for a fixed beam energy.

(ii) As the beam current is increased further, a two-stream instability develops in which the electric fields of the excited waves heat the electrons to energies comparable to the ionization energy of the neutral species. The "heated" electrons create an enhanced ionization process which results in an avalanche breakdown during the BPD.

Detailed theoretical considerations (Rowland et al., 1981) involving finite beam-plasma geometries suggest that the threshold for BPD ignition corresponds to the onset of convective instability. Quantitatively that threshold takes the form

$$\omega_p \overset{>}{=} 1.4 \; v_b/r_o \; \sqrt{\ln(R/r_c)} \tag{3}$$

where r_o and v_b are the beam radius and velocity, and R is the radius of the plasma with which the beam interacts. For the experimental conditions, r_o is taken to be controlled by the gun half-divergence angle θ, the beam velocity v_b and the superimposed magnetic field. We therefore write

$$r_o = (v_b \; \text{Sin} \; \theta)/\omega_c \; , \tag{4}$$

allowing the theoretically predicted threshold condition to be re-written as

$$\frac{\omega_p}{\omega_c} > \frac{1.4 \; (1.2)}{\text{Sin} \; \theta} \tag{5}$$

where $1.2 = 1/\sqrt{\ln \; (R/r_o)}$ has been selected as the experimental average. Equation (5) suggests that the ω_p/ω_c threshold condition is a constant, independent of B itself, and controlled only by the beam cross section through the half-divergence angle θ. Qualitatively this is in agreement with the experimental results. For a quantitative comparison, we estimate θ in the range, $5^o \overset{<}{-} \theta \overset{<}{-} 10^o$, yielding

$$9.6 \overset{<}{-} \omega_p/\omega_c \overset{<}{-} 19.3 \tag{6}$$

as the spread in values theoretically predicted for BPD ignition. This result, while sensitive to the uncertainties in θ and R/r_o (e.g., electrostatic forces and beam spreading have not been included), is taken to be in reasonably good agreement with the experimentally derived conditions (2). Inclusion of beam spreading would effectively increase θ (Linson and Papadopoulos, 1981) and improve the agreement

Table I. Abbreviated summary of beam-plasma survey

RUN #	BEAM-PLASMA STATE	ELECTRON GUN		CHAMBER CONDITION			N_e^{max}	f_c	f_p/f_c
		I_B [ma]	V_B[v]	B [g]	P [Torr]	THRUSTER			
40	THRESHOLD	37	$1.9 (10^3)$	0.9	$0.7\text{-}1.5 (10^{-5})$	ON	$3.6 (10^6)$	$2.5 (10^6)$	6.92
41	BPD	47	$1.9 (10^3)$	0.9	$0.7\text{-}1.5 (10^{-5})$	ON	$5.6 (10^6)$	$2.5 (10^6)$	8.60
48	THRESHOLD	34	$1.9 (10^3)$	0.9	$0.7\text{-}1.5 (10^{-5})$	OFF	$3.3 (10^6)$	$2.5 (10^6)$	6.6
49	BPD	45	$1.9 (10^3)$	0.9	$0.7\text{-}1.5 (10^{-5})$	OFF	$5.0 (10^6)$	$2.5 (10^6)$	8.12
57	THRESHOLD	18.5	$1.3 (10^3)$	0.9	$0.7\text{-}1.5 (10^{-5})$	OFF	$1.5 (10^6)$	$2.5 (10^6)$	4.45
58	BPD	28	$1.3 (10^3)$	0.9	$0.7\text{-}1.5 (10^{-5})$	OFF	$4.5 (10^6)$	$2.5 (10^6)$	7.71
63	THRESHOLD	7.8	800	0.9	$0.84\text{-}1.5 (10^{-5})$	OFF	$0.98 (10^6)$	$2.5 (10^6)$	3.7
64	BPD	9.9	800	0.9	$0.84\text{-}1.5 (10^{-5})$	OFF	$2.6 (10^6)$	$2.5 (10^6)$	5.9
69	THRESHOLD	6.2	800	0.9	$0.7 (10^{-5})$	ON	$3.8 (10^6)$	$2.5 (10^6)$	7.08
70	BPD	7.8	800	0.9	$0.7 (10^{-5})$	ON	$3.6 (10^6)$	$2.5 (10^6)$	6.89
81	THRESHOLD	20	$2.0 (10^3)$	1.5	$0.6\text{-}1.2 (10^{-5})$	OFF	$7.0 (10^6)$	$3.7 (10^6)$	6.65
82	BPD	30.5	$2.0 (10^3)$	1.5	$0.6\text{-}1.2 (10^{-5})$	OFF	$1.8 (10^7)$	$3.7 (10^6)$	10.7
86	THRESHOLD	12	$1.3 (10^3)$	1.5	$0.6\text{-}1.2 (10^{-5})$	OFF	$3.9 (10^6)$	$3.7 (10^6)$	4.96
87	BPD	18	$1.3 (10^3)$	1.5	$0.6\text{-}1.2 (10^{-5})$	OFF	$1.1 (10^7)$	$3.7 (10^6)$	8.34

providing even stronger arguments which deny the original notion that $\omega_p = \omega_c$ described BPD threshold.

ACKNOWLEDGMENTS

This work was supported in part by NASA/NOAA Contract No. NA79RAA04487. Support for analyses was supplemented by the Office of Naval Research under Program Element 61153N-33 in Task Area RR033-02. The authors would like to thank W. Bernstein for suggesting the experiment and helping make possible the NRL participation in the JSC experiments. We also wish to thank J.C. Holmes for his critical care in electronics design and L. Kegley for diligence and professionalism in instrument fabrication and technical assistance in experiment execution.

REFERENCES

Bernstein, W., Leinbach, H., Cohen, H., Wilson, P.S., Davis, T.N., Hallinan, T., Baker, B., Martz, J., Zeimke, R., and Huber, W., 1975, "Laboratory observations of RF emissions at ω_{Pe} and $(N + 1/2)\omega_{CE}$ in electron beam-plasma and beam-beam interactions", J. Geophys. Res., 80:4375.

Bernstein, W., Leinbach, H., Kellogg, P., Monson, S., Hallinan, T., Garriott, O.K., Konradi, A., McCoy, J., Daly, P., Baker, B., and Anderson, H.R., 1978, "Electron beam injection experiments: The beam-plasma discharge at low pressures and magnetic field strengths", Geophys. Res. Lett., 5:127.

Bernstein, W., Leinbach, H., Kellogg, P.J., Monson, S.J., and Hallinan, T., 1979, "Further laboratory measurements of the beam-plasma discharge", J. Geophys. Res., 84:7271.

Bernstein, W., Whalen, B.A., Harris, F.R., McNamara, A.G., and
    Konradi, A., 1980, "Laboratory studies of the charge
    neutralization of a rocket payload during electron beam
    emission", Geophys. Res. Lett. 7:93.
Cambou, F., Dokoukine, V.S., Ivchenko, V.N., Managadze, G.G.,
    Migulin, V.V., Nazarenko, O.K., Nesmyanovich, A.T., Pyatsi,
    A. Kh., Sagdeev, R.Z., and Zhulin, I.A., 1975, "The Narnitza
    rocket experiment on electron injection", Space Research
    XV, 491-500, Akademie-Verlag, Berlin.
Cambou, F., Lavergnat, J., Migulin, V.V., Morozov, A.I., Paton,
    B.E., Pellat, R., Pyatsi, A., Reme, H., Sagdeev, R.Z.,
    Sheldon, W.R., and Zhulin, I.A., 1978, "ARADS-Controlled or
    puzzling experiment"? Nature, 271:723.
Chen, F.F., 1965, Chapter 4, in Plasma Diagnostic Techniques,
    edited by R.H. Huddlestone and S.L. Leonard, Academic, New
    York.
Hendrickson, R.A., and Winckler, J.R., 1976, "Echo III: The study
    of electric and magnetic fields with conjugate echoes from
    artificial electron beams injected into the auroral zone
    ionosphere", Geophys. Res. Lett., 3:409.
Hess, W.N., Trichel, M.C., Davis, T.N., Beggs, W.C., Kraft, G.E.,
    Strasinopoulos, E., and Maier, E.J.R., 1971, J. Geophys.
    Res., 76:6067.
Holmes, J.C., and Szuszczewicz, E.P., 1975, "A versatile plasma
    probe", Rev. Sci. Instr., 46:592.
Holmes, J.C., and Szuszczewicz, E.P., 1981 (in press), "A plasma
    probe system with automatic sweep adjustment", Rev. Sci.
    Instr.
Jost, R.J., Anderson, H.R., and McGarity, J.O., 1981, "Measured
    electron energy distributions during electron beam-plasma
    interactions", Geophys. Res. Lett. 7:509.
Monson, S.J., and Kellogg, P.J., 1978, "Ground observations of
    waves at 2.96 MHz generated by an 8- to 40-KEV electron
    beam in the ionosphere", J. Geophys. Res., 83:121.
Rowland, H.L., Chang, C.L., and Papadopoulos, K., 1981 (in press),
    "Sealing of the beam plasma discharge", J. Geophys. Res.
Szuszczewicz, E.P., and Holmes, J.C., 1975, "Surface contamination
    of active electrodes in plasmas: Distortion of conventional
    Langmuir probe measurements", J. Appl. Phys. 46:5134.
Szuszczewicz, E.P., and Holmes, J.C., 1976, "Reentry plasma
    diagnostics with a pulsed plasma probe", AIAA Paper No.
    76-393, AIAA 9th Fluid and Plasma Dynamics Conference,
    San Diego, CA.
Szuszczewicz, E.P., 1979, "Plasma diffusion in a space-simulation
    beam-plasma-discharge", Geophys. Res. Lett., 6:201.
Winckler, J.R., Arnoldy, R.L., and Hendrickson, R.A., 1975,
    "Echo 2: A study of electron beams injected into the high-
    latitude ionosphere from a large sounding rocket", J.
    Geophys. Res., 80:2083.

PLASMA WAVES STIMULATED BY ELECTRON BEAMS IN THE LAB

AND IN THE AURORAL IONOSPHERE

R. H. Holzworth, W. B. Harbridge and H. C. Koons

Space Science Laboratory
The Aerospace Corporation
El Segundo, California 90245

Abstract

Energetic electron beams are frequently used as active probes of space plasmas. Often the assumed test particle nature of these electrons is violated when the electron beam stimulates plasma wave emissions. Such complex phenomena have been observed on rockets and satellites and are being modeled in laboratory plasmas. The large vacuum chamber at NASA Johnson Space Center in Houston, Texas has been used for modeling F-region type ionospheric plasmas. A VLF receiver has been flown into an auroral plasma and the spectra from this flight will be compared to VLF spectra obtained in the NASA/JSC laboratory chamber. The electron beam is believed to have produced beam plasma discharge (BPD) on the rocket similar to that seen in the lab. At times during the rocket flight the electron beam was operated at 4 kilovolts and the electron current modulated at 3 kilohertz from 0 to 80 milli-amps. This resulted in the beam pulsing in and out of BPD and a variety of propagating wave modes.

The laboratory VLF electric field spectra during BPD show a characteristic peak at a few kilohertz with amplitudes over 100 mV/m. This peak broadens and moves to higher frequencies as the current is increased at a fixed electron voltage. Other features of BPD in the lab as seen in the VLF spectra include appearance of the spectral peak prior to optical BPD threshold, differences between $\tilde{E}$ and $\tilde{B}$ spectra below the peak and oscillation in and out of BPD even under a steady state electron gun current on time scales of 100 ms.

INTRODUCTION

In an active experiment to study plasma dynamics in the auroral ionosphere, NASA sounding rocket 27.010 AE was launched on April 9, 1978 from Ft. Churchill, Manitoba, Canada. The rocket carried an electron accelerator and a full complement of plasma diagnostic devices including electric and magnetic receivers, particle detectors and photometers. The accelerator was mounted on the aft payload which remained attached to the rocket motor throughout the flight. The diagnostic devices were arranged on various Throw Away Detectors laterally ejected (TAD's) and a forwardly ejected payload. One important experiment performed in this flight (described in detail by Holzworth and Koons, 1981) involved the 3 kHz modulation of the electron beam current at fixed voltage and the subsequent detection of a 3 kHz signal by electric and magnetic receivers on the forward payload at distances up to several kilometers away. Furthermore, steady state gun operation at maximum current appeared to result in beam plasma discharge (BPD).

Following ejection of the forward payload at about 10 m/s and antenna deployment the electron accelerator on the aft payload was operated in a mode in which the current was modulated between I_{min} = 0 to 10 mA and $I_{max} \cong$ 80 mA at 3 kHz for 450 ms every 11 s. Every one of these accelerator modulation periods (AMP's) were detected by the forward payload in the electric VLF spectrum. In the first half of this paper the wave spectra from the rocket flight will be discussed. In particular the spectral features indicative of beam plasma discharge will be emphasized. This includes a time delay analysis which suggests that a variety of wave modes were present and the presence of a characteristic spectral enhancement near a few kHz during pulsed gun operation. A more detailed description of the 3 kHz VLF spectra from flight 27.010 may be found in Holzworth and Koons (1981).

The second half of this paper deals with experimental laboratory simulation of the ionospheric rocket observed phenomena. The experiments were conducted in the large vacuum chamber (see Bernstein et al., 1981) at the NASA Johnson Space Center in Houston, Texas. In this chamber we have conducted experiments to simulate the ionospheric phenomena seen by several rockets launched into the auroral ionosphere. Of particular interest is the beam plasma discharge (BPD) phenomena wherein the electron gun, when operated above some threshold current for a given energy, produces a vastly enhanced ionization rate in this collisionless environment. Lab studies of BPD at the NASA/JSC chamber have been conducted for some time (Bernstein, et al., 1978). These investigations have shown that the radio frequency waves above 1 MHz may not be large enough in amplitude to provide the necessary energization for BPD (Bernstein and Kellogg, 1980). In these studies relatively low resolution measurements in the VLF spectrum suggested that these very low frequency waves were responsible for the energization. Therefore the last half of this paper will characterize these low frequency electric and magnetic waves during BPD with high resolution spectrograms. Variations in gun energy and current result in differing thresholds for the BPD which are reflected in these spectra. VLF spectrograms obtained during the NASA rocket flight 27.010 AE (E||B) from Ft. Churchill in 1978 will be compared to these lab plasma spectrograms. This comparison strongly suggests that BPD phenomena were responsible for the measured spectral features from the rocket.

Rocket Instrumentation

The payload instrumentation have been described by Wilhelm et al., 1980 and by Bernstein et al., 1981. The rocket instruments relevant to this work are the electron accelerator and the wave receivers. The VLF wave receiver has been described in Holzworth and Koons (1981). Briefly, the VLF instrument includes an electric antenna consisting of a pair of spherical probes separated by 2.75 meters on rigid booms mounted $\perp$ to the spin axis and a magnetic antenna consisting of a ferrite rod with multiple windings mounted inside the forward payload. The preamps fed both broadband (up to 16 kHz) and fixed-frequency, narrowband channels. One of these narrowband filter channels was set at 3 kHz which allowed accurate determination of absolute signal amplitudes for the 3 kHz AMP's. All electronics in these receivers operated perfectly except that onboard EMI away from our range of interest caused the magnetic AGC to operate in its least sensitive mode throughout the flight. Thus the broadband magnetic spectra are considerably more noisy than the electric.

Rocket Observations

In Holzworth and Koons (1981) it was shown that a VLF signal at 3 kHz was clearly radiated during the beam modulation periods. An important point made in that paper was that the time delay between the beginning of electron accelerator modulation periods (AMP's) and the onset of the detection at the forward payload was not only measurable but variable over the flight. Fig. 1 shows these measured time delays vary from near zero to several tenths of a second. Furthermore, the point was made in that paper that the electric and magnetic signals from the AMP's were not received simultaneously but sometimes the electric signal preceded the magnetic and sometimes the other way around.

The propagation time for the 3 kHz signal to reach the forward payload is often significantly slower than for an electromagnetic wave which could not be resolved on this time scale. The apparently systematic changes in the time delays shown in this figure suggest that the signals are not simply space charge disturbances emanating from the aft payload. Since the modulation

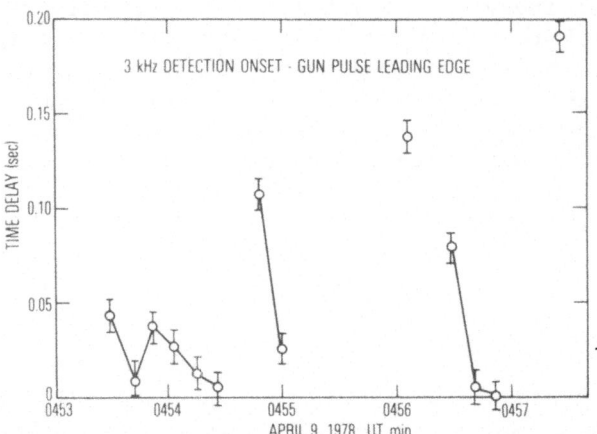

Fig. 1.   Time delay between turn on of gun AMP and detection at
           forward payload.   Data gaps are due to irregularity of
           occurrence of AMP's and interference by other VLF pheno-
           mena.

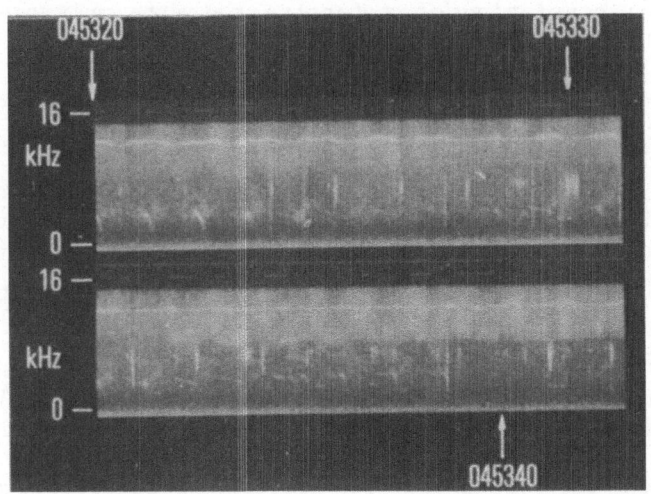

Fig. 2   A 20-second electric spectrogram from E||B.   The
          distinct broadband pulses are due to the dc gun steps at
          2 and 4 kV at the maximum current.

was between 0 and 80 mA at 4 kV, the beam was probably pulsing in
and out of BPD.

    A 20-second sequence of spectral data from the electric
receiver on the rocket is shown in Fig. 2. This spectrogram shows
several discrete broadband pulses which occurred during gun opera-

tions at 2 and 4 kV. These dc pulses were of 50 ms duration and the sequence repeats about every second. A longer pulse of 450 ms occurs near 04:53:30 UT. During these larger pulses (AMP's) the current was modulated at 3 kHz as discussed above. The spectra during AMP are very similar to the dc steps but show considerable time variability. Enlarged spectra during some of these pulses are shown in Fig. 3. Broad spectral peaks near 5 or 6 kHz are seen to extend up to the lower hybrid frequency which is a little above 7 kHz.

Laboratory Experimental Apparatus

The chamber set up used in these experiments included an electron gun on a movable cart on the floor of the chamber with a target collector for the electron beam mounted near the top of the chamber. The ac electric and magnetic antennas were mounted near the center of the chamber just outside the gun generated plasma column. Various other particle and field measuring devices were also available which are discussed elsewhere (Bernstein et al., 1978). The electron gun voltage and current were individually varied from 0 to 2 kV and 0 to 80 mA respectively.

The electric receiver consisted of a pair of crossed 2-foot dipole antennas connected to high impedance preamplifiers followed

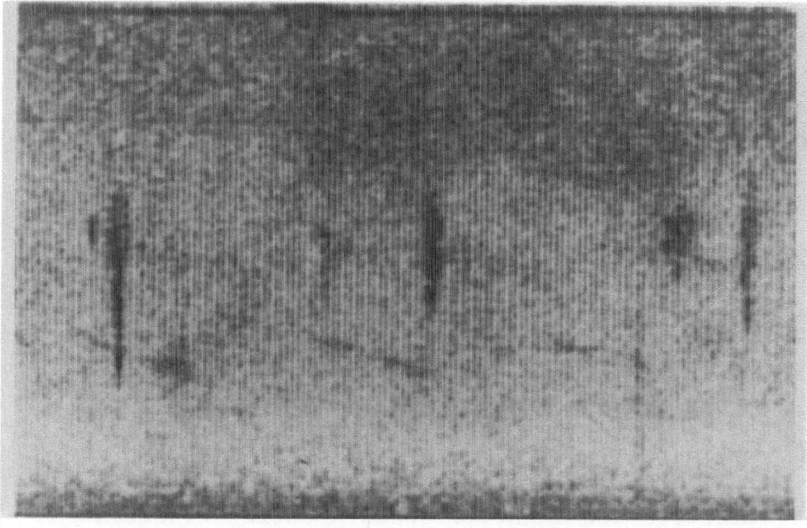

Fig. 3. Rocket observed electric spectra during pulsed mode operation; this is an expanded section of Fig. 1 starting at 04:53:31 UT.

by various attenuators and amplifiers in the chamber providing a dynamic range of 160 db above 100 nanovolts. The electric antennas were mounted on a two pulley arrangment allowing location anywhere in a plane which intersected the gun plasma column. The magnetic antenna consisted of a two foot diameter loop with 2000 turns of wire having pick off points at 20, 200 and 2000 turns. Magnetic sensitivity extended down to 0.2 picotessla at 3000 Hz. The dynamic range was 110 db. The loop was mounted on a fixed support rope about 3 meters from the gun plasma column. The vacuum chamber was cryogenically pumped at liquid helium temperatures and maintained a pressure of about a few times 10^{-6} torr throughout these experiments.

Laboratory VLF Spectral Data

Fig. 4 shows a sequence of electric and magnetic spectra during electron gun operations at 2 kilovolts and varying current from 9 mA to 48 mA. In this mode the gun was operated in a dc manner while the background plasma thruster was off. These spectra are averaged data for about one second and show the general sequence of signature variation from single particle behavior in panel a at 9 mA to a supersolid BPD in panel e. The determination of BPD onset was provided by optical instrumentation including narrowband photometers and a low level light TV system.

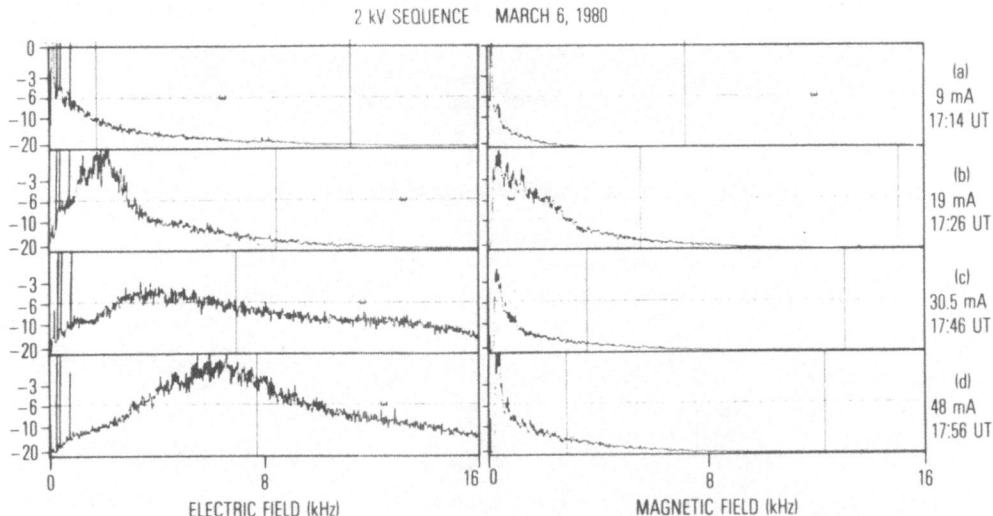

Fig. 4. Electric and magnetic spectra during electron gun operation at 2 kV. a. Single particle spectrum at 9 mA; 0db = 13.4 mV or 1.2 mγ. b. Prior to BPD threshold at 19 mA; 0db = 84.5 mV or 0.38 mγ. c. Solid BPD at 30.5 mA; 0db = 42.4 mV or 1.1 mγ. d. Supersolid BPD at 48 mA; 0db = 42.4 mV or 1.5 mγ.

Fig. 4 panel b is a spectrum from just prior to BPD conditions as determined optically. However a well developed broad peak has already appeared in the VLF electric signal near 2.2 kHz. The sharp narrowband peaks in the low frequency electric signal are 60 cycle harmonics. A sharp onset of BPD occurs within a milliamp change in the electric current. As the current is increased to 30.5 mA, a solid BPD is formed (see Fig. 4 panel c) and the VLF peak of panel b has become substantially broader and the tail more enhanced. Now the magnetic spectrum is radically different and appears more like a single particle spectrum. Finally, the current is raised to 48 mA in Fig. 4 panel d and the peak and tail are both broader and more enhanced in the electric signature.

The 2-kV BPD ignition sequence is summarized by three electric VLF spectrograms in Fig. 5. Here it is seen that the spectra also became time-variable on the scale of a few hundred milliseconds appearing to turn BPD on and off. This effect is averaged out in the one-second average spectra of Fig. 4.

As an example of another gun energy, Fig. 6 presents an ignition sequence at 800 volts. Here the BPD threshold occurred at 7.8 mA and the VLF peak progressively moved to higher frequency as the gun current was raised to the "supersolid" 30 mA 800 V level. At this energy, unlike 2 kV, the magnetic signature (not

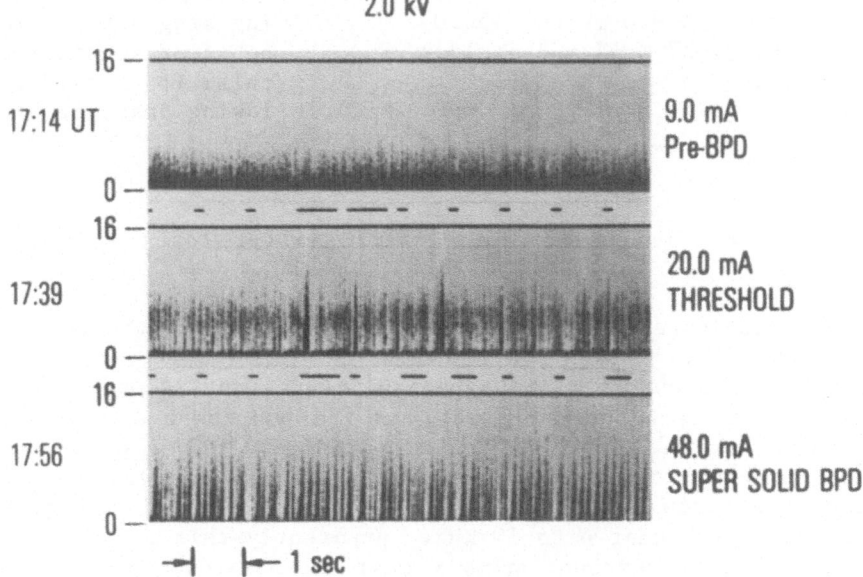

Fig. 5. Three 10-second samples of spectrograms during various gun currents at 2.0 kV.

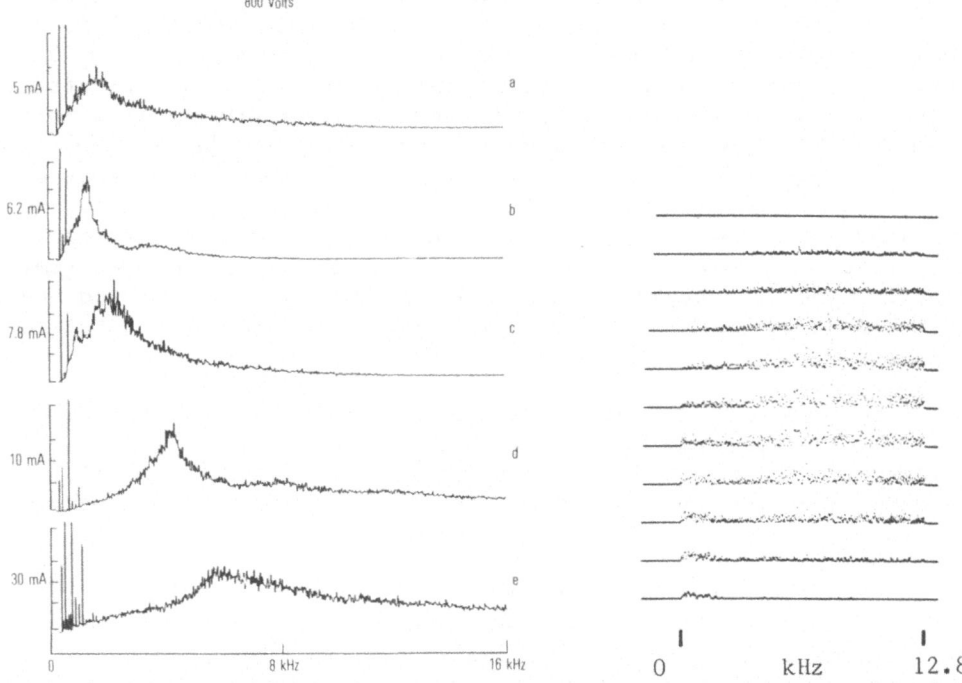

Fig. 6. Electric spectra at 800   Fig. 7. Multiple raw spectra
         V for beam currents. 0db   taken during a single
         for each panel is a. 37.8   90-ms gun pulse show-
         mV; b. 150.0 mV; c. 150.0   ing single particle
         mV; d. 37.8 mV and e. 13.4   behavior at start of
         mV.   pulse (bottom) fol-
                                                lowing onset of BPD.

shown) is very similar to the electric except for the very low
frequencies.

    To investigate the time variability we operated the gun in a
pulsed mode with 80 msec on and 250 ms off at 2 kV and 70 mA.
Fig. 7 shows a sequence of raw spectral traces during one of these
pulses with no background plasma. As the gun comes on the spectrum looks like a single particle spectrum (bottom of Fig. 7,
compare this to Fig. 4a). Then within a few milliseconds the
spectrum begins to look like a BPD spectrum (compare to Fig. 4d)
with a broad peak near 6 to 10 kHz. Apparently this is an indication that some background plasma must be built up by the beam
before the BPD onsets. DC-field strengths over 10 volts/meter
were measured near the BPD threshold condition in very burstlike
events.

Rocket/Laboratory Spectral Comparison

Figs. 2 and 3 from the rocket flight can be compared directly to Figs. 4 through 7. The rocket clearly shows broadband spectral enhancements from a few kHz up to a cutoff at the lower hybrid frequency. In this frequency range a clear peak occurs at about 5 to 6 kHz in Figs. 2 and 3. This is exactly the region of frequency space where the lab spectral peaks were seen. Furthermore the lab VLF spectral peaks were within this frequency range over a wide range of beam parameters. The time resolution of Figs. 5 and 7 are not high enough to show the exact beginning of BPD signature after gun turn on, however, Bernstein et al. (1981) have shown the BPD ignition appears within about a few milliseconds at these gun parameters. It cannot be stated absolutely therefore that the gun was pulsing in and out of BPD during the AMP's. However, it is believed that the ignition time is the time required to build a background plasma to the right density so subsequent maximum current pulses would build on previous conditions. Thus it is still expected that the 3-kHz AMP's at maximum gun voltage and current also produced BPD. Certainly the average spectra during AMP's (see Fig. 2) and during dc gun steps are similar. Unfortunately the ion gyroradius at 1 gauss is larger than the laboratory chamber for the data in Figs. 4 - 7 so ion gyrodynamics are not well modeled in the lab. However, the general spectral shape in Fig. 3 is very similar to that seen in Figs. 4c and 4d under similar conditions in the auroral ionosphere.

CONCLUSIONS

The VLF spectrum during beam plasma discharge can be characterized in general by the following:

1. The electric spectrum has a broad spectral peak typically between 1 and 16 KHz.
2. Peaks tend to move to higher frequencies or new peaks form as gun current increases.
3. The magnetic spectral signature can be considerably different from the electric especially below the peak.
4. Electric field strengths of over 100 mV/m at the peak VLF frequency have been observed in the lab.
5. Effects near threshold can be highly spiky and turbulent with quasi-dc electric spikes over 10 volts/meter.
6. The spectral peak develops prior to BPD threshold conditions as determined by optical and rf measurements.
7. Background plasma densities are required for the BPD to occur as evidenced by a few milliseconds delay in spectral change from gun turn on.

The rocket gun program operated at 0, 2 and 4 kV with maximum space charge limited currents of 0, 30 and 80 mA respectively.

There is evidence for pulses in the rocket data at both the 2-kV step, and the 4-kV steps. Unfortunately the chamber work was limited to 2 kV in these experiments. Also, the background electromagnetic noise above the lower hybrid was not modeled in the chamber. However, the general character of the VLF spectrum involving a rapid onset of a peaked spectral shape in the few kHz region and the high time variability when taken with the photometric and particle energization data reported by Bernstein et al., 1979, strongly suggest that BPD occurred on this rocket flight.

ACKNOWLEDGEMENTS

    The authors wish to acknowledge the support of W. Bernstein, the principal investigator for the flight, S. Monson who provided support for the experiment at the range and C. W. Jordan for help in constructing and integrating the payload. The spectral data analysis was assisted by M. Dazey and R. Maulfair.

    This work was supported in part by The National Oceanic and Atmospheric Administration under Contract 03-5-022-95 and in part by the Space Division of the U. S. Air Force under Contract F04701-80-C-0081.

REFERENCES

Bernstein, W., and Kellogg, P. J., 1980, Laboratory simulation of the injection of energetic electron beams into the ionosphere-ignition of the beam plasma discharge, Space Plasma Center, Rice University, Houston, Tex (preprint).
Bernstein, W., Kellogg, P. J., Monson, S. J., Holzworth, R. H., and Wahlen, B. A., 1981, Recent observations of beam plasma interactions in the ionosphere and a comparison with laboratory studies of the beam plasma discharge, elsewhere in this publication.
Bernstein, W, Leinbach, H., Kellogg, P., Monson, S., and Hallinan, T., 1978, Electron beam experiments: the beam plasma discharge at low pressures and magnetic field strengths, Geophys. Res. Lett. 5:127.
Bernstein, W., Leinbach, H., Cohen, H., Wilson, P. S., Davis, T. N., Hallinan, T., Baker, B., Martz, J., Zeimke, R., and Huber, W., 1975, Laboratory observations of RF emissions at ω_{pe} and $(n + 1/2)\,\omega_{ce}$ in electron beam-plasma and beam-beam interactions, J. Geophys. Res., 80:4375.
Bernstein, W., Leinbach, H., Kellogg, P. J., Monson, S. J., and Hallinan, T., 1979, Further laboratory measurements of the meam-plasma discharge, J. Geophys. Res., 84:7271.
Galeev, A. A., Mishin, E. V., Sagdeev, R. Z., Shapiro, V. D., and Snevekenko, I. V., 1976, Discharge in the region around a rocket following injection of electron beam in the iono-

sphere, Sov. Phys. Doklady 21:641.

Holzworth, R. H., and Koons, H., 1981, VLF emissions from a modu-
    lated electron beam in the auroral ionosphere, J. Geophys
    Res. 86:853.

Mishin, E. V., and Rushin, Yu Ya, 1978, Beam plasma discharge in
    the ionosphere: dynamics of the region in rocket envionment
    in ARAKS and Zarnit ZAZ experiments, Acad. of Sciences, USSR,
    Institute of Terrestrial Magnetism, Ionosphere and Radio Wave
    Propagation, preprint 21a, Moscow.

Wilhelm, K., Bernstein W., and, Wahlen, B. A., 1980, Study of
    electric fields parallel to the magnetic lines of force using
    artificially injected energetic electrons, Geophys. Res.
    Lett. 7:117.

STUDIES OF BEAM PLASMA INTERACTIONS IN A SPACE

SIMULATION CHAMBER USING PROTOTYPE SPACE SHUTTLE INSTRUMENTS

Peter M. Banks, W. John Raitt, William F. Denig

Physics Department and
Center for Atmospheric and Space Sciences
UMC 34, Utah State University
Logan, Utah 84322 U.S.A.

INTRODUCTION

In March, 1981, electron beam experiments were conducted in the large space simulation chamber at Johnson Space Center using equipment destined to be flown aboard NASA's Office of Space Science-1 pallet (OSS-1). Two major flight experiments were involved: The Vehicle Charging and Potential (VCAP) experiment from Utah State University and the Plasma Diagnostics Package (PDP) from the University of Iowa. Apparatus connected with VCAP included a Fast Pulse Electron Gun (FPEG), a Charge and Current Probe (CCP), a Spherical Retarding Potential Analyzer and Langmuir Probe (SRPA/LP), a Digital Command and Interface Unit (DCIU), and associated ground support equipment. The PDP included a wide variety of particle and wave analyzers packaged within a sub-satellite which will both be flown with VCAP aboard the OSS-1 payload eventually be released from the Space Shuttle on the Spacelab-2 mission for the purpose of making diagnostic plasma measurements in the ionosphere. In the present circumstances principal interest was directed towards determining the characteristic physical phenomena which occur within the plasma column created by the firing of the FPEG into the low pressure atmospheric gases of the space simulation chamber. For this purpose, the PDP was suspended approximately 15 meters above the cart-mounted FPEG in such a manner that rotations and lateral movements of the FPEG and the PDP could probe the 2-3 m diameter plasma column in ways similar to those planned for operations aboard the Space Shuttle.

In this paper we wish to give a preliminary view of the results obtained when the electron emissions were held steady over
relatively long periods of time such that steady state conditions
could be obtained with respect to the electron beam interaction
with the neutral gases and plasma of the vacuum chamber. Of particular interest to us was the plasma instability feature known as
the Beam Plasma Discharge (Bernstein, et al., 1979) which has been
found to occur when the beam current and electron energy are
larger than some critical threshold value.

Details of the scientific objectives of the VCAP and PDP experiments are given elsewhere (Neupert, 1979), while a summary of
the VCAP apparatus is given later in this volume (Raitt, et al.
1981). For the present experiments the FPEG was used in a d.c.
mode with a range of currents of 2 to 80 ma at a beam energy of
970 eV. In the chamber the FPEG was placed on a movable pan and
tilt head which allowed a variation of pitch angle of nearly $+90^{\circ}$
with respect to the local magnetic field. The FPEG and its rotation unit were mounted on a remotely controlled cart which allowed
movement of the FPEG over a distance of about 5 meters along a
magnetic meridian near the center of the chamber. Through external coils, the magnetic field intensity within the chamber could
be varied from a maximum intensity of 1.6 G to a minimum value
near zero.

The particular results given here relate to observations of
the emissions of VLF and HF noise associated with the d.c. beam.
The VLF data were gathered using an electric antenna suspended inside the chamber roughly parallel to the electron beam at a distance of approximately 8 m. The signals were monitored outside
the chamber using an HP3582 Signal Analyzer and were also recorded
on analog magnetic tape for later analysis. The HF results were
obtained with a short vertical electric dipole antenna suspended
within the chamber at a distance of about 10 m from the electron
beam. The HF signals were preamplified within the chamber and fed
to a Tektronics spectrum analyzer for photographic and video recording.

HIGH FREQUENCY RESULTS

Results obtained with the present experimental apparatus confirm the general picture of the BPD described by Bernstein, et al.
(1979). At low currents and neutral gas pressures the electron
beam propagates as a narrow feature clearly showing a classical
helix. Upon entering the BPD mode, however, a halo of extra light
surrounds the beam and the helix disappears after less than
one-half turn. Optical measurements show a general filling in of
the cylinder circumscribed by the electron helix and there is an

outward extension of optical emission beyond the beam energy electron gyroradius, presumably caused by electrons scattered by plasma waves in the column.

Within the HF band there occur strong radio emissions in two
frequency regimes: That lying below the electron cyclotron frequency and in a band extending from the plasma frequency to the
upper hybrid resonance frequency. The former emissions are regarded as originating in the electron whistler mode while the
latter are thought to be the result of intense electrostatic plasma waves existing largely within the BPD column.

Our measurements in the HF regime were designed to explore
the temporal variations of the HF emissions. To do this, we made
video recordings of the sweeps of the HF spectrum analyzer, yielding images of the spectrum at 40 msec intervals. These snapshots
show the presence of large amplitude fluctuations in the HF emissions in both the BPD and sub-BPD regimes. Although we expected
to find many dynamic features, we were not prepared for the abrupt
changes of emission modes which were found to occur. These, we
feel, represent a basic instability of the plasma column and may
have an important bearing on the possible observations of BPD from
the Space Shuttle.

To illustrate the results we have chosen to display the HF
spectrums obtained with two d.c. beam currents: 5 ma and 80 ma.
At a chamber pressure of $5x10^{-6}$ T and magnetic field intensity of
1.0 G, the former is below the BPD threshold at a pitch angle of
0°, while the latter is always above this threshold. Figure 1
shows the basic format of the results with logarithmic amplitude
given by the ordinate in 10db divisions and frequency displayed
along abcissa. The HF emissions obtained with the 80 ma beam show
a frequency spectrum with a strong cyclotron noise band up to the
electron cyclotron frequency of 2.8 MHz. In addition, a second
harmonic of the cyclotron frequency is present between 5 and 6
MHz. The plasma frequency enhancement for the 80 ma beam is beyond the scale of this figure and occurs between 30 to 40 MHz. The
5 ma beam, in contrast, shows a series of discrete HF spikes
between 1 to 3 MHz and very little emission above 6 MHz. Because
the 5 ma beam is below the BPD threshold, there is no strong enhancement of the plasma density, and plasma frequency emissions
are absent. Inspection of the time variations of these characteristic emissions at 0° pitch angle reveals that while the 80 ma
cyclotron emission flickers somewhat (5+db), there is no evidence
for its complete disappearance. However, the second harmonic component at 5 to 6 MHz does flicker strongly (30 to 40db) at a rate
of approximately 10 Hz. Likewise, the strong line components of
the 5 ma beam emissions flicker at a similar frequency.

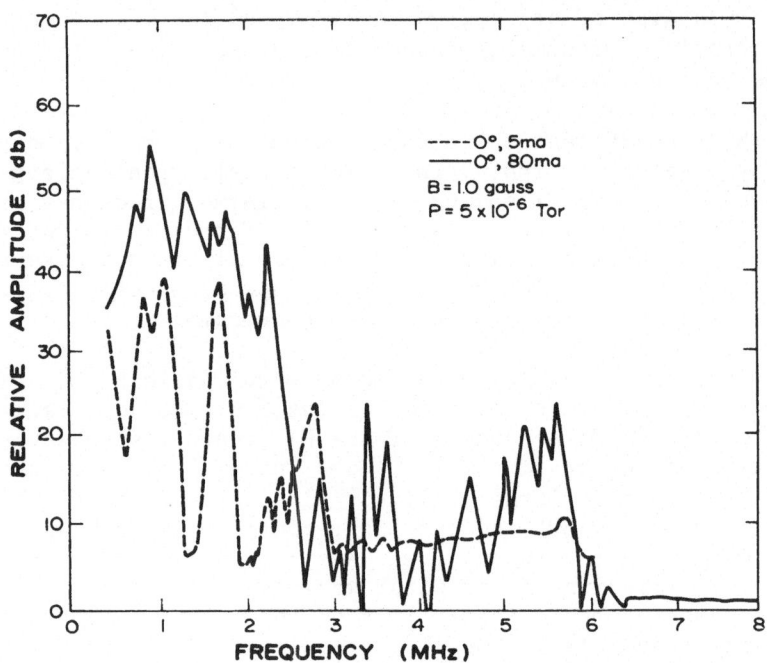

Fig. 1 HF frequency spectrums for an 80 ma beam (dashes) and a 5 ma beam (solid) at 0° pitch angle and a 1.0 G magnetic field.

The situation at 80° pitch angle differs greatly from the 0° case. Figure 2 shows a typical result, indicating that both the 5 ma and the 80 ma beams are in BPD and that both have a smooth cyclotron emission at frequencies below 2.5 MHz. Both beams have cyclotron harmonics and the 5 ma beam plasma frequency emission is visible starting at 7 MHz. The strongest flicker of emission seen in this frequency range are the cyclotron harmonics, although there do seem to be, at times, discrete lines superposed on the 7 to 9 MHz plasma frequency emissions. These results provide direct evidence for the strong dependence of BPD upon injection pitch angle.

A further view of the variations of cyclotron harmonic emission is shown in Figure 3 where we have used an extended frequency scale to include the 80 ma beam plasma frequency-upper hybrid emissions. At 0° pitch angle there seems to be more energy in the cyclotron harmonics than in the plasma frequency component. The situation is reversed at 80 degrees and substantially greater energy is present in the plasma waves. It also seems that the width of the plasma frequency spectrum is increased in this situation, but further studies are required before a definite statement can be made.

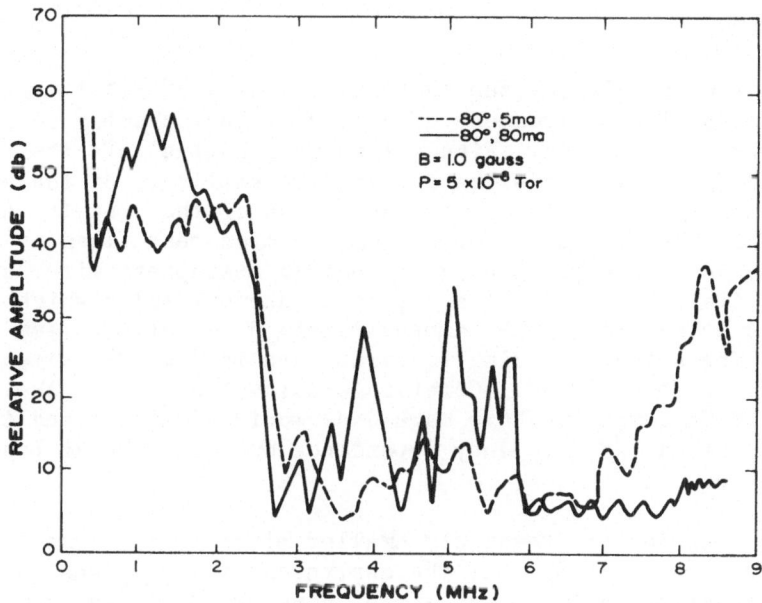

Fig. 2   HF frequency spectrums for 5 ma and 80 ma   electron   beams
at 80° pitch angle.

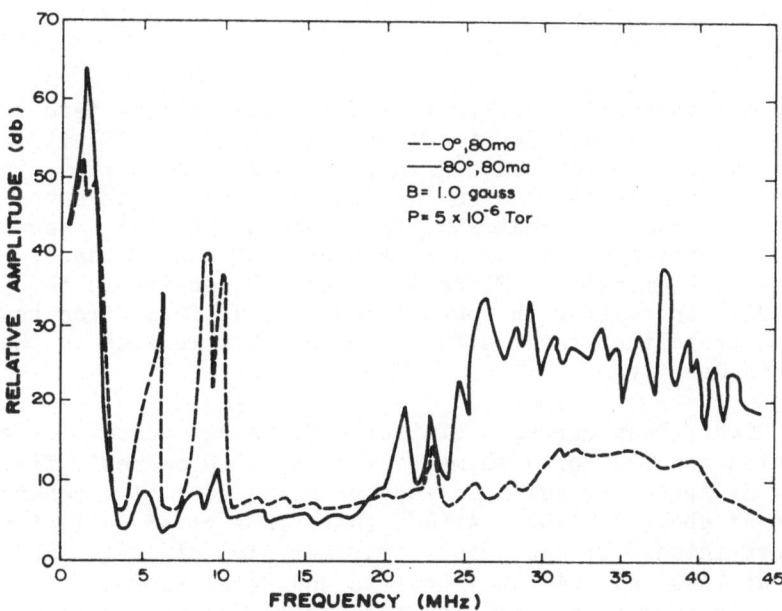

Fig.   3   HF frequency spectrums for an 80 ma electron   beam   at   0°
and 80° pitch angle.

VLF RESULTS

A different view of the temporal features of electron beam operations in the environment of a large plasma chamber is provided
by our VLF measurements. Figure 4 shows a sample of the results
expressed in a frequency-time format with amplitude of the emission
represented by intensity modulation. Each of the panels shows a
segment of VLF emission for a 5 ma electron beam fired at various
pitch angles into a previously un-ionized atmosphere with a fixed
magnetic field of 1.2 G. The top panel shows results which correspond to a transition to BPD approximately 3 sec after beam firing
began. The strong, pulsing behavior seen in the BPD segment has a
frequency of about 10 Hz. Careful examination shows that these
spikes of VLF emission have marked harmonic structure and that the
intensity has a $1/f$ frequency dependence over the 10 KHz band displayed here.

The data for progressively smaller pitch angles has many of
the features seen at 85° but the sharpness of the spikes is degraded considerably. At 0° pitch angle, however, extended periods of
"non-bursts" can be seen. Further, the frequency of the spike ap-
pearances seems to be higher at the larger pitch angles.

Another feature worthy of note is the initial transient pulse
of strong radiation seen at the moment of electron beam initiation.
At times this brief pulse exhibits a beading indicative of harmonic
structure. At other times it remains as a single featureless impulse extending upwards to 10 kHz.

A close inspection of Figure 4 also shows a long term temporal
trend in the overall level of VLF intensity. At 85° it is clear
that the sub-BPD region had a greater spread of emissions in frequency and was more intense on the average than was observed in the
BPD regime. At 0°, in contrast, there was an initial 5 sec period
of strong emission, followed by a period of less intense, spikey,
discharge-like emission. Since the electron density in the chamber
is clearly increasing throughout the period of electron beam firing, we suppose that these effects are somehow related to the average plasma density.

At higher beam currents similar effects are noted. Figure 5
illustrates results for a 40 ma beam in a 1.2 G magnetic field. At
85° a 10 Hz spikey emission is seen with the hint of a coherent VLF
emission at about 1.5 kHz. At 60° there is a brief lull of about 1
sec before intense spikes occur. Further, two discrete VLF lines
appear at 1320 and 2640 Hz, the 11th and 24th harmonics of 110 Hz.
We presently have no explanation for the occurrence of these lines,
but their presence has been noted previously and does not seem to
depend upon variations in any of the plasma parameters available
for us to adjust.

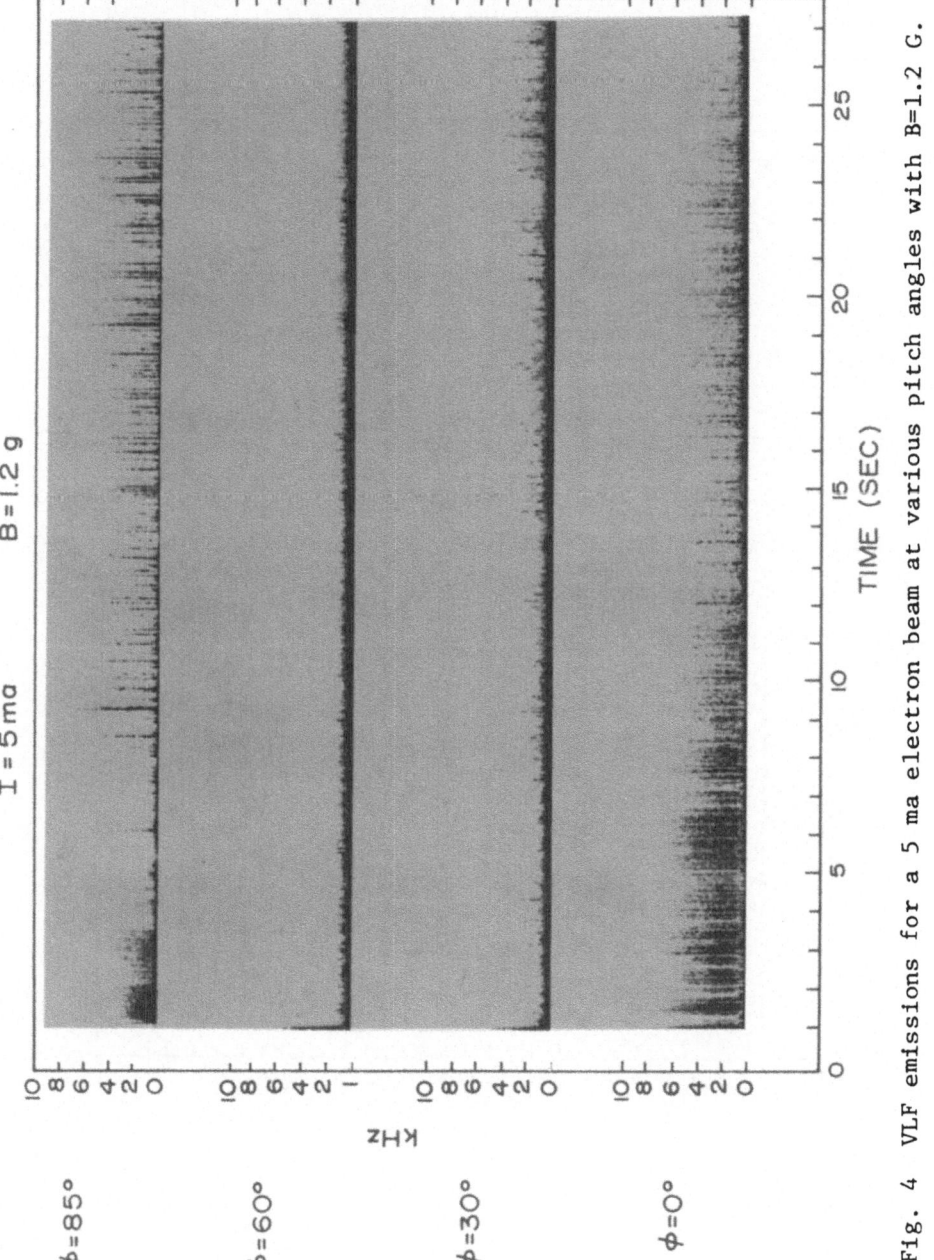

Fig. 4 VLF emissions for a 5 ma electron beam at various pitch angles with B=1.2 G.

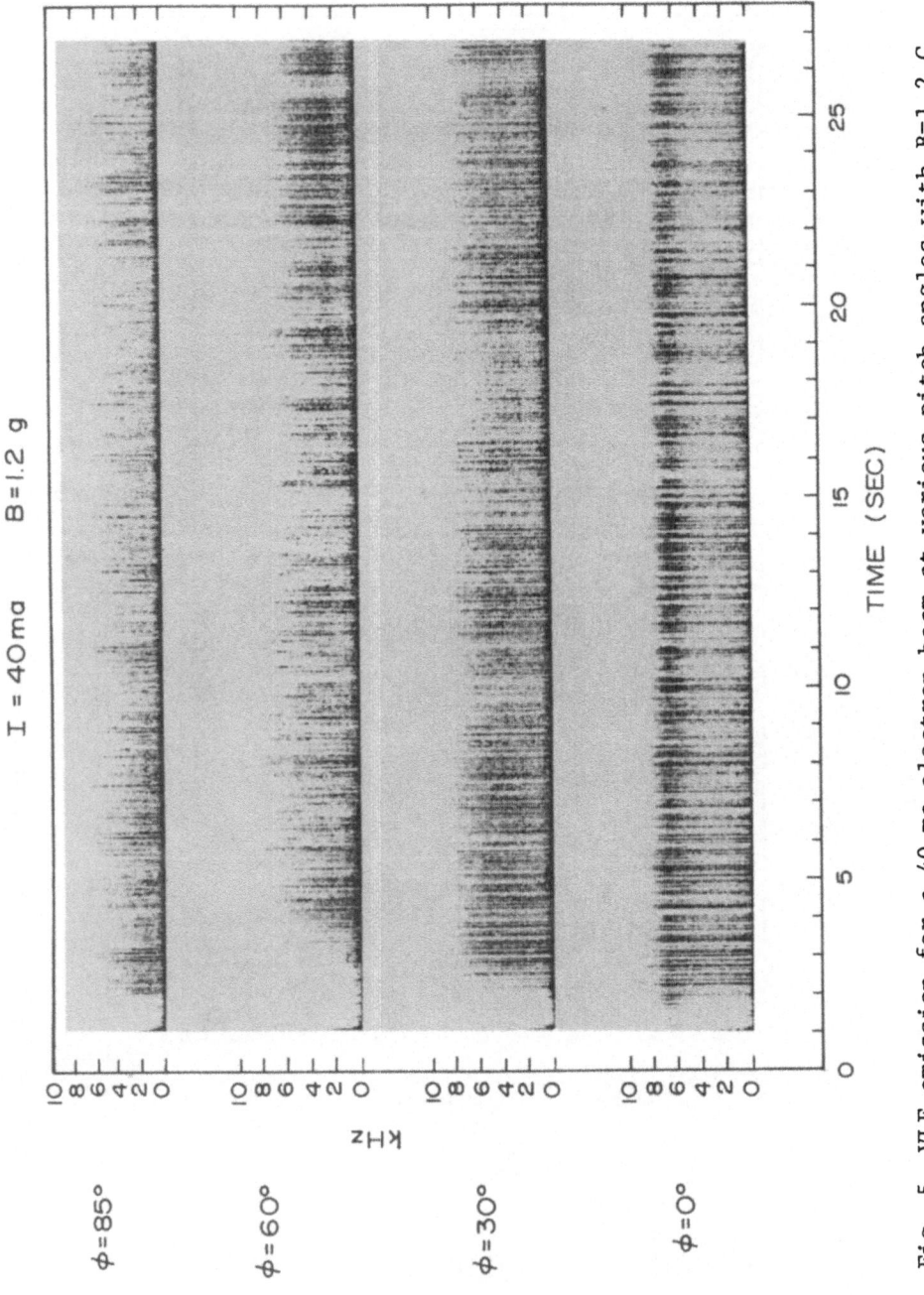

Fig. 5   VLF emission for a 40 ma electron beam at various pitch angles with B=1.2 G.

At 0° pitch angle a new emission feature is seen: a VLF band extending from about 7 to 9 kHz. Since the lower hybrid frequency is near 12 kHz, it is hard to identify this feature with the process of lower hybrid emission. More likely, it appears to result from MHD processes occurring within the plasma column established by the electron beam. It is possible that the variations of pitch angle have resulted in the attainment of a crucial alignment with respect to the column potential well and that we are seeing a resonance condition. We note, however, that the emission is seen only during the individual noise spikes; at other times, there is no detectable VLF noise.

The situation at 80 ma is shown in Figure 6 for various pitch angles and magnetic field intensities. At 30° and with B=.66 G one finds the usual spikey behavior with strong band emission at approximately 6 kHz. At the same pitch angle but with B=0.3 G a completely different situation is found. Here, there occurs only a smooth 1/f background noise with superposed discrete emissions at a variety of frequencies, all of which seem to be harmonics of 110 Hz. In the final seconds of this record, however, we note the appearance of some spikes. At 0° (bottom panel) the same behavior is seen, but the broad range of harmonics has been eliminated, leaving only a single, intense line at about 2 kHz.

Finally, it is worth considering the third panel where a negative (downward) pitch angle was employed. In this case the beam was directly impacting the floor of the chamber, with the result that there occurred plasma discharges only infrequently until 10 seconds had elapsed. Then, increasingly strong emissions were seen with the typical spikey discharge behavior and a broad emission band centered on 6 kHz.

DISCUSSION

The present results demonstrate that a basic discharge mechanism is operating within the JSC Space simulation chamber in conjunction with the operation of the FPEG electron gun. Since this discharge is seen in both BPD and non-BPD situations, and the rate of discharge increases with the beam current, it is natural to assume that it represents an electrical discharge of the plasma column.

It is also natural to assume that the rather uniform rate at which these discharges are seen is somehow related to the flickering seen at higher frequencies with respect to the electron cyclotron modes. Since the HF flickering is absent in the BPD associated electron cyclotron modes, it is possible to assume that the VLF discharge is somehow related to the sudden appearance of cyclotron spikes in the HF regime.

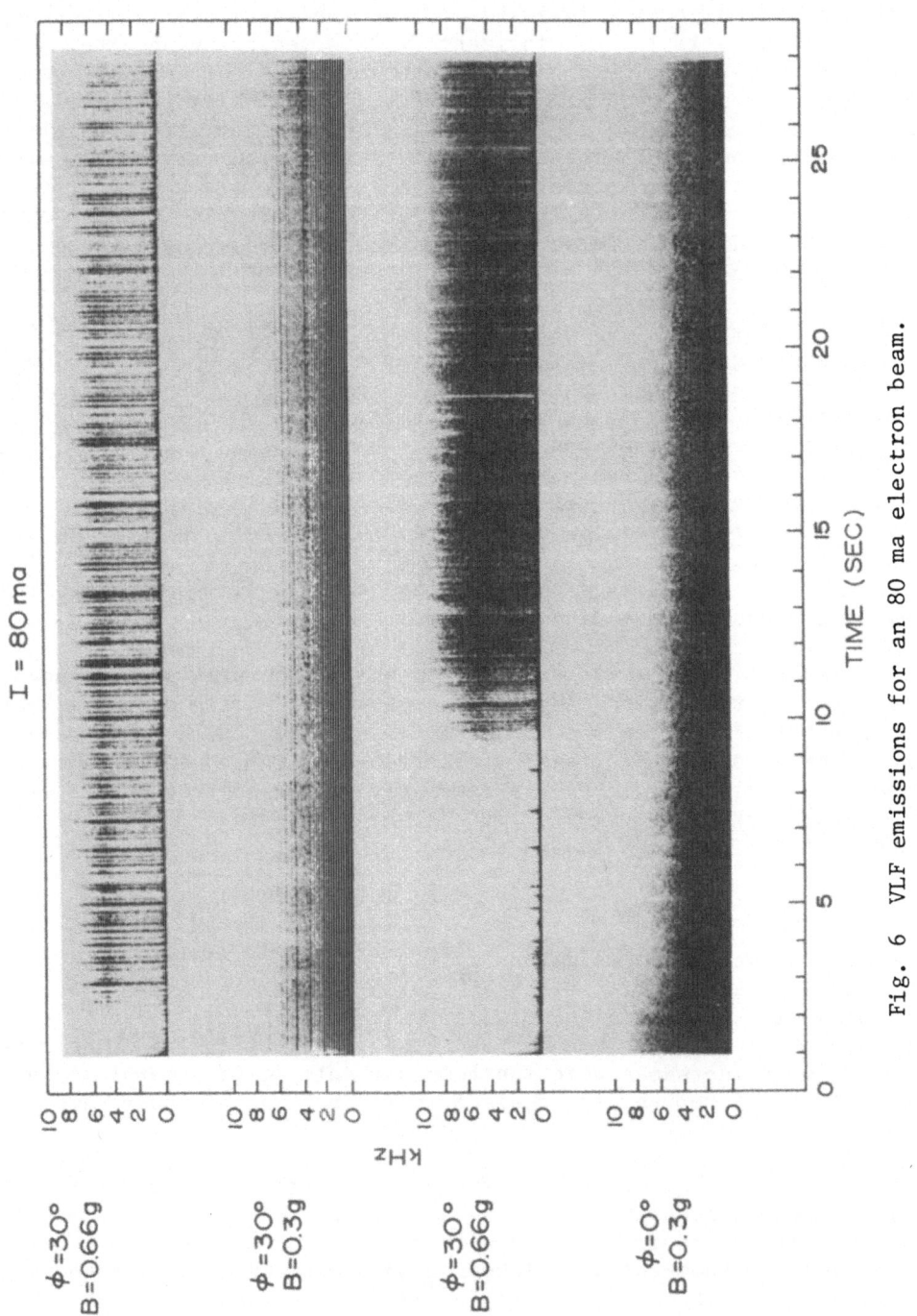

Fig. 6   VLF emissions for an 80 ma electron beam.

It is also noteworthy that the various discrete lines seen over a variety of pitch angles and magnetic field strengths are also confined to the times of occurrence of the discharges. Thus, they may represent some basic configuration of the plasma column which is driven by a combination of internal electric field, ExB, and ion inertial effects.

The discharge itself is interesting since it most likely relates to the physical constraints acting upon the electron beam produced plasma column. In particular, it is conceivable that the plasma sheaths acting at the top and bottom of the chamber undergo periodic collapse, releasing plasma column particles and energy to the walls of the chamber. If this is the case, the rate of collapse should depend directly upon the rate at which beam energy is being liberated in the plasma column. The effect such collapse has upon the BPD process and the extent to which the BPD process itself depends upon these dramatic discharges within the chamber, remains to be determined.

ACKNOWLEDGEMENTS

This research was supported through NASA grant NGR 7289 and NASA contract NAS8 34003. Discussions with Drs. S. Shawhan, H. Anderson, W. Bernstein, and J. Jost are gratefully acknowledged. The VLF spectrum analysis was done at Stanford University through the courtesy of Dr. R. Helliwell and associates.

REFERENCES

Bernstein, W., H. Leinbach, P.J. Kellog, S. Manson and T. Hallinan, 1979, Further laboratory measurements of the beam plasma discharge, J. Geophys. Res., 84:7271.

Neupert, W.M., Office of Space Science-1 Experiment Investigation Descriptions, Goddard Space Flight Center, September 1979.

Raitt, W.J., P.M. Banks and W.F. Denig, 1981, Transient effects in beam-plasma interactions in a space simulation chamber stimulated by a fast-pulse electron gun, in Artificial Particle Beams Utilized in Space Plasma Studies, B. Grandal, editor, Plenum Press.

DISCUSSION

Szuszczewicz: Were you surprised by the time-dependent
RF signatures detected by Bernstein's spectrum analyzer?

Banks: Not by the fluctuations themselves, but by the
8 to 15 Hz periodicity of the VLF and RF emissions. The
initial results indicate a discharge-like phenomenon
which seems to be a controlling influence. This, to me,
was surprising.

Papadopoulos: A general comment I would like to make
concerns measurements of waves made away from the beam.
Several modes (i e the plasma mode) are strongly localized inside the beam and are evanescent outside.
Therefore they will show as being smaller than cyclotron
and whistler waves which can propagate freely, even if
their original amplitude was many decibels higher.

Banks: For this reason one must be careful in comparing
the relative importance of processes occurring in different parts of the wave spectrum.

TRANSIENT EFFECTS IN BEAM-PLASMA INTERACTIONS IN A SPACE SIMULATION CHAMBER STIMULATED BY A FAST PULSE ELECTRON GUN

W.J. Raitt, P.M. Banks, W.F. Denig

Center for Atmospheric and Space Sciences
UMC 34, Utah State University
Logan, Utah 84322 U.S.A.

H. R. Anderson

Rice University
Houston, Texas 77001 U.S.A.

INTRODUCTION

During the 1950's and 1960's there was increased interest in the interaction of electron beams with plasma generated by ionization caused by the primary electron beam. This interest was stimulated by the need to develop special vacuum tubes to operate in the kMHz frequency region. Thus, much of the early work was performed using small plasma containers and high current density beams.

The experiments of Getty and Smullin (1963) indicated that the interaction of an energetic electron beam with its self-produced plasma resulted in the emission of wave energy over a wide range of frequencies associated with cyclotron and longitudinal plasma instabilities. This, in turn, enhanced the thermal plasma density in the vicinity of the beam, and Getty and Smullin coined the term Beam-Plasma Discharge (BPD) to describe this phenomenon.

In more recent years, the ability to operate electron guns in the unbounded conditions of space from rockets and soon from the Space Shuttle, has led to further experiments to study the BPD phenomena under conditions quite different from the earlier exper-

iments. Since the mid-1970's a number of experiments have been
made on the conditions for ignition of the BPD in a large vacuum
chamber with dimensions approximately 90 feet high and 50 feet in
diameter located at the Johnson Space Center (JSC) in Houston,
Texas. Bernstein et al (1978, 1979) have used the enhanced wave
emission from a BPD to determine empirically the dependence of BPD
on beam current, magnetic field strength, ambient gas pressure and
ambient thermal plasma density.

Recently we have had two opportunities to operate the engineering prototype of an electron gun to be flown on the Space
Shuttle using the facilities of the large JSC chamber. This electron gun has a unique characteristic in that the switch-on and
switch-off times of the electron beam are very rapid (100nS).
For this reason we refer to the instrument as a Fast Pulse Electron Gun (FPEG). This enables the starting of emission to be
quite precisely defined, thereby enabling the transient phenomena
related to the ignition phase of BPD to be studied. In May 1980,
we operated what was essentially an identical version of the
flight FPEG and obtained a number of interesting results related
to the onset time of wave energy at the instantaneous plasma frequency, the lifetime of beam-induced plasma, and the variation of
the spectrum of energetic electrons immediately following the beam
switch-on. Subsequently, in March, 1981, we returned to the
chamber to perform a series of collaborative experiments with the
University of Iowa in which we simulated the experimental arrangement for the NASA Office of Space Science (OSS-1) mission joint
USU/Iowa flight operations. For this series of tests, the FPEG
had one of its two channels modified to allow us to control the
beam current, to control the pulsing externally and to monitor the
beam current directly with a Rogowski coil. During both experimental periods, the chamber pressure was approximately 5 x 10^{-6}
Torr, further details of the general experimental arrangement are
given by Banks, et al, (1981).

In the second series of tests described here we made a systematic study of the onset of plasma frequency wave energy as a
function of pitch angle and magnetic field strength over a wide
range of frequencies from 2MHz to 40MHz. We also consolidated our
earlier measurements of particle energy spectrum changes following
beam switch-on which were very sparse due to the failure of a high
voltage chamber penetration during the May 1980 tests.

In this paper we will describe some of the transient phenomena associated with wave emission during the beam switch-on and
switch-off periods, and we will also present some results on the
changes in electron energy spectra on a time scale of tens of milliseconds following beam switch-on. The results will be discussed
in terms of the current ideas of the beam plasma discharge phenomenon, and speculation will be made as to the possibility of a

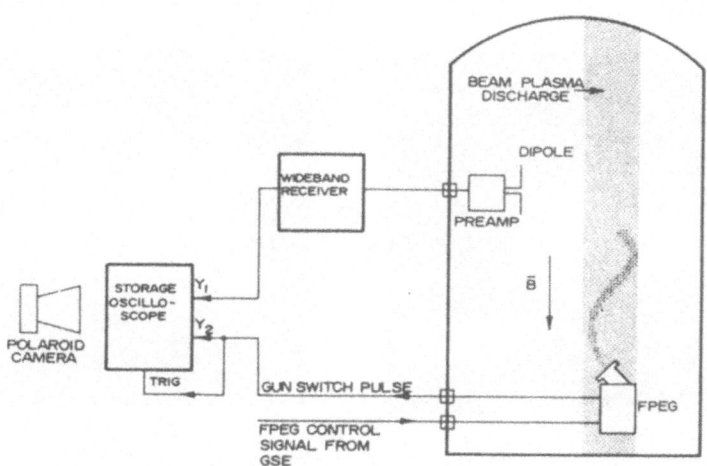

Fig. 1 Equipment configuration to study wave energy onset times.

similar discharge being produced by an electron gun operating in free space.

WAVE EMISSION STUDIES

Experimental Arrangement

In order to measure the time history of the development of the wave energy over a range of frequencies, we used the experimental set up shown schematically below in Figure 1.

The wave emission generated by the beam-plasma interaction was intercepted by a small dipole and wideband preamplifier mounted in the chamber. The signal was then sent to a wideband spectrum analyzer configured to act like a wideband receiver. The spectrum sweep was inhibited and the center frequency and bandwidth set manually. The gun switch signal was used to trigger the storage oscilloscope for a signal sweep to display both the output from the receiver and the gun switch signal.

Measurements Obtained

During our initial experimental runs in May 1980, we established that a pulse length of at least 3.3 mS for a single pulse was required to detect any plasma wave energy at a frequency near the plasma frequency corresponding to the plasma density induced by the electron beam. Further investigation using longer pulse lengths showed that for frequencies near the plasma frequency, there was a delay of 2-3 mS in the emission of wave energy.

25 MHz   2MS/DIV.   8.2 MHz   2 MS/DIV.

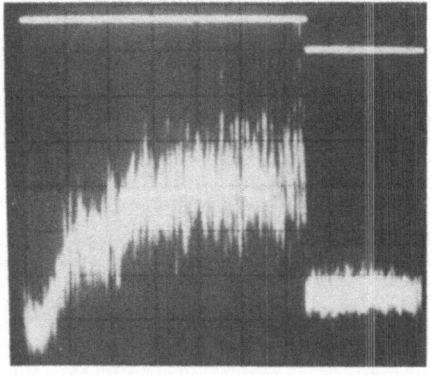 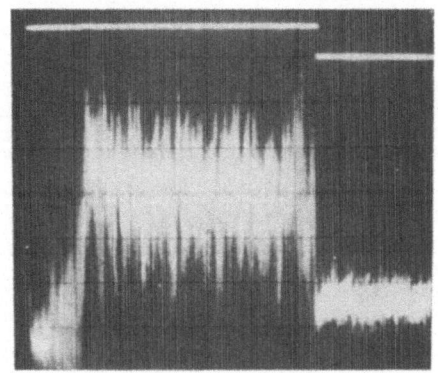

Fig. 2 Oscilloscope traces showing the onset delay of wave ener-
        gies in the steady state plasma frequency emission for an
        ambient magnetic field strength of (a) 1.1 gauss (b) 0.3
        gauss.

Examples of the effect are shown below in Figure 2 for two magnetic field strengths which resulted in a shift in the plasma frequency from near 8MHz for the lower magnetic field strength to
about 25MHz due to increased confinement of the beam-generated
plasma by the higher magnetic field.
     At frequencies below the plasma frequency resulting from an
essentially continuous beam, we observed a transient burst of wave
energy which started very nearly coincidentally when the beam
switch-on reached a peak and then died away before the electron
beam was switch off.

     During our second experimental period in March 1981, we made
a consistent set of measurements of the transient nature of the
wave energy for a variety of beam currents, magnetic field
strengths and electron beam pitch angles. Illustrated below in
Figure 3 are examples of the wave emission signature at frequencies below and near the plasma frequency resulting from a beam
pulse at the indicated pitch angle and beam current.

     In order to condense the data contained in the oscilloscope
screen photographs, we have digitized the traces and used the data
as input to a contouring program. Figure 4 shows contours of the
intensity of wave energy as a function of time and frequency for
two beam currents at the same pitch angle and magnetic field
strength.   The general shift of the frequency of characteristics
of the contour plots is consistent with plasma frequency effects
resulting from a doubling of plasma density corresponding to a
doubling of beam current.

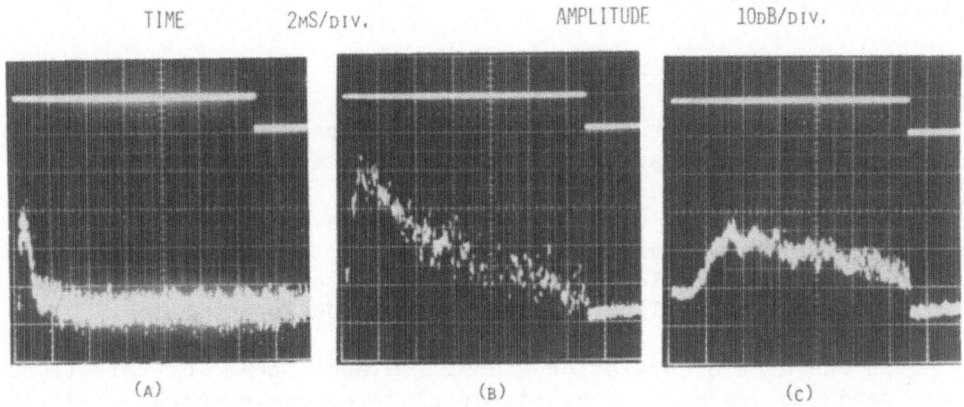

Fig. 3 Oscilloscope traces showing the transient wave energy fol-
       lowing the emission of an 80mA beam at 80° to a 1.1 gauss
       magnetic field at center frequences of a) 2MHz b) 15MHz
       c)25MHz

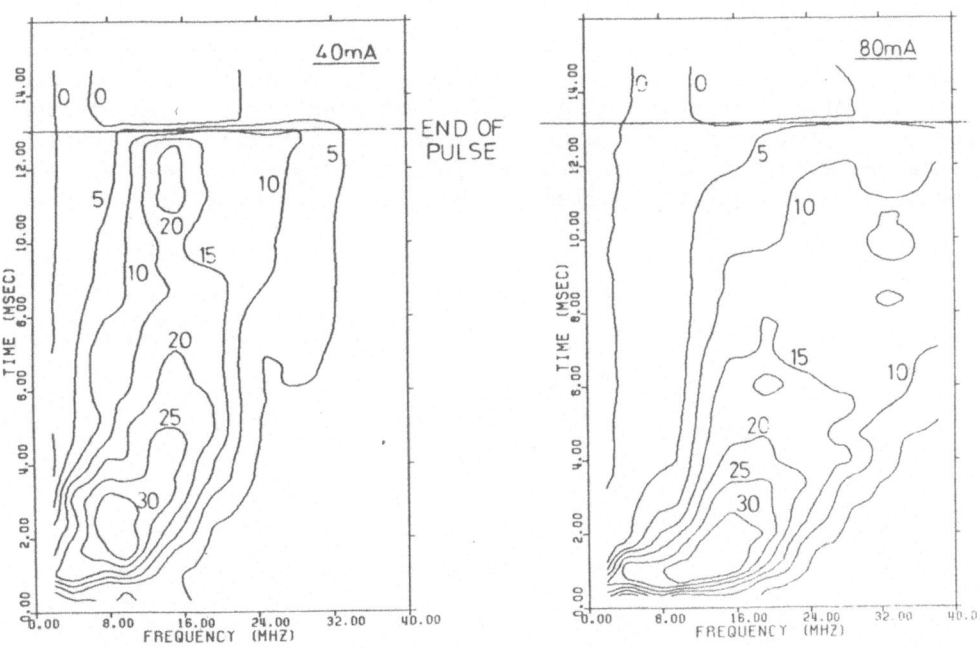

Fig. 4 Contour plots of wave energy as a function of frequency
       and time resulting from the emission of beam currents of
       a) 40mA and b) 80mA at 80° to a 1.1 gauss magnetic field.
       The contour levels are labeled in an arbitrary dB scale.

PARTICLE SPECTRUM STUDIES

Experimental Arrangement

The particle energy spectra were measured using a differential electrostatic analyzer operating over the energy range 100-1800eV for electrons. The electron gun was maneuvered on its remotely controlled cart such that when the beam was at a low pitch angle, the analyzer was directed into the beam. The beam center was then displaced approximately 60 cm to avoid saturation of the channel multiplier serving as the electron detector.

In one mode of operation a long (50 sec) beam of electrons was used to establish the steady state energy spectrum of electrons entering the analyzer. This was accomplished by sweeping the analyzer energy and feeding the channel multiplier output into a logarithmic rate meter, the output of which was displayed on an X-Y recorder.

The measurement of the transient distribution of electron energy for the same beam-analyzer geometry described above was accomplished with a different experimental arrangement shown below in Figure 5.

The FPEG was programmed to emit a series of pulses of electrons. The gun switch signal corresponding to each pulse was fed to a microcomputer which used the falling edge of the gun switch pulse to compute a new energy level for the analyzer. The switch pulse was also fed to an oscilloscope to trigger a sweep, the level of which was controlled by the signal defining the analyzer

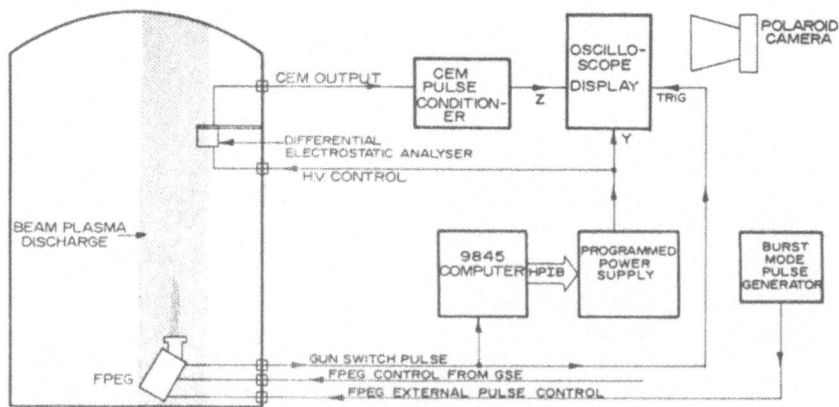

Fig. 5 Experimental configuration to study the transient development of energetic electron spectra.

energy. The intensity of the oscilloscope trace was modulated   by
the individual pulses from the channel multiplier detector in the
electrostatic analyzer. The net result was to   produce   an   image
whose density was proportional to count rate as a function of
electron energy and time.

Measurements Obtained

        The steady state electron energy spectra for continuous emis-
sion from this electron gun at four different beam currents are
shown in Figure 6. The characteristic spreading of the   electron
energy distribution above and below the beam energy for the BPD
condition (I_B 5mA) can be seen, and is in agreement with   the   results reported by Jost et al (1980).

        Two major characteristics of the initial time history of the
electron energy distribution were studied for the beam-detector
geometry described earlier. A marked delay in the   generation   of
electrons with energy below the beam energy was observed at beam
currents just above those for BPD to occur. The   change   in   the
character of this delay over the beam current range 13.0-17.5mA is

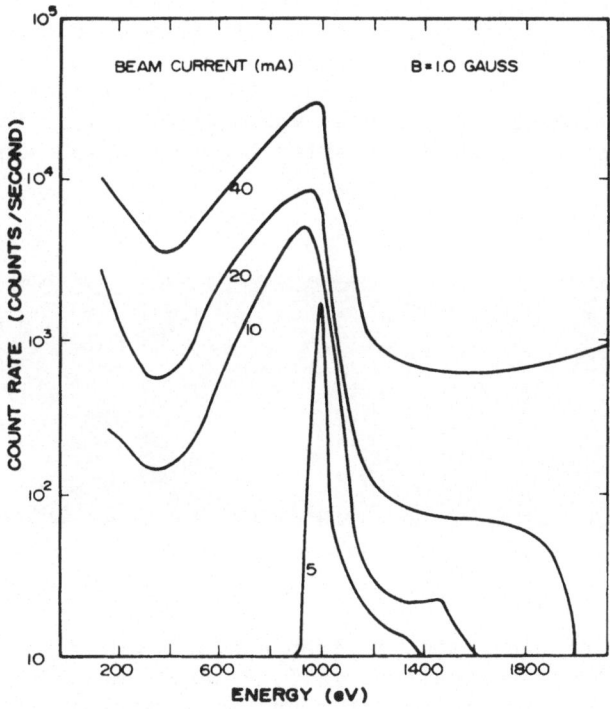

Fig. 6 Steady state energetic electron spectra for electron   beam
        currents of 5, 10, 20 and 40mA.

shown in Figure 7. It can be seen that for the lowest beam current there is an energy dependent delay in the generation of the
lower energy electrons, those near 100eV being delayed by about
12mS from the beam switch-on. The two higher beam currents show
successively shorter delays which are essentially independent of
electron energy.

Another interesting feature which can be seen in the two
higher beam currents in Figure 7 is the reduction in particle flux
at energies in the 200-600eV range which occurred about 14mS after
the beam was switched on. This effect was studied on a larger
time scale to see if it was a transient phenomenon. However, as
can be seen from Figure 8, the effect was persistent up to 80mS
after the beam switch-on. In fact, the spectrum 20mS after beam
switch-on is very similar to the steady state spectrum which also
shows the minimum between 200eV and 600eV.

DISCUSSION OF RESULTS

The individual wave energy plots of Figures 2 and 3 and the
contour plots of Figure 4 are consistent with the rapid generation
of ionization following beam emission resulting in the stimulation
of waves around a plasma frequency which changes with time as the
ionization builds up on a mS time scale. Thus, we interpret the
peak in Figure 3a as a plasma density corresponding to a plasma
frequency around 2MHz which rapidly shifted to higher densities
(frequencies) as time progressed. As the detection frequency is
increased, a threshold is reached where the decay of the wave energy following its peak value after beam switch-on slows down and

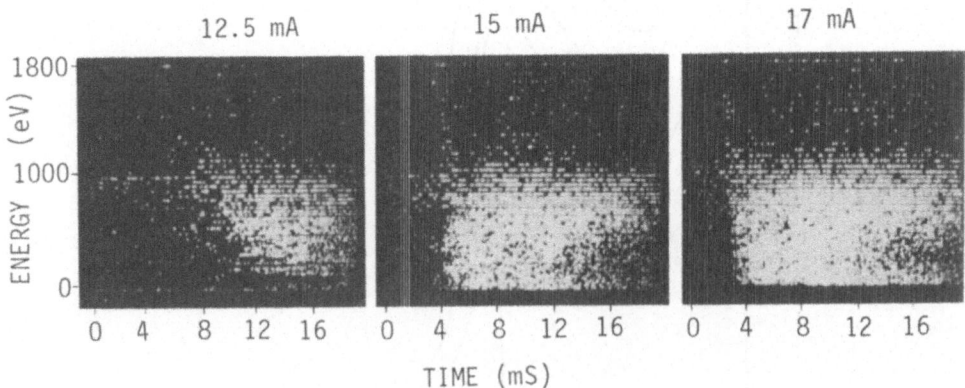

Fig. 7 Transient energetic electron energy spectra as for elec-
        tron beam currents of 12.5, 15 and 17mA. The density of
        dots is an indication of count rate which is displayed as
        a function of electron energy and time. The beam pitch
        angle was about 10°.

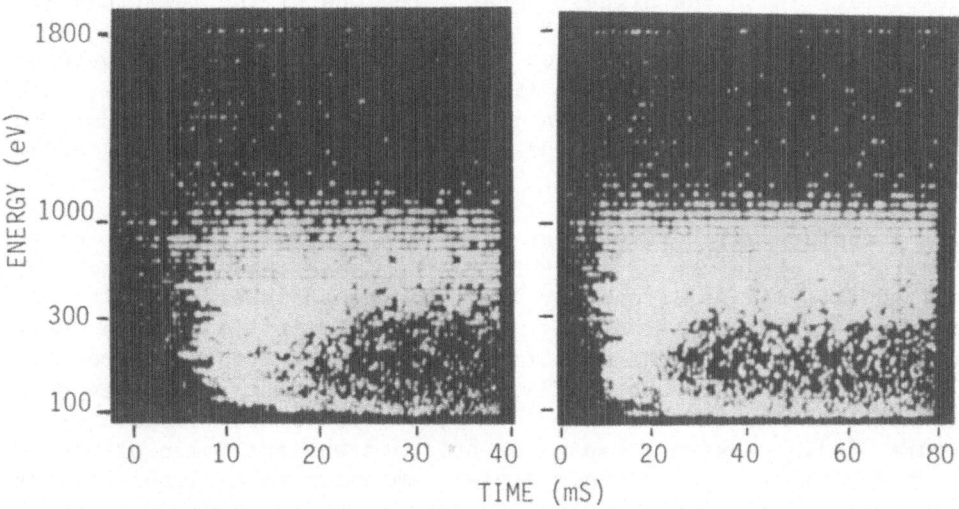

Fig. 8 Transient energetic electron spectra in the same format as
        Figure 7 for an electron beam energy of 12.5mA. The data
        is displayed on two time scales a) 0-40mS b) 0-80mS.

there is some detectable energy at the end of the pulse. This occurs at about 8MHz for the 40mA beam (Figure 4a) and about 11.5MHz
for the 80mA beam (Figure 4b). These two plasma frequencies are
consistent with a doubling of the beam-induced plasma density corresponding to the doubling of the beam current. This behavior is
interpreted as the main beam ionization still moving to higher
plasma density, but gradients in density being established with
some parts of the ionization resulting from the beam being at densities corresponding to the 8 or 11.5MHz plasma frequencies.

    At still higher detector frequencies, we see another change
in the characteristic pattern of the contours in Figure 4. At
about 13.5MHz in Figure 4a and 19MHz in Figure 4b, the onset of
received wave energy begins to show substantial delays from the
start of beam emission. These frequencies also coincide with the
wave emission being sustained until the end of the electron beam
pulse, that is, they represent steady state emission at a band of
frequencies representing the steady state distribution of plasma
density in the beam-plasma interaction region. This band of frequencies is best illustrated by the wideband spectra for a long
beam emission pulse which is shown below in Figure 8. The contour
plots clearly show an increasing delay of the onset of the plasma
frequency wave energy as the frequency is increased in this frequency region.   Eventually the detector is set to a sufficiently
high frequency that there is no detectable wave energy. The delay
associated with the upper cutoff frequency appears to be about 6mS
for the two beam currents illustrated in Figure 4.   It appears,

therefore, that the highest density regions of the beam interaction region take a significant time to become established. The results we obtained can be used to investigate instability mechanisms which result in build-up times on the few mS time scale. In all cases in this regime, when the wave energy still exists at the end of the beam pulse, we found no significant delay in the cessation of the wave emission.

A feature which is evident in Figure 9, and which is a characteristic signature of the BPD, is the broad band of frequencies emitted from the low frequency limit of the analyzer, up to the electron cyclotron frequency above which point it shows a sharp cutoff. For a 1.1 gauss magnetic field strength, the electron cyclotron frequency is about 3MHz, thus, we would expect to see this signature on the 2MHz setting for our BPD onset results (cf. Figure 3a). However, we see only a transient pulse of energy which we interpret as being momentary emission at a local plasma frequency. Since the sub-electron cyclotron frequency emission is such a clear feature of the steady state emission, we are led to the conclusion that this frequency only builds up in intensity following delay times greater than our beam pulse width of 13mS.

This result is supported by the transient particle spectra results shown in Figure 7. The relatively high flux in the 200-600eV energy range is only present for about 20mS following the beam switch-on. Thereafter, the spectrum changes its character, and shows the characteristic dip in that energy range seen in the steady state spectra (Figure 6). We believe that this change in particle spectrum is indicative of a conversion of particle energy to wave energy at and below the nominal electron cy-

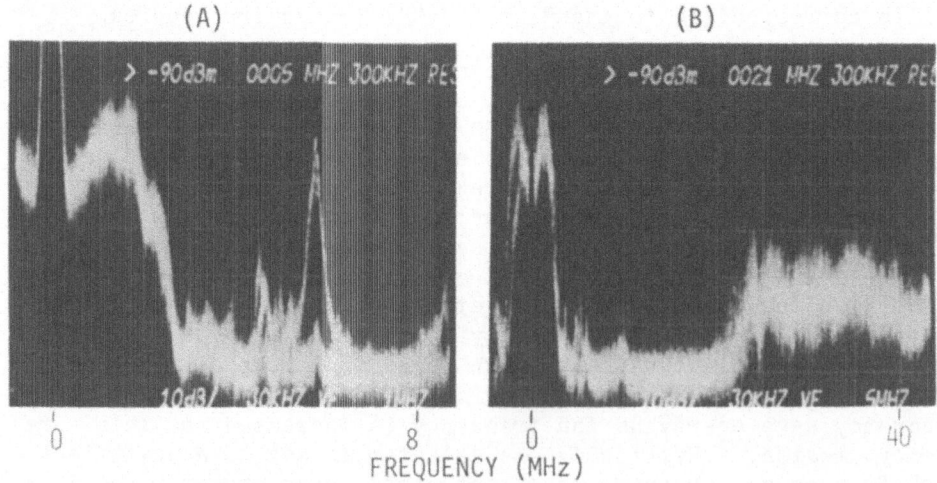

FREQUENCY (MHz)

Fig. 9 Steady state wave emission spectra on two frequency scales of a) 1MHz/div and b) 5MHz/div.

clotron frequency which occurs about 20mS after the beam switch-on for the beam current and geometry used in the particle measurements.

Unfortunately, in the limited time available for these transient tests we were not able to confirm a delay of the onset time of the wave energy near the electron cyclotron frequency. However, a prime purpose of the whole test was to work in collaboration with the University of Iowa PDP instrument, and we hope that these results will provide some pointers to the procedure of joint analysis between ourselves and the University of Iowa group.

CONCLUSIONS

We have found that for high pitch angles and beam currents above the BPD threshold, there is a transition from transient emission resulting from a changing plasma frequency to a steady emission over a range of frequencies encompassing steady state plasma frequencies. Below the steady state plasma frequency band, the onset of the emission is very close to being coincident with the beam switch-on. Within the band, however, there is an increasing delay in the onset of the emission from the low frequency end to the high frequency end.

We did not measure a transient wave emission corresponding to the steady state wave energy at and below the electron cyclotron frequency. Studies of the transient energetic electron energy spectra indicate that there is a marked depletion in flux in the 200-600eV energy range after a delay of about 20mS. We speculate that this might herald the onset of wave emission near the electron cyclotron frequency, and the fact that our wave experiment did not see it was due to an insufficient pulse width.

In all cases when there was wave emission at the end of the beam pulse, we found no significant delay before the wave emission cutoff.

The results presented in this paper represent a small fraction of the data taken by instrumentation operated by our group. We plan to study both our own data and work in collaboration with the Iowa data in the future with a view to improving our understanding of the beam plasma discharge phenomena.

We hope that our measurements and those from Iowa will enable us to establish whether BPD occurs when an electron beam is emitted from a vehicle travelling at orbital velocity. At present, we are not prepared to offer a firm opinion on this, but would like to point out that the few mS delay in establishing the steady state emission near the plasma frequency of the beam-induced plasma should be compared with the time for the electron beam to move one

Larmor radius for 1KeV electron (3.5m for 0.3 Gauss) when the gun is moving at orbital velocity. That time is 0.44mS.

ACKNOWLEDGEMENTS

We would like to thank the Space Environment Simulation Division of JSC for making their Chamber-A available to us. Also, we thank Dr. J. Jost for his help in the preparation for and execution of the tests, Drs. W. Bernstein and O.K. Garriott for allowing us to use some of the wave detection equipment. We would also like to thank Mr. J. McGarity for assistance in setting up the electrostatic analyzer and the display system. The work was supported by NASA contract NAS8-34003 and NASA grant NSG-7289.

REFERENCES

Banks, P. M., Raitt, W. J., and Denig, W. F., 1981, Studies of beam-plasma interaction in a space simulation chamber using prototype space-shuttle instruments, in Artificial particle beams utilized in space plasma studies, edited by B. Grandal, Plenum.

Bernstein, W., Leinbach, H., Kellogg, P., Manson, S., Hallinan, T., Garriott, O. K., Konradi, A., McCoy, J., Daly, P., Baker, B., and Anderson H. R., 1978, Electron beam injection experiments: The beam-plasma discharge at low pressures and magnetic field strengths, Geophys. Res. Lett., 5:7271.

Bernstein, W., Leinbach, H., Kellogg, P. J., Manson, S. J., and Hallinan, T., 1979, Further laboratory measurements of the beam-plasma discharge, J. Geophys. Res., 84:7271.

Getty, W. D., and Smullin, L. D., 1963, Beam plasma discharge: Buildup of oscillations, J. Appl. Phys., 34:3421.

Jost, R. J., Anderson, H. R., and McGarity, J. O., 1980, Electron energy distributions measured during electron beam plasma interactions, Geophys. Res. Lett., 7:509.

DISCUSSION

Jost: Our measurements of BPD onset delays were done at a low pressure in the cyclotron mode. For this case ω_{ce} is a line frequency for pre-BPD, at BPD the wide band noise below ω_{ce} abruptly appears. Using that criterion for identifying BPD, we measured the $1/I_{beam}$ time dependence for currents well above the critical current.

It seems to me that the long delays at high frequencies you are showing are associated with a slower evolution of the BPD _after_ it was achieved.

<u>Raitt</u>: When we observed at frequencies within the whistler mode band, we saw only a brief burst of energy starting with very little delay from the beam onset. We interpret this as the plasma density changing rapidly causing the plasma frequency waves to sweep in frequency and thereby sweep through our passband. We never saw any evidence for the steady emission in the whistler mode band within the 13 ms pulse used for these measurements.

DESCRIPTION OF THE PLASMA DIAGNOSTICS PACKAGE (PDP) FOR THE OSS-1 SHUTTLE MISSION AND JSC PLASMA CHAMBER TEST IN CONJUNCTION WITH THE FAST PULSE ELECTRON GUN (FPEG)

Stanley D. Shawhan

Department of Physics and Astronomy
The University of Iowa
Iowa City, Iowa   52242   USA

INTRODUCTION

The Plasma Diagnostics Package (PDP) and support systems are being readied for flight on the OSS-1 Space Shuttle Mission (STS-4 in 1982) and on the Spacelab-2 Shuttle Mission (1983 or 1984). In March 1981 in the Johnson Space Center Plasma Chamber A, the PDP was utilized to measure the state of the ambient plasma and of phenomena induced by operation of the Utah State University Fast Pulse Electron Gun (FPEG). The FPEG is an element of the Vehicle Charging and Potential (VCAP) investigation on the OSS-1 Mission. On the OSS-1 Mission, the PDP will make plasma measurements as it is swept through the FPEG beam region by the Shuttle Remote Manipulator System (RMS).

In this paper, objectives of the PDP investigations are stated, features of the PDP systems are described and measurement characteristics of the PDP instruments are listed. Sample results from the JSC Plasma Chamber Tests are presented and these results are discussed in the context of results obtained, for example, by Bernstein, et al. (1978, 1979) and by Jost, et al. (1980) and of the theory developed by Rowland, et al. (1980).

DESCRIPTION OF THE OSS-1 PLASMA DIAGNOSTICS PACKAGE (PDP)

Each of the investigations for the Shuttle OSS-1 "Pathfinder" Mission and for the Spacelab-2 Missions are described by Neupert (1979) and Clifton (1978), respectively. A brief summary of the PDP investigation objectives, instrument characteristics and on-orbit operations is given here.

Science and Technical Objectives

For the OSS-1 Mission, the primary scientific and technical objectives are as follows and these are depicted in Figure 1:

- Study the Orbiter-magnetoplasma interactions within 15 meters of the Orbiter through measurement of electric and magnetic fields, ionized particle wakes and generated waves;

- Determine the characteristics of the electron beam emitted from the Fast Pulse Electron Gun (FPEG) out to a range of 15 meters from the Orbiter and measure the resulting beam-plasma interactions in terms of fields, waves and particle distribution functions;

- Locate and measure sources of fields, electromagnetic interference (EMI) and plasma contamination in the environment of the Orbiter out to 15 meters; and

- Flight-test the systems and procedures associated with the Spacelab-2 PDP experiment with particular emphasis on operations with the Remote Manipulator System, on unlatching and relatching the PDP unit, and on evaluating the telemetry link.

The first two objectives utilize the Orbiter for carrying out active experiments in the ionosphere. Since the Orbiter itself is very large compared to an electron gyroradius, comparable to an ion gyroradius, partially conducting and partially insulating, its motion through the magnetoplasma will cause a variety of wake effects including density perturbations, currents, wave excitation and particle energization. On OSS-1 these wake phenomena are to be studied out to the 15m extent of the RMS but on Spacelab-2, when the PDP functions as an Orbiter subsatellite, the wake region out to 10km can be explored.

OSS-1 offers the only Shuttle opportunity before 1985 to perform active beam-plasma interaction studies with in-beam diagnostics. OSS-1 will carry both the VCAP Fast Pulse Electron Gun and the PDP which is articulated through the electron beam by the RMS. The Space Experiments With Particle Accelerations (SEPAC) particle injection system with some near-beam diagnostics will be flown on Spacelab-1 but the PDP is assigned for reflight on Spacelab-2. Only after 1985 with the follow-on Spacelab missions are there plans to fly the SEPAC with in-beam and subsatellite diagnostics. The JSC Plasma Chamber test of March 1981, described in the next major section, was motivated by the desire to obtain more data on beam-plasma interactions, to further describe the beam-plasma-discharge (BPD) phenomena (e.g., Rowland, et al., 1980) for identification of its occurrence on-orbit and to better understand the joint operation of the PDP and VCAP systems.

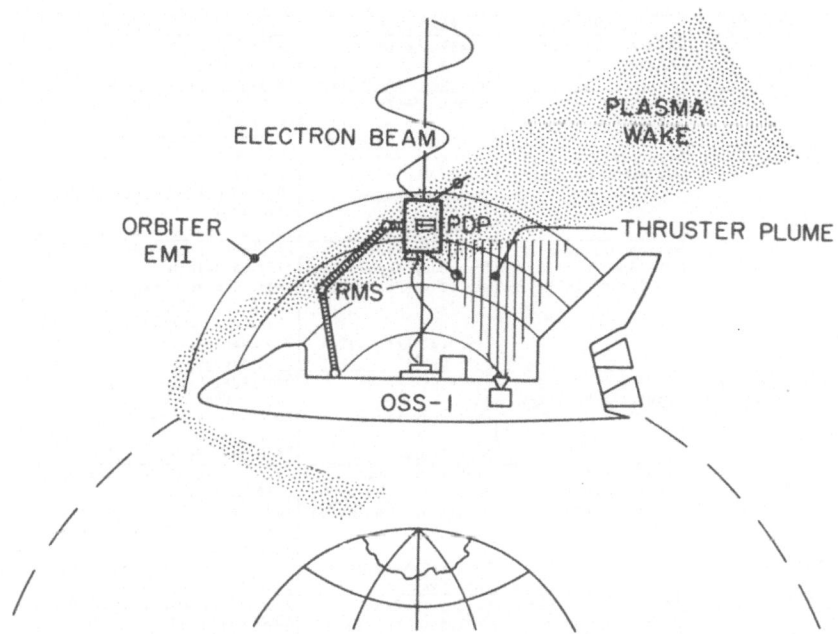

Fig. 1.   Depiction of scientific and technical objectives for the
Plasma Diagnostics Package (PDP) on the Shuttle OSS-1
Mission.   Through articulation by the Remote Manipulator
System (RMS), the PDP can be used to diagnose Orbiter-
induced plasma wakes and the Fast Pulse Electron Gun-
initiated electron beam-plasma phenomena.

    As a subsatellite of the Spacelab-2 Orbiter, the PDP will pene-
trate regions of plasma depletion created by scheduled OMS burns
exhausting 100-1000 kg of propellant.   These active plasma depletion
experiments of Mendillo and da Rosa (see Clifton, 1978) offer the
opportunity to compare ground-radar global measurements of the depletion regions with PDP in situ samples to assess chemical reaction
rates and associated fields, waves and energized plasma.   A Recover-
able PDP is being developed to continue these studies after 1985 with
active particle, wave, moving body, and chemical perturbations of the
ionospheric plasma.

PDP INSTRUMENT CHARACTERISTICS

    The PDP carries a complement of new and of flight-spare instru-
ments from previous NASA programs, to provide a comprehensive set of
fields, waves, thermal plasma and energized plasma measurements.   A
detailed listing of the measured parameters and the measurement
ranges is given in Table 1 for OSS-1.   For Spacelab-2 the electric
field dipole length will be increased to 4 meters through the use of
hinged booms.   Indicated in Figure 2 are the locations of the apera-

Table 1. PDP Scientific Instruments Performance Specifications

MEASUREMENT	TECHNIQUE	PARAMETERS	VALUE/RANGE
DC Magnetic Field	Triaxial Fluxgate Magnetometer	Dynamic Range	±12 milligauss to ±1.5 gauss each axis
		Temporal Resolution	10 samples/second each axis
DC Electric Field	1m Double Probe with Spherical Sensors	Dynamic Range	±2 mV/m to ±2 V/m (average and differential
		Temporal Resolution	20 samples/second
AC Magnetic Waves	Searchcoil Sensor; Wideband Receiver	Frequency Range	5Hz-1kHz & 0.65-10, 10-20, 20-30kHz
		Amplitude Range	100db @ 0.4db resolution; 3mγ-300γ
		Duty Cycle	12.8 seconds out of 51.2 sec.
	Searchcoil Sensor; VLF Spectrum Analyzer (IMP)	Frequency Range	16 channels 35.5 Hz to 178kHz
		Frequency Resolution	±15% bandwidth
		Amplitude Resolution	100db @ 0.4dB resolution; $3 \times 10^{-5} - 3_\gamma Hz^{-1/2}$ (peak and average)
		Temporal Resolution	0.6 sample/second each channel
		Duty Cycle	12.8 seconds out of 51.2 sec.
AC Electric and Electrostatic Waves	1m Dipole Antenna Wideband Receiver	Frequency Range	5Hz-1kHz, 0.65-10kHz, 10-20kHz & 20-30kHz
		Amplitude Range	100db @ 0.4db resolution; 3μV/m - 300 mV/m
		Duty Cycle	38.4 seconds out of 51.2 sec.
	1m Dipole Antenna VLF Spectrum Analyzer (Helios)	Frequency Range	16 channels-31.2Hz to 178kHz
		Frequency Resolution	±15% bandwidth
		Amplitude Resolution	100db @ 0.4dB resolution; $3 \times 10^{-8} - 3 \times 10^{-3} Vm^{-1} Hz^{-1/2}$ (peak and average)
		Temporal Resolution	0.6 sample/second each channel
		Duty Cycle	100%
	1m Dipole Antenna, Mid Frequency Receiver	Frequency Range	8 channels-31.6Hz to 17.8 MHz
		Frequency Resolution	±30% bandwidth
		Amplitude Resolution	70db @ 1dB resolution; 3×10^{-3} - 10 V/m (peak and average)
		Temporal Resolution	1.6 second/scan
VHF/UHF EMI Levels	Horn Antenna VHF/UHF Receiver	Frequency Range	4 channels--25-65, 65-160, 160-400, 400-800 MHz
		Frequency Resolution	±50%
		Amplitude Resolution	70db @ 1db resolution; 10^{-2} - 30 V/m; (peak and average)
		Temporal Resolution	1.6 sec/scan
S-Band Field Strength Monitor	Horn Antenna VHF/UHF Receiver + Mixer and L.O.	Frequency Range	2000-2330 MHz
		Amplitude Range	.01 to 30 V/m (peak & average)
		Temporal Resolution	1.6 sec.
Suprathermal Particles	Low Energy Proton & Electron Differential Energy Analyzer (LEPEDEA)	Energy Range	2eV-50keV in 42 steps: electrons and ions
		Energy Resolution	34%
		Field of View	$6° \times 162°$ (7 detectors)
		Flux: Electrons	$30-1 \times 10^{7}$, electrons/cm^2 sec sr eV
		Protons	$6-2 \times 10^{8}$ protons/cm^2 sec sr eV
		Temporal Resolution	1.6 sec for spectrum
	Electrometer	Flux Range	$10^{9} - 10^{14}$ elect cm^{-2} sec^{-1}
		Temporal Resolution	10 samples/second
	Retarding Potential Analyzer/Differential Ion Flux Probe	Density Range	$2 \times 10^{1} - 1 \times 10^{7}$ ions cm^{-3}
		Energy Range	0-16 eV
		Velocity Range	0-15km sec^{-1}
		Temporal Resolution	0.8 sec/scan; 51.2 sec/ analysis
Thermal Electrons	Langmuir Probe, Density	Dynamic Range	$10^{3} - 10^{7}$ electrons cm^{-3}
		Temporal Resolution	1 second sweep every 12.8 sec.
	Langmuir Probe, Density Irregularities	Scale Sizes	10 meters to 100 km
		Dynamic Range	80db @ 5db resolution; $10^{2} - 10^{8}$ cm^{-3}
Thermal Ions	Ion Mass Spectrometer	Dynamic Range	$20-2 \times 10^{8}$ ions cm^{-3}
		Mass Range	1-64 AMU @ < 1% overlap
		Temporal Resolution	1.6 seconds for mass scan
Ambient Pressure	Ionization Gauge	Pressure Range	10^{-7} to 10^{-3} torr

tures for the High Frequency Antenna, the Ion Mass Spectrometer, the
Electron Fluxmeter, the Retarding Potential Analyzer/Differential Ion
Flux Probe and the Low Energy Differential Energy Analyzer (LEPEDEA)
around the spacecraft belly. Also indicated are the locations of the
search coil magnetometer, the AC-DC electric antennas and the Langmuir Probe sensors which are supported on short booms for OSS-1.
Sample results from the JSC Plasma Chamber Tests are presented for
some of these instruments in the next major section.

PDP On-Orbit Operations

For the OSS-1 Mission, the primary PDP measurements are made
during two 6 hour periods attached to the RMS. The PDP is moved and
articulated on the RMS by a series of Automode Sequences which
specify a trajectory of locations and the PDP attitude at each location. These sequences are initiated by the crew and the crew can
take over manual control at any point. In support of the FPEG, one
sequence positions the PDP above the FPEG to measure any Orbiter-

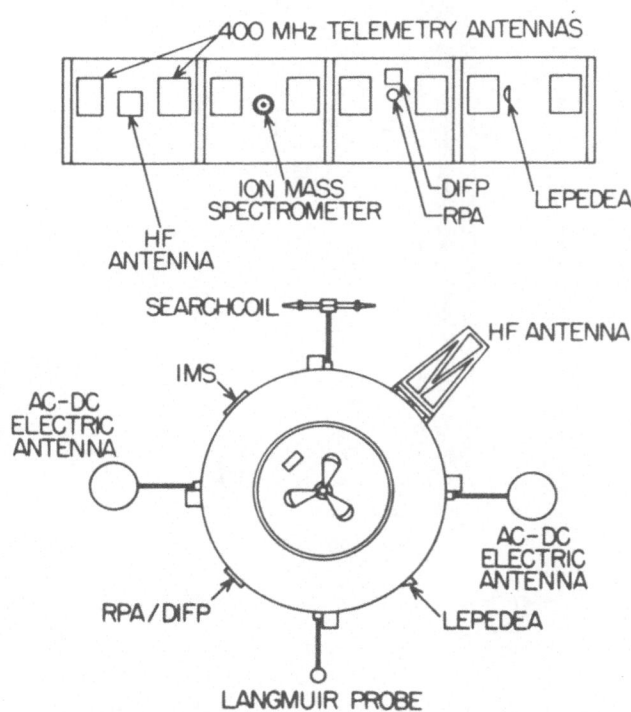

Fig. 2.   Location of PDP instrument aperatures around the belly and
          of the boom-mounted sensors protruding above the top for
          OSS-1.

generated dc magnetic fields that might affect the predicted beam
trajectory. Two sequences are set up to sweep the PDP through the
predicted beam trajectory region. The OSS-1 crew will have a display
of a few parameters from the PDP in order to judge which Automode
Sequence to use. However, the bulk of the telemetry data is tape
recorded on-board for playback after the mission. On Spacelab-2 the
bulk of the data will be relayed to the Payload Operator's Control
Center (POCC) at JSC. Therefore, realtime analysis is possible by
the PDP Team as the PDP is operated on the RMS or as the Orbiter
moves with respect to the PDP to pass the wake or the exhaust plume
past the PDP.

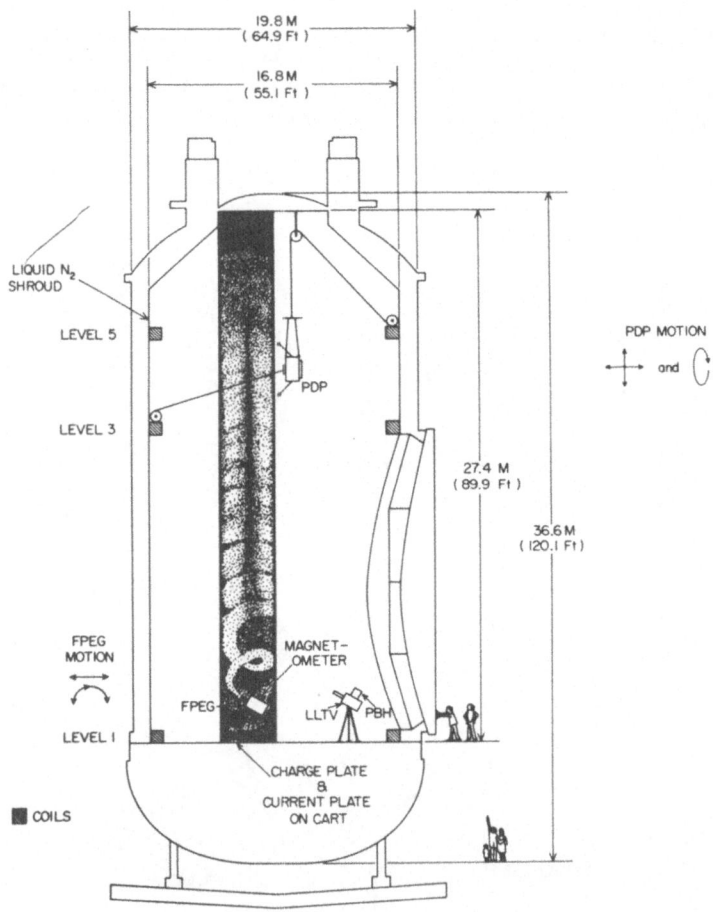

Fig. 3. Dimensions of JSC Chamber A and the relative locations of
         the PDP and FPEG systems within the chamber. The PDP can be
         rotated, moved back and forth and moved up and down. The
         FPEG can be moved across the floor.

JSC PLASMA CHAMBER TESTS

    Use of the JSC Chamber A for ionospheric-type plasma experiments
has been described by Bernstein, et al. (1978). This unique facility
has dimensions of 17m in diameter by 27m in height and the pressure
is reduced to below 10^{-5} torr in less than 8 hours pumping time.
External coils can be used to impose a magnetic field up to 2 gauss
to boost or buck the earth's field. As indicated in Figure 3, the
FPEG and other elements of the VCAP system were mounted to a moveable
cart on the chamber floor. This set-up was similar to that of Raitt,
et al. (1980). A magnetometer mounted to the FPEG was used to set
the beam pitch angle. The PDP was suspended at ~ 15 meters above the
FPEG with a rope and pulley system that allowed for it to move up/
down and in/out of the beam. In addition, it was possible to rotate
the PDP ±180°.

    Measurements were made with the PDP under variation of a wide
number of parameters including magnetic field (0.2-1.5 gauss) and
direction (up/down), chamber pressure (6 x 10^{-6} to 6 x 10^{-5} torr),
plasma (5 x 10^6 cm^{-3}) or not, FPEG pitch angle (±90°), FPEG beam current (0.5-85 ma), PDP location with respect to the beam and PDP roll
angle. Some sample results regarding fluxes, fields, wave amplitudes, and particle energy spectra under BPD and non-BPD conditions
and the variation of these parameters with distance from the beam are
presented in the next sections.

Non-BPD vs BPD Characteristics

    The data of Bernstein, et al. (1978, 1979) and the theory of
Rowland, et al. (1980) give a relation for the critical current I_c
for onset of the beam-plasma-discharge:

$$I_c > 150 \, E_b(keV)^{1.8} \, U(km/s)/(P(\mu torr)L(cm))$$

For parameters of E_b = 1, U = 1, P = 6 and L = 2000, I_c > 12.5 ma.
The beam-plasma-discharge was recognized to occur for beam currents
between 15 and 20 ma depending on the pitch angle. Distinctive characteristics between non-BPD and BPD are illustrated in Figures 4 and
5. For this case B = 1.2 gauss, the pitch angle is 30° and the gun
current starts at 16.5 ma rising to 17.5 ma at the end of a 26 second
gun pulse. In the last 4 seconds the BPD is ignited apparently due
to the increasing plasma density in the beam. As shown in Figure 4,
the electron spectrum did not change significantly (except below 10
eV which may be a charging effect) for this detector look angle (perpendicular to B) but the ion flux increased dramatically by more than
an order of magnitude and increased in average energy with a broad
peak between 10 eV and 250 eV. Electric field emissions below 10 kHz
have intensities of up to 1 V/m when the beam is on but generally
decrease after onset of BPD. Between the lower hybrid frequency and
the gyrofrequency, intensities are greater by 5-15 dB after BPD onset.

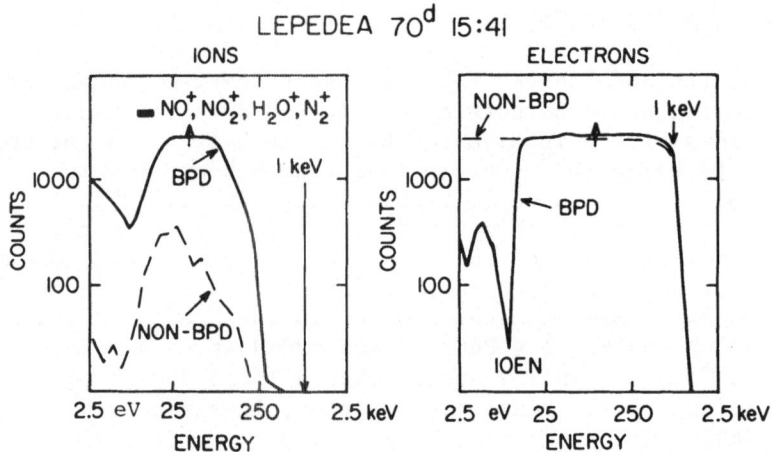

Fig. 4.   Energetic ion and electron fluxes under non-BPD and BPD con-
           ditions. Note that the electron energy distribution changes
           very little except at low energies, whereas the ion flux
           increase by an order of magnitude and becomes more energetic.
           The ion enhancement between 10 and 100 eV correlates with
           ionization potentials. The primary electron beam was 1 keV
           with a current of 16 ma for non-BPD and 17.5 ma for BPD at a
           30° pitch angle and B = 1.2 gauss.

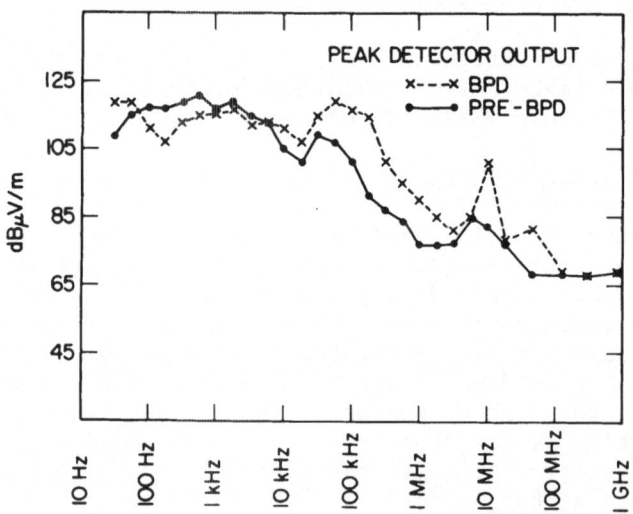

Fig. 5.   Electric field intensities for wave emissions under non-BPD
           and BPD conditions with same plasma and FPEG parameters as in
           Fig. 4. Note the enhanced emission above the lower hybrid
           frequency (~ 15 kHz) and at the plasma frequency (~ 10 MHz).
           Intensities range up to 1 volt/m.

There is also a 20 dB enhancement near the plasma frequency to fields
of ~ .1 V/m during BPD.

Also noted under these critical current conditions was the 1-5 Hz
oscillatory behavior of the plasma which is probably the same flickering phenomena reported by Bernstein, et al. (1978) and Jost, et al.
(1980). This occurred only before BPD onset. Order of magnitude variations were observed in the Langmuir Probe current and in the electron flux; ~ 1 V/m electric field fluctuations were measured and the
energetic particle spectra were modulated. Visually, it appeared as
if sections of the plasma column were torn loose and driven perpendicular to the magnetic field lines to the chamber walls.

Spatial Variation of Parameters

Spatial variations for the total electron upward flux and for the
dc electric field perpendicular to the downward magnetic field of 1.18
gauss are shown in Figure 6 for an 80 ma, 80° pitch angle beam. Two
parallel electric field values are also indicated. The upward flux is
minimum in the region of one gyroradius where the primary beam electrons are perpendicular to $\vec{B}$ and seems to go to zero near the chamber
walls. Directions of the perpendicular E-field indicate an excess of
positive charge inside of one gyroradius (beam center) and an excess

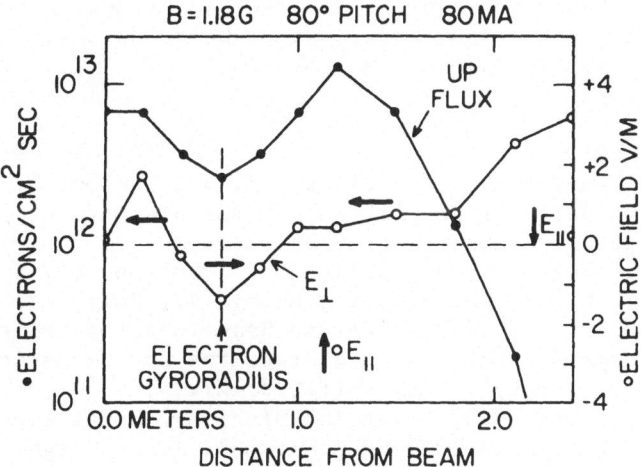

Fig. 6. Variation of the energetic electron flux and the DC electric
field with distance from the electron beam center. A 1 keV,
80 ma beam was injected into a 1.18 gauss field at an 80°
pitch angle. Note that there is a significant electron flux
outside of the illuminated column (1 gyroradius) and that
the electric field perpendicular to B reverses twice near
0.6 meters.

of negative charge outside. The upward directed field component at
1.2 meters (2 gyroradii) indicates that upward going electrons are
being decelerated. The 10 MHz plasma frequency emission peaks at one
gyroradius then decreases by 20 dB at two gyroradii.

Discussion

The energetic electron spectrum results and the electric field
wave emission results are consistent with those of Bernstein, et al.
(1978, 1979), Jost, et al. (1980) and others. Unique to this experiment with the PDP are the measurements of the energetic ion spectra,
the electron flux, the dc electric field and the ion composition of
the chamber which are yet to be processed and interpreted for the
many cases available. The PDP is limited to a temporal resolution of
~ 1.6 seconds so that onset characteristics cannot be determined.

ACKNOWLEDGEMENTS

The PDP is being developed for the OSS-1 and Spacelab-2 Shuttle
Missions through Contract NAS8-32807 with Marshall Space Flight Cen-
ter. Funding for the JSC Plasma Chamber Tests was provided by NASA/
HQ through Grant NAGW-190. I thank the contractors and staff of the
JSC Chamber A, especially Jerry Jost, for the set-up and operation of
the experiments and I thank John Raitt, Peter Banks, Bill Denig and
Brent White for our collaborative effort with the FPEG. I am espe-
cially appreciative of the PDP Co-Investigators and Engineering staff
with special thanks to Roger D. Anderson as Project Manager, and to
Marty Kerl, Terry Clausen, Harry Owens, Gerry Murphy, Dave Cramer and
Lowell Swartz for the JSC operations.

REFERENCES

Bernstein, W., Leinbach, H., Kellogg, P. J., Monson, S. J. and Halli-
    nan, T., 1979, Further Laboratory Measurements of the Beam-Plasma
    Discharge, J. Geophys. Res., 84:7271.
Bernstein, W., Leinbach, H., Kellogg, P., Monson, S., Hallinan, T.,
    Garriott, O. K., Konradi, A., McCoy, J., Daly, P., Baker, B. and
    Anderson, H. R., 1978, Electron Beam Injecton Experiments: The
    Beam-Plasma Discharge at Low Pressures and Magnetic Field
    Strengths, Geophys. Res. Lett., 5:127.
Clifton, K. Stuart (ed.), "Spacelab Mission-2 Experiment Descrip-
    tions," 1978, NASA TM-78198, Marshall Space Flight Center,
    Huntsville, Alabama.
Jost, R. J., Anderson, H. R. and McGarity, J. O., Electron Energy
    Distributions Measured During Electron Beam/Plasma Interactions,
    1980, Geophys. Res. Lett., 7:509.
Neupert, Werner M. (ed.), 1979, "Office of Space Sciences-1 Experi-
    ment Investigation Descriptions," NASA/Goddard Space Flight
    Center, Greenbelt, Maryland.

Raitt, W. J. Banks, P. M., Denig, W. F., White, A. B. and Jost, R. J., 1980, Studies of Electron-Beam Plasma Interactions In a Space Simulation Environment, preprint, Utah State University, Logan, Utah.

Rowland, H. L., Chang, C. L. and Papadopoulos, K., 1980, Scaling of the Beam Plasma Discharge, preprint, Science Applications, Inc., McLean, Virginia.

DISCUSSION

Szuszczewicz: What were the relative populations of energized ions (25 eV or greater) compared with their thermal counterparts?

Shawhan: The ion flux was about 10^8 ions/cm^2 sec eV str and for 100 eV ions (NO^+) the velocity is 10^6 cm/sec so the net density is greater than 100 cm^{-3} which is very small compared to the 10^5-10^7 cm^{-3} "ambient" densities.

Szuszczewicz: If your DC electric field measurement perpendicular to the magnetic field (or perpendicular to the beam axis) is a standard differential floating potential measurement, then there appears to be a distinct possibility that the results could be in substantial error if the sensors are in different plasma environments. Our own results have shown significant variations in plasma-electron energy populations over short perpendicular distances (1-2 meters) (e g change from a single component "cold" Maxwellian distribution to a two-component plasma-electron population with a high energy component) that result in a much greater change in floating potential than in the local plasma potential. Sensors immersed in these two environments would therefore measure potential changes which suggest a higher field than actually exists. Could you comment on this potential problem?

Shawhan: That condition is possible. We do have a Langmuir probe that extends to the same radius as that of the spheres so that as the PDP is rotated the plasma state at each sphere location can be determined. Probably we cannot correct the electric field measurement but we can know if it is not valid.

RADIAL DEPENDENCE OF HF WAVE FIELD STRENGTH IN THE BPD COLUMN

R.J. Jost,* H.R. Anderson,[#] W. Bernstein,[#] and
P.J. Kellogg[◊]

* NASA Johnson Space Center
Houston, TX   77058

[#] Rice University
Houston, TX   77001

◊ University of Minnesota
Minneapolis, MN   55455

INTRODUCTION

Several detailed beam/plasma investigation programs have been
conducted in the large vacuum chamber of the Space Environment
Simulation Laboratory (SESL) at the NASA Johnson Space Center.   In
particular, the experiments of Bernstein et al. (1979) show that at
a critical electron beam current an instability, the Beam Plasma
Discharge (BPD), between the beam and the ambient plasma develops.
The BPD onset shows a dramatic change in the light emission, RF
spectrum, and beam/plasma electron velocity distribution.   Ration-
ale for studying this instability has been generally established
in view of describing beam experiments aboard sounding rockets,
from the SESL plasma chamber work, and proposed for Shuttle missions.
It has been noted in Jost et al. (1980) that an increase in energy
by as much as a factor of two, exists for a fraction of the primary beam particles.   For the laboratory research with keV beam
energies this corresponds to an energy increase of a few keV.
Electric field strengths of volts/m must be present within the
discharge column to accelerate primary beam electrons by that
amount, however, previous work did not isolate the particular
waves in RF spectra during BPD that contained volts/m field

431

strengths. The objective of this recent experiment was to determine a quantitative value for the field strength in the plasma
frequency region of the spectrum. Determination of the plasma
wave field strengths would also help determine whether the high
frequency waves (plasma frequencies) or the low frequency waves
(cyclotron frequencies) were primarily responsible for heating the
beam electrons. Figure 1 shows a typical RF spectrum taken in the
SESL chamber during a BPD condition.

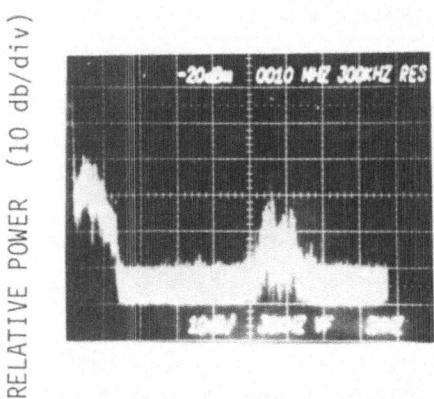

Fig. 1 Typical RF spectrum during BPD.
        Compare the amplitudes of the plasma
        frequency and low frequency waves.

It was assumed that the electrostatic field strengths for the
BPD were strongly localized in the discharge column and that previous measurements failed to probe the critical regions. This
experiment was done using a set of very simple dipole antennas
supported on a positioning device that could probe the beam column
with a spatial resolution of a few centimeters over a limited
range in three dimensions inside and outside the beam discharge
column.

EXPERIMENT DESCRIPTION

Figure 2 shows the arrangement of the antenna positioning system. Four short dipole elements were mounted on individual booms of 2 m length each. The booms were fixed to a motor hub that enabled the entire orthoganol boom system to rotate in a plane approximately perpendicular to B. With the electron beam along B at 2 m from the boom hub, the motor assembly could be rotated to position the antennas quite accurately within the beam cross section.

In addition, the entire motor-boom assembly could be position-ed along the magnetic field (i.e., along the beam). This particu-lar motion was limited to a range of 1-4 meters from the electron gun. Both motions (along and radial to the beam) were remotely controlled from outside the plasma chamber during test conditions.

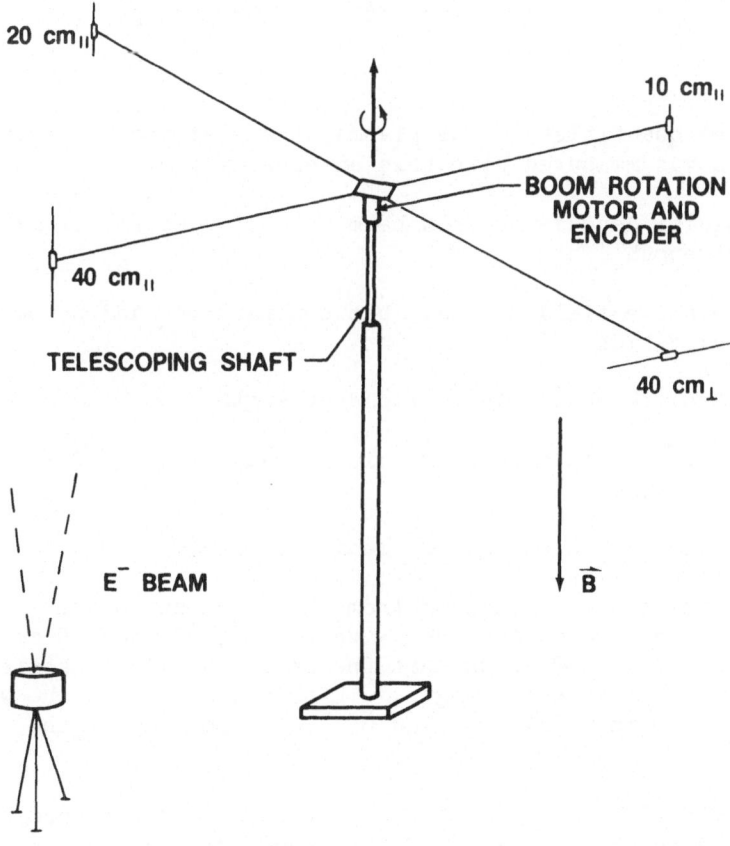

Fig. 2   Antenna positioning system for measuring
         RF electric fields as a function of radial
         distance from the beam core.

Three of the four dipole antennas, with lengths of 10, 20, and 40 cm, were oriented parallel to the magentic field while the fourth antenna, 40 cm long, was oriented perpendicular to the field. The active elements were non-insulated, straight conductors coupled at the base ends through 4 pf capacitors to a terminated coaxial cable. The coaxial cables then carried the signals through the plasma chamber wall to a standard laboratory spectrum analyzer over a total distance of about 50 meters. No amplifiers were used in this system. Further, it was assumed, based on discussions by Balmain (1965) and Grabowski (1975), that the antennas responded to the electrostatic fields and that additional asymmetrical potentials due to direct interaction with the plasma were small.

A steady state BPD configuration for the beam/plasma interaction was the primary mode investigated. An electron beam was injected parallel to B at 1500 eV energy and 40 ma beam current (BPD condition). The chamber pressure and magnetic field strength were maintained at $8(10^{-6})$ torr and 1.25 gauss respectively.

RESULTS

Several quantities of the plasma line electric field behavior for the BPD was measured with this antenna system:

1) Radial dependence from beam core of electric field strength

2) Relative field strength between parallel and perpendicular fields

3) Absolute magnitude of field strength

4) Longitudinal dependence along beam center of field strength

5) Frequency dependence on beam current

The data presented was obtained from the 40 cm antennas. A comparison between the fields measured with the 10, 20, and 40 cm antennas parallel to B shows that the signal amplitude scaled directly with the antenna length. This indicates that the wavelength of the primary oscillation was long compared to the longest antenna.

Figure 3 plots the relative electric field dependence on radial distance from the core of the beam discharge column for both the parallel and perpendicular antenna. A strong radial dependence was measured with the perpendicular antenna which showed a field decrease of 40 dB (in spectral density) within

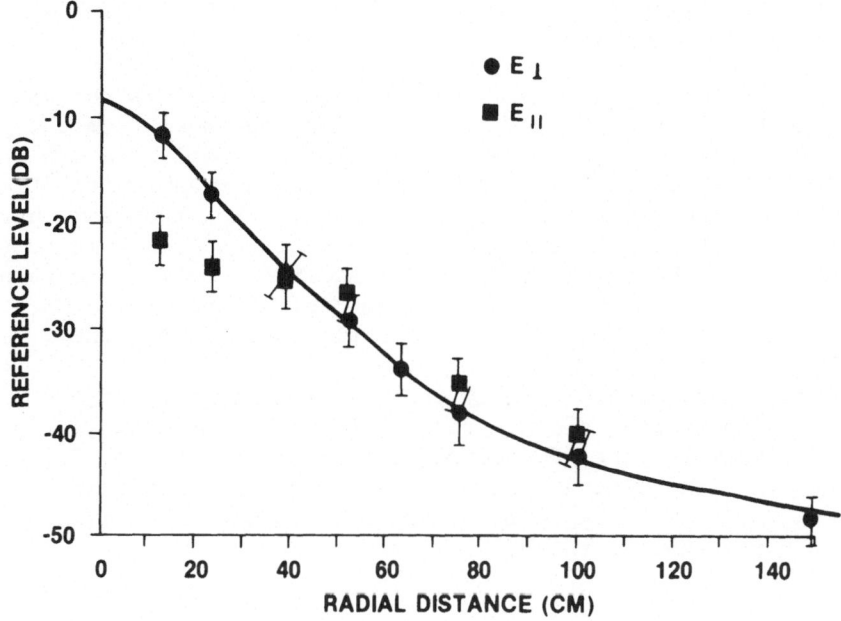

Fig. 3   Radial dependence of electric field strength
for both the perpendicular and parallel fields.

one meter from the core; the parallel antenna also had a strong
radial dependence but with a smoother decrease of about 25 dB over
the same distance.   The BPD column diameter was estimated to be
approximately one meter based on low light TV images.   Once the
antennas were moved outside the column the field strength dependence on distance from the beam was much smaller.

In this particular BPD condition at 1500 eV and 40 ma (BPD
threshold at 20 ma) the maximum field strength, measured at the
center, was approximately 10 dB greater for the perpendicular
antenna than that measured by the parallel antenna.   Qualitatively however, it appears that the perpendicular field strength is
not always greater than the parallel field inside the column.
The ratio of field strengths may depend on the antenna position
along the beam and on the beam/plasma parameters themselves.

The field strengths correspond to 10 volts/m to within a
factor of five.   This uncertainty results from difficulties in
determining the analyzer output amplitude reading, the lack of a
thorough calibration of the system, although a preliminary cali-

bration was completed, and uncertainties regarding the plasma effects on the antenna behavior.

In addition to the radial dependence, the field strength varied both as a function of distance along the beam from the gun and of the beam current. Over the 3 m range from 1 m to 4 m along the beam, an increase of 5-10 dB was observed for the field strengths in the center of the beam column. When the beam current was increased the field strength showed a corresponding increase. Qualitatively, the dependence was strongest for beam currents near the critical BPD current. As shown in Fig. 4, increasing the beam current raised both the plasma line amplitude and frequency.

These measurements clearly indicate the behavior of the buildup of ionization plasma prior to BPD ignition. Again, for this configuration the BPD occurs at a critical current, I_c, of about 20 ma. The plasma frequency wave was first detectable when $I_{beam}/I_c \approx 0.5$. With $I_{beam}/I_c \rightarrow 1$ the frequency and amplitude violently increase to the BPD levels at I_c.

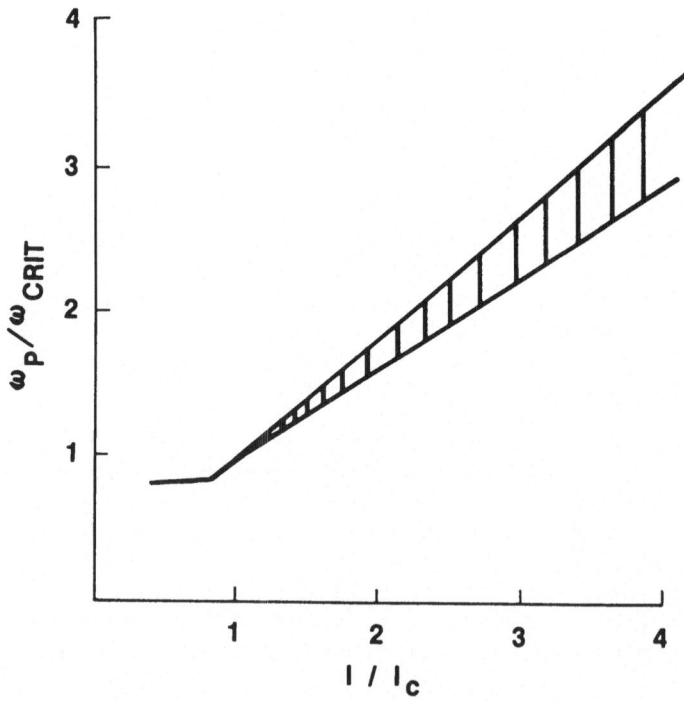

Fig. 4   Plasma wave frequency dependence on beam
current near beam core. Note the appearance
of the wave prior to BPD ignition.

SUMMARY

    Careful measurement of the plasma wave characteristics for
the BPD will provide invaluable information for describing the
instability. In summary, the findings of a recent set of RF
measurements of the BPD in the SESL plasma laboratory are:

1) The parallel and perpendicular components of the plasma
    wave electric fields inside the BPD column have compa-
    rable field strengths of the order 10 volts/m.

2) The radial dependence of the field strength is very
    strong, decreasing by as much as 40 dB within 1 meter
    from the beam center with the illumination or discharge
    column approximately 1 meter in diameter.

3) The field strength inside the column increases as a
    function of distance along the beam at least for several
    meters from the gun aperature.

4) The frequency and amplitude of the plasma wave increases
    with beam current. A particularly rapid increase in
    these parameters occurs as the beam current approaches
    the critical current.

REFERENCES

Balmain, K.G., Impedance of a short dipole in a compressible
        plasma, Radio Sci., 69D, 559, 1965.

Bernstein, W., H. Leinbach, P. J. Kellogg, S.J. Monson, and
        T. Hallinan, Further laboratory measurements of the beam
        plasma discharge, Jour. Geophys. Res., 84, 7271, 1979.

Grabowski, R., Antenna response to random electric fields due to
        thermodynamic fluctuations in plasmas, Phys. Fluids,
        18, 1387, 1975.

Jost, R. J., H. R. Anderson, and J. O. McGarity, Measured electron
        energy distributions during beam/plasma interactions,
        Geophys. Res. Lett., 7, 509, 1980.

DISCUSSION

Stenzel: The spectra you showed should be corrected by
the frequency response of the antenna which has a sheath-
plasma resonance near ω_{pe}. At large wave amplitudes the
nonlinearity of the sheath can generate the harmonics of
ω_{pe} which you showed.

Jost: We realize the difficulties of making this
measurement. The antenna elements were coupled to the
feed line through capacitors of a few pico farads which
is much smaller than the sheath capacitance in an effort
to minimize the resonance effect. However, problems do
exist and we have put an uncertainty of approximately
one order of magnitude on the measurement. Your question
on the harmonics of ω_{pe} should not affect the measurement I indicated since the antenna was located well outside the beam region.

Papadopoulos: For strongly turbulent plasmas, currents
can be generated in principle to all harmonics of ω_{pe},
with amplitudes of course falling fast with the harmonic
number. Their electromagnetic signature outside the
plasma can therefore be a real thing rather than a probe
peculiarity.

LABORATORY BEAM-PLASMA INTERACTIONS - LINEAR AND NONLINEAR

P.J.Christiansen, V.K. Jain, and J.W. Bond

Space and Plasma Physics Group
School of Mathematical and Physical Sciences
University of Sussex
Brighton BN1 9QH, UK

INTRODUCTION

Though the laboratory studies of beam-plasma interactions which we will discuss have been made within the last decade or so, the literature on this subject extends back many decades and includes early work by the progenitor of plasma physics, Langmuir (see Crawford and Kino (1961) for an historical survey).We will avoid the Herculean task of reviewing a field which has produced many hundreds of papers, and our selection of material has a high degree of arbitrariness - we therefore give advanced apologies for our many sins of omission. The reader will, however, find a number of helpful partial reviews of this field. The linear aspects are discussed by Briggs (1964), Seidel (1969), Mikhailovskii (1975), Akhiezer et al (1975), Allen and Phelps (1977), while various aspects of nonlinear phenomena are discussed by Davidson (1972), Akhiezer et al., (1975), Franklin (1977), Thornhill and ter Haar (1978), ter Haar and Tystovich (1980), Rudakov and Tsytovich (1978) and Sagdeev (1979).

The configuration of a cool plasma (often magnetized axially) penetrated by an injected electron beam is much favoured because of its experimental and (relative) theoretical simplicity. We hope to demonstrate that despite unavoidable scaling limitations, laboratory experiments can illuminate, in a controlled fashion, details of beam plasma interaction processes in a way which will never be possible in the space plasma physics. In view of the increasing interest in high frequency instabilities in the auroral zone, the injection of artificial beams into natural plasma, upstream waves in the solar wind, and the relations of, for example, type III radio bursts and electron beams of solar origin, the possibilities for interesting cross fertilisations of the two fields are extensive.

Linear Theory

In this section we outline the elements of the linear theory (see Mikhailovskii chs 3-5 for details).

Unstable electrostatic waves arising from an exchange of energy with the "free energy" beam feature are considered from two viewpoints ie as (a) kinetic and (b) hydrodynamic or fluid, instabilities. In Fig.1 we show electron distribution functions which lead to these instabilities. For kinetic instability the details of the plasma and beam distribution are important and waves grow in regions of velocity space where $\partial f/\partial_U > 0$ obeying the Cerenkov resonance condition $\omega - k_\| V_b = 0.$ Assuming perturbation of the kind of $e^{i(\omega t - \underline{k}\cdot\underline{r})}$ with $\omega = \omega_r + i\gamma$ it can be shown that

$$\gamma_{max.} = \sqrt{\frac{\pi}{8}} \frac{1}{k_\|^3 \lambda_D^3} \cdot \omega_p \left[\frac{n_b}{n_p} \frac{T_p}{T_b} k_\|^3 \lambda_D^3 \frac{mV_b}{KT_p} \cdot exp(-\tfrac{1}{2}) - exp(-\frac{1}{2k_\|^3\lambda_D^3} - \frac{3}{2}) \right] \overset{(1)}{>0}$$

ω_p = plasma frequency

In Fig. 1(c) the dispersion relation is sketched with the region of kinetically unstable waves shown. The behaviour at low k is determined by the inclusion of small but finite perpendicular wave number $k_\perp \ll k$. We also note that the spatial growth rate $k_i = -\gamma/(\partial\omega/\partial k_r)$ for $\gamma \ll \omega_r$. This quantity is important since the beam plasma instabilities in normal circumstances are convectively, rather than absolutely, unstable (Briggs 1964). In equation (1) it can be seen that in order to increase the growth rate, n_p, and V_b must be increased, and ΔU_b decreased. In the limit we have a cold electron plasma penetrated by a mono-energetic beam, and in this situation it is possible to use a fluid description of the plasma and derive linear dispersion relation

$$\varepsilon(\omega, k_\|) = 1 - \frac{\omega_p^2}{\omega^2} - \frac{\omega_{pb}^2}{(\omega - k_\| V_b)^2} = 0 , \quad n_b \ll n_p \tag{2}$$

Fig. 1(b) shows the region of instability, which is somewhat different from the kinetic case. The fluid instability is conventionally thought of as the coupling of the 'negative energy' wave on the beam (the beam has two branches $\omega - k_\| V_b = \pm \omega_{pb}$, and the slow, 'negative energy' wave takes the -ve sign) with the 'positive energy', normal plasma mode (Mikhailovskii 1975). The fluid instability arising from the coupling of negative and positive energy waves is a concept derived from travelling wave tube

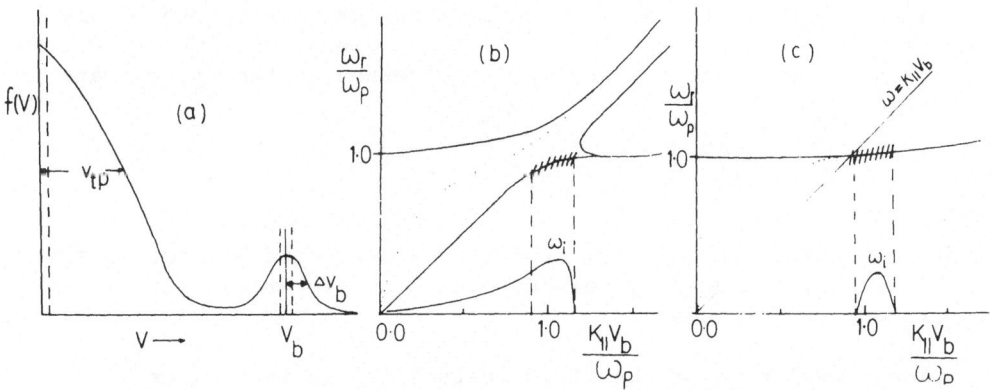

Fig.1 (a) Bump in tail (solid), cold beam plasma (dashed)
      distributions. (b) Dispersion relation in fluid limit showing
      unstable regions (hatched) on negative energy beam branch.
      (c) Kinetic dispersion, instability on plasma branch.

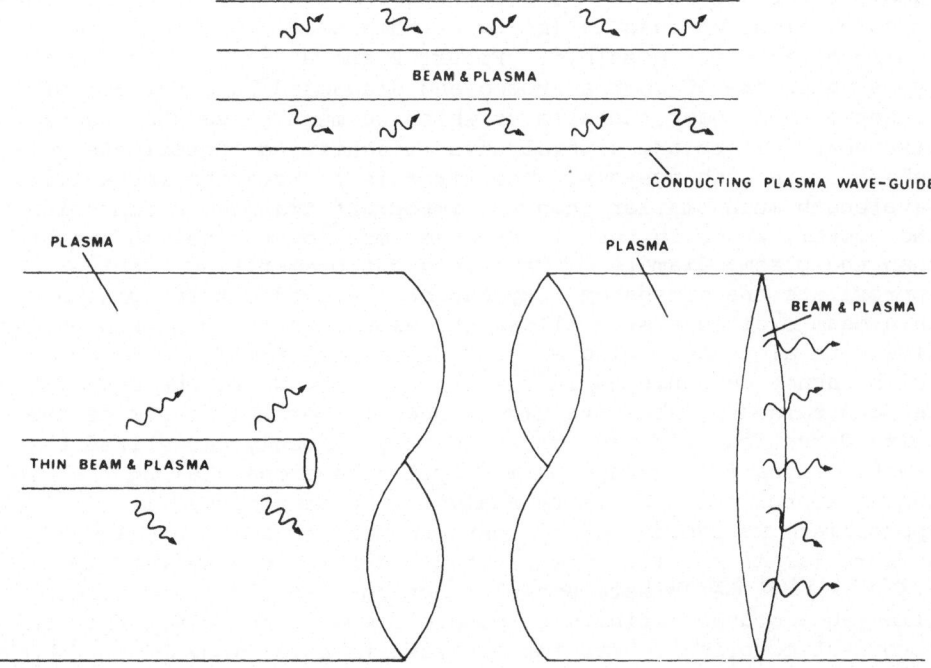

Fig.2 Typical beam plasma configurations. Upper, thin beam wave-
      guides. Lower, large diameter machines, thin (left) and filled
      beam (right) geometries. Wavy lines denote possible energy
      flows.

terminology. Note that the unstable waves propagate more slowly
than the beam i.e., there are no resonant electrons. The transition
from kinetic to fluid instability is governed by the approximate
condition

$$\frac{\Delta V_b}{V_b} \left(\frac{n_p}{n_b} \frac{V_{tp}^2}{V_b^2} \right)^{1/3} \geq 1$$

and the kinetic instability, requiring a very gentle 'bump in the
tail' is only achieved by very careful beam preparation in
laboratory experiments.

The growth rate of the fluid instability is very large,
$\gamma = 0.3\, \omega_p (n_b/n_p)^{1/3}$ even for weak beams. The group velocity of waves
is V_b rather than V_{tp} as in the kinetic case and the spatial growth
rate is also rather large, with an e-folding distance of a few
wavelengths.

Laboratory beam plasma devices come in two basic shapes,
relatively thin axially magnetized devices, with or without
provisions for separate plasma production, and rather stouter
objects with options of axial magentization. In (a) type
machines the beam fills a significant fraction of the plasma tube
cross section, whereas in (b) types thin beam or filled beam
configurations are possible. Figure 2 shows sketches of the two
types which are of course probed and diagnosed in a variety of
rather elegant ways, details of which we must leave the reader to
discover. There are several ways of achieving approximately 1-D
behaviour in such systems. The first is to have the instability
wavelength much smaller than the important transverse dimension of
the system, which in turn is governed more by the beam diameter
than the plasma diameter. Thus in a fat beam-filled machine
approximate one dimensionality can be achieved. A fat plasma
thin-beam excited system allows the experimenter to assess energy
flow both along and obliquely out of the amplifying beam region,
which cannot be achieved in the thin variant where the energy flow
is constrained by the waveguide properties of the vessel to the
axial direction. If the system is very strongly magnetized the
electron dynamics are approximately 1 dimensional (along the field)
and an approximate stability analysis can be achieved if
appropriate waveguide corrections are made at small axial wave
numbers (as in Fig.1). The excited modes in such systems $\phi \sim$
$f(r)\exp.i(\omega t - k_\parallel z - m\theta)$ have Bessel function-like radial profiles
allowing a double infinity of possible modes, but given that the
strongest coupling occurs for the lowest radial mode $\phi \sim J_0(k_\perp r)$
with $m = 0$, the effects of finite diameter are conveniently mocked
up by assuming an effective perpendicular wave number $k_\perp \sim 2.4/R$
where R is the beam radius. More realistic theoretical models are now
beginning to appear (Strangeways and Dungey 1976, Le Queau et al,
1980).

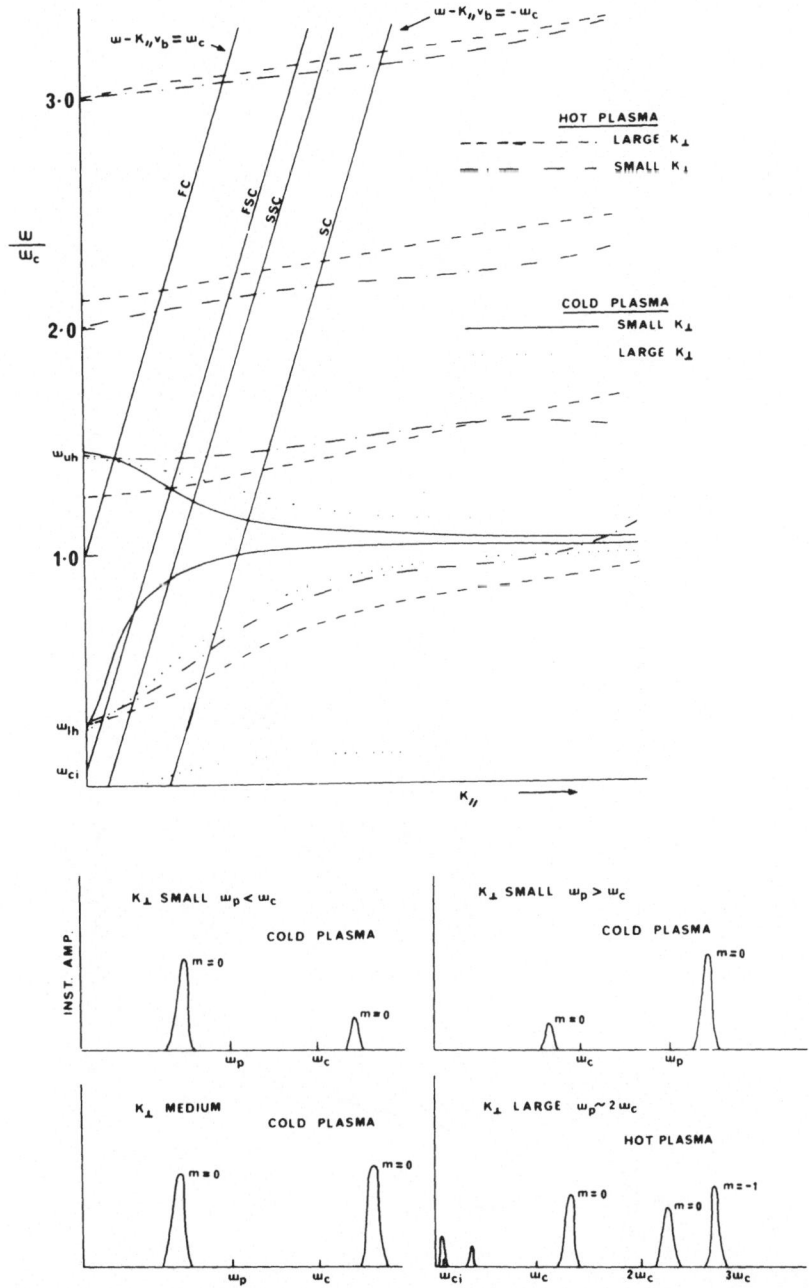

Fig. 3 Schematic dispersion for magnetized plasma ($\omega_p \sim \omega_c$) showing plasma branches in cold and hot plasma limits, for small and large $k_\perp$. Also shown are slow and fast beam (SSC, FSC), and cyclotron beam modes. Lower — schematic spectra with $m = 0$, $m = -1$ couplings displayed.

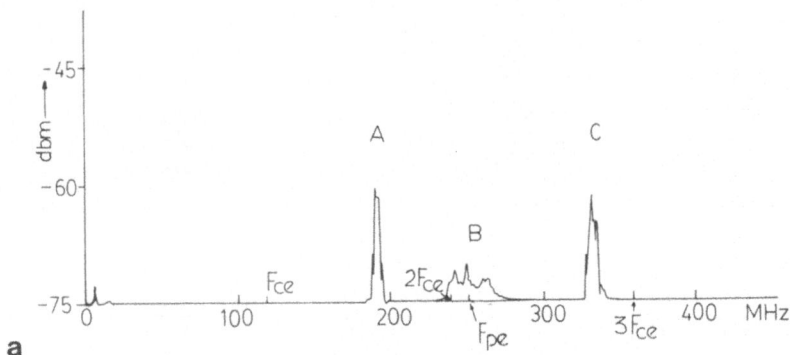

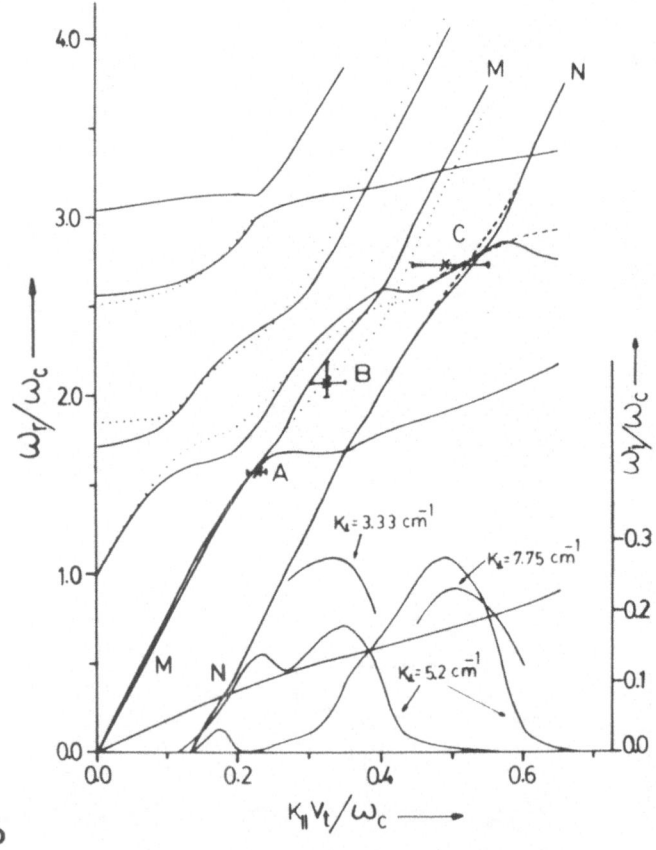

b

Fig.4 (a) Complex instability spectrum in thin beam-plasma. (b) Real
and imaginary dispersion for m = 0, -1 (MM,NN) beam modes
compared with experiment. Best fit is for $k_\perp$ = 3.33, 7.7 cm^{-1}
(dotted,dashed) as measured in experiment.

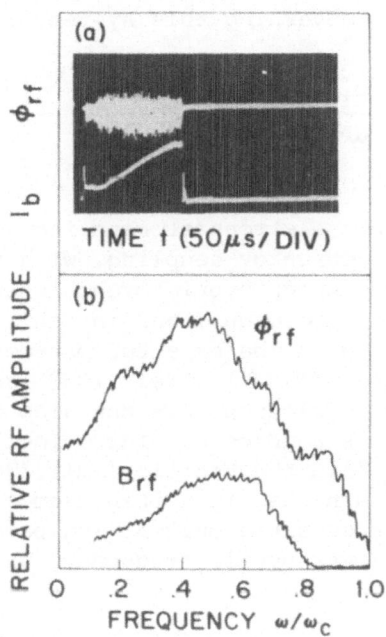

Fig.5 Broadband spectra below ω_c in large plasma, during beam
      injection ($I_b > 0$), demonstrates importance of oblique prop-
      agation. (Stenzel 1977).

The assumption of finite $k_\perp$ and the addition of magnetic field
increases the allowed number of normal modes, and additionally
introduces the possibility of unstable coupling to the so-called
fast and slow cyclotron waves on the beam $\omega = k_\parallel V_b \pm \omega_c$
Here the negative sign goes with the slow, negative energy,
cyclotron beam wave.

In Fig.3 we have attempted to summarise the possible instability
coupling between negative energy beam waves and allowed plasma
branches for the case $\omega_p \sim \omega_c$ for several values of $k_\perp$. We have
additionally included possible couplings to large $k_\perp$, hot plasma
waves, i.e., Bernstein modes, in addition to the fluid plasma
and upper hybrid branches and shown the nominal frequency spectra
to be expected in the different regimes.

The fluid equation governing this system is

$$\mathcal{E}(\omega,k) = 1 - \left[\frac{\omega_p^2}{\omega^2} + \frac{\omega_{pb}^2}{(\omega-k_{||}V_b)^2} \right] \frac{k_{||}^2}{k_{||}^2+k_\perp^2} - \left[\frac{\omega_p^2}{\omega^2-\omega_c^2} + \frac{\omega_{pb}^2}{(\omega-k_{||}V_b)^2-\omega_c^2} \right] \frac{k_\perp^2}{k_{||}^2+k_\perp^2}$$

As a rule of thumb the strongest coupling and largest growth rate occurs due to the Cerenkov coupling, with the plasma branch when $\omega_p < \omega_c$, and on the upper hybrid branch when $\omega_p > \omega_c$. The slow cyclotron coupling can also occur, and instabilities of all these kinds have been observed in the expected parameter regions for many years, e.g. Hopman et al (1968), Apel (1969) and by many others since then. (The fast cyclotron wave can also make unstable couplings if it contains a perpendicular velocity-space ring feature, or if $T_{b\perp}/T_{b||} > 1$ in the kinetic limit, (Sugaya et al, 1977). From Fig.3 it can be seen that in a thin beam system ($k_\perp >> k_{||}$) the plasma branch is accessible only at low parallel wavenumber, i.e. in the region of the lower hybrid frequency, and then with small growth rates.

In Fig.4, we give an example of a complex frequency spectrum (Jain and Christansen 1981) showing couplings to many of the available branches including some to Bernstein wave modes. In Fig. 4(b), the experimental frequencies and wavenumbers are compared with a linear stability model which takes into account observed perpendicular wavenumber variations, with satisfactory results. The less complex interaction of an electron beam with waves on the oblique plasma branch $\omega < \omega_p, \omega_c$, has been elegantly analysed both experimentally and theoretically by Malmberg and Wharton (1969).

In large diameter machines one has the opportunity of observing oblique propagation effects on beam amplified waves. In this sort of machine Stenzel has detected both strongly amplified waves propagating parallel to the beam (on the upper hybrid branch of Fig.3), and obliquely propagating waves on the plasma branch below ω_c which he terms 'whistlers'. In a large diameter system with $R >> 2\pi/k_\perp$, broadband instabilities are possible (see Fig. 5, Stenzel, 1977). At a given frequency $k_{||}$ is determined by the resonance condition $\omega - k_{||}V_b = 0$, and the corresponding value of $k_\perp$ is fixed by equation (2). If ω (and therefore $k_{||}$) is varied, $k_\perp$ also varies, and Fig.3 shows that on the oblique plasma branch $\omega < \omega_c$, lower frequency instabilities have large $k_\perp$, while at higher frequencies $k_\perp$ is small. The direction of energy flow is governed by the group velocity direction (see e.g. Maggs (1978)). This is discussed by Boswell (1975) and Stenzel in terms of the whistler resonance cone, i.e. the refractive index infinity of the electromagnetic dispersion relation, but is more easily understood in the electrostatic limit considered here.

Low frequency couplings

So far we have considered high frequency electrostatic
instabilities excited by an electron beam, whose frequencies are of
the order of electron plasma or electron cyclotron frequencies.
Weak couplings between electron beams and low frequency instabilities
in the frequency range $(\omega_{ci}, \omega_{pi})$(where ω_{ci} is the ion cyclotron
frequency and ω_{pi} is the ion plasma frequency)have also been
observed.   In the low frequency rating three basic electrostatic
oscillations are possible, namely, ion cyclotron waves, $\omega \sim \omega_{ci}$
ion acoustic waves $\omega_{ci} \ll \omega < \omega_{pi}$ and lower hybrid waves.

Though the ion wave branches are not shown on Fig.3, there is no
possibility of coupling to modes below ω_{LH}, ω_{pi} via the Cerenkov
branch (with the exception of lower hybrid waves with very large
values of k_L).   However, weak instabilities around ω_{pi} have been
studied experimentally by Vermeer et al (1967,1970) Hopman (1971)
and Bhatnagar et al (1972).   The conclusion is that unstable
couplings are possible via both the Cerenkov and slow cyclotron
couplings, the latter having low phase velocities in this frequency
range.   Slow cyclotron wave couplings to the so-called second ion
cyclotron wave ($\omega < \omega_{ci}$) with a resonance condition $\omega = \omega_{ci} \cos\theta$; $k_{\parallel} = \dfrac{\omega_{ce}}{V_b}$

and the ion acoustic wave ($\omega_{ci} \ll \omega < \omega_{pi}$)

$$\omega \approx k_{\parallel} C_s \quad , \quad k_{\parallel} = \omega_{ce}/V_b \quad , \quad C_s = \sqrt{T_e/m_i}$$

have been observed by Nyak and Christiansen (1974).

Indirect effects

These direct couplings to ion waves tend to be rather weak, but
there is a number of stronger low frequency instabilities observed
which are driven by effects indirectly related to the presence of
the electron beam.   Hopman (1971) observed unstable waves
corresponding to those of ion Bernstein waves
and the cause of this instability was identified by Yamada et al
(1977) as follows: the injected beam is not initially neutralised
and its excessive charge is compensated by movement of plasma ions
into the beam region. At the beam edges significant sheath effects
exist and give rise to a local radial electric field, E_r (see
Fig. 6). The electrons, whose larmor radii are less than the sheath
thickness,drift azimuthally with velocity $(E_r \times B_0)/B_0^2$ and a two stream
instability of ion Bernstein modes is driven by this azimuthal
(cross field) current.   Recent observations by Wall et al (1981)
and Jain (1981) show that the instability is possible over a large
range of the ratio ω_{pi}/ω_{ci}.

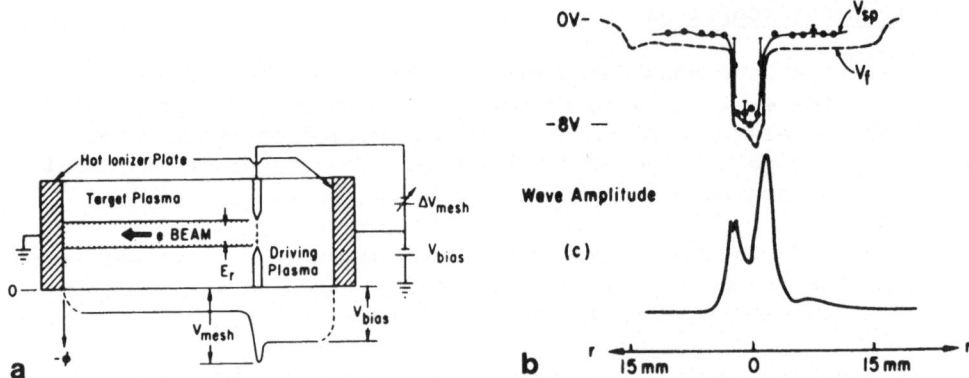

Fig. 6(a) Indirect effects in unfilled configuration. (b) Electric
fields measured at beam edges. (c) Ion Bernstein waves
concentrated at beam edges driven by $\underline{E} \times \underline{B}$ drifts (Yamada
et al 1977),

An analogous effect, giving rise to strong magnetic field
aligned current-driven ion acoustic waves has been reported by
Stenzel (1978). In situations where a compensating return current
flow from the beam injector is not possible via the machine walls,
this current is provided by a drift of slow plasma electrons anti-
parallel to the beam. If the drift exceeds the critical velocity
(Krall and Trivelpiece 1975), ion acoustic waves (and presumably
ion Bernstein waves) are driven unstable.

NONLINEAR EFFECTS

So far we have only considered the evidence for the existence
of exponentially growing instabilities in beam plasma interactions.
It is more interesting to consider the consequences of such
instabilities, i.e. when the waves have grown to a finite level
and to review some studies of non linear processes which have been
made in an attempt to understand how the free energy originally
available in the beam is redistributed to produce a final state of
equilibrium turbulence (we assume that the beam injection is
maintained for a significant length of time). Non linearities fall
into several broad categories which, however, will finally become
inextricably mixed. First there are those which occur because of
the fluid-like behaviour of the medium, for example, amplitude
dependent phase velocity effects such as shock and soliton formation,
modulational instabilities and resonant non linear coupling effects
such as wave-wave interactions obeying conservation relations

$$\sum_i \omega_i = 0 \quad , \quad \sum_i k_i = 0$$

(ω_i, k_i are the frequency and wave number of the ith wave)
in which energy may be transferred from beam-driven instabilities to
waves in other regions of velocity space where it may be reabsorbed.

Secondly, there are the non linear wave particle interactions
peculiar to plasmas and related to its discrete nature, which
include the interaction and trapping of particles nearly resonant
with large amplitude waves, and interactions of particles with non
linear beatings between two (or more) propagating waves such that

$$ U = \frac{\omega(k_1) - \omega(k_2)}{|k_1 - k_2|} $$

Finally, there is an important class of so-called 'quasilinear'
processes where the source of the instability, the beam, is
removed by the scattering in velocity space of beam electrons by
large amplitude waves in the plasma.

Though we ignore collision dependent non-linearities such as
the Luxembourg effects, there is evidence that all other processes
mentioned occur in beam plasma interactions.

Quasi-linear effects

We discuss quasi-linear effects first because though the subject
has been one of considerable theoretical interest, beginning with the
original papers by Vedenov et al (1971) and Drummond and Pines (1972)
and some controversy (see e.g. Cook 1974, Grognard 1975), quasilinear phneomena are difficult to produce in the laboratory. To
achieve a regime in which regions of velocity space initially with
$\partial f/\partial v > 0$ are gently eroded by a diffusion-like process driven by
the instability wave fields, and ultimately lead to a flattening
of the beam feature and stationary,final state, requires a weakly
unstable broad spectrum of fluctuations. Earlier we saw that the
condition for a transition from fluid instability (implying
rapidly growing narrow band instabilities) to the kinetic
instability (implying lower growth rates and broader fluctuation
spectrum) could be roughly described by the condition (O'Neill
and Malmberg 1968)

$$ \frac{\Delta V_b}{V_b} \left[\frac{n_p \, v_{tP}^2}{n_b \, V_b^2} \right]^{1/3} \sim 1 $$

This condition (which itself has been subjected to detailed
criticism by Self et al, 1971), points to the fact that a hot weak
beam (like a bump in the tail)not too much separated from the
background distribution, is required. The elegant experiments by
Roberson and Gentle (1971) have verified many aspects of the
theory's prediction, but such conditions are rarely achieved in the
majority of laboratory experiments.

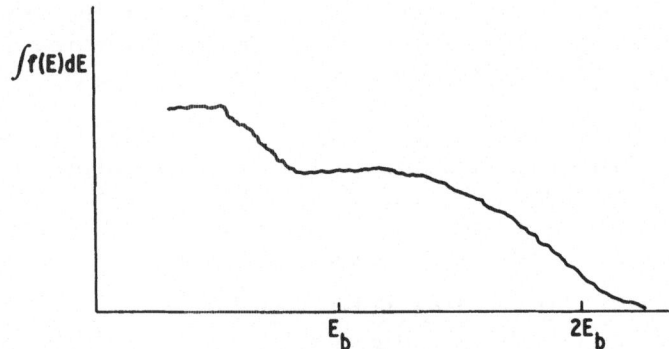

Fig.7 Integral electron energy spectrum after interaction, showing
      beam broadening and acceleration. E_b is initial beam energy.
      I_b = 7.4 mA, $f_p \sim$ 0.7 GHz, f_c = 1 GHz (Nyack et al 1974).

        This is not to say that scattering of the beam distribution
by unstable waves does not occur, but that such a process occurs
much more rapidly and more dramatically than would be predicted
in the quasi-linear framework. Strong beam scattering and
acceleration is almost always observed (see e.g.
Cabral and Hopman (1970), Nyack et al (1974)). In Fig.7 we show a
not untypical integral energy spectrum of a linear distribution,
initially quasi-monoenergetic, following the interaction. The
inferred beam scattering, including the presence of significant
numbers of electrons at twice the initial injection energy, strongly
implies some coherent process. We return to this point later.

Trapping effects

        A process which also produces saturation of the linear
instability but in almost the opposite regime to the weak
turbulence   state, is the so-called "single wave" strong
turbulence model (O'Neil et al 1971, 1972, Shapiro and Schevchenko
1971, Matisborko et al 1972). This is based on the attractive
idea that in the fluid beam plasma interaction the most rapidly
growing mode soon dominates the system so that the next stage can
be treated by considering the effects associated with a single, large
amplitude wave.

        Conceptually, what happens is as follows: the fastest growing
mode, travelling slower than the beam, has an exponentially growing
amplitude with which an electrostatic potential $\phi(t)$ is associated.
The region in velocity space which is affected by this potential has
a width $\propto \phi^{1/2}(t)$, and when this reaches the beam velocity,
i.e., when

$$ e\,\phi(t) \simeq \frac{1}{4}\,m\,(\,V_b - V_\phi\,)^2 $$

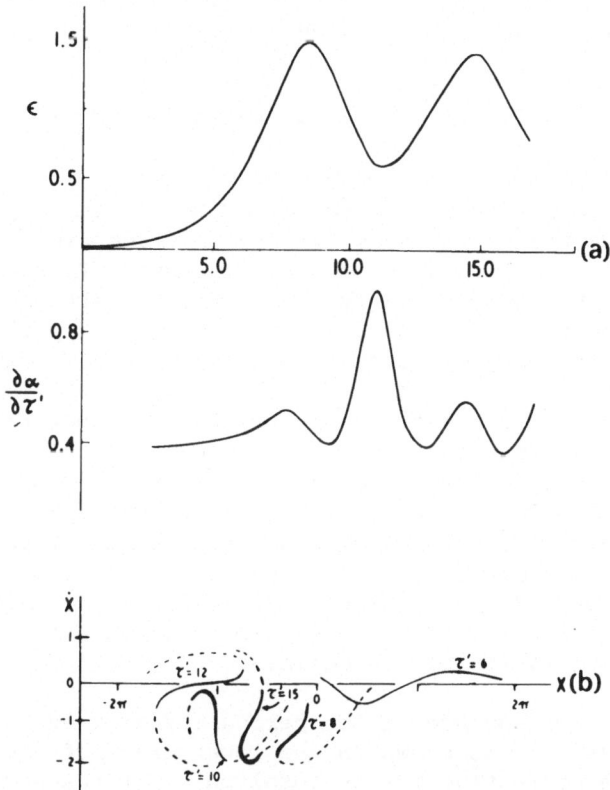

Fig. 8.   Computed prediction of single wave theory for the fastest
          growing mode for a fixed wave number k_O. (a) Top trace shows
          wave amplitude E with time τ' and lower trace the correspond-
          ing variation of $\partial\alpha/\partial\tau'$. The time τ' is in units of $\omega_O\eta^{1/3}$
          and the amplitude in units of $e\Phi/mV_b^2\ \eta^{2/3}$, where $\omega_O = k_OV_b$.
          (b) Semischematic phase space evolution of particle distri-
          bution seen in a frame moving at the original beam velocity,
          i.e., position $X = k_Ox$. Heavy lines are intended to convey
          concentration of particles and tenuous filaments have been
          partly omitted. Notice that at $\tau' \sim 8$ the particles are
          bunched in space giving rise to harmonics. The obvious
          rotation in phase space would eventually assume a double
          spiral form.

(where V_ϕ is the phase velocity of the wave) the beam is 'trapped', that is the beam electrons start to decelerate, then fall into the wave potential wells and begin to 'slosh' back and forth in the potentials, periodically exchanging energy with the wave. If the oscillation of the trapped electrons takes place at the bottom of the well, its frequency (the 'bounce frequency') is just

$$\omega_B = k \sqrt{\frac{\ell \phi}{m}} \sim \omega_p \, (n_b/n_p)^{1/3}$$

Details of such models can be found in the references given. Fig.8 shows computed results of such calculations, carried out for a one-dimensional interaction and fixed wave number. To make the transformation to the experimentally more realistic situation of a fixed frequency, the dimensionless time τ' has been multiplied by the ratio of the phase to group velocity.

The top trace of the figure shows that the amplitude grows exponentially in accordance with the linear theory up to about $\tau' = 4$, then begins to trap the beam until at $\tau' = 8.6$ the beam particles are moving backwards in the wave frame, the energy transfer to the wave having reached a maximum, so that the wave amplitude saturates. The oscillation of the particles continues until at $\tau' = 11.6$ the bulk of the particles is moving with maximum velocity in the potential well and the wave amplitude is a minimum; at $\tau' = 15$ the oscillation begins again.

A further consequence of the periodic energy exchange between waves and particles is shown in the lower trace of the figure. Here the time derivative of the total phase of the wave can be seen also to undergo periodic variations in time at twice the frequency of the wave amplitude. Since the wave number is fixed in the calculations, this can be related to variations in the wave phase velocity by

$$V_p = V_b \left(1 - \frac{\partial \alpha}{\partial \tau'} \eta^{1/3}\right), \quad \eta = n_b/n_p$$

The theory shows that $\partial \alpha/\partial \tau'$ is a quadratic function of the average particle velocity, so the phase velocity has a deep minimum when the wave amplitude is a minimum, and increases with increasing wave amplitude, though with a subsidiary minimum at amplitude maximum. Finally, Fig. 8 shows the positions of the beam electrons at several times in the trapping process in a frame moving with the beam velocity.

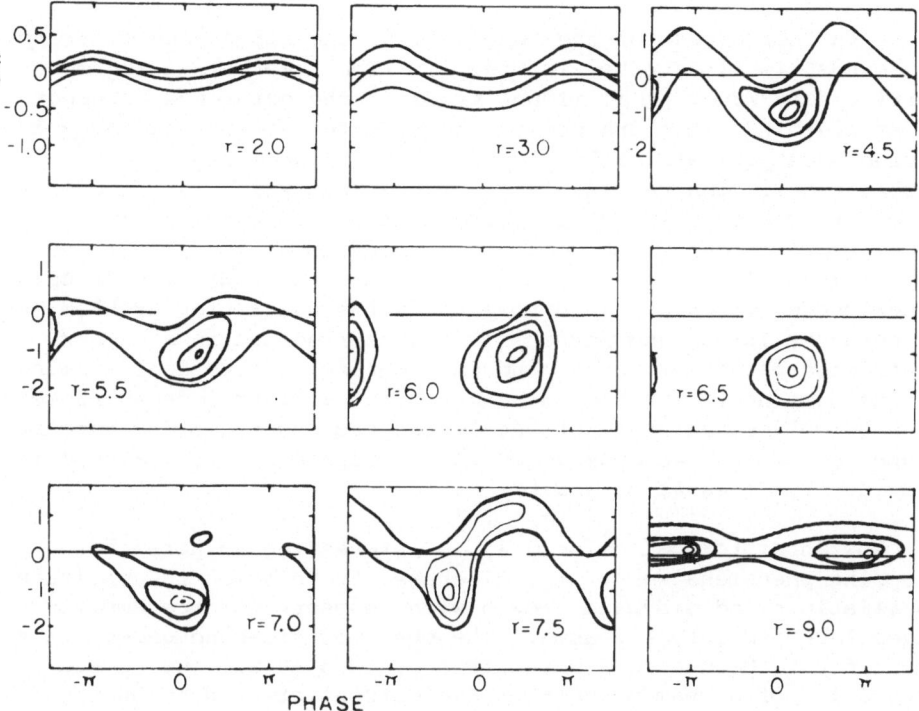

Fig.9 Experimental phase space distributions at various stages in
     trapping process. For comparison with fig. 8, $\tau = 1.4\tau'$.
     (Gentle and Lohr 1973).

Many observations of beam plasma interaction phenomena have been interpreted in terms of the single wave model (van Wakeren and Hopman 1972, Carr et al 1972, Gentle and Lohr 1972, Mizuno and Tanaka 1972, Nyack and Christiansen 1974 a,b.)

In Fig. 9 we show an experiment, due to Gentle and Lohr, of phase space rotation of the trapped particle distribution.

In Fig. 10 the phase velocity variation of nonlinear wave number shift is compared with measured amplitude variation or bounce length.

A further aspect of the interaction, the appearance of trapping sidebands, is illustrated in Fig.11. The theoretical argument initiated by Kruer et al (1969) predicts the existence of upper and lower sidebands at a frequency obeying the condition $\omega - k V_\phi \sim \pm \omega_B$ having been parametrically pumped by the trapped electron oscillations. The results, even with a beam which is premodulated in order to ensure spectral purity of the main wave, contain features which cannot be completely understood in terms of the simple model (see e.g. Nyack and Christiansen 1974). The trapping experiments, which appeared to explain the saturation problem, were, as noted by Throop and Parker (1979), carried out in a rather restricted parameter range, and in a careful theoretical examination of the pump depletion brought about by a parametric decay process (see below) emphasised that the latter process might dominate at higher densities and beam velocities. Corroborative evidence is provided by Jones et al (1976).

Now an historical aside: it is interesting to note that trapping phenomena were first discussed in relation to amplitude oscillations and sideband growth with respect to large amplitude waves launched into a plasma. The observed sideband growth is now known not to be due to a parametric process at all but to the generation of a beam of untrapped electrons created by energy exchange during the initial damping of the large wave - a non-linearly induced, linear beam plasma instability (Frankling and MacKinlay, 1976).

Non linear wave - wave interactions

Another elementary non-linear process can occur in a beam plasma system in which the situation is initially dominated by a "single" strong high frequency instability - the so-called parametric decay process. In the simplest case, i.e. that in which the strong pump (ω_1, k_1) couples two initially low amplitude linear waves (ω_2, k_2) and (ω_3, k_3), the well known conservation relations

$$\omega_1 = \omega_2 + \omega_3 \quad , \quad \underline{k}_1 = \underline{k}_2 + \underline{k}_3 \quad , \quad \varepsilon_i(\omega_i, k_i) = 0$$

are obeyed and the electric potentials of the coupled waves

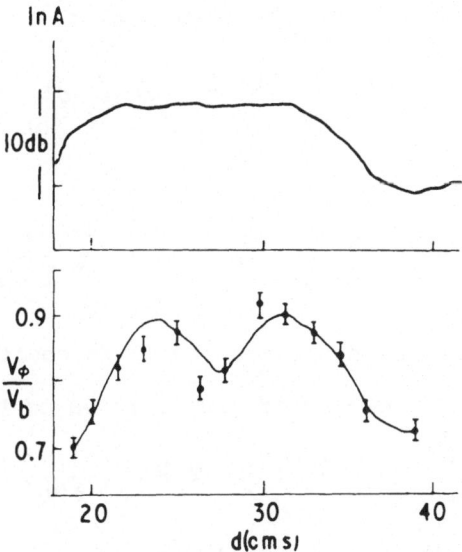

Fig.10 Observations of amplitude (upper) and phase velocity (lower)
       variations over one bounce length, as predicted in fig.8.
       Premodulated beam, I_b = 0.8 mA, V_b = 200 ev, f = =25 MHz,
       f_{ci} = 0.5 GHz. (Nyack et al 1974).

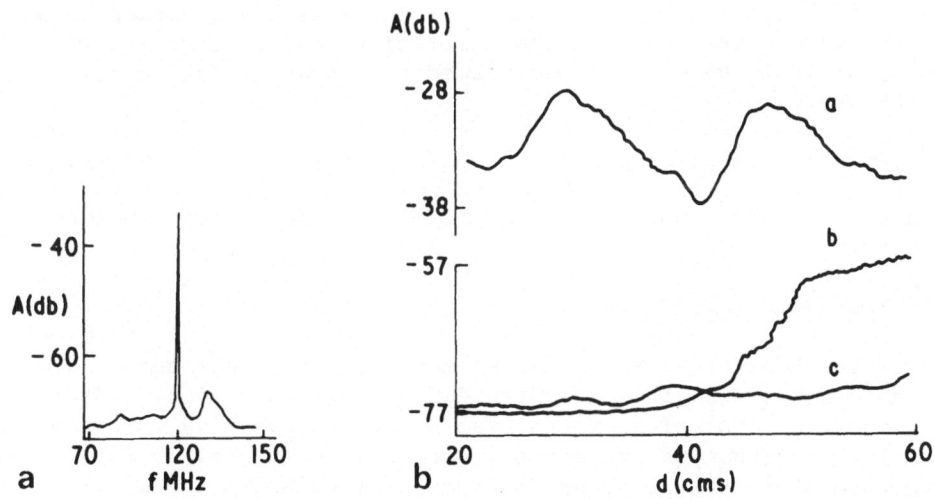

Fig. 11.   [a] Frequency spectrum of the main wave with upper and
           lower trapping sidebands. [b] Amplitude of main (premod-
           ulated) wave (a), f = 96 MHz, and upper (b) f = 132 MHz,
           and lower (c) f = 72 MHz sidebands, as a function of dis-
           tance. I_b = 1.1 mA, V_b = 250 V, f_{ce} = 500 MHz.

are governed, in the absence of damping, by :

$$\frac{\partial \phi_2}{\partial t} = -2i \Gamma_{123} \phi_1 \phi_3^* \Big/ \left(k_2^2 \frac{\partial \varepsilon}{\partial \omega_2} \right)$$

$$\frac{\partial \phi_3}{\partial t} = -2i \Gamma_{123} \phi_1 \phi_2^* \Big/ \left(k_3^2 \frac{\partial \varepsilon}{\partial \omega_3} \right)$$

where Γ_{123} is the non-linear coupling coefficient predicting Q - L non-linear growth rates of the decay products

$$\gamma_{NL} = 2 \Gamma_{123} (\phi_1 \phi_1^*)^{1/2} \Big/ k_2 k_3 \left[\left(\frac{\partial \varepsilon}{\partial \omega_2} \right) \left(\frac{\partial \varepsilon}{\partial \omega_3} \right) \right]^{1/2}$$

(see Weiland and Wilhelmsson (1978) for details.)

     That such processes occur in beam plasma interactions is not in
doubt, as can be seen in Fig.12, in which the verification of the
frequency and wave number conditions is shown for the decay of a
forward propagating beam-driven plasma wave to a backward plasma
wave and forward ion acoustic wave. It has, however, been difficult to pursue the basic three-wave scenario much further in the
laboratory despite the observation of other couplings (Hopman 1971,
Nyack and Christiansen 1975, Boswell and Giles 1977, Parker 1973).
to the stage where the coupling coefficient can be compared with
theory or that unequivocal measurements of pump depletion effects
can be observed.

     The much weaker processes of non-linear wave-wave and wave
particle interactions in the random phase approximation have not yet
been observed in laboratory beam plasma experiments for obvious
reasons.

Self Modulation and Cavitation

     It is clear that the rather elementary non-linear and quasi-
linear processes discussed so far do not individually account for
the observed saturation of the beam plasma instability, nor the
particle acceleration processes , nor could they provide for the
long-term stabilization over interaction lengths in excess of
those currently employed. The experiments do show however that in
restricted parameter regimes certain dominant processes can be
highlighted. An individual mechanism,such as the apparently
stable beam trapping process, in the longer term might well be
destroyed by the growth of trapping sidebands which subsequently
de-trap and scatter the beam distribution. Three wave processes

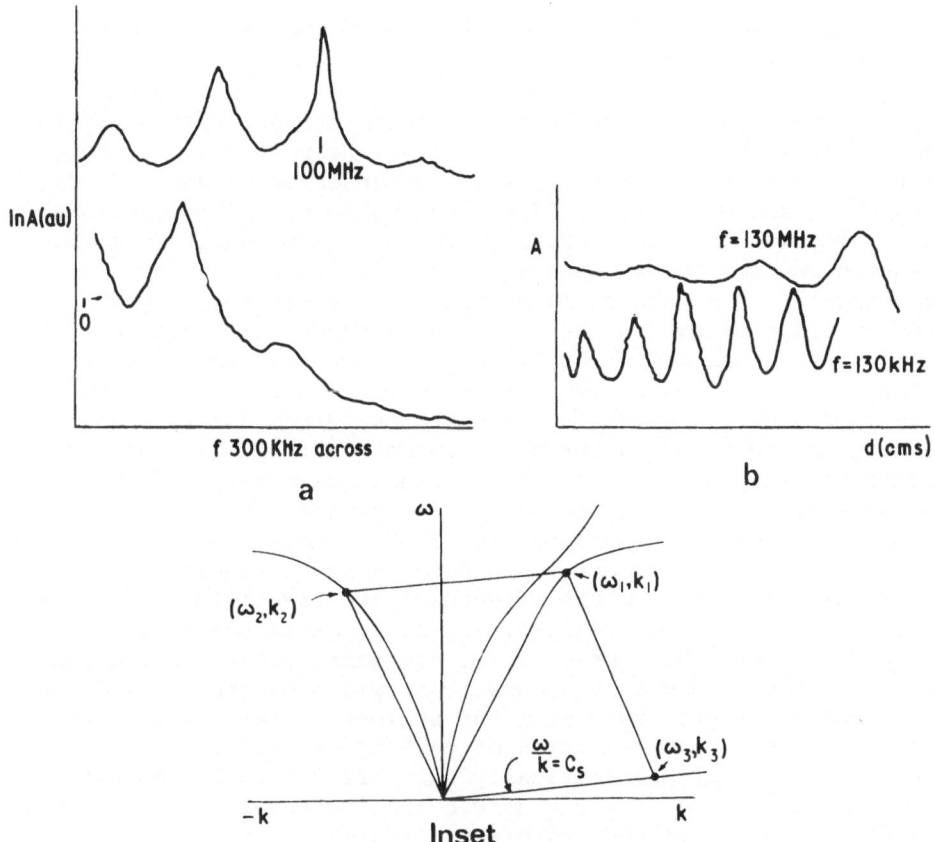

Fig. 12. (a) Frequency spectra recorded at high (upper trace) frequency showing large wave at 100 MHz, and lower (decay)
sidebands. The lower trace shows the spectrum at low
frequencies. (b) Interfermeter traces of main wave, and
low frequency wave recorded under slightly different conditions.
Inset: Schematic diagram of decay interaction. The ion
sound speed has been exaggerated for illustrative purposes.

will ultimately be dominated by higher order wave-wave and wave-wave
particle interactions as the coupled wave amplitudes increase. This
would both diffuse the beam distribution and also, by reabsorption
of the product waves, lead to heating of the background distribution.
The likelihood is of course that the various non-linear effects may
occur simultaneously, and the situation is further complicated by
finite interaction length effects due to plasma inhomogeneities
and energy transfer by both particles and waves out of the interaction region.

    In view of the intractibility of these various approaches to the
problem of explaining the observed stationary status of strong
electron (Langmuir) wave turbulence, an interesting and different
approach to the problem is offered by the so-called modulational
instability (Vedenov and Rudakov, 1965) and subsequent formation
of what have come to be known as 'cavitons' (Wong, 1977).
Theoretically, the problem is put rather clearly in the weak
turbulence approximation where it can be shown that, even
considering higher order non-linearaties, wave energy tends to
collect in a small wave number range of k   . In these regions, the
waves have high phase velocity and are therefore not in contact with
the body of the plasma distribution. This so-called 'Langmuir
condensate' (Tsytovich, 1971, 1977) cannot be dissipated and
therefore offers no prospect for the existence of an equilibrium
turbulence condition. However, it has been shown theoretically
that this small wave-number condensate is modulationally unstable,
the process being due to the effects of the so-called pondermotive
or Miller force (Gabonov and Miller, 1958; Motz and Watson,1967;
Wong and Schmidt, 1973) which acts, essentially, as a radiation
pressure force. This force tends to displace electrons (and ions
by ambipolar effects) away from the regions of large wave intensity
and hence leads to the formation of caverns or cavities (cavitons)
within the plasma. Mathematically this effect can be understood
from a consideration of the electron and ion equations of motion,
from which can be derived (Zakharov, 1972);

$$\frac{\partial^2 \underline{E}}{\partial t^2} + \omega_{pe}^2 \underline{E} + c^2 (\nabla \wedge \nabla \wedge \underline{E}) - \frac{\gamma_e T_e}{m_e} \nabla (\nabla \cdot \underline{E}) = -\frac{\partial n}{n_0} \omega_{pe}^2 \underline{E} \qquad (4)$$

and

$$\frac{\partial^2 (\partial n/n_0)}{\partial t^2} - \frac{\gamma_e T_e}{M_i} \cdot \nabla^2 \cdot \frac{\partial n}{n_0} = \frac{\nabla^2 \langle \underline{E}^2 \rangle}{8 \pi n_0 M_i^2} \qquad (5)$$

γ_e, T_e and m_e denote the electron adiabatic exponent, temperature
and mass respectively, $\underline{E}$ the wave field and M_i the ion mass with
$\partial n/n_0$ being a measure of the plasma density perturbation (is
the equilibrium number density). $\langle \cdots \rangle$ represents averaging over
the 'fast time' of $\sim 1/\omega_{pe}$ and thus takes into account the effect
of the high frequency waves acting on the average (slow time)

characteristics of the plasma. The force associated with the potential $\langle \underline{E}^2 \rangle$ in equation (5) is the ponderomotive force discussed $\overline{8\pi n_0}$ above. From these equations, Zakharov (1972) obtained the so-called non linear Schrödinger equation which (in dimensionless form) can be expressed as :

$$i \frac{\partial \underline{\varepsilon}}{\partial t} - (\underline{\nabla} \wedge \underline{\nabla} \wedge \underline{\varepsilon}) + \underline{\nabla}(\underline{\nabla} \cdot \underline{\varepsilon}) - \underline{\varepsilon} |\underline{\varepsilon}^2| = 0$$

$$(\langle \underline{E}^2 \rangle) = \frac{1}{2} |\underline{\varepsilon}^2| \qquad (6)$$

with the equivalent wave function and potential well being represented by $\underline{\varepsilon}$ and the density cavity respectively. Equations (5) and (6) are collectively known as Zakharov's equations (Thornhill and ter Haar, 1978).

Zakharov (1972,75) also demonstrated that the self modulating process could lead to singular behaviour in systems of higher than one dimension, that is, the region of enhanced field becomes progressively more peaked until it collapses into spatially very small objects. In these cavitons the wave number has increased so much that the original Langmuir condensate is now able to interact with the thermal electrons and be dissipated while stimulating suprathermal electron tails. This provides a sink for the injected energy and therefore offers the possibility of a state of equilibrium turbulence.

The significance of the often quoted strong turbulence parameter W, the ratio of the wave to particle energy densities can be seen from the full 1-D form of (6)

$$i \frac{\partial \langle \underline{E} \rangle}{\partial t} + \frac{3}{2} \frac{v_{tP}^2}{\omega_P} \frac{\partial^2}{\partial x^2} \langle \underline{E} \rangle = \frac{\omega_P}{2} \frac{\partial n}{n_0} \langle \underline{E} \rangle \qquad (7)$$

Neglecting the first term is (5) gives $\partial n / n_0 \sim \frac{\langle E^2 \rangle}{4\pi n_0 kT} = W$
and using the expression in (7) it can be seen that the non-linear term becomes important when $W \geq (k\lambda_D)^2$ and strong turbulence effects can be anticipated. Details of the behaviour to be expected for various values of W are discussed by Buchelnikova and Motochkin (1981), the reviews previously cited, and by Smith and Nicholson (1979), this last in the context of Type III radio bursts. It is interesting to note that the oscillating two-stream instability, a process closely related to self modulation (Lashmore-Davies 1975) was considered rather early by Papadopoulos et al (1974) as an explanation for the mysterious persistence of electron streams of solar origin.

For typical laboratory parameters e.g. $n_p \sim 3 \times 10^9 \, cm^{-3}$, $T_e \sim 5 \, eV$
$_b$ 10^9 cm/sec, $\delta n /n$ $10^{-2} - 10^{-1}$ the condition $W \gtrsim$
is easily fulfilled by beam-driven waves with amplitudes of a few V/cm. A further condition $(k\lambda_D)^2 < m_e/m_i$

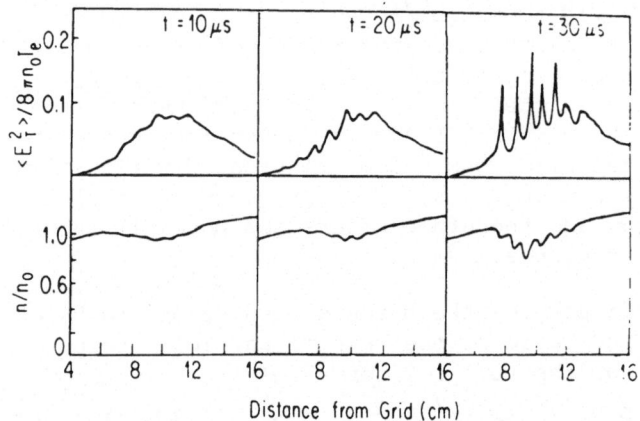

Fig. 13   Spatial profiles of caviton RF fields and density perturb-
          ations against distance from beam source. Note spiky
          profiles developing after t = 30 μs > ω_{pi}^{-1}. (Wong 1977).

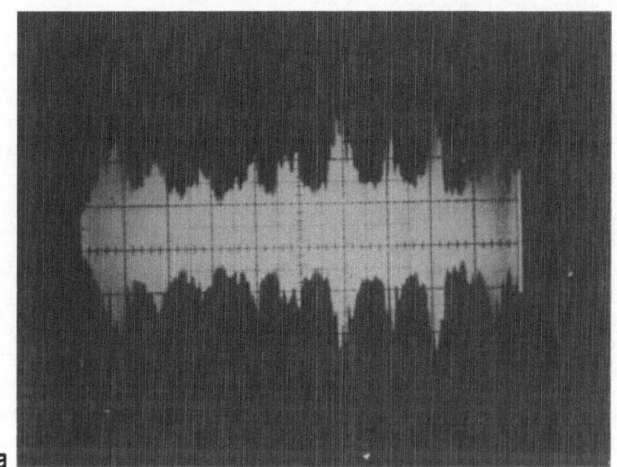

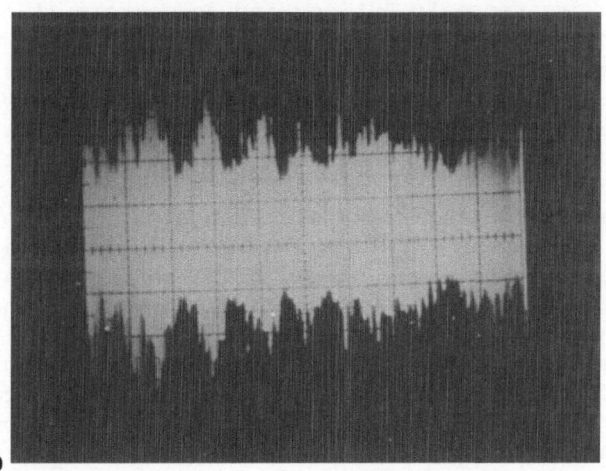

Fig. 14   Real time wave burst structure (a) scale = 2 μs/div.
          (b) 200 μs/div. Burst repetition time $\tau > f_{pi}^{-1}$ and
          $\tau < f_{ci}^{-1}$. (Christiansen et al 1981).

which forbids the occurrence of parametric decays may, however, not be obeyed, but these decays, both resonant and non resonant, can set up regions of standing wave interference between the initial pump ($\underline{k}_o$) and the backscattered electron wave (- $\underline{k}_o$) from which cavitation can proceed. This is a characteristic feature of many of the experiments considered briefly below.

Experimental studies of cavitational phenomena in the laboratory have been carried out for a number of years by the UCLA group (Wong 1977,1979) using both launched and beam excited waves in a large (essentially one-dimensional) plasma machine. In Fig.13 we show measurements of simultaneous density depletions and associated high frequency wave intensity enhancements (Wong and Quon 1975), the intensity peaks are separated by half the initial wave length of the beam-driven wave; they appear on timescales of the order of ion plasma frequencies in the system. The collapsed wave amplitudes are measured by an electron beam technique which gives peak values of $E \sim O (10's \ V/cm)$ depending on the system parameters. Electron acceleration is also observed in these experiments, up to 1.5 V_b or twice the initial beam energy. Similar results have also been reported by Ikezi et al (1976) the characteristic of both experiments is the approximately one-dimensional nature of the system, and the use of premodulation of the injected beam to produce a spectrally pure pump. Density inhomogeneities may help the initial backscatter. More recently Antipov and co-workers (1978,1979) working in a plasma that was essentially collisionless but strongly magnetzied (and therefore quasi one-dimensional) demonstrated strong self compression of beam-driven Langmuir waves with self consistent density depressions of up to 30%. The axial scale length of these depressions is of the order of a few Debye lengths and the structures, separated by $\lambda/2$ of the pump,were within experimental accuracy stationary with respect to the moving plasma. The numerical experiments of Buchelnovika and Matochkin (1981 and references therein) show many of the features demonstrated by the experiments including the production of separate thermal electron tails. Leung and Wong (1978), have also considered this effect and demonstrate that particles crossing a region of intense collapse fields can be significantly accelerated or decelerated depending on their initial phase. The calculation, though far from self consistent, certainly predicts strong beam scattering.

More recently the little considered (Dysthe et al, 1978) phenomenon of three-dimensional collapse has been demonstrated in beam plasma interactions. The results shown in Fig.14 demonstrate the well known phenomenon (see also Cabral et al 1978) that natural beam plasma interactions are extremely spikey in real time and that these quasi repetitive timescales are of the order of natural ion modes in the system. If one conducts experiments designed to examine the spatial distribution of these intensive wave field bursts, it can be seen that they are very strongly confined to the

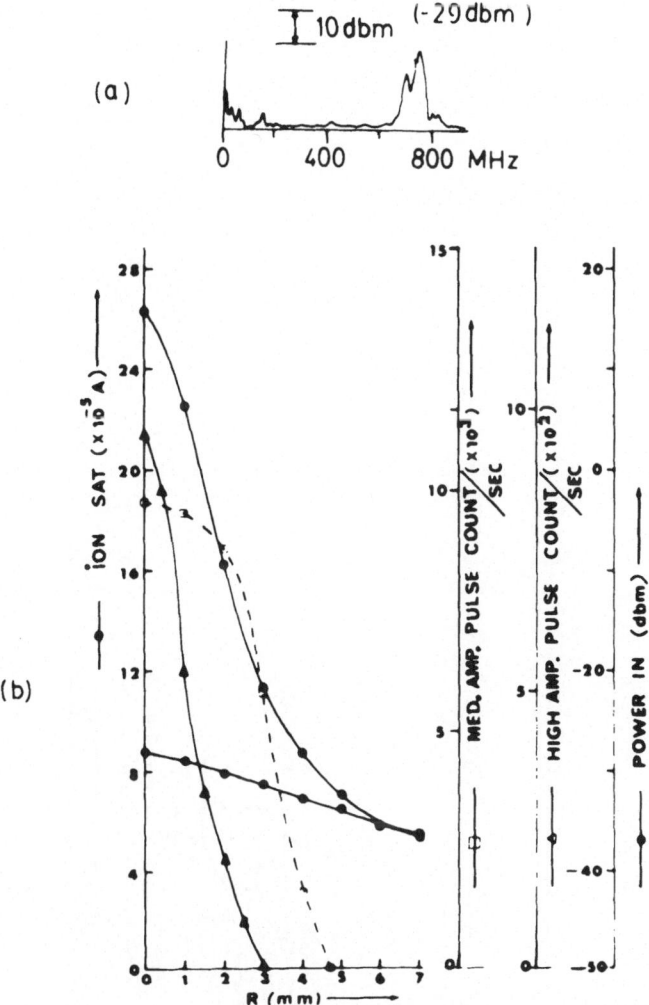

Fig. 15   (a) Frequency spectrum of upper hybrid instability.
          (b) Comparison of large and medium pulse recurrence rates
          with average power and density profile. Significant
          radial confinement is seen. (Christiansen et al 1981).

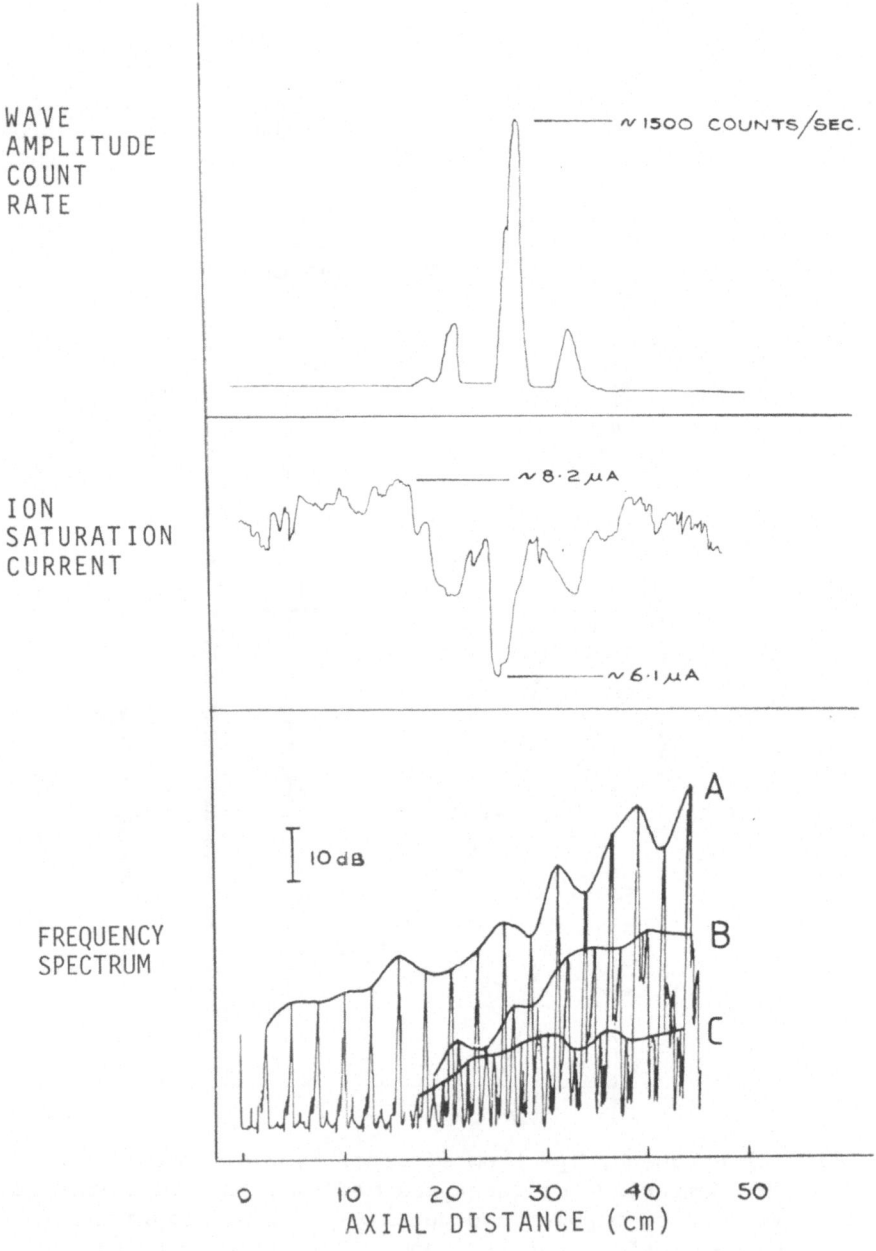

Fig. 16 Composite picture showing standing Langmuir waves, density
         variation and wave frequency spectra in an electron beam
         excited plasma. A more detailed discussion can be found
         in the text.

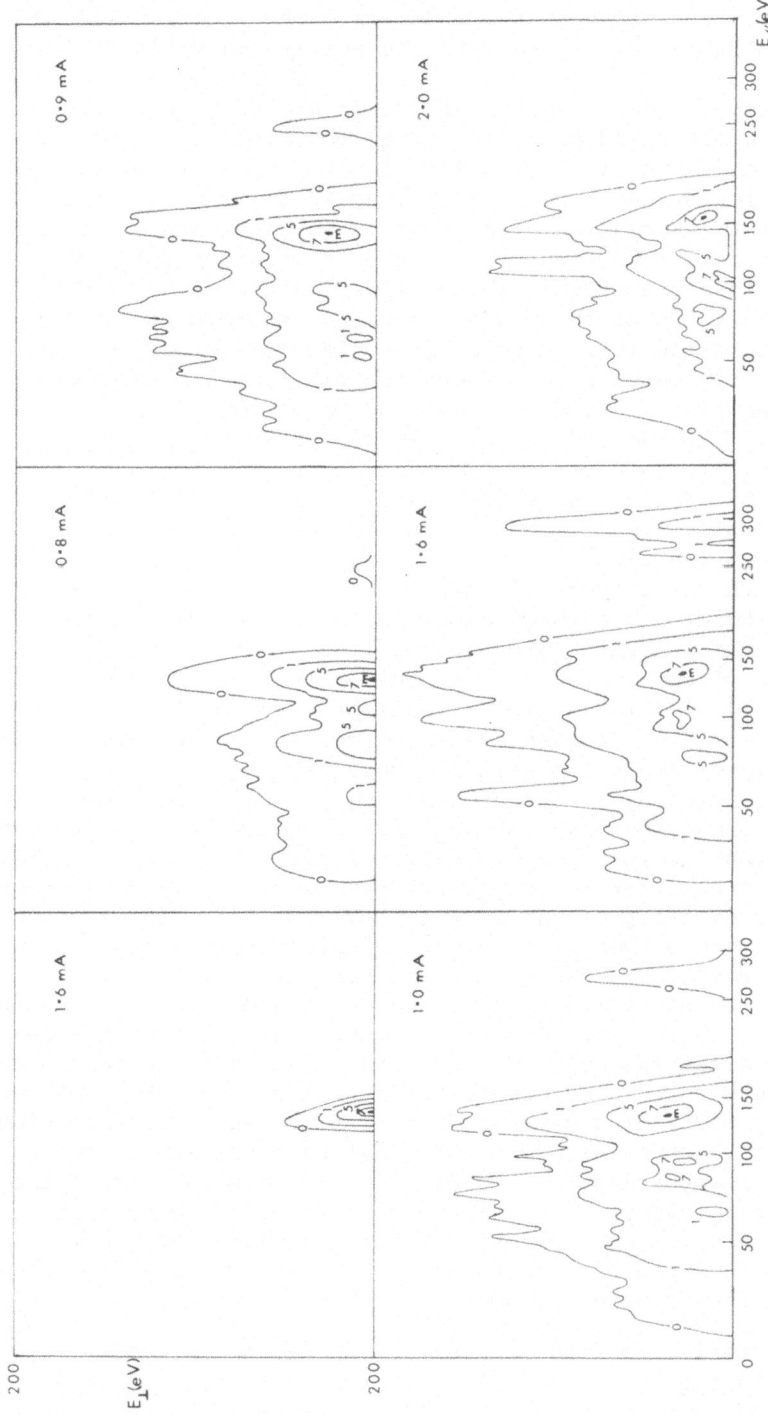

Fig. 17 Two dimensional energy scattering of beam electrons as a function of beam current. m represents maximum current and 7,5,1,0 represent 25%, 50% and 1% of maximum.

central axis of a thin beam plasma system (Christiansen et al 1981) with radial scale sizes of the order of a few electron larmor radii, implying collapse transversely to the applied magnetic field, Fig. 15

These experiments have recently been extended to a larger system and similar radial collapse was observed. A composite picture is shown in Fig. 16, which, in the top part, gives the axial location of the largest amplitude Langmuir wave spikes. The trace shows the formation of a standing wave pattern which, for different plasma conditions, was found to occur at different positions along the axis of the machine. The middle picture shows the axial variaton of the plasma density, recorded by monitoring the ion saturation current to a biased Langmuir probe, and a clear correspondence can be seen between the position and amplitude of the standing wave and the position and magnitude of the density variation. This region also displays an enhanced electron temperature. In this case, $\partial n/n_0$ can be seen to be 25%. The final part of Fig. 16 shows the axial variation of the frequency spectrum of plasma waves and consists of seventeen compressed frequency spectra from 0-200 MHz. The important point to be noted here is the increase in plasma wave activity (curves B and C) coinciding with the large amplitude "hot" region. Curve A represents the initial 'pump' plasma wave which results from the beam-driven plasma instability discussed earlier while B and C can be related to additional linear modes promoted by features in the scattered beam distribution discussed below. The effect of this localized large amplitude standing wave region on the beam electrons is clearly illustrated in Fig. 17 which shows the contours of the two dimensional variation in beam energy (related to the two dimensional velocity distribution function) for various beam currents. Except for the top left distribution, all were recorded with a neutral gas pressure of $\sim 10^{-4}$ Torr, top left distribution being taken at $\sim 10^{-6}$ Torr. which therefore effectively shows the variation of the beam energy in vacuo and is included to illustrate the scattering of the beam by plasma effects. As the beam current is increased, a significant scattering of the beam in both parallel (axial) and perpendicular directions can be seen. The peak beam scattering occurs for a beam current of 1.6mA, with the maximum parallel and perpendicular components corresponding to an approximate doubling of the original beam energy of 150eV. This significant transverse scattering and acceleration of the beam is clear evidence for the two dimensional nature of the cavitons and gives an indication of the magnitude of the localized potentials that can be developed in the plasma as a result of this highly non-linear wave phenomenon.

Acknowledgements We wish to acknowledge the Sociéte Francaise de Physique, Centre National de la Recherche Scientifique, American Institute of Physics and the D. Reidel and North Holland Publishing companies for permission to use figures.

REFERENCES

Allen, J.E. and Phelps, A.D.R., 1977, Rep. Prog. Phys. 40: 1305.

Antipov, S.V., Nezlin, M.K., Snezhkin, E.N., Trubnikov, A.S., 1978, Sov. Phys. JETP 47: 506.

Antipov, S.V., Nezlin, M.K., Snezhkin, E.N., Trubnikov, A.S., 1979, Sov. Phys. JETP 49: 797.

Apel, J.R., 1969, Phys. Fluids 12: 291.

Bhatnagar, V.P., Van Oost, G., Messiaen, A.M. and Van den Plas, P.E., 1976, Plasma Phys. 18: 525.

Boswell, R.W., 1975, Nature (London), 258: 58.

Boswell, R.W. and Giles, M.J., 1977, Phys. Rev. Lett. 39: 277.

Briggs, R.J., 1964, 'Electron Stream Interaction with Plasmas', (MIT Press: Massachusetts).

Buchelnikova, N.S. and Matochkin, E.P., 1981, Plasma Phys. 23: 35.

Cabral, J.A.C., Silva, M.E.F. and Varandas, C.A.F., 1978, 20: 21.

Cabral, J.A.C., 1971, Thesis, F.O.M. Netherlands.

Carr, W., Boyd, D., Lin, H., Schmidt, G. and Seidl, M., 1972, Phys. Rev. Lett. 28: 662.

Christiansen, P.J., Jain, V.K. and Stenflo, L., 1981, Phys. Rev. Lett. 46: 1333.

Conrad, J.R., Walsh, J.E., Diaz, C.J. and Freese, K.B., 1973, Phys. Rev. Lett, 30: 827.

Cook, I., 1974, 'Plasma Physics', ed. B.E. Keen (Inst. of Phys., London).

Crawford, F.W. and Kino, G.S., 1961, Proc. I.R.E., 49: 1767.

Drummond, W.E. and Pines, D., 1962, Nucl. Fusion Suppl. 3: 1049.

Dysthe, K.B., Mjolhus, E., Pecseli, H.C. and Stenflo, L., 1978, Plasma Phys. 20: 1087.

Fedorchenko, V.D., Mazalov, Yu.P., Bakai, A.S., Pashchenko, A.V. and Rutkevich, B.N., 1973, JETP Lett. 18: 281.

Franklin, R.N., 1977, Rep. Prog. Phys. 40: 1369.

Franklin, R.N. and MacKinlay, R.R., 1976, Phys. Fluids 19: 173.

Gabonov, A.V. and Miller, M.A., 1958, Sov. Phys. JETP 7: 168.

Gentle, K.W. and Lohr, J., 1973, Phys. Fluids 16: 1464.

Gentle, K.W. and Roberson, C.W., Phys. Fluids 14: 2780.

Grognard, R.J.M., 1975, Aust. J. Phys. 28: 731.

Hopman, H.J., 1971, Proc. 10th Int. Conf. Ionized Gases, (Oxford, Parsons) 323.

Hopman, H.J., Mattiti, T. and Kistemaker, J., 1968, Plasma Phys. 10: 1051.

Ikezi, H., Chang, R.P.H. and Stern, R.A., 1976, Phys. Rev. Lett. 36: 1047.

Jain, V.K. and Christiansen, P.J., 1981, Phys. Lett. 82A: 127.

Jones, R.W., Carr, W. and Seidl, M., 1976, Phys. Fluids 19: 607.

Krall, N.A. and Trivelpiece, A.W., 1973, 'Principles of Plasma Physics' (McGraw Hill: London).

Kruer, W.L., Dawson, J.M. and Sudan, R.N., 1969, Phys. Rev. Lett. 23: 838.

LeQueau, D., Pellat, R. and Saint Marc, A., 1980, Ann. Geophys. 36: 433.

Leung, P., Tran, M.W. and Wong, A.Y., 1978, A.P.S. Bull. 23: 844.

Maggs, J.E., 1978, J. Geophys. Res. 83: 3173.

Malmberg, J.H. and Wharton, C.B., 1969, Phys. Fluids 12: 2660.

Matsiborko, N.G., Onischenko, I.N., Shapiro, V.D. and Shevchenko, V.I., 1972, Plasma Phys. 14: 591.

Mikhailovskii, A.B., 1974, 'Theory of Plasma Instabilities' (Consultants Bureau).

Mizuno, K. and Tanaca, S., 1972, Phys. Rev. Lett. 29: 45.

Motz, H. and Watson, C.J.H., 1967, Adv. Electron. 23: 153.

Nyack, C.A., 1973, Unpublished D. Phil Thesis (Sussex).

Nyack, C.A., Christiansen, P.J. and Martelli, G., 1974, in 'Magnetospheric Physics', ed. B.M. McCormac (D. Reidel) 215.

Nyack, C.A. and Christiansen, P.J., 1974, Phys. Fluids 17: 2025.

Nyack, C.A. and Christiansen, P.J., 1974, Phys. Lett. 48A: 191.

Nyack, C.A. and Christiansen, P.J., 1975, Plasma Phys. 17: 355.

O'Neil, T.M. and Malmberg, J.H., 1968, Phys. Fluids 11: 1754.

O'Neil, T.M., Winfrey, J.M. and Malmberg, J.M., 1971, Phys. Fluids 14: 1204.

Papadopoulos, K., Goldstein, M.L. and Smith, R.A., 1974, Astrophys.J. 190: 175.

Parker, R.R. and Throop, A.L., 1973, Phys. Rev. Lett. 31.

Quon, B.H. and Wong, A.Y., 1976, Phys. Rev. Lett. 37: 1393.

Roberson, C. and Gentle, K.W., 1971, Phys. Fluids 14: 2462.

Rudakov, L.I. and Tsytovich, V.N., 1978, Phys. Rep. 40: 1.

Sagdeev, R.Z., 1979, Rev. Mod. Phys. 51: 1.

Shustin, E.G., Popovich, V.D. and Kharchenko, I.F., 1969, Sov. Phys. Tech. Phys. 14: 745.

Self, S.A., Shoucri, M.M. and Crawford, F.W., 1971, J.Appl. Phys. 42: 704.

Shapiro, V.D. and Shevchenko, C.I., 1971, J.E.T.P. 33: 555.

Shustin, E.G., Popovich, V.D. and Kharchenko, I.F., 1969, Sov. Phys. Tech. Phys. 14: 745.

Smith, D.F. and Nicholson, D.R., 1979, in 'Wave Instabilities in Space Plasmas', ed. P. Palmadesso and K. Papadopoulos (D. Reidel) 1225.

Stenzel, R.L., 1977, J. de Phys. 38: C6-89.

Stenzel, R.L., 1978, Phys. Fluids 21: 53.

Strangeways, R.J. and Dungey, J.W., 1976, Planet Space Sci. 24: 731.

Sugaya, R., Sugawa, M. and Nomoto, H., 1976, Phys. Lett. 56A: 458.

ter Haar, D. and Tsytovich, V.N., 1981, Phys. Rep. (in press).

Throop, A.L. and Parker, R.R., 1979, Phys. Fluids 22: 491.

Thornhill, S.G. and ter Haar, D., 1978, Phys. Rep. 43: 43.

Tsytovich, V.N., 1971, 'Non-linear Effects in Plasmas' (Plenum: London).

Tsytovich, V.N., 1977, 'Theory of Turbulent Plasmas' (Plenum: London).

Van Wakeren, J.M.A. and Hopman, H.J., 1972, Phys. Rev. Lett. 28: 295.

Vedenov, A.A. and Rudakov, L.I., 1965, Sov. Phys. Doklady 9: 1073.

Vedenov, A.A., Velikhov, E. and Sagdeev, R.Z., 1962, Nucl. Fusion
    Suppl. 2: 465.
Vermeer, A., Matiti, T., Hopman, H.J. and Kistemaker, J., 1967,
    Plasma Phys. 9: 241.
Wall, D.N., Edgley, P.D. and Franklin, R.N., 1981, Plasma Phys.
    23: 145.
Weiland, J. and Wilhelmsson, H., 1977, 'Coherent Non-Linear Inter-
    actions of Waves in Plasma', Pergamon.
Wong, A.Y., 1977, J. de Physique 38: 27.
Wong, A.Y., 1979, Comm. Plasma Phys. 5: 79.
Wong, A.Y. and Quon, B.H., 1975, Phys. Rev. Lett. 34, 1499.
Wong, A.Y. and Schmidt, G., 1973, Unpublished U.C.L.A. Rpt. No.
    PPG-151.
Yamada, M. and Owens, D.K., 1977, Phys. Rev. Lett. 38: 1529.
Zakharov, V.E., 1972, Sov. Phys. JETP 35: 908.
Zakharov, V.E., Mastryukov, A.F. and Synakh, V.S., 1975, Sov. J.
    Plasma Phys. 1: 339.

ELECTROMAGNETIC RADIATION FROM BEAM-PLASMA INSTABILITIES

R. L. Stenzel and D. A. Whelan

Department of Physics
University of California
Los Angeles, CA 90024

INTRODUCTION

The mechanism by which unstable electrostatic waves of an electron-beam plasma system are converted into observed electromagnetic
waves is of great current interest in space plasma physics. Electromagnetic radiation arises from both natural beam-plasma systems,
e.g., type III solar bursts (Papadopoulos, 1979) and kilometric
radiation (Maggs, 1978), and from man-made electron beams injected
from rockets and spacecraft (Grandal et al., 1980; Winckler et al.,
1975; Kawashima, 1980). In spite of numerous laboratory experiments
on beam-plasma instabilities, little attention has been paid to electromagnetic wave generation. Only the simplest process, i.e., the
conversion of electrostatic to electromagnetic waves in nonuniform
plasmas at the critical layer ($\omega = \omega_p$) is well established in both
theory (Piliya, 1966) and experiments (Stenzel et al., 1974). In
uniform plasmas it has been theoretically shown (Tsytovitch, 1977)
that the scattering of electron plasma waves off ion acoustic waves
and other electron plasma waves produces electromagnetic waves near
$\omega \gtrsim \omega_{pe}$ and $\omega \simeq 2\omega_{pe}$, respectively, where ω_{pe} is the electron plasma
frequency. In some three-wave interactions, where the electrostatic
plasma waves excited by the beam grow to very large amplitudes ($e\phi \simeq
kT_e$), the normal modes of the system are modified. The wave radia-
tion pressure creates growing density depressions (Wong and Quon,
1975; Morales and Lee, 1976) which lead to trapping of plasma waves,
i.e., formation of Langmuir solitons, which may collapse and thereby
emit an enhanced level of electromagnetic radiation (Goldman et al.,
1980). In spite of considerable theoretical effort such processes
have not yet been confirmed by observations. For type III solar
radio bursts the source location and propagation effects are not
sufficiently known to confirm the model (Lin et al., 1981). In

471

the present work the diagnostic difficulties encountered in space plasmas are overcome by using a large laboratory plasma. A finite diameter ($d \simeq 0.8$ cm) electron beam is injected into a uniform quiescent magnetized afterglow plasma of dimensions large compared with electromagnetic wavelengths ($\lambda_{em} \simeq 5$ cm, plasma diameter $D \simeq 1$ m, length $L \simeq 2$ m). Electrostatic waves grow, saturate, and decay within the uniform central region of the plasma volume so that linear mode conversion on density gradients can be excluded as a possible generation mechanism for electromagnetic waves. The arrangement closely models the injection of an intense electron beam into an unbounded space plasma. The observations indicate that upon injection of a cold beam ($V_b \simeq 500$ V, $kT_b < 1$ eV, $I_b \simeq 1$ A, $n_b/n_e \simeq 1\%$) electrostatic electron plasma waves ($\omega \gtrsim \omega_{pe} \simeq 2\pi \times 6$ GHz) rapidly grow and saturate, temporally within $\Delta t \simeq 10^3 f_{pe}^{-1} \simeq 200$ nsec, spatially within $\Delta z \simeq 10 \lambda_{||} \simeq 2$ cm.   Then, on a slower ionic time scale ($\Delta t \simeq 50 f_{pi}^{-1} \simeq 2$ μsec) density fluctuations grow along the beam with similar spatial properties as the Langmuir waves. Most important is the observation of electromagnetic waves which grow at the same slow rate as the ion density fluctuations, showing conclusively that their generation involves a mode coupling process. In the saturated state the electromagnetic waves are emitted in random bursts of short durations ($10 < \Delta t < 100$ nsec), while the ion density fluctuations exhibit a broad spectrum ($0 < \omega \lesssim \omega_{pi}$) with sub-millimeter correlation lengths. These observations are consistent with models of soliton collapse, but a strong level of harmonic emission ($\omega \simeq 2 \omega_{pe}$) is not observed. Polarization measurements of the electromagnetic field show $\vec{E} \parallel \vec{v}_b$ where $\vec{v}_b$ is the beam velocity. With movable dipole the source location for the emission has been identified to lie within the first few centimeters from the beam injection point. Beyond this region the beam is strongly scattered in velocity space such that the electrostatic waves decay to low intensities. Measurements of the axial beam current show that the injected current ($I_b \simeq 1$ A) penetrates only about $\Delta z \simeq 5$ cm into the plasma except during the ~ 200 nsec prior to saturation of the electrostatic instabilities.

EXPERIMENTAL ARRANGEMENT

The experimental setup is shown in Fig. 1. A pulsed magnetized discharge plasma is produced with a 1 m diam. oxide-coated cathode and the experiment is performed in the quiescent afterglow of typical parameters $n_e \simeq 5 \times 10^{11}$ cm^{-3}, $T_e \simeq 10 T_i \simeq 3$ eV, $B_o \simeq 20$ G, 2×10^{-4} Torr in H_2, He, Ar, Xe. A pulsed electron beam ($V_b \simeq 500$ V, $I_b \simeq 1$ A, $n_b/n_o \simeq 1\%$, rise time ~ 50 nsec) is injected from a small oxide cathode (0.8 cm diam.) along the axis of the 2 m long plasma at various pitch angles ($0 < \theta_{||} < 360°$). Plasma properties are derived from Langmuir probe, ion acoustic wave, and microwave

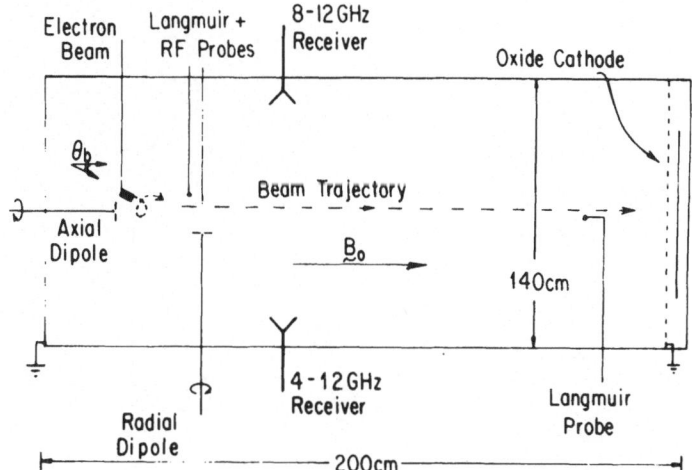

Fig. 1. Schematic diagram of the experimental setup.

interferometer measurements. The electromagnetic radiation is received either with fixed horn antennas outside the plasma column or
with movable dipole antennas inside the plasma. Coaxial rf probes
are used to detect electrostatic high and low frequency modes. The
microwave receiver consists of a broadband mixer, local oscillator,
i.f. amplifier, crystal detector, and fast oscilloscope with an overall noise figure $F \simeq 8$ dB and temporal resolution $\Delta t \simeq 10$ nsec.

OBSERVATIONS

The primary beam-plasma instability involves the excitation of
electrostatic plasma waves whose spatial amplitude distribution is
shown in Fig. 2. The initially cold beam ($kT_b \simeq 0.25$ eV) leads first
to an exponential wave growth along the beam at a spatial rate
$k_i/k_r \simeq 0.5 \ (n_b/n_o)^{1/3}$ (see insert), then to a saturation near
$z \simeq 10 \ \lambda_{||} \simeq 2$ cm when the beam distribution forms a plateau, and
finally to a wave decay when the convective loss from the finite
diameter beam-plasma volume exceeds the reduced growth rate of the
marginally unstable beam. Radially, the wave intensity decays within
a few millimeters outside the beam. Thus, intense plasma waves exist
only over limited distances in comparison with which the plasma is
uniform and unbounded. Wave reflections, which in small devices may
lead to absolute instabilities, do not exist here.

Electromagnetic radiation from the beam-plasma system is observed with microwave antennas outside the plasma (all probes removed) or with coax-fed dipoles which can be inserted radially and
axially into the plasma. Figure 3a shows the frequency spectrum as
obtained by a fast sweep of the local oscillator during one afterglow
pulse. Compared with time averaged measurements over many after-

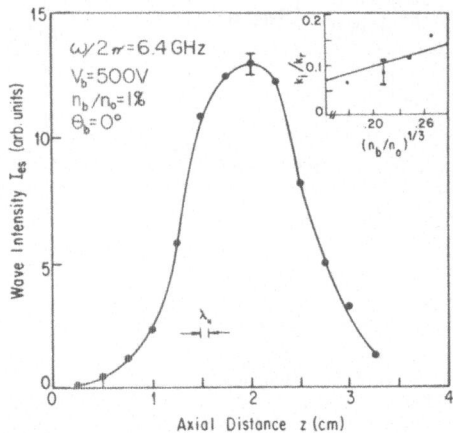

Fig. 2. Electrostatic wave intensity vs. axial density z from beam
         source. Insert shows linear growth rate at different beam
         densities normalized to the background electron density.

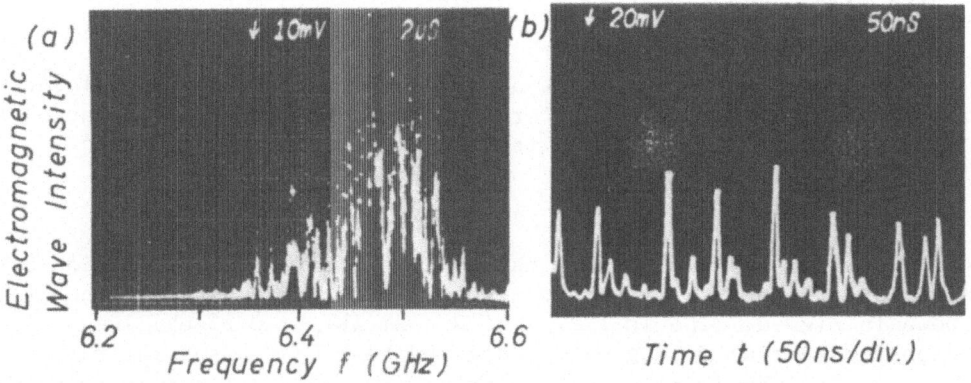

Fig. 3. (a) Frequency spectrum of the microwave emission. $\omega_{pe}/2\pi \simeq$
         5.9 GHz.
         (b) Temporal behavior of the electromagnetic radiation at
         $\omega/2\pi \simeq 6.5$ GHz showing emission in burst as short as
         $\Delta t \simeq 10$ nsec. Ar, 2×10^{-4} Torr.

glows, this technique avoids apparent spectral broadening due to
small density variations. The frequency is identified to be slightly
above the local plasma frequency ($1.1 \lesssim \omega/\omega_{pe} < 1.2$; note $\omega_{ce}/\omega_{pe} \simeq$
10^{-2}) where the most accurate density values are obtained from
electromagnetic wavelength measurements inside the plasma ($\omega_{pe}^2 =$
$\omega^2-k^2c^2$). Harmonic emission is found to be very small
$[P(2\omega_{pe}) \simeq P(\omega_{pe}) - 30$ dB$]$. Most remarkable, however, is the burst-

like emission in time (Fig. 3b) as seen at a constant receiver frequency with high time resolution (500 MHz i.f. bandwidth). Spectral width or burst duration ($\Delta t \sim \Delta f^{-1}$) indicates that the electromagnetic radiation process involves fluctuation processes on the time scale of the ion plasma frequency ($f_{pi} \simeq 25$ MHz in Ar^+).

In order to investigate the role of low frequency density fluctuations, the temporal evolution of the instability has been carefully studied (Fig. 4). With the electrostatic probe, the temporal growth and saturation of the electron plasma waves at $z \simeq 2$ cm are observed to occur within $\Delta t \simeq 500$ nsec after applying the beam current I_b.

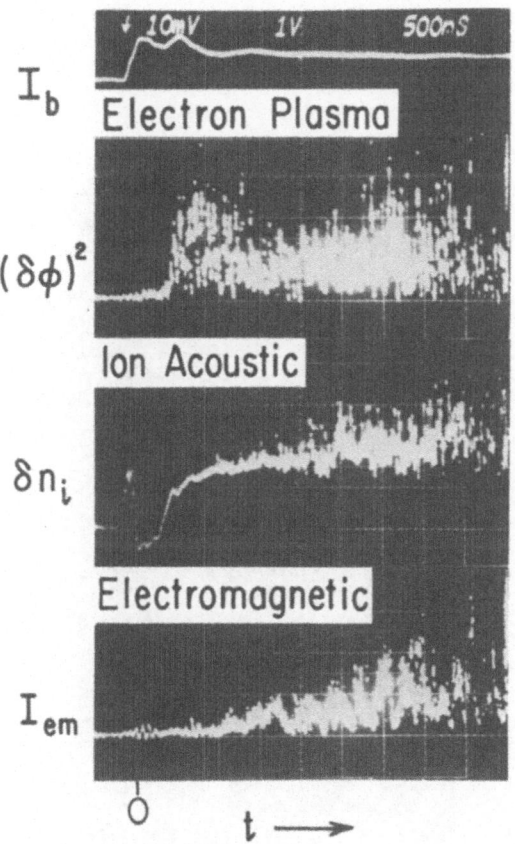

Fig. 4. Temporal growth of characteristic modes in the beam-plasma system after turn-on of the beam current I_b (top trace). Note that electromagnetic waves are generated only due to the build-up of ion acoustic turbulence in the presence of intense plasma waves. (Xe, 10^{-4} Torr).

With the same electrostatic probe, low frequency fluctuations δn_i are detected and found to grow on a slower time scale, $\Delta t \simeq 2.5$ μsec. Electromagnetic waves are generated at the same slow rate as the low frequency fluctuations, implying that there is <u>no</u> electromagnetic radiation produced by the intense plasma waves unless a high level of ion density fluctuations has built up. This behavior is confirmed by varying the ion mass as shown in Fig. 5. The growth rate of the externally observed electromagnetic radiation decreases with increasing ion mass.

The spatial distribution of the low frequency turbulence is similar to that of the electron plasma waves, i.e., the fluctuations δn_i grow along the beam, saturate and decay, axially within $z \simeq 4$ cm from the source, radially within $r \simeq 1-2$ cm from the beam. The fluctuations have been identified as ion acoustic modes by measuring both the frequency spectrum (Fig. 6a) and the cross power spectral density $C(\omega,\Delta z)$ by two-probe correlation techniques (Gekelman and Stenzel,

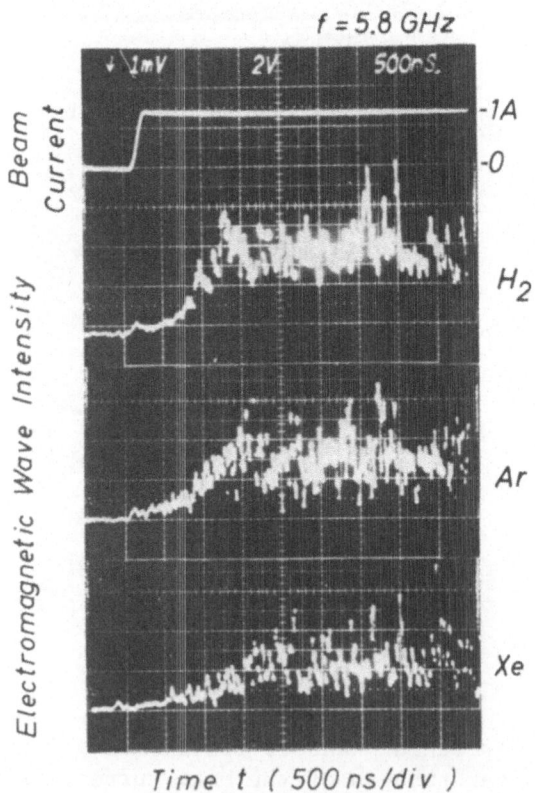

Fig. 5.   Temporal growth of the electromagnetic emission for different ion masses; hydrogen (2), argon (40), xenon (131).

1978). The spectrum falls into the range of ion acoustic waves
$(0 < \omega < \omega_{pi} \simeq 2\pi \times 25$ MHz). The dispersion $\omega(k)$ (Fig. 6b), derived from wavelength measurements at different frequency components
of the noise, follows a straight line with slope $c_s = (kT_e/m_i)^{1/2}$
where $T_e \simeq 2.8$ eV is consistent with the electron temperatures obtained from Langmuir probes. Correlation measurements are possible
only at the low frequency end of the spectrum where the wavelengths
exceed the probe dimensions (0.1 mm diam., 1.5 mm length, Debye
length $\lambda_D \simeq 0.03$ mm). Likewise, correlation functions (see insert,
Fig. 6b) are mostly taken at lower beam currents ($I_b \lesssim 0.2$ A) where
the correlation lengths [defined, for example, as the half width
of the envelope of $C(\omega,\Delta z)$] extend over a few wavelengths or millimeters. With increasing beam current, the correlation length decreases to as little as one wavelength, i.e., a state of strong turbulence is generated. While at low beam currents the sound waves
could be generated by parametric decay of Langmuir waves (Tsytovitch,

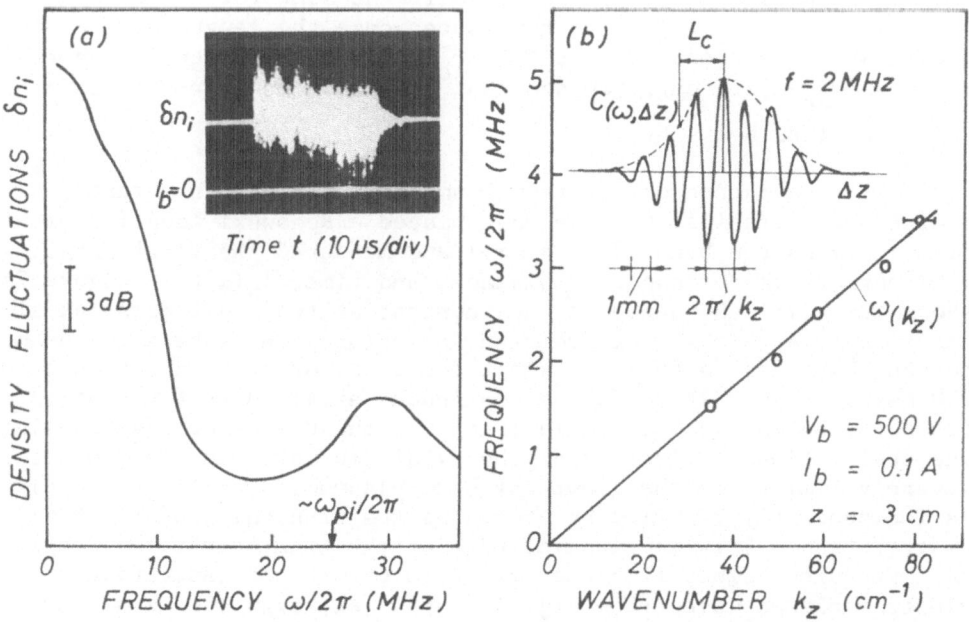

Fig. 6.   Characteristics of the low frequency turbulence generated
          by the injected electron beam. (a) Frequency spectrum,
          extending up to and beyond the ion plasma frequency, ω_{pi}.
          Insert shows noise generation only during the beam
          pulse, not at $I_b = 0$. (b) Dispersion relation of noise
          identifies low frequency modes as ion sound waves, $\omega = kc_s$.
          Insert shows typical cross power spectral function which
          yields wavenumber k_z and correlation length L_c.

1977; Stenzel and Wong, 1972), at high beam currents ($I_b \simeq 1$ A) the observed spiky turbulence is more consistent with the picture of density cavitons formed by Langmuir solitons (Morales and Lee, 1976).

The properties of the electromagnetic waves are not only inferred from remote measurements, as is common in the observation of solar radio bursts, but also from in situ measurements with movable dipole antennas (see Fig. 1). By rotating two orthogonal balanced half-wave dipoles, the polarization has been determined. Near the beam the electric field is dominantly polarized along the propagation direction of the electrostatic waves, i.e., $\vec{E} \parallel \vec{v}_b$. By moving the dipoles through the plasma the source location for the electromagnetic emission has been determined. The radiation originates from the first few centimeters beyond injection of the beam-plasma system which is the region of intense electrostatic waves. The subsequent section of the beam-plasma system, nearly 2 m long, produces a negligible contribution to the total radiation. This is independently confirmed by varying the injection angle of the beam and observing the total power received by the microwave horns in the far zone. The emission is found to be essentially the same for pitch angles $\theta = 0$ and $\theta = 90°$. Since in the latter case the length of the beam-plasma system is reduced to $L \simeq 2\pi\, r_{cb} \simeq 15$ cm (beam cyclotron radius $r_{cb} = v_b/\omega_{ce} \simeq 2.5$ cm), the dominant emission must arise from a region of length $\ell < L$.

The penetration of electron beams into plasmas is severely limited by instabilities. We have placed a Rogowski loop (17 mm diam.) around the beam (8 mm diam.) and measured the total axial beam current vs. propagation distance and time, $I_b(z,t)$. Figure 7a shows the time dependence of the current at two different axial positions. At $z \simeq 4.5$ cm from the injection point, the axial beam current has decayed to zero except for a precursor of duration $\Delta t \simeq$ 200 nsec. Figure 7b shows the continuous axial current variation for both the main pulse (solid line) and the precursor (dashed line). The observations indicate that the axial flux of beam electrons is severely limited by the growth of beam-plasma instabilities. This is independently verified by observing the broadening of the beam distribution. Only the leading edge of the beam penetrates deeply as a precursor since it occurs prior to growth and saturation of electron plasma waves (see Fig. 4, top). Although the axial current vanishes it does not imply a divergence in the current density, resulting in a build-up of space charges and subsequent current cutoff. At both the beam source and ground, a continuous current flows. But inside the plasma the current is carried only for a short distance by beam electrons and then transferred to background electrons to flow over a larger cross sectional area. Plasma waves accomplish the momentum transfer. Electromagnetic waves play a secondary role in the momentum and energy transport processes. Nevertheless, they are an integral part of a strong beam-plasma instability and a highly visible long-range phenomenon.

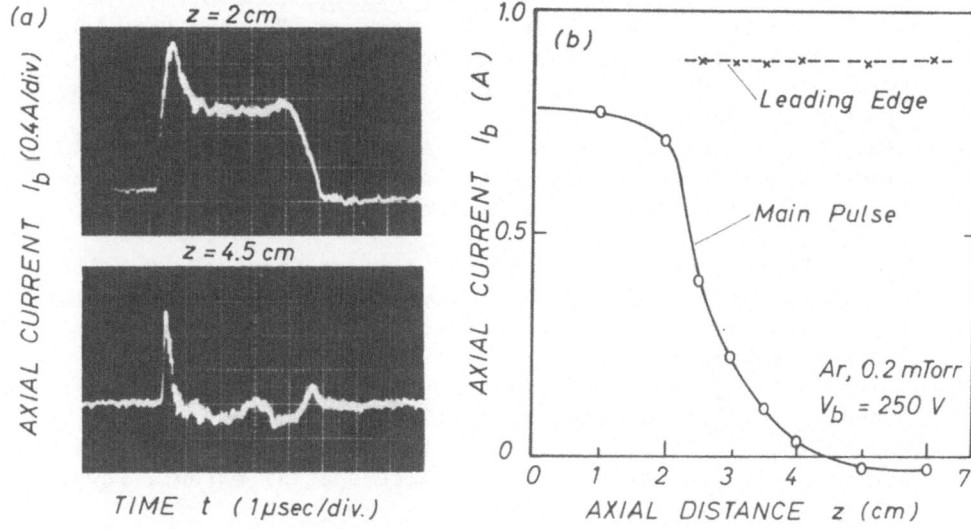

Fig. 7.   Beam current penetration into the plasma. (a) Temporal
          variation of the beam current at two different axial dis-
          tances z from the injection point. The leading edge pene-
          trates deeply while the main pulse is scattered due to the
          onset of instabilities. (b) Spatial variation of leading
          edge (precursor) and main pulse.

SUMMARY

We have identified experimentally the physical process of mode
coupling by which electromagnetic radiation is generated in an electrostatic beam-plasma instability. The observations were made under
controlled conditions in a large uniform laboratory plasma (equiva-
lent dimensions in space, 2 km $\perp$ $\vec{B}_o$, 4 km $||$ $\vec{B}_o$ at $n_e = 10^5$ cm^{-3}).

The limited penetration of the beam current into the plasma due to
instabilities has been demonstrated. These results should be relevant to beam injection experiments from rockets or satellites into
space plasmas.

ACKNOWLEDGMENTS

This work was supported by NASA grant NSG-7616 and NAGW-180.

REFERENCES

Gekelman, W., and Stenzel, R. L., 1978, ion sound turbulence in a
     magnetoplasma, Phys. Fluids, 21:2014.
Goldman, M. V., Reiter, G. F., and Nicholson, D. R., 1980, Radiation
     from a strongly turbulent plasma: Application to electron

beam-excited solar emission, Phys. Fluids, 23:388.

Grandal, B., Holtet, J. A., Trøim, J., Maehlum, B., and Pran, B., 1980, Observation of waves artificially stimulated by an electron beam inside a region with auroral precipitation, Planet. Space Sci., 28:1131.

Kawashima, N., Sasaki, S., Ushikoshi, A., and Obayashi, T., 1980, Electron beam experiments in space, in: Proceedings of the International Conference on Plasma Physics, Nagoya, Japan, Vol. I:217.

Lin, R. P., Potter, D. W., Gurnett, D. A., and Scarf, F. L., 1981, Energetic electrons and plasma waves associated with a solar type III radio burst, Univ. Calif. Berkeley Space Sciences Laboratory Report.

Maggs, J. E., 1978, Theory of electromagnetic waves on auroral field lines, J. Geomag. Geoelectr., 30:273.

Morales, G. I., and Lee, Y. C., Spiky turbulence generated by a propagating electrostatic wave of finite spatial extent, Phys. Fluids, 19:690.

Papadopoulos, K., 1979, Interplanetary type III radio bursts, Rev. Geophys. and Space Phys., 17:624.

Piliya, A. D., 1966, Wave conversion in an inhomogeneous plasma, Sov. Phys. - Tech. Phys., 11:609.

Stenzel, R. L., and Wong, A. Y., 1972, Threshold and saturation of the parametric decay instability, Phys. Rev. Lett., 28:274.

Stenzel, R. L., Wong, A. Y., and Kim, H. C., 1974, Conversion of electromagnetic waves to electrostatic waves in inhomogeneous plasmas, Phys. Rev. Lett., 32:654.

Tsytovitch, V. N., 1977, Theory of turbulent plasmas, Consultants Bureau, New York.

Winckler, J. R., Arnoldy, R. L., and Hendrikson, R. A., 1975, Echo 2: a study of electron beams injected into the high latitude ionosphere from a large sounding rocket, J. Geophys. Res., 80:2083.

Wong, A. Y., and Quon, B. H., 1975, Spatial collapse of beam-driven plasma waves, Phys. Rev. Lett., 34:1499.

DISCUSSION

Maggs: Have you tried starting out with a warm beam?

Stenzel: Not yet. We would have to use a higher energy beam and scatter it on a thin foil. This may be tried in the future.

ELECTRON BEAM INJECTION AND ASSOCIATED PHENOMENA

AS OBSERVED IN A LARGE SPACE SIMULATION CHAMBER

C. Béghin, Y. Arnal, P. Gille, D. Henry, J.L. Michau,
F.X. Sené
CRPE/CNET/CNRS - 45045 - Orléans Cédex (France)

J. Lavergnat, J.Y. Delahaye
L.G.E. - 4, avenue de Neptune, Saint-Maur 94100 (France)

J.P. Lebreton, A. Gonfalone, F. Malerba, D. Klinge
SSD/ESA - Noordwijk (The Netherlands)

B. Maehlum, J. Troim, B. Narheim
NDRE - N.2007 - Kjeller (Norway)

INTRODUCTION

Phenomena Induced by Charged Particle Beams (PICPAB) are planned to be investigated during the first Spacelab flight (FSLP) using a European payload. The PICPAB experiment (Béghin et al., 1979) consists of two accelerators of electron and ion beams and associated diagnostic instruments including wave receivers, thermal plasma probes and return current particle energy-analysers. The functional tests using prototypes of those instruments have been carried out in the large vacuum chamber SIMLES ($\sim$ 300 m^3) at CNES-SOPEMEA (Toulouse-France) in february 1980. The main purpose was to perform a simulation under conditions where the ambient neutral and ionized gas, magnetic field strength and lay-out of the different packages were as close as possible to those anticipated for FSLP mission. But such a simulation cannot be undertaken ignoring plasma physics processes which are associated with the neutralization of the gun body and of the beam itself, those problems which are precisely among scientific objectives of this experiment for FSLP mission. Obviously it was felt that the involved processes and observed phenomena could be significantly different as compared to those already identified in previous space experiments or those anticipated for the flight, mainly for two reasons :

- limited size for beam propagation

- limited reservoir of background plasma

Our feeling was even supported by the results obtained by pio-
neer works in the large NASA's chambers (Bernstein et al., 1975 and
1979) where future space accelerators were tested in the frame of the
anticipated AMPS program. It is why the SIMLES chamber experiment
was also designed for studying some unanswered questions about beam
injection in space simulation chambers, such as : what is the influence of the target used for collecting the beam ? what are the
parameters with determine the growth rate of instabilities leading
to beam plasma discharges or convecting discharges ? etc....

We are reporting here the main results obtained during this
test, for the electron beam alone ; the results related to ion beam
injection will be the subject of a further report.

After a brief description of the experiment lay-out in the chamber and of the diagnostic capabilities, we present the results
related to the DC electric field component perpendicular to the beam,
the ambient plasma response, the HF electric component, the returning
flux, and the target's and accelerator's voltage-current characteristics. Commenting those results we tentatively try to interpret them
in term of qualitative analysis of beam-plasma interaction processes,
according to the current ideas from previous theoretical and experimental works in this field (see for instance Papadopoulos review-
paper in this volume).

EXPERIMENT CONFIGURATION

The experiment lay-out is shown in fig. 1. The electron gun of
the Pierce type with a tantalum filament cathode is located inside
a metallic box containing also the high voltage generator, the ion
gun and two diagnostic instruments. This Active Package (AP), as in
flight will be grounded to the space shuttle reference, here it is
connected to the earth of the chamber through a small resistor (1Ω)
used for monitoring the net current flowing from and to the package.
Another mode of operation allows to let the package floating electrically through a high resistor ($\gtrsim 2 M\Omega$) ; in that case the AP potential is monitored as referred to the ground. Because the return
current collection surface is quite small in this mode ($\lesssim 0.5 m^2$)
the AP voltage is expected to rise up to high value in some occasions. For safety, this voltage is automatically limited to $\pm$ 200 v
by using two Zener diodes. Thus the active package can be considered
like really floating only as long as its absolute voltage with respect to the ground is lower than 200 v. One mode of gun operations
allows to vary the beam current from about 1 mA up to about 100 mA
with an energy decreasing correspondingly from 10 down to 8 kev. The
results which are presented here are obtained in pulsed mode -
20.8 ms duration - at a repetition rate of 3.75 Hz. The beam is
injected parallel to the local B field which varies in amplitude
inside the chamber from 0.2 to 0.24 Gauss.

For having a proper termination the beam was collected by a target located 3.3 m away along the beam axis. This collector is an aluminium disc of 2 m in diameter either connected to the ground through 50 Ω (current monitor) or let floating through 2 GΩ (voltage monitor) in parallel with coaxial cable capacitor (∼1.2 nF). In order to trap the secondary electrons extracted from the target under beam impact, the back of the disc was covered with a crowd of small permanent magnets in close array with opposite polarities facing each other (see Arnal, 1977). The front of the disc then offers to secondary electrons a magnetic confinement along the strong field lines of the dipoles (200 Gauss at 2 cm from the surface). Because of the close array, the field strength decreases very fastly so that 20 cm from the target-front the ambient magnetic field remains unperturbated.

At about 1.5 m from the beam axis was set up a preliminary version of the diagnostic package which is intented to be installed in the air-lock of the Spacelab (see Béghin et al., 1979).

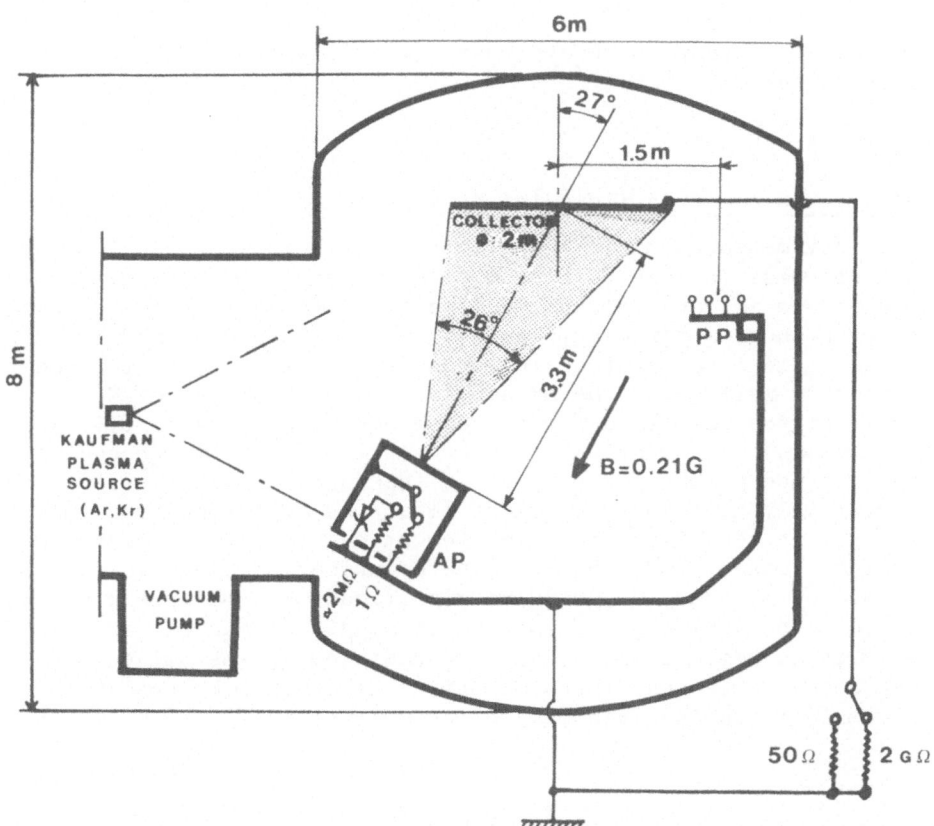

Fig. 1 Schematic representation of the experiment lay-out.

For the SIMLES test, this passive package (PP) was equipped with a very low frequency electric dipole (DC to 500 kHz), a high frequency electric dipole (100 kHz - 100 MHz), an electron density probe of the mutual-impedance oscillator type and an electron temperature probe derived from the principle of spherical electrostatic analysers (Sagalyn et al., 1963).

The pumping capability allows to reach inside the chamber a pressure as low as few 10^{-7} torr of residual gas consisting mainly of water vapor and nitrogen. In order to determine the effect of the highest expected pressure in the orbiter environment, one run has been conducted by introducing dry nitrogen inside the chamber, increasing the pressure up to 10^{-4} torr. The plasma was produced by two different manners. First, by ionization of the residual gas by the beam itself ; following current ideas that produces a quite warm (few eV) plasma column more or less confined around the beam by the B field in a cylinder of four mean Larmor radius in diameter ($\sim$ 1 m). The second manner was using a plasma thruster of the Kaufman type, injecting Argon or Krypton plasma. When this plasma source was used, the all chamber was filled with cold plasma ($<$ 1 ev) of conventional like ionospheric densities, adjusted from few 10^4 up to 10^6 e/cc. The density gradient in PP and beam vicinity was of the order of 30-40 % per meter.

EXPERIMENTAL RESULTS

Transverse DC electric field in the absence of background plasma

Direct measurements of DC electric field in weak density plasma are not easy to make (Fahleson et al., 1970) and our short double sphere dipole is certainly not well adapted for accurate and absolute determination for many reasons. It is mainly designed for VLF electric field investigation in ionospheric plasma where the sheath impedance is well below the preamplifier input impedance. Moreover, in spite of the use of carbon-coated spheres the work functions on each probe were not identical, the presence of PP metallic body near the probes was probably perturbating the equipotential surfaces and the relatively high energy flux of electrons scattered from the beam plasma column could be lightly different on one sphere due to the wake of the other. Nevertheless, the polarisation electric field external to the beam, assuming no significant plasma shielding, is expected to be very high. The order of magnitude can be easily calculated if we neglect also the field radiated by the collector. Assuming a zero-divergence beam of infinite length, the transverse DC electric field is given by Laplace's law :

$$\left|E_{\perp}\right| = \frac{i_b}{2\pi\,\varepsilon_o\,rv_{/\!/}}$$

where i_b is the beam current, ε_o is the free space permittivity, r the distance to beam axis and $V_{//}^o$ the parallel velocity of beam particles. For 8 keV beam energy one gets 0.22 v/m per mA at 1.5 m from the axis. Thus such high field could be detected with modest dipole length provided that few scattered electrons allow a resistor coupling between the probe and the ambient medium.

Figure 2 represents the voltage difference as measured by the double-sphere dipole (40 cm center to center) oriented at about 65° and located 1.5 m from the beam axis, as a function of the beam current in the case where the plasma thruster was off and the collector grounded. According to the above discussion we have intentionally plotted the voltage difference rather than the electric field deduced from some uncertain transfert function. Taking account of the limited bandwidth of the VLF receiver ($\sim$ 500 Hz), the transient rise and fall times were not measurable (<< 2 ms) ; the plotted values correspond to the flat plateau of each 20.8 ms pulses. For low current, be-

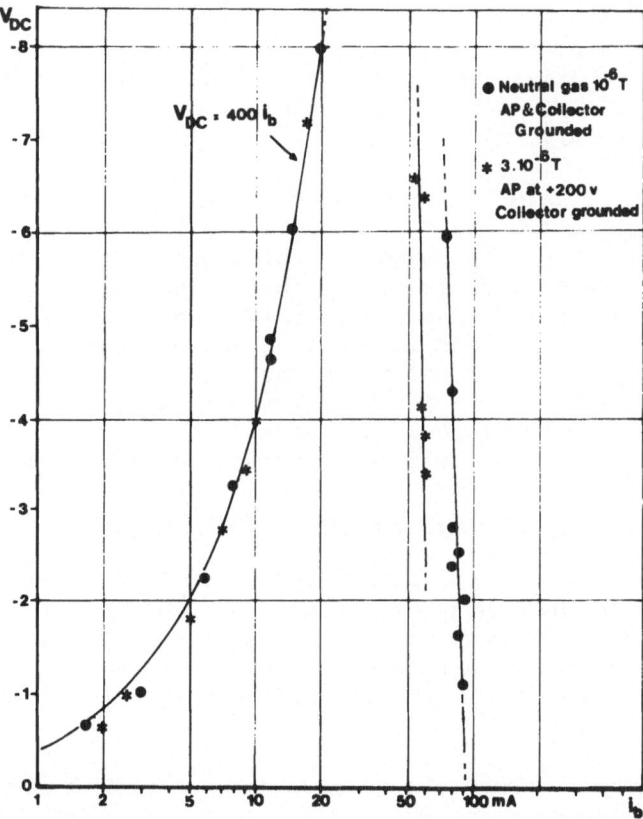

Fig. 2   DC component of potential difference (in volt) as measured by the double-sphere dipole versus the beam current.

low $\sim$ 20 mA, the received signal follows a linear law versus the beam current (i_b in logarithmic scale). If the effective length of the dipole was assumed to be 40 cm, the corresponding electric field should be :

$$|E| \stackrel{\sim}{=} 10^3 \cdot i_b$$

which is about 4.5 times larger than the above rough estimation but nevertheless shows the expected linear dependence. In this linear zone, one can see that the measured values do not seem to depend on the neutral gas pressure, at least below 3.10^{-6} torr, neither on the fact that the AP is floating or grounded.

Unfortunately no measurement is available here for i_b between 20 and 50 mA ; but it is clear that the DC field suddenly collapses when the beam current exceeds a certain threshold. This threshold is thought to be depending somewhat on the neutral gas pressure but more likely here on the potential drops at the extremities of the plasma column near the AP and the collector. As a matter of fact, when the AP is in floating mode, even for 1 mA current, its potential is observed rising up very fast (<< 1 ms) to the safety limited value of + 200 v as regard to the ground. Consequently a sharp potential drop has to exist just above the package front panel because obviously the plasma column cannot support a so strong parallel electric field as 50 v/m.

Better still this is illustrated by Fig. 3 (lower panel). Here we are considering two cases of extremely different neutral gas pressure, up to 10^{-4} torr, when the collector is floating electrically and consequently falls down to highly negative voltages. We see the same linear dependance for small beam currents but the threshold occurs around 8 mA for low neutral gas pressure, then the DC transverse electric field collapses and remains nearly zero, below the instrument sensitivity for beam currents greater than 15 mA. For high pressure the slope of the linear part is lower, indicating as expected a significantly greater perpendicular conductivity.

Transverse DC electric field in the background plasma

Now, when the external plasma-source is on, in the absence of electron beam or when the beam current is lower than few mA (see Fig.4), the double-sphere dipole has an off-set of the order of a volt which results from a combination of the work function off-sets and $V \wedge B$ field due to the plasma drift with regard to the stationary B field.

When the beam current increases, the background cold plasma density as measured at the PP location decreases significantly and more especially as the initial density is low (fig. 4, higher panel, left side). Correspondingly to the background plasma vanishing the

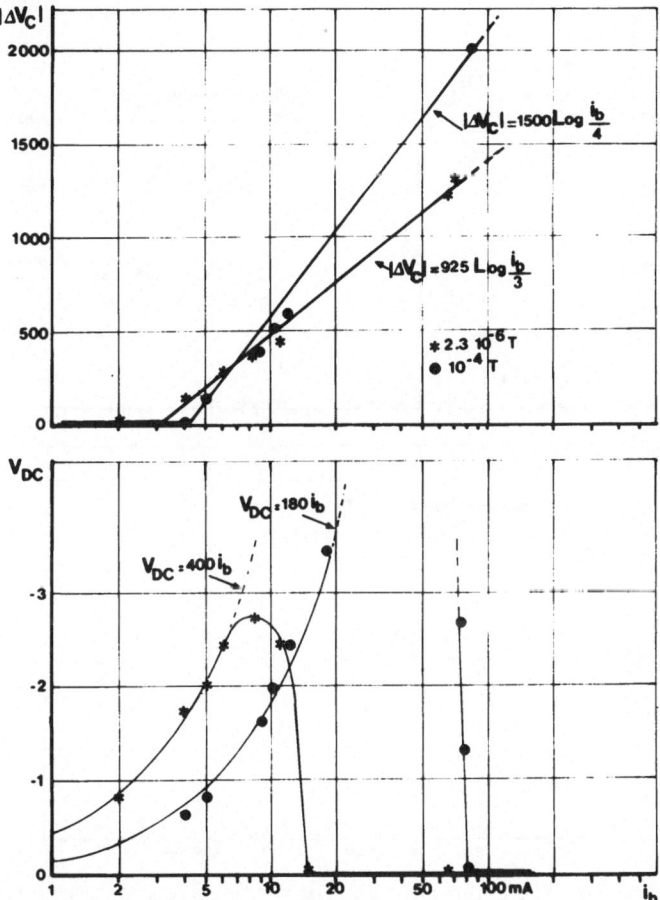

Fig. 3   Runs with floating collector under low and high gas pressure.
Lower panel : V_{DC} like in fig. 2. Higher panel : collector
potential variation (see text).

DC transverse electric field starts to recover the previous neutral
gas conditions. Then, at the time when the background cold plasma
has been totally absorbed by collector and beam neutralisation processes the DC field collapses like before.

Ambient plasma response

Like it is shown on figs. 4 and 5 the background plasma absorption process is clearly depending on the initial density ; for the
highest value of $1.6.10^6$ e/cc (fig. 5), even for 100 mA beam current
and floating collector the entire volume of the plasma chamber provides a sufficient instantaneous charged particles reservoir for
neutralisation processes.

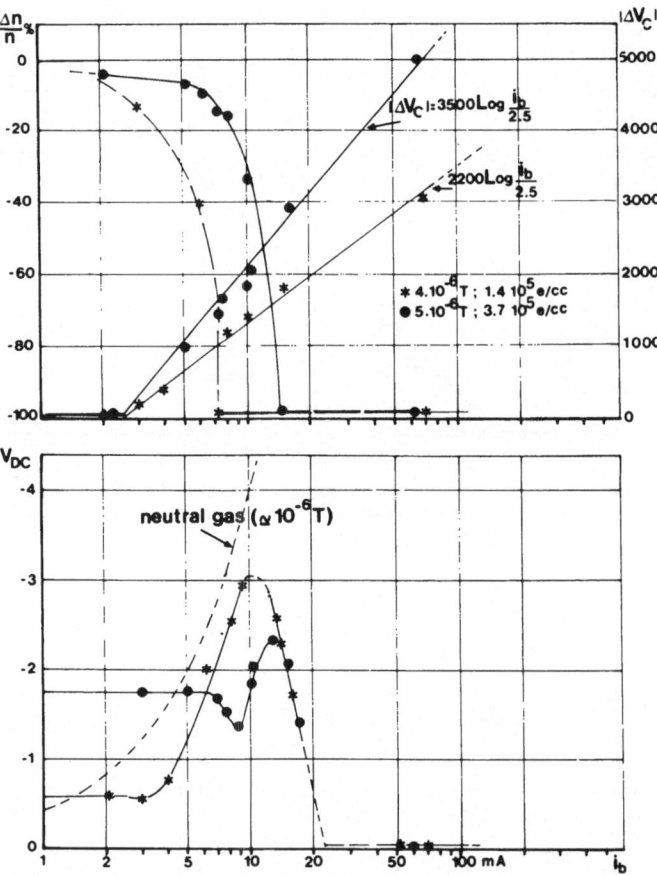

Fig. 4   Same as fig. 3 exepted that the plasma source is on.
The plasma densities are measured at PP level and the
relative electron density variations are plotted in
higher panel (left).

Here it is needed to look at the time-scale and dynamics of
the ambient plasma response. For Fig. 4 runs, when the background
plasma density is low and the collector in floating mode, 20 mA
beam-current is sufficient to drain off the chamber during the all
duration of the pulse. Some after- effects are even detected several ms after the end of the pulse; that has been identified to be
caused by the collector negative-potential reaction with the continuously injected cold plasma. But during the high density run of
Fig. 5 for high beam current the production rate of the beam-generated plasma column plus the external plasma-source production rate
are thought to exceed the losses (by recombination and neutralisation
processes). We had no possibility to measure the time-evolution of
the plasma density in the all space, which would be ideal to describe
the rise of beam plasma instabilities, nevertheless the example shown

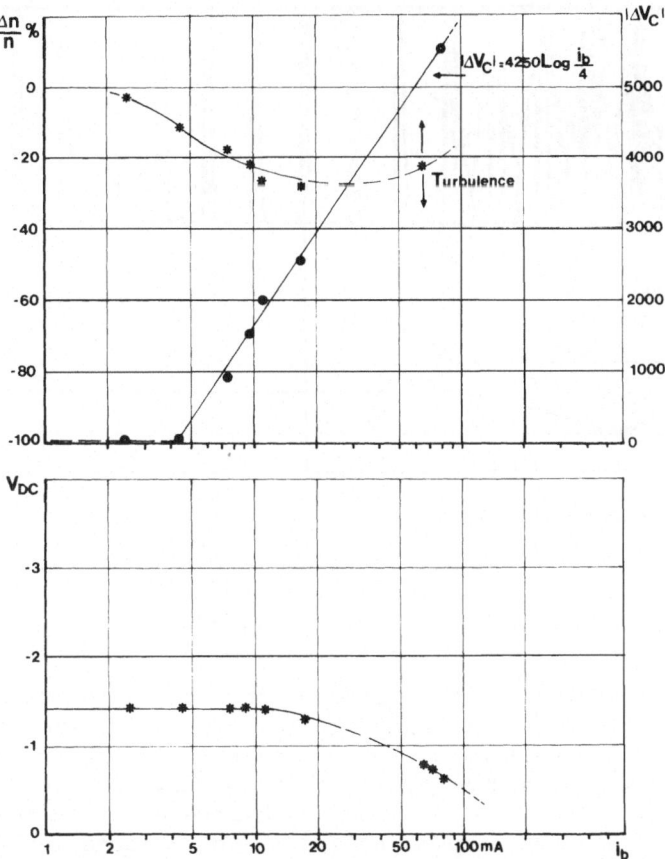

Fig. 5   Same as fig. 4, run with high ambient plasma density :
         $1.6.10^6$ e/cc. Neutral gas pressure is 9.10^{-5} torr.

in Fig. 6 allow us to foresee some instable mechanism. The instantaneous plasma-frequency of the ambient plasma as given by the mutual-impedance oscillating probe is 11.4 MHz ($n_e = 1.6.10^6$ e/cc)
during the unperturbated inter-pulse periods. One can see a fast
decrease in electron density during the first 5 ms of the pulse.
Usually, after such a transient effect, even shorter most of the time, the plasma density reaches an equilibrium (n - Δn), the relative normalized values of which were plotted in fig. 4. But here
for high beam current (70 mA) the plasma density starts to fluctuate
for a while, then grows continuously until the end of the pulse.

    In several other occasions, the fluctuations remain for the
rest of the pulse duration, like in the runs of fig. 4 for beam current just below the critical value when the DC electric field and
the electron density collapse. This indicates a turbulent instable
regime when the plasma reserve contained in the chamber and the
source production are just unable to satisfy the neutralization

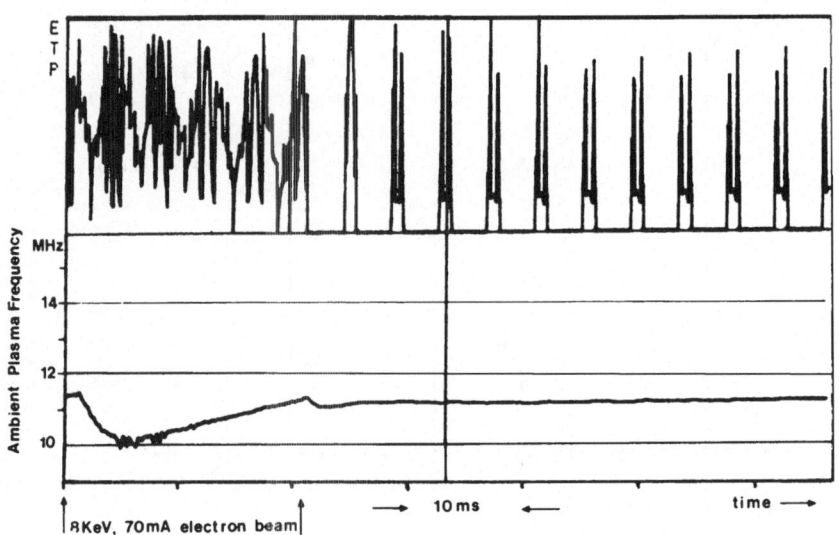

Fig. 6 Time dependent raw data given by Electron Temperature Probe
(ETP) and Mutual-Impedance Oscillating Probe in the same
run as fig. 5 for high beam current.

demand. In those cases, the system restricted to the beam and to the
beam-created plasma column has to find in itself, like in neutral
gas, the conditions of stability or instability when the beam current increases more and more. In the case shown in fig. 6, the balance reverses after 7 - 8 ms and we observe a net increase in plasma density with a tendancy to exceed the background level what is
confirmed by other examples. Unfortunately we have no measurement
above 100 mA to determine a possible threshold for the beam current
where the drastic increase in density could lead to the Beam Plasma
Discharge (BPD, see Fainberg et al., 1962, and Getty and Smullin,
1963).

In the same fig. 6 we have plotted the raw data of the Electron Temperature Probe (ETP). That is the Langmuir's positive current corresponding to the swept voltage following a saw-tooth law
applied on the spherical analyser successively with positive and
negative slope every 2 ms. In spite of noise due to several causes
(turbulence, gain commutation of step amplifiers and chopped beam
current) one can see a clear difference in the slope of the signal
between the perturbated and unperturbated conditions. During the
beam injection, the response does correspond to a quite warm plasma
(few eV) with a non-Maxwellian distribution, whereas few ms after
the end of the gun-pulse the plasma recovers the low temperature
($\sim$ 0.1 ev) of the injected flow. Taking into account of the results
obtained in several different conditions including grounded collector runs the general feature is the following : the cold ambient
plasma seems to flow by ambipolar diffusion towards the AP-beam-

collector region while warm electrons (few ev) are scattered from that region. The only difference when the collector is grounded is for low beam current; first a small increase (20% max) of ambient plasma density is observed, followed after that by a breakdown for high beam current values. In that case, the observed plasma out of the column is dominated by warm scattered electrons.

HF electric field

Intentionally we prefer avoid the word wave when we observe the close field near or inside the generation region (< 2 m) even though the radiated fields are likely of electrostatic nature in our experimental conditions. The High Frequency (100 kHz - 100 MHz) electric field as received by PP antenna at 1.5 m from the beam axis, has a remarkable reproductible behaviour in all the runs, independantly of the termination conditions. Two cases are observed : either the ambient plasma surrounding the PP has a high density, that is to say when the plasma-frequency f_p is above the electron gyrofrequency f_c ($\sim$ 600 kHz), or the reverse. Intermediate cases were not observable mainly due the instable conditions when they occur, as reported in the above section.

In the first case, always corresponding to ambient plasma runs with external thruster and beam current below the drain-off effect, the HF electric field (parallel to DC component) exhibits a broad band signal centered around the local value of f_p. This signal does exist even in the absence of beam due to the natural background of small instabilities associated with the plasma injection in the chamber. A typical example is given in Fig. 7 where one can see the shift of this plasma-frequency noise towards lower frequencies, associated with a decrease of ambient electron density due to 8 mA beam injection. Here the spectrum broadens significantly while a lower-frequency peak appears. The latter does not correspond apparently to any characteristic frequency and when it exists, its erratic feature and the lack of density measurement inside the beam region do not allow us to determine its origine.

More significant of beam-plasma interactions with the present available diagnostic tools is the second case, when the ambient plasma-frequency is lower than f_c. A constant feature is the detection of a strong peak, only during the gun-pulses, at a surprising stable frequency around 300 kHz, which is near $f_c/2$ as shown in fig. 8. The amplitude of this peak is found to be nearly proportional to the beam current from few mA up to 100 mA, and to be independant of the presence or not of external plasma as long as the drain-off effect persists. At higher frequencies the spectrum decreases in amplitude with harmonic stengthenings more or less close to $\frac{m}{2} fe$, sometimes up to high orders like in the present example when the plasma-source is on, but unable to compensate the losses during the gun-pulse.

One quite likely candidate to generate such a figure is the contra-stream instability which has been studied in several particle-beam laboratory experiments (see for instance Landauer and Muller, 1966). This could indicate that warm electrons of the plasma-column would be trapped between the AP and the target, possibly reflected back and forth by space-charge potential drops. This could be also specially favoured by the mirror-effect on our target, as described before.

Supporting this assumption is the energy-spectrum of the returning flux, as measured by the electrostatic analyser located on the top front of the AP, and reported in the next section.

Return current characteristics

One of the diagnostic instruments located in the AP is an electron electrostatic analyser covering the range from few ev up to 200 ev, energy stepped from one gun pulse to the next one. The instrument opening angle is 8 x 8 degrees centered parallel to the beam axis. A typical returning-flux spectrum is given in Fig. 9, for two values of beam current. For low current, the spectrum is in good agreement with the concept of neutral gas ionisation by the energetic electron beam. But for high current beam, the all low energy population, up to about 60 ev in this run, is absent. This suggests two possible explanations: a loss-cone distribution for electrons

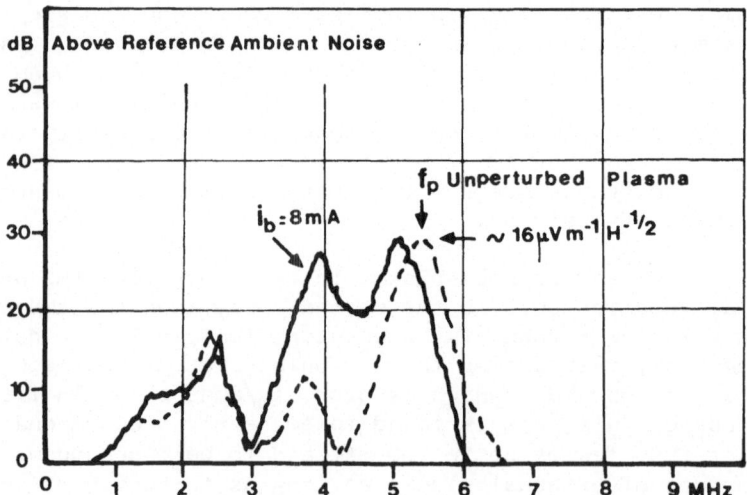

Fig. 7   HF electric field frequency-spectrum as received without
         and with low current beam, by PP high impedance Hertz's
         dipole in ambient plasma run. $n_e = 3.7.10^5$ e/cc,
         p = 5.10^{-6} torr, collector in floating mode.

60dB ≡ 200 μV m⁻¹Hz⁻¹/²

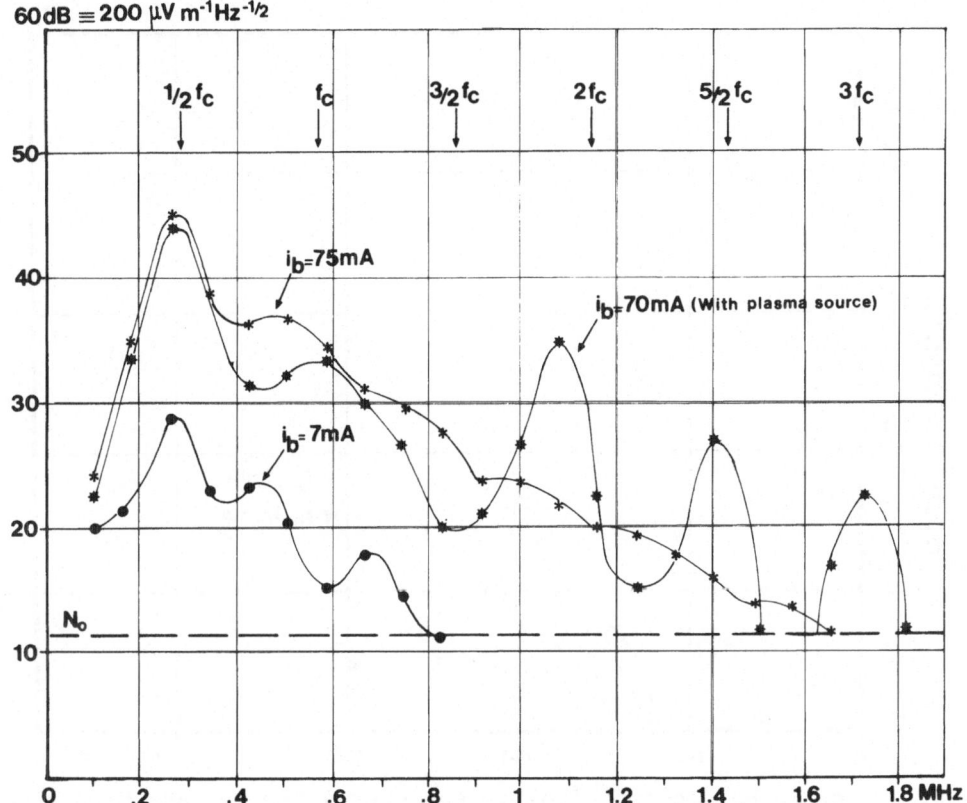

Fig. 8 HF electric field for high and low beam currents without
plasma source (p = 10⁻⁶ torr) and for high beam current
with plasma source (n_e = 3.4.10⁵ e/cc, p = 6.10⁻⁶torr.
Collector is floating in both cases, N_o is the reference
ambient noise-level without gun-pulse.

of energy lower than 60 ev, or reflection of electrons of the same
energy by space-charge potential wall. Both possibilities could
support the half-harmonic cyclotron-wave observations.

Another interesting parameter is the AP voltage (in floating
mode) versus the beam current, giving indications on neutralisation
capabilities of the return current. As already said, during the neutral gas runs, the AP in floating mode is rising up to + 200 v
(safety-saturation level) within a fraction of ms whatever the beam
current is. With background plasma the situation is quite different.
We show in Fig. 10 the AP voltage (V_{ap} evolution versus i_b for
two ambient plasma conditions. This dependance is nearly linear
up to a certain value, then the slope increases very rapidly towards
the saturation level. These results indicate clearly that the back-

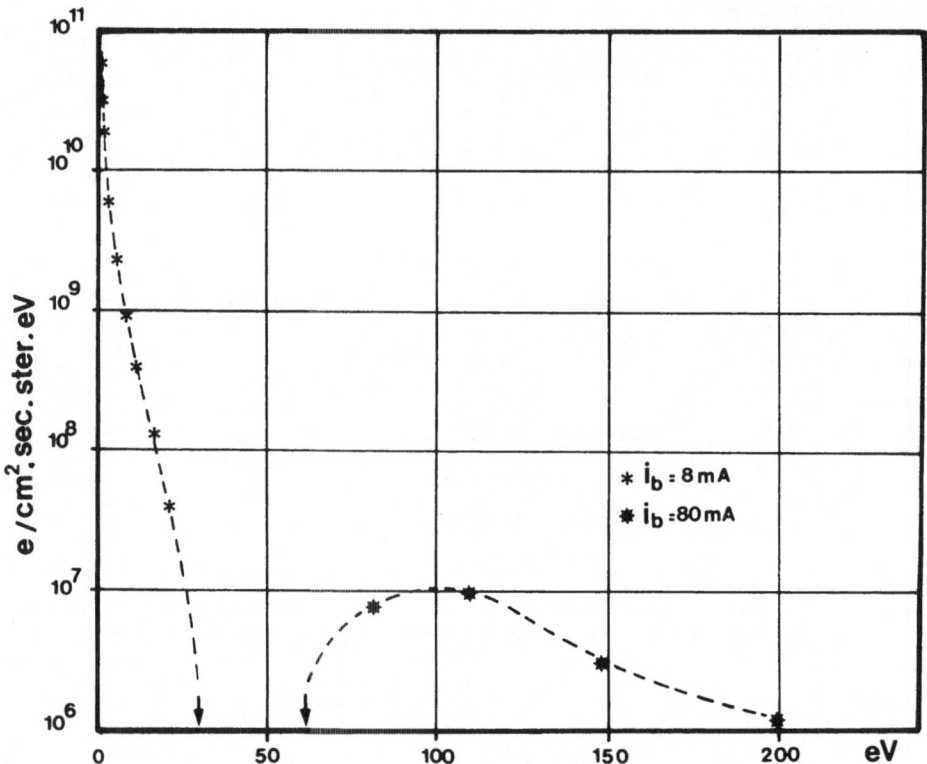

Fig. 9   Returning electron-flux versus energy for low and high beam
         currents, with ambient plasma. $n_e = 3.7.10^5$ e/cc,
         p = 5.10^{-6} torr, collector in floating mode.

ground plasma contributes to the package neutralisation much more
than the beam-created plasma-column although the latter produces
also a return current, which is measured by the electrostatic analyser even when the plasma thruster is off. The equivalent resistance
V_{ap}/i_b in the nearly linear regime depends on both the neutral gas
and the ambient plasma densities but not following a simple law.
Due to the lack of data this law has not been determined experimentally.

Finally we have plotted in fig. 10 the values of the plasma
potential as measured by a high-time-resolution Langmuir probe in
the neighbourhood of the AP. The location of this probe is such
that it avoids direct impact of beam particles, but it might be
immersed inside the beam-created plasma column 30 cm above the AP
top-front. We measure an increasing plasma-potential (referred to
the chamber ground) when the beam current grows, but much slowly
than the AP voltage (fig. 10). This indicates, like stated above,
that a sharp potential-drop does exist few cms above the AP top-front.

Collector versus plasma behaviour

The last measurement reported here is concerning the voltage
variations of the collector, in floating mode, versus the beam current for different gas pressure and plasma conditions. This could
be of great practical interest to understand the charging-up effects
of a conductive surface immersed in a plasma under high flux bombardment of energetic particles. In this particuliar experiment the
results provide also interesting data on the collector/plasma-column
coupling.

Under direct impact of the electron beam, when the collector
is floating electrically, its potential referred to the chamber
ground (V_c) can reach high negative voltages, specially in the absence of background plasma. Because the existence of a beam-generated plasma column the collector cannot really be considered as floating like in free-space, hence the potential never reaches the 8-10
kv of the beam energy during a single pulse. But the accumulated
charges in the parallel capacitor (due mainly to the cabling) stay

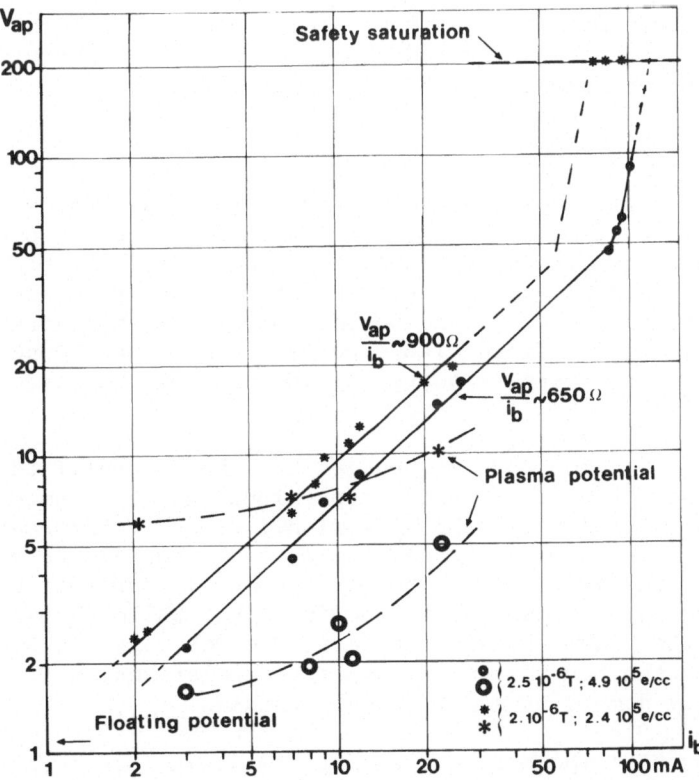

Fig. 10 Active package potential in floating mode and plasma-
         potential, both in volts, versus beam current for two
         ambient plasma conditions.

for long time after the end of the gun pulse, so that the time constant for discharging the system through 2 GΩ (for neutral gas runs)
can be much longer than the repetition time between pulses. Hence a
cumulative effect occurs and we observe that the only significant
parameter is the absolute voltage variation between V_C values during
the pulse and just after. We call $|\Delta V_C|$ this variation, which repre-
sents the dynamic response of the target impedance to the beam current. By definition, and that has been verified experimentally, for
several consecutive identical pulses $|\Delta V_C|$ shows no cumulative effect.

This value is plotted in figs. 3, 4 and 5 (top panel) for different conditions, as a function of i_b. A very similar feature appears in neutral gas runs as well as for background plasma runs :

- it exists always a threshold for the beam current, of the
order of 2-4 mA, below which $\Delta V_C \simeq 0$, exactly like expected in
free space.

- above this threshold, ΔV_C (negative for electron beam injection) increases logarithmically with i_b .

- the slope of the logarithmic law <u>increases</u> with the neutral
gas pressure and the background plasma density, in absolute contradiction with the normal DC resistivity of a plasma.

SUMMARY AND CONCLUSIONS

Those results show the complexity of the beam-plasma-collector-
gun system where non-linear processes are generated in several consecutive steps. Further theoretical and experimental works would be
needed to explain all the observed phenomena, nevertheless the observations agree quite well with the concept of a beam-generated plasma
colum aligned with B field. Two important results are concerning the
DC and the HF transverse electric field, probably out-side the plasma column.

For the DC component two regimes have been identified ; a linear one, where the DC field is proportional to the beam current,
interpreted like an radial polarisation field not compensated by
an external-plasma ambipolar diffusion, and a non-linear regime
where the DC radial field collapses. A possible explanation of this
breakdown could be connected to the fact that while the electron
density of the plasma column increase, the Debye length becomes
smaller than the Larmor radius, hence the polarisation field is
exponentially damped at short radial distance.

For the HF component, the beam-plasma-column is seen to radiate
strong cyclotron-waves near harmonics of $f_c/2$ like in two-stream
interactions of previous laboratory experiments, suggesting some
mirroring processes of warm electrons at the extremities of the
plasma column.

The results indicate that in our peculiar conditions, i.e. beam propagation distance shorter than the first node focalisation length and nearly zero pitch-angle injection, neutral gas pressure ranging from less to 10^{-6} up to 10^{-4} torr and a maximum of 100 mA electron beam of 8-10 kev energy, we never triggered the beam plasma discharge like in the NASA's chamber (Bernstein et al., 1979 although we observed in some occasions a temporal growth of the beam-created plasma density, which could be initial conditions for a such instability.

Finally, the behaviour of the AP body and of the collector potentials, versus the beam current, have been studied under several conditions, allowing us to sketch an equivalent circuit, as shown in Fig. 11. Because the AP voltage in floating mode with ambient plasma is proportional to the beam current up to a certain threshold, we think that a linear regime does exist for which the return car-

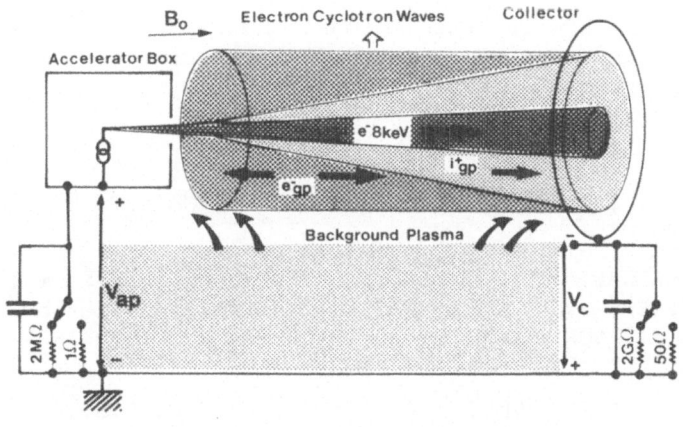

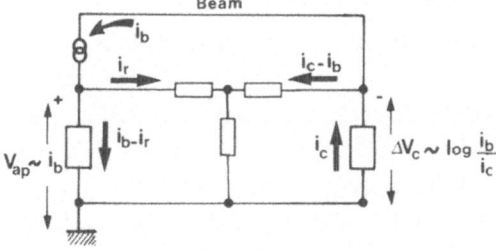

Fig. 11   Schematic representation and equivalent electric scheme of the beam-created plasma column, accelerator body, collector and ambient plasma system. e^-_{gp} and i^+_{gp} are electron-ion pairs of the beam-generated plasma assumed to diffuse along the B field and reflected back.

rent due to the ambient cold plasma is exactly proportional to the beam current. The collector-voltage/beam-current behaviour in floating mode is quite puzzling when the beam current exceeds 2-4 mA. The equivalent DC impedance of the collector for a given beam current increases with the neutral gas pressure and the ambient plasma density contrary to what expected from normal conductivity of a plasma. Further detailed analysis will be needed to understand this point (among many others) which is certainly related to the complex processes occuring in a such beam-plasma bounded system.

ACKNOWLEDGMENTS

We wish to thank many participants in this experiment and the sponsoring authorities   of CRPE/CNET/CNRS, LGE, SSD/ESA, NDRE, DRET, CNES and SOPEMEA.

We acknowledge the enthusiastic support from Dr. A. Pedersen and have greatly appreciated usefull discussions with him as also with many other competent scientists during the GEILO meeting.

The authors are indebted to their National Agencies, ESA and NASA for help and support to the PICPAB experiment as part of the joint FSLP program.

REFERENCES

ARNAL, Y., 1977, Thèse d'Ingénieur-Docteur, Note Technique, CRPE, 45.

BEGHIN, C., ARNAL, Y., BOSWELL, R., HENRY, D., PIRRE, M., BERTHELIER, J.J., LAVERGNAT, J., SYLVAIN, M., MAEHLUM, B., TROIM, J., GONFALONE, A. and MALERBA, F., 1979, ESA Journal 3, 123.

BERNSTEIN, W., LEINBACH, H., COHEN, H., WILSON, P.S., DAVIS, T.N., HALLINAN, T., BAKER, B., MARTZ, J., ZEIMKE, R. and HIBER, W., 1975, J. Geophys. Res., 80, 4375.

BERNSTEIN, W., LEINBACH, H., KELLOGG, P.J., MONSON, S.J. and HALLINAN, T., 1979, J. Geophys. Res., 84, 7271.

FAHLESON, U.V., KELLEY, M.C. and MOZER, F.S., 1970, Planet. Space Sci., 18, 1551.

FAINBERG, Ja.B., KHARCHENKO, I.F., KORNILOV, E.A., NIKOLAEV, R.M., PEDENKO, N.S. and LUCENKO, E.I., 1962, Nuclear Fusion (supplement), 3, 1101.

GELLY, W.D. and SMULLIN, L.D., 1963, J. Applied Physics, 34, 3421.

LANDAUER, G. and MULLER, G., 1966, Physics Letters, 23, 555.

SAGALYN, R.C., SMIDDY, M. and WISNIA, J., 1963, J. Geophys. Res., 68, 199.

DISCUSSION

Papadopoulos: The transverse electric field is usually
an ambipolar field created by the fact that one species
diffuses faster than the other. In this fashion its
value reflects the dominant diffusion process (i e
classical, Bohm etc). Did you attempt such a correlation
and if so what was the dominant diffusion process?

Beghin: In the regime where the transverse electric
field is observed outside the plasma column, it is not
obvious, through the available data, that any plasma
subject to ambipolar diffusion does exist. There is
some indication that only free suprathermal particles
are in this external region.

GENERAL DISCUSSION ON

ACCELERATOR EXPERIMENTS IN THE LABORATORY

Chairman: W   Bernstein

In the early 1960's, the beam plasma discharge
(BPD) had been studied in detail experimentally in the
laboratory. Prior to the first electron accelerator
experiment on rockets (Hess) the two-stream instability
was (theoretically) predicted to be the main source of
severe beam modifications. The recent laboratory
experiments, however, indicate that the BPD occurs
under near ionospheric plasma conditions. Most of the
rocket flights with accelerators have only provided
circumstantial evidence for these strong interactions.
How does one explain this apparent inability to characterize a phenomenon which is extremely obvious in the
laboratory, but apparently very subtle in space?

Firstly, the laboratory configuration is typically
characterized by:

a)   a higher ambient density, which may explain why
     the BPD is more obvious visually in the labora-
     tory than in space,

b)   a finite geometry, which may produce end effects
     important for the occurrence of the BPD,

c)   diagnostic instruments placed downstream from
     the accelerator in contrast to the space experi-
     ments where the accelerator and the instruments
     usually are placed on the same payload.

Secondly, the accelerator experiments in space are
typically characterized by:

a) the limitations of the diagnostic instruments,
    which usually were not designed to look for
    the BPD,

b) a moving accelerator whose velocity transverse
    to the geomagnetic field may preclude BPD in
    rockets as well as in the space shuttle.

The radio frequency measurements in space experiments are very important in determining when BPD occurred. If one considers electromagnetic radiation above
the plasma frequency as an indicator of BPD, it was
there on ARAKS, but not on Echo V. The broadband whistler
mode below the electron gyro frequency is a reliable
indicator of BPD in the vacuum chamber. This was present
on the E-parallel-B rocket.

On one hand it was emphasized that we really do have
a good understanding of the basic processes in the BPD,
while on the other hand it was pointed out that there is
still considerable disagreement on the details and interpretation of the laboratory studies. Thus it is not just
the space experiments that raise questions.

When the electron beam is employed as a probe in
magnetospheric physics, one wants to avoid the BPD and
retain a cold beam. Does one know enough about the BPD
to be able to avoid it? Yes, one simply uses low current
or small diameter beams.

Sometimes, however, one wants the BPD, for example
to neutralize a payload employing a particle accelerator.
One should not think simply in terms of avoiding the
BPD so that beams can be used as ideal probes. There are
plasmas throughout the universe and there are important
processes that are not well understood. Understanding
the BPD is important basic physics, thus laboratory
experiments in the unbounded laboratory of space are
needed. Furthermore, one wants to understand hot beams
since they are common in nature. The BPD is a very good
way to make hot beams.

Does the BPD play a role in natural auroras? The
intensity variation of auroral luminosity is consistent
with simple collisions. This is true, even for auroral
rays extending to high altitudes. However, there are
occasional anomalies. Thin patches observed in some
pulsating auroras, with vertical extent of less than
1 km, cannot be explained by simple Coulomb scattering.

Chapter 4: THEORETICAL ASPECTS OF THE BEAM
PLASMA INTERACTIONS

THEORY OF BEAM PLASMA DISCHARGE

K. Papadopoulos*

Science Applications, Inc.
McLean, VA 22102

ABSTRACT

Although the beam-plasma discharge (BPD) represents one of the first plasma physics phenomena, investigated as early as 1960, many of the quantitative aspects of the problem are only now beginning to emerge. This can be attributed partly to the fact that the understanding of the energization processes of the ambient plasma interacting with energetic electron beams required the theoretical development of strong turbulence, which was not accomplished till the middle 1970's, and partly to the importance of BPD, for active electron beam experiments in space. On a superficial level BPD is nothing more than an R-F discharge, with the exception that the excited waves are electrostatic and near the ambient plasma frequency (ω_e). This difference has profound consequences in the resulting absorption of the wave energy by the plasma particles. Under collisionless circumstances the energy absorption is by a few ($\sim 1\%$) electrons which are thus accelerated to high energies (~ 100's eV). The ionization process is then dominated by the suprathermal tails rather than by the heating of the mainbody of the electrons. A proper description of BPD requires the following information:

(1) The deposition rate of the beam energy to plasma waves, and the accompanying beam energy relaxation length.

(2) The ratio of wave energy heating electrons by electron neutral collisions, and creating suprathermal tails by collapse (i.e., turbulent acceleration).

---

*Permanent Address: University of Maryland, College Park, MD 20742

(3) The physics and time scale of axial and radial confinement
     of the cold, energetic electrons and beam electrons (i.e.,
     electrostatic, collisional or anomalous diffusion, vehicle
     motion, etc.).

Based on these we present a description of BPD, its properties and its
accompanying wave emission phenomena and apply the results to
laboratory situations and to beam plasma experiments from rockets
and the space shuttle.

INTRODUCTION

The beam plasma discharge (BPD) has been one of the oldest
manifestations of important collective plasma phenomena observed in
the laboratory (Kharchenko 1962; Getty and Smullin, 1963). This
type of discharge can be described qualitatively as an r-f discharge in which the high frequency oscillations are driven by a
beam plasma instability and can be electrostatic. Despite its early
identification in the laboratory, a theoretical understanding past
the linear level, is only now emerging. This can be attributed
partly to the fact that the key nonlinear phenomena by which the
energy of the beam excited waves is transferred to the ambient cold
electrons were not identified till the early 1970's (Papadopoulos,
1975; Papadopoulos and Coffey, 1974; Papadopoulos and Rowland, 1978;
Papadopoulos and Freund, 1979), and partly to the early failures of
creating hot plasmas using BPD. The interest in BPD was revived
following Bernstein's suggestion that it might be a key factor in
achieving vehicle and beam neutralization in electron beam experiments in space. This suggestion was followed by laboratory demonstration of BPD in the large vacuum facility at the NASA-Johnson
Space Center under conditions prevailing in the lower ionosphere
(Bernstein et al. 1975, 1978, 1979). It is the purpose of this
paper to present the current theoretical understanding of the BPD.
The emphasis will be on the nonlinear plasma behavior, because, as
it will be shown, is the controlling factor of the phenomenon. In
comparing the theoretical model with the experiments we will focus
first on the Johnson Space Center results several of which are
presented in this meeting. This will be followed by an analysis of
the corresponding space born accelerator results.

BPD WITH CLASSICAL ENERGY DEPOSITION

Basic Mathematical Description

The BPD can be described by a system of equations consisting of
the particle and energy transfer equations. The most general description, involves kinetic equations for the ambient plasma and
beam electrons. Such a description is not only complicated but
unnecessary. A lower level of description, treats the beam elec-

trons as a fluid, while the ambient electrons can be treated also
as one or more fluids, depending on their temperature (i.e., we
split the electron distribution function in energy bits). We
discuss first the transfer equations, by considering the ambient
electrons as one fluid characterized by one temperature T_e.
Although, as we will see, such a description is most often inade-
quate, it illustrates in a rather simple fashion the underlying
physical principles. In this case and for $n_e \ll N$ the basic
equations are:

$$\frac{d}{dt} n_e = n_b N <\sigma_i(V_b)V_b> + n_e N < \sigma_i(V_e)V_e > - \beta n_e^2 - \frac{n_e}{\tau} \qquad (1)$$

$$\frac{d}{dt} T_e = - \varepsilon_i N <\sigma_i(V_e)V_e> - H - \frac{T_e}{\tau} + \frac{Q}{n_e} S(n_e,n_b,\Omega_e \ldots) \qquad (2)$$

Equation (1) describes the rate of increase of the ambient electron
density n_e, in the presence of a neutral gas with density N, for
$n_e \ll N$. The first term on the r.h.s., of Eq. (1) gives the gas
ionization by a beam of density n_b and energy $\varepsilon_b = mV_b^2/2$ ($\sigma_i(V_b)$ is
the ionization cross section for energy ε_b). The second term corresponds to ionization by ambient electrons of temperature $T_e =
3/2 \, mV_e^2$. The third term describes electron loss from the discharge
region due to recombination (β is the recombination coefficient).
The last term, describes the electron loss rate, by using a parameter
τ, which depends on the electron confinement time in the discharge
region. The value of τ depends on the specific circumstances and
can be controlled by collisional or anomalous (i.e., Bohm) diffusion
across the magnetic field (i.e., $\tau = R^2/D$), free streaming through
ambipolar potentials along $\underline{B}$, etc.

Equation (2) describes the electron heating rate. The first
term on the r.h.s. describes the energy loss due to ionization (ε_i
is the ionization energy). The term H describes the energy loss
due to excitation of neutral atoms and heating of the ions. The
third term represents the convective energy loss, consistent with
the particle loss time τ. The last term, is a key term in the BPD
and describes the heating of electrons, due to waves excited by
beam-plasma instabilities. The factor $S(n_e,n_b,\Omega_e,\ldots)$, is a switch
on-off factor (i.e., a step function), which has the value 1, for
plasma and beam parameter conditions such that the threshold for
linear instability is exceeded, and zero, for parameters below the
linear threshold. Evaluation of the value of Q, i.e., the resulting
heating rate, requires a nonlinear calculation or particle
simulations. We will expand on these later. Before proceeding any
further, it is instructive to look at some properties of Eqs. (1)
and (2).

a.   In the absence of b-p instabilities (i.e., S = 0 at all
     times) the system described in Eq. (1) represents purely
     collisional processes. Neglecting recombination, we find

$$n_e(t) = n_o + n_b \, N \, <\sigma_i(V_b)V_b> \tau \left[1 - e(-\tfrac{t}{\tau})\right]$$

with an asymtotic value

$$n_e = n_b \, N <\sigma_i(V_b)V_b> \tag{3}$$

b.   If initially S = 0 (i.e., S(n$_o$) = 0), but due to the
     collisional increase in density, as given by Eq. (3), S
     becomes unity, the situation could lead to BPD, because
     the term QS in Eq. (2) will trigger electron heating. BPD
     will result only if $Q > n_e T_e/\tau$ for T_e such N $<\sigma_i(V_e)V_e>>1/\tau$.

c.   If initially S = 1, BPD will result as described above if
     $Q > n_e T_e/\tau$ and N $<\sigma_i(V_e)V_e> > 1/\tau$.

On the basis of the above rather general description, we can
identify the specific physical problems that have to be solved
separately for each space or laboratory situation. These are:

a.   The functional dependence of the trigger function S on
     the ambient and beam parameters. This can be accomplished
     on the basis of linear plasma physics.

b.   Determination of the functional dependence of the heating
     rate Q on the beam-plasma system parameters. This requires
     nonlinear computation of the excited wave spectrum and of
     the collisional or collective processes by which this
     energy is transferred on the plasma electrons.

c.   A determination of the loss time τ. The time τ, could be
     a function of velocity, which we have ignored in our sim-
     plified description of the electrons as a one temperature
     fluid. As will be shown in the next few sections the
     value of τ depends on whether the dominant loss is axial
     or radial, whether ambipolar, collisional, or Bohm diffusion
     dominates, and on the type of laminar electrostatic poten-
     ial sheaths that are formed at the ends to slow down elec-
     tron free streaming. These concepts parallel very closely
     the confinement physics of mirror and long solenoid fusion
     devices.

We will address below these issues as applied to some particular
situations of interest. We should, however, warn the reader at this
stage that the use of the one temperature equation for the electrons

presents several pitfalls. It can be justified only if the energy deposition from instability excited waves to electrons is dominated by electron neutral collisions, or if the collisionless heating results in a one temperature distribution. However, many experimental situations of interest are dominated by the formation of suprathermal electron tails, which end up as the controlling agent of the ionization processes. In the analysis that follows we neglect recombination and energy losses due to excitation (i.e., $\beta = 0$, $L = 0$ in Eqs. (1) and (2)). This can be justified in most but not all situations.

Energy Deposition Rate

Determination of the value of the term QS in Eq. (2), requires answering the following questions:

(a) Is the beam plasma system linearly unstable and if so to which modes?

(b) Is the instability absolute or convective, and if convective is it suppressed by the finite system size (i.e., growth length longer than the system size)?

(c) What is the amplitude of the instability generated waves at saturation?

(d) How is the wave energy transferred to the ambient plasma particles?

A large amount of work has been done over the last 20 years with respect to the linear theory of beam plasma instabilities, required to answer the first two questions. Besides the monograph by Briggs (1964) an extensive review of the subject can be found in Linson and Papadopoulos (1980). The question of absolute vs. convective instabilities has been discussed in Manickam et al. (1975, 1976) and more recently in Le Queau et al. (1980) for arbitrary beam to plasma ratios. We are not going to expand on the subject further than pointing out that the frequency range of the excited waves, is between

$$\Omega_{UH} \geq \omega \geq \max\ (\omega_e, \Omega_e)$$

$$\min(\omega_e, \Omega_e) \geq \omega \geq \Omega_{LH}$$

(4)

where Ω_{UH} and Ω_{LH} are the upper and lower hybrid frequencies. The saturation level of waves required to answer question (c), has also been the subject of extensive research as can be found in Linson and Papadopoulos (1980).

The approach we follow in this paper is to deemphasize the complex problems associated with the first three questions, while emphasizing the importance of the last question. As we will see from the analysis that follows this can provide an important insight on the dynamic behavior of BPD. We therefore, assume that a wave energy level $E_0^2/8\pi$, in the frequency range defined by (4) is given and we determine the heating rate Q.

The phase velocity of the waves excited by the beam $v_{ph} \gg V_e$. Therefore, in the absence of nonlinear processes, the transfer rate will be controlled by the electron neutral collision frequency. Namely if the wave frequency is ω_0 the heating rate will be

$$Q_c = \frac{n_e e^2 E_0^2}{2m} \frac{\nu_{en}}{(\omega_0 - \Omega_e)^2 + \nu_{en}^2} \tag{5}$$

For $\omega_0 \neq \Omega_e$ and $\omega_0, \Omega_e > \nu_{en}$, Eq. (6) can be approximated by

$$Q_c = \frac{n_e e^2 E_0^2}{2m} \frac{\nu_{en}}{(\omega_0 - \Omega_e)^2} \approx \frac{\omega_e^2}{(\omega_0 - \Omega_e)^2} \nu_{en} \frac{E_0^2}{8\pi} \approx \nu_{en} \frac{E_0^2}{8\pi} \tag{6}$$

The resulting heating will of course be bulk electron heating. The heating determined by Eqs. (5) and (6) assumes that

$$\tau \gg \frac{1}{\nu_{en}}, \quad \frac{1}{\nu_{ion}(T_e^i)} \tag{7}$$

otherwise the electrons will be lost from the system before having time to randomize their oscillatory velocity or making ionizing collisions. This is equivalent to the ionization mean free path λ_{ion} and λ_{en} such that

$$\lambda_{en}, \lambda_{ion} \ll L \tag{8}$$

where L is the axial system size. It is interesting to note that, for the Johnson S.C. experiment both $\lambda_{en}, \lambda_{ion} > L$ (the axial system size is 20m). Therefore the existence of BPD, in this experiment cannot be understood on the basis of the above considerations (Szuszczewicz et al. 1981).

An equation for $E_0^2/8\pi$ can be found by considering conservation of energy flux (Lebedev et al. 1976), as

$$\frac{d}{dt} \frac{E_o^2}{8\pi} = \eta\, n_b\, \varepsilon_b\, V_b\, \frac{1}{L} - \nu_{en} \frac{E_o^2}{8\pi} - V_g \frac{E_o^2}{8\pi} \frac{1}{L} \tag{9}$$

where $\eta \leq 1$ is the conversion efficiency of beam energy to plasma waves which can be found experimentally or theoretically from the beam relaxation length (Linson and Papadopoulos (1980)) and V_g is the group velocity of the waves. Notice that Eq. (9) includes the convective stabilization term for large V_g. We can simplify Eq. (9) by assuming that the last term is small or by including it in the switch factor S. In this case Eq. (9) can be solved so that

$$\frac{E_o^2}{8\pi} (t) = \frac{\eta}{2} n_b \varepsilon_b \frac{V_b}{L\nu_{en}} (1 - e^{-\nu_{en} t}) \tag{10}$$

For time scales of interest $\nu_{en} t \gg 1$, so that

$$\frac{E_o^2}{8\pi} = \frac{\eta}{2} n_b \varepsilon_b \frac{V_b}{L\nu_{en}} \tag{11}$$

and

$$Q_c = \frac{1}{2} \eta\, n_b \varepsilon_b \frac{V_b}{L} \tag{12}$$

Particle and Energy Loss Rates

In order to close our system of Eqs. (1), (2) and (12) we require the confinement time τ (i.e. the time the electrons remain in the BPD region). Since in all situations of interest, there is an axial magnetic field, the confinement time will be the shortest of the axial or radial confinement. The radial confinement time will be given by ambipolar diffusion

$$\tau_\perp \sim \frac{R^2}{D_\perp} \tag{13}$$

where R is the discharge radius and $D_\perp$ the transverse diffusion coefficient. The value of $D_\perp$ depends on whether classical or Bohm diffusion dominates. If classical diffusion dominates

$$D_\perp \sim D_C = \frac{V_e^2}{\nu_{en}} \left(\frac{\nu_{en}}{\Omega_e} \right)^2 \tag{14}$$

while for Bohm diffusion

$$D_{\perp} \stackrel{\sim}{\sim} D_B = \frac{V_e^2}{\Omega_e} \tag{15}$$

The parallel confinement time will depend on whether L, the discharge length is $L > \lambda_{en}$ or $L < \lambda_{en}$. For $L > \lambda_{en}$

$$\tau_{||} \stackrel{\sim}{\sim} \frac{L^2}{D_{||}} \stackrel{\sim}{\sim} \frac{L^2 \nu_{in}}{c_s} \tag{16}$$

where $c_s = (T_e/M)^{\frac{1}{2}}$ the sound speed. For $L < \lambda_{en}$

$$\tau_{||} = \frac{L}{c_s} \tag{17}$$

The appropriate time to be inserted in Eqs. (1) and (2), corresponds to

$$\tau_{||} = \min(\tau_{||}, \tau_{\perp}) \tag{18}$$

Trigger Requirements, Scaling and Observables

On the basis of the above we can write the set of equations describing the BPD for the case of classical deposition from Eqs. (1), (2) and (11), for the case that some linear instability triggered the switch factor S to become 1. Neglecting the direct ionization by beam electrons and assuming $\beta = 0$ and $H = 0$, we find

$$\frac{d}{dt} n_e = n_e N < \sigma_i(V_e)V_e > - \frac{n_e}{\tau} \tag{19}$$

$$\frac{d}{dt} T_e = - \varepsilon_i N < \sigma_i(V_e)V_e > - \frac{T_e}{\tau} + \frac{1}{2} \eta \frac{n_b}{n_e} \varepsilon_b \frac{V_b}{L} \tag{20}$$

These equations must be supplemented by the value of τ as given by Eqs. (13–18). A numerical solution of these equations for parameters of interest will be presented elsewhere. From Eqs. (19) and (20) we see that in addition to switching an instability, BPD has additional requirements. They are given by

$$\frac{1}{2} \eta \frac{n_b}{n_e} \varepsilon_b \frac{V_b}{L} > \frac{T_o}{\tau} \tag{21}$$

with the value of $T_e \stackrel{\sim}{\sim} T_o = mV_o^2/2$, determined by

$$N < \sigma_i(V_o)V_o >> \frac{1}{\tau} \tag{22}$$

Notice that even using the maximum value of $<\sigma_i(V_o)V_o> \sim 10^{-7}$ and with $\tau = \tau_{||} = L/c_s$ for the parameters of Johnson S.C. BPD experiment, we can never satisfy the inequality given by Eq. (22). This indicates clearly the necessity for axial confinement, an issue that will be addressed in the next section.

A steady state BPD based on Eqs. (19) and (20), will have density n_e given by

$$n_e = \frac{1}{2} \eta \, n_b \frac{\varepsilon_b}{\varepsilon_i + T_o} \frac{V_b \tau}{L} \tag{23}$$

and a temperature to such that

$$N < \sigma_i(V_o)V_o > = \frac{1}{\tau} \tag{24}$$

with τ determined by Eqs. (13–18). For example if the dominant loss mechanism is Bohm diffusion

$$n_e = \frac{1}{2} \eta \, n_b \frac{\varepsilon_b}{\varepsilon_i + T_o} \frac{V_b}{V_o} \frac{R^2}{L r_e} \tag{25}$$

where r_e is the electron gyroradius. The value of the temperature T_o, will be given by

$$< \sigma_i(V_o)V_o > = \frac{V_o^2}{NR^2 \Omega_e} \tag{26}$$

Similar considerations apply to any other values of τ.

Before closing this part we should note that the case of BPD described above should be associated with the following observables.
   (i)   Maxwellian electrons heated to temperatures comparable or larger than the ionization potential
   (ii) Absence of laminar potentials
   (iii) Absence of suprathermal tails
   (iv) Ionization rates proportional to N.

BPD WITH ANOMALOUS ENERGY DEPOSITION

The Role of Suprathermal Electrons

It was shown before that the triggering of BPD in the Johnson S.C. experiments of Bernstein et al (1975, 1978, 1979), cannot be explained on the basis of the considerations of the previous part.

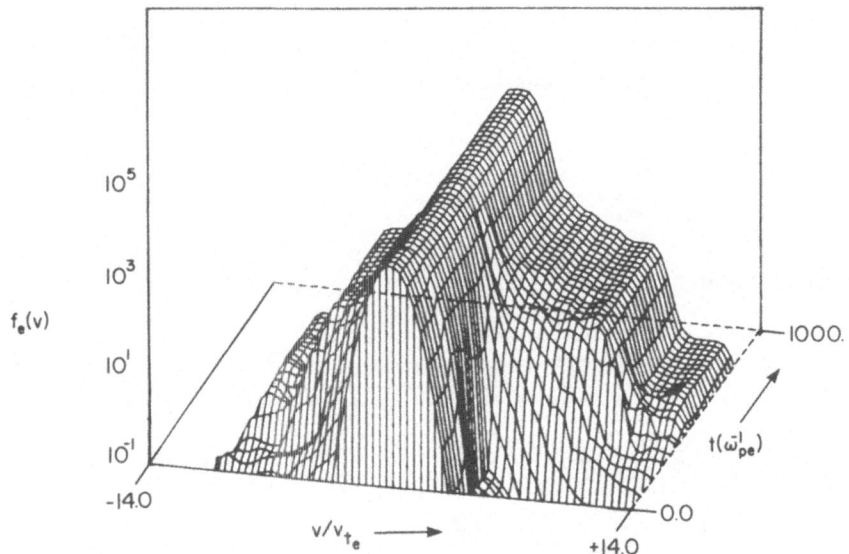

Fig. 1   Time evolution of the electron distribution function in the
         presence of large amplitude (i.e., $E_o^2/8\pi nT_e \underset{\sim}{\sim} .4$) electron
         plasma oscillations as seen in particle simulations (from
         Rowland, 1979).   The vertical scale is logarithmic; the
         phase velocity of the pump is at $v_{ph} = .50V_e$ to the left.
         The energy is deposited in nonthermal tails.

The resolution of the dilemma, as well as a major factor in our
progress in understanding BPD phenomena, comes from the recognition
of the importance of suprathermal electrons collisionlessly accelerated by the interaction of the beam with the ambient plasma. This
was first recognized in the work of Papadopoulos and Coffey (1974)
in connection with the interaction of natural auroral beams with
the ionospheric plasma. More detailed analyses were later performed
in Papadopoulos (1975), Matthews et al. (1976), Papadopoulos and
Rowland (1978) and Rowland (1979). We summarize the pertinent
results here and refer the interested reader to the above references.

When the beam plasma excited waves have frequency $\omega_o \underset{\sim}{\sim} \omega_e$ and
their amplitude exceeds a threshold given by the inequality

$$\frac{E_o^2}{8\pi nT_e} > k_o^2 \lambda_D^2 \tag{27}$$

where k_o is the typical wavenumber of the excited waves, the energy transfer of the wave-energy to the plasma is dominated by nonlinear collective effects (i.e. plasma wave collapse) and results in the formation of suprathermal electron tails. Figure 1 shows the results of particle simulations performed by Rowland (1979) which demonstrate clearly the formation of energetic electron tails, as well as the nonlinear energy transfer. For our purposes here it is sufficient to note that

(i)   The energy transfer rate is given by

$$\gamma_{NL} \sim \omega_e \left(\frac{m}{M} \frac{E_o^2}{8\pi nT} \right)^{\frac{1}{2}} \tag{28}$$

and maximizes at

$$\gamma_{NL}^m \sim \omega_e \left(\frac{m}{2M} \right)^{\frac{1}{2}} \tag{29}$$

(ii) If the ambient plasma is thermal the ratio of the suprathermal particles n_T to cold particles n

$$\alpha = \frac{n_T}{n} \sim \left(\frac{m}{M} \right)^{\frac{1}{2}} \tag{30}$$

(iii) The heating of the suprathermals can be described as a diffusion process (Manheimer and Papadopoulos (1975); Papadopoulos and Rowland (1978)).

In this case the energy transfer rate Q is given by

$$Q_w = \nu^* \frac{E_o^2}{8\pi} \tag{31}$$

$$\nu^* \sim \omega_e \left(\frac{m}{2M} \right)^{\frac{1}{2}} \tag{32}$$

Therefore, when $\nu^*/\nu_{en} > 1$ the energy Eq. (2) should be replaced or at least supplemented by an equation describing the suprathermal tails and an ionization term caused by them should be included in Eq. (1). In addition to these, the suprathermal tails have a profound effect on the electron confinement time when $L < \lambda_{en}$, such as the Johnson S.C. experiment. We address this next.

Axial Confinement for L < λ_{en}

The presence of suprathermal electrons can increase substantially the axial confinement time $\tau_{||}$, over the one given by Eq. (17) for L < λ_{en}. This can be understood by examining the physics involved in deriving Eq. (17). For L < λ_{en} electrons free stream out of the ends of the system with their thermal velocity in a time L/V_e. Quasineutrality requires an ambipolar electric field which will equalize the electron and ion fluxes. A potential $e\Phi/T_e \sim 1$ is built up at the system ends with dimensions of the order of a few λ_D, which decelerates the electrons and accelerates the ions so that the two fluxes become equal. The rate of the plasma flow is then $n_e(2e\Phi/M)^{\frac{1}{2}}/2 \sim n_e(T_e/M)^{\frac{1}{2}} \sim n_e c_s$. In some sense this potential confines half of the electrons (i.e. the ones with $mV^2/2 < e\Phi$).

Let us examine next the case where the suprathermal tails generated with temperature $T_T = mV_T^2/2$, are such that $\alpha V_T > V_e$, while L < λ_{en}. In this case balancing the electron and ion fluxes requires a potential

$$e\Phi \sim \alpha^2 \frac{M}{m} T_T \tag{33}$$

since the suprathermal electrons control the axial loss rate. Since $\alpha^2 M/m \sim 1$ and $T_T \gg T_e$, the cold electrons will be electrostatically confined and their loss will be controlled by radial diffusion. A large fraction (i.e. almost half) of the suprathermals will be also confined. The energy loss rate will be given by $\alpha n_e T_T (T_T/M)^{\frac{1}{2}}/L$ corresponding to an energy confinement time

$$\tau_E = \frac{L}{V_T} \tag{34}$$

Analytic Model and Observables

On the basis of the above considerations we generalize our equations describing BPD by including the suprathermal electrons with density $n_T \ll n_e$ and temperature $T_T = mV_T^2/2 \gg T_e$. The system of equations that replaces Eqs. (19) and (20) is

$$\frac{d}{dt} n_e = n_e N <\sigma_i(V_e)V_e> + n_T N <\sigma_i(V_T)V_T> - \frac{n_e}{\tau} \tag{35}$$

$$\frac{d}{dt} T_e = -\varepsilon_i N <\sigma_i(V_e)V_e> - \frac{T_e}{\tau} + \frac{1}{2} \frac{\nu_{en}}{n_e} \frac{E_o^2}{8\pi} \tag{36}$$

$$\frac{d}{dt} T_T = -\varepsilon_i N <\sigma_i(V_T)V_T> - \frac{1}{2} \frac{T_T}{\tau_E} + \frac{\nu^*}{n_T} \frac{E_o^2}{8\pi} \tag{37}$$

with $E_o^2/8\pi$ determined by Eq. (9) with ν_{en} replaced by $\nu + \nu^*$. For $\nu^* \gg \nu_{en}$, the only relevant equations are (35) and (37) while Eq. (36) describes the temperature of the cold component which we assume to be below the ionization potential, so that the only source term of Eq. (35) is the second term on the r.h.s. In this case

$$\frac{d}{dt} \, n_e = n_T N <\sigma_i(V_T)V_T> - \frac{n_e}{\tau} \tag{38}$$

$$\frac{d}{dt} \, T_T = - \, \varepsilon_i N <\sigma_i(V_T)V_T> - \frac{1}{2} \, \frac{T_T}{\tau_E} + \frac{\nu^*}{n_T} \frac{E_o^2}{8\pi} \tag{39}$$

The confinement time will be given by $\tau \sim \tau_\perp$, i.e. dominated by transverse losses. A proper determination of τ_E requires a kinetic description, and will be presented elsewhere. Some comments with respect to approximate values will be given in the next section. Although Eqs. (38) and (39) represent a very crude description of the systems behavior, they exhibit many properties, which help to identify the physics involved in both laboratory and space electron beam experiments. This will be clarified in the subsequent parts

In closing this part we should note that the observables associated with this situation will be

   (i)   Presence of energetic non-Maxwellian electron tails
  (ii)   Laminar potential for $L < \lambda_{en}$
 (iii)   Wave activity near ω_e, confined mainly inside the beam
  (iv)   Energetic ion fluxes.

APPLICATIONS TO SPACE AND LABORATORY SITUATIONS

The Johnson S.C. BPD Experiment

Based on the excellent results produced by this experiment and discussed in detail in Bernstein et al. (1978, 1979) as well as in many contributions of the present proceedings, we attempt to validate the theoretical framework set forth in the previous parts. Space limitations do not allow more than a brief description of the experiment and the observed scaling. An electron beam with perveance $10^{-6} AV^{-3/2}$ was injected in a large vacuum chamber (i.e. 17m x 26m). Typical operating conditions were: Beam current I (0-100mA), beam energy ε_b (.5-2keV), magnetic field B (.3-2G), path length, L(10-20m), injection angle θ (0-80°) and neutral gas composed of water vapor and N_2 with density between $3 \times 10^{10} - 10^{12}$ cm^{-3}. The main features were:

   (a)   BPD appeared at a critical current I_c, which obeyed the
         empirically found relation

$$I_c \sim \frac{\epsilon_b^{3/2}}{B^{.7}PL}$$

where P was the ambient pressure ($P < 2 \times 10^{-5}$torr).

(b)   Measurement of the electron density indicated that near
      BPD threshold (Sczuszczewicz, 1981)

$$\frac{\omega_e}{\Omega_e} \sim 6 \pm 1$$

for parallel injection.

(c)   The ambient electrons were cold (i.e. .5eV), while there
      were indications of a superimposed energetic electron
      flux component, as well as energetic ion fluxes.

(d)   Strong oscillations near the local plasma frequency (ω_e)
      were observed at the BPD threshold; they were localized
      inside the beam radius (Jost and Anderson, private
      communication 1981). The beam distribution was broadened
      by 10-15%, towards the low energy side (Jost et al., 1981).

(e)   The ignition conditions were insensitive to the injection
      angle up to $\theta \sim 60^{o}$.

(f)   There was a marked increase in ion-pair production accom-
      panied by a greatly enhanced profile in 3914 $\overset{o}{A}$ emission.

     In attempting to reconcile these facts with the model of BPD
with classical deposition we face irreconcilable difficulties. The
basic problem is that at the experimental pressures the
mean free paths for both heating and ionization processes are much
larger than the chamber length L $\sim$ 10-20m. This necessitates axial
confinement. As shown in the second part of the previous section,
axial confinement can be achieved for the case of anomalous deposition in which case most of the beam energy is depositied in
suprathermal electrons. For this to be achieved, it is required
that the frequency of the excited waves be near ω_e. An analysis
of the conditions for this to occur, including the finite size of
the beam (r_o) and the plasma R, is given in (Rowland et al., 1981).
The result is that the following condition should be satisfied

$$\omega_e^2 \geq 2 \frac{V_b^2 \cos^2\theta}{r_o^2 \ln \frac{R}{r_o}} \tag{40}$$

This result is independent of the ω_e/Ω_e ratio. In applying this result to the experiment the value of r_o should be determined including both the Larmor radius as well as the space charge term (Linson and Papadopoulos, 1980). The result is

$$r_o = 2.1 \times 10^2 \frac{I^{\frac{1}{2}}}{BV_g^{\frac{1}{4}}} \left[\ell n \frac{r_o}{r_g} + \frac{33}{K} \sin^2\theta \right]^{\frac{1}{2}} \text{ cm} \qquad (41)$$

where K is the gun pervervs in micropervs, r_g is the gun radius, I and V_g the gun current in amperes and potential in kV and the magnetic field B is in gauss. For our range of parameters with $r_g \sim 2mm$ Eqs. (40) and (41) give

$$\frac{\omega_e}{\Omega_e} \geq \frac{5 \times 10^{-1} V_g^{3/4}}{I^{1/2}} \frac{\cos\theta}{\left[1 + \frac{8.5}{K} \sin^2\theta \right]^{\frac{1}{2}}} \qquad (42)$$

Notice that for $\theta \leq 60^o$ this condition is rather insensitive to the angle θ in accordance with observation (e). For typical values of parallel injection Eq. (42) gives $\omega_e/\Omega_e \sim 5$, which is consistent with the experimental results. Note that we expect Eq. (42) to underestimate the observed ratio, since it is a necessary but not a necessary and sufficient condition. Observations (c), (d) and (f) are consistent with the ones expected from our analysis as given in the previous section.

While the condition given by Eq. (40) is a necessary condition for BPD based on anomalous absorption for the Johnson S.C. experiment it is not always a sufficient one. This can be seen from Eqs. (38) and (39), which we rewrite here for convenience

$$\frac{d}{dt} n_e = n_b N < \sigma_i (V_b) V_b > n_T N < \sigma_i (V_T) V_T > - \frac{n_e}{\tau} \qquad (38a)$$

$$\frac{d}{dt} T_T = - \varepsilon_i N < \sigma(V_T) V_T > - \frac{1}{2} \frac{T_T}{\tau_E} + \frac{\nu^*}{n_T} \frac{E_o^2}{8\pi} \qquad (39a)$$

In order to create the energetic tails

$$\frac{E_o^2}{8\pi n_T T_T} > \nu^* \tau_E \qquad (43)$$

and

$$\frac{n_T}{n_e} > \frac{1}{N<\sigma_i(V_T)V_T> \tau} \tag{44}$$

If conditions (43), (44) are less restrictive than Eq. (40), Eq. (40) will give the necessary and sufficient condition. Otherwise inequalities (43) and (44) should be added. A detailed numerical study of these requires a better description of τ_E and will be published later. We comment here on some important scalings based on (40), (43) and (44).

Equation (40) with $R/r_o = 2$ and $\cos\theta \sim 1$ can be written as

$$\frac{\omega_b^2 r_o^2}{V_b^2} \geq 2.8 \ \frac{n_b}{n_e}$$

where ω_b is the beam plasma frequency. Note that the left hand side is proportional to the beam perveance K

$$\frac{\omega_b^2 r_o^2}{V_b^2} = 3\times10^4 \ K \tag{45}$$

So that the gun perveance in micropervs (i.e. $10^{-6} \ \frac{A}{V^{3/2}}$), necessary for triggering ω_e waves will be

$$K \geq 10^2 \ \frac{n_b}{n_e} \tag{46}$$

Note that for our case where $K \sim 1$ microperv, $n_e/n_b > 10^2$ which is consistent with the observations. From Eq. (38a) we can find near BPD threshold, i.e. when the first term of Eq. (38a) is still larger than the second, that

$$\frac{n_b}{n_e} \sim \frac{1}{N<\sigma_i(V_b) V_b> \tau} \tag{47}$$

From Eqs. (40) and (47) we find that the experimental scaling

$$I_c \sim \frac{V_g^{3/2}}{B \ L \ P} \tag{48}$$

is consistent with radial Bohm loss being the dominant particle loss (i.e. $\tau \sim 1/B$) and for $L < \lambda_{ion}$, the ionization rate being proportional to L.

More details on these will appear later in the literature. We should note that an independent estimate by Leinbach (private communication), shows that ion pair production as seen from the 3914 Å emission is consistent with a 1% of electrons having been energized to above 20eV and confined. This is consistent with the numbers emerging from our Eqs. (38) and (39) and from the basic nonlinear physics which predicts $\alpha \sim n_T/n \sim (m/M)^{\frac{1}{2}}$ (see Eq. (30)). A final note, before closing this section concerns the stability of the axial confinement. Since the laminar potentials are a function of the anomalous deposition rate, one would expect large bulk oscillations at kHz range, due to the relaxation oscillations in the beam deposition rate as discussed in Papadopoulos and Freund (1979). Such oscillations have been observed in the experiment with amplitude of the order of V/m.

Application to Space Experiments

A detailed analysis of the space experiments is beyond the scope of the present paper. We constrain ourselves here to addressing some major issues as related to the analysis of the second and third parts of this paper. The confinement and finite size aspects of the problem are dominated by the transverse size R, since the axial length, which corresponds to the energy deposition length is very long. In this sense the analysis and interpretation of the space experiment is easier than the corresponding laboratory experiments. Following the analysis of the second and third parts, we can distinguish two regions in altitude corresponding to BPD with classical and anomalous energy deposition. The altitude at which the transition occurs will be given by

$$\nu^* \sim 10^{-2} \omega_e < (\nu_{en})_{max}$$

which for night time conditions corresponds to $N > 10^{11}$ or altitudes between 140-150 km. For altitudes below 140-150 km the density and electron temperature will be given by Eqs. (21-25) and will be characterized by the presence of hot electrons but not any major suprathermal tails; the associated ionization rate, as observed by the 3914 Å emission, will be proportional to the ambient pressure (i.e., N). For altitudes above 140-150 km, a transition to the BPD with anomalous dissipation occurs, characterized by the presence of energetic tails carrying most of the ambient plasma energy. An interesting observational consequence of this can be found from Eq. (39b). Since $\epsilon_i N < \sigma_i (V_T) V_T > > T_T/2\tau_E$ (i.e., no axial loss of energetic electrons since L is very long), the ionization rate

$$n_T N <\sigma_i(V_T)V_T> \sim \frac{\nu^*}{\varepsilon_i} \frac{E_o^2}{8\pi}$$

which is independent of N and therefore of altitude. This explains
the observations of the Polar 5 experiments (Grandal et al, 1980)
as well as some earlier results by the Minnesota group. More refined
computations of experimental data and comparison with the theoretical
concepts are beyond the limitations of this presentation and will
appear elsewhere.

SUMMARY AND CONCLUSIONS

We have presented above a general theory of BPD as it applies
to space and laboratory beam injection situations. A key concept
introduced above is that even when beam plasma instabilities are
excited, there are two regimes of BPD with radically different
observational properties. They were described as BPD with classical
or anomalous energy depositions. For high pressures or low altitudes
we expect the classical to dominate. For high altitudes and laboratory experiments with $L < \lambda_{en}$, no BPD will be triggered unless the
unstable waves are near ω_e and their amplitudes at saturation are
large enough to create suprathermal tails by collapsing. On the
basis of these considerations we were able to account for the details
of BPD at the Johnson S.C. experiment and resolve several mysteries
connected with space injection experiments.

ACKNOWLEDGMENTS

The author is appreciative of many discussions with Drs. Anderson
Bernstein, Chang, Kellog, Leinbach, Linson, Szuszczewicz and Yost.
The work was supported under Rice University NASA subcontract
NAS8-33777-1.

REFERENCES

Bernstein, W., Leinbach, H., Cohen, H., Wilson, P.S., Davis, T.N.
    Hallinan, T., Barker, B., Martz, J., Zeimke, R., and Huber, W.,
    1975, "Laboratory observations of RF emissions at ω_{pe} and
    $(n+1/2)\omega_{ce}$ in electron beam-plasma and beam-beam interactions,"
    JGR, 80:4375.
Bernstein, W., Leinbach, H., Kellogg, P., Monson, S.J., Hallinan, T.,
    Garriott, O.K., Konradi, A., McCoy, J., Daley, P., Baker, B.,
    and Anderson, H.R., 1978, "Electron beam experiments: the
    beam-plasma discharge at low pressures and magnetic field
    strengths," Geophys. Res. Ltrs., 5:127.

Bernstein, W., Leinbach, H., Kellogg, P.J., Monson, S.J., and Hallinan, T., 1979, "Further laboratory measurements of the beam-plasma discharge," JGR, 84:7271.

Briggs, R.J., 1964, "Electron-stream interactions with plasmas," MIT Press, Cambridge, MA.

Getty, W.D., and Smullin, L.D., 1963, "Beam-plasma discharge: buildup of oscillations," J. Appl. Phys., 34:3421.

Grandal, B.,Thrane, E.V. and Troim, J., 1980, "Polar 5 - an electron accelerator experiment within the aurora," Planet. Space Sci., 28:309.

Jost, R.J., Anderson, H.R., and McGarity, J.O., 1981, "Measured electron energy distributions during electron beam plasma inter-actions," Geophys. Res. Lett. (in press).

Kharchenko, I.F., Fainberg, Ya. B., Nikoloev, R.M., Kornilov, E.A., Lutsenko, E.I. and Pedenko, N.S., 1962, "The interaction of an electron beam with a plasma in a magnetic field,"Sov. Phys.-Tech. Phys., 6:551.

LeQueau, D., Pellat R., and Saint-Marc, A., 1980, "An investigation of the electrostatic linear.instabilities of a radially limited electron beam," A. Geoph., 36:433.

Linson, L.M. and Papadopoulos, K., 1980, "Review of the status of theory and experiment for injection of energetic electron beam in space," 1980, SAI Tech. Report No. LAPS 69.

Manheimer, W., and Papadopoulos, K., 1975, "Interpretation of soliton formation and parametric instabilities," Phys. Fl., 18:1397.

Manickam, J., Carr, W., Rosen, B. and Seidl, M., 1975, "Convective and absolute instabilities in beam plasma systems," Phys. Fl., 18:369.

Mathews, D.L., Pongratz, M., and Papadopoulos, K., 1976, "Nonlinear production of suprathermal tails in auroral electrons," J. Geophys. Res., 81:123.

Papadopoulos, K., 1975, "Nonlinear stabilization of beam plasma instabilities by parametric effects," Phys. Fl., 18: 1769.

Papadopoulos, K., and Coffey, T., 1974, "Nonthermal features of the auroral plasma due to precipitating electrons," J. Geophys. Res., 79:674.

Papadopoulos, K., and Rowland, H., 1978, "Collisionless effects in the spectrum of secondary auroral electrons at low altitudes," J. Geophys. Res., 83:5768.

Papadopoulos, K., and Freund, H., 1979, "Scaling laws for the strongly turbulent electron-beam-plasma interaction," Comments Plasma Phys., 15:113.

Rowland, H., 1980, "Strong turbulence effects on the kinetic beam plasma instability," Phys. Fl., 23:508.

Rowland, H., Chang, C.L. and Papadopoulos, K., 1981, "Scaling of the beam plasma discharge," J. Geophys. Res. (in press).

Szuszczewicz, E., Lin, C.S., Walker, D.N. and Bernstein, W., 1981, "Ignition of the beam plasma discharge and its dependence on electron density," these proceedings.

DISCUSSION

Mæhlum: What are the critical beam parameters for testing your theory? Current, energy, injection angle?

Papadopoulos: The most important parameter is the beam current. The energy deposition is basically (i e for constant perveance) linearly proportional to the current. Another important parameter is injection angle, but only for angles greater than 65-70 degrees. For smaller angles the results should be rather insensitive.

Schmerling: If an accelerator on a spacecraft is moving with a typical orbital velocity of 7 km/sec, the source is spread over about 140 m for a 20 msec pulse. How do the critical parameters for the onset of BPD scale under these circumstances?

Papadopoulos: The answer to this depends critically on the ambient neutral density (which for the shuttle can be its own atmosphere). Our preliminary estimates indicate BPD with time scale shorter than milliseconds for 200-250 km altitudes.

Lavergnat: In the anomalous energy deposition the cold plasma must be confined inside the beam in order to create a tail of the distribution function. Could you explain how this confinement occurs, and what are the basic principles of it?

Papadopoulos: The radial confinement is standard magnetic confinement, controlled either by classical diffusion or by Bohm diffusion. The axial confinement is by electrostatic sheaths. In all cases, the confinement time is at least one order of magnitude larger than the acceleration time.

ELECTRON BEAM AS A SOURCE OF ELECTROSTATIC WAVES

J. Lavergnat[*], D. Le Queau[**], R. Pellat[**], A. Roux[**],
A. Saint Marc[***]

[*] L.G.E., 4 Avenue de Neptune   94100   Saint Maur, France
[**] Centre de Physique Théorique, Ecole Polytechnique
91128 Palaiseau
[***]CESR , CNRS-Université Paul Sabatier, 31029 Toulouse

INTRODUCTION

The waves measurements performed during ionospheric electron beam injection experiments such as the ARAKS one (Lavergnat et al., 1980 ; Dechambre et al., 1980), show clearly two bands of electrostatic waves, the LF one, with frequencies lying below the electronic gyrofrequency f_c, and the HF one in the range $f_p < f < f_h$ where f_p is the electronic plasma frequency and f_h the upper hybrid frequency.

The transient nature of some of the HF observations has been explained by Lavergnat and Pellat (1979) and has been related to spatial particle bunching at the beam's front.

To interpret the continuous part of both the LF and the HF waves , we have explored the electrostatic instabilities of a radially limited electron beam flowing through a magnetized plasma and some nonlinear consequences of these instabilities.

A. Linear stability analysis (Le Queau et al. (1980),(1981))

The experimental injection situation is described by a cold electron beam of radius a flowing along the magnetic field lines through a cold magnetized infinite plasma. The ions form a motionless neutralizing background. All dissipation processes (collisional or Landau damping) are neglected. Radial spreading

of the beam due to out-of-axis injection is modelized by changing
the beam radius when retaining the same injection conditions (gun
voltage and current). This implies the decrease of beam to plasma
density ratio ε. The pertinency of these approximations are
extensively discussed in the previously cited papers.

In such a system, the role of the radially limited beam is
twofold. It is, at first, the energy source of the linearly instable waves, which may be of two kinds, depending of the resonance
line which gives rise to their unstable character : Cherenkov waves
with $\omega \simeq k \, V_b$ and slow cyclotronic waves with $\omega \simeq k \, V_b - \omega_c$.

Secondly, through the boundary conditions at the beam's surface
a denombrable set of well defined modes is picked out from the continuous spectrum of the infinite plasma electrostatic waves whose
dispersion equation is simply :

$$D^{out}(k_{//}, k_{\perp}, \omega) \equiv k_{//}^2\left(1 - \frac{\omega_p^2}{\omega^2}\right) + k_{\perp}^2\left(1 - \frac{\omega_p^2}{\omega^2 - \omega_c^2}\right) = 0$$

In the beam's interior the dispersion function of the electrostatic waves is similarly :

$$D^{in}(k_{//}, k_{\perp}, \omega, \varepsilon) \equiv k_{//}^2\left(1 - \frac{\omega_p^2}{\omega^2} - \frac{\varepsilon\omega_p^2}{(\omega - k_{//}v_b)^2}\right) + k_{\perp}^2\left(1 - \frac{\omega_p^2}{\omega^2 - \omega_c^2} - \frac{\varepsilon\omega_p^2}{(\omega - k_{//}v_b)^2 - \omega_c^2}\right)$$

Taking into account the boundary conditions at the beam's
cylindrical surface we obtain the dispersion equation of the (axi-
symmetric) linearized perturbations of the beam plasma :

$$\begin{cases} T \dfrac{J_1(T)}{J_0(T)} - U \dfrac{H_1^{(1)}(U)}{H_0^{(1)}(U)} \cdot \dfrac{\varepsilon_{\perp}^{out}}{\varepsilon_{\perp}^{in}} = 0 \\[4mm] D^{in}(k_{//}a, T, \omega, \varepsilon) = 0 \\[2mm] D^{out}(k_{//}a, U, \omega) = 0 \end{cases} \qquad (1)$$

T and U are normalized radial wave number, which describe the
radial structure of the solution in the beams interior ($\Phi \simeq J_0(T_r)$)
and the surrounding plasma ($\Phi \simeq H_0^{(1)}(U_r)$). J_0 and $H_0^{(1)}$ are respectively the zero order Bessel and Hankel functions.

These solutions may are organized with respect to the internal structure of the electrostatic potential. We shall speak of
n-nodes solutions when $j_{o,n} < T < j_{o,n+1}$ where $j_{o,n}$ is the nth zero

of Bessel function J_o. We may also encounter surface waves with $\text{Im}(T) >> \text{Re}(T)$.

Depending of the frequency range, hence of the radial behaviour of the solutions at large distance from the beam surface we also have to distinguish between :

- radiating modes ($\omega < \omega_c$ and $\omega_p < \omega < \omega_H$; $\text{Re}(U) >> \text{Im}(U)$)
- guided modes ($\omega_c < \omega < \omega_p$; $\omega > \omega_H$; with $\text{Im}(U) >> \text{Re}(U)$).

In the second case, the electrostatic energy remains confined to the beam's region.

a) <u>High frequency region</u>

Both analytical and computational study of the waves stability show that strong electrostatic turbulence is to be expected in that region.

Figure 1 shows in a particular case (zero node mode ; $\dfrac{\omega_c}{\omega_p} = .4$: $\epsilon = \dfrac{1}{2}\left(\dfrac{\omega_c}{\omega_p}\right)^2$ (which corresponds to Brillouin radius for the beam)) the essential features of the dispersion curves. α and z are nor-malised frequency and parallel wave number ($\alpha = \dfrac{\omega}{\omega_p}$; $z = \dfrac{k\,V_b}{\omega_p}$)

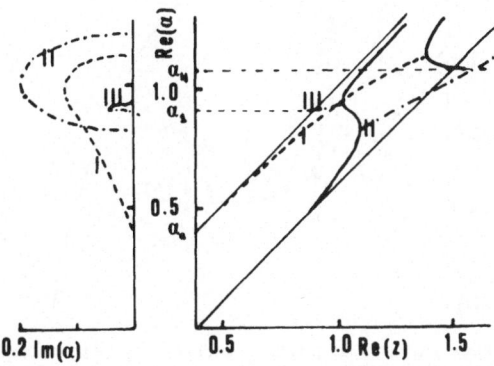

Fig. 1 : Dispersion curves for "zero node" HF modes.
I : Cherenkov ; II:Cyclotronic ; III : Surface wave

At first, it shall be pointed out that, considering the temporal growth rate, there is no essential difference between the
propagating region ($\omega_p < \omega < \omega_H$) and the guided wave region ($\omega_c < \omega < \omega_p$)

Besides the cyclotron Cherenkov (I) and cyclotron wave (II) an
unstable surface wave (III) does appear at the frequency

$$\omega_1 = \omega_c \left(1 + \frac{\omega_p^2}{\omega_c^2 + \epsilon\omega_p^2} \right)^{1/2} .$$

Using the usual Bers procedure (1972), we also have looked for
an eventual absolute character of these instabilities, in particular in the higher frequency range, where the negative parallel
group velocity of the plasma waves allows a "feed back" mechanism.
The figure 2 represents the mapping of the α complex plane into
the z complex plane correlated to the dispersion equation (1), for
the same beam, plasma and wave parameter as in fig. 1. A saddle
point appear at a frequency very close to the upper hybrid frequency

($\alpha_H = \dfrac{\omega_H}{\omega_p} \simeq 1.077$). As this saddle point proceeds from the collapse

of two branches (Cherenkov (I) and an essential branch characterized
by the divergence of Im z when $\alpha \to 1$) whose behaviour is different
when Im α goes through infinity (Im(z_I) $\to +\infty$ when Im(z_{II}) $\to -\infty$),

it is the evidence of an absolute instability.

Physically, this absolute instability is related to the reflecting efficiency of the beam's surface : when high, the energy of
the waves remains sufficiently confined to the beam's region and

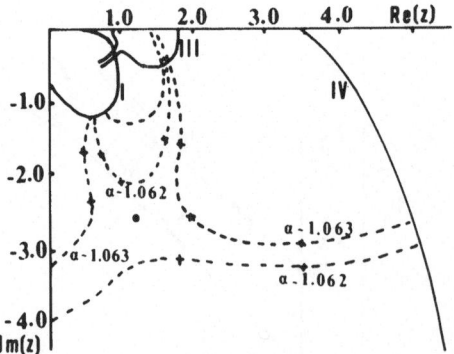

Fig. 2 : Complex mapping of α plane into 3 plane (continuous lines:
         Im(α)=0; dashed lines:lines with Re(α)=cte, indicated by
         the number)

the feed back mechanism associated with negative group velocity
can take place. When the beam's parameter ϵ is decreased below a
certain threshold, the efficiency of the feed back mechanism
becomes unsuficcient to balance the power losses out of the beam
region. Our computation shows this threshold decreases with the
nodes number of the solution.

Moreover we have shown that if an absolutely unstable character may be expected in the axial direction of the beam plasma system, the perturbation remains propagative in the radial direction.

These highly unstable High frequency waves allow a strong
electrostatic turbulence which in the frequency range $\omega_p < \omega < \omega_H$ may
be the origin of the high level emissions observed during the
ARAKS flights.

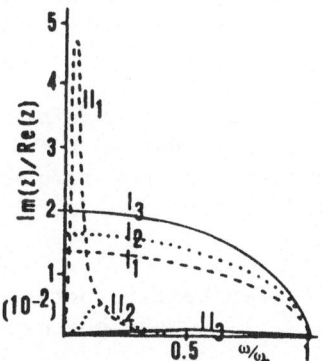

Fig. 3 : Relative spatial growth
rate of the LF instabilities

b) Low frequency region ($\omega < \omega_c$)

In figure 3, we represent the computed values of the relative
spatial growth rate for (zero-node) Cherenkov (I) and (one node)
cyclotronic (II) unstable waves, in this frequency region. 1,2,3
read for three values of the beam's parameter ϵ (normalized to plas-
ma parameter $\dfrac{\omega_c}{\omega_p}$; $\epsilon \dfrac{\omega_p^2}{\omega_c^2} = 0.5, 0.05, 0.005$). The plasma parameter
$\omega_c/\omega_p = .4$ and the gun's characteristics I = 0.5 Amp and $\nu = 15$ keV,
are choosen to fit the ARAKS conditions.

The waves are always convectively instable, in this frequency range, and the computed values of the e - folding length, which are of the order of a few kilometers, are rather large.

The radial attenuation length $\lambda_r^{-1} \sim \mathrm{Im}\, k_r \sim - (\omega_c/\omega)\, \mathrm{Im}\, k_{\parallel})$ is too small to allow these low frequency electrostatic disturbances to be observed at a large distance from the beam.

Moreover these instabilities are always convected along the beam direction, which prevents any observation when the receiver lies behind the electron gun (with respect to beam's propagation) This is in contradiction with the observed results of the ARAKS experiment.

B. <u>Nonlinear production of low frequency waves</u> (Lavergnat et al.(1980)

In view of the above linear stability analysis a very high level of high frequency electrostatic turbulence is expected, in the form of both out-of-the-beam radiated waves and of to-the-beam's region-confined solutions.

When trapping type saturation mechanisms is invoked an average electrostatic HF field of the order of a few $10\ Vm^{-1}$ may be expected. This is sufficient to prevent the Low frequency unstable waves (which have been described in the first part of this paper) to be excited.

In contrast, by nonlinear beating between two HF guided waves (say $(\omega_1,\ k_{\parallel 1})$ and $(\omega_2,\ k_{\parallel 2})$), whose energy cannot escape freely out of the beam's region, Low frequency waves can be obtained. As their frequencies $\omega = \omega_1 - \omega_2$ may be below ω_c, these forced waves can be radiated far from the beam region. This nonlinear dissipation mechanism of the electrostatic energy confined by the beam's radial inhomogeneity will play, for an outer observer, the role of a nonlinear antenna mechanism.

When two Cherenkov waves $(\omega_{1,2} \sim \omega_p,\ k_{\parallel 1,2} \sim \omega_{1,2}/V_b)$ beat together they produce an LF wave with $\omega/k \sim V_b$ which always propagate along the beam direction. But when a Cherenkov wave beats with a slow cyclotronic wave $(\omega_2 \sim \omega_p ;\ k_{\parallel 2} \sim (\omega_c + \omega_2)/V_b)$, the forced LF wave may propagate downstream as well as upstream, depending of the relative situation of ω_1 and ω_2 $(\omega/k_\parallel \sim V_b(\omega_2 - \omega_1)/\omega_c)$. This fact is well supported by observational evidences.

To perform a greater insight into the details of phenomena let us show, without entering the rather cumbersome algebra,

how the frequency spectrum of the nonlinearly forced LF waves may
be estimated.

   Consider a cylindrical beam of radius a streaming along the
ionospheric magnetic field lines. The analysis is restrained to
cold plasma hydrodynamic approximation (with motionless ions)
and axisymmetric electrostatic disturbances.
The 2nd order potential $\phi^{(2)}$ (r,k ,ω) follows in the surrounding
plasma the simple inhomogeneous wave equation

$$\epsilon_\perp \frac{1}{r} \frac{\partial}{\partial r} r \frac{\partial}{\partial r} \phi^{(2)} - k_\parallel^2 \epsilon_\parallel \phi^{(2)} = \frac{e}{\epsilon_o} n^{(2)}, \qquad (2)$$

where $\epsilon_\perp = 1 - \frac{\omega_p^2}{\omega^2 - \omega_c^2}$; $\epsilon_\parallel = 1 - \frac{\omega_p^2}{\omega^2}$ are the transverse and paral-

lel dielectric permittivities, $n^{(2)}$ is the second order density
perturbation.
$n^{(2)}$ may be computed from first order linearized quantities
(e.g $\phi_1^{(1)}$, $n_1^{(1)}$, $n_2^{(1)}$ etc.) associated to the HF unstable waves,
through the mass conservation equation and the momentum conser-
vation equations

$$n^{(2)} = -\frac{1}{i\omega} \nabla \left\{ \sum_{\substack{k_{\parallel 1} + k_{\parallel 2} = k \\ \omega_1 + \omega_2 = \omega}} \left(n_{k_\parallel 1, \omega_1}^{(1)} v_{k_\parallel 2, \omega_2}^{(1)} + n_o v_{k_\parallel, \omega}^{(2)} \right) \right\} \qquad (3)$$

$$(i\omega - \frac{e}{m} B_o \times) v_{k_\parallel, \omega}^{(2)} = - \sum_{\substack{k_{\parallel 1} + k_{\parallel 2} = k_\parallel \\ \omega_1 + \omega_2 = \omega}} (v_{k_\parallel 1, \omega_1}^{(1)} \cdot \nabla) v_{k_\parallel 2, \omega_2}^{(1)} \qquad (4)$$

   As we consider mother waves which are confined to the beam
region (r < a), so is the source term of equation (2), whose solu
tions, in the far field approximation reads simply

$$\phi^{(2)} (k_\parallel, \omega, r) = \frac{\pi}{2} \frac{e}{\epsilon_o \epsilon_\perp} H_o^{(2)} (Kr) \int_0^a r' \, dr' \, J_o (Kr') n^{(2)} (r', k_\parallel, \omega) \qquad (5)$$

where $K^2 = -k^2 \epsilon_{\parallel}/\epsilon_{\perp}$ and J_o, $H_o^{(2)}$ are the zero order Bessel and Hankel functions.

Restricting our calculations to approximations relevant to LF observations and ionospheric injection

$\omega_1 \sim \omega_2 \sim \omega_p$; $\omega_c \ll \omega_p$; $\omega \lesssim \omega_c$, we get, to the most significant order :

a) for Cherenkov - Cherenkov heating process :

$$\Phi_{c-c}^{(2)} = \sqrt{2\pi}\,\frac{e}{m}\,\sqrt{\frac{a}{r}}\,\frac{1}{V_f^2}\,(Ka)^{1/2}\,J_1(Ka)\left\{\sum_{\substack{k_{\parallel 1}+k_{\parallel 2}=k \\ \omega_1+\omega_2=\omega}} \Phi_1^{(1)}\Phi_2^{(1)}\right\}e^{i\psi} \tag{6}$$

with $K = \frac{1}{V_f}\sqrt{\omega_c^2 - \omega^2}$ and $\psi \sim Kr + \frac{\pi}{4}$ (radial wave number and phase)

b) for Cyclotron - Cherenkov heating process

$$\Phi_{c-cy}^{(2)} = \sqrt{2\pi}\,\frac{e}{m}\,\sqrt{\frac{a}{r}}\,\frac{1}{V_f^2}\left\{1 - \frac{1}{\left(1+\frac{\omega}{\omega_c}\right)} + \frac{\omega_c^2}{2\omega_p^2\left(1+\frac{\omega}{\omega_c}\right)^2}\right\}(Ka)^{1/2}\,J_1(Ka)$$

$$\times\left\{\sum_{\substack{k_{\parallel 1}+k_{\parallel 2}=k \\ \omega_1+\omega_2=\omega}} \Phi_1^{(1)}\Phi_2^{(1)}\right\}e^{i\psi} \tag{7}$$

with $\qquad K = \frac{1}{V_f}\,(1+\frac{\omega_c}{\omega})\,\sqrt{\omega_c^2-\omega^2}$ and $\psi = Kr + \frac{\pi}{4}$

It is not a priori simple to estimate the integrals present in the right hand side of the above equations, because this needs to know the radial dependencies of the HF mother waves into the beam region, as well as the fourier transform of their spatio temporal correlation function.

Nevertheless it has been shown in our linear stability analysis that the most unstable solutions are always the zero node ones. We may consequently choose as an excellent approximation radially constant first order potential.

Using as a last approximation the crude simplification of a flat frequency spectrum for the initial turbulence, the second order potentials $\Phi_{c-c}^{(2)}$ and $\Phi_{c-cy}^{(2)}$ may be exactly computed.

It can then be seen   the LF spectrum due to c-c beating is quasi flat, when the c-cy spectrum has an important slope in the upper part of the spectrum and exhibits an oscillatory behaviour in the lower part which could explain a peak structure, which has been sometimes observed by Dechambre and al. (1980) during the ARAKS experiment (see fig. 4)

These facts agree well with the experimental results (Dechambre and al. (1980) fig. 4 and 5) which show, in the outward situation of the observer with respect to the beams direction (where only waves due to c-cy beating are available) LF spectra which are peaked in the lower frequency range with slope slightly greater than - 1.5, likely due to the width of the spectral correlation function of HF turbulence.

In the inward situation (where both c-cy and c-c forced waves are present) a slightly higher potential level is observed and the LF spectra have mainly a weaker slope due to the combination of c-c and c-cy spectral superposition.

As the radial integral present on the right hand side of (5) is to be interpreted as an average of the squared HF turbulent electric field this quantity may be estimated from the observed LF electrostatic level, at a given distance from the beam's region.

With typical value of $\simeq 100\,\mu V$ observed at $r \simeq 1km$, we obtain 10 to 100 Vm^{-1} for the RMS of the HF electric field.

This is consistent with a trapping type mechanism for the saturation of the HF waves

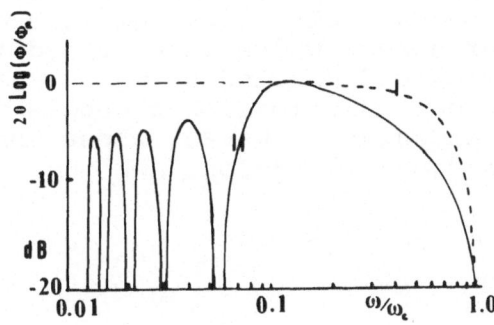

Fig. 4 : Theoretical spectra of the nonlinear forced waves ($a\omega_c/V_f=0.2$ $\omega_c/\omega_p=.4$) I:c-c beating, II : c-cy beating.

REFERENCES

Bers, A., Proceeding of the Summer School at les Houches, Gordon and
    Breach Science Publishers (1972)
Dechambre, M., Gusev, G.A., Kushnerevsky, Yu.V., Lavergnat, J.,
    Pellat, R., Pulinets, S.A., Selegei, V.V., Zhulin, I.A.,
    High frequency waves during the ARAKS experiment Ann. Geophys.
    36 : 333 (1980)
Dechambre, M., Kushnerevsky, Lavergnat, J., Pellat, R., Pulinets, S.A.
    Selegei, V.V., Waves observed by the ARAKS experiments : the
    whistler mode, Ann. Geophys. 36 : 341 (1980)
Lavergnat, J. , and Pellat, R., High frequency spontaneous emission
    of an electron beam injected into the ionosphere, J.G.R., 84A,
    12, 7223 (1979).
Lavergnat, J., Dechambre, M., Pellat, R., Kushnerevsky, Yu. V.,
    Pulinets, S.A., Waves observed by the ARAKS experiment :
    generalities, Ann. Geophys. 36 : 323 (1980)
Lavergnat, J., Le Queau, D., Pellat, R., Roux, A., Nonlinear mecha-
    nism for the production of the low frequency electrostatic
    waves, Ann. Geophys. 36 : 439 (1980)
Le Queau, D., Pellat, R., Saint-Marc, A., An investigation of the
    electrostatic linear instabilities of a radially limited
    electron beam, Ann. Geophys. 36 : 433 (1980)
Le Queau, D., Pellat, R., Saint-Marc, A., Electrostatic instabilities
    of a finite electron beam propagating in a cold magnetized
    plasma , to be published in Phys. Rev. (A) (1981).

DISCUSSION

Maggs: You said the confined HF waves were saturated by
particle trapping and then discussed radiation of
whistler waves. Is saturation caused by trapping or
radiative losses?

LeQueau: In my opinion the confined HF waves are at
first saturated by particle trapping. The obtained
electrostatic turbulence which is confined to the beam's
region is then slowly deposited by wave - wave interaction. One of these dissipation channels is the conversion into LF propagating waves, whose energy can then
flow out of this "source" region.

SPONTANEOUS EMISSION OF A CHARGED PARTICLE BEAM INSIDE A PLASMA :

COHERENT AND INCOHERENT ASPECTS

J. Lavergnat[*] and R. Pellat[**]

[*] L.G.E. 4, Av. de Neptune 94100 St-Maur-des-Fossés
France.
[**] Centre de Physique Théorique, Ecole Polytechnique
91128 Palaiseau

INTRODUCTION

Injection of charged particle beam into ionospheric plasma has revealed some unexpected behaviour of the radio waves emission (Cartwright and Kellog, 1974 ; Lavergnat et al, 1980). Two kinds of behaviour in time are observed for the wave envelopes : a 'continuous emission' with a one to one relation between the wave amplitude and the nominal beam current and a "pulsed emission" that reaches a maximum amplitude a few milliseconds after the beginning of an electron pulse and then decreases continuously to a background value. Along different experiments (Echo 1 and 2, Araks) both behaviours have been observed in two frequency ranges : in the whistler mode $\omega < \omega_b$ and in the plasma mode $\omega_p < \omega < (\omega_p^2 + \omega_b^2)^{1/2}$. Pulsed emission will be the main topic of this paper. Apart from the collective processes spontaneous emission and specially the Cerenkov emission is a good candidate to explain the observations. Singh (1973) found by a crude computation of the incoherent Cerenkov emission electric field values which are a few orders of magnitude below the observed value and no possible explanation for the pulsed character. Alekhin and Karpman (1973) are the first who suggested that a beam with an infinitely sharp front and no spread in velocity can radiate a pulsed emission. More detailed evaluations have been made since them (Lavergnat and Pellat, 1974 ; Lavergnat and Pellat, 1979) and allow to present a comprehensive study of the spontaneous emission of an electron beam (or a charged particle beam). We shall restrict ourself to a general presentation without going into cumbersome algebra but giving numerical estimations in order to compare them with the observed values. At the end of the paper we shall discuss the possibility to increase the coherent emission with the use of a charged beam as a VLF antenna.

THE ONE DIMENSIONAL BEAM : COHERENCE AND INCOHERENCE.

Discussion about the spontaneous emission is very simple in an oversimplified model : the "plane sheet beam" which allows to get a physical insight into the matter.

Let us consider a plane electron gun which emits at t = 0 a one dimensional sheet particle in a neutral unmagnetized cold plasma of density n_o. The particle has a velocity V, a charge -e and a mass m per unit area.

One can compute easily the field generated by this particle and the work done by the self field is simply $dW/dt = -e^2 V/\epsilon o$ (1)

If the gun emits now a current density I per unit area the standard computations (e.g. Bell 1968) consider that each sheet sees only its self field and the resulting 'incoherent part' becomes $\frac{dW}{dt} = IeVt/\epsilon o$. This asumption is obviously wrong : a given sheet located at x at time t sees its self field $e/\epsilon o$ and also the field created by the already emitted sheets. As a result one obtains easily the total power per unit area lost by the beam $\epsilon o\, dW/dt = IeVt + I^2 V/\omega_p^2$ (2). The first term represents the incoherent emission and the second the coherent one. The inspection of this formula shows that it is uselless to imagine one kind of emission as a limit case of the other. In order to prove this statement let us introduce in our model a velocity dispersion for instance a square distribution $((2\delta u)^{-1}$ for $|V-u| < \delta)$; the coherent potential reads

$$\Phi_c = \Phi(\delta=0)\, \sin(\omega_p\, x\, \delta/u)\, /\, (\omega_p\, x\, \delta/u)\quad (3)$$

where x is the space variable. From (3) we see that the coherent emission can be diminished at will when the velocity dispersion is increased ($\delta \nearrow$) ; during the same time the incoherent part of the emission remains constant. Such behaviour result from the under-lying description of the beam which has been considered in a fluid approximation ; it is important to note that a fluid description of the source of radiation (here the plane beam) gives nothing more than the coherent emission and is unable to produce the incoherent part. Inversely the computation of the incoherent emission by the total power radiated by the source is unable to give the local value of the electromagnetic field without some asumptions which are often debatable.

In order to make no error and to forget nothing the good way seem to compute directly the local electric field value and afterwards its quadratic average value. As we know from statistical mechanics

$$<E^2> = \int f_1 \, E_1^2 \, d\Omega_1 + \int f_{12} \, E_1 \cdot E_2 \, d\Omega_1 \, d\Omega_2 \qquad (4)$$

where angle brackets indicate a statistical ensemble average and where f_1 and f_{12} are the simple and the double (two points) distribution function, respectively of the electron population ($d\Omega_i = d^3v_i \, dr^3_i$). If we now make a cluster expansion of f_{12} ; $f_{12} = f_1 \, f_2 + C_{12}$ we get

$$<E^2> = \int f_1 \, E_1^2 \, d\omega_1 + (< E_1>)^2 + \int C_{12} \, E_1 \cdot E_2 \, d\Omega_1 \, d\Omega_2 \qquad (5)$$

For a perfect fluid (equivalent to the Vlasov approximation) one has $C_{12} = 0$, the emitted power is as in the plane beam, the sum of the coherent part and the incoherent one. In our point of view a method of computation based upon equation (5) avoids any mistake or misinterpretation.

GENERALITIES ABOUT THE SPONTANEOUS EMISSION

Standard method of computation consists of a Fourier analysing in time and space of the Maxwell equations (Poisson for the electrostatic waves) and the momentum transfer equations, the electric field (or magnetic field) being then related to the charge current (or density) of the beam. In any case one finds after an inverse Fourier transform and some straight forward integrations, a formula of the following kind :

$$A(r,t) = \sum_n \int f(k_\perp,\tau) J_n(k_\perp r_\perp) \frac{\exp i\left[(n\omega_b + k_{/\!/} v_{/\!/})(\tau - t) + k_{/\!/} r_{/\!/}\right]}{\mathscr{D}(\omega,k)}\Bigg|_{\omega = k_{/\!/} v_{/\!/} + n\omega_b} dk_{/\!/} dk_\perp d\tau$$

$$(6)$$

A represents a quantity directly related to the radiated field. ω_b is the gyrofrequency, $r_{/\!/,\perp}$ the space variables, t the time, τ a dummy variable which describes the temporal evolution of beam characteristic (velocity, current). J_n is a Bessel function of first kind which is due to the axial symmetry imposed by the magnetic field. $\mathscr{D}(\omega,k)$ is the dispersion relation.

Recalling the discussion of the previous paragraph we underline that (6) is used to compute the coherent emission of an extended source but also in the case of the emission of a test particle (in this case the distinction between coherent and incoherent is useless).

Whatever the source $|f(k_\perp)|$ or the plasma ($\mathscr{D}(\omega,k)$) equation (6) has some common proprieties we shall derive. For sake of simplicity let us restrict ourself to the Cerenkov emission (n=0)

Integration over $k_{/\!/}$ by the calculus of residues may be performed easily because $\mathcal{D}(\omega,k)$ has generally two isolated zeros $k_{/\!/} = \pm g(k_\perp)$. Equation (6) reads now

$$A = \int f(k_\perp) \, J_0(k_\perp r_\perp) \, \frac{\exp\{\pm ig(k_\perp) \, (V_{/\!/}(\tau-t)+r_{/\!/})\}}{\frac{\partial \mathcal{D}}{\partial \omega}(V_{/\!/} - Vg_{/\!/})} \, dk_\perp \, d\tau \qquad (7)$$

$Vg_{/\!/}$ is the parallel group velocity of the waves. Two main cases appear

(I) $V_{/\!/} \neq V_{g/\!/}$

It is the case of the plasma mode and the Bernstein modes. Essentially concerned by the far field emission we can use the method of stationary phase for performing the two last integrations. Net results are

(i) Over $k_\perp$ the stationary point is given by

$$r_\perp V_{g\perp} = (V_{/\!/} - V_{g/\!/}) \, (V_{/\!/}(\tau-t) + r_{/\!/}) \qquad (8)$$

$V_{g\perp}$ is the perpendicular group velocity of the waves. Physical interpretation of (8) is quite natural : it is the condition of propagation which allows the waves to reach the point $r_{/\!/}$, $r_\perp$ at the time t. In some cases (e.g. plasma mode) there is some value of t such that the stationary point is in fact a double zero. For this value the amplitude will be very important and around it will be adequatly described by an Airy function. This is the origin of the front effect delayed from the beginning of injection of the electron beam. We must emphasize, following previous remarks, that this effect is also true for the incoherent emission.

(ii) Over τ the stationary point is given by

$$\frac{dV_{/\!/}}{d\tau} = V_{/\!/}^2 \, /r_{/\!/} \qquad (9)$$

If $\frac{dV_{/\!/}}{d\tau} > 0$ the slow particles can be caught up by the faster

ones emitted a little bit later. There is consequently an increase of the emission by a large factor depending upon the sharpness of

the velocity profile. If $\frac{dV_{/\!/}}{d\tau} < 0$ there is no dynamical front.

(II) $V_{/\!/} = V_{g/\!/}$

This is the case of the whistler mode ($\omega < \omega_b$). Residues in

equation (6) are no longer simple but double, that leads to unaccept-
able secular terms ; the computation must be undertaken in an inhomogeneous plasma.

This case requires more involved algebra (Lavergnat and Pellat
1979) and is beyond the scope of this paper but at the end we find
again the same kind of behaviour : front effect (pulsed emission),
dynamical front.

APPLICATION TO ACTIVE EXPERIMENTS.

Detailed computations have been already published (Lavergnat
and Pellat 1979) and we would just summarize the results obtained
which are relevant to our purpose. The plasma mode is the more
convenient for that.

Up to now we have not discussed the actual shape of the beam
although evidently it is an important parameter. We satisfy ourselves with two models : an helicoidal beam and an hollow beam. However if we consider a longitudinal injection a full cylindrical beam
with the Brillouin radius is suitable.

In such a case, ignoring the velocity dispersion, the coherent
amplitude of the radiated plasma waves reads

$$|E| = \frac{2.51 \ I \ (\tau_o) \ \omega_p}{\epsilon_o r_B \ V_{//} (\tau_o)} \ J_1(\frac{r_B \omega_p}{V \sqrt{3}})\rho_\perp^{-5/6} \ \ Ai(0.76 \ \rho_\perp^{2/3} \ \frac{\sigma_o - \sigma}{\sigma_o}) \ .$$

$$\frac{e^{-\sigma \nu / 2\omega_p}}{(f \ (\tau_o))^{1/2}} \qquad (10)$$

where $\sigma = \omega_p \left[t - \tau_o - r_{//} / V_{//} \right]$ and $\sigma_o = 16\sqrt{3} \ \omega_p^2 / 9\omega_b^2$

τ_o solution of $1 + (r_{//} + r_\perp / \sqrt{3}) \ \frac{d}{d\tau} \ (1/V_{//}) = 0$

and $f \ (\tau_o) = \omega_p (r_{//} + r_\perp / \sqrt{3}) \ \frac{d^2}{d\tau^2} \ (1/V_{//})$, $\rho_\perp = \frac{r_\perp \omega_p}{V_{//}}$ and r_B the

radius of the beam.

Formula (10) gives a temporal evolution in agreement with the
observed one (emission delayed and pulsed). With realistic parameters it leads however to a too optimistic value in the case of
Echo experiment (1.4 V m^{-1} well above the experimental value of

3mV at the input of the receiver). Let us recall that this theoretical result assumes that the velocity dispersion has been completely removed by trapping (Lavergnat and Pellat 1974) that is certainly incorrect and remains an open question.

In Araks experiment $\dfrac{dV_{//}}{d\tau} < 0$ thus the coherent emission is very weak. Incoherent emission must be estimated by using the procedure leading to equation (5).

The finite size of the beam being negligible in comparison with the distance of observation, we have taken the particles moving along the cylindrical axis of coordinates (B axis). E_e being the emission of a test particle one gets

$$E_i^2 = \int_{r_{//}}^{V_{//}t} n(z)\, E_1^2\, dz \qquad (11)$$

where $n(z)$ is the density of the emitting particles along the z axis. Introducing the new variable $\zeta = \omega_p(t - z/V)$ (11) reads

$$E_i^2 = \frac{I}{e\omega_p} \int_0^\infty E_1^2\, d\zeta \qquad (12)$$

Here E_i represents the asymptotic value of the electric field (i.e. $t \to \infty$).

Integration of (12) is subject to the generalities previously mentioned. As in the coherent case integration over $k_\perp$ is performed by the method of stationary phase and we are faced with a double root for a particular time. But when it is licit to restrict oneself to this double root in the coherent case we have now due to the new integration in ζ, to take into account the contribution of the simple root which may be more important at intermediate distance to the beam (quoted ionospheric experiments). The final result is

$$E_i^2 = \frac{Ie}{\varepsilon_o^2}\left\{ 4.2 10^{-3} \frac{\omega p^2}{V_{//}^2} \frac{\exp(-1.97\zeta_o \nu/\omega_p)}{\nu r_\perp^2} \right. +$$

$$\left. + 1.3 10^{-2} \frac{\omega_p}{\omega_b (\nu\omega_p)^{1/2}} \left(\frac{\omega p}{V_{//}}\right)^{15/6} \frac{\exp(-\zeta_o \nu/\omega_p)}{r_\perp^{3/2}} \right\} \qquad (13)$$

with $\zeta_o = 16\sqrt{3}\, r_\perp \omega_p^3/9\, V_{//}\omega_b^2$ and ν the collision frequency.

Within the Araks conditions formula (13) leads to theoretical values around 100μV m^{-1} which are close to the experimental ones. However we are not yet able to explain the pulsed character of the emission.

Nevertheless other indications are in favour of spontaneous emission in the ionospheric experiments and for the plasma waves. Fig 1 gives the pitch angle variations of the coherent emission for an helicoidal beam and an hollow beam. Comparison with Echo I is tricky due to the unsymmetrical results around 90° obtained by Cartwright and Kellog (1974). However assuming as these authors a general shift in pitch angle, it becomes rather clear that the helicoidal model seems to explain the experimental results very

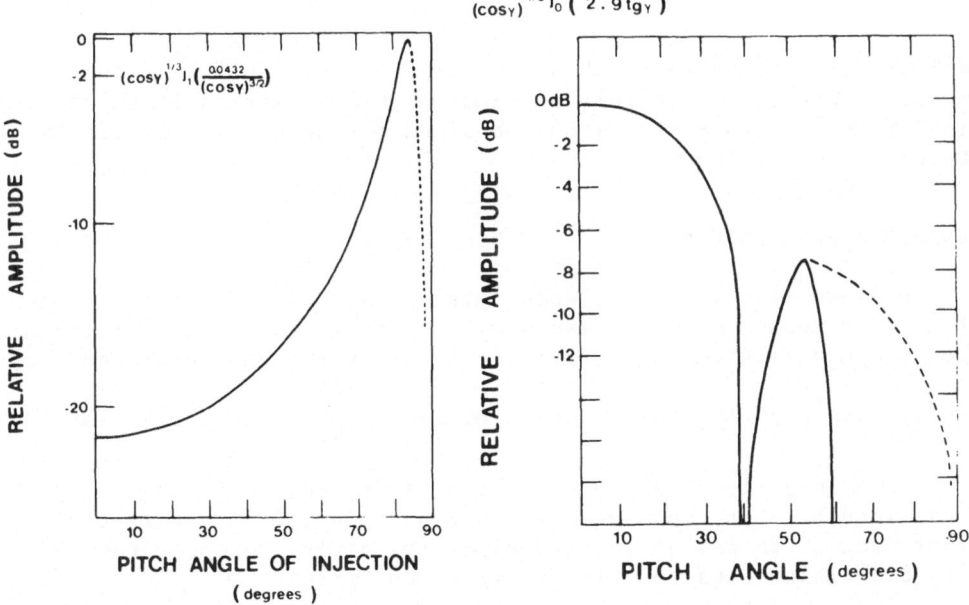

Fig. 1. Variation of the amplitude of the maximum electric field at the plasma frequency versus pitch angle. Dashed curves are the envelopes of the variations. The beam model is at the left helicoidal and at the right, hollow.

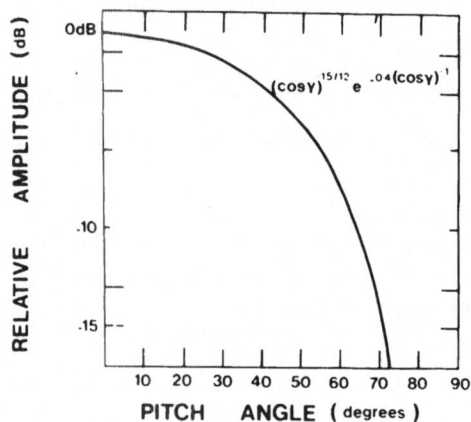

Fig. 2. Variation of the amplitude of the incoherent emission at the plasma frequency versus pitch angle of injection.

well. Fig. 2 gives the pitch angle variation of the incoherent emission. Within our numerical example this variation is quite similar to the experimental one obtained during Araks experiments (Dechambre et al 1980).

TOWARDS A BEAM ANTENNA

We have discussed with some details the spontaneous emission of an electron beam. Comparisons with experimental data have shown that the coherent emission from the front of the beam had been very likely seen when the theoretical inequality $\frac{dV_{\parallel}}{d\tau} > 0$ was satisfied.

The large level of the observed values indicates that when the conditions are met, spontaneous coherent emission is a very efficient process to radiate electromagnetic energy. Moreover the velocity dispersion does not play a disastrous part.

Rather than use the poor coherence of the front of the beam it is tempting to organize the beam in order to increase its coherence and then to radiate more efficiently.

The first idea is to modulate the intensity of the beam at the frequency which we want to radiate. For many reasons a modulation in the VLF range is the most convenient. Exact computations are very difficult to perform because of the non linearity of the problem. Nevertheless a crude estimation of the linear efficiency of this process is possible if one follows the method already presented.

Let us assume a cylindrical beam of radius r_B bearing a modulated current

$$I : I_o \left[\sin \Omega(t - \frac{r_{//}}{V_{//}}) + 1 \right] \qquad (14)$$

Disregarding the first steps of the calculus (see Lavergnat and Pellat 1979) we can estimate the braking work of the emitted waves on the beam at

$$P = \int E.J \, d^3r \qquad \text{with}$$

$$|E.J| = \frac{I_o^2}{2\pi^2 \epsilon_o r_B} \int \frac{g}{\mathcal{S}} \frac{J_1(k_\perp r_B)}{2\Omega} \sin^2 \Omega \, (t - \frac{r_{//}}{V_{//}}) \, dk_\perp \, dr_{//} \qquad (15)$$

where $g/\mathcal{S}$ expresses the properties of the surrounding plasma ($\mathcal{S}$ is the dispersion relation).

In the case of an electron beam the interaction is expected with the right polarized whistler. Integration over $k_\perp$ involves the zeros of the dispersion relation which are given in our case by

$$\frac{k^2 c^2}{\Omega^2} + \frac{\omega_p^2}{\Omega^2 (1 - \frac{\omega_b}{V_{//}k})} = 0 \qquad (16)$$

Equation (16) has a solution if and only if $V_{//} < \frac{c\,\omega_b}{2\,\omega_p}$ (17)

It becomes clear that any modulated beam does not radiate its frequency of modulation. If the inequality (17) is satisfied the power lost by the beam reads :

$$P = \frac{I_o^2 L}{8 \, \varepsilon_o \, \pi \, r_B \, \Omega} \, \frac{g}{D} \, J_1(k_\perp r_B) \qquad (18)$$
where L is the length of the beam.

As a numerical example let us take the following conditions :

$\omega_b = 6 \, 10^6$ $\omega_p = 4.5 \, 10^7$ $\Omega = 6.38 \, 10^4$ (10 kHz) $I_o = 1$ A

$V = 5 \, 10^6$ m/s.

These figures correspond to a beam, the power of which is equal to 70 W, r_B has been taken as the Brillouin radius.

Then we get by (18) $P = 10^{-1}$ L. Emitted waves have a refractive index of 70 W. It appears that when L > 10 wavelengths the power lost by the beam is equal to the initial power of the beam ! This absurd situation comes from the pure linear point of view we have adopted but it indicates clearly that a good organized beam can radiate electromagnetic waves efficiently. What is the exact output of such a process need further investigations, both theoretical and experimental.

Before to end we would mention that in some previous experiments (Lavergnat et al 1980 ; Holzworth and Koons, 1981) a modulated beam has been already used and the frequency of modulation observed on a detachable cone, but the condition (17) was not fulfill. Some preliminary computations show that it is likely not waves which have been measured but the near field.

REFERENCES

Alekhin, J.U. and V.I. Karpman, 1973. "On Cerenkov radiation by electron beam injected into the ionosphere", Cosmic Electrodynamics, 3, 406-415.

Bell T.F., 1968, "Artificial production of VLF hiss", J. Geophys. Res, 73, 4409-4415.

Cartwright, D.G. and P.J. Kellog, 1974 "Observations of radiation from an electron beam artificially injected into the ionosphere", J. Geophys. Res, 79, 1439-1457.

Dechambre, M., G.A. Gusev, Yu. V. Kushnerevsky, J. Lavergnat, R. Pellat, S.A. Pulinets, V.V. Selegei, I.A. Zhulin, 1980, "High frequency waves during the Araks experiments", Ann. de Geophys., 36, 3, 333-340.

Holzworth R.H., H.C. Koons, 1981, "VLF emission from a modulated electron beam in the auroral ionosphere", J. Geophys. Res., 86, 853-857.

Lavergnat, J., M. Dechambre, R. Pellat, Yu. V. Kushnerevsky, S.A. Pulinets, 1980, "Waves observed by the Araks experiments : generalities". Ann. de Geophys., 36, 3, 323-332.

Lavergnat, J. and R. Pellat, 1974, "Onde de neutralisation d'un faisceau d'électrons injectés dans un plasma", C.R. Acad. Sci., 278 B, 827-829.

Lavergnat, J. and R. Pellat, 1979, "High-frequency spontaneous emission of an electron beam injected into the ionospheric plasma", J. Geophys. Res., 84, 7223-7238.

Singh, R.P., 1973, "Cerenkov cyclotron power generated in the magnetosphere by natural and artificial injection of electron beams", Ann. de Geophys., 29, 227-238.

DISCUSSION

Winckler: In Echo, due to the finite rise time of the accelerating voltage the gun emits low velocity particles first, higher later. Hence the beam front steepens. In ARAKS, due to the extremely short rise time, the gun emits particles of all velocities together, so the beam front flattens.

Lavergnat: I agree completely with this comment, that is the reason why in Echo we think that one has seen the coherent emission while in ARAKS the coherent emission was too low to be observed.

THE BEAM-PLASMA DISCHARGE UNDER SPACE-LIKE CONDITIONS*

S. Cuperman and I. Roth

Department of Physics and Astronomy
Tel-Aviv University, 69978, Ramat-Aviv, Israel

I. INTRODUCTION

The beam plasma discharge was investigated in the past in
fusion research (e.g., Getty and Smullin, 1963) and more recently
in space-oriented laboratory experiments (e.g., Bernstein et al.,
1975; 1978). In the fusion research the studies were carried out
at relatively high magnetic fields (300-1000 G) and gas pressures
(>10-^{4} torr) and a short interaction range (< 1m). In the space-
oriented studies, relatively low magnetic fields ($\sim$1G) and pressures
(<10-^{4} torr) and a long interaction range ($\sim$20m) were used; these
conditions being also relevant to artificial electron beam injection
experiments in the ionosphere by rockets and spacecraft.

In the context of space physics, Galeev et al.(1976)considered
theoretically the discharge in the region around a rocket following the injection of electron beams into the ionosphere. Preliminary theoretical results for the BPD under space-like conditions
have been reported by Rowland and Papadopoulos (1979). Further
theoretical predictions and explanations of the results obtained by
Bernstein et al.(1975,1978) were recently proposed by Cuperman and
Roth (1981) and Roth and Cuperman (1981). The present paper represents an extension of the last two investigations.

Thus, we consider theoretically the BPD under conditions of
(a) low pressures and magnetic field strength and (b) configurations
having outer boundaries far removed from the edges of the beam.
Two main aspects are covered, namely: (i) the beam-plasma interaction producing the plasma instability (which leads to plasma

---

*Editor's note: This paper was submitted to, but not presented at
                  the meeting.

heating and consequently to enhanced ionization) and (ii) the competition and balancing of plasma production rate and plasma loss
outside of the interaction region. More specifically, the following problems are investigated and discussed: 1. The electrostatic
beam-plasma instability of a bounded magnetized cylindrical configuration; 2. The critical conditions for BPD; 3. Effect of the
surrounding plasma; 4. Dependence on the radius of the surrounding plasma; 5. Effect of the temperatures of the background and
surrounding plasmas; and 6. Ion-effects on the low frequency
range of the rf spectrum.

II. STABILITY ANALYSIS

a. Dispersion equation

Consider a cylindrical configuration consisting of i) a cold
neutralized electron beam (density n_b, streaming velocity V_b V_z
flowing through a background plasma (density n_1 and temperature T_1),
both of radius a and ii) a surrounding plasma (n_2 and T_2) of radius
b. The entire system is embedded in a static magnetic field $\underline{B} \equiv B_z$
(See Fig.1). The dispersion relation for electrostatic, azimuthally
symmetric modes of the form $\Psi(r)\exp[i(kz-\omega t)]$ which vanish at r=b
is (see, e.g., Davidson 1974)

$$\varepsilon_\perp^{(1)} \tau_1 a \frac{J_1(\tau_1 a)}{J_0(\tau_1 a)} = \varepsilon_\perp^{(2)} \tau_2 a \frac{I_0(\tau_2 b) K_1(\tau_2 a) + I_1(\tau_2 a) K_0(\tau_2 b)}{I_0(\tau_2 b) K_0(\tau_2 a) - I_0(\tau_2 a) K_0(\tau_2 b)} \quad (1)$$

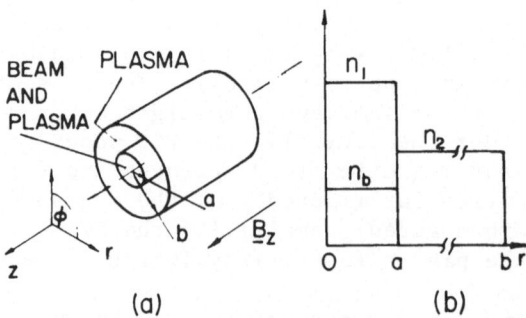

Fig. 1.   The configuration (a) and density profiles (b) considered
in this work.

where τ_j are the transverse wave numbers in the beam region (j=1) and outside it (j=2) and are given by

$$\tau_j{}^2 = (-1)^j k^2 \varepsilon_\parallel^{(j)} / \varepsilon_\perp^{(j)} . \tag{2}$$

Here $\varepsilon_\parallel^{(j)}$ (ω, k, ω_b, V_b, $\omega_{e,j}$, $\omega_{i,j}$, T_j) and $\varepsilon_\perp^{(j)}$ (ω, k, V_b, ω_b, $\omega_{e,j}$, $\omega_{i,j}$, Ω_e, Ω_i, T_j) are the parallel and perpendicular components of the dielectric constant, respectively; ω and k are the frequency and wavenumber of the disturbance, respectively; V_b is the equilibrium streaming velocity of the electron beam; ω_b, $\omega_{e,j}$ and $\omega_{i,j}$ are the plasma frequencies of the electron beam, electron plasma and ion plasma, respectively; Ω_e and Ω_i are the electron and ion cyclotron frequencies, respectively; T_j are the electron plasma temperatures; (J_0, J_1), (I_0, I_1) and (K_0, K_1) are the Bessel functions and modified Bessel functions of the first and second kind, respectively.

To proceed, we treat the case of long wavelength perturbations. Then, the Bessel and the modified Bessel functions in eq.(1) may be expanded in the small arguments to obtain the following simpler dispersion equation:

$$k^2 \varepsilon_\parallel^{(1)} + [2/a^2 \ln (b/a)] \varepsilon_\perp^{(2)} = 0 . \tag{3}$$

b. The Cold Case

In order to simplify the problem, we shall first neglect the temperature effects. Then, after some algebraic operations, eq.(3) reads

$$\{1-\omega_1^2/\omega^2+[2/k^2a^2 \ln(b/a)][1-\omega_{e,2}^2/(\omega^2-\Omega_e{}^2)-\omega_i^2/(\omega^2-\Omega_i^2)]\}$$

$$- \omega_b^2/(\omega-kV_b)^2 = 0 \tag{4}$$

where $\omega_1^2 \equiv \omega_{e,1}{}^2 + \omega_{i,1}{}^2$. Eq.(4) describes the interaction of the space-charge waves of the beam* ($k_b \simeq \omega/V_b$) and the ion ($k^{(i)}$) and electron ($k^{(e,1)}$, $k^{(e,2)}$) plasma waves. The plasma waves are given by the expression in the curly brackets equated to zero which implies

$$k^2(\omega)=[-2/a^2 \ln(b/a)]\{\omega^2[(\omega^2-\Omega_e^2)(\omega^2-\Omega_i^2)-\omega_{e,2}^2(\omega^2-\Omega_i^2)$$

$$- \omega_{i,2}^2(\omega^2-\Omega_e^2)]\}\cdot[(\omega^2-\Omega_e^2)(\omega^2-\Omega_i^2)(\omega^2 \omega_1^2)]^{-1} . \tag{5}$$

---

* In the long wavelength approximation the cyclotron mode of the beam ($k_b = (\omega + \Omega_e)/V_b$) does not participate in the interaction.

The plasma modes $k^{(i)}$, $k^{(e,1)}$ and $k^{(e,2)}$ described by eq.(5) together with the space-charge wave of the beam are sketched in Fig.2.

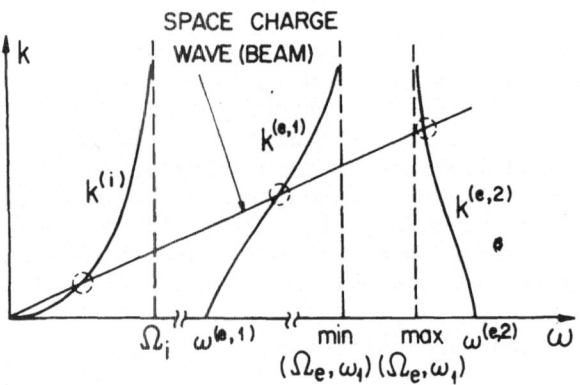

Fig. 2. Dispersion diagram for the plasma waves ($k^{(i)}$, $k^{(e,1)}$ and $k^{(e,2)}$) (eq.5) and the beam space charge wave ($k = \omega/V_b$).

In this paper we concentrate on the modes $k^{(e,1)}$ and $k^{(i)}$ which can be treated consistently by the aid of eq.(4) which was derived for long wavelength perturbations.

Thus, looking for a solution described by complex wave numbers ($k \equiv k_r + ik_i$) and real frequencies ($\omega \equiv \omega_r$), we set $kV_b = \omega(1+\alpha)$, $|\alpha| \ll 1$, insert it into eq.(4) and obtain the following equation for α:

$$\alpha^3 - (q/2p)\alpha^2 + \varepsilon/2p = 0 \tag{6}$$

where

$$p = [2V_b^2/a^2\omega_1^2 \ell n(b/a)][1-\omega_{e2}^2/(\omega^2-\Omega_e^2)-\omega_{12}^2/(\omega^2-\Omega_i^2)] \tag{7}$$

and

$$q = \omega^2/\omega_1^2 - 1 + [2V_b^2/a^2\omega_1^2 \ell n(b/a)] \cdot [1-\omega_{e2}^2/(\omega^2-\Omega_e^2)-\omega_{12}^2/(\omega^2-\Omega_i^2)]. \tag{8}$$

Denoting $\alpha \equiv x + iy$, inserting into eq.(6), separating the real and imaginary parts, provides the following two equations for y and x:

$$y^2 = - x^2 + (\varepsilon x/p)^{\frac{1}{2}} \tag{9}$$

and

$$x^3 - 0.5(q/p)x^2 + (1/16)(q/p)^2 x - (1/16)(\varepsilon/p) = 0 . \tag{10}$$

Close to origin one has

$$x \simeq \varepsilon \, p/q^2 , \quad y^2 \simeq \varepsilon/|q| . \tag{11}$$

The maximum growth rate is found to be

$$y_{max} = (3^{\frac{1}{2}}/2) \, (\varepsilon/2p_m)^{\frac{1}{2}} \tag{12}$$

where $p_m \equiv p \, (\omega_m)$, ω_m being the solution of the following equation

$$\omega_m^2 - \omega_1^2 + [2V_b^2/a^2 \ell n (b/a)][1 - \omega_{e,2}^2/(\omega^2 - \Omega_e^2) - \omega_{i,2}^2/(\omega - \Omega_i^2)] = 0. \tag{13}$$

For the electron plasma mode, $k^{(e,1)}$, assuming $\omega >> \Omega_i$, $\omega_{i,2}$ from (9)-(13), one obtains the following specific results:
   i) low frequency and wavenumber range:

$$k_i^{(e,1)} \simeq (\omega/V_b) \, [(n_b/n_1) / (1 - \delta^{(e,1)})]^{\frac{1}{2}} \tag{14}$$

$$k_r^{(e,1)} \simeq (\omega/V_b)[1 + (n_b/n_1)\delta^{(e,1)}/(1 - \delta^{(e,1)2})] \tag{15},$$

$$\delta_1^{(e,1)} \equiv [2V_b^2/a^2\omega_1^2 \ell n(b/a)] (1 + \omega_{e,2}^2/\Omega_e^2) . \tag{16}.$$

   ii) maximum growth rate:

$$k_{i,max}^{(e,1)} \simeq (3^{\frac{1}{2}}/2)(\omega_b^2/2\omega_1^2)^{1/3}(\omega_m^{(e,1)}/V_b) \, \cdot$$
$$\cdot \, [a^2\omega_1^2 \ell n \, (b/a)/2V_b^2]^{1/3} /\{1 - \omega_{e,2}^2/[(\omega_m^{(e,1)})^2 - \Omega_e^2]\}^{1/3} \tag{17}$$

$$k_r^{(e,1)} \simeq (\omega_m^{(e,1)}/V_b) \{1 + 0.5[\omega_b^2 a^2 \ell n \, (b/a)/4V_b^2]^{1/3}\} \tag{18}$$

where

$$[\omega_m^{(e,1)}]^2 \simeq \{\omega_1^2 \Omega_e^2 - [2V_b^2/a^2 \ell n \, (b/a)](\omega_{e,2}^2 + \Omega_e^2)\}$$

$$[\Omega_e^2 + \omega_1^2 - 2V_b^2/a^2 \ell n \, (b/a)]^{-1} . \tag{19}$$

iii) threshold condition for instability

$$(\omega_1^2 a^2/V_b^2) > [2/\ell n(b/a)] \ [1 + \omega_{e,2}^2/\Omega_e^2] \tag{20}$$

or, in terms of the quantities $\varepsilon \equiv n_b/n_{p1}$, $K \equiv a^2 \omega_b^2/V_b^2$ (which is proportional to the perveance of the beam and therefore character-izes the beam) and $\Gamma \equiv 0.5 \ell n(b/a)$

$$K^{(e,1)} > (\varepsilon/\Gamma) \ (1 + \omega_{e,2}^2/\Omega_e^2) \ . \tag{20'}$$

For the ion plasma mode, $k^{(i)}$, assuming $\omega \ll \Omega_e$, ω_1, from (9)-(13), one finds :

i) low frequency and wavenumber range :

$$k_i^{(i)} \simeq (\omega/V_b)[(n_b/n_1)/(1 - \delta^{(i)})]^{\frac{1}{2}} \tag{21}$$

$$k_r^{(i)} \simeq (\omega/V_b)[1 + (n_b/n_1) \ \delta^{(i)}/(1 - \delta^{(i)})^2] \tag{22}$$

$$\delta^{(i)} = [2V_b^2/a^2\omega_1^2 \ell n(b/a)][1 + \omega_{e,2}^2/\Omega_e^2 + \omega_{i,2}^2/\Omega_i^2] \ . \tag{23}$$

ii) maximum growth rate :

$$k_{i,max}^{(i)} \simeq (3^{\frac{1}{2}}/2)(\omega_b^2/2\omega_1^2)^{1/3} (\omega_m^{(i)}/V_b)$$

$$/[1 + \omega_{e,2}^2/\Omega_e^2 - \omega_{i,2}^2/(\omega_m^{(i)} - \Omega_i^2)]^{1/3} \tag{24}$$

$$k_r^{(i)} \simeq (\Omega_i/V_b)\{1 + (\omega_b^2 a^2 \ell n(b/a)/4V_b^2)^{1/3}\} \tag{25}$$

where

$$[\omega_m^{(i)}]^2 \approx \Omega_i^2 \ \{1 - [2V_b^2/a^2 \ell n(b/a)](\omega_{i,2}^2/\Omega_i^2)$$

$$\cdot [\omega_1^2 - 2V_b^2/(1 + \omega_{e,2}^2 /\Omega_e^2)a^2 \ell n(b/a)]^{-1}\} \ . \tag{26}$$

iii) the threshold condition for instability:

$$(\omega_1^2 a^2/v_b^2) > [2/\ell n(b/a)][1+\omega_{e,2}^2/\Omega_e^2 + \omega_{i,2}^2/\Omega_i^2] \tag{27}$$

or, alternatively

$$K^{(i)} > (\varepsilon/\Gamma)(1 + \omega_{e,2}^2/\Omega_e^2 + \omega_{i,2}^2/\Omega_i^2). \tag{27'}$$

Inspection of equations (9) – (27) indicates the following:

1.   For given <u>physical parameters</u> $(n_b/n_1, \omega_{e,1}/\Omega_e$ and $\omega_{e,2}/\Omega_e)$ and <u>geometry</u> (Γ), there exists a critical K-value, K_c, above which instability can occur.

2.   An increase in the magnetic field lowers the critical value K_c for instability. This is due mainly to the corresponding decrease in the radius of the beam, a, and therefore, in the factor $1/\Gamma \equiv 2/\ell n(b/a)$ in the r.h.s. of the conditions (20') and (27'). In addition to this, a decrease in the factors $1+\omega_{e2}^2/\Omega_e^2$ and $1+\omega_{e2}^2/\Omega_e^2 + \omega_{i2}^2/\Omega_i^2$ with increasing B-value also occurs. This B-dependence is observed in the experiments of Bernstein et al. (1975) as reflected in the empirical relation for the critical for B.P.D., $I_c \sim B^{-0.7}$.

3.   The effect of the plasma surrounding the beam is to increase the threshold for instability; this is indicated by the factors $1+\omega_{e2}^2/\Omega_e^2$ and $1+\omega_{e2}^2/\Omega_e^2+\omega_{i2}^2/\Omega_i^2$ appearing in the conditions (20')and (27'). In fact, this is exactly what has been observed in the experiments of Bernstein et al (1980), where <u>filling the chamber with ambient plasma required increased critical currents for instability.</u>

4.   The condition (27') for the ion mode is more stringent than the condition (20') for the electron mode; this being due to the fact that the factor $1+\omega_{e2}^2/\Omega_e^2 + \omega_{i2}^2/\Omega_i^2 \equiv 1+(m_i/m_e+1)(\omega_{e2}^2/\Omega_e^2)$ may be much larger than the factor $1+\omega_{e2}^2/\Omega_e^2$ in the r.h.s of (20'). Thus, in the case of relatively low beam perveance, the ion mode can become unstable only when the density of the surrounding plasma is very low or the magnetic field is very strong. Obviously, if the ion mode is excited, the electron mode is also excited.

5.   When both ion and electron modes are excited, the later is dominant. Indeed, from (14) and (21) one finds

$$k_{i,max}^{(1,e)}/k_{i,max}^{(i)} \simeq min(\Omega_e, \omega_1)/\Omega_i. \tag{28}$$

6.   In the low frequency range, $\omega \sim \Omega_i$, the ion mode can be dominant. Indeed, for $\omega_{e1}/\omega_b \gg 1$

$$(k_i^{(i)}/k_i^{(1,e)})_{\omega \approx \Omega_i} \approx k_{i,max}^{(i)}/k_{i,\omega \approx \Omega_i}^{(1,e)} \approx (\omega_{el}/\omega_b)^{1/3} > 1 \qquad (29).$$

c.   <u>Temperature Effects On The Electron Plasma Mode</u>

To account for the temperature effects of the background and ambient plasmas too, the following expressions are utilized for the dielectric functions [See,e.g., Krall and Trivelpiece, 1973]:

$$\varepsilon_{\parallel}^{(1)} = 1-\omega_b^2/(\omega-kV_b^2)-(\omega_1^2/\omega^2)(1 + 3k^2 v_{th,1}^2/\omega^2)$$

$$\varepsilon_{\parallel}^{(2)} = 1-(\omega_{e,2}^2/\omega^2)(1+3k^2 v_{th,2}^2/\omega^2)$$

$$\varepsilon_{\perp}^{(1)} = 1-\omega_b^2/[(\omega-kV_b)^2-\Omega_e^2]-[\omega_1^2/(\omega^2-\Omega_e^2)^2] \cdot$$

$$\cdot [1 + k^2 v_{th,1}^2 (3\omega^2-\Omega_e^2)/(\omega^2-\Omega_e^2)]$$

$$\varepsilon_{\perp}^{(2)} = 1-[\omega_{e,2}^2/(\omega^2-\Omega_e^2)][1+k^2 v_{th,2}^2(3\omega^2-\Omega_e^2)/(\omega^2-\Omega_e^2)^2]$$

$$(30)$$

where, $v_{th,j}=(kT_j/m)^{1/2}$ (j=1,2), are the thermal velocities of the background (1) and surrounding plasma (2), respectively. After a lengthy calculation one obtains the following generalization of the cold-plasma results:
threshold condition:

$$K^{(e,1)} > (\varepsilon/\Gamma)(1+\omega_{e,2}^2/\Omega_e^2)/(1+v_{th,1}^2/V_b^2) \qquad (31)$$

growth rate and real wave number:

$$k_i^{(e,1)} \approx (\omega/\Omega)[(n_b/n_1)/(1-\delta^{(e,1)}+ \bar{\delta}^{(e,1)})]^{1/2} \qquad (32)$$

$$k_r^{(e,1)} \approx (\omega/\Omega)[1+(n_b/n_1)\delta^{(e,1)}+3v_{th,1}^2/V_b^2)/(1-\delta^{(e,1)}+\bar{\delta}^{(e,1)})^2]$$

$$(33)$$

where $\delta^{(e,1)}$ is given by eq. (16) and

$$\bar{\delta}^{(e,1)} = 3v_{th,1}^2 / V_b^2 + [2/\ell n(b/a)](\omega_{e,2}^2 v_{th,2}^2 / \Omega_e^4 a^2) . \tag{34}$$

Inspection of the results (31)-(34) indicates the following effects due to finite plasma temperatures:

1. The threshold for instability is lowered by the factor $(1+v_{th,1}^2/V_b^2)^{-1} \approx 1-v_{th,1}^2/V_b^2$;

2. The growth rate, $k_i^{(e,1)}$ is decreased by a factor $[1+\bar{\delta}^{(e,1)}/(1-\delta^{(e,1)})]^{-\frac{1}{2}} \approx 1-0.5\bar{\delta}^{(e,1)}/(1-\delta^{(e,1)})$.

d.   Effect of Finite Interaction Range.

In laboratory experiments the interaction length is determined by the length of the drifting tube, L. The simplest conditions on the perturbations (potentials) at the ends are $\phi(z=0)=\phi(z=L) = 0$. These conditions may affect the instability in several ways, namely by i) determining the maximum wavelength which the system can accommodate, i.e. setting a lower k_r-limit ($\lambda_{max}/2=L$; $k_{min}=2\pi/\lambda_{max}$); ii) producing a discretization of the wave spectrum, $k=nk_{min}=n(\pi/L)$, n=1,2 ...; iii) in the case of convecting instabilities (considered here), determining the extent to which the amplitude of the wave can grow; and iv) under favorable conditions producing new unstable modes, resulting from the coupling of forward and backward (reflected) waves.

We consider here only the effect the finite interaction length L might have on the instability threshold. The following results are obtained,

$$K^{(e,1)} > (\epsilon/\Gamma)(1+\omega_{e,2}^2/\Omega^2-\pi^2 V_b^2/L^2\Omega_e^2)/(1-\pi^2 V_b^2/L^2\Omega_e^2)$$

$$+ (\pi^2 a^2/L^2)/(1-\pi^2 V_b^2/L^2\Omega_e^2) \tag{20''}$$

and

$$K^{(i)} > (\epsilon/\Gamma)[1+\omega_{i,2}^2/\Omega_i^2-\pi^2 V_b^2/L^2\Omega_i^2]/(1-\pi^2 V_v^2/L^2\Omega_i^2) \tag{27''}$$

In the case $L>>\pi V_b/\Omega_e$ the expression (20'') reads

$$K^{(e,1)} > (\epsilon/\Gamma)\{1+\omega_{e,2}^2/\Omega_e^2(1+\pi^2 V_b^2/\Omega_e^2 L^2)\}+\pi^2 a^2/L^2 \tag{20'''}$$

and, if $L>>\pi V_b/\Omega_i$, expression (27'') becomes

$$K^{(i)} > (\epsilon/\Gamma)[1+\omega_{i,2}^2/\Omega_i^2+(\pi^2 V_b^2/L^2\Omega_i^2)(\omega_{i,2}^2/\Omega_i^2)] \tag{27'''}$$

The last expressions reduce to the $L=\infty$ case treated previously.

They indicate, however, that under certain circumstances, the finite L-effect raises the threshold for instability above that obtained for the L=∞ case.

IV.   THE BEAM-PLASMA DISCHARGE

An increase in the plasma density implies an ionization rate which is larger than that of plasma loss outside of the system. Assuming the loss of plasma out to the ends to be more important than that in the radial direction, we may write the following balance equation, determining the critical conditions (for which the two rates just balance):

$$\sigma V_b N_b n_o = 2\pi a^2 n_p V_i \quad , \tag{35}$$

where $\sigma \propto E_b^{-c}$ (E_b-beam energy) is the ionization cross section, V_b the beam velocity, N_b the total number of the electrons in the cylindrical interaction region of radius a and length L, n_o the density of neutral particles, n_p the ambient plasma density and V_i the thermal velocity of an ion (electrons and ions are coupled electrostatically).Since $n_o \propto P$(P-the ambient neutral pressure), $V_b \propto E_b$, $V_i \approx$const. and $N_b \cong \pi a^2 L\, n_b$, solving for n_p in (35) one obtains

$$n_p \propto n_b \, E_b^{0.5-c} \, PL \ . \tag{36}$$

Next, the "anomalous" feature of the BPD has to be considered. The anomality merely indicates that rather than being due to collisions between beam electrons and neutral particles the enhanced ionization is produced by plasma-electrons which were heated by the waves generated through the beam-plasma instability. Thus, the balance equation (35) or its equivalent form, eq.(36) has to satisfy simultaneously also the condition required for the beam plasma instability to be set up, which is the threshold condition (20'). Combining the two conditions and using the definition of the current ($I_b \equiv \pi a^2 n_b e V_b$) yields

$$I_c \propto (E_b^{2.4-c}/PL\Gamma)(1+\omega_{p,2}^2/\Omega_e^2)/(1+v_{th,1}^2/V_b^2) \tag{37}$$

Eq.(37) gives the critical dependence of the beam current on the beam energy, neutral pressure, magnetic field and interaction range in a beam-plasma-discharge of the type considered in this paper.

As a particular case, consider the Bernstein et al. (1978) experiment. Here, the base pressure ($\sim 10^{-6}$ torr) consisted of 30% water and 70% air. The working pressures were achieved by addition of dry nitrogen; for the B dependence experiments a total pressure of 4×10^{-6} torr was used so that the neutral background consisted of 7.5% water, 17.5% air and 75% nitrogen. Also, the beam energies used for the B dependence measurements were E_b=1KeV and 1.5KeV ($v_{th,1}/V_b$<<1). Thus, we use $\sigma_{eff} \propto E_b^{-0.7}$ (Brown, 1966), i.e. c=0.7. Concerning the dependence on the magnetic field: the radius of the beam, a, varies as B^{-1}, that is, $a_1B_1=a_2B_2$, etc. Then, for fixed external radius, b, one may write Γ as $\Gamma(B_2)=0.5[\ell n(b/a_1)+\ell n(B_2/B_1)]$; if, for example $B_2/B_1=2$, one has $\Gamma(B_2)/\Gamma(B_1)=1.7$ (1.3) if $b/a_1=$ =2.7(10). Thus, $1/\Gamma \propto B^{-d}$, 0.5$\lesssim$d$\lesssim$1. Thus, eq.(37) converts to

$$I_c \propto (E_b^{1.7}/P~L~B^{-d})~(1+\omega_{p,2}^2/\Omega_e^2)~(1+v_{th,1}^2/V_b^2) \qquad (38)$$

Eq. (38) reproduces satisfactorily the empirical expression found in the experiment. (An experimental error bar of 10% is reported by the authors). It shows the correct dependence on E_b,P,L, and B in the case of pure neutral ambient gas (i.e., $\omega_{e,2}^2$=0); it indicates an increase in the critical current in the case of presence of an ambient plasma, prior to the injection of the beam.

V.   SUMMARY

     The relation between the critical current, accelerating voltage, the magnetic field, the neutral pressure and the interaction range found in recent experiments on space-oriented beam-plasma discharge was reproduced analytically. In order to carry this out, the threshold condition for electrostatic instability in a magnetized bounded beam-plasma system and the balance equation for the rate of plasma production and that of loss through ends were derived and solved simultaneously.
     The principal scaling of the threshold criterion with variations in the magnetic field is determined by the variations in the radius of the beam-plasma region.
     The presence of a surrounding plasma increases the stability threshold, i.e. larger critical currents are required for the instability to occur. On the other hand, the growth rates are enhanced by the ambient plasma. On the contrary, the temperature effects of the background plasma are to decrease the threshold critical values as well as the growth rates.

REFERENCES

Bernstein, W., and Kellogg, P. J., 1980, Communicated.
Bernstein, W., Leinbach, H., Cohen, H., Wilson, P.S., Davis, T.N.,
     Hallinan, T., Baker, B., Martz, J., Zeimke, R. and Huber, W.,
     1975, J. Geophys, Res., 80: 4375.
Bernstein, W., Leinbach, H., Kellogg, P., Monson, S., Hallinan, T.,
     Garriott, O. K., Konradi, A., McCoy, J., Daly, P., Baker, B.,
     Anderson, H. R., 1978, Geophys. Res. Letters., 5:127.
Brown, S.C., 1966, Basic Data of Plasma Physics, The M.I.T.Press,
     Mass.

Cuperman, S., and Roth, I., 1981, Geophys. Res. Letters, 8:117.
Davidson, R. C.,1974, Theory of Nonneutral Plasma, Benjamin,
        Reading, Mass. .
Getty, W. D. and Smullin, L. D., 1963, J. Appl. Phys., 34:3421.
Krall and Trivelpiece, 1973, Principles of Plasma Physics, McGraw
        Hill Book Company.
Rowland, H. L., and Papadopoulos, K., 1979, Communicated.
Roth, I., and Cuperman, S., 1981, TAUP 901-81, Department of
        Physics and Astronomy, Tel Aviv University.

PLASMA WAVES GENERATED BY RIPPLED, MAGNETICALLY FOCUSED ELECTRON

BEAMS SURROUNDED BY TENUOUS PLASMAS*

S. Cuperman and F. Petran

Department of Physics and Astronomy
Tel Aviv University
Ramat Aviv, Israel

ABSTRACT

The electrostatic instability and the corresponding unstable wave spectrum of a neutralized rippled cylindrical electron beam (finite radius) surrounded by a tenuous plasma (large radius) at rest is investigated. The rippling is due to the imbalance of the centrifugal (defocusing) and Lorentz (focusing) forces acting on the fluid element of beam.

The analysis is carried out for both long and short wavelength surface perturbations. Closed analytical expressions for the growth rates are derived and discussed. It is found that the growth rates are proportional to the relative ripple amplitude and to the effective beam plasma frequency. Because of the "reduction" effect due to the finiteness of the beam radius, the growth rates are larger in the short wavelength limit. The presence of the surrounding plasma is such as to increase the growth rates of the instability.

The generation of unstable surface waves in the presence of a rippled beam-edge (and, correspondingly, a modulated beam particle density) places this process in the class of parametric instabilities.

I. INTRODUCTION

Electron beams are currently used in laboratory and space physics for purposes such as transport of energy, mapping, generation

Editor's Note: This paper was submitted to, but not presented, at the meeting.

559

of waves, basic studies of beam-plasma interactions, etc. As a rule, theoretical treatments assume a laminar regime for the equilibrium configurations, i.e., no dependence along the externally produced magneto-static field (See, e.g., Briggs, 1964).

There are, however, many situations in which, because of the imbalance of the focusing (Lorentz) and defocusing (centrifugal) forces on the element of beam, the beam is rippled rather than laminar (See, e.g. Bernstein et al., 1975). The amplitude of the ripple is determined by the slope at the exit from the gun. The rippled equilibria of electron beams in vacuum have been recently investigated by Mahaffey et al. (1975, 1976); Mahaffey and Trivelpiece (1977); Theiss et al. (1975, 1977). The stability of these equilibria was considered by Cuperman and Petran, (1981a) (nonneutral beams) and (1981b)(neutralized beams). However, in the last two works the beams were considered to be surrounded by vacuum and enclosed by coaxial metallic containers of relatively small radia ($\gtrsim$ beam radia). While these conditions are appropriate for laboratory experiments (including electronic tubes), they do not hold for space explorations.

Thus, in this paper we investigate the electrostatic instability and the corresponding unstable wave spectrum of magnetically focused neutralized rippled electron beams under spacelike conditions. That is, we treat the case of electron beams of finite radius surrounded by a tenuous plasma of relatively large radius (practically, infinite extension) at rest.

The paper is organized as follows: In Sect. II we sketch the derivation of the equilibrium conditions for the rippled configurations considered in the paper. In Sec. III we derive the dispersion relation for the electrostatic surface waves. This dispersion relation is then separately solved and discussed for the cases of long and short wavelength (Sec. IV). A summary and discussion are given in Sec. V.

II. GENERAL EQUATIONS AND EQUILIBRIUM

Consider a configuration consisting of a cold rippled cylindrical electron beam (average radius b) which is surrounded by a cold plasma at rest (radius a, a>>b). The electron beam is neutralized by positive ions (protons) at rest. The entire system is embedded in an axial static magnetic field, B_z. In the present analysis we consider small induced azimuthal and axial motions and, therefore, neglect the self magnetic fields. Then, in cylindrical polar coordinates with azimuthal symmetry, the motion and Poisson equations describing systems as defined above are respectively, (the unlabelled quantities refer to electrons)

$$\frac{\partial n}{\partial t} + \frac{1}{r}\frac{\partial}{\partial r}(nv_r r) + \frac{\partial}{\partial z}(nV_z') = 0 \quad , \tag{1}$$

$$\frac{\partial v_r}{\partial t} + v_r \frac{\partial v_r}{\partial r} + V'_z \frac{\partial v_r}{\partial z} - \frac{v_\theta^2}{r} = - \frac{e}{m} (E_r + \frac{v_\theta B_z}{c}) , \tag{2}$$

$$\frac{\partial v_\theta}{\partial t} + v_r \frac{\partial v_\theta}{\partial r} + V'_z \frac{\partial v_\theta}{\partial z} - \frac{v_r v_\theta}{r} = \frac{e}{m} \frac{v_r B_z}{c} , \tag{3}$$

$$\frac{\partial v_z}{\partial t} + v_r \frac{\partial v_z}{\partial r} + V'_z \frac{\partial v_z}{\partial z} = - \frac{e}{m} E_z , \tag{4}$$

$$\frac{1}{r} \frac{\partial}{\partial r} (rE_r) + \frac{\partial E_z}{\partial z} = 4\pi \Sigma_j q_j n_j \quad (j=e,p) . \tag{5}$$

Here $V'_z = V_z + v_z$, V_z is the constant axial streaming velocity in vacuum and v_z, v_r, v_θ are the self generated velocity components; B_z is the uniform external magnetic field; $E_i (i=r,z)$ are the components of the electric field; and $n(r,z)_j$, q_j are the particle density and charge, respectively.

Equilibrium

For the specific case considered here one has to distinguish between the beam region (1) and the surrounding region (2). Thus, one has

Region 1 $(0 \leq r \leq b)$

$V_{z,1} \equiv V_z \neq 0$, $v_{z,1} \equiv v_z \neq 0, v_{r,1} \equiv v_r \neq 0, v_{\theta,1} \equiv v_\theta \neq 0; n_b = n_{p1}$.

Region 2 $(b \leq r \leq a)$

$V_{z,2} = 0, v_{z,2} = v_{r,2} = v_{\theta,2} = 0;$ $n_{e2} = n_{p2}$.

We now consider configurations with radially uniform densities and relatively small ripple amplitudes (See Fig.1a and Fig.1b). Moreover, we assume the constancy of the axial streaming velocity of the beam and look at the case in which $n_{p1} >> n_{p2}$, i.e., dense neutralized electron beams surrounded by tenuous plasmas at rest.

The equilibrium edge of the rippled beam can be found by solving the "envelope" equation for the outermost electrons as follows:

First, from eqs. (2), (3) and (5) one obtains the following expression for the equilibrium rotation velocity, $\dot{\theta}(\dot{\theta} \equiv v_\theta/r)$(see, e.g., Davidson, 1974, Mahaffey, et al.,1976)

$$\dot{\theta}^{\pm}(r) = (\omega_c/2)(1 \pm r_o^2/r^2) \tag{6}$$

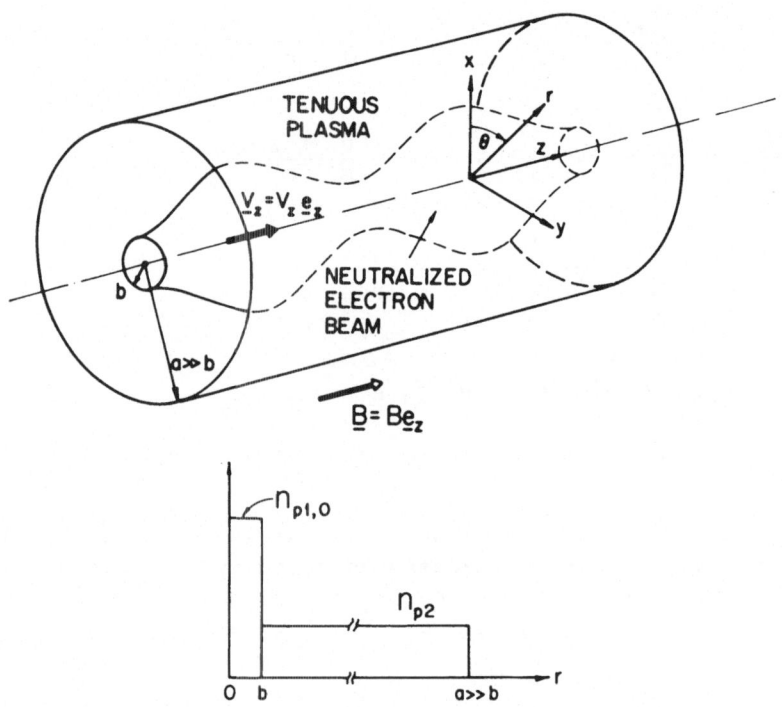

Fig. 1.   Equilibrium configuration (a) and density profiles (b)

where $\omega_c = eB_z/mc$ is the electron cyclotron frequency, r_o is the laminar (i.e. $\partial/\partial_z = 0$) equilibrium beam radius and $\dot{r}$ is the instantaneous beam radius. Thus, $\dot{\theta}$ is a constant only in a given cross section of the beam; it depends, however, on the radius of that cross section, which varies along the beam (z direction). The + and − signs in (6) correspond to the fast and slow modes of beam rotation, respectively.

Next, from eq. (2), in a reference frame moving with the beam, the following radial equation of motion is obtained

$$\ddot{r} - r\dot{\theta}^2 + \omega_c r\dot{\theta} = 0 \quad . \tag{7}$$

For small ripple amplitude, $|\delta|$ ($r = b = r_o + \delta$, $|\delta| \ll r_o$), one obtains the following linearized equation

$$\ddot{\delta} + \omega_c^2 \delta = 0 \quad . \tag{8}$$

For a beam which enters the interaction region with $r = r_{min}$ (See Fig. 1) the solution of Eq. (8) (after converting to the rest frame) is

$$b = r_o - \delta_o \cos k_S z, \quad k_S \equiv \omega_c/V_z \quad . \tag{9}$$

Thus, the net restoring force (centrifugal less magnetic) in the radial equation (7) produces the periodic motion of the edge electrons [according to eq. (9)] and determines the actual rippled equilibrium shape of the entire beam. The amplitude of the ripple is determined by the slope at the entry plane: $\delta_0 = v_r(z=0)/V_z k_S$. The modulation of the beam radius in conjunction with the condition of radially uniform particle density leads to the axial modulation of the beam particle density and, therefore, of the electron plasma frequency, namely

$$\omega_{p1}(z) = \omega_{p1,0}[1 + (\delta_0/r_0)\cos k_S z] \quad , \tag{10}$$

where $\omega_{p1,0}^2 = 4\pi n_{p1,0} e^2/m_e$ is the square of the laminar electron beam plasma frequency.

III. THE DISPERSION RELATION

For the small ripple amplitude case treated here, we consider the waves to have space varying wave numbers as given by a uniform beam dispersion relation but modulated according to the local value of the plasma frequency, eq. (10) (See, e.g., Tien, 1977; Cuperman and Petran, 1981a).

Thus, for the uniform beam we look for electrostatic perturbations of the form

$$\tilde{S}(r,z,t) = \tilde{S}(r) \exp[i(\omega t - k_z z)] \tag{11}$$

($\tilde{S} = \tilde{n}, \tilde{v}_j, \tilde{E}_j$; $j = r,\theta,z$). Upon substitution of eq. (11) into the linearized motion and field equations for the neutralized laminar beam-plasma system and identification of the exponents, one obtains the following electrostatic wave equations for the regions 1 and 2, respectively

$$\frac{1}{r}\frac{\partial}{\partial r}\left(r\frac{\partial \tilde{E}_z}{\partial r}\right) - \bar{\tau}^2 \tilde{E}_z = 0 \tag{12}$$

with

$$\bar{\tau}^2 = \tau^2 k_z^2 \tag{13}$$

and

$$\tau_1^2 = (1 - \omega_{p1,0}^2/\Omega^2)/[1 - \omega_{p1,0}^2/(\Omega^2 - \omega_c^2)] \quad , \tag{14}$$

$$\tau_2^2 = (1 - \omega_{p2}^2/\omega^2)/[1 - \omega_{p2}^2/(\omega^2 - \omega_c^2)] \tag{15}$$

$$\Omega \equiv \omega - k_z V_z \quad . \tag{16}$$

For surface waves, $\tau^2 > 0$, the solution of eq. (12) inside the beam which remains finite at $r=0$ is

$$\tilde{E}_{z,1} = A_0 I_0(\bar{\tau}_1 r)\exp i(\omega t - k_z z) \quad , \quad r \leq b \quad , \tag{17}$$

where I_0 is the modified Bessel function of the first kind of order zero. In the region surrounding the beam the perturbed field solution which vanishes at large distances, a>>b, is

$$\tilde{E}_{z,2} \equiv B_0 K_0(\bar{\tau}_2 r)\exp[i(\omega t - k_z z)] \qquad b<r\leq a, \qquad (18)$$

where K_0 is the modified Bessel function of second kind and order zero.

Upon matching the field solutions at $r = b = r_0$, one obtains the following dispersion relation

$$\tau_1[I_0'(\bar{\tau}_1 r_0)/I_0(\bar{\tau}_1 r_0)][1-\omega_{p1,0}^2/(\Omega^2-\omega_c^2]$$

$$= \tau_2[K_0'(\bar{\tau}_2 r_0)/K_0(\bar{\tau}_2 r_0)][1-\omega_{p2}^2/(\omega^2-\omega_c^2)] \qquad . \qquad (19)$$

where the prime sign indicates derivatives with respect to the argument.

Notice that the condition for surface waves, $\tau^2>0$, defines the following regimes of parameter range:

Region 1:   Regime I: $\omega_{p1,0}^2/\Omega^2>1$: $\omega_c^2<\Omega^2<\omega_{p1,0}^2$ $\qquad (20)$

Regime II: $\omega_{p1,0}^2/\Omega^2<1$:

IIA: $\omega_c^2/\Omega^2>1$: $\omega_{p1,0}^2<\Omega^2<\omega_c^2$ $\qquad (21)$

IIB: $\omega_c^2/\Omega^2<1$: $\Omega^2>\omega_{p1,0}^2+\omega_c^2$ $\qquad (22)$

Region 2:   Regime I: $\omega_{p2}^2/\omega^2>1$: $\omega_c^2<\omega^2<\omega_{p2}^2$ $\qquad (23)$

Regime II: $\omega_{p2}^2/\omega^2<1$:

IIA: $\omega_c^2/\omega^2>1$: $\omega_{p2}^2<\omega^2<\omega_c^2$ $\qquad (24)$

IIB: $\omega_c^2/\omega^2<1$: $\omega^2>\omega_{p2}^2+\omega_c^2$ $\qquad (25)$

These consistency conditions have to be used for the selection of the actual solutions.

IV. SOLUTION OF THE DISPERSION RELATION

A. Long Wavelength Perturbations

By using the series expansions for the Bessel functions in eq. (19) and retaining only the leading terms one obtains the following dispersion equation describing the laminar ($\partial/\partial_z=0$) neutralized beam-plasma configuration:

$$(1-\omega_{p1,0}^2/\Omega^2) = (1-\omega_{p2}^2/\omega^2)(1-1/k_z^2 R) \qquad (26)$$

with $(\bar{\tau}_z r_0 \ll 1)$

$$R \equiv -0.5(\tau_2 r_0)^2 \ln(\bar{\tau}_2 r_0) > 0. \tag{27}$$

Solving for Ω provides

$$\Omega^2 \simeq k_z^2 R \omega_{p1,0}^2 / (1 - \omega_{p2}^2/\omega^2) \quad . \tag{28}$$

From (28) and (16) one obtains

$$(k_z)_s^f = (\omega/V_z) \mp \bar{R}^{\frac{1}{2}}(\omega_{p1,0}/V_z)(1 - \omega_{p2}^2/\omega^2)^{-\frac{1}{2}} \equiv k_L \quad . \tag{29}$$

The quantity $\bar{R}^{-\frac{1}{2}} \equiv R^{\frac{1}{2}}(\omega/V_z)$ represents the reduction factor due to the finite beam geometry. Eqs. (29) describe two space-charge surface waves (f-fast, s-slow) that are supported by the laminar beam. Inspection of eq. (29) indicates the existence of two distinct frequency ranges, namely:

1. $\omega < \omega_{p2}$. In this range k_z is complex and

$$Re k_z \equiv k_{z,r} = \omega/V_z \tag{30}$$

and

$$Im k_z \equiv k_{z,i} = \bar{R}^{-\frac{1}{2}}(\omega_{p1,0}/V_z)(\omega_{p2}^2/\omega^2 - 1)^{-\frac{1}{2}} \tag{31}$$

These solutions indicate an instability of the beam-plasma type which is due to the coupling of the negative energy wave of the beam (slow wave) and the positive energy wave of the plasma.

2. $\omega > \omega_{p2}$. In this range, the two surface waves, k_z^f and k_z^s [eq. (29)] supported by the laminar beam are stable. The parameter range in which consistent solutions exist (i.e., eqs. (26) and (20)-(25) are satisfied simultaneously) is given by the inequalities $\omega_c^2 < \Omega^2 < \omega_{p1,0}^2$; they correspond to the case of relatively weak magnetic field.

B. Short Wavelength Perturbations

Using asymptotic expansions for the Bessel functions in eq. (19) and retaining only the leading terms provides the following dispersion relation for the laminar beam-plasma configuration in the short wavelength limit:

$$\tau_1[1 - \omega_{p1,0}^2/(\Omega^2 - \omega_c^2)] = -\tau_2[1 - \omega_{p2}^2/(\omega^2 - \omega_c^2)] \quad . \tag{32}$$

To make progress, in the following, we consider two different cases, namely the relatively high frequency range $(\omega^2 \gg \omega_c^2)$ and low frequency range $(\omega^2 \ll \omega_c^2)$.

1. $\omega^2 \gg \omega_c^2$. In this range, eq. (32) becomes

$$\tau_1[1 - \omega_{p1,0}^2/(\Omega^2 - \omega_c^2)] \simeq -\alpha \tag{33}$$

where

$$\alpha \equiv 1 - \omega_{p2,0}^2/\omega^2 \quad . \tag{34}$$

After some algebra, eq. (33) reads

$$\Omega^4(1-\alpha^2) - \Omega^2[2\omega_{p1,0}^2 + \omega_c^2(1-\alpha^2)] + \omega_{p1,0}^2(\omega_{p1,0}^2 + \omega_c^2) = 0 \tag{35}$$

and provides the following solution:

$$\Omega_{\pm}^2 = \{\omega_{p1,0}^2 + 0.5\omega_c^2(1-\alpha^2) \pm [\alpha^2\omega_{p1,0}^4 + 0.25(1-\alpha^2)^2\omega_c^4]^{\frac{1}{2}}\}$$

$$\cdot(1-\alpha^2)^{-1}. \tag{36}$$

As in the case of long wavelength perturbations, we check also here the conditions for consistent solutions satisfying eqs. (32) and (20)-(25). We distinguish two situations:

1a. $\omega_{p2}^2 < \omega^2$. The consistent parameter range is found to be determined by the conditions $\omega_{p2}^2 < \omega_c^2 << \omega^2 < \Omega^2 < \omega_{p1,0}^2$, $0 << \alpha \lesssim 1$. Then, expanding in the small parameter $(\omega_c^4/\omega_{p1,0}^4)[(1-\alpha^2)^2/4\alpha^2]$ provides *

$$\Omega_{\pm}^2 \simeq \omega_{p1,0}^2/(1\mp\alpha) + \omega_c^2/2 \quad . \tag{37}$$

Since for consistency $\Omega^2 < \omega_{p1,0}^2$, it follows that the only physical solution is Ω_-^2. Then by (37) and (16) one has

$$(k_z)_s^f = (\omega/V_z) \mp (\omega_{p1,0}^2/V_z)[1/(1+\alpha) + \omega_c^2/2\omega_{p1,0}^2]. \tag{38}$$

1b. $\omega_{p2}^2 > \omega^2$. In this case $\alpha < 0$ and the consistency conditions are $\omega_c^2 << \omega^2 < \omega_{p2}^2 < \omega_{p1,0}^2, 0 < \Omega^2$. Then Ω_-^2 given by eq. (37) represents a consistent solution in the range $1 < \omega_{p2}^2/\omega^2 < 2$.

2. $\omega^2 << \omega_c^2$. For this range eq. (32) reads

$$\tau_1[1 - \omega_{p1,0}^2/(\Omega^2 - \omega_c^2)] = -\beta(1 + \omega_{p2}^2/\omega_c^2)^{\frac{1}{2}} \tag{39}$$

where

$$\beta \equiv (1 - \omega_{p2}^2/\omega^2)^{\frac{1}{2}} \quad . \tag{40}$$

The consistency conditions required in this case are $\omega_{p2}^2 < \omega^2 << \omega_c^2 < \Omega^2 < \omega_{p1,0}^2$. To make further progress, we neglect the term ω_{p2}^2/ω_c^2 on the r.h.s. of eq. (40) and solve it for Ω^2 to obtain

$$\Omega_{\pm}^2 = \{\omega_{p1,0}^2 + 0.5\omega_c^2(1-\beta^2) \pm [\beta^2\omega_{p1,0}^4 + 0.25(1-\beta^2)^2\omega_c^4]^{\frac{1}{2}}\} \tag{41}$$

$$\cdot(1-\beta^2)^{-1}.$$

It can be easily shown that only Ω_-^2 represents a consistent solution. Then one finds $(0 < \beta < 1)$

$$\Omega_-^2 = \omega_{p1,0}^2/(1+\beta) + \omega_c^2/2 \quad , \tag{42}$$

which leads to

$$(k_z)_s^f = (\omega/V_z) \mp (\omega_{p1,0}/V_z)[1/(1+\beta) + \omega_c^2/2\omega_{p1,0}^2]^{\frac{1}{2}} \quad . \tag{43}$$

We notice that in the short wavelength limit the surface modes imply a relatively weak magnetic field and are stable in the laminar beam case.

___

* Since $\omega_c^2 << \omega^2 < \omega_{p1,0}^2$, the second term on the r.h.s. of eq. (37) may be neglected to obtain $\Omega_-^2 = \omega_{p1,0}^2/(2 - \omega_{p2}^2/\omega^2)$ which is the equation for the beam plasma system in the absence of a static magnetic field in in the short wavelength limit.

C. The Rippled Beam

In the case of the laminar beam the axial electric field $\tilde{E}_z$ represents a plane wave in the axial direction (z). The appropriate wave (Helmholtz) equation is

$$\partial^2 \tilde{E}_z / \partial z^2 + k_z^2 \tilde{E}_z = 0 \quad . \tag{44}$$

It describes the behavior of two independent space-charge waves (f,s) with $(k_z)_s^f$ given by (29) (long wavelength perturbations) or (38) and (43) (short wavelength perturbations).

The rippled beam equilibrium is characterized by a beam radius modulation whose axial dependence is given by eq.(9) and by a corresponding beam-plasma frequency modulation given by eq. (10). Then for small ripple amplitudes one may assume for $\tilde{F}_z$ an axial space dependence given by $\tilde{E}_z \sim \exp\{ik_z(z)z\}$, $k_z(z)$ being the local (modulated) wave number obtained by an expansion in Taylor series about the laminar beam value, $(k_z)_s^f \equiv k_{z,L}$: (See, e.g., Tien, 1977, Cuperman and Petran 1981a)

$$k_z(z) = k_{z,L} + (\partial k_z / \partial b)_{b=r_0} (b-r_0) \; - \; - \; - \quad . \tag{45}$$

To first order one obtains the following results:

Long wavelength limit:

$$\{k_z(z)\}_s^f = (k_z)_s^f [1 \mp \frac{\omega_{p1,0} \bar{R}^{\frac{1}{2}} \delta' \cos k_s z}{V_z (1-\omega_{p2}^2/\omega^2)^{\frac{1}{2}}(k_z)_s^f}] \tag{46}$$

and, from here,

$$[\{k_z(z)\}_s^f]^2 \simeq [(k_z)_s^f]^2 (1+2\varepsilon_s^f \cos k_s z) \quad , \tag{47}$$

where

$$\varepsilon_s^f = \mp \delta' \bar{R}^{\frac{1}{2}} (\omega_{p1,0}/V_z)/(1-\omega_{p2}^2/\omega^2)^{\frac{1}{2}}(k_z)_s^f \quad . \tag{48}$$

Short wavelength limit , $\omega^2 >> \omega_c^2$:

$$\{k_z(z)\}_s^f = (k_z)_s^f [1 \mp \frac{\omega_{p1,0} \delta' \cos k_s z}{V_z (1+\alpha)^{\frac{1}{2}}}] \tag{49}$$

$$\varepsilon_s^f = \mp \delta' (\omega_{p1,0}/V_z)/(1+\alpha)^{\frac{1}{2}}(k_z)_s^f \quad . \tag{50}$$

Short wavelength limit, $\omega^2 << \omega_c^2$:

$$\{k_z(z)\}_s^f = (k_z)_s^f \{1 \mp \frac{\omega_{p1,0} \delta' \cos k_s z}{V_z (1+\beta)[1/(1+\beta)+\omega_c^2/2\omega_{p1,0}^2]^{\frac{1}{2}}} \} \quad , \tag{51}$$

$$\varepsilon_s^f = \mp \, \delta' (\omega_{p1,0}/V_z)/(1+\beta)[1/(1+\beta)+\omega_c^2/2\omega_{p1,0}^2]^{\frac{1}{2}}(k_z)_s^f \quad . \tag{52}$$

Thus, with $k_z(z)$ given by eqs. (46)-(52) the wave equation for the rippled beam reads

$$\partial^2 \tilde{E}_z / \partial z^2 + k_L^2 (1 + 2\epsilon \cos k_S z) \tilde{E}_z = 0 \tag{53}$$

This is Mathieu's equation, the solution of which predicts wave-resonances for $k_L/k_S = \pm 1/2, \pm 1, \ldots$ Thus, according to Floquet's theorem, the general solution of the wave equation in a periodic medium is given by a superposition of waves propagating in opposite directions (forward and backward waves). The continuous spectrum of wave numbers of the laminar configuration, $k_{z,L}$ is now replaced by a Brillouin type spectrum consisting of side bands at k_L, $k_L \pm k_S$, $k_L \pm 2k_S$, etc.

For small ϵ-values, i.e., small relative ripple amplitudes and ω not too close to ω_{p2} (long wavelength) or $\omega_{p2}/2^{\frac{1}{2}}$ (short wavelength), one may obtain an analytical solution of Mathieu's eq. (53) by a small-perturbation resonant coupling method.

Thus, using the exponential decomposition for $\cos k_S z$ and the Fourier transform for $\tilde{E}_z(z)$ in eq. (53), multiplying the resulting equation by $\exp(-ik_z' z)$ and integrating over z, by the ortogonality conditions one obtains an equation relating the mode $\tilde{E}_z(k_z)$ and the modes $\tilde{E}_z(k_z \pm k_S)$. A second equation relating the mode $\tilde{E}_z(k_z - k_S)$ and the modes $\tilde{E}_z(k_z)$ and $E_z(k_z - 2k_S)$ is easily obtained by multiplying eq. (53) by $\exp[-i(k_z' - k_S)z]$ and following the same procedure. Next, within the resonant two-mode coupling assumption we neglect off-resonant modes and are left with two equations for $\tilde{E}_z(k_z)$ and $\tilde{E}_z(k_z - k_S)$, with $k_L \approx k_S/2$. Thus, a forward wave and a backward wave couple to produce an unstable standing wave. Solving for complex k one obtains ($k_z = k_r + ik_i$)

$$k_r = k_S/2 \quad , \tag{54}$$

$$k_i = \pm 0.5 (\epsilon^2 k_L^2 - \Delta^2)^{\frac{1}{2}} \quad , \tag{54'}$$

where $\Delta = k_S - 2k_L$ is the mismatch from perfect resonance. The wave spectrum is a band spectrum, the width of which is given by

$$(\tfrac{1}{2} - \epsilon/4) \leq k_L/k_S \leq (\tfrac{1}{2} + \epsilon/4) \quad . \tag{55}$$

The maximum growth rate corresponds to perfect matching ($\Delta = 0$) and has the value

$$k_{i,max} = \pm \epsilon k_S/2 \quad . \tag{56}$$

We notice that we considered the resonant coupling of the modes $\tilde{E}_z(k_z)$ and $\tilde{E}_z(k_z - k_S)$. In a quite similar manner one can treat the resonant coupling of the modes $\tilde{E}_z(k_z)$ and $\tilde{E}_z(k_z + k_S)$ to find identical results.

Finally, upon substitution in eq. (56) of the various expressions derived previously for ϵ, one obtains the following explicit results for the maximum growth rates:

long wavelength limit, $\omega > \omega_{p2}$:

$$k_{i,max} = 0.5\delta'\bar{R}^{\frac{1}{2}}(\omega_{p1,0}/V_z)/(1-\omega_{p2}^2/\omega^2)^{\frac{1}{2}} \qquad (57)$$

short wavelength limit, $\omega^2 >> \omega_c^2$:

$$k_{i,max} = 0.5\delta'(\omega_{p1,0}/V_z)/(1+\alpha)^{\frac{1}{2}} \qquad (58)$$

short wavelength limit, $\omega^2 << \omega_c^2$:

$$k_{i,max} = 0.5\delta'(\omega_{p1,0}/V_z)/(1+\beta)^{\frac{1}{2}}[1/(1+\beta)+\omega_c^2/2\omega_{p1,0}^2]^{\frac{1}{2}}. \qquad (59)$$

Notice that, since $k_r = k_S/2 = \omega_c/2V_z = k_L$ [k_L given by (29), (38) or (43)], only fast-fast wave coupling is possible in the parameter ranges considered here. Also, since ω_c and V_z are prescribed by the equilibrium conditions, both k_r and ω are uniquely determined.

Thus, we found that neutralized rippled electron beams can support unstable long and short wavelength electrostatic surface waves; the growth rate is proportional to the relative ripple amplitude, $\delta' \equiv \delta_0/r_0$, and the effective beam plasma frequency. Physically, due to the modulation (rippling) of the beam, two natural oscillations of the laminar system, $\tilde{E}_z \sim \exp[-ik_Lz]$ and $\tilde{E}_z \sim \exp[ik_Lz]$, couple and produce amplification, the required energy for amplification being supplied by the rippled beam. This instability belongs to the class of parametric instabilities.

V. SUMMARY

We investigated the electrostatic instability of neutralized, magnetically focused rippled electron beams of finite radius surrounded by tenuous plasmas of infinite extension. The following results were found:

In the long wavelength limit two types of instability (extending over different frequency ranges) exist. In the range $\omega < \omega_{p2}$, an instability of the beam-plasma type occurs due to the interaction between the beam electrons and the surrounding plasma electrons at the beam-plasma interface. In the range $\omega > \omega_{p2}$ a parametric type instability produced by the coupling of a fast forward wave and a fast backward wave due to the rippling (modulation) of the beam is present; the energy source for the parametric instability is provided by the rippled beam. The growth rate of the parametric instability is smaller than that of the beam-plasma surface mode by a factor $\sim\delta'(<<1)$ representing the relative ripple amplitude; notice, however, that it occurs for higher frequencies than those corresponding to the beam-plasma mode.

In the short wavelength limit, surface waves which are stable for the laminar beam may become unstable in the rippled beam case; thus, if resonant conditions for parametric instability are fulfilled, standing waves with growing amplitudes are produced. The growth rates of the parametric instability in the short wavelength limit are larger than those corresponding to the long wavelength limit; this is due to a larger value of the effective beam-plasma frequency.

The presence of a surrounding plasma leads in all cases to an enhancement of the growth rates.

REFERENCES

Bernstein, W., Leinbach, H., Cohen, H., Wilson, P.S., Davis, T.N., Hallinan, T., Baker, B., Martz, J., Zeimke, R. and Huber, W., 1975, J. Geophys. Res., 80:4375.

Briggs, R.I., 1964, Electron-Stream Interaction with Plasmas, M.I.T. Press, Cambridge, MA

Cuperman S. and Petran F., 1981, J. Plasma Phys., In Press.

Cuperman S. and Petran, F., 1981b, TAUP 891-81, Tel Aviv University, Israel.

Davidson, R.C., 1974, Theory of Non-neutral Plasmas, Benjamin, Reading, MA.

Mahaffey, R.A. Batchelor, D.B. and Trivelpiece, A.W. 1976, J. Appl. Phys. 47:4464.

Mahaffey, R.A., Goldstein, S.A., Davidson, R.C. and Trivelpiece, A.W. 1977, Phys. Rev. Lett. 35:1439.

Mahaffey, R.A. and Trivelpiece, A.W. 1975, Phys. Fluids, 20:469.

Theiss, A.J., Mahaffey, R.A. and Trivelpiece, A.W. 1975, Phys. Rev. Lett. 35:1436.

Theiss, A.J. Mahaffey, R.A. and Trivelpiece, A.W. 1977, Phys. Fluids 20:785.

Tien, P.K. 1977, Rev. Mod. Phys. 49:361.

Chapter 5: NEUTRALIZATION OF A CHARGED BODY IN
           SPACE

CHARGE NEUTRALIZATION AS STUDIED EXPERIMENTALLY AND THEORETICALLY

Lewis M. Linson

Science Applications, Inc.
La Jolla, California   92037

INTRODUCTION

The problem of vehicle neutralization is simply stated. The electrostatic potential, ϕ, of an isolated sphere of radius a meters emitting negative charge at a rate of I amperes will charge up at the rate of

$$\frac{d\phi}{dt} = 9 \, \frac{I}{a} \text{ kV/}\mu\text{s} \, . \tag{1}$$

If there is no return current, a meter-size body emitting as little as a 10 mA electron beam would be charged up to 9 kV potentials in 0.1 ms. Such large potentials would inhibit the electron beam from leaving the vicinity of the vehicle. A one ampere beam would charge up a vehicle in microseconds.

In order to avoid charging to high positive potentials, the vehicle must attract a return current equal to the emitted current. A measure of the return current that can be provided by an ambient plasma of number density n_p (cm^{-3}) and electron temperature, T_e, is given by the total flux of electrons across an imaginary surface of area S (m^2);

$$I_o = \frac{1}{4} \, en_p\bar{c}S = \frac{n_p}{10^5} \left(\frac{T_e}{1600^o}\right)^{1/2} S \quad mA \tag{2}$$

where $\bar{c}$ is the mean thermal velocity ($8kT_e/\pi m$) and k is Boltzman's constant. For effective collecting areas of order 1 m^2 and ionospheric densities less than 10^6 cm^{-3}, this current is less than 10 mA.

For electron beams significantly greater than I_o, either the effective collecting area, S, must be increased or the ambient plasma in the vicinity of the vehicle must be modified so as to increase n_p and/or T_e. The effective collecting area can be increased by modifying particle orbits due to a positive vehicle potential or causing anomalous transport across the ambient magnetic field by turbulent electric and magnetic fields. The plasma can be modified by heating the plasma or energizing electrons, or by ionizing ambient neutrals either directly by beam electrons, beam-plasma discharge, or by a vehicle-induced discharge. The time constant associated with these processes is unclear, but unless it is less than a few microseconds for a 1 A beam, the possibility exists of the vehicle acquiring at least a transient voltage of order kilovolts.

This fundamental problem was recognized in the 1960's when the first rocket-borne electron beam experiments were planned. Theoretical approaches to this neutralization problem concentrated initially on two nonoverlapping altitude regimes. Below about 100 km altitudes, the mean-free-path for ionization of the ambient neutral atmosphere by primary beam electrons is less than or comparable to the scale size of the vehicle. It was recognized that beam-produced ionization would be very important and modify the nearby plasma density. In this low altitude regime dominated by collisionally-produced ionization, models were developed to investigate both the initial potential rise of the vehicle during the initial charge-up phase as well as the drop to a steady-state potential sufficient to attract a return current equal to the emitted current, I. The return current in these models, to be discussed in more detail below, was assumed to be provided by the beam-produced ionization alone.

Above 200 km altitudes, the mean-free-path for ionization by the primary beam electrons is many kilometers or larger and thus far greater than the vehicle size. It was concluded that ionization of the ambient neutrals is unimportant and that the source of the return current would have to be the ambient ionospheric plasma. The potential, ϕ, on the vehicle would have to rise so as to increase the effective current-collection area, S, in Eq. (2). Three theoretical models were developed to calculate the potential to which the vehicle must rise in order to attract to it a neutralizing return current equal to the emitted current I. The first high-altitude rocket-borne electron beam experiment (Hess et al., 1971; Davis et al., 1971) was also carried out. Hess et al. (1971) solved the neutralization problem by deploying a collector with a sufficiently large collecting area (530 m^2) such that the potential on the vehicle would not be expected to rise very much.

In the 1970's, many rocket-borne electron beam experiments have been conducted. The general conclusion is that the rockets did not charge significantly and the beam electrons left the vicinity of the rocket at a substantial fraction of the energy imparted to them by the electron gun. Indeed no experimenter has interpreted his results as implying that the rocket charged up and prevented the primary electrons from leaving the rocket due to a lack of vehicle charge neutralization. The overwhelming evidence is that the emitted electron beam leaves the vicinity of the rocket with little loss of energy throughout the altitude range from below 100 km up to 350 km. This general conclusion seems to be independent of gun voltage between 1-40 kV, the injected pitch angle, the background electron density, or whether the beam is injected upwards or downwards. It is true that many of the observables such as the indicated vehicle potential, locally enhanced plasma density, plasma heating, secondary electrons, wave characteristics, and light emission do vary with these parameters. Thus, the physical processes taking place are strongly dependent on the injection conditions.

As Winkler (1980) has remarked, positive vehicle potentials accompanying electron beam emission have proven difficult to measure. The characteristics of various wave emissions and secondary electron spectra indicate that collective effects frequently play an important role. The measurement of enhanced plasma density and temperatures and intensified light emissions indicate that the plasma in the vicinity of the vehicle has been greatly modified. We shall discuss in more detail below the processes that produce these plasma modifications.

There are two generically different ways to ionize ambient neutrals to increase the local plasma density. Ionization can take place solely due to the presence of the beam in which case the return current would depend on the electron gun voltage but not significantly on the potential of the vehicle. The second source for plasma modification is provided by ambient or secondary electrons accelerated to the vehicle at positive potential. In this case the return current would depend on the potential of the vehicle but not significantly on the electron gun voltage. The determination of which of these processes is dominant under various conditions is experimentally unclear and currently an area of active research.

Another important occurrence that took place in th last decade was the rediscovery of the beam-plasma dishcarge (BPD) in parameter ranges of ambient neutral density, magnetic field intensity, and beam length 2 to 3 orders of magnitude different than the values of these parameters when the beam-plasma discharge phenomena was originally discovered in the early 60's. The revelance of this strong beam-plasma interaction to space experiments and its importance for vehicle neutralization will be discussed below.

GENERAL TIME-SCALE CONSIDERATIONS

In the previous section we have indicated that the charging time for instantaneous beam turn on is in the submillisecond time range and may be as short as the microsecond range for large currents. If the beam turn-on is slower, then it will control the charging rate.

In a steady-state situation long after beam turn-on, there are two possible sources of the neutralizing return current; the ambient ionospheric electrons and electrons produced by ionization of neutrals. There are various time scales associated with setting up a possible steady state and there will be transient effects on these time scales. We discuss some of the important characteristic time scales and how they relate to the charging problem.

The electron gyrofrequency and plasma frequency in the ionosphere are $> 10^6$ s^{-1} so that the corresponding time scales are shorter than a microsecond. The ionospheric Debye length is in the millimeter range, much smaller than the body size. For $I > I_o$, the amount of charge emitted in 1 μs exceeds the charge contained in a Debye sheath ($\sim 10^{-8}$ C). Thus, a steady-state cannot be achieved on the microsecond scale. In order to collect additional current, the spatial distribution of ions will have to be modified. Two independent time scales are associated with the time it takes a thermal ion to move a body scale and the time it takes the vehicle to move a body scale. Because both velocities are ~ 1 km/s, these characteristic times are of order one millisecond for meter-size bodies. These times are long compared to the charging time and thus transient effects will be very important if the source of return current is the ambient plasma.

The ionization cross section for atmospheric gases (N_2, O_2, NO) maximizes at around 100 eV at about 3×10^{-16} cm^2. For energetic electrons in the range $0.5 < V_g < 40$ keV, the ionization cross section can be approximated by $\sigma_i = 10^{-16} V_g^{-0.83}$ cm^2. The ionization time, τ_i, for each electron to make an electron-ion pair can be approximated by

$$\tau_i = (n_n \sigma_i v)^{-1} \sim \frac{5.3 \times 10^6}{n_n} V_g^{0.33} \text{ s} \qquad (3)$$

where n_n is the atmospheric neutral particle concentration in cm^{-3}. The ionization time minimizes for about 200 eV electrons such that

$$\tau_i \gtrsim \frac{4 \times 10^6}{n_n} \text{ s.} \qquad (4)$$

At altitudes ~ 100 km where n_n is $> 10^{13}$ cm^{-3}, the ionization time is less than 1 μs and short compared to the charging time. Thus ionization will play an important role in controlling the transient phase. At altitudes $\gtrsim$ 190 km where $n_n \lesssim 10^{10}$ cm^{-3}, the ionization time approaches the millisecond time-scale or longer, long compared to the charging time.

These arguments strongly suggest that above 190 km, the time to reach a steady-state (if it exists) is long compared to a millisecond and transiently the vehicle must charge up to a high potential before it can collect a neutralizing return current. There presently are no models that are appropriate for the charging problem above 190 km that are valid in the steady-state and include the effects of ionizing collisions, or that treat the time-dependent charging problem including the potential build-up and the transition from times short compared to τ_i to times long compared to τ_i.

The status of theoretical models is different below 190 km. Considering first the low altitude end, two detailed models have been developed for altitudes below ~ 120 km. At these altitudes, where the neutral number density is high, collisions with neutrals dominate the electron energization, transport, and energy decay processes. At intermediate altitudes, plasma instabilities are very important and collisions with neutrals serve as a damping mechanism to limit wave growth and ionizing collisions provide a source of new plasma.

Experimentally, there have been no observations of the charging of rockets with sufficient time resolution to detect transient charging on time scales less than 1 ms. Hence, the role of ionizing collisions can definitely be considered to be important for interpreting observations below 190 km. If the vehicle is surrounded by its own atmosphere at comparable number densities of 10^{10} cm^{-3} (~ 3 x 10^{-7} Torr at 300°C) due to gas venting, outgassing, or sputtering, then ionizing collisons will also be important for experiments conducted at higher altitudes.

LOW-ALTITUDE REGIME

Electron beam experiments in the low-altitude regime have been proposed and carried out as a means of studying the excitation mechanisms of atmospheric species. As discussed above, the source of the neutralizing return current is the ionization of the atmospheric neutrals. Ionization by both the primary beam electrons and secondary electrons are important.

Initial Transients

Linson (1970) considered the time dependence of the electrostatic potential of an electron-emitting body traveling at 7 km/s

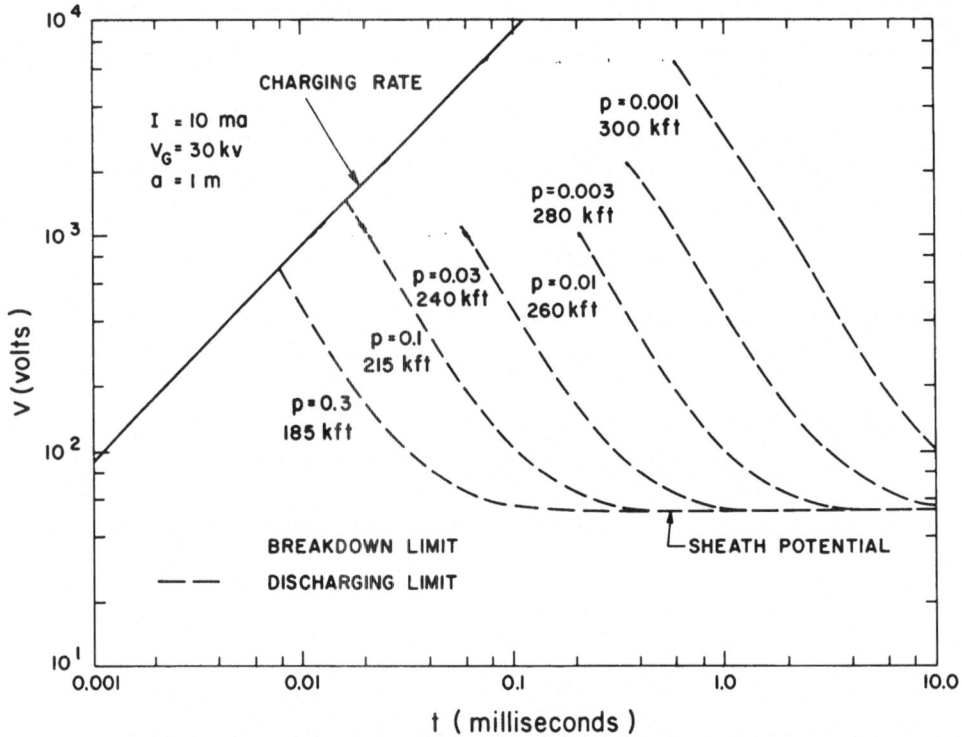

Fig. 1.   Upper limit to the voltage of a 1 meter size vehicle which
          is emitting a 30 kV, 10 mA electron beam as a function of
          time for several altitudes. p is the pressure in Torr. At
          altitudes above 215 kft, breakdown is likely to limit the
          voltage of the vehicle before sufficient current can be
          drawn through the ionized column. After a sufficient time,
          the vehicle voltage approaches the sheath potential. From
          Linson (1970).

at altitudes below 90 km. Because of the low altitude and high ambient neutral densities, the problem is collisionally dominated,
the earth's magnetic field is unimportant, and the ambient ionosphere is irrelevant and plays no role in limiting the potential
rise. Figure 1, taken from Linson (1970), shows a representative
upper limit to the vehicle potential for a 30 kV, 10 mA beam. Qualitatively, the vehicle potential initially rises at the charging rate
given by Eq. (1) until sufficient ionization is produced by the primary beam to provide a neutralizing return current. At the upper
altitude range the electric field due to the large potential is sufficiently high to break down the ambient air. In the approach to
steady-state, the column of ionization surrounding the primary beam
is treated as a nonuniform conductor through which a self-consistent
electric field attracts the required current. Asymptotically, the

vehicle potential drops to the sheath potential in the ionized column corresponding to an anode drop. According to this model, for rockets traveling at slower velocities the dashed curves would move down and to the left. The sheath potential is independent of the ambient neutral density, depends mostly on the gun voltage and vehicle velocity, and only weakly on the other gun parameters.

Baum et al. (1975) have developed a different model for a similar discharge process wherein the source of neutralizing electrons are the secondaries created by the primary beam electrons. They note that the inertia of the electrons between collisions can be important when the collisional mean-free-path is comparable to the body size. The presence of an ambient ionosphere is neglected, the magnetic field serves to confine all the ionization to a cylindrical volume with a several meter diameter, and there is no further ionization of the ambient neutrals by the secondaries. Their model, applied to the 90-120 km altitude regime, predicts that the vehicle potential initially charges rapidly to above a kilovolt and then oscillates, finally settling to a low positive potential (which asymptotically approaches zero) within a few electron-neutral collision times or $\sim$ 20 μs at $n_n = 5 \times 10^{12}$ cm^{-3}. The absolute magnitude of the initial charging and the frequency of the initial oscillations are increasing functions of the rate at which the beam current is turned on. The oscillations and quantitative results predicted by this model are strongly suspect for two reasons: 1) the plasma column has no shielding effect on the electric field which is taken to be a pure Coulomb field varying inversely with the square of the distance from the vehicle; and 2) neither quasi-neutrality nor Poisson's equation is used. The neglect of the self-consistent electric fields resulting from the space charge produced by the variations in electron density in the plasma column omits the strong electrostatic forces that result in plasma oscillations or tend to maintain quasi-neutrality. These forces would greatly modify the density variations that must accompany the calculated oscillations.

Both of the above models are attempts at treating the time-dependent charging process in the high-density regime where collisions are important. Both models predict that transient variations in voltage take place on time scales fast compared to 1 ms. There have been no observations of charging of rockets with sufficient time resolution to detect such rapid variations.

Asymptotic State

As discussed above, at sufficiently low altitudes where the neutral density is high, a steady-state should be set up on time scales short compared to a millisecond. The potential on the vehicle is just sufficient to collect the required return current

from the ionization created by the primary beam. The level of ionization available to neutralize the vehicle is determined by a balance
between the ionization production rate and some loss rate. In
Linson's model, the principal potential drop takes place in a sheath
near the vehicle and the principal loss mechanism is the drag of the
neutrals on the electrons due to their relative velocity with respect
to the vehicle. Baum et al. (1975) examined only the initial transients, included no plasma loss mechanism, and their solution asymptotically approached zero due to the infinite build up of plasma.

The Air Force Geophysics Laboratory (formerly AFCRL) has conducted a series of rocket-borne high-power electron beam experiments
in the 80-140 km altitude range. These artificial auroral experiments in the Excede Program were designed to investigate the detailed
production and loss processes of various excited electronic and vibrational states resulting in optical and infrared emissions due to
the energy deposited by the primary and secondary electrons. O'Neil
et al. (1978) have applied Langmuir probe theory in order to obtain
estimates of the expected steady-state potential on the Precede
rocket. The source of return current electrons is assumed to be the
secondary electrons in the vicinity of the vehicle whose density is
estimated by balancing the primary beam production rate with an estimate of the loss rate due to recombination. The result yields an
expression for the vehicle potential

$$\phi \; \propto \; I \; V_g^{5/2}/n_n. \tag{5}$$

which gives values increasing from 1.4 V at 100 km to 28 V at 120 km
altitude. These values are in good agreement with experimental results obtained as a function of altitude (or n_n) in the 105-120 km
range during the Precede experiment. This agreement is perhaps surprising because the modeling takes no account of the earth's magnetic
field even though the mean-free-path for collisions, $\sim$ 2-40 m, is far
greater than the electron gyroradius, $\sim$ 3 cm. The dependence on current and voltage predicted by Eq. (5) was not tested.

The AFGL analyses of their experiments have been based solely
on single particle collisions. No consideration is given to possible
collective effects between the primary beam and the background plasma
it creates. These strong beam-plasma interactions produce significant modifications of the primary electron distribution and the
energy distribution of the secondary electrons leading to significantly enhanced secondary electron production rates and greatly modified excitation rates. The resulting beam-plasma discharge (BPD) is
expected in the altitude and gun parameter regime where these experiments have taken place as we shall see below. A detailed analysis
of the AFGL observational data in the light of the almost certain
presence of a BPD would provide valuable information of the effects
it produces in rocket electron beam experiments.

HIGH ALTITUDE REGIME

The earliest models of the vehicle neutralization process were developed in the 1960's prior to the first rocket electron beam experiments. These models were developed for the high altitude regime where the mean-free-path for ionization is many kilometers. For this reason, ionization of ambient neutrals was neglected and the ambient ionosphere was assumed to be the sole source of the neutralizing return current. Linson (1969) summarized these early theoretical treatments that attempted to calculate the positive potential to which an isolated body in the ionosphere must rise in order to attract to it a current I significantly greater than I_O given by Eq. (2). The neglect of secondary ionization would appear to be justified for sufficiently high altitude and for vehicles that do not carry their own neutral atmosphere.

Beard and Johnson (1961) neglected the effect of the earth's magnetic field and thus essentially solved the spherical space-charge-limited flow problem. This model predicts

$$\frac{I_{BJ}}{I_O} = \left(\frac{n_p \bar{c}}{2.5 \times 10^{12}}\right)^{-4/7} a^{-8/7} \left(\frac{\phi}{40}\right)^{6/7} \tag{6}$$

where $n_p \bar{c}$ is in $cm^{-2}s^{-1}$ consistent with Eq. (2) and ϕ is the vehicle potential in volts. This expression represents an upper limit to the amount of current that can be collected to a positively-charged body (neglecting the production of ionization) because the earth's magnetic field restricts the ability of electrons to be collected transverse to the magnetic field.

The earth's magnetic field has a major effect on the treatment of the current collection process. The gyroradius of the ambient ionospheric electrons in the earth's magnetic field is ~ 3 cm which is small compared to a typical body dimension a. The magnetic field affects the current collection from the ambient plasma because the mean-free-path for collisions is much larger than a. The body size and magnetic field strength define a characteristic potential ϕ_O such that the gyroradius of an electron with kinetic energy $e\phi_O$ is equal to the characteristic body size, or

$$\phi_O = \frac{m\Omega_e^2 a^2}{2e} = 178a^2 \left(\frac{B}{0.45}\right)^2 \text{ V} \tag{7}$$

with B in gauss and a in meters. One would expect that the current that the vehicle could attract from the inosphere would not increase substantially above I_O until the vehicle potential was large compared to ϕ_O.

Parker and Murphy (1967) investigated this problem and, based

on the conservation of single particle constants of motion in static
electric and magnetic fields, defined a critical radius r_0 in terms
of the potential ϕ on the body. If a particle originates at a radial
distance greater than r_0 transverse to the magnetic field and rigor-
ously conserves its constants of motion, it is dynamically forbidden
to reach the surface of the body. Those electrons that originate at
a radial distance less than r_0 are dynamically allowed to reach the
body but whether they do so depends on their particular orbit. Thus,
Parker and Murphy (1967) obtain what they consider to be an upper
limit to the current that can reach the vehicle,

$$\left(\frac{r_0}{a}\right)^2 = \frac{I_{PM}}{I_0} \leq 1 + 2\left(\frac{\phi}{\phi_0}\right)^{1/2} . \tag{8}$$

As an example, if the vehicle were approximated by a sphere of
radius $a = 1$ m, then the cross-sectional area of the vehicle transverse to the magnetic field is 3.14 m^2. If the ionospheric density
were $n_p = 10^5$ cm^{-3}, then the current I_0 defined by Eq. (2) is 6.3
mA due to current collection in both directions parallel to the magnetic field. If the electron gun were emitting 200 mA, the two expressions given by Eqs (6) and (8) produce $\phi_{BJ} = 2.3$ kV and $\phi_{PM} = 42$
kV, respectively. Smaller current collection areas or larger beam
currents produce larger values for ϕ_{BJ}, ϕ_{PM}, and ϕ_{PM}/ϕ_{BJ}.

Consider the configuration of ionization in an hypothesized
steady-state surrounding a positively-charged spherical body of radius
a as indicated schematically in Fig. 2. The vehicle is emitting an
electron beam parallel to the magnetic field $\vec{B}$. There is a boundary
far from the body at very low potential, indicated by the heavy curve
labeled $\phi \sim 0$, that separates the region of high potential devoid of
ions from the ambient ionospheric plasma. Because of the reduced
mobility of the electrons transverse to the magnetic field, this high
potential region is elongated parallel to $\vec{B}$. The dimension of this
high potential region parallel to $\vec{B}$ is dominated by space-charge-
limited flow parallel to $\vec{B}$ and the motion of the vehicle across the
magnetic field.

The source of neutralizing current is provided by the ionosphere.
Due to the small electron gyroradius, this current flows parallel to
the magnetic field. Ambient electrons on magnetic field lines within
r_0 of the field line that passes through the center of the vehicle
have direct access to it (Parker and Murphy, 1967) and form a portion
of the neutralizing current. The dashed boundary is a representation
of the region within which the large electric fields can self-consist-
ently modify electron orbits allowing them to contact the vehicle.

Ambient electrons on magnetic field lines that enter the region
of high potential at distances greater than r_0 are strongly affected
by the large electric fields. Any dissipation in the form of self-

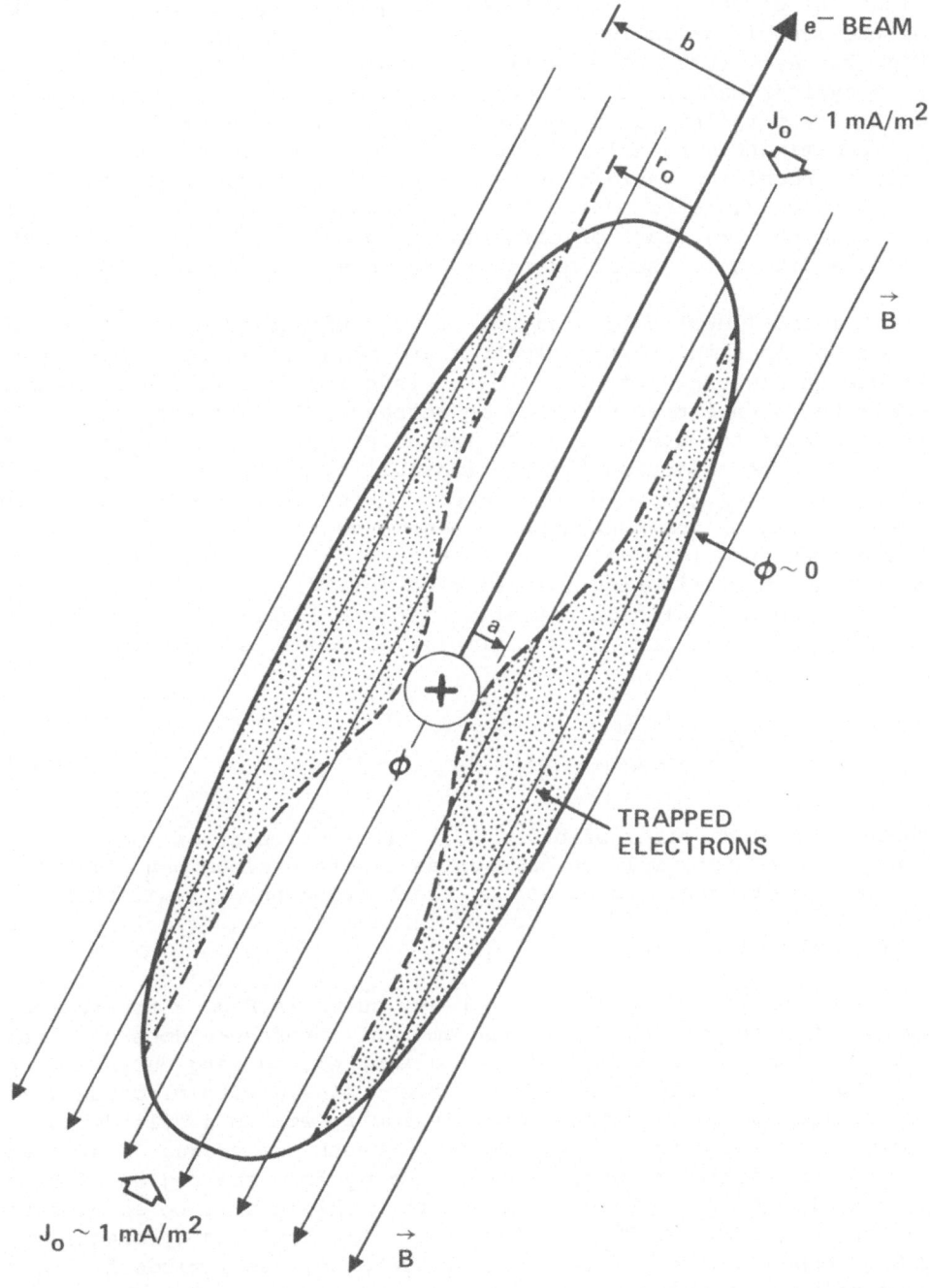

Fig. 2. Schematic of region of large positive potential surrounding a spherical body of radius a.

collisions, collisions with neutrals, or collective interactions will cause the electrons to be trapped (or quasi-trapped) in the vicinity of the vehicle as indicated schematically by the dotted region in Fig. 2. As a result there will be a nonneutral cloud of electrons in a cylindrical shell surrounding the vehicle. Due to the high conductivity parallel to the magnetic field, the electric field, $\vec{E}_r$, will be predominantly radial transverse to $\vec{B}$. The electron cloud will rotate around the vehicle in $\vec{E}_r \times \vec{B}$ flow. This configuration is known to be diocotron unstable. The growth of nonlinear waves can increase the transport of electrons across the magnetic field and allow additional return current to reach the vehicle.

Linson (1969) suggested that this configuration of a nonneutral electron cloud in crossed electric and magnetic fields may be unstable at high frequency ($\omega \sim \Omega_e$). This instability leads to small-scale turbulence that produces an increase in the transport of electrons across the magnetic field enhancing the return current to the vehicle. The criterion for the instability to be present is $\omega_p^* \gtrsim \Omega_e/2$ or $n_p^* \gtrsim 6 \times 10^3$ cm^{-3} where ω_p^* is the electron plasma frequency based on the nonneutral electron concentration n_p^* in the cloud. Because the required value is lower than typical ionospheric concentrations, this criterion is easily satisfied. The resulting asymptotic expression at high potential for the increase in return current to the vehicle according to a simplified model is given by

$$\frac{I}{I_0} \sim \frac{2\phi/q_c\phi_0}{[\ell n(2\phi/q_c\phi_0) - 1]} \tag{9}$$

where q_c is a constant of order 0.2. This calculation indicates that the presence of turbulence in such an electron cloud may allow the collection of current approaching the Beard-Johnson limit that neglects the restricting effect of the earth's magnetic field in the high potential region.

We should caution that the heavy curve in Fig. 2 separating the region of high potential from the ambient ionosphere does not imply that the ionosphere outside is not disturbed. Strong currents are being drawn parallel to $\vec{B}$ in the ambient plasma with electron drift velocities approaching the thermal velocity within a cylinder of radius b as shown in Fig. 2. These currents can produce disturbances in the ambient plasma many kilometers away from the region of high potential. Ions in the vicinity of the vehicle will be accelerated outward and a 100 eV O$^+$ ion can travel more than two gyroradii or 250 m perpendicular to $\vec{B}$. These disturbances can include fluctuations in density, heating of plasma, and wave emissions. This broad region of disturbance is consistent with observations of such effects seen tens and hundreds of meters away from electron-emitting rockets as has been reported by Cartwright et al. (1978) and Jacobsen (1981).

The above models indicate that at high altitude rockets should charge up to high potentials for $I \gg I_0$. There is little experimental evidence that this has occurred. Two important effects have not been included in the above models: the motion of the vehicle across magnetic field lines and the ionization of neutrals. We will not address the first effect here and the effect of the presence of neutrals will be addressed in the next section. We look to laboratory experiments for tests of the theoretical concepts embodied in the models.

Some laboratory experiments have been conducted in the Johnson Space Flight Center (JSC) large tank and by the Japanese Group on the charging of an isolated electron gun. The general result has been that when the gun is isolated and is emitting more than a few milliamps the gun charges up to the accelerating voltage in the kilovolt range and cuts off the electron beam (Bernstein, 1979; Kawashima, 1981). There is no source of return current to allow the continued emission of energetic electrons. The Japanese Group has been able to discharge the electron gun by firing a MPD arc thus surrounding the gun in a high density plasma. This plasma provides a source of return current until the plasma decays and the gun charges back up.

The Japanese have also carried out a series of experiments by charging up a spherical ball in a magnetized plasma in a large vacuum tank and investigated the current-voltage characteristics as a function of background neutral density, ambient plasma density and magnetic field strength (Kaneko et al., 1979; Kawashima, 1981). At low voltages, below 100 V, the results of these experiments were generally in agreement with the above models. Without a magnetic field, the potential distribution surrounding the sphere deviated from a vacuum field due to the space charge of the unneutralized electrons in the high potential region. The current collected was limited by spherical-space-charge flow as expected. In the presence of a magnetic field, the current collected could exceed the Parker-Murphy limit but was limited by the Beard-Johnson spherical-space-charge flow.

However, at voltages somewhat greater than 100 V the return current would seem to increase. The source of the enhanced current was ionization produced by the energetic electron cloud surrounding the charged sphere. In these experiments there was no electron gun and therefore there was no beam-produced ionization. The picture that emerged was that at sufficently high voltage a cross-field discharge could be induced. The effect of ionization is more pronounced when the ambient plasma density is low ($\sim 10^4$ cm^{-3}) than when it is high ($10^6 - 10^8$ cm^{-3}). It was concluded that both the trapping of electrons in the vicinity of the sphere as well as neutrals emitted due to electron bombardment are important processes (Kaneko et al., 1979).

DISCHARGE MODELS

Vehicle-Induced Discharge

Returning to Fig. 2, we see that the combination of a positive potential on a vehicle which attracts electrons and the earth's magnetic field which restrains electrons results in a quasi-trapped cloud of unneutralized electrons rotating around the vehicle with a velocity $\vec{E}_r \times \vec{B}/B^2$. If the vehicle potential approaches 100 V, the radial electric field will be of order 100 V/m and the kinetic energy of the rotating electrons will approach 15 eV sufficient to ionize ambient neutrals.

Cartwright et al. (1978) proposed such a description in order to explain a number of observations made on the Echo II rocket experiment near 250 km altitudes. They found that the background plasma density was enhanced, the background electrons were heated to higher than 1 eV, and there was no evidence of an ion-free region around the rocket during gun pulses. Their observations show that significant ionization was taking place near the rocket. This ionization supplies the return current necessary to keep the rocket potential from rising above the inferred 60-80 V. Cartwright et al. calculated that the ionization created by the primary beam is insufficient to provide the necessary return current. Thus, they conclude that return current is produced by a vehicle-induced discharge. The region surrounding the rocket in Fig. 2 is not devoid of ions when ionization of neutrals is taking place. As Cartwright et al. remark, the creation of secondary electrons greatly relaxes the restrictions discussed by Linson (1969).

The model described above is analogous to a Penning discharge with the rocket acting as an anode and the ionospheric plasma as a cathode. For a fixed voltage, the current is proportional to the ambient neutral density. Zhulin et al. (1976) suggested such a model in order to provide an explanation for observations made during the Zarnitza I rocket experiment (Cambou et al., 1975). These observations consisted of a visible glow surrounding the rocket and radar reflections at 45 MHz implying a plasma of density $> 10^7$ cm^{-3} surrounding the rocket. It was recognized that such an enhanced electron density produced by a "glow" discharge would provide the source for the neutralizing currents.

In order for a discharge to exist, a mechanism is needed to energize ambient electrons sufficiently in order to ionize ambient neutrals, and, in order to ignite and sustain the discharge, the ionization time must be short compared to the time for the electrons to be lost from the discharge region. These requirements provide a basis for developing quantitative estimates as to the conditions under which a discharge can be struck.

Galeev et al. (1976) indicated that there are at least two mechanisms for igniting such a discharge: a) the discharge driven by the electric field of the rocket as discussed above; and b) a beam-plasma discharge that we will discuss below. In the former case the electric field due to the potential surrounding the rocket provides the source of energy to energize the electrons. Galeev et al. developed a semi-quantitative conceptual model of such a self-sustained discharge. The electron loss rate is taken to be due to the drift of electrons across magnetic field lines in the direction of the electric field due to collisions with neutrals. The two key assumptions of this model are that all the electrons created in the volume surrounding the rocket are collected by the rocket, and the size of this volume is determined by the potential on the rocket. A feature of their self-sustained discharge model is that the potential is independent of the ambient plasma density, temperature, magnetic field strength and size of the rocket. They find that the potential varies as

$$\phi \ \sim \ 2\left(\frac{10^{12}}{n_{n}}\right)^{2/3}\left(\frac{I}{0.1}\right)^{2/3} \ kV. \tag{10}$$

This result indicates that the potential on the rocket would have to rise to many tens of kilovolts at altitudes greater than 190 km and currents $\gtrsim$ 100 mA in order to sustain the discharge and neutralize the vehicle. This fact throws into doubt the effectiveness of this mechanism operating at high altitudes without some modification to their model.

Beam-Plasma Discharge

Galeev et al. (1976) also considered the conditions under which a beam-plasma discharge (BPD) could be struck. The BPD can be visualized as an rf break-down in which the high-frequency electric fields are created by a beam-plasma instability. The mechanism which causes such a discharge is the following. When a strong electron beam interacts with a weakly ionized plasma, intense plasma oscillations are excited. Thermal electrons are heated in the fields of these oscillations up to energies of the order of the ionization energy. When these collide with neutrals, they produce secondary electrons. In this way, if there is a sufficient quantity of neutral gas, a cascade-like process can develop, increasing the plasma density, and igniting a discharge.

The BPD has been studied recently in the large tank at the Johnson Space Center (Bernstein et al., 1978; Bernstein et al., 1979). These experiments were done in weak magnetic fields (B ~ 1 Gauss) and in the density regime $n_n \sim 10^{10} - 10^{12}$ cm^{-3} similar to space experiments and the beam path length, L, was varied between 10 and 20 m. In the voltage regime $0.5 < V_g < 2.5$ kV, Bernstein et al. (1979)

determined an empirical relationship for a critical current necessary
to strike the BPD,

$$I_c \quad \propto \quad \frac{V_g{}^{3/2}}{B^{0.7} n_n L} \quad , \tag{11}$$

for $n_n < 10^{12}$ cm $^{-3}$. Many of the phenomena associated with the BPD
are discussed in several other papers in these proceedings. The
importance of BPD for vehicle neutralization is that it produces an
enhanced plasma density, greater than 10^7 cm^{-3} in the laboratory
experiments, that provides a source of neutralizing current.

In recent laboratory studies (Bernstein et al., 1980) it was
found that the potential of an isolated rocket electron gun payload
was reduced when the emitted electron beam ignited the BPD indicating
that it lead to a more efficient neutralization process. Figure 3
is a schematic of the gun floating potential as a function of gun
emission current (Bernstein, private communication). The gun was
isolated in the JSC tank and allowed to charge up to ϕ = 400 V where
the potential was clamped by a diode. When the critical current
I_c = 14 mA and 30 mA for V_g = 1 kV and 1.6 kV, respectively, a BPD
was struck and the potential on the isolated gun dropped to about
50 V and 100 V, respectively. At these higher currents the BPD
plasma supplied the neutralizing return current. The points indicated by a circled cross designate equal plasma production rates as
determined optically. The interpretation is that a gun potential of
100 V was necessary to collect 50 mA from the same plasma that could
supply 30 mA of neutralizing current when the gun potential was 50 V.

The laboratory experiments have provided a wealth of information
regarding the phenomena associated with the BPD. Papadopoulos (1981)
has recently developed a theory of the BPD that successfully predicts
the scaling relationships shown in Eq. (11), explains a number of
associated measurements, and has applied it to the space environment.
This theory is contained in these proceedings and we will not discuss
it further here. The principal factors differentiating the space
experiments from the laboratory experiments are the motion of the
rocket or satellite in the former case and the presence of boundaries
in the latter.

Lyachov and Managadze (1977) have analyzed the scattered electron measurements on the Zarnitza II rocket experiment, the periods
of telemetry blackout, and the variations of optical glow in the
vicinity of the rocket. Based on a detailed analysis of the variations of these quantities with the geometry of the beam injection
(injection pitch angle and phase angle around the rocket), they
conclude that a BPD is struck when the beam interacts with neutral
gases emitted from the still-hot rocket motor. They infer that the
neutral density in the exhaust gas cloud is ~ 10^{12} cm^{-3} thus allowing
the BPD to be struck even at high altitudes where the ambient neutral

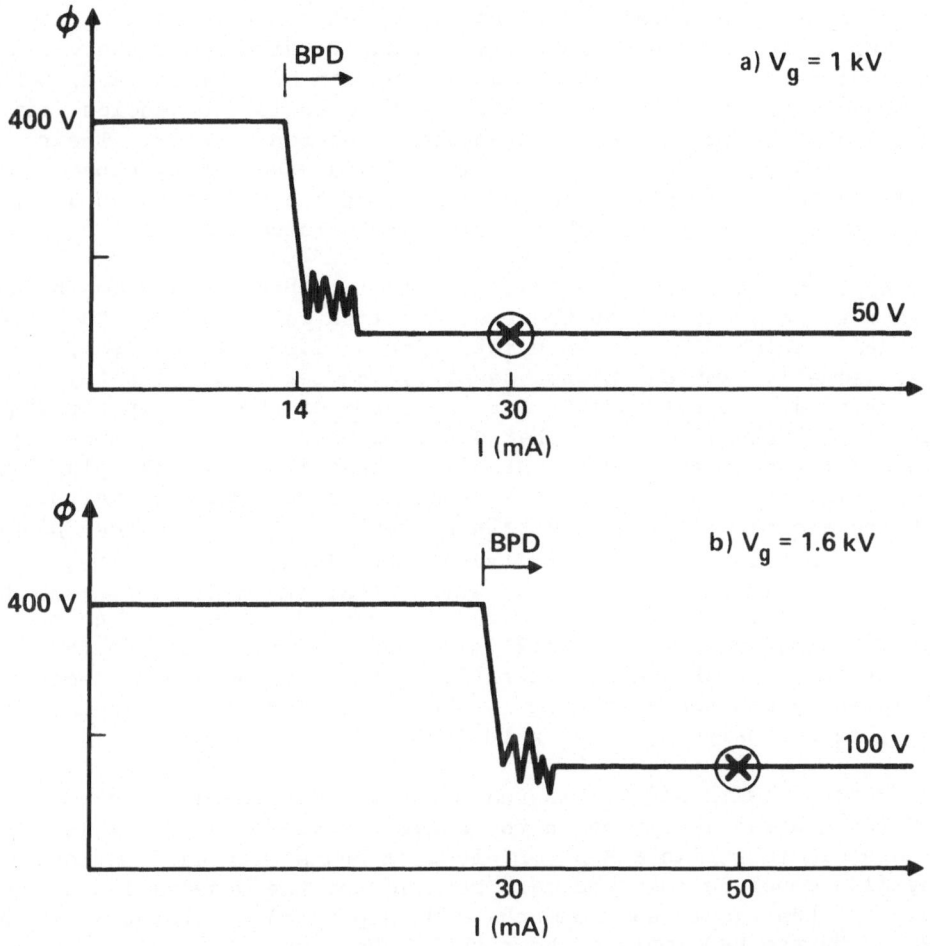

Fig. 3. Schematic of gun floating potential as a function of gun
         emission current for two applied gun voltages: a) $V_g = 1$ kV;
         b) $V_g = 1.6$ kV.

density is too low to sustain a discharge. They also infer that the
plasma density in the beam where the discharge was occurring probably
exceeded 10^9 cm^{-3}! In support of their conclusion, a discharge based
on the potential on the rocket body would, in principle, show less
sensitivity to the orientation of the beam injection.

     Mishin and Ruzhin (1978; 1980) have recently carried out a de-
tailed theorectical analysis of the BPD as applied to the ARAKS and
Zarnitza II rocket experiments. They claim a large degree of success
in explaining many details of a variety of observed phenomena including the luminosity surrounding the rocket, radar echos from the
plasma surrounding the rocket, and the amplitude variations of the

rocket-transmitted telemetry signals received on the ground. In particular, they claim that their theoretical results model the variations shown by the observations as a function of beam voltage, beam divergence, geometry of injection, observed radio frequencies, time-history of the beam pulses, and ambient electron density. Their approach and basic physics is similar to that outlined by Papadopoulos (1981) but the analysis is rather complicated and depends on a large combination of parameters that merit careful study.

Applying the criterion that the electron heating and ionization rates must be greater than the electron loss rate leads to the condition under which BPD can can exist. For a given beam current, voltage, geometry, ambient plasma density, and beam pulse length, there is a critical neutral density for the onset of BPD. It is the magnitude of the neutral density that distinguishes a BPD from other types of beam-plasma interactions. Strong interactions may take place with the heating of electrons and the creation of suprathermal tails, but if there are insufficient neutrals present, a BPD cannot take place. For the conditions of the Zarnitza II experiment, ($V_g \sim 10$ kV, $I \sim 0.5$ A), Mishin and Ruzhin determine that the minimum neutral density for ignition of the BPD is 2-$4 \times 10^{10} cm^{-3}$. Likewise, at too high a neutral density, the collisions of electrons with neutrals produce only a weak plasma instability that does not heat electrons sufficiently and the discharge is shut off. For these rocket experiments, this density is ~ 2-$3 \times 10^{13} cm^{-3}$.

Hence, Mishin and Ruzhin reach the rather general conclusion that for neutral densities in the range $3 \times 10^{10} \lesssim n_n \lesssim 3 \times 10^{13}$ it is possible to strike a BPD for electron beams emitted from rockets. They also conclude that the main behavior of the observables in the region of the rocket environment can be explained, at least qualitatively, by the BPD ignition near the rocket; and the electron-produced discharge current is capable of ensuring rocket neutralization at small rocket potential, i.e., $\phi \lesssim 1$ kV, small compared to the gun voltage.

Figure 4 summarizes the altitude and density range over which BPD is to be expected. The dashed line at $n_n \sim 2 \times 10^{11} cm^{-3}$ is roughly the transition between bulk collisional heating of the plasma by the wave electrostatic fields and the creation of suprathermal tails which takes place predominantly at lower neutral densities. The importance of these two regimes and the different signatures of observables is discussed by Papadopulos (1981). Figure 4 also shows that the pressure range in which the laboratory experiments have been conducted in the large JSC tank corresponds to the density range in which BPD has been observed. It is also indicated in Fig. 4. that the high-power electron beam experiments conducted by AFGL satisfied all of the criterion necessary to strike a BPD. The bulk electron

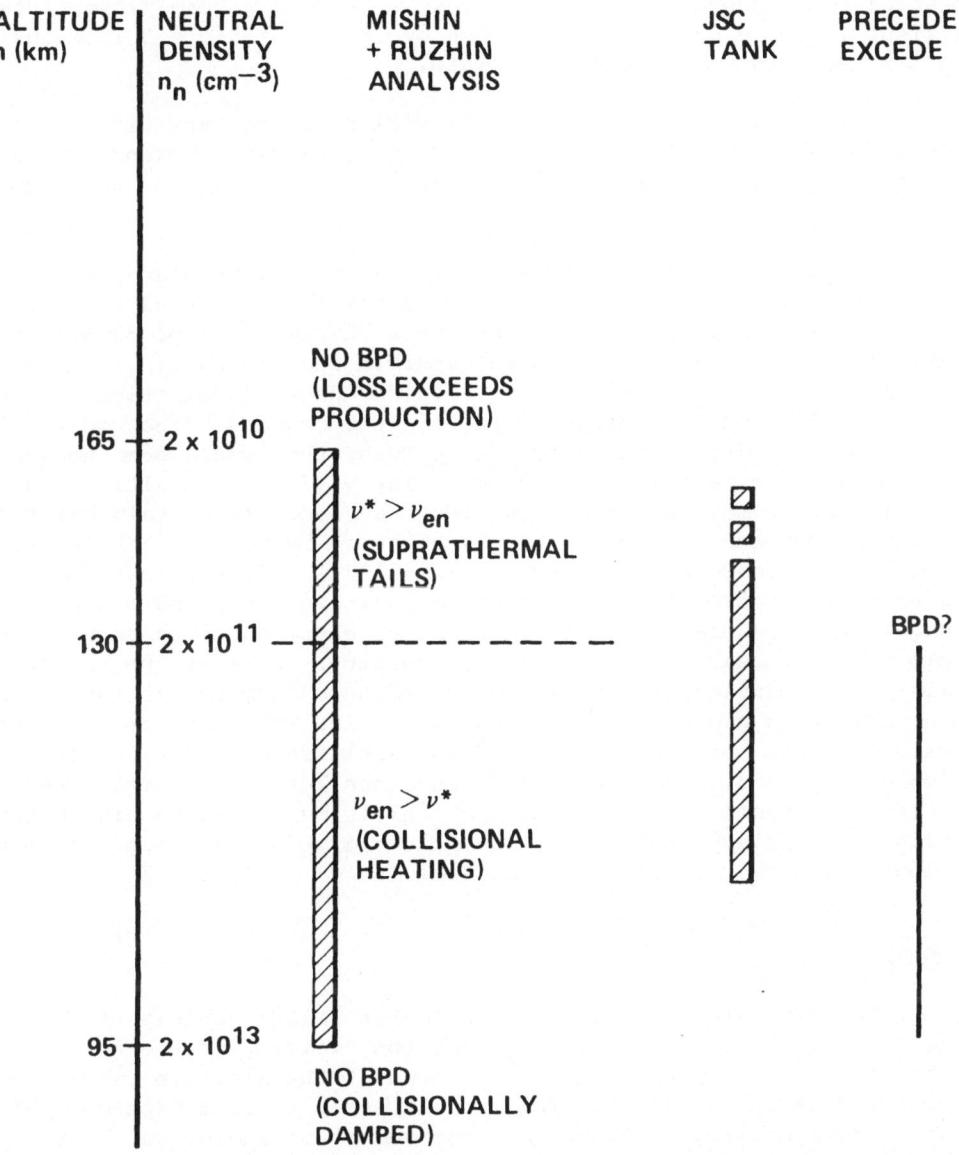

Fig. 4. Altitude and neutral density range over which BPD can be
expected. The first striped column indicates the range expected from the detailed analysis of Mishin and Ruzhin (1978;
1980). The second striped column indicates the parameter
range over which BPD has been observed in the large tank at
the Johnson Space Center (JSC). The vertical line on the
right indicates the altitude range over which the Air Force
Geophysics Laboratory has conducted high-power rocket-borne
electron beam experiments emitting current far in excess of
the critical current necessary to strike BPD given by Eq. 11.

heating rates should have been dominated by strong beam-plasma interactions in those experiments and not by single particle collisions
as has been assumed in their analyses.

If a BPD is struck, the vehicle will be in contact with a relatively dense plasma that can supply the neutralizing current. This
plasma source will likely keep the vehicle potential rise to no more
than a hundred volts or so.

In summary, motivated by the results of various high power, high
altitude rocket experiments conducted since 1969, several theoretical
models have been proposed that rely on a discharge to produce secondary electrons as a source for the vehicle neutralization. These
theoretical models generally have required neutral densities in
excess of 10^{10} cm^{-3} in order to provide the required electron multi-
plication that drives the discharge. Thus, they would seem to apply
only to those experiments conducted below ~ 170-190 km altitude unless it is assumed that there is always a cloud of neutrals emitted
by the rocket and accessible to the electron beam. The laboratory
experiments conducted by Kaneko et al. (1979) indicate that both the
effects of electron trapping around a positively-charged probe and
outgasing of neutrals by the probe due to electron bombardment contribute to increasing the ionization density. None of the theoretical models examines the contribution of secondary ionization to the
neutralization process at altitudes above 190 km. None of the theoretical models applicable above 120 km altitude consider the time-
dependent potential rise and decay in times less than 1 ms. Theoretical arguments indicate that the vehicle must rise to high potentials, perhaps approaching the gun potential, and may trap, at least
temporarily, the primary electron beam.

SUMMARY

The overwhelming evidence from rocket flight experiments is
that the emitted electron beam leaves the vicinity of the rocket
with little loss in total energy throughout the altitude range from
below 100 km up to 350 km. This result seems to be independent of
gun voltage between 1-40 kV, the injected pitch angle, the background electron density, and whether injected upward or downward.
It is true that many of the observables such as, indicated vehicle
potential, locally increased plasma density, plasma heating, energization of secondary electrons, emitted-wave characteristics, and
light emission, do vary considerably with these parameters. Thus,
the physical processes taking place are strongly dependent upon the
injection conditions. We have discussed some theoretical models of
some processes believed to be important for the vehicle neutralization problem under various conditions.

The general pattern that emerges as a result of the measurements is that below altitudes ~ 125 ± 5 km the vehicle potential rarely rises more than several tens of volts. (An exception is the observation of Kaneko et al. (1979) of a transient voltage ~ 200 V lasting less than 10 ms at low altitudes.) Whatever the mechanism of secondary electron production and energization, the high neutral densities ($\gtrsim$ 5 x 10^{11}cm^{-3}) provide sufficient ionization to limit the necessary potential rise to attract to the rocket a neutralizing return current in the steady state.

In the altitude range between 125 ± 5 km and 175 ± 15 km theoretical models indicate that various discharge mechanisms can operate to produce sufficient neutralizing ionization. At higher altitudes, there are no theoretical models developed that include the ionization of the ambient neutrals, but theoretical work suggests that it is impossible to strike a discharge in space for $n_n < 10^{10}$ cm^{-3}. The measurements of vehicle potential above 125 km altitude suggest that the vehicle potentials are limited to values generally reported as tens of volts to hundreds of volts. The two exceptions are the ARAKS rocket for which there are some suggestions of 3 kV potentials and the Norwegian Polar 5 mother-daughter experiment which may have charged up to $\leq$ 1 kV.

None of the experimental observations indicate that the vehicle potential rises dramatically at altitudes greater than 175 ± 15 km as might be expected if a discharge mechnism were shut off. A possible explanation for this behavior may be that rockets carry along with them their own atmosphere having neutral densities ~ 10^{10} cm^{-3}. Aside from this suggestion, there is insufficient theoretical work to explain the neutralization of rockets at high altitudes.

Lastly, there are a number of indications from a variety of experiments that the current collection process is nonsteady. Indeed, oscillations have been seen in the 10-400 Hz range both in rocket experiments (Echo IV, Polar 5, Japanese Rocket series) and the laboratory (Bernstein et al., 1980). The suggestion occurs that there is no steady-state discharge process. Ionization times of neutrals, ion transit times across the beam, electron transport times, and electron collision times with neutrals all have time scales in the millisecond range and could contribute to the observed oscillations.

ACKNOWLEDGEMENTS

The author is appreciative of many useful discussions with Drs. W. Bernstein and K. Papadopoulos. This work was supported by Rice University subcontract (NASA prime contract) No. NAS8-3377-1.

REFERENCES

Baum, H., Bien, F., and Tait, K., 1975, An analysis of transient
    vehicle charging in the Excede experiment, Report RR-65, Aero-
    dyne Res., Inc., Bulington, Mass.
Beard, D. B. and Johnson, F. S., 1961, Ionospheric limitations on
    attainable satellite potential, J. Geophys. Res., 66:4113.
Bernstein, W., 1979, Private communication.
Bernstein, W., Leinback, H., Kellogg, P. J., Monson, S. J. and
    Hallinan,T., Further laboratory measurements of the beam
    plasma discharge, 1979, J. Geophys. Res., 84:7271.
Bernstein, W., Leinback, H., Kellogg, P. Monson, S., Hallinan, T.,
    Garriott, O. K., Konradi, A., McCoy, J., Daly, P., Baker, B.,
    and Anderson, H. R., 1978, Electron beam experiments: the
    plasma discharge at low pressures and magnetic field strengths,
    Geophys. Res. Ltrs., 5, 127.
Bernstein, W., Whalen, B. A., Harris, F. R., McNamara, A. G., and
    Konradi, A., 1980, Laboratory studies of the charge neutraliza-
    tion of a rocket payload during electron beam emission, J. Geo-
    phys. Ltrs., 7:93.
Cambou, F., Dokoukine, V. S., Ivchenko, V. N., Managadze, G. G.,
    Migulin, V. V., Nazarenko, O. K., Nesmyanovich, A. T., Pyatsi,
    A. Kh., Sagdeev, R. Z. and Zhulin, I. A., 1975, The Zarnitza
    rocket experiment on electron injection, Space Research XV,
    491.
Cartwright, D. G., Monsoon, S. J., and Kellog, P. J., 1978, Heating
    of the ambient ionosphere by an artifically injected electron
    beam, J. Geophys. Res., 83:16.
Davis, T. N., Hallinan, T. J., Mead, G. O, Trichel, M. C., and W. N.
    Hess, 1971, Artificial aurora experiment: ground-based optical
    observations, J. Geophys. Res., 76:6082.
Galeev, A. A., Mishin, E. V., Sagdeev, R. Z., Shapiro, V. C., and
    Shevchenko, V. I., 1976, Discharge in the near rocket region
    during injection of electron beams into the ionosphere, Doklady,
    Academy NAUK, USSR, 231:71.
Hess, W. N., Trichel, M. C., Davis, T. N., Beggs, W. C., Kraft, G. E.,
    Stassinopoulos, E., and Maier, E. J. R., 1971, Artificial aurora
    experiment: experiment and principal results, J. Geophys. Res.,
    76:6067.
Jacobsen, T. A., 1981, Observations of plasma heating effects in the
    ionosphere by a rocket borne electron accelerator, these pro-
    ceedings.
Kaneko, O., Sasaki, S., and Kawashima, N., 1979, Active experiments
    in space by an electron beam (revised), Preprint, Institute of
    Space and Aeronautical Science, University of Tokyo, Komaba,
    Meguro-ku, Tokyo.
Kawashima, N., 1981, Experimental studies of the neutralization of
    a charged vehicle in space and in the laboratory in Japan,
    these proceedings.

Linson, L. M., 1969, Current-voltage characteristics of an electron-
     emitting satellite in the ionosphere, J. Geophys. Res., 74:2368.
Linson, L. M., 1970, Electrostatic potential of a current-emitting
     vehicle below the ionosphere, RR296, Avco Everett Res. Lab.,
     Everett, Mass.
Lyachov, S. B., and Managadze, G. G., 1977, Beam plasma discharge
     near the rocket (Zarnitza II Experiment), Academy of Sci.,
     USSR, Space Research Institute, (1K1) Moscow, Preprint 310.
Mishin, E. V., Ruzhin, Yu. Ya., 1978, Beam-plasma discharge during
     electron beam injection in ionosphere; dynamics of the region
     in rocket environment in ARAKS and Zarnitza-2 experiments,
     Preprint No. 21, (a & b), Academy of Sci. USSR, Institute of
     Terrestrial Magnetism, Ionosphere and Radio Wave Propagation,
     (IZMIRIN) Moscow.
Mishin, E. V., and Ruzhin, Yu. Ya., 1980, The model of beam-plasma
     discharge in the rocket environment during an electron beam
     injection in the ionosphere, Ann. Geophys., 36:423.
O'Neil, R. R., Bien, F., Burt, D., Sandock, J. A. and Stair, A. T.,
     Jr., 1978, Summarized results of the artificial auroral experi-
     ment, Precede, J. Geophys. Res., 83:3273.
Papadopoulos, K., 1981, Theory of beam-plasma discharge, these pro-
     ceedings.
Parker, L. W., and Murphy, B. L, 1967, Potential buildup on an elec-
     tron emitting ionospheric satellite, J. Geophys. Res., 72:1631.
Winkler, J. R., 1980, The application of artificial electron beams
     to magnetospheric research, Rev. Geophys. and Space Physics,
     18:659.
Zhulin, I. A., Kopaev, I. M., Koshelets, T. E., and Moskalenko, A. M.,
     1976, On rocket charge neutralization in Zarnitza experiments,
     preprint, Ismirin, Moscow.

EXPERIMENTAL STUDIES OF THE NEUTRALIZATION

OF A CHARGED VEHICLE

IN SPACE AND IN THE LABORATORY IN JAPAN

Nobuki Kawashima

Institute of Space and Astronautical Science
4-6-1, Komaba, Meguro-ku, Tokyo, Japan

ABSTRACT

Electron beam expriments in space and laboratory
that have been done in Japan are reviewed with an
emphasis on vehicle charging and charging neutralization.

1. Introduction

The injection of high energy electron beam in space
has a wide variety of use in atmospheric research,
magnetospheric research and plasma physics.
Artificial reproduction of natural phenomena in space
by a high intensity electron beam injection offers
a new experimental technique and opens a new scientific
regime in space physics.

Objectives of the experiment are:

i) Study of beam-atomospheric interaction such as the
    excitation of artificial aurora and airglow,

ii)   Study of beam-plasma interaction such as beam
      plasma instabilities, non-linear phenomena, plasma
      heating et al and

iii) Tracing of magnetic field configuration and
      detection of the electric field in the magneto-
      sphere.

A number of rocket experiments have been performed
and their results are summarized in a review paper by
Winkler (1980). In this electron beam experiment in
space, the charging phenomenon is a very important
problem and we cannot avoid to pass through for the
investigation along the above objectives. It is also
practically important in respect of the satellite
charging in geostationary orbit (Deforest, 1972).

In Japan, we have performed a few rocket experiments
and laboratory studies of the charging and charging
neutralization. Here, we will summarize the results
obtained so far in space and laboratory (Kaneko, 1979,
Kaneko et al 1979, Yamori et al 1979, Kawashima et al
1980).

First, the rocket experiment results are reviewed
in chapter 2 with an emphasis on vehicle charging and
charging neutralization. Laboratory simulation experiments are reviewed in Chapter 3 including a large
vacuum chamber experiment using SEPAC hardwares.

2.   Electron Beam Experiments in Space on-board Japanese
     Rockets and Satellite

Electron beam experiments have been performed
using Japanese sounding rockets and satellites. The
results of satellite experiment is described in detail
in a separate paper in this symposium so that here is
mentioned briefly (See Page   of this proceeding).

2-1 Rocket Experiments

2-1-1 Experiment Configuration

A list of rocket experiments that have been done
is shown in Table 1. Experiments started with a low

Table 1
List of electron beam experiments on Japanese rockets

ROCKET	LAUNCHING TIME	MAXIMUM BEAM ENERGY & CURRENT	DIAGNOSTICS
K-10-11	1975.9.24 14:00 LST	300 V 3.7 mA CW	LANGMUIRE PROBE (BIAS-0.5V +3V)
K-10-12	1976.1.18 14:20 LST	200 V 2 mA C.PULSE (100 mS)	FLOATING PROBE LANGMUIRE PROBE (BIAS -12 V + 5V) RECEIVERS VLF (0 30 kHz) HF (1 MHz 10MHz)
K-9M-57	1976.8.31 04:55 LST	3 kV 100 mA PULSE (180 mS)	FLOATING PROBE (RANGE + 10V -100V) LANGMUIRE PROBE (BIAS + 10V -50V) PHOTOMETERS 3914 A 5577 A RECEIVERS VLF (0 10 MHz)
K-9M-58	1977.1.16 21:45 LST	5 kV 350 mA PULSE (130mS)	FLOATING PROBE (RANGE + 10V -200V) LANGMUIRE PROBE (BIAS + 10V -50V) PHOTOMETERS 3914 A 5577 A 8446 A
K-9M-61	1978.1.27 20:00 LST	2 kV 35 mA C	FLOATING PROBES (RANGE +10V -400V) LANGMUIRE PROBE (BIAS +10V -100V) ELECTRON ENERGY ANALYSER
K-9M-66	1979.1.21 18:06 LST	1 kV 1 mA Max	ELECTRON ENERGY ANALYSER RETURN CURRENT COLLECTORS
K-9M-69 INT. COLABORA-TION WITU USU)	1980.1.16 12:00 LST	1 kV 30 mA Max	WAVE DETECTOR FLOATING PROBE CHARGE PROBE LANGMUIRE PROBE (TETHERED MOTHER AND DAUGHTER)
S-520-2 (INT. COLABORA-TION WITH USU)	1981.1.29	1 kV 30 mA	WAVE DETECTOR FLOATING PROBE CHARGE PROBE LANGMUIRE PROBE (TETHERED MOTHER AND DAUGHTER)

power electron beam experiment in K-10-11 and 12.
High power experiment was done in K-9M-57 and 58 using
300 W and 1.5 kW electron beam, respectively.   Since
phenomenon associated with high power electron beam is
too violent to analyse the charging phenomena, the beam
power has been reduced to an intermediate level of tens
of Watts (1 ~ 2 keV, 10 ~ 30 mV).

K-9M-69 and S-520-2 were intended to perform a
mother-daughter rocket experiment with a tethered wire,
though they were partially successful.

Diagnostics were mostly floating probe and Langmuir
probe.   The floating potential of the rocket with
respect to the space potential is measured by either
the output of the floating probe or the shift of the
Langmuir V-I characteristic of Langmuir probe.   The
V-I characteristic usually shifts toward negative when
the vehicle potential rises positive.   The floating
potentials determined by both methods agree with
each other fairly well.   The spatial distribution of
the floating potential was measured in K-9M-61 and
S-520-2 deploying a multiprobe.

2-1-2   Experimental Results

Typical experimental results of the floating probe
measurement are shown in Figs. 1 and 2.   Fig. 1 shows
examples of low power beam experiment of K-10-11, 12
and K-9M-61, while Fig. 2 shows a high power experiment results.   It is clear that the floating potential
shows a violent variation for a high power beam.   This
indicates a strong discharge is excited when a high
power electron beam is emitted.   For low power beam
experiments, data are very stable and reproducible.
Generally, the floating potential increases as the
beam current is increased and the density of space
plasma decreases (i.e. altitude decreases).

The electron temperature is measured by Langmuir
probe and a plasma heating by the electron beam is
observed in K-10-12 and K-9M-58 (Fig. 3).   It should
be noted that the Langmuir characteristic has a high
energy tail forming a two component plasma.

For a high power electron beam operation, a large
amount of plasma production is observed as can be seen
in the increase of the ion saturation current in the

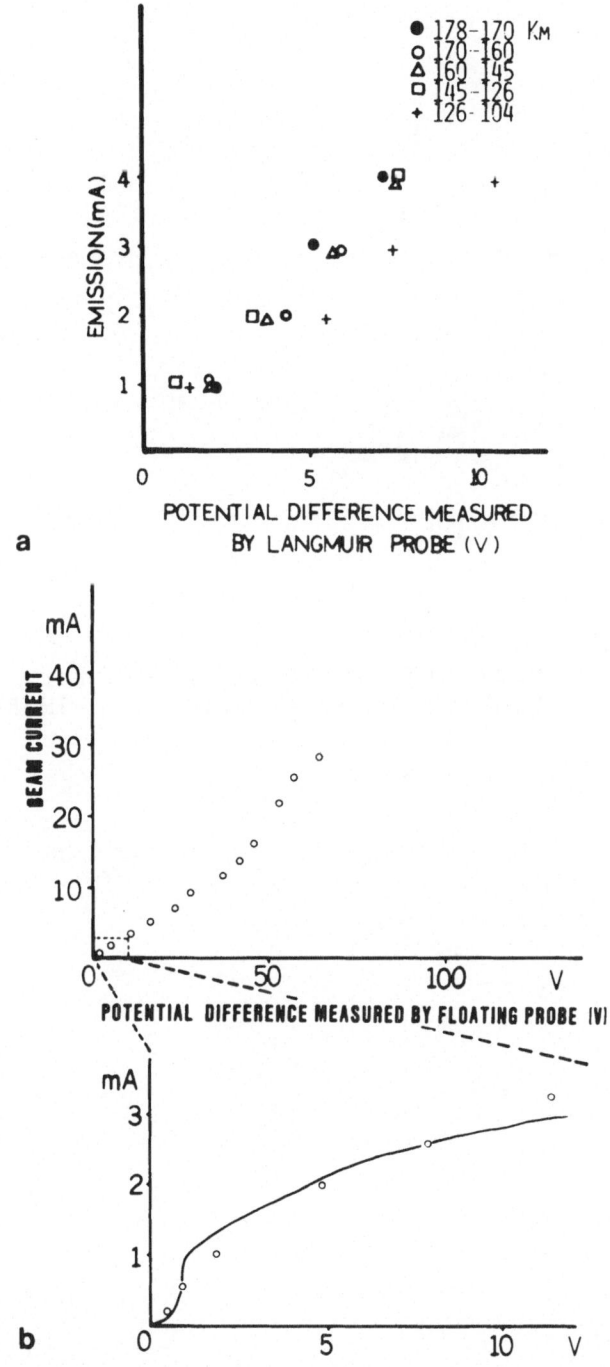

Fig. 1 Potential shift of rocket measured by Langmuir probe for low power experiments. (a) K-10-12; (b) K-9M-61.

FLOATING VOLTAGE

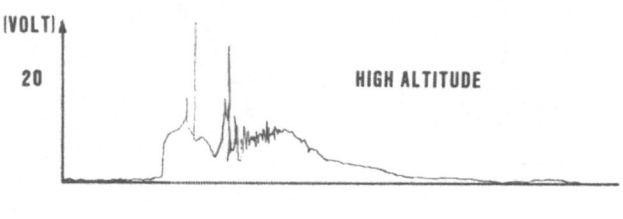

BEAM ON

Fig. 2   Potential shift of rocket for
         high power experiment of K-9M-58

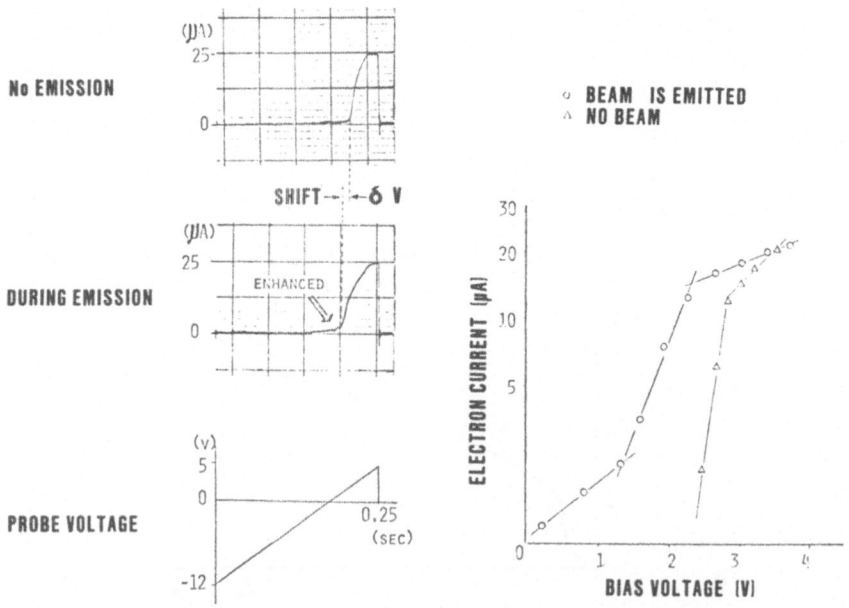

Fig. 3   Electron temperature increase of
         ambient plasma by the electron
         beam emission (K-10-11)

Langmuir characteristic in Fig. 4. It should be noted that neither the beam accelerating voltage nor the beam current is affected by this plasma production. This implies that the plasma production is not due to a simple discharge between the cathode and the anode, but it is caused by the electron beam - neutral gas interaction.

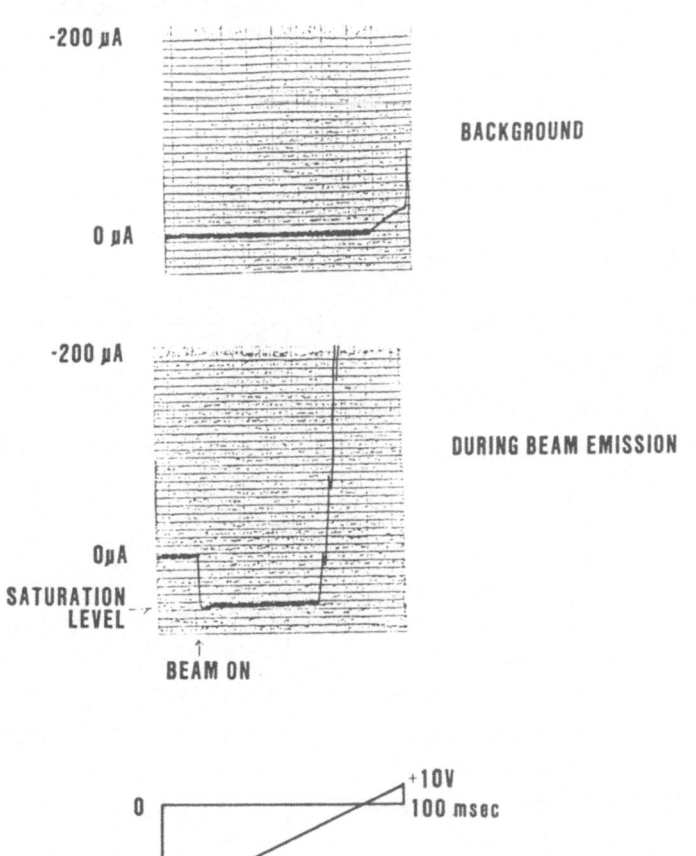

Fig. 4 Increase of the plasma density by
the beam plasma discharge (K-9M-58)

The spatial distribution of the potential is measured by a multi-probe in K-9M-61 and the result is shown in Fig. 5. Though the spatial resolution is not good, it can be seen from this figure that the sheath size increases as the beam current is increased and as the beam density is decreased.

The floating potential is dependent on an angle between the axis of the rocket and the geomagnetic field. In K-9M-61, the precession of the rocket axis was fortunately fairly large. When the rocket axis was parallel to the geomagnetic field, the floating potential was maximum compared with that in other rocket configuration. This implies that plasma electrons are confined in the magnetic field so that they can come into the rocket surface only along the magnetic lines of force to compensate the emitted electron beam. Consequently, the cross-sectional area of the rocket viewed along the magnetic lines of force is important determinig the balance of the emitted electron beam and return current.

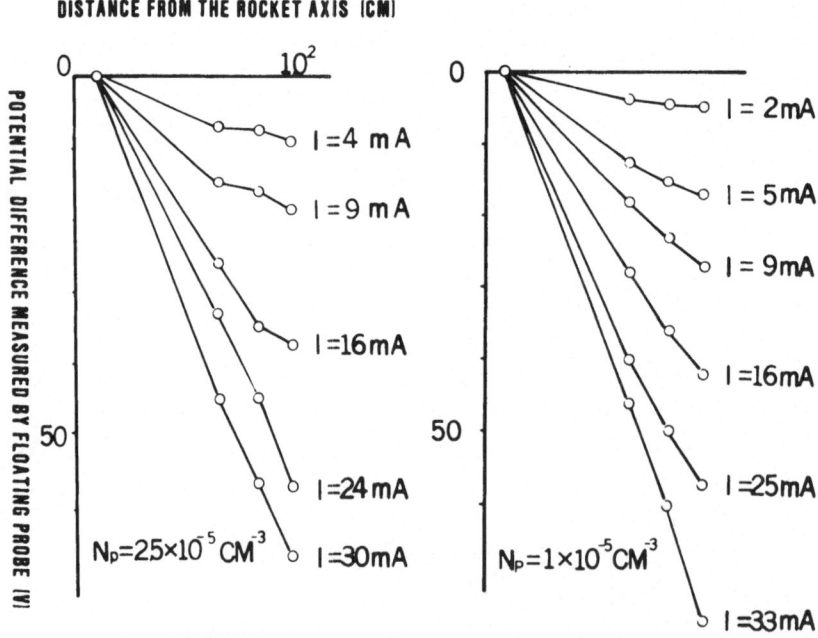

Fig. 5   Potential distribution around the
         rocket during the beam emission
         (K-9M-61)

The relative direction of the electron beam with respect to the geomagnetic field is also important. When the electron beam is emitted perpendicular to the geomagnetic field, it returns directly back to the rocket surface. Then the resulting charging is expected to be much smaller for a perpendicular injection. This was verified in K-10-12 and K-9M-61.

Summarizing the results obtained in these rocket experiment, Fig. 6 displays the floating potential versus the beam current normalized by the electron saturation current I of space plasma, which is given as

$$I_o \equiv J_o \, S$$

J_o : thermal electron flux

$$(1/4 \; n_e \; e \; \sqrt{\frac{8kTe}{\pi \; n}} \;)$$

S : crosssectional area of the metal surface.

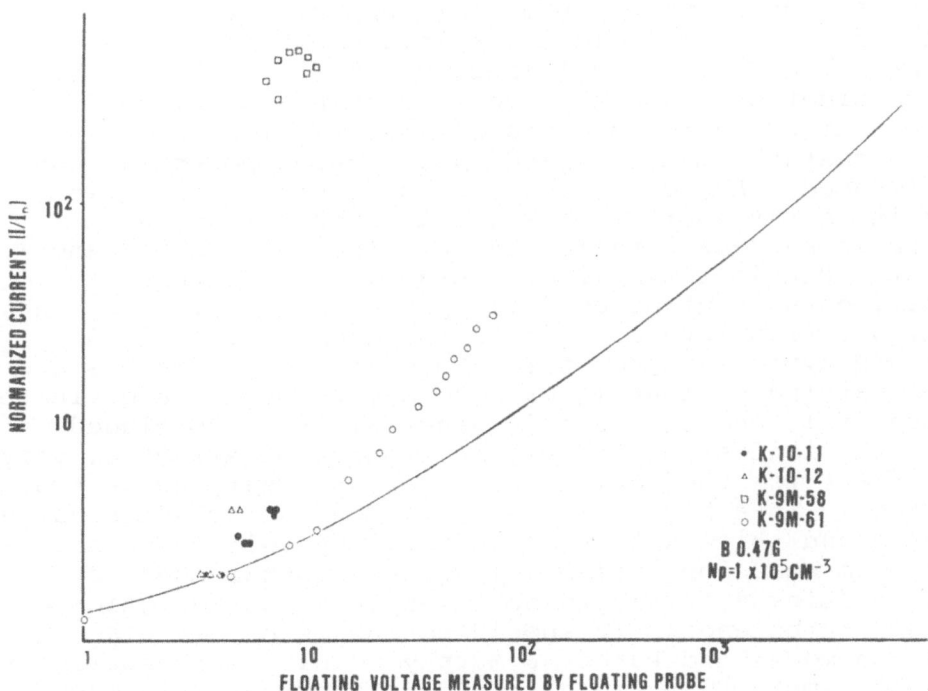

Fig. 6   Summary of floating voltage versus normalized current in beam emission experiment in rocket

It is natural that when the beam current is below I_o, the potential rise is small and the rocket body follows Langmuir characteristic. The floating potential-beam current (ϕ - I) characteristic remains in the transition region of Langmuir probe

$$I \equiv I_o \exp \frac{e (\phi - \phi_s)}{KTe}$$
where ϕ : rocket potential
ϕ_s : space potential

For the beam current higher than I_o, the sheath around the rocket should expand. There seems to exist two critical currents I_1 and I_2 which characterize the balance of charging neutralization. So long as the beam current is between I_o and I_1, the ϕ - I characteristic follows Beard & Johnson Model (Beard & Johnson, 1961) based on a simple space - charge - limited characteristic without the magnetic field, that is, I/I_o c $_{-}$ $\phi^{6/7}$. In Fig. 6, $I_1 \sim 3 I_o$ in K-9M-61 experiments. Since the expansion of the sheath is not large, the effect of the magnetic field can be neglected. Above I_1, the characteristic tends to depart from Beard & Johnson model in a sense opposite to that with the magnetic field taken into consideration such as Parker & Murphy model (Parker & Murphy, 1967). It is considered due to the increase of plasma by the beam ionization in the sheath. This mitigates the space charge limitation and the floating potential can be reduced. Above I_2, the discharge effect comes in and it is far apart from Beard & Johnson model as shown in K-9M-58 data in Fig. 6. The value of I_1 and I_2 are functions of various parameters. In rocket experiments, neutral density, plasma density, outgasing, configuration with respect to the geomagnetic field etc are typical parameters. Some of the parameters are time varying so that it is not easy to obtain a definite value of I_1 and I_2. But in general, it is concluded that the ionization and plasma production effect is very important for the vehicle charging and charging neutral- ization. In a higher current, the beam plasma discharge occurs and this reduces the charge-up voltage making the beam emission easier. There is another model by Linson (Linson, 1969) which takes into consideration of plasma turbulence. It should be inbetween Beard & Johnson model and Parker & Murphy model. Our experimental results imply that the discharge effect comes in much earlier than plasma turbulence effect so that the latter is masked by the discharge accompanied with the increase of plasma by ionization.

2-2   Satellite Experiment

   Electron beam experiment on a Japanese satellite
" JIKIKEN (EXOS-B) " is described in detail in a
separate paper in this proceeding.

3.   Laboratory Experiments

3-1   General

   Experiments in space using rocket or satellite
are restricted in various respect, for example,
diagnostic means are limited, experiment time is limited,
parameters such as plasma density or neutral density
are time varying, unknown or uncontrollable parameters
exist such as outgasing etc.

   Laboratory experiments have advantages over space
experiments in these respects, though it is done in a
limited (bounded) space and cannot fully simulate
the situation in space.

   Laboratory experiments shall be performed in two
respect:

 i)   to simplify the situation as much as possible and
      to find out physical basis governing the charging
      and charging neutralization phenomena comparing
      with theoretical analysis and
ii)   to simulate the actual configuration as much as
      possible to find out and study the configuration
      peculiar problems.

We have been doing laboratory experiments along these
two lines.

3-2   Charging of a Spherical Body Placed in Plasma

3-2-1   Experimental Setup

   A spherical body is immersed in a plasma produced
by a quiescent plasma source produced by either a glow
discharge or ECR plasma. The former is a quiet
non-magnetized plasma and the latter magnetized.

In the former, the density varies from 10^4 to $10^8/cm^3$ with $T_e \simeq 2 \sim 5$ eV and the density of the latter plasma varies from 10^7 to $10^8/cm^3$ with $T_e \simeq 1 \sim 2$ eV and $B \simeq 500$ gauss. The background neutral gas pressure is $10^{-5} \sim 5 \times 10^{-4}$ Torr.

A high voltage is applied on the sphere to simulate the charging due to the beam emission and the voltage-current characteristic is taken.

3-2-2 Experimental Results

In the non-magnitized plasma, the voltage-current characteristic agrees with Beard & Johnson model (Fig. 7) until a discharge occurs (in Fig. 7, it is up to 200 V). The dimension of the sheath and the potential distribution are also in a good agreement with the model.

In the magnetized plasma, when the charging voltage is low ($\phi < \phi_c$: $\phi_c \sim 100$ V in Fig. 8), the V-I characteristic follows Parker & Murphy model, that is, $I/I_0 \sim 1 + (4\phi/\phi_0)^{1/2}$

$$\text{where } \phi_0 = \frac{m\omega_{ce} a^2}{2e}$$

ω_{ce} : electron cyclotron frequency
a   : radius of the sphere

as shown in Fig. 8.
In this regime, no significant ionization occurs.
When the applied voltage is above ϕ_c, a significant ionization takes place and the V-I characteristic departs from Parker & Murphy's limitation approaching and going over Beard & Johnson limit. (Fig. 9)
ϕ_c is dependent on the background neutral density and plasma density. ϕ_c increases as the plasma density is increased and the background neutral density is decreased. The bright halo formation around the sphere can be observed visually and it is displayed schematically in Fig. 10.

3-2-3 Charging Experiments Using A Simulated Model

This experiment is to simulate and study the physical aspect of charging problem which is peculiar

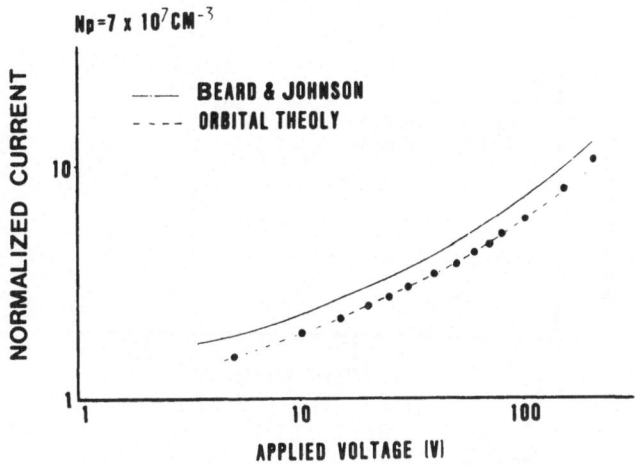

Fig. 7 (a)   Voltage - Current characteristic
of a sphere placed in a non-
magnetized plasma for low
floating voltage

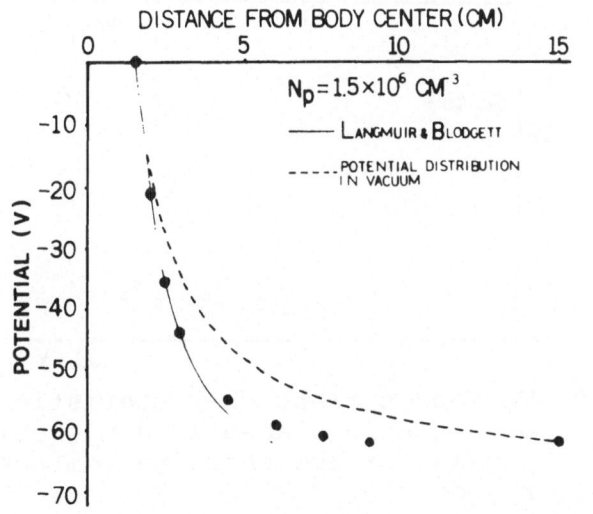

Fig. 7 (b)   Spatial potential distribution

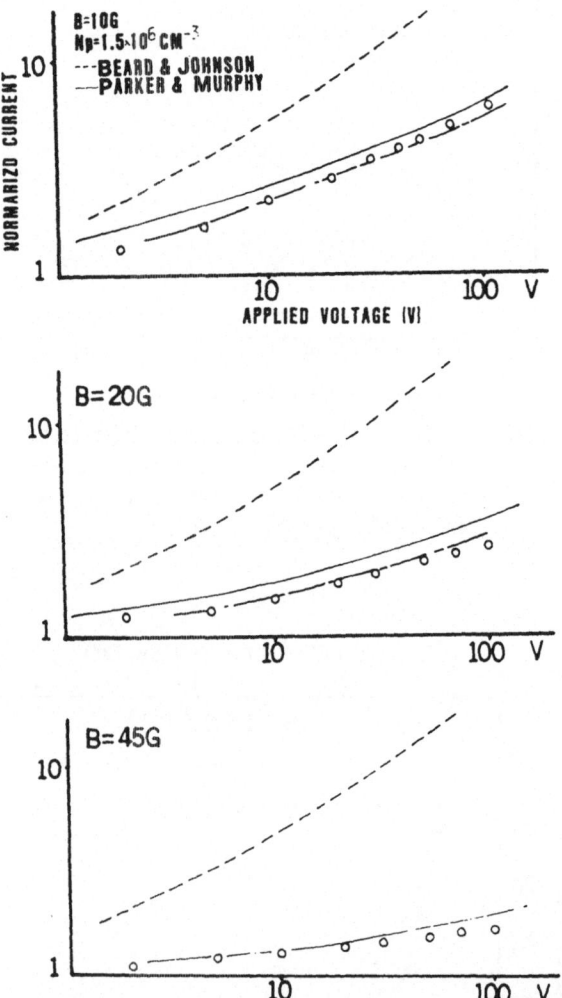

Fig. 8 Voltage-current characteristic
of a sphere placed in a magnetized
plasma for low charging voltage
region

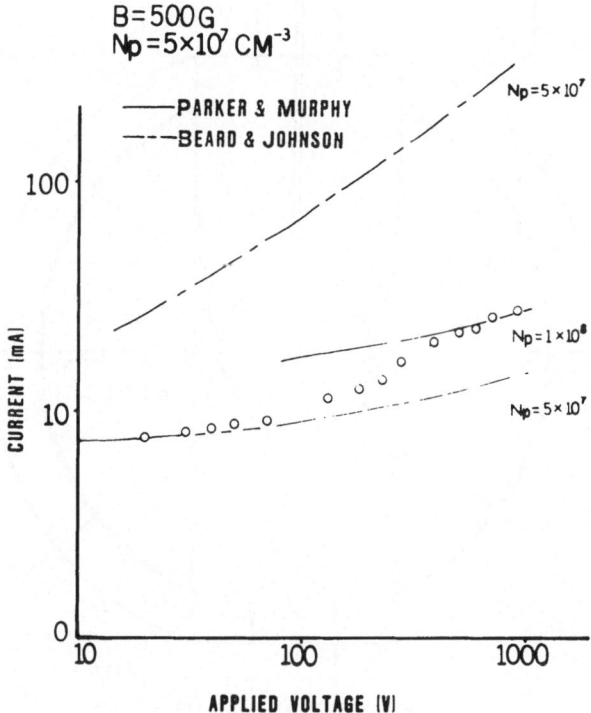

Fig. 9 Voltage-current characteristic
showing a departure from Parker
& Murphy model for higher charging
voltage

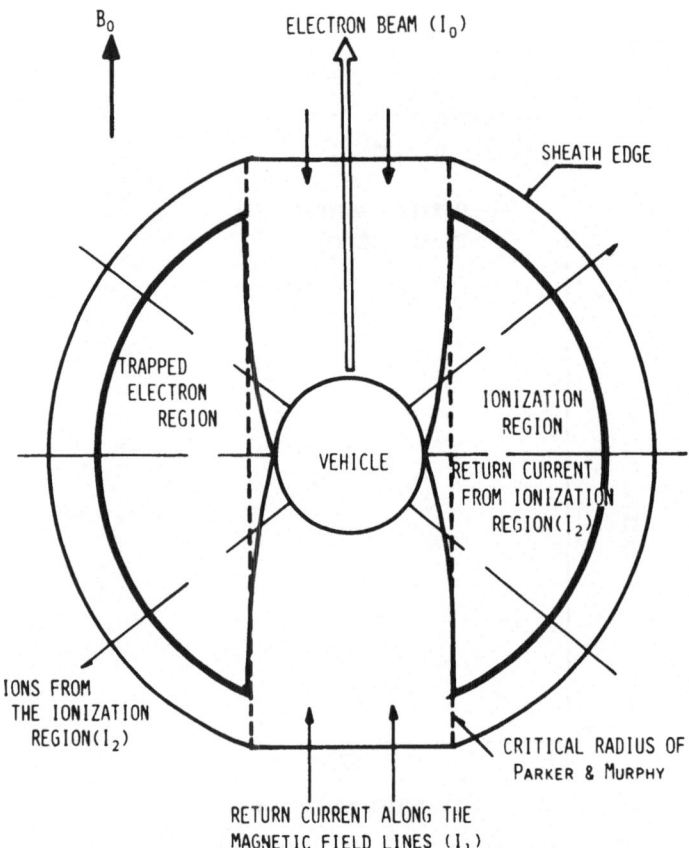

Fig. 10   Schematic illustration of a
           bright halo formation around
           a sphere in a magnetized plasma
           when the charging voltage is
           high

to the configuration. We have performed several types
of experiments simulating rocket, satellite and space
shuttle. Here we will show some typical experiments
of space shuttle experiment.

3-3-1 Experiments Using Space Shuttle Model

3-3-1-1 Return Current Collection Experiment

     Space shuttle has only a limited area of conductive
surface which will be useful for the return current
collection. Possible return current collectors on
the shuttle is shown in Fig. 11. The main engine
nozzles are the largest, and others have not a large
contribution. The objective of this experiment is to
verify how much of return current flows in each part
of conductive surface and whether it is proportional
to the surface area.

     A 17 cm long shuttle model with metal surfaces at
the corresponding places was used in this experiment.
Plasma flow from an coaxial plasma gun was used.
Its plasma parameters were $n \sim 10^{11}/cm^3$, $T_e \sim 5$ eV
flow velocity ~ 8 km/sec and plasma duration ~ 100
μsec. The plasma density was so chosen to keep the
parameter S of Debye length (λ_d)/Scale of Shuttle
(L) $\equiv$ S should be kept invariant ($S \sim 10^{-3}$).

     Fig. 12 shows that how return current is
distributed over the shuttle surface for various values
of angle of attack θ. From this figure, it is clear
that the contribution of the main engine nozzles is
the largest even in the case of the nozzles in the
wake (θ = 0°). Though the effective current collecting
area relative to the main engine is 3 to 5 times the
actual conductive area for hinges and wheels, the
return current collected by the main engine is dominant
so long as the engine nozzle is facing, in some way,
to the plasma (Fig. 13).

     From Fig. 13, it should be noted that the relative
importance of the main engine for the return current
collection increases as the ambient plasma density
is increased. It may be explained that the sheath
thickness increases as the plasma density is decreased
and this in turn increases the effective return current
collecting area of hinges and wheels.

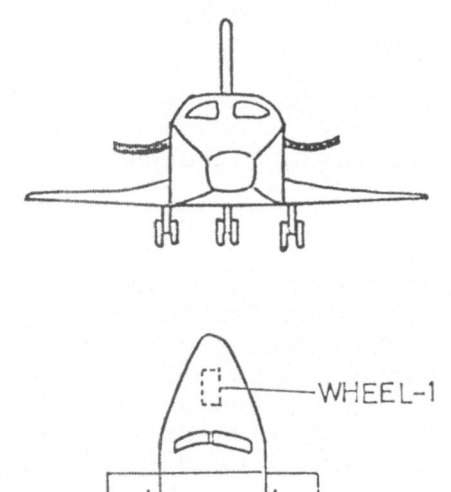

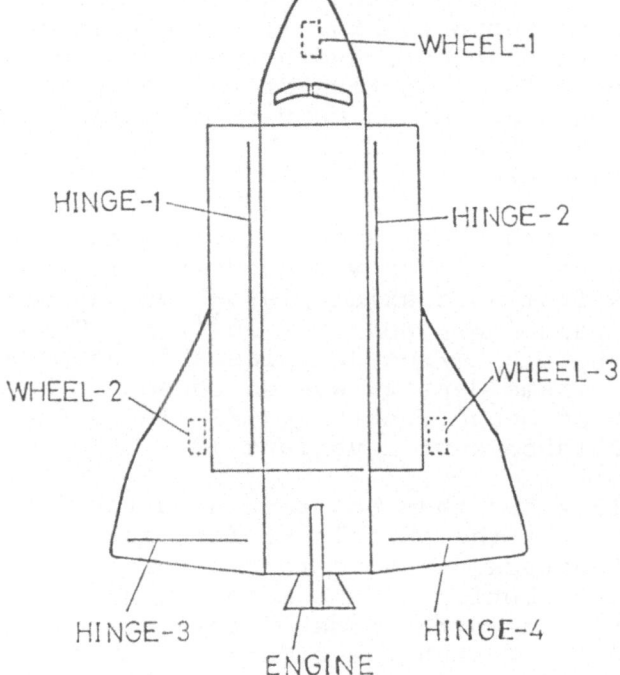

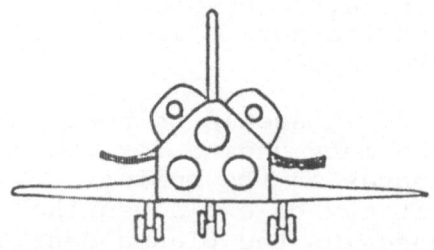

Fig. 11   Possible return current
          collectors on the space shuttle

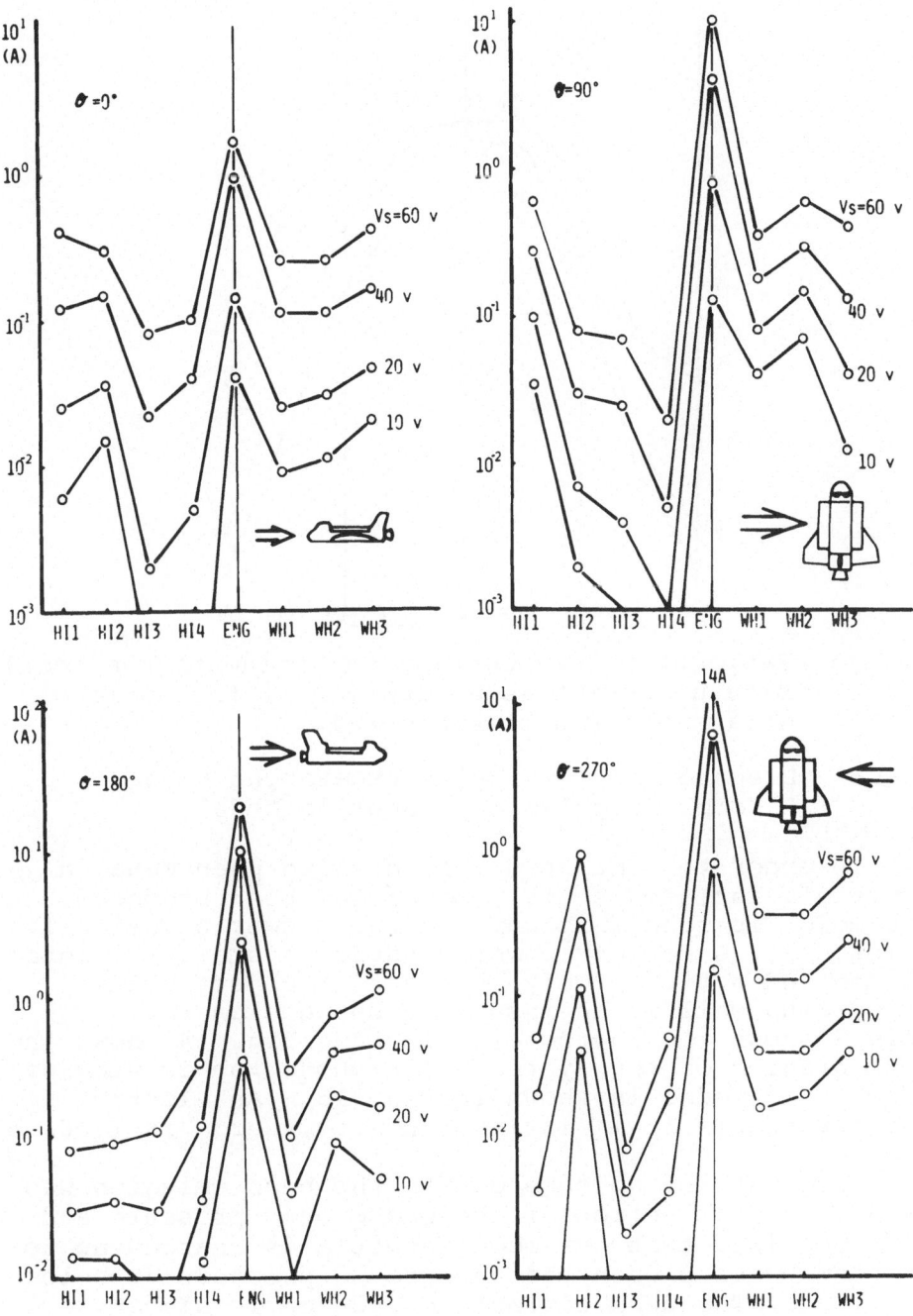

Fig. 12   Return current distribution
           over the shuttle surface for
           various values of angle of
           Attack θ

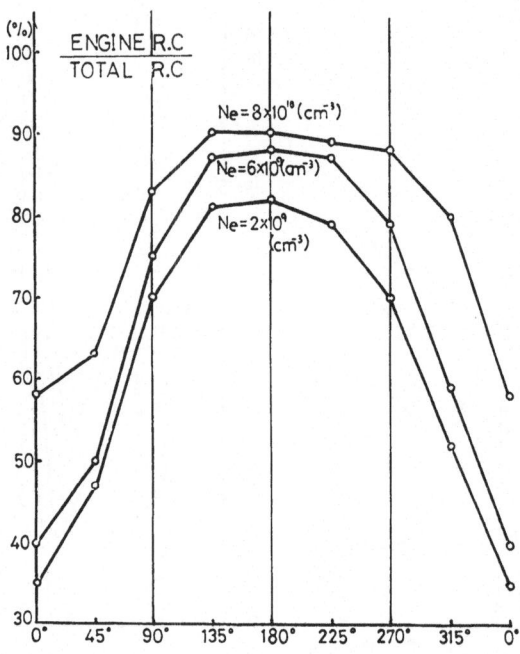

Fig. 13   Ratio of return current to Engine to the total
          return current as a function of the angle of
          attack and the plasma density

3-3-1-2   Electron Beam Emission Experiment Using
          A Completely Isolated Shuttle Model

     In order to simulated the charging phenomena, a
completely isolated shuttle model has been produced.
This model is free from any external power and signal
cables.   It is battery powered having a small telemeter
and command system.   Only interactive object with
surrounding wall is an insulated suspension wire.
Table 2 shows its characteristics and Fig. 14 shows the
block diagram of the system and a photographic view of
the shuttle model is shown in Fig. 15.   An electron
beam current is 10 mA maximum and its energy is 1,000 eV.

     The charging voltage due to the beam emission was
measured as a function of the background pressure and
it is obtained that the model shuttle is charged up to
the beam voltage when the background pressure is below
10^{-5} Torr.   The beam current dependence is sliest,
though there is a tendency that the charging is hore
suppressed for higher beam current.   This implies that
the plasma production by the beam is non-linear process
and the plasma production rate increases more than
linearly.

Table 2. Specification of space shuttle
          model completely isolated from
          the wall

(1) length   50 cm

(2) surface   brass covered with
                       insulator

(3) beam energy   0   1   keV

(4) beam current   0   100 mA

(5) beam duration   5 msec

(6) telemeter system

         analog signal 105   110 MHz

         discrete signal

                   135 MHz   155 MHz

(7) bsttery   Ni-Cd

(8) system start   laser beam

                        first -on

                        second -off

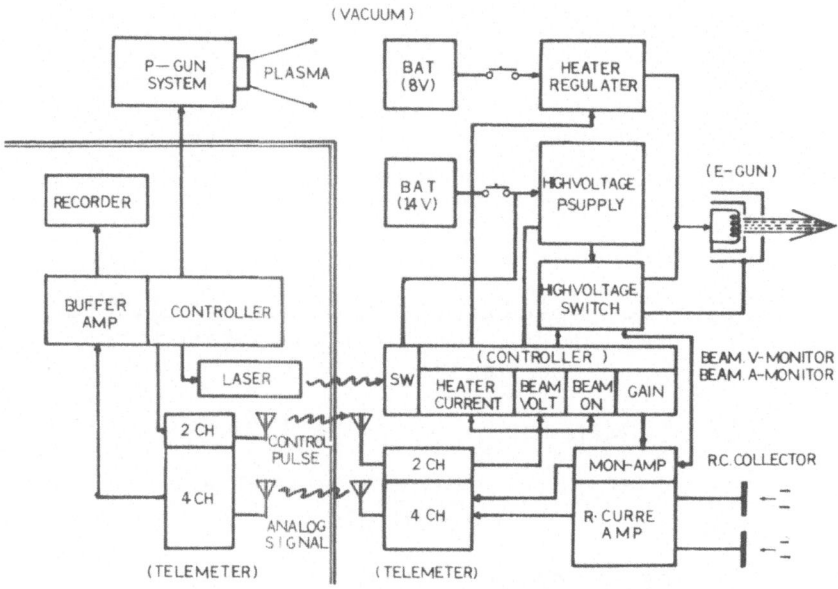

Fig. 14   Block diagram of a laboratory
space shuttle experiment completely
isolated from the chamber wall

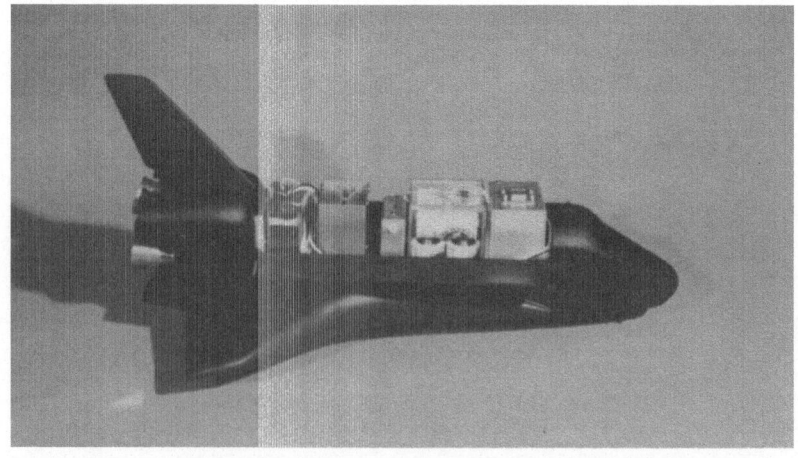

Fig. 15   Photograph of model space shuttle
with subsystems

The return current distribution is shown in Fig. 16. When the background pressure is high, the return current is coming onto the shuttle surface uniformly so that the contribution of the main engine nozzle is the largest. On the other hand when background pressure is low, the plasma production by the beam has a negligible contribution to the return current. Consequently, the return current distribution is such as shown in Fig. 16 and most electrons emitted from the gun comes back to the most adjacent conductive surfaces. The fraction of the current flowing into the main engine nozzle is extremely small because it is in the shadow with respect to the electron gun.

Further experiment is now under way to simulate the shuttle experiment more realistically producing a plasma flow by a plasma gun.

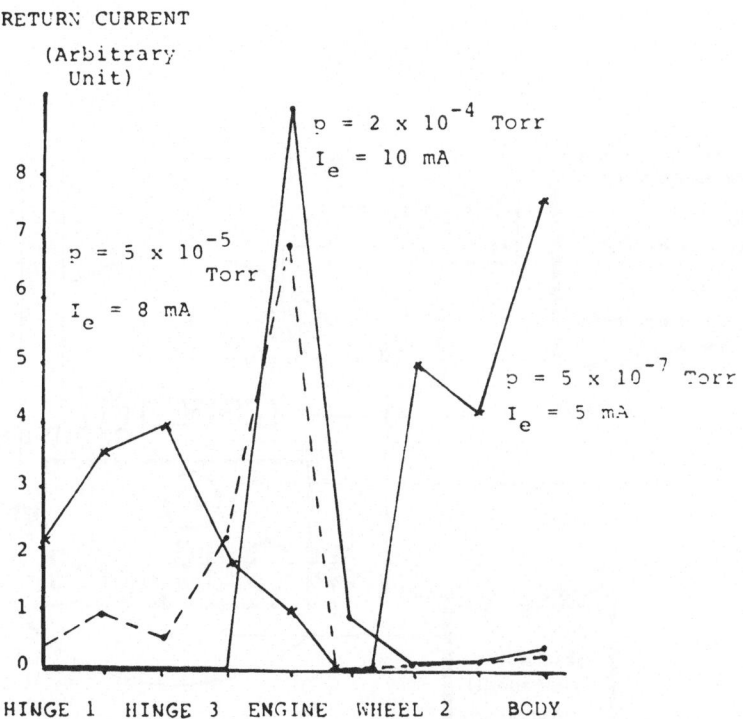

Fig. 16   Return current distribution on the shuttle for various values of the background presure

3-4   Large Space Chamber Experiment Using SEPAC
      Instruments

      A high power particle beam experiment on Spacelab 1
SEPAC (Space Experiment with Particle Accelerators) has
a capability of injecting a 12 kW (7.5 KeV, 1.6 A)
electron beam in space (Ref. to p.   of this
proceeding).

      We have done already twice large chamber (8 m in
diameter 20 m high) experiments using SEPAC engineering
model and protomodel (Obayashi et al, 1978, 1980).
In this autumn, the final test of SEPAC flight model is
scheduled. Though the prime objective is to test SEPAC
instrument operation in vacuum, some scientific experiments were done on charging. Fig. 17 shows a schematic
layout of the experiment setup. SEPAC instruments were
located on a similar size simulated pallet, which was
set floating from the wall.

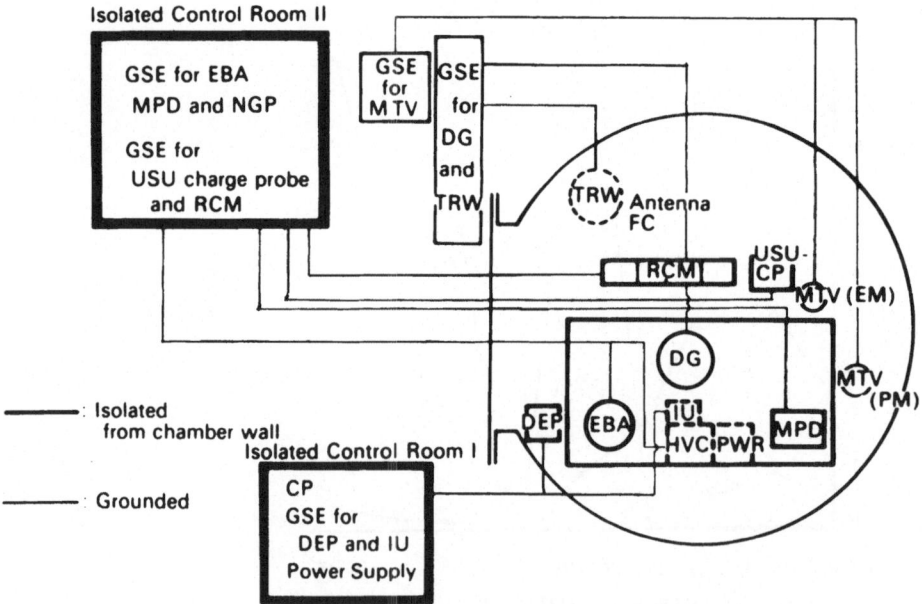

Fig. 17   Schematic layout of the SEPAC
          experiment setup in a large
          vacuum chamber (8 m ∅ 20 m h)

The charging voltage of the pallet with SEPAC instruments was measured directly between the pallet and the chamber wall. The profile of the electron beam was monitored by a high sensitivity TV camera (MTV) as shown in Fig. 18.

When the beam current is more than 10 mA, the pallet potential goes up to the voltage corresponding to the beam accelerating voltage (Fig. 19).

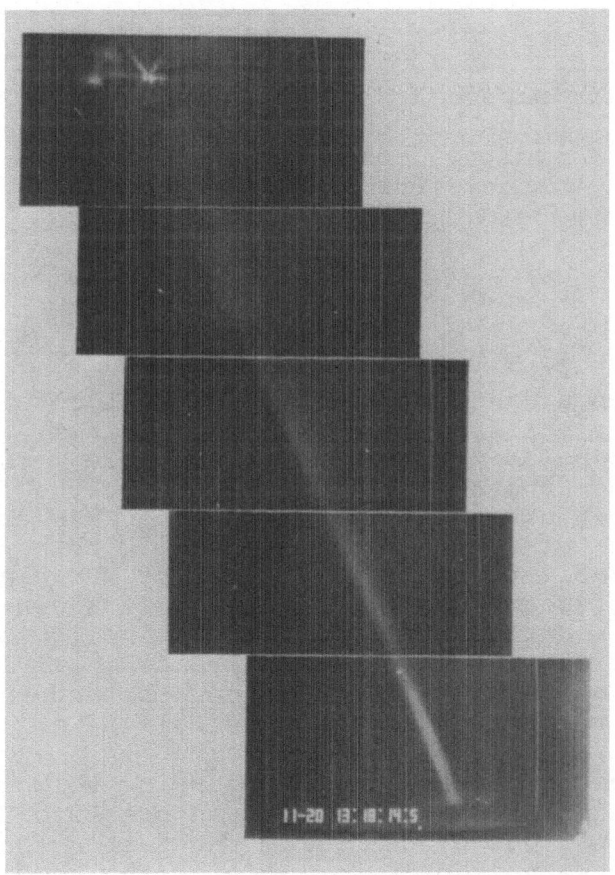

Fig. 18   TV image of the electron beam
          profile in a large vacuum chamber

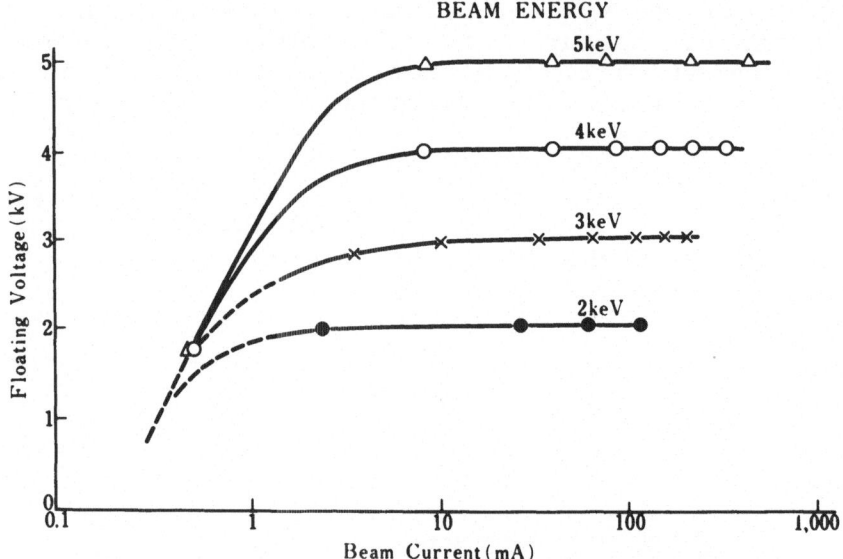

Fig. 19   Dependence of floating voltage on
          beam current. The floating
          voltage coincides with the beam
          acceleration voltage when the
          beam current is large enough.

When MPD arcjet is fired, the pallet potential goes
down abruptly but soon recovers (Fig. 20(a)). The
effectiveness of MPD plasma lasts about 20 ~ 30 msec.
This is rather short but it may be due to the finiteness
of the chamber size. Plasma produced by MPD arcjet dies
away when the plasma reaches the wall. The dependence
of the neutralization effective time by MPD plasma as
a function of beam current and energy is shown in Fig.
20(b). The neutralization effective time is shorter as
the beam current and energy is increased. It is natural
because the increase of the beam current and the beam
energy requires more charge to neutralize.

The subsequent drop of the charging voltage is
caused by the rise of the background pressure due to the
remnant neutral gas ejected from the Fast Acting Gas
Value(FAV) Reservoir of MPD. This is the same effect
as the Neutral Gas Plume generator (NGP) of SEPAC.
Though the amount of the neutral gas ejected from FAV
of MPD is small, the pressure around the pallet increases

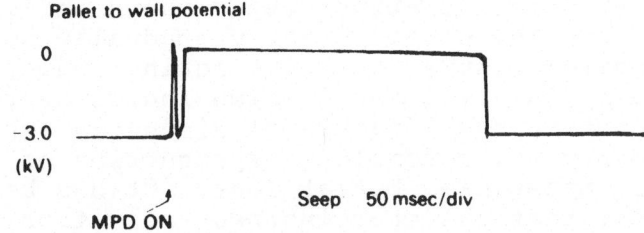

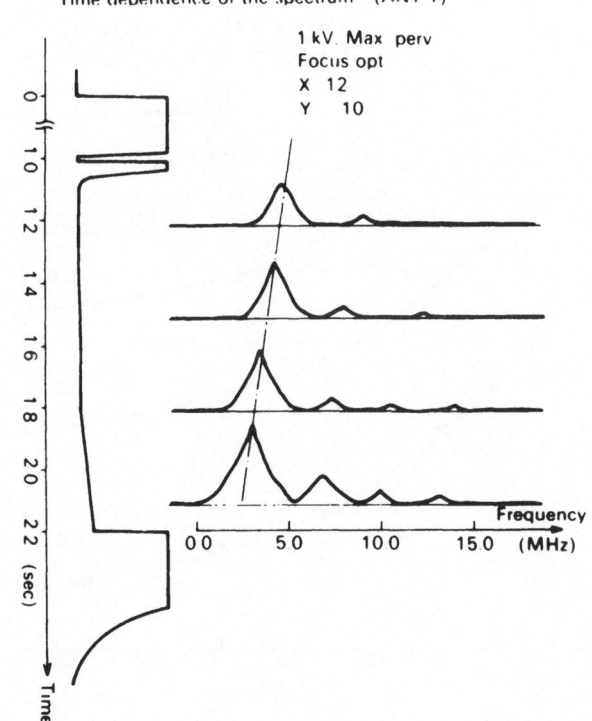

Fig. 20 (a)   Time variation of the
               pallet-to-wall potential
               in a joint operation of the
               electron beam and MPD
               arcjet plasma

Fig. 20 (b)   Neutralization time by MPD
               plasma as a function of the
               beam current and energy

considerably (~ 10^{-5} Torr) because the chamber is finite.
Plasma created by the ionization of this neutral gas by
the electron beam contributes to the charging neutralization.   During this period, a very strong emission of
plasma wave possibly at Upper Hybrid Frequency is
observed with harmonic structure as shown in Fig. 21
indicating that the plasma density gradually decreases
until the pallet resumes charging again.   The recovery
of charging is fairly abrupt, which indicates that the
plasma production has a nature of discharge (BPD).
In this experiment, a complete reproduction of experiment
in space is impossible, but at least, it has been shown
qualitatively that the charging neutralization by MPD
plasma and NGP gas should be effective.   We must wait
until SL-l SEPAC experiment in 1983 for quantitative
conclusion.

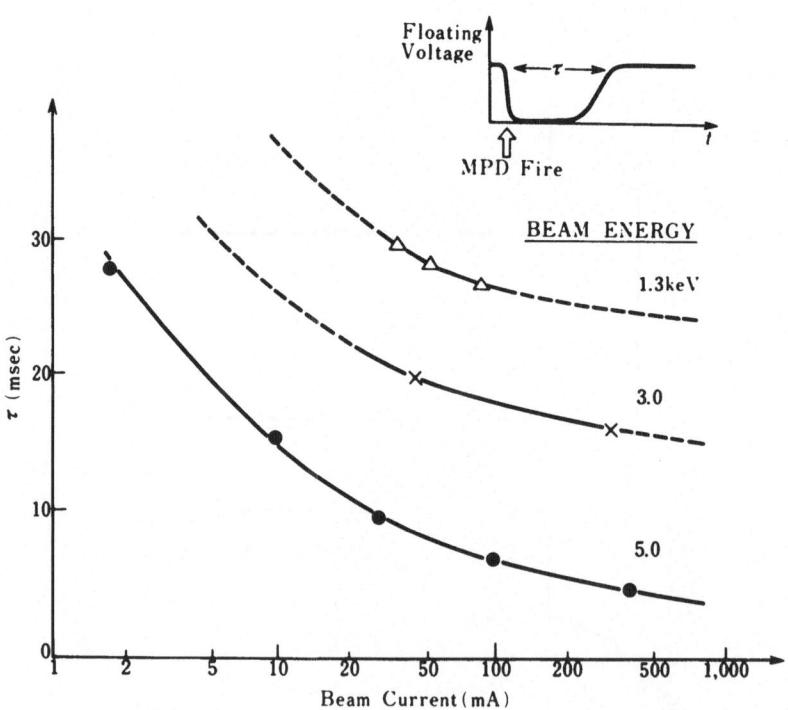

Fig. 21   Plasma wave generation in the
          electron beam-FAV neutral gas
          (MPD) interaction (1 keV 80 mA)

4. Summary and Discussion

Rocket and laboratory experiments of vehicle charging by the beam emission performed in Japan are reviewed. It is shown that in both rocket and laboratory experiments, the charging process obeys a simple probe theory or space charge theory so long as the beam current is low so that the charging voltage is low. As the beam current increases the ionization effect becomes significant and the voltage - current characteristic departs from the theory in such a way that the charging voltage is supressed. For a much higher current, a discharge phenomenon occurs resulting in an abrupt drop of the charging voltage. In laboratory experiment of studying the configuration peculiar problems related to the space shuttle charging it has been studied how effective each conductive surface is depending on its configuration with respect to the plasma flow. It has been concluded that the contribution other than the main engine nozzle is not expected much. In the experiment using SEPAC instruments in a large vacuum chamber, the system is charged up to the beam voltage easily if the surrounding vacuum pressure is below 10^{-5} Torr. This implies that the space shuttle will be charged up soon to the beam energy as long as the surrounding environment is a good vacuum. In order for an intense electron beam to be emitted from the shuttle, a fairly large amount of plasma shall be present or created around the shuttle in order to increase the effective conductive surface area to collect sufficient return current.

Rocket experiments are very difficult in a sense that diagnostic means are limited, time is limited and parameters are changing. Laboratory experiment is important to supplement these aspects of rocket experiments. A more intimate coordination with space experiment and laboratory experiment is required.

REFERENCES

Beard, D.B. and Johnson, F.S., 1961, J. Geophys, Res. 66, 4113.
De Forest, S.E., 1972, J. Geophys, Res. 77, 651.
Kaneko, O., Kawashima, N. and Sasaki, S., 1979, ISAS RN 85.
Kaneko, O., 1979, PhD Thesis of Univ. of Tokyo.

Kawashima, N., Sasaki, S., Yamori, A., Obayashi, T. and
      Kaneko, O., 1980, Space Res. and Advances in Space
      Exploration (in press).
Linson, L.M., 1969, J. Geophys, Res. 74,2368.
Obayashi, T., Kuriki, K., Kawashima, N., Nagatomo, M.,
      Ejiri, M. and Sasaki, S., 1978, ISAS Report No. 568.
Obayashi, T., et al, 1980, ISAS Report No. 577.
Parker, L.W. and Murphy, B.L., 1967, J. Geophys, Res. 72,
          1631.
Yamori, A., Kawashima, N. and Sasaki, S., 1979, ISAS RN
      111.
Winckler, J.R., 1980, Rev. of Geophys, Space Phys. 18,
      659.

ELECTRIC FIELD OBSERVATIONS OF TIME CONSTANTS RELATED TO CHARGING

AND CHARGE NEUTRALIZATION PROCESSES IN THE IONOSPHERE

N. C. Maynard[1], D. S. Evans[2], J. Trøim[3]

[1]Laboratory for Extraterrestrial Physics, Goddard Space
Flight Center, Greenbelt, MD 20771
[2]Space Environment Laboratories, NOAA, Boulder, CO 80302
[3]Norwegian Defense Research Establishment
N-2007 Kjeller, Norway

INTRODUCTION

The accumulation and neutralization of charge in space plasmas
has been a subject of interest for many years. Spacecraft charging
has been the topic of special conferences (see Finks and Pike, 1979)
and a dedicated satellite mission (SCATHA). Charge separation or
regions of excess change become a source of electric fields, often
of a transient nature, in the plasma medium. Also, vehicles in
space that emit charged particles must be neutralized by current
from the plasma medium. The charging of payloads by accelerator
experiments has been a concern since their inception by Hess et
al. (1971).

The Polar 5 experiment (Maehlum, et al. 1980) with its "mother-
daughter" configuration provided a unique opportunity to observe
the effects of spacecraft charging by an electron accelerator. The
electric field instrument on the mother payload observed this
charging and discharging process at distances from 10 to 100 m.
Jacobsen and Maynard (1980) concluded that the daughter payload,
which contained the accelerator, charged to between several hundred
volts and 1 kV.

An earlier rocket flight, Polar 3, observed transient events
which are thought to involve the "natural" accretion and neutralization of space charge. Electric field, energetic particle and
cold plasma density instruments recorded over 50 transient events
over a period of about 40 seconds north of the main auroral arc

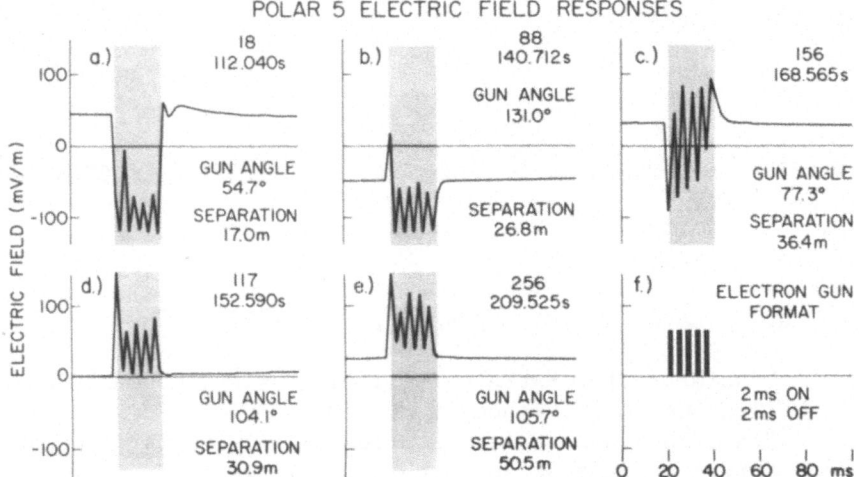

Figure 1. Five examples of the electric field responses
at the mother payload to the firing of the electron gun
on the daughter payload. The electron gun format is
shown in panel f. The shaded area represents the interval
over which the gun was fired. The pulse number, the
elapsed time, the angle of the gun to the magnetic field
and the separation distance between the mother and the
daughter payloads are given for each pulse.

(see Maynard et al. 1977 and Evans et al. 1977). We will, review
the Polar 5 electric field results, present the Polar 3 transients,
and look at these phenomena from the space charge perspective.

ARTIFICIAL PHENOMENA - POLAR 5

The 20 ms Polar 5 electron accelerator pulse consisted of 5
sub-pulses, each 2 ms long and separated by 2 ms. A 0.1A beam of
10 keV electrons was emitted in this pulsed mode every 410 ms. The
electric field response at the mother payload to these pulses has
been described by Jacobsen and Maynard (1980). An overview of the
whole program including a description of the background geophysical
conditions is found in Maehlum et al. (1980). The payload was
launched over a quiet arc on February 1, 1976 at 19:28:55 U.T.
(about 20:29 local time).

The electric field experiment was a single axis double probe
instrument (Aggson, 1969) using cylindrical antennas deployed in the
spin plane. The baseline or separation distance between the mid-
points of the sensors was 8.7 m. The background electric field was

relatively small in magnitude throughout the flight (Maehlum et al. 1980). The electric field response to each electron gun pulse was variable and dependent on several factors. The most significant of these was that the measurement was of only one component rather than the complete electric field vector. The resulting character of each signal was therefore a function of the attitude of the mother payload as well as the attitude or orientation of the electron gun on the daughter payload. In addition, the electron gun format (see panel f of Figure 1) meshed with the sampling period of the electric field data (.002048S) such that a variety of responses were possible. Another complicating factor was that the daughter spin rate was almost exactly twice the mother spin rate.

The electric field responses over a complete mother spin cycle have been presented by Jacobsen and Maynard (1980). Some clearly delineate all five electron gun sub-pulses while others show little variation relative to the sub-pulses. Figure 1 displays five responses at different separation distances and where the five sub-pulses are clearly evident. Panels b, d and e illustrate that the response to the first sub-pulse was generally of greater magnitude and sometimes of different direction than the response to the remaining sub-pulses. The uniform, magnitude of the negative excursions in panels a and b are at least partially a result of saturation. Note that magnitudes of over 100 mV/m were observed at distances of 50 m (panel e).

Figure 2 shows a plot of the amplitudes of the remainders of the responses (without the larger initial values) over the whole flight. The dc terms have been suppressed. Values in the early part of the flight are limited by saturation.

Large electric field responses were seen at over 100 m separation distances, even though the Debye length for the medium is of the order of centimeters. The Debye length for plasma shielding is derived for potentials less than kT of the ambient plasma and should not be used for higher potentials. The electric field variation with distance is nearer that of free space, especially for the initial sub-pulse response. Some shielding is evident for the remainder of each gun pulse (cf. Figure 2). Cartwright and Kellogg (1974) saw "quasi-dc electric fields" at distances of hundreds of meters from the ECHO-1 accelerator payload, but interpreted their mV potentials as Cherenkov radiation.

Jacobsen and Maynard (1980) did a systematic study of all the electric field responses and data from a retarding potential analyser. They concluded that the daughter payload had to have charged to a potential of several hundred volts to 1 kV and that, to first order, the charge configuration was a dipole consisting

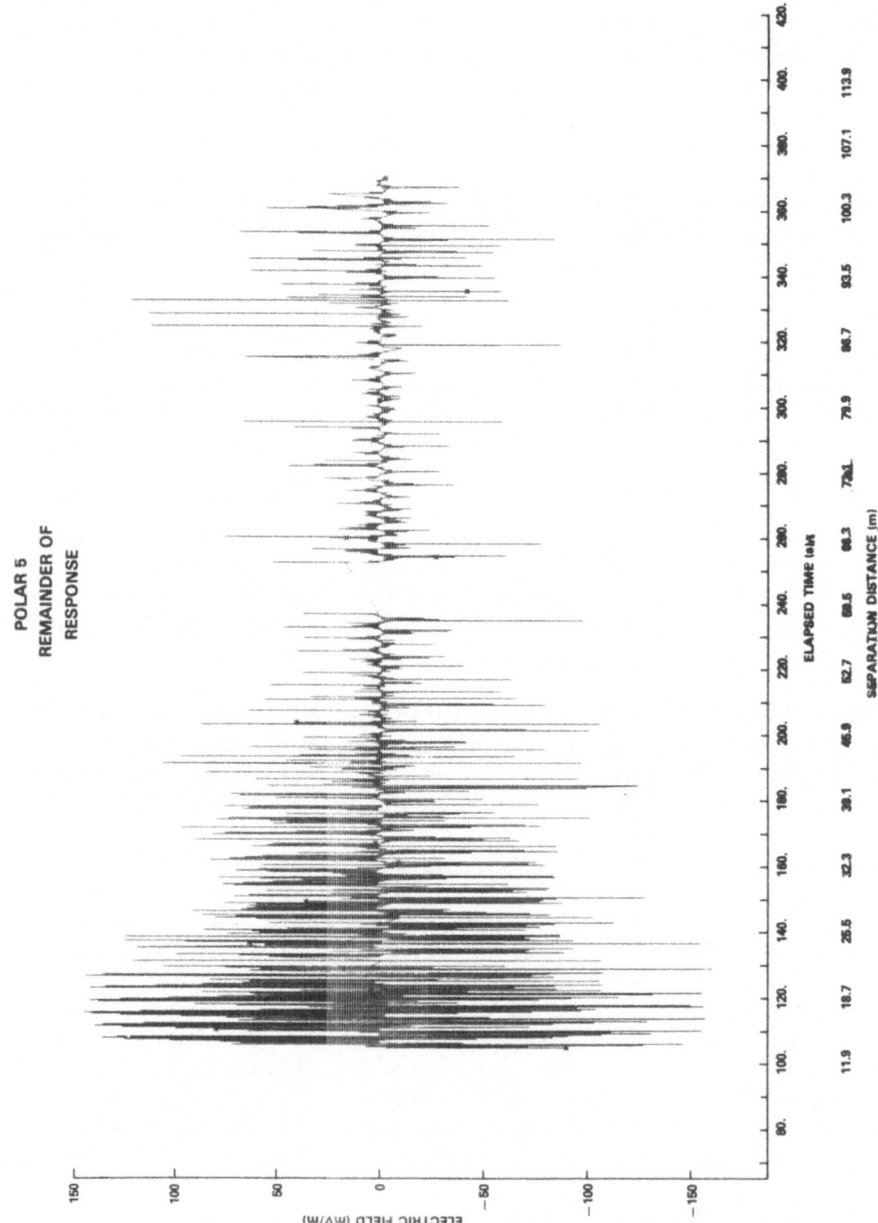

Figure 2. The maximum amplitude of the remainder of the electric field responses after the initial value has been removed as a function of time and the separation distance (from Jacobsen and Maynard, 1980).

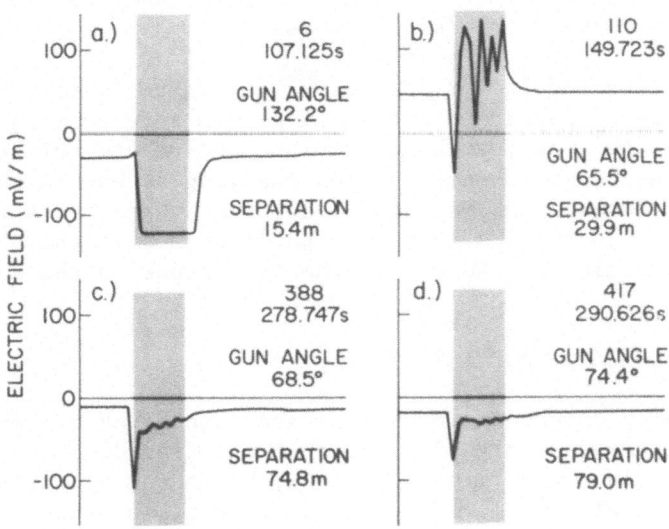

Figure 3. Four examples of electric field responses to the electron accelerator pulses which have observably slower decay rates. (See Figure 1 caption and text.)

of the positively charged daughter payload and the negatively charged beam. This did not completely explain the data; however, the qualitative nature of single axis measurements in a rapidly varying situation precluded a more quantitative approach.

The question of accretion and neutralization rates can be further explored in the examples in Figure 3. Example a shows a saturated response that recovers several milliseconds after end of the electron gun pulse. Examples b, c and d indicate a time constant of several milliseconds to return to normal after the last sub-pulse. A few responses were observed with long lasting after effects (sometimes more than 10 ms; see Figure 6 in Jacobsen and Maynard, 1980); however, these phenomena, which may be related to secondary electrons, have not been considered here. The general case is that accretion of charge occurs within the time period between telemetry samples (about 2 ms). Figures 1a and 3a show a small response prior to a saturated response which is interpreted as occurring as the beam is turned on, an indication that the charging time is at least of the order of tens of μs. The neutralization time varies between a value somewhat less than the sample

period and several ms. The difference between the initial responses
and the remainder of the responses indicates that partial shielding
is set up on a time scale of a few ms.

A complicating factor in the charge neutralization process is
that a source of particles from beam plasma interactions is thought
to have been necessary to have provided the return current that
prevented the daughter payload from charging to even higher potentials (cf. Jacobsen and Maynard, 1980; and Grandal et al. 1980).
This source of particles must have participated in the shielding
and neutralization of the charge that did occur on the daughter
payload.

NATURAL PHENOMENA - POLAR 3

The Polar 3 sounding rocket payload contained a comprehensive
set of instruments designed to measure electron densities, electron
temperatures, auroral light emissions, DC and AC electric fields,
and the energetic particle population in regions nearby and over
discrete auroral arcs. The payload was launched from Andenes,
Norway on 27 January, 1974 over a bright, well defined discrete
aurora. With the exception of a Langmuir probe and a portion of
an experiment which measured the greater than 20 keV component in
the energetic particle population, all the instruments operated
satisfactorly.

A region of enhanced energy influx of precipitating electrons
occurring between 150 s and 230 s of flight time corresponded to
the approximently 70 km wide discrete auroral arc that was the
target of the rocket launch. The gradual increase in energy flux
after about 340 s of flight time had the spatial dimensions,
spatial gradients, electron energy spectrums, and pitch angle distributions (field aligned) that would be appropriate to "inverted-
V" events often observed by satellite instrumentation. The rocket
re-entered the atmosphere at about 390 s before the entire
"inverted-V" region could be traversed. The analysis and interpretation of the data obtained over the auroral arc and "inverted-
V" portions of this flight have been published (Maynard et al. 1977;
Evans et al. 1977).

The purpose of this section is to describe and discuss a series
of 50 or more short lived events that were observed between 237 s
and 273 s of flight time in the region of otherwise quite modest,
and uninteresting, energy fluxes located poleward of the bright arc
and equatorward of the "inverted-V". These episodes were of such
a curious and unexpected nature that we have, on occasion, referred
to them as "glitches" although there is no evidence of any instrumental malfunction. As is clear from the examples given below,
these events appear simultaneously, or in close time association,

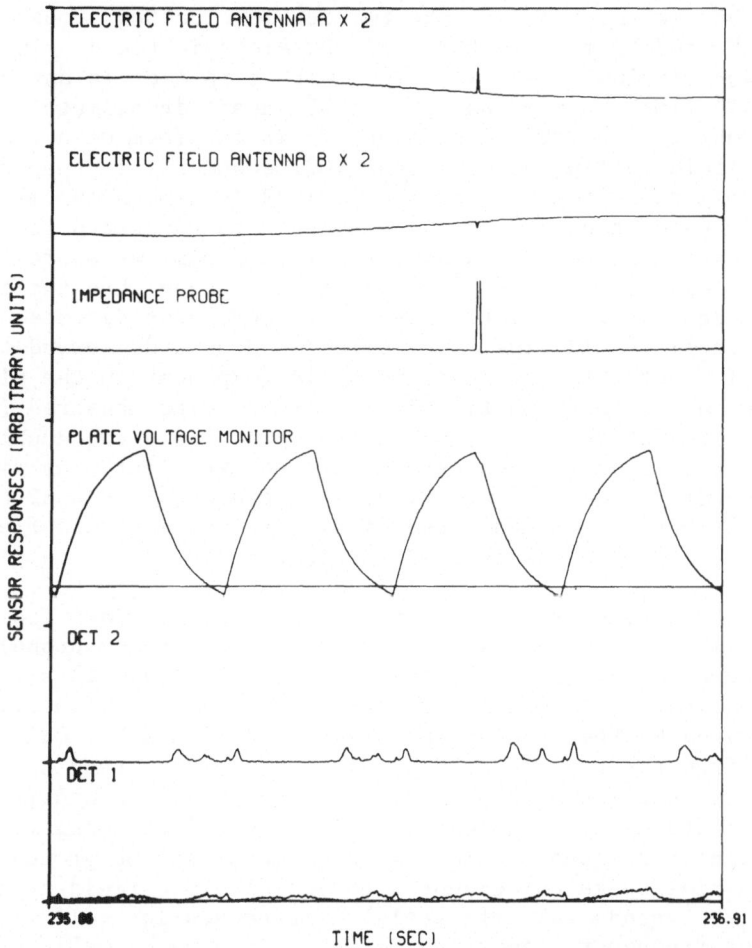

Figure 4. Raw data from the electric field, impedance probe and auroral particle instruments for the first of the Polar 3 transient events (see text). In this and the next three figures the scales have been kept identical for each instrument.

on many, although not all, of the payload experiments. Often the nature of the event, as viewed by a specific instrument was so far outside the designed operating range of that instrument that little quantitative information can be inferred beyond the fact that some-thing very unusual occurred. For this reason, the data displays are of the raw sensor responses rather than converted to physical quantities.

Figure 4 displays data from the very first of these unusual events encountered at 236.6 s into the flight. The top two panels

display the raw responses of the two double floating probe
"antennas" used to measure the local DC electric field. In this
case a two axis instrument was flown with both axes (a and b)
located 45° from the spin axis, but 180° apart in azimuth. The
event as observed by this instrument is in the form of a transient
electric field lasting no more than .002 s which corresponds to
only 2 m motion on the part of the payload (a dimension smaller
than the span of the antennas). With perhaps two exceptions, each
of the events observed was associated with a similar short-lived
electric field transient. Moreover, in each case that the transient
was observed, the transient always had a component directed parallel
to the magnetic field line (i.e. directed downwards toward the
earth). In contrast, the electric field component in the plane
perpendicular to the magnetic field that could be measured (the
short duration of the transient together with only two double probes
prevented a full vector measurement of the transient) was distributed randomly in direction. Electric field magnitudes of over
100 mV/m both parallel and perpendicular to the magnetic field were
observed. Many events saturated the instrument.

The third panel from the top displays the raw data from the
on-board impedance probe (also known as a capacitance probe) which
was designed to measure the local electron density. This instrument is basically an oscillator where the frequency of oscillation
is determined by the free space capacitance of a spherical probe
extended from the tip of the payload. Because the free space
capacity is determined in part by the local electron density, the
oscillator frequency is a measure of this electron density. It is
the oscillator frequency which is telemetered to the ground. The
large positive spike coincident with the electric field transient
(this spike extends well off scale) corresponds in reality to the
oscillator frequency attempting to shift to a value well outside its
designed range of operation. For this reason the data cannot be
converted to a meaningful measure of local electron density and
all that can be said is that the local plasma environment was outside the range of temperatures and densities normally encountered
in the auroral ionosphere.

The time for the impedance probe to resume normal operation
is about .008 s in this example. It is also noteworthy that the
instrument senses a perceptively different electron density after
resuming operation than just before the event. In every case of
a transient in the DC electric field, there is a simultaneous loss
of ability of the impedance probe to measure the local electron
density. The time required by the probe to recover normal operation after these events ranges from about .010 s, as in Figure 4,
to about .100 s in other examples. Moreover, the impedance probe
data often displays a variation occurring before it ceases proper
operation as well as a period of recovery after resuming proper
operation. This is shown more strikingly in later examples.

The fourth panel in Figure 4 displays the monitor of the voltage between the analyzer plates of the "high energy" detectors in the auroral particle experiment. This instrument contained 4 identical particle detectors, each a cylindrical-electrostatic analyzer channel-multiplier-detector combination. The 4 detectors were mounted in pairs, one pair viewing upwards at 45° with respect to the rocket spin axis, the other viewing downwards at 135°. The analyzer plates for one detector in each pair were swept through a voltage profile such that positive ions from about 1500 eV down in energy, through zero, and then electrons from zero up to 16,000 eV would be measured. At this point a relay switch, controlling the charging and discharging of a capacitor bank, reversed the sweep. The second detector in each pair was also swept in a similar fashion except that the energy limits were about 200 eV for positive ions and 3000 eV for electrons. No ion data or electron data from energies below 100 eV from these detectors were regarded as valid. The energy sweep rates were 3.5 full cycles per second (7 energy spectrums per second) and the sweeps were synchronized to the PCM telemetry system. The count rates from the four detector systems were telemetered at a rate of 488 samples per second each. The direct monitor of the "high energy sweep voltage" was telemetered at the same rate so that each "high energy" detector count rate could be directly associated with a specific particle species and energy. The normal waveform is an exponential increase to a maximum corresponding to an electron energy of 16,000 eV followed by an exponential decrease, through zero potential difference between the plates, to a maximum "proton energy" of 1500 eV. The baseline drawn on this panel corresponds to the zero energy level of this monitor. It should be noted that there is a discontinuity in the monitor waveform occurring simultaneously with the electric field transient and the disturbance in the proper operation of the impedance probe.

The bottom two panels in Figure 4 display the responses of two of the four particle detectors in the auroral particle instrument. Both detectors 1 and 2 are the upward viewing detectors measuring precipitating electrons. Detector 1 is sweeping in electron energy up to 3000 eV while detector 2 is sweeping in energy up to 16,000 eV ("high energy detector"). The count rate pattern displayed by detector 2 is quite typical of a modest intensity auroral precipitation. The outer count rate peaks in this pattern correspond to the knee in the electron differential-directional energy spectrum that is usually observed in auroral electron precipitation. In Figure 4 these peaks convert to a knee located at 4000 eV electron energy. The two inner count rate peaks correspond to the enhanced fluxes of secondary and backscatter electrons at energies below 1000 eV (the second of these peaks is difficult to see because the energy analyzer sweeps upwards through this energy range so rapidly). The detector 1 response is entirely consistent with the detector 2

response given the difference in the range of the energy sweep and
the rate at which the two analyzers sweep through their respective
ranges. It should be noted that there is no feature in the particle
data coincident with the disturbance registered by the other
instruments. This speaks against a significant level of electrical
interference as explaining any anomolous data.

Figure 5 shows a 1.05 s long period of data during which there
were 4 transient events in the DC electric field data. The third
panel from the top displays data from the 6300 Å photometer in the
payload. There is a short lived, .002 s long, increase in the
photometer output associated with each of the electric field
transients. The other slower variations in the photometer response
are due to the instrument scanning past nearby aurora as the rocket
rotated. There were very modest or no transient variations in the
4278 Å photometer at these times, even though the two photometers
were mounted together and viewed the same direction. Because of the
long time constant for emission, we can not interpret the 6300 Å
photometer responses as light emission from atomic oxygen. Suggested explanations for this response include a contribution from a
nearby O_2^+ spectral line (half-width of the filter was 50 Å), or a
response to an electrical transient within or without the payload.

The plate voltage monitor shown in Figure 5 displays clear
discontinuities associated with two of the transient events but
nothing so easily seen in association with the other two. As was
the case in Figure 4, these two observed discontinuities occurred
when the energy sweep neared its maximum where the voltage between
the analyzer plates was about 800 volts. This lead to speculation
that there might be some sort of electrical breakdown between the
analyzer plates which, in turn, could cause anomalous behavior
on the part of other experiments on board the rocket. However,
there is no real correlation between the occurance of transients
and the magnitude of the voltage between the analyzer plates.

The nature of the discontinuities appearing on the plate
voltage monitor is quite strange and bears some discussion. The
voltage waveforms impressed between the analyzer plates were
generated by alternately charging and discharging capacitors
through a resistor which was switched by a relay between various
outputs of a high voltage supply. The voltage to the plates was
balanced with respect to ground rather than grounding one plate
and impressing a voltage waveform to the other plate. The combination of resistor and capacitor values were chosen so that the
minimum charging (or discharging) current would be large compared
to those currents collected by the analyzer plates either from
energetic auroral particles or from thermal plasma entering the
electrostatic analyzers from the auroral ionosphere. It was
calculated that the current from thermal electrons entering the
detector system would be only a few $x10^{-8}$ A. Discontinuities

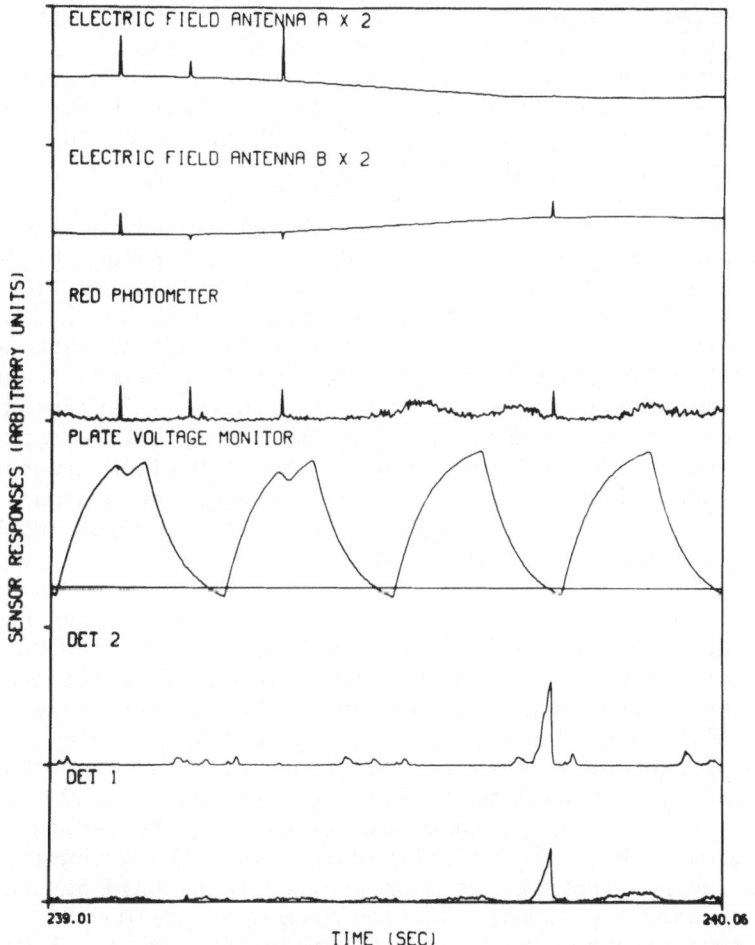

Figure 5. A second example of the Polar 3 transient events. Data from the 6300 Å photometer has been substituted for the impedance probe data. The scales are identical to these in Figure 4 (see text).

in the plate voltage waveform identical to those shown in Figure 5 were produced in the laboratory by imposing a momentary (.01 s) 5×10^{-6} A current load between the analyzer plates. This current could exceed the current that was charging the capacitor, especially when the charging phase was nearly complete, and thus result in a momentary discharge of the capacitor. If the transient current load were impressed just after the discharge cycle started (e.g. coincident with the second of the four transient shown in Figure 5) the effects would be nearly impreceptible as the initial discharge current would be large compared to 5×10^{-6} A. If the excessive current load occurred when the discharge cycle was nearly complete and the potential between the analyzer plates was close to zero, the effect would also be difficult to see as the potential would simply be driven somewhat closer to zero than would otherwise be the case. In fact, careful examination of the plate voltage monitor for the fourth transient in Figure 5 shows that the discharge cycle was completed at a point slightly closer to zero potential between the plates than normal.

Thus, every feature associated with the discontinuities in the plate voltage waveform was duplicated in the laboratory by imposing a current load to the analyzer plates of some 50 to 100 times greater than the currents that were anticipated to be collected from the ionospheric plasma. This, in turn, would suggest that at the instant of the discontinuity in the plate voltage sweep (together with the transient in the DC electric fields and malfunction in the impedance probe) the product $nT^{1/2}$ (n the ionospheric electron density, T the electron temperature) was 50–100 times greater than the normal auroral ionosphere. While this would be unusual, to say the least, such conditions would easily explain why the impedance probe oscillator would cease to operate momentarily.

The bottom two panels in Figure 5 again show charged particle data from detectors 1 and 2. There were no features in the auroral electron data associated with the first 3 electric field transients. However, the fourth one, which occurred when the detectors were measuring electrons of energies less than 1 keV, was accompanied by intense fluxes of low energy electrons. Indeed, the count rate of the detector 2 channel multiplier reached nearly $2x10^5$ counts per second, a value higher than observed during the period when the payload was above the bright auroral arc. The intensity of these low energy electrons is much greater than could be accounted for by secondary electron production from the very modest energetic electron precipitation that existed during this event. Finally, it is noteworthy that these intense electron fluxes return abruptly to normal levels exactly coincident with the electric field transient. It is as though these transients that appear on many instruments mark the termination of an event of greater duration.

Figure 6 gives further support to the hypothesis that a common feature to these events is the presence of large fluxes of low energy electrons. Here, for 3 consecutive energy sweeps of the particle detectors an ever increasing intensity of low energy electrons is observed. Then an electric field transient occurs which apparently terminated the event in as much as the large electron fluxes had disappeared by the time the spectrometers again studied that energy range.

If should be noted that in Figure 6 it is the detector 3 and detector 4 responses that are plotted. These two detectors were mounted to view downward so as to measure upcoming electrons (detector 4 is the "high energy" spectrometer). A comparison bet-

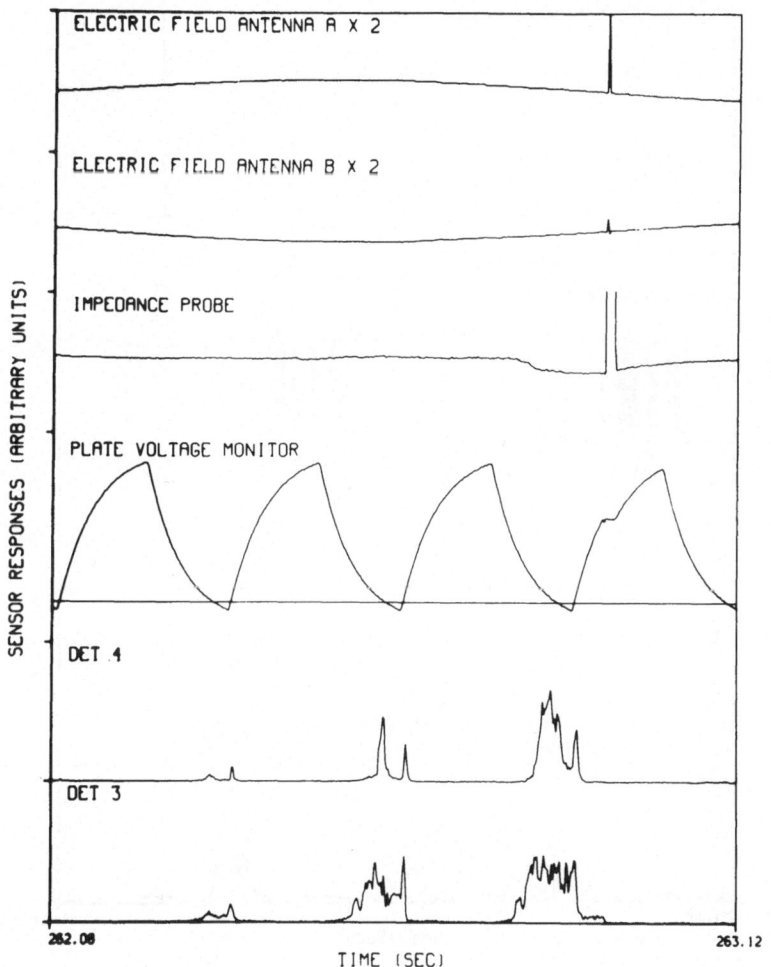

Figure 6. An examples of a longer duration Polar 3
transient in which the particles are clearly restricted
to low energeies. In this case the downward looking
detectors are plotted in place of the upward looking
detectors.

ween these two detectors and those two which measured downgoing
electrons showed that at times during the event in Figure 6 (which
lasted for 0.7 s in the low energy electrons) the upcoming electron
intensities exceeded the downgoing.

Figure 7 is the final example to be shown. It also was the
final episode of this sort as well as being the most intense and
longest lived. The electron spectrometer data show that for 4
sweeps there were increasingly intense fluxes of low energy electrons.
The fluxes which were observed during the final sweep were the
largest ever encountered by an electron spectrometer of this design

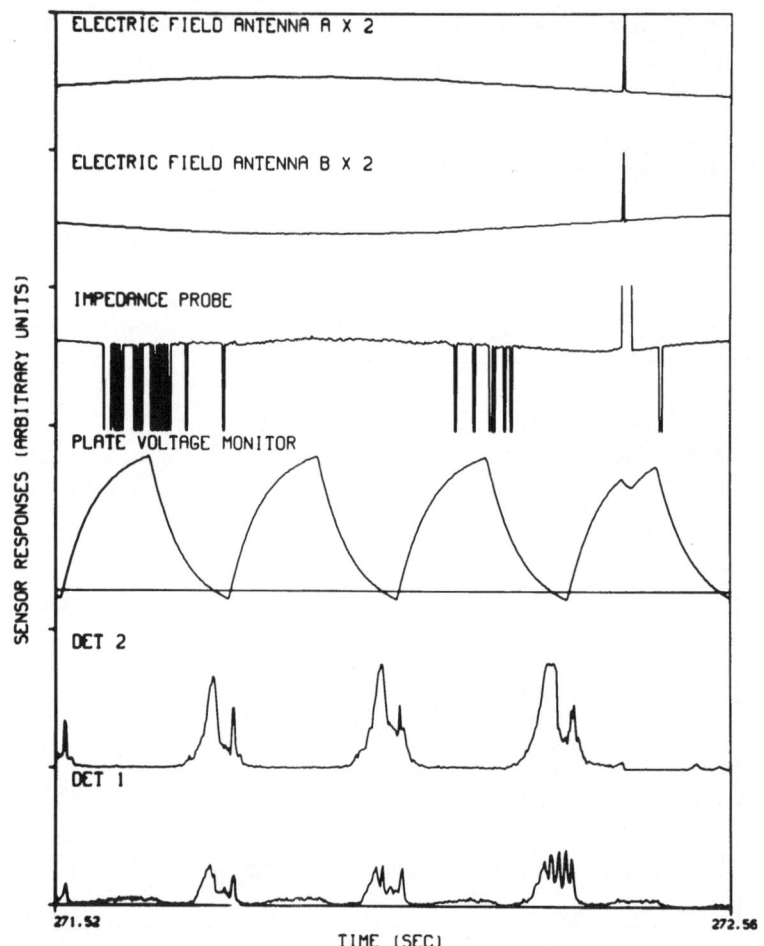

Figure 7. The last and longest duration Polar 3
transient. The low energy of the particles is
clearly emphasized.

over the course of many rocket flights over auroras. A rough conversion of the detector 2 response to a measure of the differential-
directional electron intensities that existed during the final
phases of this event yielded a value of nearly 10^{11} electrons cm^{-2}
sec^{-1}ster^{-1}keV^{-1} at an energy of 0.5 keV. Flux values of this
magnitude are truely astounding.

As was invariably the case, the low energy electron event
terminated with the DC electric field transient and the oscillator
dropout on the part of the impedance probe. The negative going
transients seen in the impedance probe data during this event are
not explained. This was the only event during which this instrument exhibited such behavior. A comparison between the detector

1 and detector 2 responses shows that these two detectors did not
agree with one another when they were measuring the same electron
energy at nearly the same time. Outside of the 30 s period
containing these events, the 4 individual detectors were <u>always</u> in
excellent agreement with one another. While it is tempting to use
a small number of instances of disagreement like this as proof that
there were detector malfunctions, we believe that there is enough
independent evidence that something physical was actually happening
and that the disagreements are the result of the transient nature
of the event.

After the termination of the episode shown in Figure 7, all on
board instruments returned to uneventful and flawless operation for
the remainder of the flight.

The nature of these events may be summarized as follows. An
event (or episode) is associated with the presence of intense fluxes
of low energy electrons as detected by the auroral electron spectrometer. During this phase, there are no observable responses on
the part of the photometers or DC electric field instruments on the
payload. The impedance probe often detects an enhancement of the
local electron density during this phase of the event. The event,
as defined by intense low energy electron fluxes, terminates very
abruptly, coincident with a transient (.002 s long) in the local
electric field. The termination is also accompanied by a dramatic
perturbation in the local ionosphere as inferred by the oscillator
dropout in the impedance probe and from the discontinuities in the
electrostatic analyzer plate voltage wave form. Immediately after
the termination of an event, all instrument responses return to a
normal level with the exceptions of the impedance probe and plate
voltage monitor which require .008 s or more to recover.

Given the dynamic behavior of these low energy fluxes and the
fact that no event lasted as long as the rotation period of the
rocket, the determination of complete pitch angle and energy
distributions cannot be made. However, instantaneous comparisons
between the detector pair viewing precipitating electrons and the
pair viewing upgoing electrons gives some information about the
angular distribution of these electrons. As a general rule, the
upgoing and downcoming electron intensities tended to differ by no
more than a factor of 3 or so from one another. There were notable
exceptions, however, emphasising both upgoing and downcoming
intensities.

If the count rate variation in the low energy event shown in
Figure 6 is due to the spectrometer sweeping in energy (as opposed
to a purely temporal variation) then the e-folding energy of the
electron spectrum below 1000 eV (a measure of the temperature of the
electrons) can be estimated as about 200 eV. A similar estimate
for the low energy event in Figure 7 based upon the detector 2 data

where enhanced fluxes were observed up to energies of a few keV, yields an e-folding energy of about 700 eV.

Although we can not settle on a location of the source of these particles or a generation mechanism, it is of interest that each of these events produced a space charge electric field which terminated the event. The fact that the parallel electric field was always directed downward would be consistant with a region of excess electrons located in the vicinity of but below the payload. The accretion of this change must occur within a time less than 2 ms, as must its dissapation, since it is never seen in more than one telemetry sample per event. On two events a measurement of the voltage from one antenna to the spacecraft was made on the next telemetry word occurring 128 μs later (these measurements were commutated so that the observation was by chance) showed the continued presence of the transient. Hence, the measurements indicate that the discharge time constant is greater than 200 μs.

DISCUSSION

Although the two types of phenomena described here are of totally different origins, the electric field response to the space charge that was created in each case provides a common thread. The magnitudes in each case are similar and the durations, while shorter in the natural case, are of the same order of magnitude. Based on the Polar 5 results, the large magnitude of the electric field from Polar 3 would indicate that the observed space charge was probably within a few kilometers or less of the payload.

The time constant for shielding of small charges should be at the plasma frequency or tenths of μs according to the derivation of the Debye length (cf. Jackson, 1962). Cole (1960) predicted that charges in the ionosphere would reach equilibrium with a time constant of the order of a few μs, based on arguments involving Pedersen conductivity. Processes involved in the two cases presented in this paper require time constants of the order of ms.

McCoy and Konradi (1979) have recently performed measurements of the sheath effects from charging a 1 by 10 m panel to high voltages in the thermal vacuum test chamber at the NASA Johnson Space Center. The plasma density, flow direction, and magnetic field strength were controlled to simulate low Earth orbit. The flat "array" was biased in many modes up to $\pm$ 4 kV. They concluded that the current collection is space charge limited and that the dimensions of the sheath can be derived by equating the random thermal current to the space charge limited current. The resulting expression for spherical geometry and the thick sheath approximation is

$$(d + a) = r_o = 137 \frac{(aV)^{3/7}}{N_o^{2/7}E^{1/7}} \tag{1}$$

where d is the sheath thickness, a is the radius of the body, r_o is the outer radius of the sheath, V is the potential of the body, N_o is the ambient plasma density and E is the energy in eV of the ambient plasma. Note that the expression is independent of mass indicating that the electron and ion sheaths will be of the same dimensions if their temperatures are the same.

The electon density observed by Polar 5 varied from just under 10^6 to near 2×10^4 cm^{-3} during the times when the electron gun was operating (Maehlum et al. 1980). If we assume a nominal value of 10^{11} m^{-3} for N_o, a plasma energy of 1 eV, a daughter radius of 0.5 m, and a potential of 500 V on the daughter, then the resulting value for the sheath radius r_o is 1 m. Unfortunately this is at least 1 to 2 orders of magnitude too low if we assume the sheath extends to the limit of the electric field response.

If the sheath dimensions are picked to be between 50 and 100 m which is not unreasonable relative to the electric field measurements, then a qualitative estimate of the neutralization time would be the transit time for ions across the sheath. The kinetic velocity of a 1 eV proton is about 14 km/s. Thus it would traverse the distance in 4 to 8 ms, assuming freedom of movement across magnetic field lines, which is the order of the decay times observed on Polar 5. A higher energy or smaller sheath would result in the smaller times inferred from the Polar 3 data. The geometry of the Polar 3 situation is obviously very different, and the space charge is probably very dispersed.

The above discussion is admittedly simplistic. It does show that a new treatment is necessary to account for the significant electric fields at large distances from regions of space charge that have a higher potential than the plasma energy, and for the relatively long neutralization time constants that we observe. The dynamics of the ionosphere or the charging of spacecraft can involve active and complicated process of which we know very little.

ACKNOWLEDGEMENTS

We would like to thank K. Måseide and B. Maehlum for helpful discussions. Dr. Maehlum was the project scientist for both rocket payloads. The research was jointly sponsored by the Royal Norwegian Council for Scientific and Industrial Research and the National Aeronautics and Space Administration.

REFERENCES

Aggson, T. L., 1969, Probe measurements of electric fields in space, in Atmospheric Emissions, ed. B. M. McCormac and A. Omholt, Van Nostrand, New York.

Cartwright, D. G., and Kellogg, P. J., 1974, Observations of radiation from an electron beam artificially injected into the ionosphere, J. Geophys. Res., 79, 1439.

Cole, K. D., 1960, A dynamo theory of the aurora and magnetic disturbance, Australian J. Phys., 13, 484.

Evans, D. S., Maynard, N. C., Maehlum, B., Trøim, J., and Egeland, A., 1977, Auroral vector electric field and particle comparisons. 2. Electrodynamics of an arc, J. Geophys. Res., 82, 2235.

Finke, R. C., and Pike, C. P. editors, 1979, Spacecraft Charging Technology - 1978, NASA Conference Publication 2071 and AFGL-TR-79-0082, NASA Science and Technical Information Office, Washington, D.C.

Grandal, B., Thrane, E. V., and Trøim, J., 1980, POLAR 5 - An electron accelerator experiment within an aurora. 4. Measurements of the 391.4 nm light produced by an artificial electron beam in the upper atmosphere, Planet. Space Sci., 28, 309.

Hess, W. N., Trichel, M. G., Davis, T. N., Beggs, W. C., Kraft, G. E., Stessinopoulis, E., and Maier, E. J. R., 1971, Artificial auroral experiment: Experiment and principle results, J. Geophys. Res., 76, 6067.

Jackson, J. D., 1962, Classical Electrodynamics, John Wiley and Sons, New York.

Jacobsen, T. A., and Maynard, N. C., 1980, POLAR 5 - An electron acceleration experiment within an aurora. 3. Evidence for significant spacecraft charging by an electron accelerator at ionospheric altitudes, Planet. Space Sci., 28, 291.

Maehlum, B. N., Måseide, K., Aarsnes, K., Egeland, A., Grandal, B., Holtet, J., Jacobsen, T. A., Maynard, N. C., Søraas, F., Stadsnes, T., Thrane, E. V., and Trøim, J., 1980, POLAR 5 - An electron accelerator experiment within an aurora. 1. Instrumentation and geophysical conditions, Planet. Space Sci., 28, 259.

Maynard, N. C., Evans, D. S., Maehlum, B., and Egeland, A., 1977, Auroral vector electric field and particle comparisons. 1. Premidnight convection topology, J. Geophys, Res., 82, 2227.

McCoy, J. E., and Konradi, A., Sheath effects observed in a 10 m high voltage panel in simulated low earth orbit plasma, in Spacecraft Charging Technology - 1978, edited by R. C. Finke and C. P. Pike, p. 315, NASA Conference Publications 2071 and AFGL-TR-79-0082, NASA Science and Technical Information Office, Washington, D.C.

MEASUREMENTS OF VEHICLE POTENTIAL USING A MOTHER-DAUGHTER TETHERED ROCKET

P. Roger Williamson, William F. Denig, Peter M. Banks,
W. John Raitt

Physics Department and
Center for Atmospheric and Space Sciences
UMC 34, Utah State University
Logan, Utah 84322 U.S.A.

N. Kawashima, K. Hirao, K.I. Oyama, S. Sasaki

Institute of Space and Aeronautical Science
Komaba, Meguro-Ku
Tokyo, Japan

INTRODUCTION

When plans were being made in 1969 to perform the first electron
beam experiment in space a question was raised concerning the amount
of neutralization current which could be collected from the ambient
ionospheric plasma when a beam of electrons was emitted from an
isolated vehicle such as a sounding rocket or satellite. Although
the first such experiment was a success and more than 20 similar
flights have been made since that time the original question of the
amount of neutralization current available from the ambient
ionospheric plasma is still unanswered.

In the first artificial aurora experiment (Hess et al., 1971) a
500 mA beam of 10 keV electrons was emitted and optical observations
from the ground showed conclusively that the beam was propagating to
distances far away from the vehicle. In experiments since 1969 beams
with energies up to 40 keV and, in other cases, with currents as high
as 7 amperes have been used. In some experiments evidence for
vehicle charging has been found although in general the degree of
charging is small enough that the primary investigation has not been
affected. It is uncertain whether or not vehicle charging was a
factor in experiments which failed to achieve the desired results.

The early experiments demonstrated that electron beams could be emitted from a sounding rocket in contradiction to the theory which predicted a much smaller return current than actually observed. It is apparent that other mechanisms usually dominate experiments performed in the altitude region normally accessible to sounding rockets. Linson (1969) has reviewed two models of charge collection for a spherical satellite and developed an intermediate model based on the onset of plasma turbulence. These models, which formed the basis of the early controversy on the probable success or failure of the artificial aurora experiment, have never been tested. The experimental conditions required to test these theories of charge collection in the ambient ionospheric plasma are not reproduced in the previous electron beam experiments in space. The electron beam produces a disturbed, hot plasma in the neighborhood of the vehicle and beam-plasma interactions as well as the enhanced ionization produced by the beam alter the environment of the vehicle to such an extent that the amount of charge collected from the ambient ionosphere in previous experiments cannot be determined.

A test of charge collection theories requires an accurate measurement of both the emitted current and the potential of the vehicle relative to the ambient plasma. Vehicle potentials have been determined in the past from Langmuir probe data, retarding potential analyzers, electrostatic analyzers and floating potential probes deployed on booms. In each case the measurement is affected by the surrounding disturbance at the time the electron beam is emitted.
EXPERIMENT

The Tethered Payload Experiment (TPE) is designed to provide a reference point in the ambient plasma far away from the disturbed region around the vehicle by deploying one section of the rocket payload (a daughter) away from the main payload (the mother section) with an insulated wire tether attached between them to provide a relative measurment of potential. The daughter acts as a "ground" reference point for measurement of the mother vehicle potential. Measurements of the relative potential between the daughter and mother can be made by measuring the potential difference between the end of the tether and the chassis ground at the mother end of the payload using a high input impedance voltage monitor. When a low impedance is switched onto the tether the voltage of the mother and daughter is very nearly the same relative to the ambient plasma and current in the tether can be measured.

The experimental objectives of the first TPE were several fold. The accurate measurement of vehicle potential with high time resolution was to be performed using the mother-daughter tether. In addition, the reference point supplied by the daughter would provide a measurement of the mother vehicle potential when perturbed by an electron beam emitted from the mother. A swept power supply which could be switched into series with the tether at the mother end of the wire could be used to measure the current voltage characteristic

of the daughter. Finally, the stability of the system was observed
to determine the usefullness of the TPE technique for future
experiments.

INSTRUMENTATION

Instruments on the mother and daughter sections of the payload
are shown in Figure 1. The daughter section contained a charge probe
(DCP) which measured the surface charge on a sheet of teflon,
a flat plate Langmuir probe (DLP), a flat plate floating probe (DFP)
of the same size as the DLP but mounted on the opposite side of the
payload.

The mother section contained a similar set of instruments
including a flat plate Langmuir probe (MLP) and a flat plate floating
probe (MFP). The tether deployment mechanism in the mother held
several hundred meters of teflon insulated, stainless steel wire
composed of 7 strands 0.08 mm in diameter.   The tether was wound on
a spool located at the forward end of the mother payload. Black
marks at one meter intervals were detected with an optical device in
the end of the deployment mechanism.

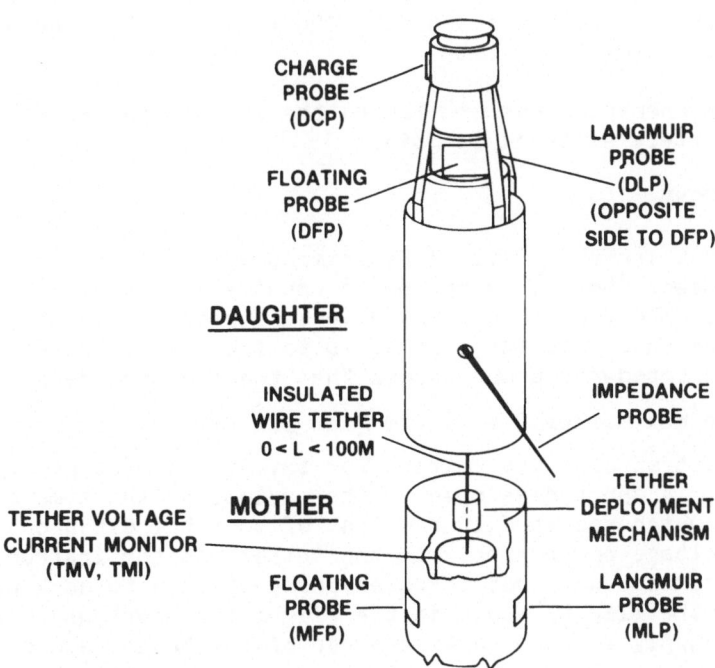

Fig. 1.   Tethered Payload Experiment (TPE) Instrumentation.
          An insulated wire tether is attached between the mother
          and daughter sections of the payload.

The DCP sensor was a thin sheet of teflon 0.14 mm thick covered by a thin sheet of polyimide material 0.025 mm thick. The backing of the dielectric was isolated from the vehicle by a charge amplifier with a very long time constant compared to the flight time so that changes in the vehicle potential which resulted in charging of the dielectric could be observed in the charge displacement on the dielectric surface. The known capacitance of the dielectric surface and the measured charge on the dielectric yield a measure of the vehicle potential if the sheath size at the surface of the dielectric is small and the potential of the surface is the same as the ambient plasma.

The DLP flat plate (4 cm x 6 cm) was mounted flush with the surface of the daughter and swept linearly from -10 V to +10 V with a sawtooth wave shape at a frequency of 4 Hz. The gain of the probe current amplifier was switched on alternate sweeps between high and low gain.

The DFP plate was identical to the DLP plate and located on the opposite side of the daughter. The gain of the DFP was switched at the same time as the DLP gain.

The MLP was similar to the DLP except that after separation of the daughter the sweep was set for -100 V to +100 V. The sweep frequency was 8 Hz.

The MFP operation was similar to the DFP. Both the MLP and MFP were alternately switched in gain.

FLIGHT HISTORY

The first Tethered Payload Experiment was launched from the Kagoshima Space Center, Japan, on 16 January 1980 at twelve noon local time. The two stage K-9M-69 vehicle attained an apogee of 328 km at 291 seconds into the flight. Data from an ionosonde and from the onboard impedance probe showed that the electron density was very high with a maximum of 3×10^6 cm^{-3} at 328 km.

The electron beam was emitted for several seconds prior to separation. A short developed in the payload at the time of separation which resulted in the loss of both the electron beam and the high voltage power supply in the tether monitor instrument. The other instruments were not affected and continued to work well throughout the flight. Late in the flight the short was removed from the power supply and one voltage sweep of the tether was obtained.

The vehicle was spin stabilized at 2 rps and was not despun before separation. The daughter was separated at 106 seconds into the flight with an initial separation velocity of 0.60 m/s and left smoothly with very little coning motion. The optical measurement of black marks on the tether indicated a minimum separation of 38 m was obtained but saturation of the optical detector by sunlight may have occured after that time so that the length may have been somewhat greater than as measured. The tether system was stable throughout the flight and a positive confirmation of tether integrity was obtained at 73 km on the downleg during a sweep of the tether power supply which produced a voltage sweep of the daughter payload.

A second TPE flight on sounding rocket S-520-2 was also made from the Kagoshima Space Center. The second flight on 23 January 1981 included a forward looking CCD camera of low resolution to view the separation of the daughter and to aid in determining the stability of the tether. The deployment was successful and images showing the daughter moving away from the mother with the tether attached were obtained. Data on the vehicle potential were not obtained on this flight because of a power supply problem and the data reported here are from the K-9M-69 flight.

VOLTAGE MODE MEASUREMENTS

After separation vehicle potentials on both the mother and daughter were found to change as a result of saturation in the electron current collection for both the DLP and the MLP. Maximum negative potentials during the flight were -5 V for the daughter and -9 V for the mother. Data taken at 124 seconds into the flight are shown in Figure 2. The tether is in the voltage mode with a 100 megohm resistor between the end of the tether and the mother chassis ground. Thus, the potentials of the two sections of the payload are completely independent (assuming they are sufficiently far apart so that there is no coupling through the plasma).

The tether voltage as shown is positive when the daughter is relatively positive with respect to the mother. Since neither of the payloads are driven positive, a positive tether voltage indicates an equal but negative mother vehicle potential. A negative tether voltage corresponds to a negative daughter potential. Several of the telemetry channels were switched in gain. Full scale values for the data shown in Figures 2 and 3 are given in Table 1.

A comparison of the peaks in the TMV data with the Langmuir probes shows that the Langmuir probe electron current has saturated and the continuing voltage sweep applied to the probe is producing a negative offset in the vehicle potential. The different sweep rates in the DLP and MLP result in some overlap in the data. The mother potential is equal to the negative of the tether voltage when the DLP is in the ion collection portion of the sweep. At this time the daughter potential is very near the plasma potential (the floating potential of both mother and daughter is assumed to be the same and is not included in these measurements since it is small compared to the potentials of primary interest).

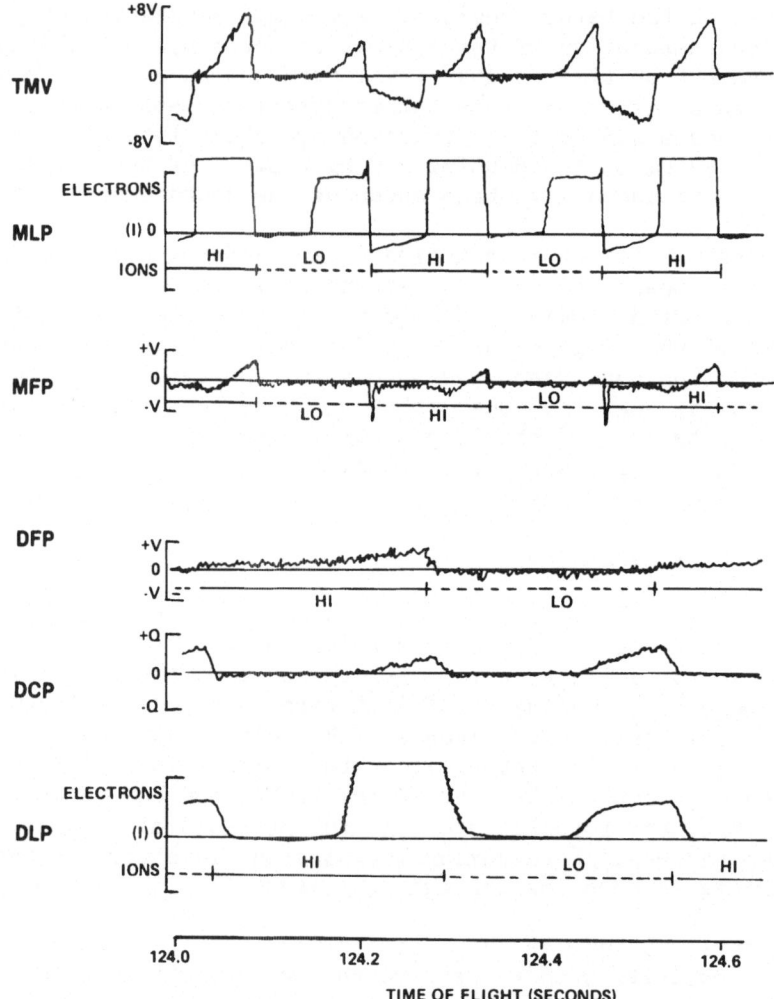

Fig. 2. Voltage Mode Data. Tether voltage (TMV) is the potential
difference between the daughter and mother payloads. The
input impedance for the tether voltage monitor is
100 megohms.

Peaks in both the DFP and MFP are about one-half the potentials
measured through the tether. This difference is not surprising since
the probes are flush with the surface of the payloads. The charge
probe data follow the daughter potential very well at this time. The
vehicle potential cannot be determined from the Langmuir probe data
since at the time the probe sweep is near zero the vehicle potential
is also near zero. The ratios for the daughter and mother surface to
probe areas were 200:1 and 400:1, respectively. The ratio for the

mother may be larger than 400:1 since the motor was not included in
the calculation because the surface was painted.

Some modulation of the vehicle potentials is associated with the
roll rate. Variations in the peak values of the TMV data are
produced by a combination of overlap in the DLP and MLP sweeps and by
the roll rate modulation. In Figure 2 the modulation is clearly seen
in the DCP data. This effect is more difficult to observe in the
other channels because they are gain switched at a rate comparable to
the roll rate.

CURRENT MODE MEASUREMENTS

The tether was switched during the flight into different modes
every 8 seconds. In the current mode the mother and daughter are at
the same potential since the current in the tether and the small load
resistor of 10 ohms result in a voltage drop of less than 1 mV. Data
at 134 seconds are shown in Figure 3 when the tether was in the
current mode. The TMI current is positive when the mother is
negative. Ions are collected at the daughter and photoelectrons
emitted. The DFP and DCP are evidence that the daughter potential is
driven negative along with the mother although the DLP is in the ion
collection mode. The DLP does not produce an offset in the vehicle
potential because the collecting area is enlarged to include the
mother skin and the motor attached through the tether. The DLP also
sweeps only to +10 V as compared to the +100 V on the MLP. The
decrease in the DLP current which is observed in the low gain mode is
a result of the MLP sweep driving the entire system negative so that
although the DLP probe is sweeping positive relative to the daughter
skin the DLP plate potential is actually decreasing until the MLP
sweep retrace occurs and the system potential recovers.

Table 1. Full scale values for data in Figures 1 and 2.

Channel	Gain	Full scale
DCP	*	3 nC (3 V)
DFP	HIGH	2 V
	LOW	40 V
DLP	HIGH	20 μA
	LOW	100 μA
MFP	HIGH	6 V
	LOW	60 V
MLP	HIGH	20 μA
	LOW	100 μA
TMV	*	8 V
TMI	*	100 μA

* Not switched in gain

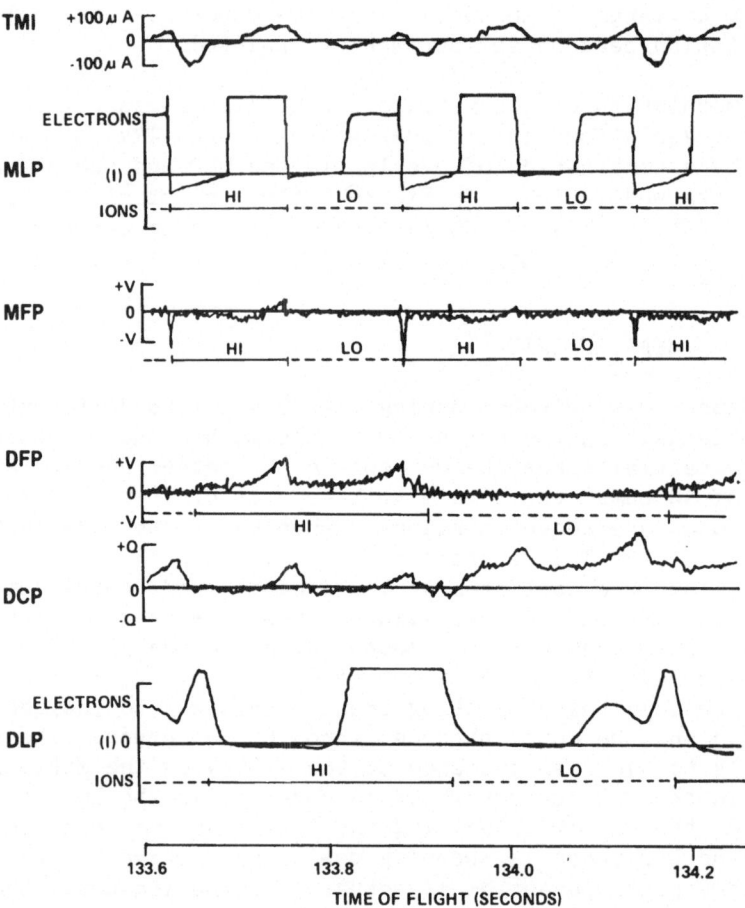

Fig. 3. Current Mode Data. A 10 ohm resistor is connected between the daughter and mother payloads at the mother end of the tether.

DISCUSSION

The vehicle charging produced by the Langmuir probes is a direct measurement of the probe voltages which can be used for a given collection area to probe area ratio. For the daughter, with a ratio of 200:1, probe voltages of more than 5 volts always produced a change in the daughter potential relative to the ambient plasma. For the mother, where the area ratio was 400:1 or greater, charging was not significant until more than 10 volts was applied to the probe. Saturation of the probe current occurs gradually and the interpretation of the current–voltage characteristic is difficult unless the changing vehicle potential is known by some measurement other than from the Langmuir probe data.

The accuracy of the potential measurements depends upon the decoupling of the mother and daughter vehicles and upon knowledge of the potential of the system relative to the plasma. Measurements of the plasma temperature were made during the flight so that the system floating potential can be estimated. However, for the measurements of relative potential this offset from the plasma potential is not important. The mother-daughter potential difference measured during times when both mother and daughter Langmuir probes were in the ion collection mode (and thus not affecting the vehicle potentials) was within the noise level of the telemetry. A small offset from the VxB potential was present but very nearly equal to the noise level of 100 mV. Frequency response of the tether voltage monitor was limited only by the telemetry response as can be seen from the step change which occurs during the Langmuir probe retrace. The time for the vehicle potential to recover when the probe voltage switches from +V to -V is less than the time response of the TMV channel (3 ms).

CONCLUSION

The Tethered Payload Experiment has shown that a fast, accurate measurement of vehicle potential can be made using an insulated wire tether connected between the mother and daughter sections of a rocket payload. Langmuir probes with area ratios of less than 400:1 may produce changes in the vehicle potential if probe voltages of more than 5 volts are applied in the electron collection mode.

ACKNOWLEDGEMENTS

The Tethered Payload Experiments were cooperative efforts by the Institute of Space and Aeronautical Science (ISAS) of the University of Tokyo and Utah State University (USU). With the exception of the charge probes and the tether current-voltage monitor all other systems, the integration, testing and the sounding rocket flights were supported by ISAS including the design and development of the tether deployment system. We are indebted to the personnel of ISAS and the Kagoshima Space Center for their many efforts in these experiments. We thank especially Professor T. Obayashi for his encouragement and advice during the early stages of planning the TPE flights. Support at USU was provided by NASA grant NSG-6027.

REFERENCES

Hess, W.N., M.G. Trichel, T.N. Davis, W.C. Beggs, G.E. Kraft, E. Stassinpoulos, and E.J.R. Maier, Artificial aurora experiment: experiment and principal results, J. Geophys. Res., 76, 6067-6081, 1 Sept. 1971.

Linson, Lewis M., Current-voltage characteristics of an electron-emitting satellite in the ionosphere, J. Geophys. Res., 74, 2368-2375, 1 May 1969.

GENERAL DISCUSSION ON

NEUTRALIZATION OF A CHARGED BODY IN SPACE

Chairman: L Linson

Data from a rocket flight showing apparent ion depletions, indicate that the charging effect may disturb the ambient plasma at a distance of more than one kilometer from the vehicle transverse to the geomagnetic field. Others have observed large electric fields 50-100 m from the accelerator after the end of beam injection, thus indicating that the ionosphere is severely perturbed. The overall balance of charge (emitted electrons versus ions) in the ionosphere as a whole was discussed with the recognition that there is still much that one does not understand.

The general question of outgassing was discussed, in particular for the "mother"-"daughter" payload configuration. In the absence of the beam or a hot rocket motor, the outgassing is not believed to be important at altitudes around 200 km. In the presence of an electron beam, however, the possibility of enhanced outgassing due to the bombardment of the rocket surface by the return current electrons was suggested.

The following questions were singled out as being of vital importance for a better understanding of the charging and neutralization processes and should therefore be addressed in the planning of future work:

1) What is the relative importance of the beam plasma discharge (BPD), the $(EXB)/B^2$ discharge and other discharge mechanisms?

2) How can the initial, transient charging be explained?

3) What is the mechanism for the oscillatory behaviour (flickering) of the charging?

4) How can reliable measurements be made of the
floating potential of the spacecraft?

5) How is charge neutralization achieved at altitudes above 200 km where the neutral density
is so low?

6) What is the size of the sheath around electron
emitting rockets and how is it determined?

7) What is the importance of outgassing?

Chapter 6:   FUTURE PLANS

SPACE EXPERIMENTS WITH PARTICLE ACCELERATORS (SEPAC)

T. Obayashi, N. Kawashima, K. Kuriki,
M. Nagatomo, K. Ninomiya, S. Sasaki,
A. Ushirokawa, I. Kudo[1], M. Ejiri[2],
W.T. Roberts[3], R. Chappell[3], J. Burch[4],
and P. Banks[5]

    The Institute of Space and Astronautical
    Science, 4-6-1 Komaba, Meguro, Tokyo, JAPAN
[1] Electrotechnical Laboratory
    1-4 Umezono, Sakuramura, Ibaraki, JAPAN
[2] National Institute of Polar Research
    1-9-10 Kaga, Itabashi, Tokyo, JAPAN
[3] NASA, MSFC, Alabama, USA
[4] SWRI, San Antonio, Texas, USA
[5] Utah State University, Logan, Utah, USA

ABSTRACT

Space Experiments with Particle Accelerators (SEPAC)
is a project of performing an active experiment on the
Spacelab 1 in 1983 using electron and plasma beams
injected into space plasma. General description,
present status and future program are described.

1. Scientific Objectives

Scientific objectives of SEPAC are;

i) to study the vehicle charging in space and charging
neutralization by using plasma and neutral gas
plume,

ii)   to study the beam-plasma interaction in space,
      in particular,
        wave excitation in VLF to HF frequency range
      in the interaction of the electron beam with
      space plasma and plasma heating resulting from
      the non-linear beam-plasma interaction,

iii)   to study the beam atmospheric interaction exciting
       artificial aurora and airglow and

 iv)   to trace the magnetic field configuration of the
       magnetosphere and detect the field aligned electric
       field.

2.   SEPAC Hardwares

     SEPAC consists of the following subsystems :

  i) Electron Beam Accelerator (EBA)
 ii) Magnetoplasma Arcjet (MPD)
iii) Neutral Gas Plume Generator (NGP)
 iv) Power Supply for EBA and MPD (PWR)
  v) Diagnostic Package (DG)
 vi) Monitor TV (MTV)
vii) Control and Data Management System (CD)

     The block diagram is shown in Fig. 1.

     Each subsystem has the following characteristics :

  i) Electron Beam Accelerator (EBA)

      It consists of Electron Gun (EBA), Gun Power Supply
      (GPS) and High Voltage Converter (HVC). EBA and
      GPS is integrated in one unit.

      Beam energy : 0 ~ 7.5 keV Variable
      Beam current : 0 ~ 1.6 A
      Maximum Output power : 12.5 kW
      Average Output power : ~ 1 kW
      Perveance : 2.5 x 10^{-6} (A/V$^{2 \cdot 3}$)
      Pulse width : 1 msec ~ 1 sec (High Power)
                    1 msec ~ CW   (Low Power)
      Deflection Capability : ± 30°
      Focusing : Magnetic
      Beam Diameter : 20 mm at the exit
      Cathode : impregnated cathode

ii)   Magnetoplasma Arcjet (MPD)

It consists of MPD-AJ and Capacitor Bank (CAP) integrated in one unit together with NGP below.

Energy stored : 2 kJ
Discharge Pulse Width : 1 msec
No. of ion-electron
   pairs produced per shot : 10^{19} pairs
Plasma density
   1 m from MPD : $10^{12}/cm^3$
   15 m from MPD : $10^{9}/cm^3$
Plasma flow velocity : 2 x 10^4/sec
Beam spread : $\pm$ 20°
Electron Temperature : 3 ~ 5 eV
Ion Temparature : ~ Te
Gas : Argon

iii) Neutral Gas Plume Generator (NGP)

It is integrated in MPD subsystem.

Gas : Nitrogen
Pulse width : 0.1 sec
Ejected Velocity :   ~ 400 m/sec
Ejected Number of
  Molecules/shot   : 10^{23} molecules/shot

iv)   Power Supply for EBA and MPD (PWR)

It consists of Battery Package (BAT) for EBA and Charger (CHG) for MPD and BAT.

Charger for MPD : 500 V 1 kW
Battery Package for EBA : 450 V 4 AH
                 Endurable to 15 C 1 sec discharge

v)   Diagnostic Package (DG)

It consists of several instruments as described below :

v-1) Photometer (PHO)

Zenith : controlled by gimbal system within $\pm$ 60°

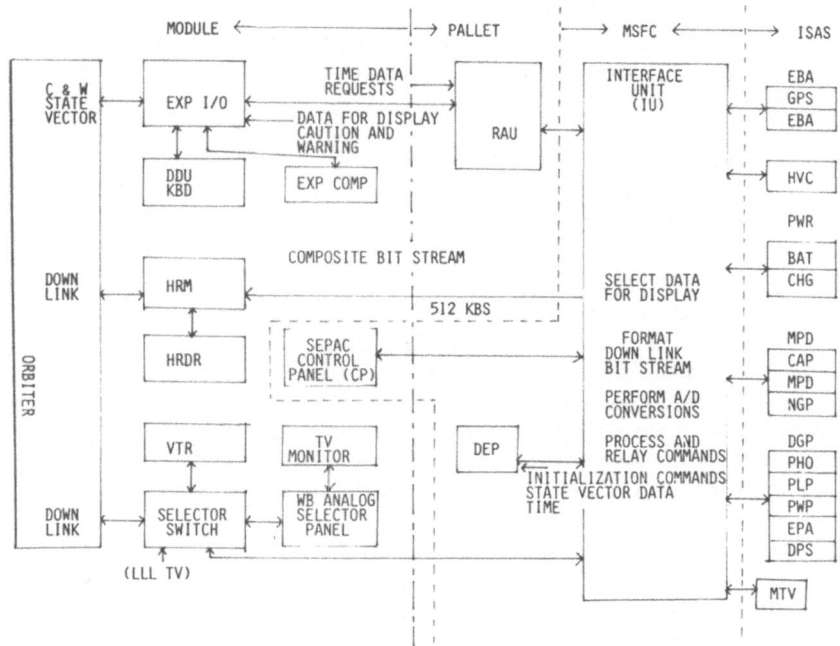

Fig. 1(a)   System Block Diagram

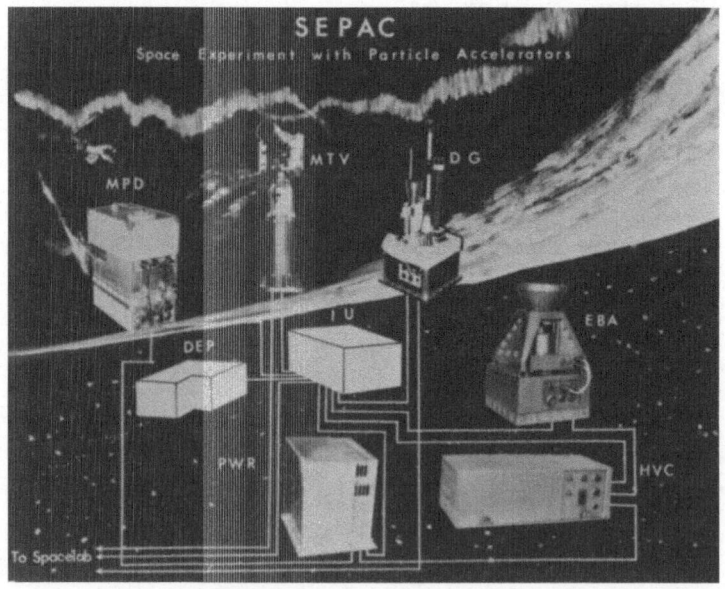

Fig. 1 (b)   Schematic Layout of SEPAC with
              Flight Model Hardwaves

                Field of view : determined by iris from
                        9° to darkness
                Filter : one is selected from 3914Å, 5577Å
                        and 6300Å filters
                Measuring range : 10^{-13} ~ 5 x 10^{-10} W
                                (power of incident light)
                Sampling rate : 1 kHz

v-2) Plasma probe : Langmuir Probe (PLP-LP)

        Measuring range :
        * electron density 10^4 ~ 10^8 els/cm^3
        * electron temperature 600 ~ 5000°K
        * current   0.01 ~ 10 A

V-3) Particle Analyser : Vacuum Gauge (EPA-V)

        Measuring range : 5 x 10^{-4} ~ 5 x 10^{-8} Torr
        Sampling rate   : 1 kHz

v-4) Plasma Probe : Floating Probe (FLP-FP)

        Measuring range : -8000 ~ 8000 V (potential
                        of each probes)
        Sampling rate : 1 kHz

v-5) Plasma Wave Package (PWP)

        Measuring range: 750 Hz ~ 10 kHz or 100 kHz
                        ~10 MHz (step mode)
                        400 Hz ~ 4 MHz or 4 MHz ~
                        7.5 MHz (wide band)

v-6) Particle Analyser : Electron (EPA-E)

        Measuring range :
        * energy   100 eV ~ 15 keV
        * flux rate 10^{-10} ~ 2 x 10^{-4} A/cm^2 (current
                        collector) and 2 x 10^6 ~ 10^{10}
                        els/cm^2.sec. str. keV
                        (channeltron)
        Resolution
        * energy $\Delta E/E$ ≈ 0.18
        * solid angle 4 x 10°
        Mode
        * fixed mode
        * scan mode   320 msec/1 scan

vi)   Monitor TV

This is a high sensitivity TV camera to monitor
the near field and far field views of the electron
beam profile in space.

Field of view   : 28.7° x 21.7°
Sensitivity   : 0.01 ~ 10^5 lux
Spectral response   : 3900 ~ 7000 Å
Frame rate   : 30 Hz

vii) Control and Data Management (CD)

It consists of Dedicated Experiment processor
(DEP), Interface Unit (IU), and Control Panel (CP).
Its functions are;

Control of SEPAC Hardwares by sending a
sequence of commands,

Control of SEPAC instrument and scientific
data,

PCM   : 512 kbps
Analog : 4.2 MHz
TV Video signal

Communication with the Spacelab Experiment
Computer (EC),

Display of SEPAC status and data on DDU
(Data Display Unit) of EC,

Control of IU and DEP through DDU key
board of EC, and

Management of data coming from Space
Shuttle through EC.

All SEPAC instruments other than CP are located
on the pallet. SEPAC is controlled by the
payload crew through the DDU of EC and CP of
SEPAC. The layout on the SL-1 is shown in Fig. 2.
CD subsystem is developed in NASA MSFC and other
subsystems are developed in Japan. Total weight
is about 400 kg, and the average electric power
consumption is about 1 kW.

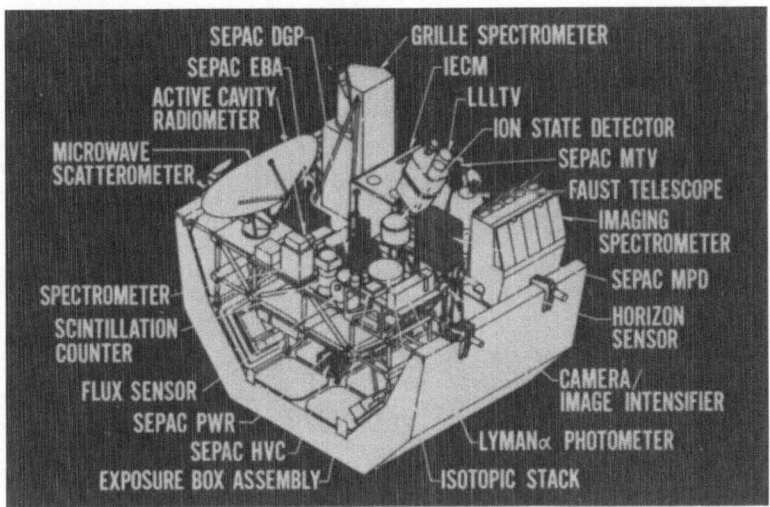

Fig. 2   SEPAC on SL-1 pallet

SL-1 experiment is now scheduled in June, 1983.
All harwares are completed fabrication and a final
system test and large vacuum chamber test is scheduled
before integration.

3. SEPAC Operation

   SEPAC Operation consists of various FO's. A list
of FO's is shown in Table 1. They are classified as

  i) Test modes
 ii) Low power electron beam experiment modes
iii) High power electron beam experiment modes
 iv) Plasma beam experiment mode
  v) Battery charging
 vi) Passive modes

   A brief description of each experiment FO is given
below :

FO-5 Low Power Electron Beam Experiment 1

   This experiment is to study the shuttle charging
effect and beam-plasma interaction effects such as wave
excitation due to beam-plasma instabilities in a low
power beam mode.

   FO-5A is a CW mode and FO-5B is a pulse mode with
a pulse width of 100 msec.

Table 1.   SEPAC Functional Objectives

F.O.-1	T-0	SEPAC System Checkout
2	T-1	EBA Firing Test (Level I)
3	T-2	MPD Firing Test
4	T-3	EBA Firing Test (Level II)
5	A-1	Electron Beam Experiment 1
		(Low Power (1 - 5 keV),CW and Pulse)
6	A-2	Electron Beam Experiment 2
		(Low Power (1 - 5 keV),EBA/NGP)
7	A-3	Electron Beam Experiment 3
		(Low Power (1 - 5 keV,EBA/MPD)
8	A-4	Plasma Beam Production
9	A-5	Artificial Aurora Excitation
		(High Power (3-7.5 keV) 0.5 sec ON 1.5 sec
		OFF 3 pulses in series Every 15 sec,
		EBA/MPD/NGP)
10	A-6	Equatorial Aerochemistry
11	A-7	Electron Echo Experiment
		(High Power (7.5keV, 1.6 A) 0.5 sec ON/
		Every 15 sec, EBA/MPD)
12	A-8	E//B Experiment
		(High Power (1 - 7 keV, 0.08 - 1.0 A)
		100 msec/Every 1 sec)
13	P-1	Passive Experiment
14	P-2	Passive Experiment (IESO20 Support)
15	CFR	SEPAC System Deactivation
16	CHG	Battery Charging

FO-6   Low Power Electron Beam Experiment 2

A simultaneous operation of EBA and NGP is performed in a low beam power mode to study the charging neutralization effect by the neutral gas plume. The beam propagation profile is monitored by MTV and by a payload specialist taking a still photograph.

FO-7   Low Power Electron Beam Experiment 3

A simultaneous operation of EBA and MPD is performed in a low beam power mode to study the charging neutralization effect by the MPD-Arcjet plasma. The beam profile is monitored by MTV and by a payload specialist taking a still photograph.

FO-8   Plasma Beam Experiment

Physics of plasma beam produced by the MPD-Arcjet in space is studied. Airglow excitation and plasma beam propagation and spreading in a magnetized space plasma are measured by a photometer (PHO). Joint experiment with LLLTV (AEPI (INS 003)) is scheduled.

FO-9   Aritificial Aurora Excitation

This Experiment is to excite artificial aurora with a high power electron beam injection. MPD-AJ or NGP is used to neutralize the electric charge-up due to the electron beam emission. Joint experiment with AEPI is scheduled and ground observation may be coordinated.

FO-10 Equatorial Chemistry

This Experiment is a support experiment for AEPI (INS 003). The electron beam is injected in a high power mode along almost horizontal geomagnetic field lines and the artificial airglow is monitored by the LLLTV.

FO-11 Electron Echo Experiment (Fig. 3)

This experiment is to verify the electron beam reflection by the magnetic mirror field effect. The experiment is performed at South Atlantic Anomaly where the geomagnetic field is very low. The electron beam injected at the shuttle altitude

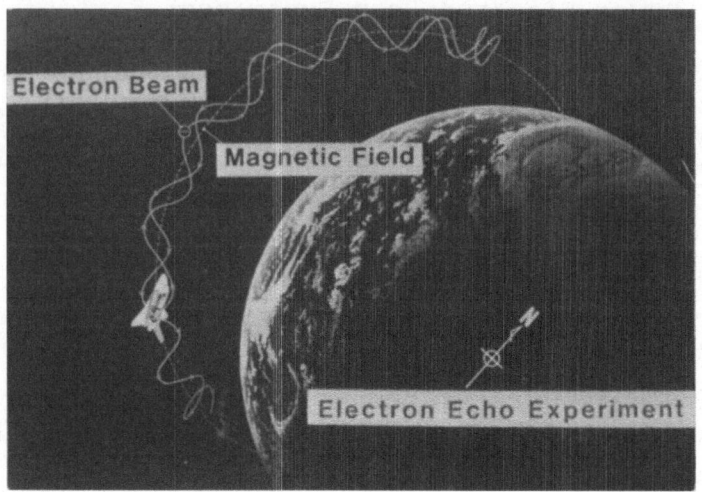

Fig. 3   Schematic illustration of Electron
Echo Expriment

is reflected at a rather high altitude in the
opposite hemisphere by the mirror, not by the
reflection due to the atmospheric scattering. The
artificial airglow excited below the shuttle by
the reflected electron beam is observed by the LLLTV.

FO-12 E//B Experiment

This experiment searches for the existence of
E//B field in the auroral zone. The electron beam
is injected upward. Reflected beam from the E//B
region precipitates into the atmosphere at the
altitude of 100 km, thereby producing artificial
auroras. High power electron beam and MPD-Arcjet
is used.

Typical operation sequence of each FO is shown
in Fig. 4(a). First, the communication with the
Experiment Computer (EC) of SL-1 is established
(ECAS software load). It takes a certain time to
configure SEPAC instruments such as software load,
switch on, parameter change, et al. The cathode
heater warm-up time usually requires 10 minutes and
during this period, the attitude aquirement is
done. During the last two minutes of the heater

Test Mode
Low Power Electron Beam Mode
High Power Electron Beam Mode   with Plasma (MPD)
                                     with Gas Plume (NGP)
Plasma Beam Mode
Passive Mode

Typical FO

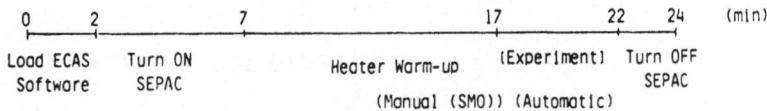

Load ECAS   Turn ON   Heater Warm-up   [Experiment] Turn OFF
Software   SEPAC   SEPAC
                                           (Manual (SMO)) (Automatic)

Fig. 4(a)   Typical SEPAC Operation Sequence

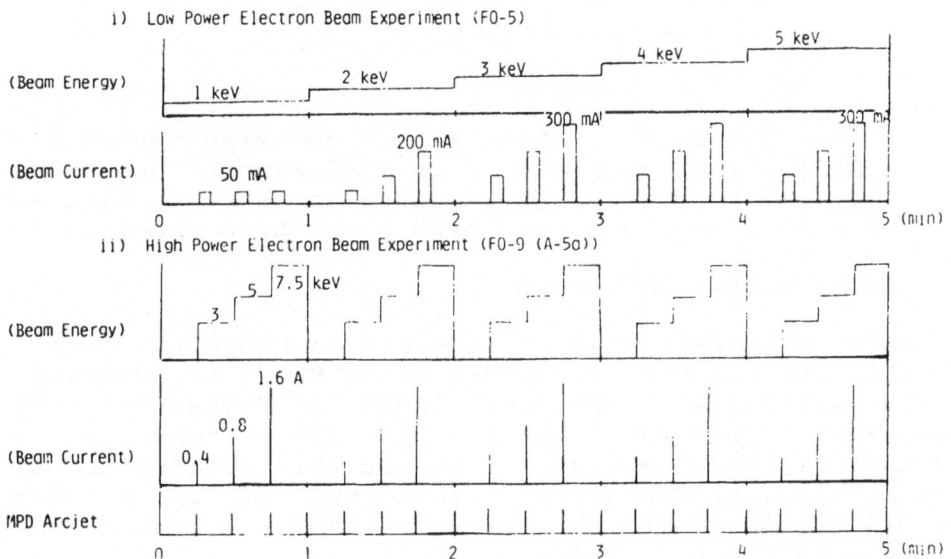

Fig. 4(b)   Typical SEPAC Experiment Sequence

Table 2.   SEPAC follow-on Mission

Level I   The same unit as SL-1 is used
          but its location shall be at higher
          place on the pallet and isolated
          from other payload.

Level II   Upgrading of Diagnostic
           Instruments

    i)   MTV
         Focusing/Zooming
         Capability
    ii) New Diagnostic Package or RMS
         Wave Detectors
         Plasma Probes
         Energy Analyser

Level III
    i)   Accelerators
         Electron Gun   10 keV 2.5 A
         MPD-AJ   5 kJ

    ii) Tethered Satellite
         Diagnostic Package

    iii) Colored High Sensitivity
         TV (SIT or CCD)

warm-up time, the SEPAC can be operated manually
(SMO). The experiment time is 5 minutes and
typical beam emission sequence is shown in Fig. 4(b),
for both low power and high power beam expeirments.

4.   SEPAC Follow-on Mission

SEPAC follow-on mission is based on PROPOSAL FOR A
CONTINUING SCIENTIFIC PROGRAM USING SPACE EXPERIMENTS
WITH PARTICLE ACCELERATORS: SEPAC FOR FUTURE SPACELAB
MISSIONS BY TATSUZO OBAYASHI Nov. 1978. Though the
actual scale and evolution are not well defined because
it is strongly dependent on the result of SL-1 experiment, the following is a general outline of the SEPAC
follow-on concept (Table 2).

It is generally agreed that SEPAC shall have a chance
of reflight without any major modification of SL-1
configuration. (Level I).

The location of Diagnostic Package shall be improved to avoid the effect of other instruments. In Level II, the grade-up of diagnostic instruments and MTV will be done. Diagnostic instruments may be either on RMS or on a tethered satellite. The combination of the tethered satellite and SEPAC will become a very interesting experiment.

The upgrading of the electron beam and plasma beam will be done in level III, to at least 25 kW level (10 kV, 215 A).

The SEPAC follow-on mission will be done in close coordination with other spacelab payloads and SEPAC will work as a core instrument for future active experiment in space.

THE NORWEGIAN PROGRAM USING PARTICLE ACCELERATORS IN
SPACE

Bernt N. Maehlum

Norwegian Defence Research Establishment
P O Box 25
N-2007 Kjeller Norway

The Norwegian accelerator program was originally
based on the auroral physics tradition. We hoped to
conduct controlled simulation experiments to understand
the auroral processes better. Now the program is based
on a combination of auroral and general plasmaphysics
interest.

The program is centered on five problems.

1) How is a beam fast electrons scattered in the
    ionosphere? We started theoretical studies
    of this problem in 1962, and several theoretical
    studies have been conducted after that time,
    but no good experimental verification of the
    theoretical results has been conducted.

2) How is the energy transferred from the beam
    electrons to the ionospheric plasma by linear
    and non-linear processes?

3) How does the atmosphere respond to the dis-
    sipated energy for various beam currents and
    at various altitudes?

4) How fast does the modified plasma regions in
    the ionosphere recover to normal? Similar
    studies cannot be performed by natural
    phenomena, since abrupt cut-offs in the
    energy input to the ionospheric plasma seldom
    occurs in nature.

5) Also as part of the project are studies of the vehicle neutralization in the ionosphere.

In order to conduct these studies we are using a "mother"-"daughter" concept. There are a multitude of advantages using this configuration, some of which are listed below:

1) The EMI generated by the accelerators and the unwanted vehicle charging for some of the instruments is negligible.

2) Since the distance between the two payloads increases slowly with time both in a direction normal to and along the geomagnetical field lines we are in a position to study the two-dimensional cross-section of the ionospheric disturbance and of the scattered electron beam.

3) The "mother" enters the disturbed plasma region a time delay of Δt seconds after the "daughter". This delay increases slowly with flight time, and we can study the build-up and the decay of the disturbance.

4) The accelerator launches pulses alternatively up and down, and the "mother" is always located below the "daughter" payload. Hence, we can study the effects in front and at the rear side of the accelerator.

5) The "mother" serves at a remote platform from which the optical and electric field environment of the "daughter" can be observed. Also, angle-of-arrival measurements of the electron beam generated waves can be recorded.

6) By measuring the charging of the "mother" and "daughter", and the electric field near the "mother" we obtain a detailed picture of the potential distribution.

The Norwegian program in this field consists of 4 rockets and a satellite experiment. Christian Beghin has reviewed our satellite involvement, and I will concentrate on the rocket experiments here.

The POLAR 6-rocket which is due for launch in October this year, carries a pulsed electron accelerator "daughter" payload. The electron energy is 10keV, and the current is stepped between 0 and 100mA.

The "mother" carries a series of diagnostic instruments to monitor the effects of the beam at various altitudes and beam currents.

An array of spectrometers monitor the energy and pitch angle distribution of the scattered beam electrons and the energized ionospheric electrons in the beam and in the "halo".

Several plasma probes measures the amplitude of the disturbance as well as small-scale irregularities inside the disturbed region by Norwegian groups and by Martin Friedrich, Technical University of Graz. In addition, the waves set up by the electron beam is observed from 0 to a few Megahertz.

Photometers are mounted on the "daughter" to monitor the H_β and other emissions generated by the beam.

Finally, the vehicle charging and neutralization is studied by probes on the "daughter", by potential difference measurements between the two payloads through a connecting wire and by DC-electric field measurements on the "mother". The last measurement is conducted by Nelson C Maynard, Goddard Space Flight Center.

During the POLAR 6 launch the EISCAT installation will be in operation, although it is somewhat uncertain whether the echo from the plasma disturbance near the rocket will be concealed by the echo from the rocket body.

Let me finally say a few words about our other proposed rockets, the MAIMIK, which in 1984 and 85 will carry an accelerator which can eject beam currents up to a few Amperes. By this we hope to be able to trace the disturbance by EISCAT to monitor the ionospheric decay time. In addition, several new diagnostics instruments will be included with on-board data processors to improve the time resolution of the observations.

THE ARTIFICIALLY INJECTED CHARGED PARTICLES
AS A TOOL FOR THE MEASUREMENT OF THE ELECTRIC FIELD
IN THE MAGNETOSPHERE

Michel Pirre

CNRS/CNET, Centre de Recherche en Physique
de l'Environnement
45045 - ORLEANS Cédex (France)

INTRODUCTION

The knowledge of the electric field in the magnetosphere is essential to understand processes such as the magnetospheric convection or the acceleration of charged particles. But, contrary to the magnetic field which plays also an important role in this understanding, it is very difficult to measure it and especially its parallel component to the magnetic field. Up to now most of the measurements use double probe techniques (FAHLESON et al., 1970 ; MOZER et al., 1979). The difficulties of this measurement are due to the low values of the electric field which does not permit to neglect the perturbations of the electric potential of the natural plasma due to the presence of a spacecraft, mainly if these perturbations are not symmetric. These dissymetries are due for example to the wake effect, to the non equipotentiality of the spacecrafts and to the photoelectrons clouds surrounding these space crafts at high altitude (MOZER et al., 1979). These difficulties sometimes lead to a skepticism on the results of the electric field measurements and often on the existence of the parallel component to the magnetic field (MOZER, 1980). Another independent technique is therefore necessary to compare both set of results leading to a better confidence on these results.

Artificially injected electrons used as tracers of the electric field can be chosen to this end. Such a technique has already been used on the GEOS satellites to deduce the perpendicular component of the electric field (MELZNER et al., 1978) from the measurement of the perpendicular drift of the electrons after one gyration around the magnetic field. The comparison of the results with those obtained by a double probe technique shows the interest of simultaneous measurements by both techniques (MELZNER et al., 1979). In

677

this paper we will show in a first part how it is possible to extend
this technique to the measurement of the parallel component to the
magnetic field.

The detailed description of both the energy spectrum and the
pitch angle distribution of downward flowing and upward flowing
natural charged particles has also proved to be useful to the measurement of the parallel component of the electric field or more exactly of the parallel potential differences along the magnetic field.
To this end, nevertheless it is clear that the use of artificially
injected charged particles is also of a very great interest to loca-
lize the acceleration regions and to measure the electric potential
at each point of the magnetic field line. Barium ions jets followed
by optical techniques from the ground (HAERENDEL et al., 1976) have
indeed allowed the discovery of an acceleration region at 7600 km altitude of about 7 KV. Other techniques (WIHLHELM,1977 ; PIRRE, 1981)
can lead to a detailed description of the electric potential on the
whole magnetic field line, in measuring the echo time delay of artificially injected charged particles reflected by the potential drop,
as a function of the particles energy at the injection. Such techniques are very difficult to bring about a successful issue mainly because the rendezvous between the returning fluxes and the detectors
are difficult to obtain (WILHELM et al., 1980). In this paper we will
show in a second part what is the best technique to measure downward
parallel electric field by use of artificially injected ions beams.

MEASUREMENT OF THE ELECTRIC FIELD IN THE VICINITY OF A SPACECRAFT

Principle of the Measurement

PIRRE et al. (1979) have given the principle of the measurement
of both perpendicular and parallel components of the electric field
by a technique using artificially injected electrons from a spacecraft and detected aboard the same spacecraft, after they have performed a large number of gyrations around the magnetic field. Here
below is a brief recall of this principle.

Figure 1 shows the movement in the plasma frame of reference of
both the spacecraft and the injected particles, projected in the
plane P perpendicular to the magnetic field B, at different times
between injection and detection. The upper left figure shows the
injection, as the lower right one shows the detection. The rendezvous exists if the particles and the spacecraft reach the point R
at the same time t given by :

$$t = \frac{2\, r_b (\alpha + n\pi)}{V_p} = \frac{2\, r_b \sin \alpha}{V_{sp}} = \frac{2\, m\, V_{//}}{q\, E_{//}} \qquad (1)$$

where r_b is the gyroradius of the particles, V_p and V_{sp} are the
projected velocities in the plane P of, respectively the electron
and the spacecraft, α is the angle between these two velocities,
m and q are the mass and charge of the electrons, $V_{//}$ is their

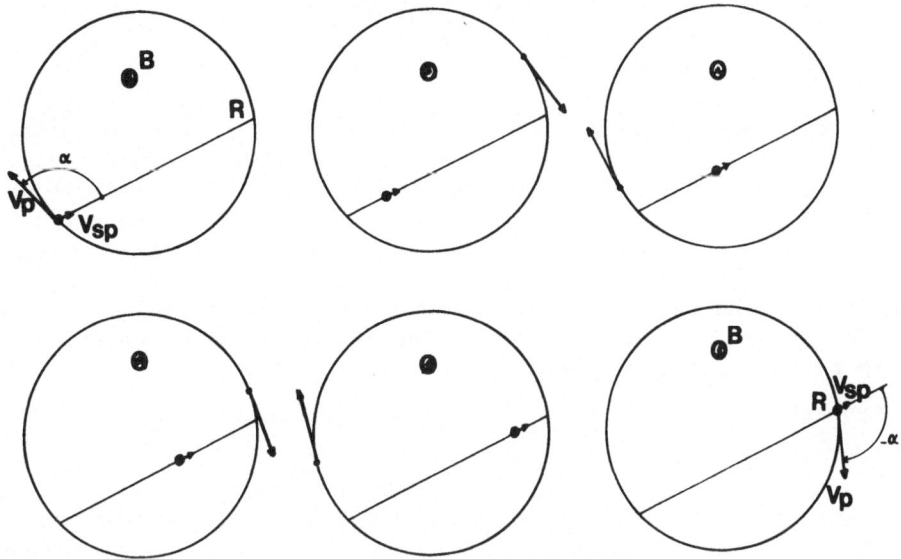

Fig. 1. Rendezvous between the injected electrons and the satellite

parallel velocity to the magnetic field and $E_{//}$ is the parallel component of the electric field. Therefore, for a given velocity V_0
of the particles at the injection, the measurement of the parallel
electric field is achieved by the measurement of the pitch angle γ
$(V_{//} = V_0 \cos\gamma)$ and of the time t when a rendezvous occurs, i.e.,
if α satisfies the relation ① . To this end the direction of injection must be changed until this rendezvous is obtained. It has been
shown (PIRRE et al., 1979) that it exists a large number of solutions to the rendezvous problem ($2 \times n_{MAX}$, where n_{MAX} is the maximum number of gyration). This fact allows to inject the particles in
two fixed opposite directions in the plane P, while simply changing
the pitch angle of injection, to be sure to have a rendezvous in most
cases, assuming nevertheless that the particles can be detected in
any azimuth around the magnetic field. The latter requirement upon
the detectors is severe, but the difficulty is decreased by the fact
that the pitch angle of detection as well as the injection one have
just to be varied of a few degrees (5° maximum).

Parallel Electric Field Minimum Theoretical Value Measurable

After previous measurements (MOZER et al., 1979) the parallel
component of the electric field is certainly at least one order of
magnitude lower than its perpendicular component in most cases. Our
proposed technique has therefore to be able to measure such low
fields.

After ① the maximum time of rendezvous which allows the measurement of the minimum value $E_{//MIN}$ of $E_{//}$ is

$$t_{MAX} = \frac{2r_b}{V_{sp}} = \frac{2mV_{//}}{q\, E_{//MIN}} \qquad\qquad ②$$

After this later relation, low values of $E_{//}$ can be measured if the parallel velocity $V_{//}$ of the injected electron is small, i.e. if γ is close to 90°. Let ϵ be $\epsilon = 90° - \gamma$. ② can be written :

$$E_{//MIN} = \epsilon\, E_\perp$$

where $E_\perp$ is the perpendicular component of the electric field in the spacecraft frame of reference. The accuracy on the measurement of the pitch angle could be 1°.

Therefore let us assume that $\epsilon = 1°$. This leads to :

$$E_{//MIN} = 0.02\, E_\perp$$

The corresponding time of rendezvous depends on the energy W of the electrons at the injection and on $E_\perp$, it is easy to show that if $W = 3$ Kev , and $E_\perp = 100$ mv/m, this time is $t_{MAX} = 3.6$ ms if $E_\perp = 500$ mv/m , t_{MAX} is 0.72 ms.

These values can be obtained only if the angle of injection α is 90°. For any α we find

$$E_{//} = \frac{\epsilon\, E_\perp}{\sin\alpha} \qquad \text{and} \qquad t = t_{MAX} \times \sin\alpha$$

Limitations Due To Spacecraft Environment

The low value of the minimum parallel electric field measurable seems to indicate that the proposed technique is very suitable to measure this field. But, obviously many limitations to this measurement are expected, mainly because , parallel electric fields are expected to exist in regions where natural background of precipitating electrons are high, cold plasma density is low and where a high level of turbulence exists.

As a consequence of natural background being high we must inject high enough beam intensities, but because cold plasma density is small, too high intensities would lead to spacecraft charging. This is in general not permissible for the good functioning of the other experiments aboard the spacecraft. So let us assume in one hand a natural background of $10^9/cm^2$s.ster.kev and the following characteristics of the detectors :
- geometrical factor : 0.5 cm^2 x 0.03 ster
- energy resolution : $\Delta W = W/10$
- integration time : $\Delta t = 100$ μs

This leads to a natural background of $N_B = 150$ counts. Artificial electron fluxes must therefore be larger than $\sqrt{N_B} = 12$ counts to be detected.

Table 1 : Count rate N, minimum parallel electric field measurable $E_{//min}$, and time t of rendezvous for two values of the angle α of injection in the plane perpendicular to B; α = 5° and α = 85°; also given are the distance $d_{//}$ between the electrons and the spacecraft along the magnetic field after one gyration and $d_{\perp}$ the maximum distance between the electrons and the magnetic field line crossing the spacecraft (see fig. 2) These values are computed for two altitudes : ∿ 1 Re (B = 6000Y) and ∿ 3 Re (1000Y) and three values of the perpendicular electric field.

W = 3 Kev ε = 1°		B = 6000 Y $E_{\perp}$ (mv/m)			B = 1000 Y $E_{\perp}$ (mv/m)		
		25	100	500	10	40	200
α= 85°	N	50	200	1020	3	13	68
	E //MIN (mv/m)	0.5	2	10	0.2	0.8	4
	t (ms)	14.5	3.6	0.72	36	9	1.8
α = 5°	N	550	2200	11220	37	152	746
	E //MIN (mv/m)	5.5	22	110	2.2	9	44
	t (ms)	1.3	0.32	0.06	3.2	0.8	0.15
	d // (m)	3.4	3.4	3.4	20	20	20
	d⊥ /m	61	61	61	366	366	366

In the other hand let us assume the cold plasma density to be $n = 1$ cm^{-3} , the temperature to be = 3 eV and the collecting surface of the spacecraft to be S = 10 m^2 then the collected current is about I = 1 µA. We will assume therefore that the injected intensity I_i cannot be superior to a few tenth of µA. The following calculations are made assuming I_i = 0.25 µA, and the opening angle of the beam = 0°.5.

After the calculations of PIRRE et al. (1979) it is possible to compute the count rate of the detectors, the characteristics of which are those defined above, to be

$$N = \frac{15314 \times E_\perp (V/m) \times B(gauss)}{\sin^2\alpha \, \cos\alpha \, W^{3/2}(Kev)}$$

This formula is not available if α is very close to 0° or 90°.

Table 1 gives the count rate for different conditions as well as the minimum parallel electric field measurable and the time of rendezvous, if W = 3 Kev and ε = 1°. We can notice that the computations have been made assuming more propitious characteristics of the injected beam and detectors and a better accuracy on the attitude than in the previous paper (PIRRE et al., 1979), which were maybe too pessimistic. The results show that the needed count rates are obtained in most cases.

Nevertheless, we have to take into account others limitations to these measurements. In fig. 2 we have shown the trajectory of the particles just after the injection or just before the detection. We

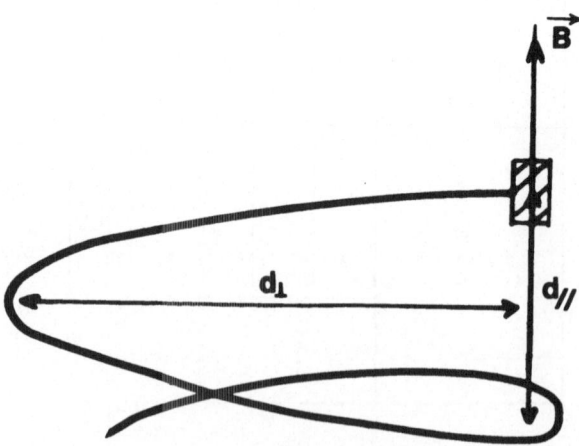

Fig. 2.   Trajectory of the electrons just after the injection

can see that this trajectory could be very much different from the one expected if the particles spend too much time in the disturbed region around the spacecraft. The values of $d_{//}$ and $d_\perp$ was seen fig. 21 have been computed and are shown in table 1 for W a s Kev and $\epsilon = 1°$. It is obvious that they decrease as W decreases. We can see therefore that too low values of W are prohibited. In all the cases the errors introduced by this disturbed region will have to be tentatively evaluated although the characteristics of the region are not very well known.

The last limitation is due to the high level of turbulence measured in the acceleration regions (MOZER et al., 1979). Indeed high frequency turbulence could lead to a broadening of the beams, the returning intensities of which would then be too low to be detected. This could limit the measurement to short times of rendezvous and consequently to very large parallel electric fields only. Such effects will have to be taken into account in future calculations. Nevertheless turbulence is not always necessary to explain parallel electric fields. Indeed such fields can be necessary to maintain the quasi neutrality between energetic particles of different charges and pitch angle distribution functions. Recent simulations (CHIU and SCHULZ, 1978) indicates the possible existence of a 0.5 mV/m parallel electric field at 2500 km, which is measurable by the proposed technique when the perpendicular component of the electric field in the spacecraft frame of reference, is not too large. To end this chapter we also have to say that we will have to be aware of the fact that the injection of electrons could modify the equilibrium between the space charges which allows the existence of a parallel electric field.

MEASUREMENT OF THE ELECTRIC POTENTIAL ALONG THE MAGNETIC FIELD LINES

Measurements of parallel acceleration of barium ions jet (HAERENDEL et al., 1976) as well as magnetic field aligned pitch-angle distributions of downward flowing electrons and upward flowing ions, detected for example at S3-3 orbit (GHIELMETTI et al., 1978 show that upward parallel electric field certainly exists at ~ 1 Re altitude. But, after GHIELMETTI et al. (1979 ,observed downward flowing magnetic field aligned ions at these altitudes are mainly due to upward flowing ions from the conjugate hemisphere and not to downward parallel electric field above but close to the satellite. They conclude that downward parallel electric fields are therefore an unfrequent phenomenon at altitudes ranging between 2000 km and $\sim$ BRe.

In the other hand, low altitude measurements of magnetic field aligned downflowing ions fluxes (REME and BOSQUED, 1971 ; HULTQVIST et al., 1971 ; GALPERIN et al., 1976 ; HULTQVIST and BORG, 1978 could be explained by 1 to 2 Kv potential differences associated with downward parallel electric field. GALPERIN et al. (1976) claim moreover, that magnetic field aligned ions fluxes could not exist at 400-500 km, where they are detected, if they had been accelerated at altitudes higher than 1500 km owing to charge exchange collisions which isotropozed the fluxes. Besides, downward electric field have

been measured at low altitude (~ 2000 km) at S3.3 satellite orbit
(MOZER, 1980). We can therefore conclude that if parallel electric
fields exist, they are localized at low altitude.

Artificially injected ions beams from ionosphere is a tool
to prove the existence or not of such downward parallel electric field. Moreover, if it exists, the measurement of the echo time
delay as a function of the energy can lead to the knowledge of the
electric potential at each point of the magnetic field line. To this
end, we must be sure to be able to detect the fluxes if they are
reflected by a potential drop. As seen in the introduction such experiments using electrons (to measure upward parallel electric field)
have shown that the rendezvous between returning electron fluxes and
spacecraft are difficult. This is mainly due to the very narrow horizontal section of these fluxes. Contrary to the electrons, the ions
fluxes are spread by charge exchange, thus the rendezvous conditions
are less difficult to be fulfilled. Nevertheless this spreading
leads to very low returning fluxes which can be separated from the
natural background only if the ions are not natural ones. To maximize the returning fluxes, it has been shown that the lithium ions must
be chosen (PIRRE, 1981) because first they are light ions, and echo
times are thus reduced. Secondly, their charge exchange cross section
is smaller than the ionizing cross section of the lithium atom, contrary to deuterium ion for example. Consequently lithium ions returning fluxes are less spread than the deuterium ones and therefore
these fluxes are the largest, especially at low energy (~ 1 Kev).
Fig. 3 shows a typical example of the returning fluxes injected and
detected at 200 km as a function of the horizontal distance from the
center of the beam. The injected intensity is assumed to be 10 mA
as the energy is 1 Kev and 10 Kev. To detect these low fluxes the
spacecraft must not be too far from the center of the returning
beam (i.e. no more than ~ 10 km) to receive detectable fluxes (10^3
to 10^4 ions/cm^2.s). After PIRRE (1981) we can expect a 30 s echo time
delay if the parallel electric field is small (~ 1 mV/m) but not too
far from the earth (~ 1000–2000 km). Consequently we can see that
the above condition can be fulfilled using rockets, the horizontal
velocity of which will have to be matched with the horizontal drift
of the beam due to the perpendicular electric field. This latter
parameter can be measured by ground technique. Indeed, a realistic
incertitude of ± 300 m/s on the real drift velocity of the particles
would assure the rocket to be at 9 km from the center of the returning
beam. The use of satellites would imply a large number of subsatel-
lites to be sure to detect the returning fluxes, which is certainly
unrealistic.

CONCLUSION

We have discussed two methods of measuring the parallel electric field using artificially injected charged particles. The first
method uses electrons to measure the parallel electric field in the
vicinity of the spacecraft. It has been shown that a very good accuracy on this measurement can be achieved ($\sim 2\%$ of the perpendicular

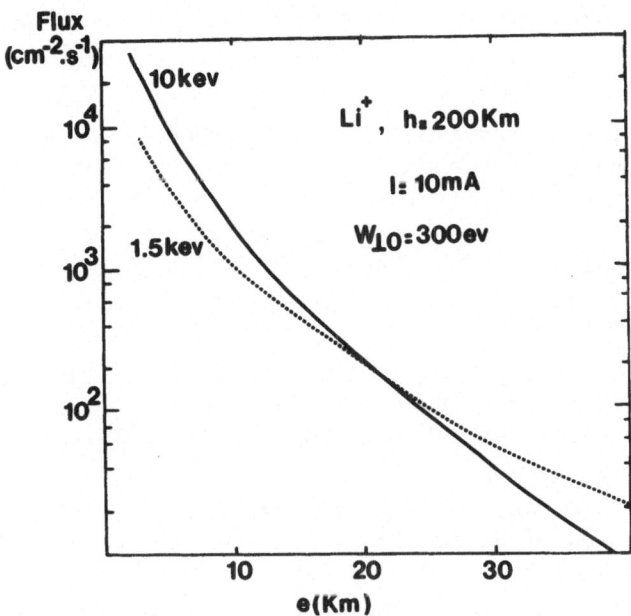

Fig. 3. Returning fluxes of lithium ions as a function of the
distance from the center of the beam. The altitude of
injection and detection is 200 km. The injection intensity
is 10 mA as the energy is 1 Kev or 10 Kev as indicated.

component). But, many limitations due in particular to the disturbed
regions surrounding the spacecraft and to high level of turbulence
could reduce significantly the accuracy of these measurements. Nevertheless it is clear that such a technique is very powerful because,
even if the parallel electric fields are too low to be measured owing
to the broadening of the fluxes by turbulence, per contra the technique can be used to study this turbulence.

The second method uses ions to measure electric potential along
the magnetic field lines if the associated parallel electric field
is directed downward. It has been shown that if parallel electric
fields are predominantly directed upward at high altitude, downward
parallel electric field can exist at lower altitude. In this case
lithium ions have to be chosen to maximize the returning fluxes and
therefore to increase the probability of detection. Moreover, rockets
have been shown to be more suitable than satellites to make these
measurements.

REFERENCES

CHIU, Y.T. and M. SCHULZ, "Self consistent particle and parallel electric field distributions in the magnetospheric ionospheric
auroral region," J. Geophys. Res., 83, 629, 1978.

FAHLESON, U.V., KELLEY, M.C. and MOZER, F.S., "Investigation of the operation of a D.C. electric field detector", Planet Space Sci., <u>18</u>, 1551, 1970.

GALPERIN, Y.I., R.A. KOVRAZHKIN, Y.N. PONOMAREV, J. CRASNIER, J.A. SAUVAUD, "Pitch-angle distributions of auroral protons", Ann. Geophys., <u>32</u>, 109, 1976.

GHIELMETTI, A.G., R.G. JOHNSON, R.D. SHARP, E.G. SHELLEY, "The latitudinal, diurnal and altitudinal distribution of upward flowing energetic ions of ionospheric region", Geophys. Res. Lett., <u>5</u>, 59, 1978.

GHIELMETTI, A.G., R.D. SHARP, E.G. SHELLEY, R.G. JOHNSON, "Downward flowing ions and evidence for injection of ionospheric ions into the plasma sheet", J. Geophys. Res., <u>84</u>, 5781, 1979.

HAERENDEL, G., E. RIEGER, A. VALENZUELA, H. FOPPL, H.C. STENBAECK-NIELSEN, E.M. WESCOTT, "First observation of electrostatic acceleration of barium ions into the magnetosphere", SCLOSS ELMAU, 115, 1976.

HULTQVIST, B., H. BORG, W. RIEDLER, P. CHRISTOPHERSEN, "Observation of magnetic field aligned anisotropy for 1 and 6 Kev positive ions in the upper atmosphere", Planet Space Sci., <u>19</u>, 279, 1971

HULTQVIST, B. and H. BORG, "Observations of energetic ions in inverted V events, Planet Space Sci., 26, 673, 1978.

MELZNER, F., G. METZNER, D. ANTRACK, "The GEOS electron beam experiment S 329", Space Sci. Instr., <u>4</u>, 45, 1978.

MELZNER, F., G. GEIGER, G. HAERENDEL, R. GRARD, K. KNOTT, A. PEDERSEN, "Simultaneous measurements of plasma drifts and quasistatic electric fields with the GEOS satellite", 1st International Conf. on IMS results, MELBOURNE, 1979.

MOZER, F.S., "On the lowest altitude S3.3 observations of electrosatic shocks and parallel electric fields", Geophys. Res. Let., <u>7</u>, 1097, 1980.

MOZER, F.S., C.A. CATTELL, M. TEMERIN, R.B. TOBBERT, S. VONGLINSKI, M. WOLDORFF, J. WYGANT, "The DC and AC electric field, plasma density, plasma temperature, and field aligned current experiments on the S3.3 satellite", J. Geophys.Res., <u>84</u>, 5875, 1979

PIRRE, M., "Interaction between an artificially injected ion beam and the neutral atmosphere, application to the parallel electric field measurement", Ann. Geophys., in Press, 1981.

PIRRE, M., M. HAMELIN, T.R. SANDERSON, G.L. WEBB, "A method of measuring the total DC electric field in the vicinity of a spacecraft using artificially injected charged particles", Ann. Geophys., <u>35</u>, 177, 1979.

REME, H., J.M. BOSQUED, "Evidence near the auroral ionosphere of a parallel electric field deduced from energy and angular distributions of low energy particles", J. Geophys. Res., 76, 7683, 1971.

WILHELM, K., "Remote sensing experiments for magnetospheric electric fields parallel to the magnetic field", J. Geophys., 43, 731, 1977.

WILHELM, K., W. BERNSTEIN, B.A. WHALEN, "Study of electric fields parallel to the magnetic lines of force using artificially injected energetic electrons", Geophys. Res. Let., 7, 117, 1980.

GENERAL DISCUSSION ON

FUTURE EXPERIMENTS

Chairman: C. Beghin

The session on future plans and recommendations was
opened by E R Schmerling, who stressed the need for well
organized team operation on future satellite accelerator
experiments with carefully selected diagnostic instruments, and a close cooperation between the experimenta-
lists and the theoreticians in the planning and development phase. Improved data system studies are needed and
continued supporting laboratory activity in this field
is recommended.

The number of STS flights devoted to space plasma
physics studies and other near space studies is anticipated not to exceed one every 12 to 18 months.

No facilities for free flyers are planned for the
future other than the "Plasma Diagnostic Package".
Individual proposals for additional, ad hoc free flyers
will be considered in the usual way.

After Dr Schmerling's introduction, the audience
was invited to present reviews of their future activity
in the field. Only a small number of the presentations
was later submitted for publication. The keywords of
the presentations are given in Table 1.

General discussion

The general discussion focused on three topics:

1)   Does the true value of the laboratory experi-
     ments lie in the plasma physics studies per
     se, or do they contribute to a better under-
     standing of the observations obtained in space?

Table 1   List of planned experiments using artificial
particle beams in space

Experiment	Presented by
FPEG (Fast Pulse Electron Gun) to be flown on OSS1 in 1982. Studies of vehicle potential and charging.	P M Banks
Tethered payloads to be flown on a sounding rocket in 1982. Studies of vehicle charging with a time resolution in the sub-microsecond range.	J Raitt
Tethered balloon current generator to be flown on STS after 1984.	P R Williamson
PDP (Plasma Diagnostic Package), to be flown on OSS1 in 1982 and reflown as a sub-satellite on SL2 in 1984. A new, recoverable sub-satellite will later be developed.	S D Shawhan
SEPAC (Space Experiments with Particle ACcelerators), will be flown on SL1 in 1983. Comprehensive plasma experiment for investigating the charging of STS and several beam-plasma interaction processes.	N Kawashima
PICPAB (Phenomena Induced by Charged PArticle Beams), to be flown on SL1 in 1983. Includes ion and electron accelerators for studies of beam-plasma interaction processes.	C Beghin
Particle spectrometer to be flown on SL1 in 1983. Monitors the artificial particle beam effects.	K Wilhelm
Series of "mother-daughter" rockets to be launched from Norway 1981-84. Studies of the atmospheric modifications generated by the beam.	B N Maehlum
Echo 6, scheduled for 1982. New member of the series of accelerator rockets used for investigating the distant magnetospheric configuration.	J R Winckler

SCEX, planned launched in 1982.   P J Kellogg
Studies of atmospheric effects of
an electron beam.

TEBPP. Proposed for STS in 1986-87.   H R Anderson
Investigations of atmospheric effects
of particle beams.

AMPTE, satellite to be launched in 1984. G Haerendel
Studies of geomagnetic field configu-
ration by chemical releases in the
outer part of the magnetosphere._____

2) Can natural plasma processes in space be
   simulated in the laboratory?

3) Does the space vehicle borne accelerators
   contribute to a deeper understanding of the
   natural space plasma processes, or are they
   mainly of non-geophysical significance?

It was generally agreed that the plasma chambers
are valuable supplements to the space experiments, and
the participants expressed some concern about the lack
of availability for large plasma chambers in the future
after the close-down of the Johnson Space Center
chamber. A few large chambers are still available in
Japan and France, but no priority is given to space
plasma studies in these.

The required chamber size for conducting studies
of plasma instabilities in the vicinity of a charged
particle beam was the topic for some discussion. It
was agreed that the scaling problem could possibly be
overcome for studies of beam plasma discharges and
other types of beam instabilities if the size of the
chamber is 2-3 meters or larger.

Some questions were raised to whether the results
obtained in the plasma tanks could be directly applied

in the interpretation of beam experiment results from
the upper atmosphere. In fact, the laboratory experiments may be of less value if not proper account is
taken to coordinate the activity with the relevant
studies conducted in space. It was stressed by the
audience that such coordination was necessary, and it
was recommended that plasma theoreticians should be
consulted and take active part in the planning and
conducting of the space plasma simulation studies in
the laboratory.

Another, equally important aspect of the utilization
of the laboratory plasma chamber was stressed by several
participants. It is well known that some of the diagnostics instruments used on accelerator rockets have
behaved in a manner which cannot be explained. Hence,
it is important to perform simulation studies of upper
atmosphere conditions in the chamber using the flight
instruments. Also, the anticipated interference from
the accelerator should be investigated before flight in
the laboratory plasma. The importance of the space plasma
studies will be significantly enhanced by conducting
such pre-flight investigations.

The importance of the SEPAC gas plume as a tool for
increasing the effective neutralizing return current was
discussed. It is at present not quite clear what the
effective area will be, but the experiment will provide
important information.

A general question as to whether the particle
accelerators on space vehicle really can provide new
information on the natural plasma processes was discussed at some length. Particle beams have been utilized as probes for investigating the geomagnetic and
electric field configuration in the distant magnetosphere, but it is quite obvious that the interaction
processes between the beam and the turbulent fields in
the magnetosphere are not well known. In the ionosphere
the interaction is even less predictable, particularly
when high-current beams are utilized which creates non-
linear effects. Studies of these processes from rockets
and satellites should therefore be given priority.

The use of incoherent scatter radars as remote
sensors for the modifications in the ionospheric plasma
during and after the particle injection was questioned
by some participants. The reflected signal from the
rocket body may drive the receivers in saturation, but
it was realized that valuable information may be

obtained if the saturation problem might be overcome.

The beam spread and geometry near the rocket has
not yet been properly measured, but it is quite clear
that future rocket and satellite experiments should try
to monitor these parameters. Also, it is not quite
clear how the beam characteristics could be monitored
in the vicinity of the beam without modifying the beam.
In fact, the geometrical size of the probe should be
considered carefully before conducting such investigations.

Finally, the audience agreed that the laboratory,
satellite and rocket experiments are valuable comple-
ments. In the future one should continue to encourage
the team effort, with space and laboratory experimenta-
lists along with theoretical plasma physicists, in the
planning and interpretation of the experiments.

PARTICIPANTS

Anderson, H R   Dept of Space Physics, Rice University,
                  Houston, TX 77001, USA

Banks, P M   Physics Dept, Utah State University,
                  UMC 41, Logan, Utah 84322, USA

Beghin, C   CRPE, La Source, 45045 Orleans Cedex,
                  France

Bering, E   University of Houston, Physics Depart-
                  ment, Houston, Texas 77004, USA

Bernstein, W   Department of Space Physics and
                  Astronomy, Rice University,
                  Houston, TX 77001, USA

Boström, R   Uppsala Ionosfäreobservatorium,
                  Uppsala, Sweden, S-75590

Christiansen, P   Physics Dept, University of Sussex,
                  Falmer, Brighton BN1 9QM, England

Duprat, G R   Herzberg Institute of Astrophysics,
                  National Research Coucil of Canada,
                  Ontario K1A 0R6, Canada

Egeland, A   Dept of Physics, University of Oslo,
                  Blindern, Oslo 3, Norway

Folkestad, K   EISCAT, Ramfjordmoen, 9027 Ramfjord-
                  botn, Norway

Gough, P   Physics Department, University of
                  Sussex, Falmer, Brighton BN1 9QN,
                  England

Grandal, B   Norwegian Defence Research Establish-
                      ment, P O Box 25, N-2007 Kjeller,
                      Norway

Gundersen, A   Royal Norwegian Council of Scientific
                      and Industrial Research, P O Box 309,
                      Blindern, Oslo 3, Norway

Haerendel, G   Max Planck Institute, Inst Extra-Terr
                      Phys, 8046 Garching bei München,
                      West Germany

Hallinan, T J   Geophysical Institute, Univ of Alaska,
                      College, Alaska 99735, USA

Holme, N   Norwegian Defence Research Establish-
                      ment, P O Box 25, N-2007 Kjeller,
                      Norway

Holtet, J   Dept of Physics, University of Oslo,
                      Blindern, Oslo 3, Norway

Jacobsen, T A   Norwegian Defence Research Establish-
                      ment, P O Box 25, N-2007 Kjeller,
                      Norway

Johnstone, A   Mullard Space Sci Lab, Holmbury,
                      St Mary, Dorking, Surrey, RH5 6NT,
                      England

Jost, R J   NASA, L B Johnson Space Center,
                      Houston, TX 77058, USA

Kawashima, N   Inst of Space & Aeronautical Sci,
                      University of Tokyo, 4-6-1 Komaba,
                      Meguro-Ku, Tokyo, Japan

Kellogg, P J   University of Minnesota,
                      School of Physics and Astronomy,
                      116 Church St SE, Minneapolis,
                      Minn 55455, USA

Kintner, P   School of Electrical Engineering,
                      Cornell Univ, Ithaca, NY 14853, USA

Klinge, D   Space Science Department of ESA/ESTEC,
                      Zwarteweg 62, 2200 AG Noordwijk,
                      The Netherlands

Kofsky, I L   PhotoMetric Inc. 422 Marrett Road,
                     Lexington, Massachusetts 02173, USA

Koons, H C   A6/2447, The Aerospace Corp,
                     P O Box 92957, Los Angeles, CA 90009,
                     USA

Landmark, B   Royal Norwegian Council for
                     Scientific and Industrial Research,
                     P O Box 309, Blindern, Oslo 3, Norway

Lavergnat, J   LGE, 4 Ave de Neptune,
                     94100 Saint-Maur des Fossé,
                     France

Lebreton, J P   Space Science Department of ESA/ESTEC,
                     Zwarteweg 62, 2200 AG Noordwijk,
                     The Netherlands

Le Queau, D   Centre de Physique Théorique de l'Ecole
                     Polytechnique, Palaiseau, Cedex,
                     France, 91128

Linson, L M   Science Applications, Inc,
                     1200 Prospect St, P O Box 2351,
                     La Jolla, CA 92037, USA

Mæhlum, B N   Norwegian Defence Research Establish-
                     ment, P O Box 25, N-2007 Kjeller,
                     Norway

Maggs, J E   Institute of Geophysics, UCLA,
                     Los Angeles, CA 90024, USA

Måseide, K   University of Oslo, Blindern,
                     Oslo 3, Norway

Maynard, N C   Goddard Space Flight Center,
                     Greenbelt, MD 20771, USA

Papadopoulos, K   Science Applications, Inc.,
                     1710 Goodridge Drive McLean,
                     Virginia 22102, USA

Pedersen, A   Space Science Department of ESA/ESTEC,
                     Zwarteweg 62, 2200 AG Noordwijk,
                     The Netherlands

Pendleton, W R   University of Stockholm,
                      Department of Meteorology, S-10691,
                      Stockholm 50, Sweden

Pirre, M   CRPE/CNRS - 3 A -
                      Avenue de la Recherche Scientifique
                      45045 Orleans Cedex, France

Raitt, W J   Center for Atmospheric and Space
                      Sciences, Utah State University,
                      Logan, Utah 84322, USA

Saint-Marc, A J   CESR, 11 Ave Colonel Roche,
                      31029 Toulouse, France

Schmerling, E R   Space Plasma Physics,
                      Solar Terrestrial Div, NASA SC-7,
                      Washington DC 20546, USA

Shawhan, S D   Dept of Physics & Astronomy,
                      University of Iowa, Iowa City,
                      Iowa 52242, USA

Soliz, P   EOARD, 223/231 Ole Marylebone Road,
                      London NW1 5TH, England

Stenzel, R L   Dept of Physics, UCLA, Los Angeles,
                      CA 90024, USA

Szuszczewicz,E P   Ionospheric Diagnostics Section,
                      Space Science Div/Code 4187,
                      Naval Research Lab, Washington
                      DC 20375, USA

Trøim, J   Norwegian Defence Research Establish-
                      ment, P O Box 25, N-2007 Kjeller,
                      Norway

Trulsen, J   Auroral Observatory, University of
                      Tromsø, P O Box 953,
                      N-9001 Tromsø, Norway

Wilhelm, K   Max Planck-Institut für Aeronomie,
                      D-3411 Katlenburg-Lindau 3,
                      West Germany

Williamson, P R   Space Plasma Physics Office/SC-7
NASA Headquarters,
Washington DC 20546, USA

Winckler, J R   School of Physics and Astronomy,
University of Minnesota,
116 Church Street SE, MN 55455,
USA